T5-CUN-741

Methods in Enzymology

Volume 254

ONCOGENE TECHNIQUES

METHODS IN ENZYMOLOGY

Methods in Enzymology

Volume 254

Oncogene Techniques

EDITED BY

Peter K. Vogt

THE SCRIPPS RESEARCH INSTITUTE
LA JOLLA, CALIFORNIA

Inder M. Verma

THE SALK INSTITUTE
SAN DIEGO, CALIFORNIA

ACADEMIC PRESS

San Diego New York Boston London Sydney Tokyo Toronto

This book is printed on acid-free paper. ♾

Academic Press, Inc.
A Division of Harcourt Brace & Company
525 B Street, Suite 1900, San Diego, California 92101-4495

United Kingdom Edition published by
Academic Press Limited
24-28 Oval Road, London NW1 7DX

International Standard Serial Number: 0076-6879

International Standard Book Number: 0-12-182155-2

PRINTED IN THE UNITED STATES OF AMERICA
95 96 97 98 99 00 MM 9 8 7 6 5 4 3 2 1

Table of Contents

Section I. Cells

Section II. Molecular Clones

Section III. Gene Function

Section IV. Protein–Protein Interactions

Section V. Protein–DNA Interactions

Contributors to Volume 254

Article numbers are in parentheses following the names of contributors.
Affiliations listed are current.

STUART A. AARONSON (13), *Derald H. Ruttenberg Cancer Center, Mount Sinai Medical Center, New York, New York 10029*

VICTOR ADLER (37), *American Health Foundation, Valhalla, New York 10595*

LEILA ALLAND (36), *Albert Einstein College of Medicine, Bronx, New York 11046*

CHRISTOPHER P. AUSTIN (25), *Department of Genetics, Harvard University Medical School, Boston, Massachusetts 02115*

LIDIA AVERBOUKH (20), *Division of Cell Growth and Regulation, Dana-Farber Cancer Institute, Boston, Massachusetts 02115*

PAUL L. BARTEL (16), *Department of Molecular Genetics and Microbiology, State University of New York at Stony Brook, Stony Brook, New York 11794*

PETR BARTUNEK (4), *Institute of Molecular Pathology, Vienna A-1030, Austria*

DAVID BAUER (20), *Max Plank Society, Humboldt University, Max Delbruck Center for Molecular Medicine, 13122 Berlin-Buch, Germany*

MARTIN BENTZ (22), *Deutsches Krebsforschungszentrum, D-69120 Heidelberg, Germany*

HARTMUT BEUG (4), *Institute of Molecular Pathology, Vienna A-1030, Austria*

T. KEITH BLACKWELL (41), *Center for Blood Research and Department of Pathology, Harvard Medical School, Boston, Massachusetts 02115*

ELIZABETH M. BLACKWOOD (15), *Department of Biology, University of California, San Diego, La Jolla, California 92093*

DAVID BOETTIGER (6), *Department of Microbiology, University of Pennsylvania, Philadelphia, Pennsylvania 19104*

AMY BRIOT (8), *Department of Pathology and Curriculum in Genetics and Molecular Biology, Lineberger Comprehensive Cancer Center, University of North Carolina at Chapel Hill School of Medicine, Chapel Hill, North Carolina 27599*

GEORGES CALOTHY (5), *Unité de Recherche Associée, 1443 du CNRS, Institut Curie, Section de Biologie, Centre Universitaire 91405 Orsay, France*

LEWIS C. CANTLEY (35), *Division of Signal Transduction, Beth Israel Hospital, and Department of Cell Biology, Harvard Medical School, Boston, Massachusetts 02115*

KEITH CAULEY (43), *Signal Pharmaceuticals, San Diego, California 92121*

CONSTANCE CEPKO (25), *Department of Genetics, Harvard University Medical School, Boston, Massachusetts 02115*

JIN-HONG CHANG (27), *Department of Microbiology and Cancer Center, University of Virginia, Charlottesville, Virginia 22908*

JIN CHEN (17), *Department of Microbiology and Immunology, Vanderbilt University School of Medicine, Nashville, Tennessee 37232*

JONATHAN CHERNOFF (12), *Fox Chase Cancer Center, Philadelphia, Pennsylvania 19111*

PAUL J. CHIAO (31), *Department of Surgical Oncology, M. D. Anderson Cancer Center, Houston, Texas 77030*

PATRICIA CRISANTI (5), *Unité Propre 9035 du CNRS, Université Paris VI, 75270 Paris, France*

CARLO M. CROCE (21), *The Jefferson Cancer Institute, Jefferson Cancer Center, and Department of Microbiology and Immunology, Jefferson Medical College of Thomas Jefferson University, Philadelphia, Pennsylvania 19107*

JAMES A. DECAPRIO (7), *Dana-Farber Cancer Institute and Harvard Medical School, Boston, Massachusetts 02115*

CAMILLE DILULLO (6), *Department of Anatomy, Philadelphia College of Osteopathic Medicine, Philadelphia, Pennsylvania 19131*

ANDRZEJ A. DLUGOSZ (1), *Laboratory of Cellular Carcinogenesis and Tumor Promotion, National Cancer Institute, Bethesda, Maryland 20892*

ROBERT N. EISENMAN (15), *Division of Basic Sciences, Fred Hutchinson Cancer Research Center, Seattle, Washington 98104*

EMELYN R. ELDREDGE (31), *Molecular Biology and Virology Laboratory, The Salk Institute, San Diego, California 92037*

PAULA J. ENRIETTO (30), *Department of Microbiology, State University of New York at Stony Brook, Stony Brook, New York 11794*

DONNA M. FEKETE (25), *Department of Biology, Boston College, Boston, Massachusetts 02167*

STANLEY FIELDS (16), *Department of Molecular Genetics and Microbiology, State University of New York at Stony Brook, Stony Brook, New York 11794*

FRÉDÉRIC FLAMANT (14), *Laboratoire de Biologie Moléculaire et Cellulaire, Centre National de la Recherche Scientifique, Institut National de la Recherche Agronomique, Ecole Normale Superiéure de Lyon, 69364 Lyon, France*

R. E. K. FOURNIER (9), *Division of Basic Sciences, Fred Hutchinson Cancer Research Center, Seattle, Washington 98104*

CHRISTOPHER C. FRANKLIN (37), *Department of Medicine, Division of Hematology/Oncology, University of Alabama at Birmingham, Birmingham, Alabama 35294*

FRED H. GAGE (2), *Department of Neurosciences, University of California, San Diego, La Jolla, California 92093*

GERALD GISH (34), *Samuel Lunenfeld Research Institute, Mount Sinai Hospital, Toronto, Ontario, Canada M5G 1X5*

ADAM B. GLICK (1), *Laboratory of Cellular Carcinogenesis and Tumor Promotion, National Cancer Institute, Bethesda, Maryland 20892*

HIROSHI HANDA (40), *Faculty of Science and Biotechnology, Tokyo Institute of Technology, Yokohama 227, Japan*

MICHAEL J. HAYMAN (4), *Department of Microbiology, State University of New York at Stony Brook, Stony Brook, New York 11794*

JEAN-MICHEL HEARD (14), *Laboratoire Retrovirus et Transfert de Génétique, Institut Pasteur, 76724 Paris, France*

GEOFFREY G. HICKS (17), *Department of Microbiology and Immunology, Vanderbilt University School of Medicine, Nashville, Tennessee 37232*

JEFFREY T. HOLT (23), *Department of Cell Biology, Vanderbilt University, Nashville, Tennessee 37232*

JUN-ICHIRO INOUE (33), *Department of Oncology, Institute of Medical Science, University of Tokyo, Tokyo 108, Japan*

YURIJ IONOV (18), *California Institute of Biological Research, La Jolla, California 92037*

MICHIYASU ITOH (33), *Department of Analytical Chemistry, Setsunan University, Osaka 573, Japan*

STEFAN JOOS (22), *Deutsches Krebsforschungszentrum, D-69120 Heidelberg, Germany*

YASUNORI KADOWAKI (11), *3rd Internal Medicine, University Hospital, University of Tokyo, Tokyo 113, Japan*

HARUMA KAWAGUCHI (40), *Department of Applied Chemistry, Faculty of Science and Technology, Keio University, Yokohama 227, Japan*

LAWRENCE D. KERR (42), *Departments of Microbiology and Immunology and Cell Biology, Vanderbilt University School of Medicine, Nashville, Tennessee 37232*

ANN MCNEILL KILLARY (9), *Division of Laboratory Medicine, Hematopathology Program, M. D. Anderson Cancer Center, University of Texas, Houston, Texas 77030*

KARLA KOVARY (28), *Departamento de Bioquimica, Instituto de Biologia, Universidade do Estado do Rio de Janeiro, 20551-013 Rio de Janeiro, Brasil*

ANDREW S. KRAFT (37), *Department of Medicine, Division of Hematology/Oncology, University of Alabama at Birmingham, Birmingham, Alabama 35294*

WILHELM KREK (7), *Dana-Farber Cancer Institute and Harvard Medical School, Boston, Massachusetts 02115*

MARK A. LABOW (24), *Roche Research Center, Hoffmann-La Roche Incorporated, Nutley, New Jersey 07110*

HARTMUT LAND (3), *Imperial Cancer Research Fund, London WC2B 3PX, United Kingdom*

LOUISE LAROSE (34), *Polypeptide Hormone Laboratory, McGill University, Montreal, Quebec, Canada*

PENG LIANG (20), *Division of Cell Growth and Regulation, Dana-Farber Cancer Institute, Boston, Massachusetts 02115*

PETER LICHTER (22), *Deutsches Krebsforschungszentrum, D-69120 Heidelberg, Germany*

NIKOLAI LISITSYN (19), *Cold Spring Harbor Laboratory, Cold Spring Harbor, New York 11724*

KUN PING LU (31), *Molecular Biology and Virology Laboratory, The Salk Institute, San Diego, California 92037*

WAYNE T. MATTEN (29), *ABL-Basic Research Program, NCI-Frederick Cancer Research and Development Center, Frederick, Maryland 21702*

MICHAEL MCCLELLAND (18), *California Institute of Biological Research, La Jolla, California 92037*

TORU MIKI (13), *Laboratory of Cellular and Molecular Biology, National Cancer Institute, Bethesda, Maryland 20892*

JOVAN MIRKOVITCH (39), *Swiss Institute for Experimental Cancer Research, CH-1066 Epalinges, Switzerland*

SHIGEKI MIYAMOTO (43), *Molecular Biology and Virology Laboratory, The Salk Institute, San Diego, California 92186*

MARY PAT MOYER (10), *Department of Surgery, The University of Texas Health Science Center at San Antonio, San Antonio, Texas 78284*

HEIKO MULLER (20), *Max Plank Society, Humboldt University, Max Delbruck Center for Molecular Medicine, 13122 Berlin-Buch, Germany*

HEIDI OKAMURA (36), *Cell Biology and Genetics Program, Memorial Sloan-Kettering Cancer Center, New York, New York 10021*

LUIS F. PARADA (26), *Molecular Embryology Section, Mammalian Genetics Laboratory, ABL-Basic Research Program, NCI-Frederick Cancer Research and Development Center, Frederick, Maryland 21702*

ARTHUR B. PARDEE (20), *Division of Cell Growth and Regulation, Dana-Farber Cancer Institute, Boston, Massachusetts 02115*

SARAH J. PARSONS (27), *Department of Microbiology and Cancer Center, University of Virginia School of Medicine, Health Sciences Center, Charlottesville, Virginia 22908*

TONY PAWSON (34), *Samuel Lunenfeld Research Institute, Mount Sinai Hospital, Toronto, Ontario, Canada M5G 1X5*

MIGUEL ANGEL PEINADO (18), *Institut of Recerca Oncologica, Hospital Duran i Reynals, 08907 Barcelona, Spain*

MANUEL PERUCHO (18), *California Institute of Biological Research, La Jolla, California 92037*

BERNARD PESSAC (5), *Unité Propre 9035 du CNRS, Université Paris VI, 75270 Paris, France*

BENJAMIN POULOSE (8), *Department of Pathology and Curriculum in Genetics and Molecular Biology, Lineberger Comprehensive Cancer Center, University of North Carolina at Chapel Hill School of Medicine, Chapel Hill, North Carolina 27599*

LYNN J. RANSONE (32), *Signal Pharmaceuticals, Incorporated, San Diego, California 92121*

JASODHARA RAY (2), *Department of Neurosciences, University of California, San Diego, La Jolla, California 92093*

HEATHER K. RAYMON (2), *Department of Neurosciences, University of California, San Diego, La Jolla, California 92093*

MARILYN D. RESH (36), *Cell Biology and Genetics Program, Memorial Sloan-Kettering Cancer Center, New York, New York 10021*

CHERYL ROBINSON-BENION (23), *Department of Cell Biology, Vanderbilt University, Nashville, Tennessee 37232*

MARKUS ROHRWILD (20), *Department of Cell Biology, Harvard Medical School, Boston, Massachusetts 02115*

MICHAEL ROSHON (17), *Department of Microbiology and Immunology, Vanderbilt University School of Medicine, Nashville, Tennessee 37232*

H. EARL RULEY (17), *Department of Microbiology and Immunology, Vanderbilt University School of Medicine, Nashville, Tennessee 37232*

ELIZABETH F. RYDER (25), *Department of Molecular Biology, Massachusetts General Hospital, Boston, Massachusetts 02114*

JACQUES SAMARUT (14), *Laboratoire de Biologie Moléculaire et Cellulaire, Centre National de la Recherche Scientifique, Institut National de la Recherche Agronomique, Ecole Normale Superiéure de Lyon, 69364 Lyon, France*

CHRISTINA SCHERER (17), *Department of Biology and Center for Cancer Research, Massachusetts Institute of Technology, Cambridge, Massachusetts 02139*

STEVEN A. SCHICHMAN (21), *The Jefferson Cancer Institute, Jefferson Cancer Center, and Department of Microbiology and Immunology, Jefferson Medical College of Thomas Jefferson University, Philadelphia, Pennsylvania 19107*

MARY ANN SELLS (12), *Fox Chase Cancer Center, Philadelphia, Pennsylvania 19111*

STEVEN M. SHAMAH (38), *Department of Microbiology and Molecular Genetics, Harvard Medical School, and Dana-Farber Cancer Institute, Boston, Massachusetts 02115*

RANDY SHEN (34), *Samuel Lunenfeld Research Institute, Mount Sinai Hospital, Toronto, Ontario, Canada M5G 1X5*

ER-GANG SHI (17), *Department of Microbiology and Immunology, Vanderbilt University School of Medicine, Nashville, Tennessee 37232*

YUZURU SHIIO (33), *Department of Oncology, Institute of Medical Science, University of Tokyo, Tokyo 108, Japan*

CATHERINE T. SIGAL (36), *Cell Biology and Genetics Program, Memorial Sloan-Kettering Cancer Center, New York, New York 10021*

ZHOU SONGYANG (35), *Division of Signal Transduction, Beth Israel Hospital, and Department of Physiology, Tufts University, Boston, Massachusetts 02111*

PETER STEINLEIN (4), *Institute of Molecular Pathology, Vienna A-1030, Austria*

CHARLES D. STILES (38), *Department of Microbiology and Molecular Genetics, Harvard Medical School, and Dana-Farber Cancer Institute, Boston, Massachusetts 02115*

MICHAEL STRAUSS (20), *Max Plank Society, Humboldt University, Max Delbruck Center for Molecular Medicine, 13122 Berlin-Buch, Germany*

WILLIAM M. SUTHERLAND (27), *Department of Anatomy and Cell Biology, University of Virginia, Charlottesville, Virginia 22908*

TAMAR TENNENBAUM (1), *Laboratory of Cellular Carcinogenesis and Tumor Promotion, National Cancer Institute, Bethesda, Maryland 20892*

LINO TESSAROLLO (26), *Molecular Embryology Section, Mammalian Genetics Laboratory, ABL-Basic Research Program, NCI-Frederick Cancer Research and Development Center, Frederick, Maryland 21702*

THEA D. TLSTY (8), *Department of Pathology and Curriculum in Genetics and Molecular Biology, Lineberger Comprehensive Cancer Center, University of North Carolina, School of Medicine, Chapel Hill, North Carolina 27599*

GEORGE F. VANDE WOUDE (29), *ABL-Basic Research Program, NCI-Frederick Cancer Research and Development Center, Frederick, Maryland 21702*

INDER M. VERMA (43), *Molecular Biology and Virology Laboratory, The Salk Institute, San Diego, California 92186*

TADASHI WADA (40), *Faculty of Bioscience and Biotechnology, Tokyo Institute of Technology, Yokohama 227, Japan*

AMY K. WALKER (30), *Department of Microbiology, State University of New York at Stony Brook, Stony Brook, New York 11794*

CHRISTOPHER WALSH (25), *Department of Neurology, Beth Israel Hospital, Boston, Massachusetts 02115*

PETER WARTHOE (20), *Danish Cancer Society, Division of Cancer Biology, 2100 Copenhagen, Denmark*

HAJIME WATANABE (40), *Faculty of Bioscience and Biotechnology, Tokyo Institute of Technology, Yokohama 227, Japan*

WENDY C. WEINBERG (1), *Laboratory of Cellular Carcinogenesis and Tumor Promotion, National Cancer Institute, Bethesda, Maryland 20892*

JOHN WELSH (18), *California Institute of Biological Research, La Jolla, California 92037*

MICHAEL H. WIGLER (19), *Cold Spring Harbor Laboratory, Cold Spring Harbor, New York 11724*

DOUG WILLIAMSON (17), *Department of Microbiology and Immunology, Vanderbilt University School of Medicine, Nashville, Tennessee 37232*

TADASHI YAMAMOTO (11), *Department of Oncology, Institute of Medical Science, University of Tokyo, Tokyo 108, Japan*

STUART H. YUSPA (1), *Laboratory of Cellular Carcinogenesis and Tumor Promotion, National Cancer Institute, Bethesda, Maryland 20892*

Preface

The past two decades have produced powerful evidence for the genetic basis of cancer. The idea that genes determine the growth behavior of a cell is now almost axiomatic, and abnormal growth is generally accepted to reflect the action of abnormal genes. The transformation of a normal cell to a cancer cell starts with changes in growth regulatory genes, and in the course of tumor progression further escape from normal growth control is caused by additional alterations in genes that direct cell multiplication and cell survival. Somatic mutations, and to a lesser extent germ line mutations, are the root cause of cancer; current attempts to develop effective and specific therapies for the disease target these heritable changes in cancer cells.

The genetic view of cancer has its origins in virology. Studies on tumor viruses had shown that viral genomes could carry individual genes that, expressed in host cells, are both necessary and sufficient for the induction of oncogenic transformation. In retroviruses, such oncogenes became particularly interesting because they turned out to be recent acquisitions from cellular genomes, pieces of host genetic information that were mutated, transduced, and expressed as part of the viral life cycle. Retroviral oncogenes constitute the bridge between virus-induced tumors and tumors of all other etiologies: a cellular oncogene activated by viral transduction is a mere special example of the general phenomenon of genetic alterations that can convert important and useful growth regulators of the cell into driving forces of unbridled growth.

The working definition of an oncogene is rather broad. It includes any gene that has the potential of becoming a constitutive growth stimulator and determinant of oncogenic cellular properties. The oncogenicity of these genes is correlated with a gain of function. There is an equally important category of cancer genes that contribute to tumorigenesis through a loss of function—the tumor suppressor genes. In contrast to the growth stimulatory oncogenes, tumor suppressors normally function as attenuators and as inhibitors of growth. The normal versions of both oncogenes and tumor suppressor genes serve in diverse regulatory systems of the cell. Most oncogenes code for components of signal transduction pathways that convert an extracellular stimulus into a programmed pattern of gene expression. A functional relatedness of tumor suppressors is less apparent; some domains of tumor suppressor gene action include cell surface properties, signal transduction, gene transcription, DNA repair, and checkpoints for cell division. The methods compiled in this volume are applicable to work with oncogenes as well as with tumor suppressors.

Techniques are the backbone of a field. They determine what can be done and what questions it makes sense to ask. But in an area of research that extends from animals to cells to molecules, the technical repertoire is very wide. It cannot be compressed into a single, comprehensive volume. In making the necessary selections, we chose techniques that reflect recent developments and that play important roles in cutting-edge research. We limited established methods, like cell culture or transformation and tumorigenesis assays, for which instruction is readily available in existing publications. We also excluded techniques such as the generation of transgenic and of knockout animals which require so much ancillary detail as to make adequate coverage impossible within one or even a few chapters. Fortunately, available publications again take care of this instructional need.

The chapters assembled in this book attempt to strike a balance between two goals: to expound the principles behind a method and to offer information that allows adoption and reproduction of the technique in the laboratory. The volume may be treated as a "cookbook," but it should also be seen as a base on which to build future developments that will advance cancer research, lead to new discoveries and insights, and result in successful therapies.

This volume could not have been put together without the able and dedicated work of Susan K. Burke and Pat McClintock. They formed the communications link between editors, authors, and publisher. They acted as organizers and were the guardians against tardiness. We thank them most sincerely.

Peter K. Vogt
Inder M. Verma

METHODS IN ENZYMOLOGY

Volume I. Preparation and Assay of Enzymes
Edited by Sidney P. Colowick and Nathan O. Kaplan

Volume II. Preparation and Assay of Enzymes
Edited by Sidney P. Colowick and Nathan O. Kaplan

Volume III. Preparation and Assay of Substrates
Edited by Sidney P. Colowick and Nathan O. Kaplan

Volume IV. Special Techniques for the Enzymologist
Edited by Sidney P. Colowick and Nathan O. Kaplan

Volume V. Preparation and Assay of Enzymes
Edited by Sidney P. Colowick and Nathan O. Kaplan

Volume VI. Preparation and Assay of Enzymes (*Continued*)
Preparation and Assay of Substrates
Special Techniques
Edited by Sidney P. Colowick and Nathan O. Kaplan

Volume VII. Cumulative Subject Index
Edited by Sidney P. Colowick and Nathan O. Kaplan

Volume VIII. Complex Carbohydrates
Edited by Elizabeth F. Neufeld and Victor Ginsburg

Volume IX. Carbohydrate Metabolism
Edited by Willis A. Wood

Volume X. Oxidation and Phosphorylation
Edited by Ronald W. Estabrook and Maynard E. Pullman

Volume XI. Enzyme Structure
Edited by C. H. W. Hirs

Volume XII. Nucleic Acids (Parts A and B)
Edited by Lawrence Grossman and Kivie Moldave

Volume XIII. Citric Acid Cycle
Edited by J. M. Lowenstein

Volume XIV. Lipids
Edited by J. M. Lowenstein

Volume XV. Steroids and Terpenoids
Edited by Raymond B. Clayton

Volume XVI. Fast Reactions
Edited by Kenneth Kustin

VOLUME 73. Immunochemical Techniques (Part B)
Edited by JOHN J. LANGONE AND HELEN VAN VUNAKIS

VOLUME 74. Immunochemical Techniques (Part C)
Edited by JOHN J. LANGONE AND HELEN VAN VUNAKIS

VOLUME 75. Cumulative Subject Index Volumes XXXI, XXXII, XXXIV–LX
Edited by EDWARD A. DENNIS AND MARTHA G. DENNIS

VOLUME 76. Hemoglobins
Edited by ERALDO ANTONINI, LUIGI ROSSI-BERNARDI, AND EMILIA CHIANCONE

VOLUME 77. Detoxication and Drug Metabolism
Edited by WILLIAM B. JAKOBY

VOLUME 78. Interferons (Part A)
Edited by SIDNEY PESTKA

VOLUME 79. Interferons (Part B)
Edited by SIDNEY PESTKA

VOLUME 80. Proteolytic Enzymes (Part C)
Edited by LASZLO LORAND

VOLUME 81. Biomembranes (Part H: Visual Pigments and Purple Membranes, I)
Edited by LESTER PACKER

VOLUME 82. Structural and Contractile Proteins (Part A: Extracellular Matrix)
Edited by LEON W. CUNNINGHAM AND DIXIE W. FREDERIKSEN

VOLUME 83. Complex Carbohydrates (Part D)
Edited by VICTOR GINSBURG

VOLUME 84. Immunochemical Techniques (Part D: Selected Immunoassays)
Edited by JOHN J. LANGONE AND HELEN VAN VUNAKIS

VOLUME 85. Structural and Contractile Proteins (Part B: The Contractile Apparatus and the Cytoskeleton)
Edited by DIXIE W. FREDERIKSEN AND LEON W. CUNNINGHAM

VOLUME 86. Prostaglandins and Arachidonate Metabolites
Edited by WILLIAM E. M. LANDS AND WILLIAM L. SMITH

VOLUME 87. Enzyme Kinetics and Mechanism (Part C: Intermediates, Stereochemistry, and Rate Studies)
Edited by DANIEL L. PURICH

VOLUME 88. Biomembranes (Part I: Visual Pigments and Purple Membranes, II)
Edited by LESTER PACKER

VOLUME 89. Carbohydrate Metabolism (Part D)
Edited by WILLIS A. WOOD

Volume 90. Carbohydrate Metabolism (Part E)
Edited by Willis A. Wood

Volume 91. Enzyme Structure (Part I)
Edited by C. H. W. Hirs and Serge N. Timasheff

Volume 92. Immunochemical Techniques (Part E: Monoclonal Antibodies and General Immunoassay Methods)
Edited by John J. Langone and Helen Van Vunakis

Volume 93. Immunochemical Techniques (Part F: Conventional Antibodies, Fc Receptors, and Cytotoxicity)
Edited by John J. Langone and Helen Van Vunakis

Volume 94. Polyamines
Edited by Herbert Tabor and Celia White Tabor

Volume 95. Cumulative Subject Index Volumes 61–74, 76–80
Edited by Edward A. Dennis and Martha G. Dennis

Volume 96. Biomembranes [Part J: Membrane Biogenesis: Assembly and Targeting (General Methods; Eukaryotes)]
Edited by Sidney Fleischer and Becca Fleischer

Volume 97. Biomembranes [Part K: Membrane Biogenesis: Assembly and Targeting (Prokaryotes, Mitochondria, and Chloroplasts)]
Edited by Sidney Fleischer and Becca Fleischer

Volume 98. Biomembranes (Part L: Membrane Biogenesis: Processing and Recycling)
Edited by Sidney Fleischer and Becca Fleischer

Volume 99. Hormone Action (Part F: Protein Kinases)
Edited by Jackie D. Corbin and Joel G. Hardman

Volume 100. Recombinant DNA (Part B)
Edited by Ray Wu, Lawrence Grossman, and Kivie Moldave

Volume 101. Recombinant DNA (Part C)
Edited by Ray Wu, Lawrence Grossman, and Kivie Moldave

Volume 102. Hormone Action (Part G: Calmodulin and Calcium-Binding Proteins)
Edited by Anthony R. Means and Bert W. O'Malley

Volume 103. Hormone Action (Part H: Neuroendocrine Peptides)
Edited by P. Michael Conn

Volume 104. Enzyme Purification and Related Techniques (Part C)
Edited by William B. Jakoby

Volume 105. Oxygen Radicals in Biological Systems
Edited by Lester Packer

Volume 106. Posttranslational Modifications (Part A)
Edited by Finn Wold and Kivie Moldave

VOLUME 107. Posttranslational Modifications (Part B)
Edited by FINN WOLD AND KIVIE MOLDAVE

VOLUME 108. Immunochemical Techniques (Part G: Separation and Characterization of Lymphoid Cells)
Edited by GIOVANNI DI SABATO, JOHN J. LANGONE, AND HELEN VAN VUNAKIS

VOLUME 109. Hormone Action (Part I: Peptide Hormones)
Edited by LUTZ BIRNBAUMER AND BERT W. O'MALLEY

VOLUME 110. Steroids and Isoprenoids (Part A)
Edited by JOHN H. LAW AND HANS C. RILLING

VOLUME 111. Steroids and Isoprenoids (Part B)
Edited by JOHN H. LAW AND HANS C. RILLING

VOLUME 112. Drug and Enzyme Targeting (Part A)
Edited by KENNETH J. WIDDER AND RALPH GREEN

VOLUME 113. Glutamate, Glutamine, Glutathione, and Related Compounds
Edited by ALTON MEISTER

VOLUME 114. Diffraction Methods for Biological Macromolecules (Part A)
Edited by HAROLD W. WYCKOFF, C. H. W. HIRS, AND SERGE N. TIMASHEFF

VOLUME 115. Diffraction Methods for Biological Macromolecules (Part B)
Edited by HAROLD W. WYCKOFF, C. H. W. HIRS, AND SERGE N. TIMASHEFF

VOLUME 116. Immunochemical Techniques (Part H: Effectors and Mediators of Lymphoid Cell Functions)
Edited by GIOVANNI DI SABATO, JOHN J. LANGONE, AND HELEN VAN VUNAKIS

VOLUME 117. Enzyme Structure (Part J)
Edited by C. H. W. HIRS AND SERGE N. TIMASHEFF

VOLUME 118. Plant Molecular Biology
Edited by ARTHUR WEISSBACH AND HERBERT WEISSBACH

VOLUME 119. Interferons (Part C)
Edited by SIDNEY PESTKA

VOLUME 120. Cumulative Subject Index Volumes 81–94, 96–101

VOLUME 121. Immunochemical Techniques (Part I: Hybridoma Technology and Monoclonal Antibodies)
Edited by JOHN J. LANGONE AND HELEN VAN VUNAKIS

VOLUME 122. Vitamins and Coenzymes (Part G)
Edited by FRANK CHYTIL AND DONALD B. MCCORMICK

VOLUME 123. Vitamins and Coenzymes (Part H)
Edited by FRANK CHYTIL AND DONALD B. MCCORMICK

VOLUME 124. Hormone Action (Part J: Neuroendocrine Peptides)
Edited by P. MICHAEL CONN

VOLUME 125. Biomembranes (Part M: Transport in Bacteria, Mitochondria, and Chloroplasts: General Approaches and Transport Systems)
Edited by SIDNEY FLEISCHER AND BECCA FLEISCHER

VOLUME 126. Biomembranes (Part N: Transport in Bacteria, Mitochondria, and Chloroplasts: Protonmotive Force)
Edited by SIDNEY FLEISCHER AND BECCA FLEISCHER

VOLUME 127. Biomembranes (Part O: Protons and Water: Structure and Translocation)
Edited by LESTER PACKER

Volume 128. Plasma Lipoproteins (Part A: Preparation, Structure, and Molecular Biology)
Edited by JERE P. SEGREST AND JOHN J. ALBERS

Volume 129. Plasma Lipoproteins (Part B: Characterization, Cell Biology, and Metabolism)
Edited by JOHN J. ALBERS AND JERE P. SEGREST

Volume 130. Enzyme Structure (Part K)
Edited by C. H. W. HIRS AND SERGE N. TIMASHEFF

Volume 131. Enzyme Structure (Part L)
Edited by C. H. W. HIRS AND SERGE N. TIMASHEFF

Volume 132. Immunochemical Techniques (Part J: Phagocytosis and Cell-Mediated Cytotoxicity)
Edited by GIOVANNI DI SABATO AND JOHANNES EVERSE

Volume 133. Bioluminescence and Chemiluminescence (Part B)
Edited by MARLENE DELUCA AND WILLIAM D. MCELROY

VOLUME 134. Structural and Contractile Proteins (Part C: The Contractile Apparatus and the Cytoskeleton)
Edited by RICHARD B. VALLEE

Volume 135. Immobilized Enzymes and Cells (Part B)
Edited by KLAUS MOSBACH

Volume 136. Immobilized Enzymes and Cells (Part C)
Edited by KLAUS MOSBACH

Volume 137. Immobilized Enzymes and Cells (Part D)
Edited by KLAUS MOSBACH

Volume 138. Complex Carbohydrates (Part E)
Edited by VICTOR GINSBURG

Volume 139. Cellular Regulators (Part A: Calcium- and Calmodulin-Binding Proteins)
Edited by ANTHONY R. MEANS AND P. MICHAEL CONN

Volume 140. Cumulative Subject Index Volumes 102–119, 121–134

VOLUME 141. Cellular Regulators (Part B: Calcium and Lipids)
Edited by P. MICHAEL CONN AND ANTHONY R. MEANS

Volume 142. Metabolism of Aromatic Amino Acids and Amines
Edited by SEYMOUR KAUFMAN

Volume 143. Sulfur and Sulfur Amino Acids
Edited by WILLIAM B. JAKOBY AND OWEN GRIFFITH

Volume 144. Structural and Contractile Proteins (Part D: Extracellular Matrix)
Edited by LEON W. CUNNINGHAM

Volume 145. Structural and Contractile Proteins (Part E: Extracellular Matrix)
Edited by LEON W. CUNNINGHAM

Volume 146. Peptide Growth Factors (Part A)
Edited by DAVID BARNES AND DAVID A. SIRBASKU

Volume 147. Peptide Growth Factors (Part B)
Edited by DAVID BARNES AND DAVID A. SIRBASKU

Volume 148. Plant Cell Membranes
Edited by LESTER PACKER AND ROLAND DOUCE

Volume 149. Drug and Enzyme Targeting (Part B)
Edited by RALPH GREEN AND KENNETH J. WIDDER

Volume 150. Immunochemical Techniques (Part K: *In Vitro* Models of B and T Cell Functions and Lymphoid Cell Receptors)
Edited by GIOVANNI DI SABATO

Volume 151. Molecular Genetics of Mammalian Cells
Edited by MICHAEL M. GOTTESMAN

Volume 152. Guide to Molecular Cloning Techniques
Edited by SHELBY L. BERGER AND ALAN R. KIMMEL

Volume 153. Recombinant DNA (Part D)
Edited by RAY WU AND LAWRENCE GROSSMAN

Volume 154. Recombinant DNA (Part E)
Edited by RAY WU AND LAWRENCE GROSSMAN

Volume 155. Recombinant DNA (Part F)
Edited by RAY WU

Volume 156. Biomembranes (Part P: ATP-Driven Pumps and Related Transport: The Na,K-Pump)
Edited by SIDNEY FLEISCHER AND BECCA FLEISCHER

Volume 157. Biomembranes (Part Q: ATP-Driven Pumps and Related Transport: Calcium, Proton, and Potassium Pumps)
Edited by SIDNEY FLEISCHER AND BECCA FLEISCHER

Volume 158. Metalloproteins (Part A)
Edited by JAMES F. RIORDAN AND BERT L. VALLEE

Volume 159. Initiation and Termination of Cyclic Nucleotide Action
Edited by JACKIE D. CORBIN AND ROGER A. JOHNSON

Volume 160. Biomass (Part A: Cellulose and Hemicellulose)
Edited by WILLIS A. WOOD AND SCOTT T. KELLOGG

Volume 161. Biomass (Part B: Lignin, Pectin, and Chitin)
Edited by WILLIS A. WOOD AND SCOTT T. KELLOGG

Volume 162. Immunochemical Techniques (Part L: Chemotaxis and Inflammation)
Edited by GIOVANNI DI SABATO

Volume 163. Immunochemical Techniques (Part M: Chemotaxis and Inflammation)
Edited by GIOVANNI DI SABATO

Volume 164. Ribosomes
Edited by HARRY F. NOLLER, JR., AND KIVIE MOLDAVE

Volume 165. Microbial Toxins: Tools for Enzymology
Edited by SIDNEY HARSHMAN

Volume 166. Branched-Chain Amino Acids
Edited by ROBERT HARRIS AND JOHN R. SOKATCH

Volume 167. Cyanobacteria
Edited by LESTER PACKER AND ALEXANDER N. GLAZER

Volume 168. Hormone Action (Part K: Neuroendocrine Peptides)
Edited by P. MICHAEL CONN

Volume 169. Platelets: Receptors, Adhesion, Secretion (Part A)
Edited by JACEK HAWIGER

Volume 170. Nucleosomes
Edited by PAUL M. WASSARMAN AND ROGER D. KORNBERG

Volume 171. Biomembranes (Part R: Transport Theory: Cells and Model Membranes)
Edited by SIDNEY FLEISCHER AND BECCA FLEISCHER

Volume 172. Biomembranes (Part S: Transport: Membrane Isolation and Characterization)
Edited by SIDNEY FLEISCHER AND BECCA FLEISCHER

Volume 173. Biomembranes [Part T: Cellular and Subcellular Transport: Eukaryotic (Nonepithelial) Cells]
Edited by SIDNEY FLEISCHER AND BECCA FLEISCHER

Volume 174. Biomembranes [Part U: Cellular and Subcellular Transport: Eukaryotic (Nonepithelial) Cells]
Edited by SIDNEY FLEISCHER AND BECCA FLEISCHER

Volume 175. Cumulative Subject Index Volumes 135–139, 141–167

VOLUME 176. Nuclear Magnetic Resonance (Part A: Spectral Techniques and Dynamics)
Edited by NORMAN J. OPPENHEIMER AND THOMAS L. JAMES

Volume 177. Nuclear Magnetic Resonance (Part B: Structure and Mechanism)
Edited by Norman J. Oppenheimer and Thomas L. James

Volume 178. Antibodies, Antigens, and Molecular Mimicry
Edited by John J. Langone

Volume 179. Complex Carbohydrates (Part F)
Edited by Victor Ginsburg

Volume 180. RNA Processing (Part A: General Methods)
Edited by James E. Dahlberg and John N. Abelson

Volume 181. RNA Processing (Part B: Specific Methods)
Edited by James E. Dahlberg and John N. Abelson

Volume 182. Guide to Protein Purification
Edited by Murray P. Deutscher

Volume 183. Molecular Evolution: Computer Analysis of Protein and Nucleic Acid Sequences
Edited by Russell F. Doolittle

Volume 184. Avidin–Biotin Technology
Edited by Meir Wilchek and Edward A. Bayer

Volume 185. Gene Expression Technology
Edited by David V. Goeddel

Volume 186. Oxygen Radicals in Biological Systems (Part B: Oxygen Radicals and Antioxidants)
Edited by Lester Packer and Alexander N. Glazer

Volume 187. Arachidonate Related Lipid Mediators
Edited by Robert C. Murphy and Frank A. Fitzpatrick

Volume 188. Hydrocarbons and Methylotrophy
Edited by Mary E. Lidstrom

Volume 189. Retinoids (Part A: Molecular and Metabolic Aspects)
Edited by Lester Packer

Volume 190. Retinoids (Part B: Cell Differentiation and Clinical Applications)
Edited by Lester Packer

Volume 191. Biomembranes (Part V: Cellular and Subcellular Transport: Epithelial Cells)
Edited by Sidney Fleischer and Becca Fleischer

Volume 192. Biomembranes (Part W: Cellular and Subcellular Transport: Epithelial Cells)
Edited by Sidney Fleischer and Becca Fleischer

Volume 193. Mass Spectrometry
Edited by James A. McCloskey

Volume 194. Guide to Yeast Genetics and Molecular Biology
Edited by Christine Guthrie and Gerald R. Fink

VOLUME 228. Aqueous Two-Phase Systems
Edited by HARRY WALTER AND GÖTE JOHANSSON

VOLUME 229. Cumulative Subject Index Volumes 195–198, 200–227

VOLUME 230. Guide to Techniques in Glycobiology
Edited by WILLIAM J. LENNARZ AND GERALD W. HART

VOLUME 231. Hemoglobins (Part B: Biochemical and Analytical Methods)
Edited by JOHANNES EVERSE, KIM D. VANDEGRIFF AND ROBERT M. WINSLOW

VOLUME 232. Hemoglobins (Part C: Biophysical Methods)
Edited by JOHANNES EVERSE, KIM D. VANDEGRIFF AND ROBERT M. WINSLOW

VOLUME 233. Oxygen Radicals in Biological Systems (Part C)
Edited by LESTER PACKER

VOLUME 234. Oxygen Radicals in Biological Systems (Part D)
Edited by LESTER PACKER

VOLUME 235. Bacterial Pathogenesis (Part A: Identification and Regulation of Virulence Factors)
Edited by VIRGINIA L. CLARK AND PATRIK M. BAVOIL

VOLUME 236. Bacterial Pathogenesis (Part B: Integration of Pathogenic Bacteria with Host Cells)
Edited by VIRGINIA L. CLARK AND PATRIK M. BAVOIL

VOLUME 237. Heterotrimeric G Proteins
Edited by RAVI IYENGAR

VOLUME 238. Heterotrimeric G-Protein Effectors
Edited by RAVI IYENGAR

VOLUME 239. Nuclear Magnetic Resonance (Part C)
Edited by THOMAS L. JAMES AND NORMAN J. OPPENHEIMER

VOLUME 240. Numerical Computer Methods (Part B)
Edited by MICHAEL L. JOHNSON AND LUDWIG BRAND

VOLUME 241. Retroviral Proteases
Edited by LAWRENCE C. KUO AND JULES A. SHAFER

VOLUME 242. Neoglycoconjugates (Part A)
Edited by Y. C. LEE AND REIKO T. LEE

VOLUME 243. Inorganic Microbial Sulfur Metabolism
Edited by HARRY D. PECK, JR., AND JEAN LEGALL

VOLUME 244. Proteolytic Enzymes: Serine and Cysteine Peptidases
Edited by ALAN J. BARRETT

VOLUME 245. Extracellular Matrix Components
Edited by E. RUOSLAHTI AND E. ENGVALL

VOLUME 246. Biochemical Spectroscopy
Edited by KENNETH SAUER

VOLUME 247. Neoglycoconjugates (Part B: Biomedical Applications)
Edited by Y. C. LEE AND REIKO T. LEE

VOLUME 248. Proteolytic Enzymes: Aspartic and Metallo Peptidases
Edited by ALAN J. BARRETT

VOLUME 249. Enzyme Kinetics and Mechanism (Part D: Developments in Enzyme Dynamics)
Edited by DANIEL L. PURICH

VOLUME 250. Lipid Modifications of Proteins
Edited by PATRICK J. CASEY AND JANICE E. BUSS

VOLUME 251. Biothiols (Part A: Monothiols and Dithiols, Protein Thiols, and Thiyl Radicals)
Edited by LESTER PACKER

VOLUME 252. Biothiols (Part B: Glutathione and Thioredoxin; Thiols in Signal Transduction and Gene Regulation)
Edited by LESTER PACKER

VOLUME 253. Adhesion of Microbial Pathogens
Edited by RON J. DOYLE AND ITZHAK OFEK

VOLUME 254. Oncogene Techniques
Edited by PETER K. VOGT AND INDER M. VERMA

VOLUME 255. Small GTPases and Their Regulators (Part A: Ras Family)
Edited by W. E. BALCH, CHANNING J. DER, AND ALAN HALL

VOLUME 256. Small GTPases and Their Regulators (Part B: Rho Family)
Edited by W. E. BALCH, CHANNING J. DER, AND ALAN HALL

VOLUME 257. Small GTPases and Their Regulators (Part C: Proteins Involved in Transport) (in preparation)
Edited by W. E. BALCH, CHANNING J. DER, AND ALAN HALL

VOLUME 258. Redox-Active Amino Acids in Biology (in preparation)
Edited by JUDITH P. KLINMAN

VOLUME 259. Energetics of Biological Macromolecules (in preparation)
Edited by MICHAEL L. JOHNSON AND GARY K. ACKERS

VOLUME 260. Mitochondrial Biogenesis and Genetics, Part A (in preparation)
Edited by GIUSEPPE M. ATTARDI AND ANNE CHOMYN

VOLUME 261. Nuclear Magnetic Resonance and Nucleic Acids (in preparation)
Edited by THOMAS L. JAMES

VOLUME 262. DNA Replication (in preparation)
Edited by JUDITH L. CAMPBELL

VOLUME 263. Plasma Lipoproteins (Part C: Quantitation) (in preparation)
Edited by WILLIAM A. BRADLEY, SANDRA H. GIANTURCO, AND JERE P. SEGREST

Section I

Cells

[1] Isolation and Utilization of Epidermal Keratinocytes for Oncogene Research

By ANDRZEJ A. DLUGOSZ, ADAM B. GLICK, TAMAR TENNENBAUM, WENDY C. WEINBERG, and STUART H. YUSPA

Introduction

Keratinocytes have been widely used as target cells for testing the activity of oncogenes in epithelial neoplasia. Many experimental studies have utilized cultured mouse skin keratinocytes, where *in vitro* results can be analyzed in the context of a substantial experience in carcinogen-induced mouse skin tumors.[1] More recent experiments have employed keratinocytes derived from human skin, oral cavity, or cervix, where results can be directly extrapolated to cancers or warts originating in the corresponding epithelia.[2–6] Several laboratories have utilized hamster or rat keratinocytes[7–10] in analyses of oncogenes.

The contribution of various oncogenes and tumor suppressor genes to the development of epithelial neoplasia has been reviewed.[11] In experimental epidermal carcinogenesis, oncogenic *ras* recombinant constructs have received considerable attention because this gene family is frequently mutated in mouse and human skin cancers. Studies of mouse keratinocytes have also revealed the biological consequences of expressing v-*fos* and derivatives, *neu*, mutant p53, and transforming growth factor α (TGF-α). These studies have focused on the participation of oncogenes as mediators of premalignant progression and malignant conversion. The effects of onco-

[1] S. H. Yuspa and A. A. Dlugosz, *in* "Physiology Biochemistry and Molecular Biology of the Skin" (L. A. Goldsmith, ed.), p. 1365. Oxford University Press, New York, 1991.
[2] S. P. Banks-Schlegel and P. M. Howley, *J. Cell Biol.* **96,** 330 (1983).
[3] J. S. Rhim, T. Kawakami, J. Pierce, K. Sanford, and P. Arnstein, *Leukemia* **2,** 151S (1988).
[4] C. D. Woodworth, S. Waggoner, W. Barnes, M. H. Stoler, and J. A. DiPaolo, *Cancer Res.* **50,** 3709 (1990).
[5] S. L. Li, M. S. Kim, H. M. Cherrick, J. Doniger, and N. H. Park, *Carcinogenesis* (*London*) **13,** 1981 (1992).
[6] B. J. Aneskievich and L. B. Taichman, *J. Invest. Dermatol.* **91,** 309 (1988).
[7] G. T. Diamandopoulos and M. F. Dalton-Tucker, *Am. J. Pathol.* **56,** 59 (1969).
[8] R. D. Storer, R. B. Stein, J. F. Sina, J. G. DeLuca, H. L. Allen, and M. O. Bradley, *Cancer Res.* **46,** 1458 (1986).
[9] S. A. Bayley, A. J. Stones, and C. G. Smith, *Exp. Cell Res.* **177,** 232 (1988).
[10] K. Yamanishi, F. M. Liew, Y. Hosokawa, S. Kishimoto, and H. Yasuno, *Arch. Dermatol. Res.* **282,** 330 (1990).
[11] S. H. Yuspa, *Cancer Res.* **54,** 1178 (1994).

genic DNA viruses have also been tested in keratinocytes, and particular emphasis has been directed toward the expression of simian virus 40 (SV40) and specific subtypes of human or bovine papilloma viruses (HPV, BPV) or their transforming sequences. Other studies have tested the biological activity of herpes, adeno-, and Epstein–Barr viruses (EBV) in cultured human keratinocytes.[6,12–15]

Because skin keratinocytes are easily isolated and culture conditions are well defined, many oncogene studies have been performed on primary or low passage cells, a quality making keratinocyte studies unique and highly valid with regard to *in vivo* biology. Several nonneoplastic mouse[16,17] and human[18,19] keratinocyte continuous cell lines have been developed. Although these cell lines have extended the keratinocyte model to laboratories less familiar with primary culture techniques, they should not be considered normal target cells since chromosomal and genetic aberrations have been detected in the parental lines. Benign neoplastic mouse[20,21] and immortalized human[3,22] keratinocyte cell lines have also been used in oncogene studies, particularly for the analysis of premalignant progression.

While details of the techniques employed in studies testing the influence of oncogenes on keratinocytes are outlined below, several general conclusions have become evident from the results. Oncogenes which appear to function early in the transformation process (e.g., *ras*, HPV E6 and E7, SV40 T antigen) alter the terminal differentiation program of keratinocytes, allowing transformed cells to grow in an environment which triggers terminal differentiation of normal cells.[2,23–26] Oncogenes which function later in

[12] A. Razzaque, O. Williams, J. Wang, and J. S. Rhim, *Virology* **195,** 113 (1993).
[13] C. W. Dawson, A. B. Rickinson, and L. S. Young, *Nature* (*London*) **344,** 777 (1990).
[14] R. Fahraeus, L. Rymo, J. S. Rhim, and G. Klein, *Nature* (*London*) **345,** 447 (1990).
[15] J. S. Rhim, G. Jay, P. Arnstein, F. M. Price, K. K. Sanford, and S. A. Aaronson, *Science* **227,** 1250 (1985).
[16] S. H. Yuspa, B. Koehler, M. Kulesz-Martin, and H. Hennings, *J. Invest. Dermatol.* **76,** 144 (1981).
[17] B. E. Weissman and S. A. Aaronson, *Cell* **32,** 599 (1983).
[18] M. L. Goldaber, J. Kubilus, S. B. Phillips, C. Henkle, L. Atkins, and H. P. Baden, *In Vitro Cell Dev. Biol.* **26,** 7 (1990).
[19] P. Boukamp, R. T. Petrussevska, D. Breitkreutz, J. Hornung, A. Markham, and N. E. Fusenig, *J. Cell Biol.* **106,** 761 (1988).
[20] J. R. Harper, D. R. Roop, and S. H. Yuspa, *Mol. Cell. Biol.* **6,** 3144 (1986).
[21] G. P. Dotto, J. O'Connell, G. Patskan, C. Conti, A. Ariza, and T. J. Slaga, *Mol. Carcinog.* **1,** 171 (1988).
[22] P. Boukamp, E. J. Stanbridge, D. Y. Foo, P. A. Cerutti, and N. E. Fusenig, *Cancer Res.* **50,** 2840 (1990).
[23] S. H. Yuspa, A. E. Kilkenny, J. Stanley, and U. Lichti, *Nature* (*London*) **314,** 459 (1985).
[24] J. B. Hudson, M. A. Bedell, D. J. McCance, and L. A. Laiminis, *J. Virol.* **64,** 519 (1990).
[25] M. S. Barbosa and R. Schlegel, *Oncogene* **4,** 1529 (1989).
[26] R. Schlegel, W. C. Phelps, Y. L. Zhang, and M. Barbosa, *EMBO J.* **7,** 3181 (1988).

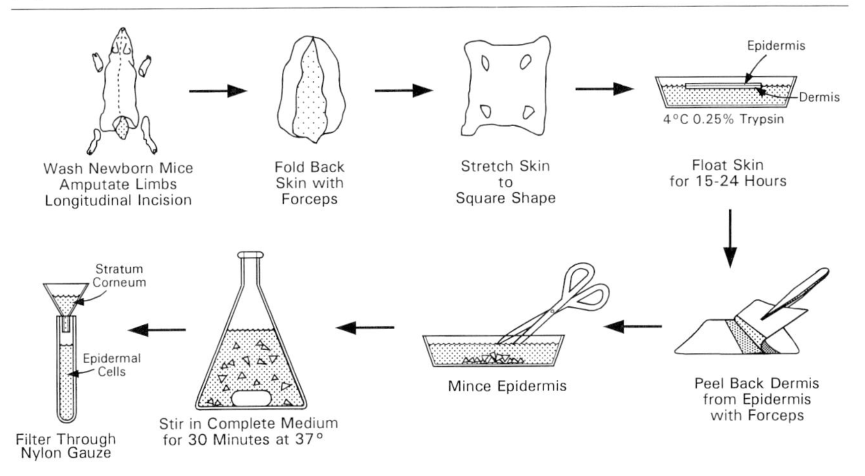

FIG. 1. Protocol for isolating primary keratinocytes from newborn mouse skin. See text for additional details.

neoplastic progression (e.g., *fos*, mutant p53, *ras*) profoundly influence keratinocyte growth control.[20–22,27]

Culture Techniques

Murine Keratinocytes

Newborn mouse epidermis yields a large number of cells (5–10 × 10^6/epidermis), with a 30–40% plating efficiency. Mice are sacrificed by CO_2 narcosis 1–4 days postpartum (prior to the appearance of hair), soaked in Betadine for 5 min, rinsed twice in 70% ethanol (v/v), and kept on ice. Using an aseptic technique, limbs and tails are amputated, a longitudinal incision is made from tail to snout, and skin is peeled off the carcass using forceps (Fig. 1). Skins are stretched out with the dermis facing down in the bottom of 150-mm culture dishes, and stored at 4° until all skins have been removed. It is essential that the edges of each skin be flattened out, otherwise the skin may not float properly. Using forceps, each skin is carefully floated on the surface of an 0.25% trypsin solution (w/v) at 4°, with the epidermis facing upward, for 15–24 hr. Skins are individually transferred to a sterile surface and stretched with the epidermis facing down, and the epidermis is separated from the dermis using forceps, minced, and stirred in standard medium (see below), with the calcium adjusted to 1.4 m*M*, to release keratinocytes. To remove cornified sheets, the cell suspension is

[27] M. Reiss, V. F. Vellucci, and Z. L. Zhou, *Cancer Res.* **53,** 899 (1993).

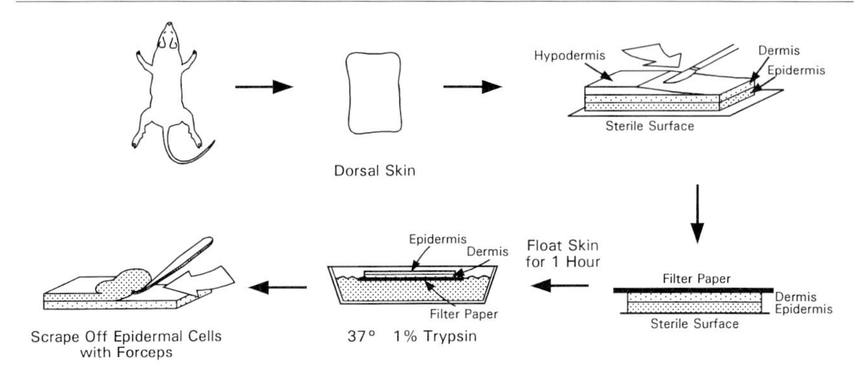

FIG. 2. Isolation of primary keratinocytes from adult mouse skin. After scraping keratinocytes off the dermis with forceps, cells are resuspended in standard medium, with Ca^{2+} adjusted to 1.4 m*M*, and filtered through nylon gauze as in Fig. 1. See text for additional details.

filtered through a sterile, 100-μm mesh nylon gauze (Nytex 157 mesh, Martin Supply Co., Baltimore, MD), and cells are plated at a density of $2–4 \times 10^5/cm^2$ in standard culture medium with Ca^{2+} adjusted to 0.1–0.3 m*M*. After 16–24 hr, cells are washed with Ca^{2+}- and Mg^{2+}-free phosphate-buffered saline (PBS), and switched to standard medium containing 0.05 m*M* Ca^{2+} (see below).[28] When epidermis from transgenic mice is utilized, intact skins can be stored overnight at 4° in standard medium containing 1.4 m*M* Ca^{2+}, while genotyping is being performed.

Adult mouse epidermal cells may be isolated from two sources: haired dorsal skin and hairless tail and ear skin. For dorsal skin, mice must be in the resting phase of the hair cycle, which can be determined by the absence of hair growth for 24–48 hr after clipping. Adult mice are sacrificed by CO_2 narcosis and dorsal skin is depilitated with Nair (Carter Products Inc., New York, NY). The entire mouse is washed in Betadine and 70% ethanol as described earlier. Dorsal and ventral skin is removed surgically and placed epidermis side down on a sterile surface (Fig. 2). The hypodermis is removed by vigorous scraping with a scalpel—a procedure essential for the successful separation of epidermis from dermis. The skin is cut into 3- to 4-cm^2 pieces, and autoclaved filter paper (Whatman No. 1, Wardstone, England) is placed on the exposed surface of the dermis as a support. The skin and filter paper are inverted onto the surface of 1% trypsin in PBS (2.5% trypsin stock, Bio-Whittaker, Walkersville, MD, Cat. No 17-160H) for 1 hr at 37°. Skins are then removed to a dry sterile surface, the epidermis

[28] H. Hennings, D. Michael, C. Cheng, P. Steinert, K. Holbrook, and S. H. Yuspa, *Cell* (*Cambridge, Mass.*) **19,** 245 (1980).

is scraped off the dermis with curved forceps, and the keratinocytes are filtered and resuspended as for newborn skin. The approximate yield from dorsal adult mouse skin is 5–10 × 10^6 cells/skin. Alternatively, nonhairy tail or ear skin is washed in Betadine and 70% ethanol.[29] A longitudinal incision is made through the tail length and skin is removed using forceps. The two skin flaps composing the ear must be separated carefully. The trypsinization of tail or ear skin requires 1.5 hr of flotation on 1% trypsin at 37° and yields approximately 10 × 10^6 keratinocytes per tail and 6 × 10^6 per pair of adult ears.

Adult keratinocytes do not attach well to culture dishes, and coating of plastic dishes with 30 μg/ml of collagen type I (Vitrogen 100, Celtrix Laboratory, Palo Alto, CA, Cat. No. PC0701), collagen type IV, or fibronectin is essential for the proper attachment and spreading of primary cells. To achieve maximal attachment and growth of adult and newborn keratinocytes, dishes with a keratinocyte extracellular matrix can be prepared from confluent cultures of newborn keratinocytes. Culture dishes previously plated with newborn keratinocytes are incubated with 0.025 *M* NH_4OH/0.5% Triton X-100 (v/v) in PBS for 1–5 min at room temperature, followed by several washes with 1 m*M* EDTA in PBS. Dishes can be stored with PBS at 4° for at least 3–4 weeks until used for the plating of freshly isolated keratinocytes.

Human Keratinocytes

Human keratinocytes can be prepared from fetal, adult, and newborn sources.[30,31] Newborn foreskin is a common source. The foreskin is spread, and the muscular and mucosal layers are completely removed from the hypodermis by scraping vigorously with a scalpel. The tissue is floated on 0.25% trypsin at 4° for 18–24 hr. The epidermis is then separated from the dermis by forceps, minced, and stirred in medium. The dissociated single cell suspension is filtered through Nytex gauze and centrifuged, and pelleted cells are resuspended and plated. Each foreskin yields 2–4 × 10^6 viable cells. Skin tissue specimens can also be obtained from surgical procedures, amputations, or suction blisters—a routine procedure used for allografts and autografts in clinical applications.[32] When small biopsies are obtained, floating on 0.25% trypsin for 1–2 hr at 37° is sufficient for the proper separation of epidermis from dermis. The development of defined medium

[29] T. Tennenbaum, S. H. Yuspa, and J. Kapitulnik, *J. Cell. Physiol.* **143,** 431 (1990).

[30] J. G. Rheinwald and H. Green, *Cell (Cambridge, Mass.)* **6,** 331 (1975).

[31] A. R. Haake and A. T. Lane, *In Vitro Cell Dev. Biol.* **25,** 592 (1989).

[32] M. R. Pittelkow and R. E. Scott, *Mayo Clin. Proc.* **61,** 771 (1986).

for optimal growth of human keratinocytes enables the growth of cells at clonal densities, prolongs the life span, and increases the growth rate without the need of 3T3 feeder layers (see below). Primary human keratinocyte cultures can be subcultured repeatedly using a dilute trypsin/EDTA solution, 0.1%/0.02%, in PBS.

In an effort to develop a cultured skin system that will mimic the characteristics of normal skin, cultures of primary keratinocytes are layered on a dermal equivalent. Dermis-like structures can be composed of organic deepidermized dermis kept at $-80°$ until use.[33] Alternatively, collagen gels or a combination of collagen lattices and fibroblasts can be utilized as collagen gels.[34] The mixture of collagen, fibroblast, and complete culture medium is cast in petri dishes. After 3 days, gelation and contraction of the dermal equivalent are completed, and keratinocytes are layered on top at the air–medium interface, resulting in a three-dimensional organ culture system.[34] These reconstituted skin equivalents stratify and differentiate similarly to normal skin with compartmentalized expression of markers specific to the different epidermal strata and polarized expression of basement membrane components.[34] This system enhances the ability to study the invasive characteristics of oncogene-transduced human keratinocytes and considers the influence of oncogenes on keratinocyte–fibroblast interactions.[21] This model was also reported to support the propagation of oncogenic human papilloma virus *in vitro*.[35]

Culture Media

In early studies, keratinocytes derived from mouse epidermis grew poorly *in vitro*. The discovery that Ca^{2+} is a key regulator of keratinocyte differentiation enabled the development of improved culture media. Proliferation of murine epidermal keratinocytes *in vitro* requires an extracellular Ca^{2+} concentration <0.10 m*M*, with higher Ca^{2+} levels inducing terminal differentiation.[28,36] Since serum is required for the optimal growth of murine keratinocytes, it must first be depleted of Ca^{2+} using either a chelating resin or dialysis. Fetal bovine serum (FBS) is passed through a column of Chelex 100 chelating resin (200–400 mesh, sodium form, Bio-Rad Laboratories,

[33] N. Basset-Seguin, J. F. Culard, C. Kerai, F. Bernard, A. Watrin, J. Demaille, and J. J. Guilhou, *Differentiation (Berlin)* **44,** 232 (1990).

[34] E. Bell, S. Sher, B. Hull, C. Merrill, S. Rosen, A. Chamson, D. Asselineau, and L. Dubertret, *J. Invest. Dermatol.* **81,** 2s (1982).

[35] S. C. Dollard, J. L. Wilson, L. M. Demeter, W. Bonnez, R. C. Reichman, T. R. Broker, and L. T. Chow, *Genes Develop.* **6,** 1131 (1992).

[36] S. H. Yuspa, A. E. Kilkenny, P. M. Steinert, and D. R. Roop, *J. Cell Biol.* **109,** 1207 (1989).

Richmond, CA, Cat. No. 142-2842), followed by 0.2 μm filter sterilization. Standard growth medium is prepared using Ca^{2+}-free Eagle's minimum essential medium (EMEM) with Earle's balanced salt solution, nonessential amino acids, and L-glutamine (Bio-Whittaker, Cat. No. 06-174D), supplemented with 8% Chelex-treated FBS, 0.25% penicillin–streptomycin solution (10,000 units/ml penicillin G sodium and 10 mg/ml streptomycin sulfate in 0.85% saline, GIBCO, Grand Island, NY, Cat. No. 600-514OAG), and the appropriate volume of $CaCl_2$ (280 mM stock) for a final concentration of 0.05 mM Ca^{2+}. The amount of $CaCl_2$ is determined for each batch of Chelex-treated serum by measuring the residual Ca^{2+} concentration in complete standard growth medium using atomic absorption spectrophotometry and adding sufficient $CaCl_2$ to a final concentration of 0.05 mM. Serum lots should be screened due to variability in supporting keratinocyte growth. When grown in standard medium and replenished three times a week, primary murine keratinocytes can be maintained as a proliferating monolayer for up to 2 weeks. In general, these cultures cannot be successfully passaged. Mouse epidermal keratinocytes can also be grown in a serum-free medium containing multiple defined supplements and bovine pituitary extract.[37] This medium has been reported to support clonal growth, long-term cultivation (25–30 population doublings), and repeated subculturing of murine keratinocytes.[37] Another serum-free nutrient-enriched medium has been developed for use with murine keratinocytes which does not require reduced levels of extracellular Ca^{2+}.[38]

Successful long-term cultivation of human keratinocytes was first achieved by growing cells on a feeder layer of irradiated mouse 3T3 fibroblasts in medium supplemented with 20% serum, hydrocortisone, and epidermal growth factor (EGF).[30] Serum-free media have subsequently been developed which support the growth of human keratinocytes for several passages, without the requirement of a feeder layer. Most of these are modifications of a formulation (MCDB 153) originally developed by Boyce and Ham,[39] containing reduced levels of extracellular Ca^{2+} and several additives. Cells derived from newborn or adult human skin are routinely grown in an enriched MCDB 153 medium containing 74 ng/ml hydrocortisone, 5 μg/ml insulin, 6.7 ng/ml triiodothyronine, 5 ng/ml EGF, 50 μg/ml bovine pituitary extract, and 0.09 mM Ca^{2+} (Keratinocyte-SFM, GIBCO BRL, Gaithersburg, MD, Cat. No. 320-7005PJ). Clonetics Corporation (San

[37] F. Bertolero, M. E. Kaighn, M. A. Gonda, and U. Saffiotti, *Exp. Cell Res.* **155,** 64 (1984).
[38] R. J. Morris, K. C. Tacker, J. K. Baldwin, S. M. Fischer, and T. J. Slaga, *Cancer Lett.* **34,** 297 (1987).
[39] S. T. Boyce and R. G. Ham, *J. Invest. Dermatol.* **81,** 33s (1983).

Diego, CA) offers a MCDB 153-based medium which can also be used, designated KGM (Cat. No. CC-3001). Primary human keratinocytes can be passaged several times when grown in these media.

Transfection and Selection Methods

Introduction of Foreign DNA

Foreign DNA can be introduced into keratinocytes using several methods, including calcium phosphate- and lipid-mediated transfection and retroviral infection. Because of the differentiating effect of extracellular Ca^{2+} on keratinocyte cultures,[28] the calcium phosphate method has been modified.[40] Murine cultures are maintained in standard 0.05 m*M* Ca^{2+} growth medium (see earlier), and culture medium is changed 1 day prior to transfection. Four hours prior to transfection, cultures are washed twice with PBS and incubated with 4.5 ml of standard growth medium containing 0.01 m*M* K^+ (prepared from K^+- and Ca^{2+}-free EMEM, Bio-Whittaker, Cat. No. 04-544D) per 60-mm dish of cells. DNA is diluted in 1 ml 0.25 *M* $CaCl_2$ and is precipitated with 1 ml of 2× HEPES-buffered saline, pH 7.04 (1 ml of 1 *M* HEPES, pH 7.04, 30 μl of 1 *M* $NaHPO_4$, and 2.5 ml of 2.5 *M* NaCl in 5 m*M* HEPES, pH 7.3, in 20 ml total volume), adding 3 drops at a time with gentle agitation. Precipitate (0.9 ml/60-mm dish) is added, and the dish is rotated to evenly distribute the precipitate. After an additional 4 hr, the cells are washed once with 3 ml PBS, treated for 3 min with 2 ml of 25% dimethyl sulfoxide (DMSO) in 0.01 m*M* K^+ medium, washed three times with 5 ml PBS, and incubated for 16 hr in fresh 0.01 m*M* K^+ medium before washing and replenishing with standard medium. Up to three plasmid constructs have been introduced in combination[41]; a total of 20 μg/60-mm dish of CsCl-purified plasmid is added using sheared salmon sperm DNA or noncoding plasmid vector as carrier DNA. Transfection of DNA for transient gene expression can be performed on primary cultures from newborn or adult mice within the first 6 days of culture with similar results. Cell lines are plated in standard medium ($1–2 \times 10^6$/60-mm dish) 1–2 days prior to transfection.

Several lipid-based methods of transfection have become available. These methods require less plasmid and are less laborious. Murine keratinocytes are washed once with serum-, Ca^{2+}-, and antibiotic-free EMEM and are covered with 2 ml of a lipid–DNA complex/60-mm dish for 6 hr [2 μg DNA and 6 μl LipofectAMINE (GIBCO BRL, Cat. No. 18324-012)

[40] J. R. Harper, D. A. Greenhalgh, and S. H. Yuspa, *J. Invest. Dermatol.* **91,** 150 (1988).

[41] T. Kartasova, D. R. Roop, K. A. Holbrook, and S. H. Yuspa, *J. Cell Biol.* **120,** 1251 (1993).

preincubated for 30–45 min in standard medium without serum or antibiotic], then washed and allowed to recover in standard medium for at least 12 hr before further experimentation. Best results are obtained if primary cultures are confluent at the time of lipofectamine-mediated transfection. Other liposome-mediated protocols have also been successfully applied for human keratinocyte transfections.[42] Both stable and transient gene expression can also be achieved following electroporation of human and mouse keratinocyte cell suspensions.[24,43,44] Human foreskin keratinocytes can be efficiently transfected with lipid- and calcium phosphate-mediated methods, but Polybrene-mediated transfection is recommended.[44] DNA (0.5–3 μg) is added to keratinocytes in the presence of 10–20 μg/ml Polybrene to increase the adsorption of DNA to the cell surface. After 6 hr, the DNA/Polybrene mix is removed, the uptake of adsorbed DNA is facilitated with a 3-min treatment with 25–30% DMSO, and fresh culture medium is added.[44,45]

Because of the higher efficiency of viral infection, retroviral vectors are better suited for establishing mass cultures of primary keratinocytes expressing exogenous DNA.[46] A replication-defective retrovirus restricts genetransfer to cultured cells and therefore simplifies the evaluation of *in vivo* tests of recipient cells. Moloney-based retroviral systems have been successfully employed for transducing DNA into keratinocytes.[47,48] Cultured mouse keratinocytes are washed with PBS and incubated with virus, at a multiplicity of infection of 1 : 1, in a total volume of 0.5 ml/60-mm dish in standard medium in the presence of 4 μg/ml Polybrene, tilting culture trays every 15 min to ensure that the cells remain moist.[47] After 1–1.5 hr, 2.5 ml of standard medium is added and replaced with fresh virus-free standard medium after 48–72 hr. Hair follicles can be infected in suspension following the same procedure but in a 50-ml tube, swirling every 15 min.[49] Murine keratinocytes can be infected with multiple retroviruses in combination. Human keratinocytes can be transduced by retroviral vectors using a

[42] X. F. Pei, J. M. Meck, D. Greenhalgh, and R. Schlegel, *Virology* **196,** 855 (1993).

[43] M. Reiss, M. M. Jastreboff, J. R. Bertino, and R. Narayanan, *Biochem. Biophys. Res. Commun.* **137,** 244 (1986).

[44] C. K. Jiang, D. Connolly, and M. Blumenberg, *J. Invest. Dermatol.* **97,** 969 (1991).

[45] J. S. Rhim, J. B. Park, and G. Jay, *Oncogene* **4,** 1403 (1989).

[46] J. A. Garlick, A. B. Katz, E. S. Fenjves, and L. B. Taichman, *J. Invest. Dermatol.* **97,** 824 (1991).

[47] D. R. Roop, D. R. Lowy, P. E. Tambourin, J. Strickland, J. R. Harper, M. Balaschak, E. F. Spangler, and S. H. Yuspa, *Nature* (*London*) **323,** 822 (1986).

[48] O. Danos and R. C. Mulligan, *Proc. Natl. Acad. Sci. U.S.A.* **85,** 6460 (1988).

[49] W. C. Weinberg, D. Morgan, C. George, and S. H. Yuspa, *Carcinogenesis* (*London*) **12,** 1119 (1991).

similar procedure, with Polybrene concentrations differing between investigators.[46,50]

Plasmid Choice and Selection of Stably Expressing Cells

Many constitutively active, virally derived promoters are highly expressed in human and mouse keratinocyte cultures, including those from CMV, SV40, and RSV[41,44,51]; however, expression driven by viral promoters is commonly modulated by extracellular Ca^{2+}.[51] A genomic promoter sequence has been defined within the mouse c-*ras*^{Ha} gene which demonstrates strong promoter activity in murine keratinocytes with less Ca^{2+} responsiveness.[52] Plasmid constructs encoding specific exogenous sequences regulated by the metallothionine gene promoter have also been used in mouse keratinocytes.[41] Promoter sequences from genes expressed specifically according to the differentiation state which have been utilized in mouse cell transfections include the 6000-bp promoter sequence of the human keratin 5 gene[53] and Ca^{2+}-inducible regulatory elements 3′ to the human keratin 1 coding sequence.[54] Regulatory sequences from the human keratin 1, bovine keratin 10, and human keratin 14 genes have been used to target the expression of several oncogenes and growth factors or their receptors to specific subpopulations within the epidermis of transgenic mice.[55–57]

For stable transfected keratinocyte cell lines, commonly used selectable markers include the genes for neomycin and hygromycin resistance in a 1:10 ratio with the sequence of interest. Stable transfectants are difficult to isolate from primary cultures of mouse keratinocytes. Primary mouse keratinocytes are very sensitive to these selecting agents (6 μg/ml G418, 0.5 μg/ml hygromycin), and resistant cells grow poorly at clonal density. Greater success has been achieved with established murine keratinocyte cell lines, which require from 40 to 200 μg/ml G418 and 3 to 8 μg/ml hygromycin to eliminate nonrecipients of these selectable markers.

[50] C. D. Woodworth, H. Wang, S. Simpson, L. M. Alvarez-Salas, and V. Notario, *Cell Growth Differ.* **4,** 367 (1993).

[51] J. I. Lee and L. B. Taichman, *J. Invest. Dermatol.* **92,** 267 (1989).

[52] R. Neades, N. A. Betz, X. Y. Sheng, and J. C. Pelling, *Mol. Carcinog.* **4,** 369 (1991).

[53] C. Byrne and E. Fuchs, *Mol. Cell. Biol.* **13,** 3176 (1993).

[54] C. A. Huff, S. H. Yuspa, and D. Rosenthal, *J. Biol. Chem.* **268,** 377 (1993).

[55] D. A. Greenhalgh, J. A. Rothnagel, X. J. Wang, M. I. Quintanilla, C. C. Orengo, T. A. Gagne, D. S. Bundman, M. A. Longley, C. Fisher, and D. R. Roop, *Oncogene* **8,** 2145 (1993).

[56] S. Werner, W. Weinberg, X. Liao, K. G. Peters, M. Blessing, S. H. Yuspa, R. L. Weiner, and L. T. Williams, *EMBO J.* **12,** 2635 (1993).

[57] R. Vassar, M. E. Hutton, and E. Fuchs, *Mol. Cell. Biol.* **12,** 4643 (1992).

Transfected human keratinocytes require 100–800 μg/ml G418[21,58,59] and 20 μg/ml hygromycin[24] for effective selection.

Markers of Neoplastic Transformation

In Vitro Markers

In cultures of normal mouse keratinocytes, raising extracellular Ca^{2+} from 0.05 to >0.10 m*M* triggers irreversible growth-arrest coupled to the expression of multiple epidermal differentiation markers. Ca^{2+} also induces differentiation of human keratinocytes at higher concentrations (0.3–2 m*M* Ca^{2+}). Induction of differentiation markers in response to Ca^{2+} is typically delayed or inhibited in keratinocytes transduced with certain oncogenes; moreover, in some cases, markers not normally expressed in epidermal cells are induced aberrantly (Table I).[60–75] In addition, formation of a cross-linked cornified cell envelope and detachment of cells from the culture substratum, which mark the final stage of squamous differentiation, are characteristically inhibited in neoplastic keratinocytes (Table I). The acquired resistance to Ca^{2+}-mediated differentiation has been used to select for oncogene-altered keratinocytes *in vitro*. Oncogenes are introduced into cultures of proliferating keratinocytes using techniques outlined earlier.

[58] C. D. Woodworth, S. Cheng, S. Simpson, L. Hamacher, L. T. Chow, T. R. Broker, and J. A. DiPaolo, *Oncogene* **7,** 619 (1992).
[59] L. Pirisi, S. Yasumoto, M. Feller, J. Doniger, and J. A. DiPaolo, *J. Virol.* **61,** 1061 (1987).
[60] C. Cheng, A. E. Kilkenny, D. Roop, and S. H. Yuspa, *Mol. Carcinog.* **3,** 363 (1990).
[61] M. Reiss, D. DiMaio, and T. A. Zibello, *Cancer Commun.* **1,** 75 (1989).
[62] T. S. Hronis, M. L. Steinberg, V. Defendi, and T. T. Sun, *Cancer Res.* **44,** 5797 (1984).
[63] N. Sheibani, J. S. Rhim, and B. L. Allen-Hoffmann, *Cancer Res.* **51,** 5967 (1991).
[64] X. F. Pei, P. A. Gorman, and F. M. Watt, *Carcinogenesis* (*London*) **12,** 277 (1991).
[65] C. Missero, C. Serra, K. Stenn, and G. P. Dotto, *J. Cell Biol.* **121,** 1109 (1993).
[66] C. Missero, E. Filvaroff, and G. P. Dotto, *Proc. Natl. Acad. Sci. U.S.A.* **88,** 3489 (1991).
[67] M. Diaz-Guerra, S. Haddow, C. Bauluz, J. L. Jorcano, A. Cano, A. Balmain, and M. Quintanilla, *Cancer Res.* **52,** 680 (1992).
[68] D. R. Henrard, A. T. Thornley, M. L. Brown, and J. G. Rheinwald, *Oncogene* **5,** 475 (1990).
[69] X. F. Pei, I. M. Leigh, and F. M. Watt, *Epithelial Cell Biol.* **1,** 84 (1992).
[70] L. Pirisi, K. E. Creek, J. Doniger, and J. A. DiPaolo, *Carcinogenesis* (*London*) **9,** 1573 (1988).
[71] P. Kaur and J. K. McDougall, *J. Virol.* **62,** 1917 (1988).
[72] M. Darmon, C. Delescluse, A. Semat, B. Bernard, J. Bailly, and M. Prunieras, *Exp. Cell Res.* **154,** 315 (1984).
[73] J. Taylor-Papadimitriou, P. Purkis, E. B. Lane, I. A. McKay, and S. E. Chang, *Cell Differ.* **11,** 169 (1982).
[74] C. Agarwal and R. L. Eckert, *Cancer Res.* **50,** 5947 (1990).
[75] E. K. Parkinson, P. Grabham, and A. Emmerson, *Carcinogenesis* (*London*) **4,** 857 (1983).

TABLE I
ONCOGENE-INDUCED PHENOTYPIC ALTERATIONS IN CULTURED EPIDERMAL KERATINOCYTES

Oncogene	Differentiation marker(s)	Species	Refs.
	Reduced or Delayed Expression of Differentiation Markers		
v-*ras*^{Ha}	Keratins K1 and 10	Mouse	60
BPV 1	Transglutaminase	Mouse[a]	61
HPV 18 E6/E7	K1, involucrin	Human	58
EBV-LMP, EJ *ras*^{Ha}	Involucrin	Human[b]	13
SV40	Involucrin	Human	62
HPV 16	Involucrin	Human	63, 64
E1A	Transglutaminase	Mouse	65
E1A	Transglutaminase	Mouse[c]	66
	Appearance of "Aberrant" Keratins		
v-*ras*^{Ha}	K8, K18	Mouse	60
T24 c-*ras*^{Ha}	K8 K18	Mouse[d]	67
v-*ras*^{Ha}	K19	Human	68
HPV 16	K18, K19	Human	69, 70
HPV 18	K19	Human	71
SV40	K8, K18, K19	Human	62, 72, 73
SV40 large T Ag	K7	Human	74

Inhibition of Squamous Differentiation

Oncogene	Differentiation stimulus	Species	Refs.
v-*ras*^{Ha}, v-*ras*^{Ki}	Ca^{2+}	Mouse	23
HPV 16	Suspension	Human	63
HPV 16, 18	Ca^{2+} + serum	Human	26
SV40	TPA	Human	75
SV40	Ca^{2+} ionophore	Human	2
HPV 18	Ca^{2+}, TPA	Human	71
EBV-LMP, EJ *ras*^{Ha}	Ca^{2+} ionophore	Human[b]	13

[a] BALB/MK cell line.
[b] SCC12F cell line.
[c] PAM212 cell line.
[d] MCA3D cell line; other studies used nonimmortalized, early passage epidermal keratinocytes.

When cultures are subsequently grown in medium with an increased Ca^{2+} level to induce terminal differentiation, "Ca^{2+}-resistant" foci emerge, representing the clonal expansion of single, oncogene-transduced keratinocytes (Table I). Oncogene-transduced keratinocytes are also resistant to other differentiation stimuli, including the phorbol ester TPA, Ca^{2+} ionophores, and growth in suspension (Table I). In addition to alterations in their ability to terminally differentiate, cultured keratinocytes expressing a variety of different oncogenes exhibit abnormalities in growth regulation (Table II).[76-83]

Growth in a semisolid medium is a sensitive assay in testing the anchorage-independent growth of malignant fibroblasts. However, in cultures of human and mouse keratinocytes, growth in suspension may induce the differentiation of malignant cell lines[63,84] as well as normal keratinocytes, and growth in semisolid medium generally represents a later stage of malignant conversion or invasion.[85,86] Keratinocytes (usually 3×10^5) are plated in complete medium (MCDB 153 for human and EMEM for mouse keratinocytes) containing 1.5% methylcellulose with standard Ca^{2+} concentrations. Cells are incubated at 37° for 4 days. The suspension is diluted in complete medium and plated on 60-mm dishes. The dishes are kept in culture for 10 days and subsequently are fixed with 3% formaldehyde and stained with Giemsa. Colonies with more than 16 cells are counted.[85]

In Vivo Markers

Two types of *in vivo* systems have been employed to analyze the tumorigenic properties of keratinocyte recipients of specific oncogenes: (1) subcutaneous injections and (2) skin grafting of the test cells onto the dorsal epidermis of an adult athymic or syngeneic adult or newborn mouse host (Table III).[87-93] With subcutaneous injections, only epithelial cells are intro-

[76] M. L. Steinberg and V. Defendi, *Proc. Natl. Acad. Sci. U.S.A.* **76,** 801 (1979).
[77] Y. Barrandon, J. R. Morgan, R. C. Mulligan, and H. Green, *Proc. Natl. Acad. Sci. U.S.A.* **86,** 4102 (1989).
[78] M. Durst, R. T. Dzarlieva-Petrusevska, P. Boukamp, N. E. Fusenig, and L. Gissmann, *Oncogene* **1,** 251 (1987)
[79] M. W. Appleby, I. M. Greenfield, T. Crook, E. K. Parkinson, and M. A. Stanely, *Oncogene* **4,** 1323 (1989).
[80] J. P. Falco, W. G. Taylor, P. P. DiFiore, B. E. Weissman, and S. A. Aaronson, *Oncogene* **2,** 573 (1988).
[81] A. B. Glick, M. B. Sporn, and S. H. Yuspa, *Mol. Carcinog.* **4,** 210 (1991).
[82] J. A. Pietenpol, R. W. Stein, E. Moran, P. Yaciuk, R. Schlegel, R. M. Lyons, M. R. Pittelkow, K. Munger, P. M. Howley, and H. L. Moses, *Cell (Cambridge, Mass.)* **61,** 777 (1990).
[83] M. Sebag, J. Henderson, J. Rhim, and R. Kremer, *J. Biol. Chem.* **267,** 12162 (1992).
[84] N. E. Fusenig, S. M. Amer, P. Boukamp, and P. K. Worst, *Bull. Cancer* **65,** 271 (1978).
[85] J. G. Rheinwald and M. A. Beckett, *Cell (Cambridge, Mass.)* **22,** 629 (1980).
[86] N. H. Colburn, W. F. Vorder Bruegge, J. R. Bates, R. H. Gray, J. D. Rossen, W. H. Kelsey, and T. Shimada, *Cancer Res.* **38,** 624 (1978).
[87] G. P. Dotto, R. A. Weinberg, and A. Ariza, *Proc. Natl. Acad. Sci. U.S.A.* **85,** 6389 (1988).
[88] J. L. Brissette, C. Missero, S. H. Yuspa, and G. P. Dotto, *Mol. Carcinog.* **7,** 21 (1993).
[89] D. A. Greenhalgh, D. J. Welty, A. Player, and S. H. Yuspa, *Proc. Natl. Acad. Sci. U.S.A.* **87,** 643 (1990).
[90] M. S. Lee, J. H. Yang, Z. Salehi, P. Arnstein, L. S. Chen, G. Jay, and J. S. Rhim, *Oncogene* **8,** 387 (1993).

TABLE II
ONCOGENE-INDUCED ALTERATIONS IN GROWTH REGULATION OF EPIDERMAL KERATINOCYTES *in Vitro*

Oncogene	Effect	Species	Refs.
SV40	Immortalization	Human	76
E1A	Immortalization	Human	77
HPV 16	Immortalization	Human	59, 78
HPV 18	Immortalization	Human	71
HPV 18 E6/E7	Immortalization	Human	24
Ad12-SV40	Immortalization	Human	15
Polyomavirus large T Ag	Immortalization	Rat	9
v-*fos*, EJ ras^{Ha}	Immortalization	Mouse	79
SV40	Reduced serum requirement	Human	2, 76
HPV 16	Reduced serum requirement	Human	70
v-*fos*, EJ ras^{Ha}	Reduced serum requirement	Mouse	79
v-ras^{Ha}, v-ras^{Ki}, v-*mos*, v-*erbB*, v-*fms*	EGF-independent	Mouse[a]	80
v-*fgr*	EGF/insulin-independent	Mouse[a]	80
v-ras^{Ha}	EGF-independent	Human	68
v-ras^{Ha}	TGFα expression increased	Mouse	81
c-ras^{Ha} (protooncogene)[b]	Hyperproliferation	Rat[c]	10
HPV 16, 18; SV40	Less responsive to growth inhibition by TGF-β	Human	82
E1A	Less responsive to growth inhibition by TGF-β	Mouse[d]	66
p53 (mutant)	Less responsive to growth inhibition by TGF-β	Mouse[a]	27
ras^{Ha}	Less responsive to growth inhibition by vitamin D_3	Human[e]	83

[a] BALB/MK keratinocyte cell line cultured in a defined medium.
[b] Overexpressed.
[c] FRSK cell line.
[d] PAM212 cell line.
[e] HPK1A cell line.

duced. Approximately 10^6–10^7 cells are injected in 0.1 ml PBS into the interscapular region of the host.[19] Most malignant keratinocytes can grow subcutaneously, whereas papilloma or benign tumor cells do not form tumors[94] (Table III).

[91] D. A. Greenhalgh and S. H. Yuspa, *Mol. Carcinog.* **1,** 134 (1988).
[92] C. M. Kim, J. Vogel, G. Jay, and J. S. Rhim, *Oncogene* **7,** 1525 (1992).
[93] E. Finzi, A. Kilkenny, J. E. Strickland, M. Balaschak, T. Bringman, R. Derynck, S. Aaronson, and S. H. Yuspa, *Mol. Carcinog.* **1,** 7 (1988).
[94] J. E. Strickland, D. A. Greenhalgh, A. Koceva-Chyla, H. Hennings, C. Restrepo, M. Balaschak, and S. H. Yuspa, *Cancer Res.* **48,** 165 (1988).

TABLE III
In Vivo EFFECTS OF ONCOGENES ON KERATINOCYTES

Oncogene	Keratinocyte recipient	Species	Transplant type	Phenotype	Refs.
ras	Primary	Mouse	Graft	Papilloma	47
		Mouse	Graft	Carcinoma[c]	87, 88
	RHEK-1[a]	Human	Subcutaneous	Carcinoma	15
	PA-PE[b]	Mouse	Graft	Carcinoma	20
	p117[b]	Mouse	Graft	Carcinoma	21
v-*fos*	Primary	Mouse	Graft	Normal skin	89
	RHEK-1	Human	Subcutaneous	Carcinoma	90
	308/SP1[b]	Mouse	Graft	Carcinoma	91
E1a	308/SP1	Mouse	Graft	Papilloma	91
c-*myc*	308/SP1	Mouse	Graft	Papilloma	91
p53mut	p117	Mouse	Graft	Dysplastic papilloma	21
neu	p117	Mouse	Graft	Dysplastic papilloma	21
fes, fms erbB, src	RHEK-1	Human	Subcutaneous	Carcinoma	3
HIVtat	RHEK-1	Human	Subcutaneous	Carcinoma	92
TGF-α	308/SP1	Mouse	Graft	Papilloma	93
v-*fos*/v-*ras*	Primary	Mouse	Graft	Carcinoma	89
v-*fos*/HPV18	Primary	Human	Subcutaneous	Carcinoma	42
HPV	Primary	Human	Subcutaneous	Nontumorigenic	25, 70, 71
HPV	Primary	Human	Subcutaneous graft under skin flap[d]	Dysplastic Epithelium	4
SV40	Primary	Human	Subcutaneous	Nontumorigenic	2
HHV-6	RHEK-1	Human	Subcutaneous	Carcinoma	12

[a] Continuous human cell line, contains Ad12-SV40.
[b] Papilloma cell lines, contain activated c-*ras*^{Ha}.
[c] Phenotype dependent on high v-*ras*^{Ha} expression.
[d] Cultured epithelial sheet implanted intact between dorsal musculature and skin.

Skin reconstitution or grafting is a more widely used method since it has the advantage of allowing the study of cells at intermediate stages of premalignant progression. In some studies the graft site on the host is prepared by inserting a ground glass disk (26 $\times$ 3 mm) between the skin and thoracic wall to induce a capsule of granulation tissue.[84] The disk is first inserted at an incision at the base of the tail and moved up to the center of the back. After 3–4 weeks the disk is removed and a silicone grafting chamber is held in place using surgical clips. The grafting chamber (Renner GMBH, 6701 Dannstadt-Schauerheim 1, Riedstrasse 6 Germany) consists of an upper hat-shaped dome with a central 3-mm hole and a lower

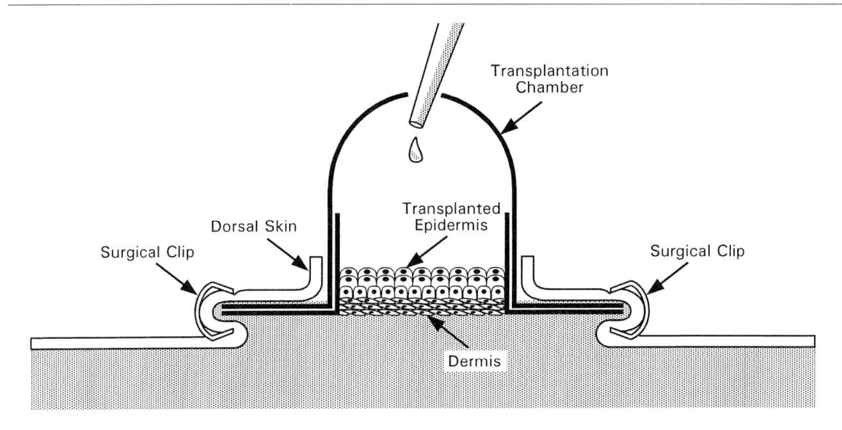

FIG. 3. Schematic cross section of silicone grafting chamber held in place under dorsal skin with surgical clips.

chamber which fits into the dome that is an open cylinder with a broad rim[95] (Fig. 3). More recent studies have shown that the granulation bed is not required for successful transplantation.[95,96] In this method the assembled chamber is put in place immediately before adding the test cells. The dorsal epidermis of anesthetized mice is wiped with Betadine and 70% ethanol. An approximately 1-cm-diameter circle of dorsal epidermis is removed, using curved scissors, and the wound is treated with an antibiotic spray (Polysporin, Burroughs-Wellcome, Research Triangle Park, NC). The assembled grafting chamber is inserted directly, with the rim of the chamber underneath the host skin, and is held in place with surgical clips. In both methods the graft chambers are removed 1 week after the addition of the transplanted cells, and the grafts are allowed to heal.

Two types of cell preparation techniques have been employed: (1) transplantation of an organotypic culture directly to the graft site, or (2) transplantation of suspensions of keratinocytes and fibroblasts. In the first method, either mouse or human keratinocytes are plated in petri dishes on a thin collagen gel prepared within the lower part of the grafting chamber. Dermal fibroblasts can also be plated directly into the collagen gel prior to plating the keratinocytes. After 2 days in culture, the top of the grafting chamber is added and the culture is transplanted to the host.[97] Alternatively,

[95] J. E. Strickland, A. A. Dlugosz, H. Hennings, and S. H. Yuspa, *Carcinogenesis* (*London*) **14,** 205 (1993).

[96] W. C. Weinberg, L. V. Goodman, C. George, D. L. Morgan, S. Ledbetter, S. H. Yuspa, and U. Lichti, *J. Invest. Dermatol.* **100,** 229 (1993).

[97] A. Bohnert, J. Hornung, I. C. Mackenzie, and N. E. Fusenig, *Cell Tissue Res.* **244,** 413 (1986).

suspensions of epidermal cell lines or primary epidermal cultures and primary dermal fibroblasts[96] can be combined, pelleted, and the thick slurry of cells added directly to the grafting chamber in place on the host, through the hole in the top dome.[95] Routinely 5–6 × 10^6 epidermal cells are combined with 8 × 10^6 fibroblasts. It is important to use dermal fibroblasts that have been cultured for at least 1 week to avoid generation of hair in the graft[96] and to remove any epithelial cells that could modulate the behavior of the test cells. Aseptic techniques must be used at all times in preparation of cells and during grafting since transplanted cells usually do not survive if the graft site becomes infected. Growth of transplanted tumor cells is usually apparent by 1 week after dome removal.

In addition to histological characterization, graft tumors can be analyzed using a series of markers that are associated with malignant progression of epidermal tumors. Tumors generated from oncogene-transduced grafted mouse keratinocytes have a similar pattern of marker expression as chemically induced mouse skin tumors.[98] Specific polyclonal antibodies have been developed which allow *in situ* localization of protein markers.[99] Keratins 1 and 10 are expressed in normal epidermis and well-differentiated papillomas, whereas keratin 13 is expressed during premalignant progression, often in cells that have lost expression of keratins 1 and 10.[100] Keratin 8 is expressed only in highly dysplastic papillomas and in squamous carcinomas.[100,101] Additionally, the $\alpha6\beta4$ integrin is a useful marker for the basal compartment which expands during the premalignant progression of mouse papillomas.[100] TGF-β1 and TGF-β2 immunostaining is also lost during the premalignant progression of chemically induced and grafted mouse skin tumors.[102] Human tumors also have a similar loss of keratins 1 and 10, and appearance of keratin 8.[103] However, expression of markers in grafted oncogene-transduced keratinocytes has not been characterized. In general, sections from tissues frozen in OCT (Miles, Elkhart, IN) give the best results in immunohistochemistry, although Carnoy's fixative (60% ethanol:30%

[98] T. Tennenbaum, S. H. Yuspa, A. Grover, V. Castronovo, M. E. Sobel, Y. Yamada, and L. M. De Luca, *Cancer Res.* **52,** 2966 (1992).

[99] D. R. Roop, C. K. Cheng, L. Titterington, C. A. Meyers, J. R. Stanley, P. M. Steinert, and S. H. Yuspa, *J. Biol. Chem.* **259,** 8037 (1984).

[100] T. Tennenbaum, A. K. Weiner, A. J. Bélanger, A. B. Glick, H. Hennings, and S. H. Yuspa, *Cancer Res.* **53,** 4803 (1993).

[101] F. Larcher, C. Bauluz, M. Diaz-Guerra, M. Quintanilla, C. J. Conti, C. Ballestin, and J. L. Jorcano, *Mol. Carcinog.* **6,** 112 (1992).

[102] A. B. Glick, A. B. Kulkarni, T. Tennenbaum, H. Hennings, K. C. Flanders, M. O'Reilly, M. B. Sporn, S. Karlsson, and S. H. Yuspa, *Proc. Natl. Acad. Sci. U.S.A.* **90,** 6076 (1993).

[103] I. M. Leigh, P. E. Purkis, A. Markey, P. Collins, S. Neill, C. Proby, M. Glover, and E. B. Lane, *in* "Skin Carcinogenesis in Man and in Experimental Models" (E. Hecker, E. G. Jung, F. Marks, and W. Tilgen, eds.), pp. 179–191. Springer-Verlag, New York, 1993.

chloroform : 10% v/v acetic acid) or ethanol fixation (70%) is also compatible with keratin immunostaining. Expression of the hair follicle-associated enzyme γ-glutamyl transpeptidase also occurs in squamous carcinomas, and this can be detected with a histochemical stain.[104] It is also useful to analyze cell proliferation in the tumors by immunohistochemical detection of bromodeoxyuridine (BrdU) incorporated into DNA.[105] Changes in number and the distribution of labeled nuclei are generally associated with premalignant progression and malignant conversion.[100,102] For BrdU staining, tumor-bearing animals are injected 1 hr before sacrifice with 250 μg/g of BrdU in 0.9% saline. Tissues fixed in either 70% ethanol or frozen in OCT can be used. Anti-BrdU antibodies are commercially available and suppliers provide the methods for use (Becton-Dickinson, San Jose, CA, Cat. No. 7580, and Zymed, San Francisco, CA, Cat. No. 93-3943).

Acknowledgment

The authors thank Margaret Taylor for secretarial assistance.

[104] C. M. Aldaz, C. J. Conti, F. Larcher, D. Trono, D. R. Roop, J. Chesner, T. Whitehead, and T. J. Slaga, *Cancer Res.* **48,** 3253 (1988).

[105] H. S. Huitfeldt, A. Heyden, O. P. F. Clausen, E. V. Thrane, D. Roop, and S. H. Yuspa, *Carcinogenesis (London)* **12,** 2063 (1991).

[2] Generation and Culturing of Precursor Cells and Neuroblasts from Embryonic and Adult Central Nervous System

By JASODHARA RAY, HEATHER K. RAYMON, and FRED H. GAGE

Introduction

The mechanisms for generation of the diversified cell types of the central nervous system (CNS) from the homogeneous population of neuroepithelial cells are not well understood. Proliferating cells can be found in different brain regions at specified times during development. In general, it is thought that this developmental wave of neurogenesis ends with a departure from the cell cycle and subsequent cellular differentiation. Observations have shown that the immature CNS is not the only site for cellular proliferation and neurogenesis; the adult brain may in fact retain this capacity, albeit at a lower level.[1-3] Diversified neuronal phenotypes (e.g., projection neurons, interneurons) arise from neuroblasts (precursor cells for neurons) which in turn arise from neural precursor cells. The precursor cells in the CNS can be stem cells or progenitor cells. In analogy to the hematopoietic system, it has been proposed that CNS stem cells are the original multipotent cell type, with extended self-renewing capacity, and the ability to give rise to progenitor cells.[4] The latter cell type, which has a limited life span, also has self-renewing properties and can give rise to either neuroblasts or glioblasts (precursor cells of neurons and glia, respectively).

The mechanism(s) for production of neuroblasts from precursor cells and their differentiation to diversified neuronal cell types can be better understood by studying individual cells in culture, where the environment can be easily manipulated. Since the establishment of long-term neuroblast cell cultures has been achieved only recently, previous studies have used immortalized neuronal cell lines.[5-7] The immortalization process arrests cells at specific stages of development and halts their terminal differentiation. Cells at these intermediate stages of development can be propagated indefinitely. Although immortalized cells offer a number of advantages, these cells do not always represent their primary counterparts. Thus, studies have devised methods to grow precursor cells which give rise to neuroblasts and glioblasts from embryonic and adult CNS *in vitro*.[1,8-11] In addition, long-term cultures of neuroblasts have been established from both embryonic and adult rat CNS.[1,2,12,13] The parameters important for generation and maintenance of neural cells in culture are discussed in this chapter and the strategies and techniques to culture both immortalized and primary precursor cells and neuroblasts are detailed.

[1] B. A. Reynolds and S. Weiss, *Science* **255,** 1707 (1992).
[2] L. J. Richards, T. J. Kilpatrick, and P. F. Bartlett, *Proc. Natl. Acad. Sci. U.S.A.* **89,** 8591 (1992).
[3] H. A. Cameron, C. S. Woolley, B. S. McEwen, and E. Gould, *Neuroscience* **56,** 337 (1993).
[4] D. J. Anderson, *Neuron* **3,** 1 (1989).
[5] C. L. Cepko, *Trends Neurosci.* **11,** 6 (1988).
[6] C. L. Cepko, *Annu. Rev. Neurosci.* **12,** 47 (1989).
[7] U. Lendhal and R. D. G. McKay, *Trends Neurosci.* **13,** 132 (1990).
[8] D. L. Stemple and D. J. Anderson, *Cell (Cambridge, Mass.)* **71,** 973 (1992).
[9] T. J. Kilpatrick and P. F. Bartlett, *Neuron* **10,** 255 (1993).
[10] B. A. Reynolds, W. Tetzlaff, and S. Weiss, *J. Neurosci.* **12,** 4564 (1992).
[11] A. L. Vescovi, B. A. Reynolds, D. D. Fraser, and S. Weiss, *Neuron* **11,** 951 (1993).
[12] J. Ray, D. A. Peterson, M. Schinstine, and F. H. Gage, *Proc. Natl. Acad. Sci. U.S.A.* **90,** 3602 (1993).
[13] J. Ray and F. H. Gage, *J. Neurosci.* **14,** 3548 (1994).

Neurogenesis during Development: Development of Neurons in Different Brain Regions/Choice of Age

The mammalian CNS is generated from the neural ectoderm which invaginates to form the neural tube. The neural tube consists of undifferentiated neuroepithelial cells which are in different phases of the mitotic cycle. Once a cell has gone through its final division it leaves the lining of the proliferative ventricular zone and migrates to its position within an individual brain region. In general, development of the mammalian brain proceeds in a caudal to rostral fashion, with phylogenetically older parts of the brain appearing earlier during ontogenesis. In most brain regions, neurogenesis takes place during a discrete prenatal period in which all the cells from that region are born. For example, the cells of the ventral mesencephalon are produced between E11 and E15, with the peak of cell division occurring between E11 and E13.[14] In contrast, neurogenesis takes place before and after birth in brain regions such as the cortex, hippocampus, and cerebellum. In the hippocampus, pyramidal neurons are produced during the prenatal period, whereas neurogenesis of the granule cells of the dentate gryus starts before birth and continues into adulthood.[3,15]

The choice of age for culturing cells derived from different brain regions depends on the cell of interest. Early embryonic time points (E10 in the mouse) have been used to culture undifferentiated neuroepithelial cells from the telencephalon and mesencephalon.[9,16] Additionally, striatal primordia from E13.5 to E14.5 rodent (the time point when neurogenesis begins in this region[17]) has been used to isolate stem or precursor cell populations.[1,11,18] In addition to isolating undifferentiated neuroepithelial cells *in vitro,* conditions for establishing neuroblast cell cultures derived from different regions of embryonic rat brain have been established. Establishment of short-term cultures of neuroblasts from the cerebral hemispheres (E13) or spinal cord (E14) has been described.[19,20] We have established long-term cultures of neuroblasts derived from different embryonic brain regions. Cells were isolated during periods of neurogenesis from the hippocampus,[12] spinal cord,[13] and ventral mesencephalon (VM)[21] at E16–18, E14–16, and E14, respectively. Based on differences in the development

[14] J. M. Lauder and F. E. Bloom, *J. Comp. Neurol.* **155,** 469 (1974).
[15] S. A. Bayer, *J. Comp. Neurol.* **190,** 87 (1980).
[16] M. Murphy, J. Drago, and P. F. Bartlett, *J. Neurosci. Res.* **25,** 463 (1990).
[17] S. A. Bayer, *Int. J. Dev. Neurosci.* **2,** 163 (1984).
[18] E. Cattaneo and R. McKay, *Nature* (*London*) **347,** 762 (1990).
[19] J. C. Deloulme, J. Baudier, and M. Sensenbrenner, *J. Neurosci. Res.* **29,** 499 (1991).
[20] C. Gensburger, G. Labourdette, and M. Sensenbrenner, *FEBS Lett.* **217,** 1 (1987).
[21] H. K. Raymon, J. Ray, D. A. Peterson, and F. H. Gage, *Soc. Neurosci. Abstr.* **19,** 654 (1993).

of cells in discrete brain regions, it is important to determine the time of neurogenesis for the brain region and the cell type of interest.

Factors That Influence Generation and Survival of Precursor Cells and Neuroblasts

Neural tissue is composed of not only neurons and glial cells but also other nonneuronal cells, connective and vascular tissues. When neural tissue is removed from the brain, dissociated, and transferred to culture conditions, the cells no longer retain their three-dimensional structures due to loss of anchorage, physiological connections with other cells, and the humoral environment. *In vitro* culture conditions have been developed to provide artificial substratum for anchorage and to imitate the humoral environments through the use of culture medium and gas exchange. Exogenous factors like substratum, culture medium, gaseous environment, and other parameters play an important role in the survival, proliferation, and differentiation of precursor cells and neuroblasts. Prior to establishing cell cultures, all the parameters for generating an optimum environment for the cells should be considered.

Effects of Substratum

The composition of the substratum is important for the adhesion, survival, proliferation, and differentiation of neural cells. Neural cells can be cultured with or without precoated substratum in serum-containing medium. Components in serum facilitate the attachment of cells to uncoated plastic. However, culturing in serum-free medium requires the presence of substratum for efficient cell attachment. Polyornithine (PORN), poly-L-lysine (pLL), poly-D-lysine (pDL), and PORN/laminin are the most common substratum used for neural cultures. Attachment and clustering of cells to substratum are variable depending on the composition and pH of the buffer used in the coating procedure. Neuronal cells tend to form clumps when plated on dishes precoated with pLL dissolved in water, whereas cells cultured in dishes precoated with pLL in 0.1 *M* boric acid–NaOH buffer, pH 8.4, remain mostly isolated.[19] In addition to survival, proliferation, and differentiation, composition of the substratum is important for the fate choice of a precursor cell. For example, subclones of neural crest stem cells established on fibronectin (FN) generate neurons when plated on FN/pDL but only generate astrocytes when plated on FN-coated substratum.[8]

Methods

Methods for Coating Culture Substrata

All solutions and buffers used for coating dishes should be sterile, and the procedure is usually done in a laminar flow hood. The most commonly used methods are outlined next.

Coating with Poly-D-lysine

Although the method for coating with pDL[22] is described here, a similar method can be used to coat plates with pLL.

1. Dissolve pDL (Sigma, St. Louis, MO) in sterile water or in sterile 0.1 *M* boric acid–NaOH buffer, pH 8.4, to make a stock solution of 1.0 mg/ml. Filter through a 0.22-μm filter (Millipore, Bradford, MA). Store in small aliquots at $-20°$. Dilute the stock solution in appropriate buffer or water to make a solution of the desired concentration. The concentrations of pDL may vary from 10 to 50 μg/ml.
2. Add enough pDL solution into the plates to cover the surface and allow to incubate at 37° for 2–24 hr. Wash plates with sterile water three to four times and allow to dry. Plates can be used for several weeks.

Coating Plates with Polyornithine

1. Make a 1- to 10-mg/ml stock solution of polyornithine hydrobromide (Sigma) in sterile water or in 0.1 *M* boric acid–NaOH buffer, pH 8.4.[23] Filter through a 0.22-μm filter. Store aliquots at $-20°$. Dilute the stock to make a 10-μg/ml solution. Add enough PORN to entirely cover the surface of the dish and incubate at room temperature for 24 hr.
2. Wash with sterile water two to three times, cover with enough water and store at $-20°$ sealed in plastic bags for future use. If coating with laminin, wash one to two times with phosphate-buffered saline (PBS).
3. Alternatively, after coating with PORN, the dishes can be air dried at room temperature in a laminar flow hood. Before plating the cells, wash plates two to three times with water followed by a wash with the medium.

Coating with Laminin

1. Make a 5-mg/ml stock solution of laminin (mouse or rat, GIBCO/BRL, Gaithersberg, MD) in sterile PBS. Store in small aliquots at $-80°$. Laminin should not be freeze-thawed more than once or twice.

[22] B. H. Juurlink, *in* "Protocols for Neural Cell Cultures" (S. Fedroff and A. Richardson, eds.), p. 49. Humana Press, Totowa, NJ, 1992.

[23] B. Rogister and G. Moonen, *in* "Protocols for Neural Cell Cultures" (S. Fedroff and A. Richardson, eds.), p. 10. Humana Press, Totowa, NJ, 1992.

2. Add a solution of 5 μg/ml laminin in PBS to PORN-coated plates. Incubate at 37° for 24 hr. Store the plates sealed in plastic bags at −20°.

3. Alternatively, wash plates two to three times in PBS, cover with PBS, and store in plastic bags at −20°. Do not let the plates dry out. The plates can be stored for 1–2 months.

Tissue Procurement

The following procedure is used in the authors' laboratory for obtaining tissues from fetal rat CNS. Deeply anesthetize timed pregnant Fisher 344 rats with the following anesthesia cocktail: ketamine (44 mg/kg), acepromazine (4.0 mg/kg), and Rompun (0.75 mg/kg). Remove uterine horns by cesarean section and place immediately on ice. Remove embryos from their individual sacs and place in sterile Dulbecco's PBS (PBS-D). Measure the crown–rump length to verify the embryonic age.

Dissection of Tissues from Different Regions of Brain

The clean dissection of tissues is extremely important in obtaining pure cell cultures. The presence of connective tissue can increase the nonneuronal cell population which will eventually overtake the cultures. Dissection procedures for a few regions of the rat CNS are described next.

Hippocampus

Long-term hippocampal cell cultures are derived from rats between E16 and E18. The brain from each embryo is removed and placed in sterile Dulbecco's PBS-D. Using forceps to stabilize the brain, the cortex on one side is peeled back from the midline and laid out flat. The hippocampus, lying just underneath the cortex, is removed with a pair of erridectomy scissors and a sharp forcep, using the scissors to cut out the hippocampus and the forceps to hold onto the tissue piece. The procedure is repeated on the contralateral side. Once dissected free from the embryo, any meningeal membranes or blood vessels that remain attached to the tissue are removed with sharp forceps.

Spinal Cord

Spinal cord cell cultures are established from rats between E14 and E16. The embryo is placed on its side in sterile PBS-D. An initial cut is made lateral to the spinal canal with a pair of erridectomy scissors. Subsequent cuts are made on the opposite side and at the level of the cervical and lower lumbar regions. Once freed from the embryo, residual connective tissue is removed from the spinal cord. At the earlier time points, the connective tissue is firmly attached to the spinal cord, making it difficult to remove without damage to the cord.

Ventral Mesencephalon

Long-term ventral mesencephalon (VM) cell cultures are established from E14 rats. The embryo is placed on its side in sterile PBS-D. The soft skull covering the mesencephalic flexure is peeled away. The block of tissue lying immediately dorsal to the mesencephalic flexure is removed by making two cuts on either side of the flexure and a third cut above the flexure. Mesenchymal tissue is carefully removed from the tissue piece with sharp forceps. Ventral mesencephali from 25 to 30 embryos are pooled to increase the cell yield following dissociation.

Medium for Neuronal Cell Cultures

The composition of the medium influences the survival, proliferation, and differentiation into neuronal cells *in vitro*. Neuronal cultures are generally maintained at pH 7.2–7.6 and at the appropriate osmolarity. Synthetic base media containing different amounts of inorganic salts, vitamins, amino acids, and buffering agents have been designed to optimize the survival of cultured neuronal cells.[24,25] The most commonly used base medium for culturing neuronal cells is Dulbecco's modified Eagle's medium (DMEM), often combined (1 : 1, v/v) with Ham's F12 medium. Cells are grown in this medium either in the absence (serum-free) or in the presence of serum [usually 10% fetal bovine serum (FBS), but horse serum has also been used]. One disadvantage of growing cells in serum is that the components of the serum are undefined, making analysis and interpretation of results difficult. When exogenous factors are added to serum-containing media, it is not clear whether the effects are due to the added factors or to an interaction of the factor with substances present in the serum. The alternative to using serum is to supply the cells with known amounts of ingredients that support cell survival. A number of serum-free media supplements containing hormones, transport proteins, and vitamins have been developed; N2[25] is the most common (Table I). To make N2 medium, DMEM : F12 medium containing L-glutamine is filtered through a 0.22-μm filter (Corning, Inc., Corning, NY) and the N2 supplement is added to the medium. This medium can be stored for 1 month at 4°. The N2 supplement

[24] R. P. Saneto and J. De Vellis, *in* "Neurochemistry: A Practical Approach" (A. J. Turner and H. S. Bachelard, eds.), p. 27. IRL Press, Oxford, 1987.

[25] J. E. Bottenstein and G. Sato, *Proc. Natl. Acad. Sci. U.S.A.* **76,** 514 (1979).

TABLE I
COMPOSITION OF N2 MEDIUM[a]

Components	Final composition	Concentrations of stock solution	Amount added/100 ml medium
Ham's F-12/DMEM with 2.5 m*M* L-glutamine, 3.1 g/liter glucose[b]	50% (v/v) of each	100%	100 ml
Insulin (bovine)[c]	5 μg/ml	2.5 mg/ml in 10 m*M* HCl	0.2 ml
Transferrin (human, iron-free)[d]	100 μg/ml	20 mg/ml in water	0.5 ml
Mixture of progesterone/ putresceine hydrochloride/ sodium selenite (Na_2SeO_3) (PPN)[e]	20 n*M*/100 μ*M*/30 n*M*	4 μ*M*/20 m*M*/6 μ*M*	0.5 ml

[a] Adapted from Bottenstein and Sato.[25]
[b] Irvine Scientific (Santa Ana, CA).
[c] Sigma; store desiccated below −20°.
[d] Sigma; store desiccated at 0–5°.
[e] PPN is made of 4 μ*M* progesterone (Sigma), 20 m*M* putrescine dihydrochloride (Sigma), and 6 μ*M* Na_2SeO_3 (Sigma) in DMEM. This sterile solution when diluted 1:200 yields the correct final concentration. To make 50 ml PPN solution, add 161.5 mg putrescine; 0.25 ml Na_2SeO_3 solution (1.2 m*M* stock, filter sterilized, stored at −70°); 0.125 ml progesterone (1.6 m*M* solution in 100% ethanol, must be made fresh); 50 ml DMEM, filter sterilize, store in small aliquots at −70°.

can be made and stored as described in Table I or a 100× N2 supplement can be purchased from GIBCO/BRL.

Cell cultures can be preenriched for neuronal cells. Dissociated neuronal and nonneuronal cells attach differentially to substratum in the presence of serum. Using this property, methods have been developed to obtain cultures enriched in neurons.[26,27] Dissociated, single cells are plated in serum-containing medium on uncoated plastic plates for 2–3 hr. Nonneuronal cells attach to plastic substratum faster and more firmly than the neuronal cells. The best enrichment of neuronal cells by this method depends on the serum batch and the duration of preplating.

At the present time, there are no specific media formulations for growing precursor cells, neuroblasts, or neurons. Cells have been grown both in serum-containing[2,9] and serum-free media.[12,13,24,25] Mitogenic growth factors [epidermal growth factor (EGF) or basic fibroblast growth factor (FGF-2)] have been used for culturing precursor cells and neuroblasts. The addition of

[26] G. R. Hanson, P. L. Iversen, and L. M. Partlow, *Dev. Brain Res.* **3,** 529 (1982).
[27] S. D. Skaper and S. Varon, *Dev. Biol.* **98,** 257 (1983).

growth factors (EGF or FGF-2) to the base medium in the absence or presence of serum can have profound effects on the *in vitro* proliferation and differentiation of precursor cells or neuroblasts. The presence of serum is essential for the differentiation of precursor cells by FGF-2, indicating that other factors in collaboration with FGF-2 can influence the fate choice of these cells.[9,11]

Preparation of Primary Cultures

1. Dissect out the regions of interest from embryonic or postnatal rat brains and transfer to 35-mm petri dishes containing Ca^{2+}, Mg^{2+}-free Hanks' balanced salt solutions, pH 7.2 (HBSS; Irvine Scientific, Santa Ana, CA), or PBS-D. Remove the meninges from the surface of the tissue by gently pulling with sharp forceps and then transfer the tissues to a 15-ml centrifuge tube (Corning Inc., Corning, NY).

2. Add ~5 ml HBSS or PBS-D to the tube and resuspended the tissues by tapping the tube gently and wash three to four times by centrifugation at 1000*g* for 3 min.

Tissue Dissociation

Mechanical Method[12]

1. After the final wash, resuspend the tissue in 1 ml culture medium and triturate gently with a large to medium bore (1.0–1.5 mm) Pasteur pipette to break the tissue pieces (about 20×) and with a medium to fine bore (0.5–1.0 mm) Pasteur pipette (10×) to make a single cell suspension. Avoid forming bubbles. Wash tissues one to two times with the culture medium to remove debris and make a final single cell suspension by trituration with a fine bore Pasteur pipette (5–10×).

2. Make an appropriate dilution of the cell suspension and mix an aliquot of the suspension with trypan blue (Sigma). Cell viability is determined by the exclusion of trypan blue by living cells. Inject ~15 μl of cell suspension slowly to fill the hemocytometer chamber by capillary action. Count cells not stained by trypan blue and determine the total cell number. Adjust the cell suspension to appropriate cell density and plate onto prepared dishes. Although the cell attachment can vary considerably between cultures, a minimum density is needed for the survival of neural cells.

Enzymatic Dissociation

a. Trypsin Digestion[28]

1. Place tissue explants (or small pieces of tissues) in HBSS or L-15 medium containing 0.25% trypsin (Worthington Biochemicals, Freehold, NJ). The trypsin solution should be made up fresh or from a stock solution stored at −70°.

2. Incubate the tissues at 37° for 30–45 min with gentle rotation or occasional shaking. Wash three to four times with plating medium. Resuspend tissues in 1 ml plating medium and triturate with a fire-polished large to medium bore Pasteur pipette (5×). Avoid forming bubbles by not squirting the last few microliters of the cell suspension. Allow the fragments to settle.

3. Remove the cell suspension to a new tube and add 1 ml medium to the tissue chunks left behind in the first tube. Triturate the chunks as described in step 2. Combine the cell suspensions. The remaining chunk of tissues can be removed by filtration through a nylon mesh (pore size 15 μm, Nitex, TETKO, Inc., Des Plains, IL) or by passing progressively through 18- to 24-gauge hypodermic needles. Triturate with a fine bore Pasteur pipette to make an even single cell suspension. Plate onto prepared tissue culture plates.

b. Papain–Protease–DNase (PPD) Digestion[29,30]

1. Prepare PPD solution as follows:
 0.01% (w/v) papain (Worthington Biochemicals, Freehold, NJ)
 0.1% (w/v) neutral protease (Dispase, grade II, Boehringer Mannheim, Indianapolis, IN)
 0.01% (w/v) DNase I (Worthington Biochemicals)

Dissolve in HBSS (without Ca^{2+}, Mg^{2+}). Supplement with 12.4 mM $MgSO_4$. Filter sterilize through a 0.22-μm filter. Store in 5-ml aliquots at −20°.

2. Put dissected tissues in HBSS or PBS-D and wash three times with 5 ml buffer. Remove the final wash by aspiration and add 5 ml PPD solution. Resuspend tissue and incubate for 15 min at 37°.

3. Triturate five times with a fire-polished, medium bore (1.0–1.5 mm) Pasteur pipette; incubate 30 min at 37°. Triturate ten times and incubate 30 min at 37°. Triturate an additional five to ten times with a fire-polished, fine bore (0.5–1.0 mm) Pasteur pipette. Centrifuge at 1000g, 3–5 min.

[28] M. I. Johnson and R. P. Bunge, *in* "Protocols for Neural Cell Cultures" (S. Fedroff and A. Richardson, eds.), p. 26. Humana Press, Totowa, NJ, 1992.

[29] R. M. Landon and R. J. Robins, this series, Vol. 124, p. 412.

[30] H. K. Raymon and F. Leslie, *J. Neurochem.* **62,** 1015 (1994).

TABLE II
RETROVIRAL VECTORS AND ONGOGENES USED FOR IMMORTALIZATION OF NEURAL CELLS[a]

Retroviral vectors[b]	Oncogenes	Refs.
pZIP-Neo SV (X)	Avian myc	31
LTR-"oncogene"-3′ss-neo-SVori-pBRori-LTR	Adeno E1a 12s	31, 32
	Adeno E1a 13s	31, 32
	Polyoma large T antigen	31, 32
	SV40 large T antigen	31, 32
	SV40 T tsA58 (temperature-sensitive)	33, 34
pneoMLV	Avian myc	31
LTR-"oncogene"-SVpro-neo-pBRori-LTR		
pLJ	Mouse N-myc	31, 35
LTR-"oncogene"-pBRori-SVpro-neo-LTR		

[a] Adapted from Ryder *et al.*[31]
[b] LTR, long terminal repeat; 3′ss, 3′ splice site; SVori, SV40 origin of replication; pBRori, pBR origin of replication; and SVpro, SV40 early promoter and origin of replication.

4. Remove the PPD solution by aspiration, add 5 ml DMEM/F-12-DNase solution [0.01% (w/v) DNase in DMEM/F-12]. Incubate for 15–30 min at 37°, triturate, and centrifuge as described in step 3.
5. Aspirate the medium and resuspend cells in the culture medium (2 ml). Count cells using the trypan blue exclusion method. Adjust the cell concentrations and plate onto prepared plates.

Generation of Immortalized Neural Cell Cultures

Before the development of methods for long-term culturing of neuroblast cells, which give rise to neurons in culture, establishment of such cultures posed a problem due to lack of or slow proliferation of neuronal cells and rapid growth of nonneuronal cells in cultures which eventually outgrew the neurons. As an alternative, immortalized neuronal cell lines have been generated either by transduction of oncogenes (Table II)[31–35] into precursor cells or neuroblasts[5–7] or by fusion of tumor lines with primary cells.[36] A description of procedures for immortalization of cells by both methods is presented next.

[31] E. F. Ryder, E. Y. Synder, and C. L. Cepko, *J. Neurobiol.* **21,** 356 (1990).
[32] C. Evrard, E. Galiana, and P. Rouget, *EMBO J.* **5,** 3157 (1986).
[33] P. S. Jat and P. A. Sharp, *Mol. Cell. Biol.* **9,** 1672 (1989).
[34] K. Frederiksen, P. S. Jat, N. Valtz, D. Levy, and R. McKay, *Neuron* **1,** 439 (1988).
[35] P. F. Bartlett, H. H. Reid, K. A. Bailey, and O. Bernard, *Proc. Natl. Acad. Sci. U.S.A.* **85,** 3255 (1988).
[36] H. J. Lee, D. N. Hammond, T. H. Large, and B. H. Wainer, *Dev. Brain Res.* **52,** 219 (1990).

1. Make a single cell suspension from dissected tissues. Plate 5×10^4 to 2×10^5 cells/cm^2 onto PORN/laminin-coated plates in DMEM + 10% FBS and incubate at 33°.

2. The oncongenes most commonly used for immortalization are *myc* oncogenes,[31,32] the SV40 large T antigen,[31] or a temperature-sensitive hybrid of the SV40 T antigen (SV40 T tsA58) (Table II).[33,34] Cells infected with SV40 T tsA58 grow at 33° (permissive temperature) but not at 39° (nonpermissive temperature). The retroviral vector construction and creation of packaging lines have been described in detail elsewhere.[37,38] The retroviral vectors contain the neomycin-resistant gene for easy selection of stably infected cells. Incubate the packaging cell line overnight in DMEM + 10–20% FBS. Collect the conditioned medium and filter the medium through a 0.45-μm filter.

3. The day after plating, incubate the cells for 2 hr with conditioned medium collected from the packaging cell lines (step 2) in the presence of Polybrene (8 μg/ml; Sigma). Replace the virus containing medium with fresh medium and incubate at appropriate temperatures. After 1–2 days, start the selection by growing cells in the presence of 200 μg/ml G418 (neomycin analog Genticine, GIBCO). Change medium every 3–5 days.

4. Pick colonies (usually appear in 3–4 weeks) separately using cloning rings and expand. For long-term storage, $\sim 10^6$ cells/cryovial (Nunc, Naperville, IL) in DMEM + 10–30% FBS + 10% dimethyl sulfoxide (DMSO) or 10% glycerol can be frozen in liquid nitrogen.

Generation of Immortalized Neuronal Cell Lines by Fusion with Tumorogenic Lines

Somatic cell fusion methods have been used to immortalize neuronal cells and generate cell lines.[36] The advantage of this method is that postmitotic cells can be immortalized.

1. Make a single cell suspension from fetal or adult tissues (septal or hippocampal) as described earlier. For postnatal tissue, centrifuge cells at the buoyant density to enrich for viable, single cells. Resuspend cells in phytohemagglutinin (PHAP, Difco Laboratories, Detroit, MI). This treatment enhances the adhesion of cells to the neuroblastoma cells, N18TG2. This is a HPRT-deficient mutant cell line and the neuroblastoma cells that did not fuse with primary cells can be eliminated from cultures by

[37] C. L. Cepko, E. F. Ryder, C. P. Austin, C. Walsh, and D. M. Fekete, this series, Vol. 225, p. 933.

[38] M. D. Kawaja, L. J. Fisher, M. Schinstine, H. A. Jinnah, J. Ray, L. S. Chen, and F. H. Gage, *in* "Neural Transplantation" (S. B. Dunnett and A. Bjorklund, eds.), p. 20. Oxford University Press, New York, 1992.

supplementing the medium with HAT (100 μM hypoxanthine, 0.4 μM aminopterin, and 16 μM thymidine). The neuroblastoma line is usually grown in DMEM + 10% FBS.

2. Layer hippocampi-PHAP cells on logarithmically growing neuroblastoma cells. After 15 min, remove the unattached cells by aspiration. Add 50% polyethylene glycol (PEG 1000, KochLight) in DMEM (v/v) to initiate cell fusion. Remove the solution after 40 sec and, after 60 sec, wash cells and incubate the cultures in DMEM + 10% FBS overnight. Replate cells in HAT-containing medium to select for growth. Remove aminopterin after 4–8 weeks and hypoxanthine and thymidine 4 weeks later.

3. Cells can be cloned, subcloned, and maintained as described earlier.

Immortalized hippocampal and septal cell lines have been generated by this method from embryonic and young mice.[36] Phenotypically stable hybrid cell lines express neuronal markers and exhibit electrophysiological properties.

Culturing of Precursor Cells from Embryonic or Adult Brains Using Growth Factors

Neural precursor cells give rise to neuroblasts or glioblasts under appropriate culture conditions. Culturing of precursor cells from fetal or adult brain with mitogenic growth factors (EGF or FGF-2) has been described[1,2,9–12] Although EGF has been used in serum-free culture conditions, FGF-2 has been used in 10% FBS-containing medium. EGF-generated clonal cultures gave rise to both neurons and astrocytes, but FGF-2-generated cultures produce either pure neuronal clones or clones containing both neurons and astrocytes.[9,11,16]

Embryonic Tissues[10]

Suspend striatum (E14 mouse) in DMEM/F-12 medium and mechanically dissociate the tissue as described earlier. Plate cells on PORN-coated glass coverslips in a 24-well plate at a density of 2500/cm^2 in DMEM/F-12 medium containing the N2 supplement (the concentrations of insulin and putrescine are 25 μg/ml and 60 μM instead of 5 μg/ml and 100 μM, respectively). Add EGF (Collaborative Research) to the N2 medium at 20 ng/ml prior to use. However, the addition of EGF at the time of plating is not required and can be added up to 7 days *in vitro* (DIV). Change medium after 10–14 DIV and then every 2–4 days thereafter. Using this method, cells have been maintained up to 25 days in culture.

Adult Brain[1]

1. Dissect the striatum from mouse brain (3–18 months old), cut 1-mm coronal sections, and transfer into normal artificial cerebrospinal fluid (aCSF) containing 124 m*M* NaCl, 5 m*M* KCl, 1.3 m*M* $MgCl_2$, 2 m*M* $CaCl_2$, 26 m*M* $NaHCO_3$, and 10 m*M* glucose (pH 7.35, ~280 mosmol) aerated with 95% O_2–5% CO_2 at room temperature.

2. After 15 min, transfer the tissues to a spinner flask with a magnetic stirrer (Bellco glass) containing low Ca^{2+} aCSF (composition same as aCSF except it contains 3.2 m*M* $MgCl_2$ and 0.1 m*M* $CaCl_2$, and is aerated at 32–35° with 95% O_2–5% CO_2) and 1.33 mg/ml trypsin (9000 benzoyl-L-arginine ethyl ester units/mg), 0.67 mg/ml hyaluronidase, and 0.2 mg/ml kynurenic acid. After 90 min, transfer tissue sections to normal aCSF, incubate for 5 min, and then triturate.

3. Transfer tissues to DMEM/F-12 medium containing 0.7 mg/ml ovomucoid (Sigma) and triturate with a fire-polished Pasteur pipette. Plate cells onto uncoated 35-mm plates at 1000 cells/plate and culture in serum-free medium containing 20 ng/ml EGF as described for embryonic cells. Cells initially attach on the substratum, but by 6–8 DIV the proliferating cell clusters detach and grow in floating spheres. The plating of cells on PORN-coated substratrum inhibits cell proliferation.

4. If it is necessary for cells to be attached to the substratum, proliferating cell clusters can be transferred with a large-bore Pasteur pipette and plated onto PORN-coated plates for 1 hr for adhesion and growth.

Short-Term Cultures of Neuronal Cells

A variety of media (both serum-containing and serum-free) and culture conditions have been used for culturing neuronal cells for a short period of time. Under these culture conditions cells survive for 1–4 weeks. Briefly, fetal tissues are washed in HBSS and then dissociated as described earlier. A single cell suspension is adjusted to the desired concentration and plated. Approximately, 2×10^5–1×10^6 cells are plated/35-mm dishes.[39,40] Cells can be grown in DMEM/F-12 medium containing 10% FBS for 2 weeks. For serum-free conditions, cells are initially plated in the presence of DMEM/F-12 medium containing 10% FBS. After 18 hr of incubation, the medium is changed to serum-free medium and the cells are grown for 2–3 weeks.[40] Cells can be directly plated on PORN/laminin-coated plates in serum-free

[39] K. Unsiker, H. Reichert-Preibsch, R. Schmidt, B. Pettmann, G. Labourdette, and M. Sensenbrenner, *Proc. Natl. Acad. Sci. U.S.A.* **84,** 5459 (1987).

[40] R. S. Morrison, A. Sharma, J. De Vellis, and R. A. Bradshaw, *Proc. Natl. Acad. Sci. U.S.A.* **83,** 7537 (1986).

medium with the N2 supplement.[41] Neuronal cells survive for 1–2 weeks under these culture conditions. In addition, neuronal cells have been cocultured on a feeder layer of astrocytes.[42] Under these culture conditions, neurons survive for about 4 weeks.

Long-Term Cultures of Neuroblasts

Utilizing the proliferative property of FGF-2, long-term neuroblast cultures have been generated from both embryonic and adult CNS of rats.

Embryonic Tissues[12]

Long-term neuroblast cultures have been generated from embryonic hippocampus, VM, locus coeruleus, septum, and spinal cord. Neuroblast cultures generated from hippocampus have survived for 8 months in culture and have been passaged 9 times. The cultures have been generated 35 times from hippocampus with equal consistency.

1. Place the dissected tissues in a 15-ml polypropylene centrifuge tube and wash three to four times with 5 ml PBS-D each time. Resuspend tissues in 1 ml N2 medium containing 20 ng/ml FGF-2 (human recombinant, Syntex-Synergen Consortium or purified bovine, R&D Systems) (N2 + FGF-2). Make a single cell suspension by mechanical dissociation (hippocampus) or by using the PPD solution (VM or spinal cord) as described earlier.
2. Count viable cells in a hemocytometer and plate $\sim 1 \times 10^6$ cells/T75 flasks (coated with PORN/laminin) in 8 ml N2 + FGF medium. The medium is changed every 3–4 days.
3. Cells can be passaged when the cultures are 70–80% confluent. During culture and cell growth, cells detach from the substratum and float in the medium. More floating cells are noted when the cell density is >80%. The floating cells can be collected by centrifugation of the culture medium at 1000*g* for 3 min and replated on PORN/laminin-coated plates.

To passage cells by trypsinization, add 1 ml ATV trypsin (Irvine Scientific, Santa Ana, CA) to a culture flask. Roll the flask to spread trypsin and immediately hit the flask gently on the side to dislodge the cells. The trypsinization process usually takes <1 min. Add 5 ml N2 medium, transfer

[41] P. Walicke, W. M. Cowan, N. Ueno, A. Baird, and R. Guillemin, *Proc. Natl. Acad. Sci. U.S.A.* **83,** 3012 (1986).

[42] G. Banker, *Science* **209,** 809 (1980).

TABLE III
IMMUNOCYTOCHEMICAL MARKERS FOR NEURAL CELLS

Cell Type	Marker	Refs.
Precursor cells	Nestin	44
	A2B5	45
Neurons	Neuron-specific enolase	46
	Neurofilament (68, 150, 200 kDa)	47
	Microtubule-associated protein (MAP2)	48
	Tau	49
	Synaptophysin	50
	Tyrosine hydroxylase (dopaminergic neurons)	51
	Glutamate decarboxylase (GABAergic neurons)	52
	Choline acetyltransferase (cholinergic neurons)	53
Astrocytes	Glial fibrillary acidic protein	54

the cell suspension to a 15-ml centrifuge tube, and spin at 1000*g* for 3 min. After removing the supernatant by aspiration, wash cells once by resuspending in 1 ml N2 medium. Remove the wash medium and make a single cell suspension by trituration using medium- to fine-bore Pasteur pipettes. Plate cells onto PORN/laminin-coated culture plates.

4. For long-term storage, remove the cells from culture flasks by trypsinization as described in step 3. After removing trypsin, wash cells in 3–5 ml N2 medium once and then resuspend in N2 + FGF-2 + 10% DMSO (freezing medium). Cells from one T75 flask (80–90% confluent) can be resuspended in 2–3 ml of the freezing medium. Cells are slowly frozen in liquid nitrogen. To reculture, thaw cells quickly in a 37° water bath and transfer to a centrifuge tube containing 5 ml N2 medium. Centrifuge the cell suspension at 1000*g* for 3 min and remove the supernatant containing DMSO by aspiration. Resuspend cells in 1 ml N2 + FGF-2, make a single cell suspension, and plate onto a PORN/laminin-coated culture plate.

Adult Tissues

1. Dissect the brain region of interest[43] and remove the meninges and connective tissues. After washing with PBS-D, suspend the tissues in 5 ml PPD solution and incubate at 37° for 30 min with occasional shaking. Dissociate tissues by mechanical trituration as described earlier. Centrifuge at 1000*g* for 3 min and remove the supernatant by aspiration.

2. Resuspend cells in 1 ml DMEM/F-12 + 10% FBS, triturate to make a single cell suspension, and wash cells two to three times. Resuspend cells in 1 ml DMEM/F12 + 10% FBS + 20 ng/ml FGF-2, make a single cell

[43] P. W. Coates, J. Ray, and F. H. Gage, unpublished data.

suspension, and plate onto uncoated plastic plates. Incubate at 37° overnight.

3. After 24 hr, replace serum-containing medium with N2 + FGF-2 medium. Initially, feed cultures every 3–4 days by replacing half of the medium. The concentration of FGF-2 in the replacement medium is 20 ng/ml. After the cultures are ~20–30% confluent, feed cultures with fresh N2 + FGF-2 medium by replacing the medium completely. This procedure has been done more than 20 times in the authors' laboratory using hippocampi of adult rats and 6 times with substantia nigra. Cultures of hippocampal cells have been maintained for one year through 30 passages.

Analysis of Neural Cell Cultures

The nature of cells present in cultures can be analyzed by morphological and phenotypical criteria. For morphological analysis, both phase-contrast and electron microscopy have been used to characterize the neuronal nature of cells. The expression of neuronal markers by cells in culture can be analyzed by immunocytochemistry using antibodies against neuron-specific antigens. The antigenic markers expressed by precursor cells, astrocytes, and mature neurons are listed in Table III.[44–54]

Conclusion

The major objective for culturing cells from the CNS is to study the morphologic, phenotypic, and electrophysiological properties of cells in a controlled environment. Methods for culturing precursor cells, which give rise to neuroblasts and glioblasts, using the mitogenic growth factors EGF and FGF-2 have been established. Although it is easier to grow long-term cultures of astrocytes, the procedure for long-term culturing of neuroblasts

[44] K. Frederiksen and R. D. G. McKay, *J. Neurosci.* **8,** 1144 (1988).
[45] M. C. Raff, R. H. Miller, and M. Noble, *Nature* (*London*) **303,** 390 (1983).
[46] D. E. Schmechel, P. J. Marangoe, A. P. Zis, M. W. Brightman, and F. K. Goodwin, *Science* **199,** 313 (1978).
[47] M. J. Carden, J. Q. Trojanowski, W. W. Schlaepfer, and V. M.-Y. Lee, *J. Neurosci.* **7,** 3789 (1987).
[48] C. G. Dotti, G. A. Banker, and L. I. Binder, *Neuroscience* **23,** 121 (1987).
[49] P. Litman, J. Barg L. Rindzoonski, and I. Ginzburg, *Neuron* **10,** 627 (1993).
[50] T. L. Fletcher, P. Cameron, P. De Camilli, and G. Banker, *J. Neurosci.* **11,** 1617 (1991).
[51] J. Engele, C. Pilgrim, M. Kirsch, and I. Reisert, *Brain Res.* **483,** 98 (1989).
[52] J. R. Cooper, F. E. Bloom, and R. H. Roth, *in* "The Biochemical Basis of Neuropharmacology," 4th ed., p. 367. Oxford University Press, New York, 1982.
[53] B. Knusel, P. P. Michel, J. S. Schwaber, and F. Hefti, *J. Neurosci.* **10,** 558 (1990).
[54] A. Bignami, L. F. Eng, D. Dahl, and C. T. Uyeda, *Brain Res.* **43,** 429 (1972).

and neurons has only recently been described. This chapter provides an overview of methods available for the culturing of precursor cells and neuroblasts. The choice of a particular procedure for culturing cells depends on the specific cell type of interest.

Acknowledgments

The authors thank Mary Lynn Gage for help in the preparation of the manuscript. The work in authors' laboratory has been supported by grants from NIH (PO1 AG10435) and Hollfelder Foundation.

[3] Transformation of Primary Rat Embryo Cells

By HARTMUT LAND

Carcinogenesis is a multistep process involving the activation of oncogenes and the loss or inactivation of tumor suppressor genes. First evidence that multiple genetic lesions cooperate in this process came from the observation that single oncogenes are unable to fully transform nonestablished rat cells. Instead, transfection of certain combinations of distinct cellular oncogenes (e.g., *ras* and *myc*) or cellular and viral oncogenes (e.g., *ras* and adenovirus E1A) is required to induce tumorigenic conversion.[1,2] Since then a large number of oncogene pairs have been shown to cooperate using secondary rat embryo cells as a simple and convenient cell system.[3,4] Moreover, this system also has been very useful to reveal the oncogenic properties of oncogenes which otherwise cannot be easily monitored in mammalian cells *in vitro*. In such cases (e.g., *myc*, see Amati *et al.*[5]) the rat embryo cell cotransformation assay provides an invaluable tool to score the biological activity of these genes easily. The preparation of rat embryo cells and the cotransformation assay are described in detail next.

Materials and Reagents

Two pairs of scissors and surgical forceps.
Two pairs of curved watchmaker's forceps.

[1] H. Land, L. F. Parada, and R. A. Weinberg, *Nature* (*London*) **304,** 596 (1983).
[2] H. E. Ruley, *Nature* (*London*) **304,** 602 (1983).
[3] H. E. Ruley, *Cancer Cells* **2,** 258 (1990).
[4] T. Hunter, *Cell* (*Cambridge, Mass.*) **64,** 249 (1991).
[5] B. Amati, M. W. Brooks, N. Levy, T. D. Littlewood, G. I. Evan, and H. Land, *Cell* (*Cambridge, Mass.*) **72,** 233 (1993).

One small curved scalpel.

PBS: NaCl, 10 g/liter; KCl, 0.25 g/liter; KH_2PO_4, 0.25 g/liter; Na_2HPO_4; 1.43 g/liter Tris–saline: Tris base, 24 m*M*; NaCl, 0.8 g/liter; KCl, 0.38 g/liter; Na_2HPO_4, 0.1 g/liter; D-glucose, 1 g/liter; phenol red, 0.015 g/liter; penicillin, 100 U/ml; streptomycin, 0.1 g/liter.

Trypsin: 0.25% trypsin in Tris–saline.

Versene: 0.54 m*M* EDTA in phosphate-buffered saline (PBS).

Dulbecco's modified Eagle's medium with 4500 mg/liter glucose, 584 mg/liter glutamine, and 110 mg/liter sodium pyruvate (DMEM).

Fetal bovine serum (FBS).

Transfection buffer: (2X) 50 m*M* HEPES, 280 m*M* NaCl, 1.5 m*M* Na_2HPO_4, adjusted to pH 7.1 with 0.5 *N* NaOH. Filter sterilize and freeze in aliquots. Do not use if older than 3 months. The pH is critical for the quality of the precipitate since the optimal pH range for the formation of the $CaPO_4$–DNA coprecipitate is extremely narrow. It is best to test the efficacy of a series of buffers between pH 6.95 and 7.15 since pH meter readings may not always be accurate.

$CaCl_2$: 2.5 *M* stock solution.

Method

Preparation of Rat Embryo Cells

Timed-pregnant inbred rats (e.g., Fisher or Wistar strains) should be sacrificed at days 13.5–14 of gestation. (The day when the vaginal plug is found is defined as day 1.) For the dissection two pairs of sterile scissors and surgical forceps are required. The dissection is best carried out in the animal facility and several animals should be handled in a single batch. Soak the lower abdomen with ethanol. Lift the fur in the lower abdominal area with forceps and make a V-shaped incision to expose the muscle layer below the dermis. Change to a new pair of instruments. Then open the abdominal cavity and identify V-shaped uteri below the intestine. These are easily recognizable because the implanted fetuses are arranged like beads on a string. Cut proximal to the chiasm and pull up uteri with forceps. Remove all fat tissue, then cut each branch distal to the last fetus and transfer uteri into a petri dish containing sterile PBS. All further precedures should be carried out in a laminar flow hood.

The fetuses should be retrieved from the uteri with two fine curved watchmaker's forceps. First, each uterus should be transferred into a new dish containing PBS. Then, tear the uterus wall on the light-colored side of each bulge. Usually the fetuses will be forced out due to the internal pressure. Collect the fetuses in a 9-cm dish with PBS, wash, and transfer to another 9-cm dish with PBS. Use one pair of forceps together with a small curved edge scalpel to remove head, viscerae, and liver, if the latter is visibly red.

Transfer the prepared torsos from a total of five pregnant rats into a 6-cm dish without PBS. Mince tissue carefully with the scalpel, add 5 ml of prewarmed trypsin/versene (1:1 mixture), and incubate for 20 min at 37°. To stop proteolysis, add 5 ml of DMEM containing 10% FBS. To remove most of the trypsin, take up the suspension with a 10-ml wide mouth pipette. At this step disaggregation should be avoided so that the tissue can settle to the bottom of the pipette. Collect the tissue in a dish and discard as much of the trypsin/medium mix as possible. Add 5 ml of fresh DMEM with 10% FBS to the tissue and disintegrate the tissue into single cells by pipetting up and down forcefully through a 5-ml narrow mouth pipette.

Transfer the cell suspension into a sterile tube and count living (dye excluding) cells after staining with trypan blue (0.9 ml PBS, 0.1 ml cells, 1 ml trypan blue solution, Sigma). Use a hemocytometer for counting. Plate at least 3×10^7 living cells in 30 ml DMEM containing 10% FBS onto one 15-cm dish. Two hours after plating the primary culture the cells should be approximately 50–70% subconfluent. On average the fetuses from a single rat yield two to three 15-cm dish cell cultures. Trypsinize and freeze cells after 3–4 days in culture in DMEM containing 40% FBS and 10% DMSO. From each 15-cm dish culture, prepare six freezing vials containing 1 ml of cell suspension each. For greater consistency of results it is advisable to pool the cells from several dishes before freezing. Freeze cells slowly and store in liquid nitrogen.

Transfection of Secondary Rat Embryo Fibroblasts

One day before transfection thaw the cells and dilute in 30–50 ml of prewarmed DMEM with 10% FBS. Determine the cell concentration, e.g., by Coulter counter, and seed $1.2–1.5 \times 10^6$ cells per 9-cm dish into 10 ml DMEM with 10% FBS. This procedure should produce dishes with approximately 10^6 attached cells at the time of transfection. Most of these cells can be characterized as fibroblasts. One vial of cells usually will be sufficient to prepare six 9-cm dishes. It is important not to passage the cells any further before transfection since the transfection efficiency will drop significantly.

Two hours before the transfection the cells are refed with 5 or 10 ml of fresh medium. It is important that the pH of the medium has fully equilibrated before adding the calcium phosphate–DNA coprecipitate.

For cotransfection of one 9-cm dish of rat embryo fibroblasts with combinations of oncogenes or potential transforming genes, most commonly 5–10 μg of each expression plasmid is mixed with 0.5 μg of a plasmid conferring drug resistance against Geneticin (G418) or puromycin. The DNAs are brought up to a final volume of 250 μl with H_2O. Then 250 μl of the transfection buffer is added and the precipitate is formed by mixing in 18–25 μl of 2.5 *M* $CaCl_2$. The appropriate amount of $CaCl_2$ must be determined by titration. The precipitate is added dropwise into the medium. It should be very fine and should not contain clumps which will lead to insufficient transfection efficiencies. The precipitate should be left on the cells for 16–20 hr. This will increase the transfection frequencies so that a glycerol shock becomes unnecessary.

Ten to 20 μg of DNA will form an abundant precipitate and yield efficient expression of the transfected genes. However, the addition of more than 20 μg of DNA per 9-cm plate should be avoided since the transfection of cells is saturated and transformation frequencies often decrease. Where titrations of expression constructs are required, DNA and enhancer concentrations must be kept constant by adding the appropriate amount of empty expression vector. Otherwise the results obtained may be affected by alterations in transfection efficiencies or by enhancer competition effects.

After transfection the precipitate is replaced with 10 ml of DMEM with 0.5% FBS. The cells are allowed to recover in this medium for 2–4 hr, after which they are refed with DMEM containing 10% FBS.

Focus and Colony Assays

The day after transfection, the cultures are split 1:6 and are subdivided into three cultures maintained in DMEM with 10% FBS to score the appearance of transformed foci. Three further cultures are plated in the same medium supplemented with the appropriate drug in order to monitor transfection efficiencies and/or potential toxicity of the transfected constructs. Geneticin (G418) is used at 500 μg/ml, while puromycin is used at 1.25 μg/ml. Puromycin is the preferable drug since selection is fast (2 days) and its application is relatively inexpensive. The resulting foci and colonies are scored 10–13 days after plating. The cultures are refed every 3–4 days. The focus assays are refed with DMEM containing 5% FBS to avoid lifting of the monolayer from the substratum.

In most cases it is best to score the number of foci in monolayers without staining the cultures. The foci can be most easily scored based on their higher light refraction. However, sometimes it is necessary to distinguish certain foci of transformed cells from clumps of normal cells by microscopic inspection. Colonies of transfected drug-resistant cells can be counted after staining for 30 sec with a 2% (w/v) crystal violet solution in 20% (v/v) methanol. The data from a given transfection can be given as the cumulative number of foci or colonies of the respective triplicate of plates. This number represents the transformation or transfection frequency per 5×10^5 transfected cells.

If required the transfected cells can also be seeded directly into semisolid media such as soft agar or methylcellulose. Colonies of transformed cells can be scored or isolated after 10–14 days. To test for tumorigenicity, the cells from a single transfected monolayer culture can be pooled and injected subcutaneously into nude (*nu/nu*) mice. The formation of tumors can be observed 7–21 days after injection.[1]

[4] Avian Hematopoietic Cell Culture: *In Vitro* Model Systems to Study Oncogenic Transformation of Hematopoietic Cells

By HARTMUT BEUG, PETER STEINLEIN, PETR BARTUNEK, and MICHAEL J. HAYMAN

Introduction

Hematopoietic cells are produced throughout the lifetime of an individual from a small set of stem cells. The progeny of these stem cells make decisions involving self-renewal and differentiation to give rise to the cells of the erythroid, myeloid, and lymphoid lineages.[1] These cells either proliferate, to expand a certain compartment, or differentiate along a lineage-specific pathway. Through a delicate balance between proliferation and differentiation, the numbers of the cells within various lineages are controlled and homeostasis is maintained. A crucial role in this process is played by cytokines, which regulate the proliferation and differentiation of progenitors as well as the function of mature cells.[2,3]

Leukemias and lymphomas are diseases in which the delicate equilibrium between proliferation and differentiation has been disturbed. This results in dramatic changes in the phenotypic spectrum of hematopoietic cells found in the body, usually characterized by the hyperproliferation of immature or partially mature hematopoietic cells at the expense of mature, functional ones.[3] In order to understand the mechanisms underlying these

[1] J. E. Till and E. A. McCulloch, *Biochim. Biophys. Acta* **605,** 431 (1980).
[2] D. Metcalf, *Nature (London)* **339,** 27 (1989).
[3] C. L. Sawyers, C. T. Denny, and O. N. Witte, *Cell (Cambridge, Mass.)* **64,** 337 (1991).

oncogenic changes it is necessary to identify the controls that operate during normal hematopoiesis and then determine how these are disturbed in the various disease states.

Our current understanding of how normal and abnormal hematopoiesis is controlled stems mainly from work on murine and human hematopoiesis. However, it has been difficult to reconstruct the different oncogenic alterations causing leukemias and lymphomas seen *in vivo* by respective *in vitro* experiments. For instance, most of the genetic alterations that occur during development of the Friend disease in mice are known at the molecular level: (i) the constitutive activation of the erythropoietin receptor by the Friend virus glycoprotein gp55, (ii) constitutive overexpression of the transcription factors spi-1(Pu-1) and fli-1 by insertional activation of the respective promotors by the retrovirus, and (iii) loss or mutation of the p53 tumor suppressor gene. Despite this knowledge, the actual contributions of these individual changes to the leukemic phenotype and the way they may cooperate are largely unknown. A similar scenario applies for human leukemias: the abnormal fusion proteins generated by chromosomal translocations have been characterized in detail, but their contribution to the leukemic phenotype has remained obscure.

The major reasons for this deficiency are methodological. Normal and/or oncogene-transformed primary mouse bone marrow cells are difficult to analyze because of the short *in vitro* life span of such cells (7–15 generations). Oncogene action in mouse hematopoiesis is therefore either studied *in vivo* (transgenic mice) or using immortalized cell lines. Because of their inherent genetic instability, the latter easily accumulate genetic and epigenetic changes, making it very difficult to distinguish between oncogene action and other changes due to "cellular progression." Primary human bone marrow cells are also problematic due to their limited availability and, more importantly, to the lack of efficient and practical (e.g., nonhazardous) gene transfer systems that would allow stable transformation of the primary cells.[4]

Avian hematopoietic cells provide a unique model system which overcomes several of these problems. First, because of their rather long *in vitro* life span (30–50 generations[5]), normal and oncogene-transformed avian hematopoietic cells can be grown into mass cultures (10^{10} to 10^{12} cells per clone). Second, avian retroviruses can be used to introduce efficiently oncogenes into avian bone marrow cells and thus recapitulate *in vitro* the erythroid, myeloid, and lymphoid leukemias seen *in vivo* after inoculation

[4] T. Graf and H. Beug, *Cell (Cambridge, Mass.)* **34,** 7 (1983).
[5] H. Beug and T. Graf, *Exp. Cell Res.* **107,** 417 (1977).

of chicks with the same viruses.[6] The avian system is also ideally suited to study aspects of oncogene cooperation during multistep oncogenesis since retroviruses containing two cooperating oncogenes can be used to infect the bone marrow cells. Finally, a variety of conditional oncogenes are available (the function of which can be turned on and off at will); the respective conditionally transformed hematopoietic progenitors are particularly suited for studying aspects of normal hematopoiesis.[7,8] This chapter first reviews the various techniques that are available to transform hematopoietic progenitors from several lineages using specific oncogenes or their combinations. Because a major impetus for obtaining transformed hematopoietic cells is their use as models for normal differentiation, methods to induce these cells to differentiate are also described. The second part of this chapter describes the techniques developed for the growth and *in vitro* differentiation of normal chicken hematopoietic cells.

Transformed Avian Hematopoietic Cells

Hematopoietic progenitors present in bone marrow are commonly believed to possess a fixed program of terminal differentiation. That is, they undergo a limited number (5–10) of cell divisions and at the same time become increasingly mature. Transformation in this system is therefore operationally defined as the ability of a cell to undergo sustained proliferation in the absence of differentiation (termed self-renewal) under conditions where normally it would either differentiate or die. In the chicken system, a large number of oncogenes are capable of transforming cells from the erythroid, myeloid, and certain lymphoid lineages as well as multipotent progenitors.[4,6,9–11] Basically, three major criteria have to be fulfilled in order to ensure successful transformation of a hematopoietic cell by an oncogene-carrying retrovirus. (i) The target cell has to be cultured under conditions allowing cell divisions to ensure successful integration and subsequent expression of the retroviral genes. (ii) The retrovirus has to be of sufficiently high titer to allow transformation of the potentially rare target cells. (iii) Since oncogenic transformation rarely renders the transformed cell independent of all growth/differentiation factors, the specific growth require-

[6] H. Beug and T. Graf, *Eur. J. Clin. Invest.* **19,** 491 (1989).

[7] J. Knight, H. Beug, J. Marshall, and M. J. Hayman, *Oncogene* **2,** 317 (1988).

[8] H. Beug, A. Leutz, P. Kahn, and T. Graf, *Cell (Cambridge, Mass.)* **39,** 579 (1984).

[9] T. Graf and H. Beug, *Biochim. Biophys. Acta* **516,** 269 (1978).

[10] T. Graf, K. McNagny, G. Brady, and J. Frampton, *Cell (Cambridge, Mass.)* **70,** 201 (1992).

[11] J. Larsen, S. Meyer, P. Steinlein, H. Beug, and M. J. Hayman, *Oncogene* **8,** 3221 (1993).

ments of the transformed cell have to be met for the transformed cell to be successfully propagated.

The most successful method of infection has been found to be cocultivation of the bone marrow cells with virus-producing, growth-arrested fibroblasts. This method is effective for all the different oncogenes tested to date, as well as for the simultaneous infection of cells with two or three different retroviral vectors.[12–16] On the other hand, the growth requirements of the transformed cells are quite unique, being specific for the different cell lineages and sometimes dependent on the particular oncogene used. Therefore, after describing the general methods of how to cultivate these rather delicate primary cells, the methods used for retroviral infection by cocultivation as well as for selection and propagation of transformed clones will be outlined. The specific growth requirements of the different cell types will then be listed.

Avian Hematopoietic Cell Culture

Components of Tissue Culture Media

General

A major problem with cell culture of avian primary bone marrow cells is that success is entirely dependent on the availability of tested batches of several important medium components. Procedures for such tests are given below. However, since generally available established cell lines are unsuitable as test cells in such assays, setting up of the techniques described below requires testing several batches of each component on various types of transformed primary cells that have to be generated for this purpose. As with most complex tissue culture techniques,[17] the purity requirements for the water used to make up the media are stringent and comparable to those required for defined media.[18]

[12] H. Beug, M. J. Hayman, M. B. Raines, H. J. Kung, and B. Vennström, *J. Virol.* **57,** 1127 (1986).

[13] M. Zenke, P. Kahn, C. Disela, B. Vennström, A. Leutz, K. Keegan, M. J. Hayman, H. R. Choi, N. Yew, J. D. Engel, and H. Beug, *Cell* (*Cambridge, Mass.*) **52,** 107 (1988).

[14] M. Zenke, A. Munoz, J. Sap, B. Vennström, and H. Beug, *Cell* (*Cambridge, Mass.*) **61,** 1035 (1990).

[15] C. Glineur, M. Zenke, H. Beug, and J. Ghysdael, *Genes Dev.* **4,** 1663 (1990).

[16] S. Fuerstenberg, I. Leitner, C. Schroeder, H. Schwarz, B. Vennström, and H. Beug, *EMBO J.* **11,** 3355 (1992).

[17] R. G. Ham and W. L. McKeehan, this series, Vol. 58, p. 44.

[18] J. Bottenstein, I. Hayashi, S. Hutchings, H. Masui, J. Mather, D. B. McClure, S. Ohasa, A. Rizzino, G. Sato, G. Serrero, R. Wolfe, and R. Wu, this series, Vol. 58, p. 94.

Growth Medium

The Dulbecco's modified Eagle's medium (DMEM) used in a modified form by the authors has been published,[19] but crucial information is missing. In the following, a simple procedure for its preparation is described. Tissue culture grade chemicals (e.g., Sigma) should be used throughout.

Stock solution A: Dissolve 5 mg biotin in 50 ml doubly distilled H_2O. Store at $-20°$ in 10-ml aliquots.

Stock solution B: Weigh out 3 g L-methionine, 6.3 g L-phenylalanine, 1.8 g L-alanine, 17 g glycine, 9.5 L-threonine, 10.5 g L-isoleucine, 2.5 g L-proline, 9.3 g L-valine, 3 g L-aspartic acid, and 7.5 g L-glutamic acid; dissolve in $\approx$1800 ml doubly distilled H_2O by warming to 60° and stirring until all solids are dissolved (may take several hours). Aliquot 200 ml/bottle and store at $-20°$ in 200-ml aliquots.

To prepare 10 liters of modified DMEM medium, add in this order into a suitable container: Commercially available DMEM powder (tested batch, high glucose, i.e., 4.5 g/liter, with L-glutamine, without sodium bicarbonate, package for 10 liters), 25 g $NaHCO_3$, 10 ml solution A, 200 ml solution B. Dissolve 50 mg hypoxanthine in 125 ml doubly distilled H_2O + 3.8 ml 1 *M* HCl, (heat to 60° if necessary). Dissolve to completion in $\approx$9.5 liters doubled distilled H_2O, adjust pH to 7.3 (with 1 *N* HCl or NaOH), bring to 10.0 liters, and sterile filter.

To prepare 2× concentrated DMEM, add 15 g instead of 25 g $NaHCO_3$ to above ingredients and dissolve in 5 liters instead of 10 liters doubly distilled H_2O. This medium must be shock-frozen (using dry ice or liquid nitrogen) and stored at $-20°$. After thawing, bring to pH < 7.2 by gassing with CO_2 and stir until all turbidity has gone. Thawed 2× medium should be used within a few days, but can be refrozen as above.

Medium Components

Fetal calf serum (FCS) batches should be tested for optimum growth of transformed primary erythroblasts. The best batches are usually also suitable for culture of normal erythroid progenitors. With chicken serum, several (three to five) batches should be first tested on established and primary transformed erythroblasts. If the same batch is to be used for normal erythroid progenitors, additional tests are required. These tests are described at the end of this section. Sera should be stored frozen in aliquots, and storage for more than 1–2 weeks at 4° should be avoided.

Detoxified bovine serum albumin (BSA) is another crucial ingredient.

[19] T. Graf, *Virology* **54,** 398 (1973).

It has to be prepared as outlined below and then tested for nontoxicity on similar cells as the chicken serum. We have found that commercially available preparations of detoxified BSA are usually unsuitable, particularly for normal erythroid progenitors.

1. Dissolve 4 g dextran T40 (Pharmacia) in 4 liters doubly distilled H_2O in a suitable glass beaker, then slowly add 40 g activated charcoal with stirring. Stir gently for 30 min (this allows binding of the dextran to the activated charcoal).

2. Add 200 g BSA (Sigma, Fraction V or equivalent) in several portions with slow stirring. Continue slow stirring until all BSA is dissolved.

3. Adjust to pH 3.0 using concentrated (10 *N*) HCl. Strong stirring and the addition of the HCl in small portions are required to avoid denaturation of the BSA.

4. Incubate for 30 min in a 56° water bath. Occasionally stir the contents. Remove from the water bath and rapidly cool on ice to 4°.

5. Centrifuge at 6000*g* (e.g., Sorvall GSA rotor, 250-ml tubes, 12,000 rpm) for 1 hr at 4°.

6. When the centrifuge stops, immediately take the tubes from the rotor and carefully decant the supernatants. The pellets are not "firm" and some of the charcoal will remain in the supernatant.

7. Filter the BSA through a double-fluted paper filter to remove most of the remaining charcoal.

8. Adjust solution to pH 5.5 with 1 *N* NaOH, then remove residual charcoal by filtration (three filters are used: a glass fiber prefilter, a 1.2-μm nitrocellulose filter, and a 0.45-μm nitrocellulose filter). The BSA is now lipid free.

9. Add 400 ml analytical grade mixed bed resin ion exchanger (Serdolit, MB-1, analytical grade, 20–50 mesh, Serva 40701) to the BSA to remove toxic ions. Any other type of ion exchanger has to be pretested since it may render the BSA toxic. Agitate suspension in the cold room overnight on a rotary shaker (do not use a magnetic stirrer as this pulverizes the resin).

10. Remove the ion-exchange beads with a porous glass filter (<0.2 mm) attached to a vacuum flask.

11. Sterile filter again if necessary. Measure the volume of the BSA at this stage. Concentrate to less than 25% of this measured volume, using the Amicon spiral membrane concentrator (PA 2000) with a 2 liter reservoir plus filter cartouches with a molecular size exclusion of 10,000 Da (S1Y10) or equivalent. Methods that cannot concentrate the BSA in less than 6 hr are unsuitable. The BSA can also be concentrated by lyophilization. It is then pulverized and redissolved in doubly distilled H_2O.

12. The final concentration of BSA should be 20%. Sterile filter the BSA, shock-freeze in 50- or 10-ml aliquots, and store at $-20°$.

Generally Used Tissue Culture Media

Standard Growth Medium

This medium consists of modified DMEM (see above) supplemented with tested batches of fetal calf serum (8%), chicken serum (2%), 10 m*M* HEPES–HCl, pH 7.2, and standard amounts of penicillin and streptomycin.

Complex Growth Medium (CFU-E Medium)

The CFU-E medium consists of 82 ml modified DMEM, 21.4 ml doubly distilled H_2O, 15 ml fetal calf serum (tested batch), 6 ml chicken serum (tested batch), 300 μg/ml iron-saturated chicken ovotransferrin [conalbumin, (Sigma C 0880), from 15 mg/ml stock in Hanks' balanced salt solution (HBSS)], 16 μl 1 *M* mercaptoethanol in doubly distilled H_2O, 4.6 ml detoxified BSA (tested batch), and 4.2 ml 5.6% (w/v) sodium bicarbonate.[20] This medium has to be stored frozen and should be used within a week of thawing. For testing of medium supplements, see below.

Semisolid Medium (Methocel)

The method of choice to clone hematopoietic cells and to isolate transformed clones is culture in semisolid medium. Methylcellulose (Methocel) is highly superior to soft agar since the colonies can be isolated and dispersed much easier. Tested batches of Methocel are commercially available for mammalian cells, but their usefulness for avian hematopoietic cells has not been assessed. Since the available batches of Methocel are highly variable, testing of such batches for supporting the growth of transformed erythroid and myeloid colonies is required (see below). Methocel is prepared according to Graf[19]; the essential features are listed here.

1. Sterilize 10 g Methocel by spreading it into a thin layer in a 145-mm plastic petri dish and UV irradiating it in a sterile environment (e.g., a laminar flow hood) for 30–60 min. If a more (or less) viscous Methocel is desired, the amount of methocel can be varied between 8 and 12 g.
2. Weigh a sterile, 1-liter Erlenmeyer flask (1000 ml).

[20] K. Radke, H. Beug, S. Kornfeld, and T. Graf, *Cell* (*Cambridge, Mass.*) **31,** 643 (1982).

3. Add 225 ml sterile doubly distilled H_2O and bring to a boil on a bunsen burner or hot plate; immediately afterward add the sterilized Methocel powder in small portions with vigorous shaking. Avoid stirring.

4. Cover the Erlenmeyer flask with sterile aluminum foil and swirl the flask vigorously until no more lumps are visible.

5. Bring once again to a boil with vigorous shaking, then immediately remove from the heat and rapidly cool to 30–40° by shaking in ice–water.

6. Add 250 ml 2× DMEM (freshly thawed, containing no undissolved material).

7. Add sterile doubly distilled H_2O until the total weight (minus the weight of the Erlenmeyer flask) equals 503 g.

8. Stir the Methocel overnight in the cold room (use a large stirring bar and a strong magnetic stirrer to provide good mixing).

9. Shock-freeze and store at −20° in 50- to 100-ml aliquots.

10. It is recommended to prepare two to four times of this amount. Before freezing, remove a 50-ml aliquot for batch testing.

Cell Counting

For successful culture of transformed and normal bone marrow cells, the monitoring of cell proliferation by frequent determination of cell numbers is indispensable. For this reason, the use of an electronic cell counter (Coulter counters or the new CASY-1 of Schärfe System Inc., FRG) is preferable to the use of counting chambers, which is just too time-consuming. In addition, it is desirable to obtain size distributions of the cells since these allow to judge many parameters of the bone marrow cultures (e.g., state of health, presence of differentiated cells, or contaminations by other cell types) in a routine fashion.

Bone Marrow Preparation

General Procedure

Preparation of normal bone marrow cells from 3- to 7-day-old SPAFAS chicks is done according to Graf.[19] The method is briefly outlined here.

1. For infection with retroviral oncogenes, bone marrow from 2- to 14-day-old chicks is recommended. Older chicks may give severely reduced transformation efficiencies.

2. Sacrifice the chick (gassing with CO_2 is recommended) and wet the legs and body feathers with 70% ethanol. Cut away the skin, make an incision at the hip, and dislocate the leg at the hip joint. Carefully remove

the muscle with a pair of scissors and a sharp scalpel. Obtain the tibia in the same fashion. If bones from several chicks are prepared, store bones submersed in modified DMEM plus antibiotics.

3. Remove cartilage from both ends of the bone.

4. Dispense 15 ml of the cold (4°) complex growth medium (should be at pH 7.2; i.e., orange in color, otherwise gas with CO_2) in a 50-ml conical plastic centrifuge tube. Fill a 10-ml syringe with a No. 1 needle with the same medium, insert needle into the bone cavity, and flush the bone marrow into the tube. Repeated flushing of the medium through the bone may be required to obtain a good yield of bone marrow cells. Use a wide bore plastic pipette to suspend cells clumps.

5. Filter the cell suspension through a suitable commercial cell strainer (e.g., Falcon 2350). Centrifuge at 500–700*g* (1500 rpm in a Heraeus Labofuge or equivalent) for 5 min at room temperature.

6. Carefully remove the lipid layer from the top of the supernatant. Wash once more in 10 ml of the complex growth medium.

7. Determine the cell number in an electronic counter and adjust to $>20 \times 10^6$ cells/ml in the complex growth medium; $1–3 \times 10^8$ cells can be expected from a 1-week-old chick.

To enrich for target cells for transformation, i.e., immature cells capable of division, the bone marrow can be fractionated using 1.070 g/cm^3 Percoll (Pharmacia); $50–100 \times 10^6$ bone marrow cells in 7–8 ml of complex growth medium are underlayed with 2 ml of a mixture of 47 ml Percoll solution [9 parts Percoll, 1 part 10× concentrated phosphate-buffered saline (PBS), pH 7.3] and 53 ml standard growth medium in 12-ml conical plastic centrifuge tubes. After centrifugation for 1500–2000 rpm (500–750*g*) at room temperature for 5–10 min (no brake), the interphase containing a white to pink layer of immature cells is retrieved with a Pasteur pipette, diluted with complex growth medium, and spun down again after removing an aliquot for counting.

Generation of Virus-Producing Fibroblasts

Chicken Embryo Fibroblasts (CEFs)

Primary chicken CEFs are prepared according to standard procedures.[19] Cells can be frozen in liquid nitrogen and thawed 2 days before use; freshly prepared cells should be used after 2–3 days of culture.

DNA

For adequate transfection efficiency, it is crucial that the DNA used be as pure as possible. Either cesium chloride-purified DNA or DNA purified using a recent version of the "Qiagen" DNA purification kit can be used. Retroviral vector plasmids should either contain a well-transforming oncogene or a selectable marker [e.g., the neomycin (neo) or hygromycin resistance markers, see below].

Transfection Method[21]

A modification of the method described in Frykberg *et al.*[22] and Chen and Okayama[23] is recommended. Since the procedure has been significantly modified further to obtain high transfection efficiency (which is crucial because of the limited life span of the fibroblasts), it is briefly outlined here.

1. To accommodate the varying viability and growth rates of cells used, seed 3, 5, and 8 $\times$ 10^5 freshly prepared CEFs into 60-mm dishes containing 5 ml standard growth medium the day before transfection and allow to grow overnight.
2. Wash nearly confluent dishes once with PBS and once with transfection medium (standard growth medium lacking chicken serum) and add 5 ml transfection medium/dish.
3. Mix 10 μg retroviral construct DNA + 1 μg RAV-1 or RAV-2 helper virus DNA in 219 μl doubly distilled H_2O (final volume) and add 31 μl 2 *M* $CaCl_2$ (in Eppendorf tube). Mix well and add the DNA/Ca^{2+} solution dropwise with vortexing to 250 μl of 2$\times$ HBS (280 m*M* NaCl, 50 m*M* HEPES, 1.5 m*M* Na_2HPO_4, pH 7.12) in an Eppendorf tube (or 5-ml Falcon tube). [The pH of the HBS is critical for obtaining a precipitate of the right particle size. If there are problems with transfection efficiency, a few HBS solutions spanning the range between pH 7.1 and pH 7.2 should be assayed for transfection efficiency, e.g., in transient transfection assays.]
4. Leave 30 min at room temperature, add calcium phosphate/DNA suspension to cells, and swirl to evenly distribute the precipitate.
5. Grow cells overnight, then wash once with standard growth medium, add fresh medium, and grow for another 24 hr.
6. Trypsinize cells, suspend in standard growth medium, spin down, count, and seed for marker selection. Note that calcium phosphate-treated cells are hard to trypsinize; check the progress of trypsinization using a microscope.
7. For G418 selection, seed approximately 1 $\times$ 10^6 transfected cells per 100-mm dish, grow overnight, and add G418 (800 μg/ml final concentra-

[21] F. L. Graham and J. D. van der Eb, *Virology* **52,** 456 (1973).

[22] L. Frykberg, S. Palmieri, H. Beug, T. Graf, M. J. Hayman, and B. Vennström, *Cell (Cambridge, Mass.)* **32,** 227 (1983).

[23] C. Chen and H. Okayama, *Mol. Cell. Biol.* **7,** 2745 (1987).

tion). Set up negative and positive controls (mock-transfected cells and cells transfected with an empty *neo* vector. After 3–4 days, the G418-sensitive cells will start to die; a massive outgrowth of G418 resistant-cells can be expected after about 1 week.

8. As soon as G418-resistant cells have grown out and the G418-sensitive control cells have disintegrated, the concentration of G418 should be reduced to 400 mg/ml and eventually omitted entirely. This is even more important for selection with hygromycin, which should be removed as soon as the control cells have died, otherwise, the resistant cells are also irreversibly damaged.

Transformation of Bone Marrow Cells by Cocultivation

Cocultivation

To infect bone marrow (BM) cells by cocultivation with virus-producing fibroblasts, a semiconfluent culture (2×10^5 cells per 35-mm dish, $5–8 \times 10^5$ cells per 60-mm dish) of such cells in standard growth medium is treated with 10 μg/ml mitomycin C (Sigma, dissolved at 2 mg/ml in 50% DMSO in PBS) for 1–2 hr, washed three times in growth medium, incubated for 30–60 min at 37°, and washed again three times. Such dishes are preferably used directly for cocultivation, but can also be trypsinized and used after allowing 4–8 hr for the cells to readhere to the plastic. Then 25×10^6 freshly prepared BM cells or 2×10^6 immature cells (purified using Percoll 1.070 g/cm^3) are suspended in 5 ml of standard or complex growth medium and incubated with the fibroblasts for 24–48 hr. It is advisable to suspend the BM cells with a Pasteur pipette (although this will dislocate some virus-producing fibroblasts) and transfer the cells to a new dish without CEF after 24 hr. The efficiency of infection can be increased by adding growth factors to the cocultivation mix (see below).

Infection of Bone Marrow Cells with Two or More Retroviruses

The possibility of growing normal hematopoietic progenitors in combinations of specific growth factors (see below) enhances the efficiency of retroviral infection to an extent allowing infection of the same progenitor with two or three retroviral vectors, if suitable selection methods are applied.[16] Typically, a transforming retrovirus can be combined with a retrovirus expressing another test gene in combination with a selectable marker. The use of vectors expressing different selectable markers, e.g., the neomycin and hygromycin resistance genes, allows the generation of transformed cells expressing two different test genes.

To infect BM cells with several different retroviruses, culures of the fibroblasts producing the different viruses are trypsinized and equal numbers of cells are mixed to form an almost confluent culture and are allowed to adhere overnight. The mixed cultures are then treated with mitomycin C as described earlier. Depending on the age of the fibroblast cultures (old cells produce less virus than freshly infected ones) and on the type of experiment (fibroblasts containing nontransforming genes should be in a 3- to 5-fold excess, fibroblasts producing strongly transforming oncogenes can be used in reduced amounts) the ratios between the different fibroblast types can be varied. The mixed fibroblast layers are then incubated with (preferably density purified) BM cells in complex growth medium containing the appropriate growth factors.

Selection for Transformed or Infected Cells

If a transforming oncogene is used, the infected bone marrow cells are seeded into Methocel containing medium appropriate for the oncogene to be analyzed (see below) and screened for growth of transformed colonies. Typically, infected cells ($1–5 \times 10^6$ BM cells or $0.5–1 \times 10^6$ density purified cells) suspended in 20–50 μl of standard or complex growth medium are added to 3–4 ml of Methocel medium in 12-ml round-bottom polypropylene tubes (Greiner, FRG, 184261). After mixing by vigorous shaking, the mixture is poured into 35-mm bacterial petri dishes (Falcon 1008), a precaution that prevents the excessive growth of macrophages or stromal cells in the Methocel cultures. Macroscopic, transformed colonies (0.5 to 1 mm in diameter) appear after 5–14 days. Special care has to be taken to avoid any evaporation of medium during this period; the use of moist chambers [245 $\times$ 245 rectangular culture plates (Nunc) containing sterilized Kleenex pads wetted with sterile water] is recommended even in incubators with moisture control.

To select for G418 resistance, a 300-mg/ml stock solution of G418 is prepared in Hanks' balanced salt solution. Activated charcoal is added (about one-fifth of the solution volume), mixed well, and allowed to stand for 30 min. The activated charcoal is removed by centrifugation (2000*g*, 5 min) and filtration through a 0.45-μm membrane filter. Then, 1 *M* HEPES, pH 7.3, is added to a final concentration of 0.1 *M*. The actual concentration required for successful selection of G418-resistant colonies has to be determined in pilot experiments since it depends on the cell type used and on the time the transformed colonies require for growing out. For such pilot experiments, G418-resistant, transformed cells are mixed with normal BM cells and compared to BM cells alone after seeding into Methocel containing a series of G418 concentrations between 1.5 and 3 mg/ml. The goal should be a concentration that allows undisturbed outgrowth of the transformed, resistant cells into macroscopic colonies but is able to kill all normal cells by the time the colonies reach their full size. (The reasons why the G418 concentrations required in Methocel are much higher than in liquid medium are unknown.)

An identical procedure is applied for hygromycin selection, except that there is no purification of the hygromycin stock solution (Calbiochem) and determination of the correct concentration is even more important than for G418. If double selections are to be performed, both antibiotics are added together in somewhat reduced concentrations as compared to single selection. It is advisable to use several different concentrations in each experiment.

Isolation and Expansion of Transformed Clones

Transformed colonies should be allowed to grow to diameters between 0.2 and 0.5 mm, but should be isolated before the Methocel medium becomes exhausted (visible by acidification or unhealthy appearance of the colonies in dense cultures). Individual colonies are isolated by drawn-out Pasteur pipettes (inner diameter of tip 0.1 to 0.5 mm, angle of drawn-out tip 45–80°) under an inverted microscope and placed in 96-well tissue culture plates containing 100 μl of a suitable growth medium (depending on the cell type, see below). At least 20 (use up to 100 for less well transforming oncogenes) colonies should be picked to ensure a sufficient number of healthy transformed clones with a long *in vitro* life span (see below). Try to pick the largest colonies first since these grow into long-lived clones with an enhanced probability.[5] It is important to avoid extensive alkalinization (blue-red color) of the medium during this procedure (e.g., by returning the microwell plates to the CO_2 incubator at regular intervals).

Culture of the picked colonies has to be done by avoiding any possible evaporation of medium. Cultures are fed daily, first by two additions of 50 μl of medium, then by partial medium changes (take out 70 μl and add 100 μl with multichannel pipettes). Even for cell types apparently factor independent in mass culture, it may be advisable to supplement the medium with appropriate growth factors until the transformed colonies have grown out to about 10^5 cells per well.

Screening of Transformed Clones for Various Parameters

As a rule, clones of transformed avian hematopoietic cells show clonal heterogeneity, differing from each other in growth rate, extent of spontaneous differentiation, and *in vitro* life span. To allow the characterization of the effect of a given oncogene, 20 to 100 clones should be screened for these parameters and several (5–10) long-lived clones with properties spanning this range of phenotypes should then be grown into mass cultures. If temperature-sensitive (ts) oncogene mutants or conditional versions of oncogenes are used, it is likewise necessary to screen clones for a maximum phenotypic effect after shutoff of the oncogene function. A basic screening method that can be adapted to the specific requirements of the experiment in question is outlined below.

1. Prepare microwell plates containing the test media (referred to as assay plates; e.g., two identical plates with differentiation medium to be placed at 36° and 42° to screen for a ts-kinase oncogene or two plates with and without a given hormone or growth factor to screen for factor-dependent growth or a ligand-dependent transforming gene). Store microwell plates in a CO_2 incubator until needed.

2. Start screening procedure when the largest colonies have filled about half of the well. With multichannel pipettes, suspend the contents of each well from the original plate (referred to as the master plate) and transfer between one-fourth (one to two screening conditions) to one-half (three to four screening conditions) of the cell suspension to a 96-well plate with conical bottoms (e.g., Greiner 651180). Refeed master plate with at least 100 μl of fresh growth medium.

3. Spin down cells at 500*g* for 5 min at 4° in a centrifuge equipped with a rotor accepting 96-well plates (slow acceleration, no brakes). Remove supernatant with micropipette tip applied to a suction tube, taking care to avoid disturbance of the pellets. Keep empty wells always covered (e.g., by the lid of the microwell dish) to avoid drying out of the wells. Wash once in 100 μl of a medium lacking any component to be assayed (complex growth medium lacking chicken serum is usually suitable).

4. Resuspend cells in a volume large enough to remove 20-μl aliquots for each screening condition (e.g., 70 μl for three assay plates). Distribute the cells to the assay plates in 20-μl aliquots. Place assay plates in moist chambers and cultivate for the appropriate time periods. If feeding of the plates during incubation is required, this is best done by adding 50 μl per well from extra aliquots of the test media kept frozen until use.

5. Screen by appropriate assays (visual inspection, various qualitative and quantitative assays, see below). Design assay in a fashion that the result is available soon since the screening procedure has to be finished before the cells grow too dense in the master plate.

6. Keep cells in the master plate healthy by daily exchange of most of the medium. Removal of cells should only be done on those wells that acidify the medium in less than 24 hr (or are obviously too dense).

Isolate 5–20 clones showing the desired phenotype and expand [very dense clones can be transferred into 35-mm dishes, the others should be transferred to the 17-mm wells of 4-well plates (Nunc 1-76740, preferable to 24-well plates because of easier handling and moisture control)]. When expanded to a dense 100-mm dish, freeze cells in liquid nitrogen in two to three aliquots. Keep an aliquot (35 mm dish) in culture for life span assessment.

Freezing and Thawing Avian Hematopoietic Cells

If primary avian hematopoietic cells are frozen and thawed according to standard protocols,[17,19] no or inadequate survival is sometimes observed. The following hints may be useful in achieving maximum recovery of primary avian cells after freezing, which is important because of the fixed life span of these cells.

1. Do not freeze cells that have slowed down in growth rate due to beginning senescence or that have been exposed to unfavorable culture conditions (acidification, lack of growth factors, etc.).

2. Freeze cells at a concentration of at least 10–20 $\times$ 10^6 cells/ml. If the cell number is limiting, freeze 0.5 or 1-ml aliquots instead of the standard 1.5–2 ml.

3. Suspend cells in the appropriate volume of undiluted, ice-cold fetal calf serum. Growth factors can be added at this stage for those cells that have problems reviving despite using other precautions. Within 1–2 min, add 10% of analytical grade dimethyl sulfoxide (DMSO) in five portions with continuous, gently swirling of the cell suspension. Dispense into prechilled freezing vials and freeze immediately according to standard procedures. Transfer to a liquid nitrogen tank within 1 month, otherwise viability can be severely reduced.

4. Thaw ampoules rapidly in a 37° water bath. When there is only a small amount of ice left, put on ice. Mix with a fivefold volume of growth medium containing necessary growth factors, spin down at 500*g* for 5 min. Resuspend at $>5 \times 10^6$ cells of prewarmed growth medium (should have orange color, alkaline media can damage cells), add growth factors, and place in CO_2 incubator. The whole procedure should be completed within 10–15 min. For particularly sensitive cells (E26 myeloblasts, normal erythroid progenitors), cells should be thawed in the morning and cultivated at very high cell density (10–20 $\times$ 10^6 cells) during the first 6–8 hr. If the cells start to acidify the medium or look healthy after 8 hr, dilute to 3–5 $\times$ 10^6, otherwise cultivate overnight before dilution.

Determination of Clonal *in Vitro* Life Span

A general feature of primary cells is their limited *in vitro* life span (30 to 40 generations in mass cultures of chicken fibroblasts). On a clonal basis, *in vitro* life spans are much more variable, with most clones having only short *in vitro* life spans, whereas rare clones can have much longer ones (see Beug and Graf,[5] and references therein). Since this phenomenon not only applies to chicken fibroblasts but also to hematopoietic cells of various lineages (H. Beug and T. Graf, unpublished observations), one prerequisite for obtaining clones suitable for molecular analysis is the identification of those which have an extended *in vitro* life span. Since a long *in vitro* life span is a stable property of a given clone, the effort to screen for such clones is justified by the extended possibilities for experimentation such clones provide. This is particularly relevant due to the observation that important features of clones such as efficient *in vitro* differentiation are lost five or more generations before senescence-associated loss of proliferation potential becomes obvious (H. Beug, unpublished).

To screen hematopoietic cell clones for *in vitro* life span, they are subjected to daily serial passages in 35-mm dishes, keeping cell densities between 1 and 3×10^6 cells/ml, and monitoring the number of passages accrued until senescence. Senescence is evident by a slowing down of proliferation, an increase in cell size heterogeneity, and an accumulation of irregular shaped as well as apoptotic cells. One week after the onset of serial passage, clones still exhibiting a normal doubling time (<24 hr) are subjected to a rescreen of their desired features, i.e., their ability to differentiate, to respond to growth factors, etc. Those clones which essentially retain their characteristics from the original screen are, as a rule, the ones to select for. Nevertheless, to avoid false positives it is important to determine the life span of such well-differentiating clones by serial passage until senescence occurs.

Primary Avian Hematopoietic Cells from Different Lineages: Transformation with Retroviral Oncogenes

Erythroid Cells

Generation of Transformed Mass Cultures and Clones

The ease with which chick erythroid progenitors can be transformed by oncogenes depends mostly on the oncogene (or oncogene combination) chosen. While retroviruses in which the primary transforming oncogene (v-*erbB*, v-*sea*, v-*fps*, v-*src*, v-*yes*, v-*fms*, v-Ha-*ras*, v-*mil*) is combined with the v-*erbA* oncogene will transform erythroblasts under a relatively wide range of conditions,[24] these oncogenes alone require special medium conditions in Methocel, during propagation of the clones, and during mass culture.[25] Since these special conditions, however, are also beneficial for v-*erbA*-containing clones, only those methods will be described here.

To generate transformed erythroblasts, BM cells are cocultivated with the respective virus-producing fibroblasts in the complex growth medium for 48 hr as described. For strongly transforming retroviruses (v-*erbB*, retroviruses containing c-*erbA*), no exogenous growth factors are required except 1 μg/ml insulin. For the other oncogenes, the efficiency of transformation can be largely increased by the addition of 2–5% anemic serum (containing a variety of avian growth factors in addition to erythropoietin) during cocultivation. The addition of 100 ng/ml of avian stem cell factor (SCF[26]) is even more efficient in increasing transformation efficiency. The preparation of these agents is described below.

After cocultivation, cells are cultivated in complex growth medium plus 1 μg/ml insulin until the transformed cells grow out into mass cultures. Alternatively, the infected BM cells are centrifuged through Percoll (density 1.072 g/cm^3), and the interphase cells are seeded at several concentations (ranging from 0.3 to 2 $\times$ 10^6 cells per 3–4 ml of methocel) into Methocel supplemented with the components of complex growth medium plus 1 μg/ml insulin. A range of cell concentrations is required to avoid insufficient numbers of transformed colonies on the one hand and premature acidification of the cultures by an excess of macrophages and/or transformed colonies on the other. For certain weakly transforming oncogenes (v-*src*, v-*mil*, certain v-*erb*B- and v-*sea* mutants), it can be helpful to add 5 ng/ml of mammalian recombinant TGFα (a ligand for avian c-*erbB* expressed in certain erythroblast progenitors, see below), but screening of the resultant clones for their transformed state is then absolutely required. Because of the large volume of the additions, it may be advisable to use a Methocel with higher than standard viscosity (see earlier discussion). Methocel cultures should be fed with 0.5–1 ml of the complex growth medium (supplemented with 0.5–1 mg/liter extra sodium bicarbonate if the cultures tend to become acidic) at days 5, 8, and 10.

After 5–10 days, compact, white to light pink colonies of transformed

[24] T. Graf, N. Ade, and H. Beug, *Nature* (*London*) **275,** 496 (1978).

[25] H. Beug, P. Kahn, G. Doederlein, M. J. Hayman, and T. Graf, *in* "Modern Trends in Human Leukaemia VI" (R. Neth, R. Gallo, M. Greaves, and K. Janka, eds.), p. 284. Springer-Verlag Berlin, 1985.

[26] M. J. Hayman, S. Meyer, F. Martin, P. Steinlein, and H. Beug, *Cell* (*Cambridge, Mass.*) **74,** 157 (1993).

cells appear in the cultures. Colonies should be allowed to grow to a size of 0.2–0.5 mm in diameter, but only if this is possible without acidification of the cultures. Colonies are then isolated, placed in 96-well tissue culture plates containing 100 μl of complex growth medium, and cultivated as described earlier. For most viruses (except those containing v-*erb*A), it is advisable to supplement the medium with either anemic serum or 100 ng/ml of avian SCF until the transformed colonies have grown out to about 10^5 cells.

Screening Procedures for Transformed Erythroid Cells

Erythroid cells transformed by nonconditional tyrosine kinase oncogenes should be screened for their ability to spontaneously differentiate. This is done (i) by visual inspection, checking for the presence of oval, flat erythrocytes in the cultures; and (ii) by staining for hemoglobin, using acid benzidine staining (see Graf *et al.*,[24] and references therein). Since the described procedures lead to high background staining in the media recommended here, the cells should be washed once in pure DMEM using conical 96-well plates, resuspended in the same medium, and transferred to normal 96-well plates (see earlier discussion) before adding the acid benzidine staining mix.

Methods for Analyzing Differentiation in Erythroid Cells

Anemic Serum. Anemic serum is prepared from young adult chickens (2–3 months old; older animals cause problems because of insufficient activity) by the following procedure[27]:

1. Phenylhydrazine solution: Rapidly dissolve 4 g phenylhydrazine hydrochloride (PHZ, Merck, p.a.) in 400 ml doubly distilled H_2O and adjust to pH 7.0 by adding 2 *N* NaOH. The mixture is highly unstable and generates toxic HCl gas, therefore work as fast as possible and immediately shock-freeze 10-ml aliquots in dry ice or liquid nitrogen. Store at $-20°$ until needed.
2. Thaw quickly an aliquot of PHZ, then keep on ice and use within 15 min after thawing.
3. Inject im (breast muscle) 2.0 ml/kg on day 1. Inject similarly 1.0 ml/kg im on day 2 and 0.5 ml/kg on day 3. If more than one-third of the animals have died after the second injection, omit the third injection or reduce the amount of phenylhydrazine injected on day 3 by half.

[27] E. Kowenz, A. Leutz, G. Döderlein, T. Graf, and H. Beug, *in* "Modern Trends in Human Leukemia VII" (R. Neth, R. C. Gallo, M. F. Greaves, and H. Kabisch, eds.), p. 199. Springer-Verlag, Heidelberg, 1987.

4. Bleed the animals on day 4 by heart puncture. Success of this procedure is entirely dependent on the availability of needles of the right length (chicks older than 3 months require special needles longer than 8 cm).

5. Allow the serum to clot at room temperature for 1 hr in 50-ml Falcon tubes. Loosen the clot from the tube wall with a sterile plastic spatula. Spin for 30 min at 4° and 5000–10,000*g* (8000–10,000 rpm, Sorvall centrifuge, Rotor HS 4). This spin is also performed if sera have not yet clotted. Retrieve supernatant (serum), and spin serum again in tabletop centrifuge at 2000*g* (4000 rpm). Freeze in aliquots at −70°, and keep a small aliquot for testing.

Anemic Serum Test.[27] As test cells, clones of erythroblasts transformed with either ts v-*sea*[28] or human c-*erbB*/EGFR plus v-*myb*EEA[29] that are growing well are suitable. The latter cells are more easily used since suitable cell populations are more readily accessible and have a higher *in vitro* life span. A third alternative is the use of 4- to 6-day-old, SCF-induced normal erythroid progenitors (see below); these do not require retroviral transformation or a shift to 42°, but their use is dependent on an unlimited supply of avian SCF (see below).

1. Prepare complex growth medium without chicken serum (Epotest medium), supplemented with 10 ng/ml human insulin (transformed cells) or 1 ng/ml insulin (normal SCF progenitors).

2. Dispense 10, 5, and 2 μl of the anemic sera to be tested into triplicate wells of a 96-well tissue culture plate. As positive and negative controls, use an anemic serum of known activity and normal chicken serum, respectively, at the same concentrations. Add 100 μl of the Epotest medium to all wells. Allow for respective empty wells as blanks if the hemoglobin assay is used to evaluate the test.

3. Seed test cells (20–40,000 for EGFR–ts-v-*mybEEA*, 30–60,000 for ts-v-*sea*, and 20–40,000 for normal SCF progenitors). Suitable cell concentrations have to be determined by pilot experiments since they depend on the test cell batch.

4. Incubate at 42° (ts v-*sea*, EGFR–ts-v-*mybEEA*) or 37° (normal SCF progenitor cells). The incubation time is dependent on the test cell batch available. It is crucial for the sucess of the assay to determine the optimal incubation time for each test cell batch. This has to be done by pilot experiments, incubating test cells for 40, 48, 60, and 72 hr and selecting the

[28] J. Knight, M. Zenke, C. Disela, E. Kowenz, P. Vogt, J. D. Engel, M. J. Hayman, and H. Beug, *Genes Dev.* **2,** 247 (1988).

[29] H. Beug, G. Doederlein, and M. Zenke, *in* "Nuclear Processes and Oncogenes" (P. A. Sharp, ed.), p. 53. Academic Press, San Diego, CA, 1992.

time where the difference between anemic and normal sera is maximal. By visual inspection, this time point is evident by oval, but still dividing, cells in the wells containing the anemic serum standard and mostly apoptotic cells (discernible by condensed chromatin under phase contrast) in the control wells. It is thus advisable to freeze potential assay cells in many small aliquots in liquid nitrogen, thaw one test ampoule, and use for the just-described pilot experiments.

5. Evaluate the test by measuring the thymidine incorporation (EGFR–ts-v-*myb*EEA cells, 2–3 days; ts-v-*sea* cells, 1–2 days) or by the photometric hemoglobin assay (ts-v-*sea* cells, 2–3 days; normal SCF progenitors, 3 days). Both assays are described next. Tests with a less than 5- to 10-fold difference between standard anemic serum and control serum should be repeated with a different test cell batch.

Thymidine Incorporation Assay. This assay is based on the fact that differentiating erythroblasts go through three to give cell divisions while accumulating hemoglobin in the presence of differentiation hormones (Epo/anemic serum, insulin) whereas they wihdraw from the cycle after 24 hr in the absence of such factors. It is performed according to Leutz *et al.*[30] with the following modifications. Add 0.8 μCi of tritiated thymidine (in 10 μl of modified DMEM) to the wells of a 96-well test plate ready for assay. Incubate for 3 hr at respective temperature. Harvest with a suitable manual (Skatron) or automated cell harvester (TOMTEC, plus 96-well scintillation counter Wallac Microbeta 1450), making sure that the machine background (no cells are added to the well) is below 500 cpm.

MTT Assay (Promega). This assay measures the amount of metabolically active, viable cells and may be helpful with certain cell types where thymidin incorporation gives uninterpretable results. The assay should be performed according to the instructions of the manufacturer, with the following alterations: Add 15 μl of the MTT solution to 100 μl of the cell suspension, incubate for 4 hr in the CO_2 incubator, add 130 μl of solubilization fluid to the wells, incubate overnight in the dark, mix contents of the wells with multichannel pipettes, and measure in an enzyme-linked immunosorbent assay (ELISA) photometer at a measuring wavelength of 590 nm versus a reference wavelength of 690 nm. Blanks receive the same medium as the experimental wells, but without cells.

Hemoglobin Assay. This assay, described in Kowenz *et al.*,[27] measures the amount of hemoglobin present in viable cells by a photometric procedure. Because of possible problems with the availability of that reference, the method is briefly described below.

[30] A. Leutz, H. Beug, and T. Graf, *EMBO J.* **3,** 3191 (1984).

1. With a multichannel pipette, resuspend the cells in a 96-well test plate ready for assay. Transfer to a 96-well plate with conical-bottomed wells, and spin the plate for 5 min at 4° in a special 96-well plate rotor (no brake).

2. Carefully aspirate the supernatants. Cell pellets will be mostly visible and terminally differentiated cells will form a red cell pellet. Suspend cells in 130 μl Hanks' balanced salt solution + 1% detoxified BSA (see above) and spin down as before. Wash the pellets once more in this buffer.

3. After careful aspiration of the supernatant, add 30 μl sterile H_2O to each well to lyse the cells (including wells for blanks that did not receive cells). Resuspend the pellets by holding each side of the plate to a vortex mixer. Incubate for 20–30 min at 4°.

4. To prepare the staining reagent, dissolve *o*-phenylenediamine (Kodak) in 0.1 *M* citrate/phosphate buffer, pH 5.0, at 0.5 mg/ml, then add 1 μl concentrated (33%) hydrogen peroxide per ml of above solution. This reagent has to be made fresh immediately before use.

5. Using a multichannel pipette, add 125 μl of the staining reagent to each well (at 10- to 30- sec interval per row). Cover the plate with aluminum foil and incubate for exactly 5–20 min (monitor the color reaction during this time; 20 min is the maximum to achieve the yellow color of a positive sample).

6. In the meantime, take a suitable flat-bottom 96-well plate and dispense 25 μl 8 *N* H_2SO_4 to each well. Stop the reaction by transferring the samples row by row (again at 10- to 30-sec intervals) to the plate with H_2SO_4. The color of the samples intensifies.

7. Determine optical density in an ELISA photometer (a measuring wavelength of 492 nm vs a reference wavelength of 620 nm). In each assay, use extracts containing a known number of chick red blood cells from peripheral blood as standards.

Histological Plus Hemoglobin Staining of Cells Cytocentrifuged onto Slides. This test allows analysis of the respective proportion of immature, partially mature, and mature cells in a given cell preparation as well as the content of apoptotic and lysed cells. It should therefore always be performed in parallel with the quantitative assay to judge the relevance of the quantitative data obtained. Cytocentrifugation and staining are performed and evaluated according to Beug *et al.*[31,32] It is very important first to stain a test slide (a cytospin prepared from normal bone marrow, stored in a dry

[31] H. Beug, S. Palmieri, C. Freudenstein, H. Zentgraf, and T. Graf, *Cell (Cambridge, Mass.)* **28,** 907 (1982).

[32] J. A. Schmidt, J. Marshall, M. J. Hayman, P. Ponka, and H. Beug, *Cell (Cambridge, Mass.)* **46,** 41 (1986).

dark place no longer than 4 weeks) to check that the benzidine solution and the histological stains are still working.

Differentiation Assays of Erythroblasts Transformed with Conditional Oncogenes

Clones to be screened for an oncogene-induced, temperature-sensitive phenotype are aliquoted to assay plates containing (i) Epotest medium with 3% high titer anemic serum plus 10 ng/ml insulin and (ii) 3% normal chicken serum plus the same amount of insulin (see above). After 2–3 days at 42°, clones are visually inspected for healthy, oval, erythrocyte-like cells in the presence of anemic serum/insulin, but dead cells in the absence of hormone. If required, they are then processed for the hemoglobin assay. A thymidine assay is not usually required in these screens. Clones with the best score in visual inspection plus the largest difference in hemoglobin content are selected for further use.

Clones expressing other conditional oncogenes can be screened by analogous methods. For instance, erythroblasts expressing exogenous avian or human c-*erbB* are screened at 37° (i) for ligand-dependent transformation in anemic serum/insulin-containing media containing or lacking TGFα or EGF and (ii) for differentiation ability in media containing or lacking anemic serum/insulin after withdrawal of TGFα or EGF.[33,34]

For quantitative monitoring of differentiation in pretested clones, cells are seeded at 1.5×10^6/ml in Epotest medium containing or lacking anemia serum plus insulin (35-mm dishes with 2 ml of medium). Cell numbers and cell size distributions are determined daily and cells are kept at cell densities between 1 and 2×10^6/ml by the addition of fresh medium and/or removal of the cell suspension. Thymidine incorporation assays are performed on three to four 100-μl aliquots at days 1 and 2, hemoglobin assays on similar aliquots at days 2 and 3. (For more slowly differentiating cells, the times of assay should be varied accordingly.) Cytospin preparations are made at days 2, 3, and 4. These combined assays allow to characterize clones transformed with conditional (e.g., ts) oncogenes after successful initial screening and assessment of the *in vitro* life span.[28] They are also quite useful in determining the effect of additional genes on erythroid differentiation in such ts oncogene-transformed clones. For further information, see Beug *et al.*[13–16, 34–38]

[33] K. Khazaie, T. J. Dull, T. Graf, J. Schlessinger, A. Ullrich, H. Beug, and B. Vennström, *EMBO J.* **7,** 3061 (1988).

[34] C. Schroeder, L. Gibson, C. Nordström, and H. Beug, *EMBO J.* **12,** 951 (1993).

[35] C. Schroeder, L. Gibson, and H. Beug, *Oncogene* **7,** 203 (1992).

[36] J. Larsen, H. Beug, and M. J. Hayman, *Oncogene* **7,** 1903 (1992).

Myeloid Cells

Transformation of Macrophages with v-myc-Containing Viruses

Transformation of macrophage-like cells by avian retroviruses containing the v-*myc* oncogenes (MC29, CM-II, OK-10, and MH-2,[9]) can be performed in Methocel containing the additions of standard growth medium. Purification of bone marrow by density is not necessary. To generate v-*myc*-transformed macrophage clones, between 0.5 and 2×10^6 unpurified BM cells are seeded per 3–4 ml of standard growth medium Methocel after infection by cocultivation with virus-producing fibroblasts. Since the target cells (monocytes and their progenitors[39]) divide readily in standard growth medium, infection can also be done by fibroblast supernatants.

The diffuse, large colonies are picked after 5–8 days and can be propagated and expanded in standard growth medium. The addition of cMGF (see below) is beneficial for growing colonies up, but is not essential. Clones should be screened for rapid growth, healthy appearance, and, most importantly, for prolonged growth without spontaneous withdrawal from the cell cycle. This "regression" phenomenon, which probably reflects spontaneous differentiation of the transformed clones, is accompanied by an increase in cell size and medium acidification, at the same time the cells become more strongly adherent and cease to proliferate. In addition to selection for clones that do not regress readily, the cells should not be allowed to become too sparse ($<1 \times 10^6$ cells/ml) nor should they be exposed to spent, acidified medium for prolonged time periods since both conditions can induce "regression" even in selected clones that are growing well.

Transformation of Myeloblasts (and Cells Differentiating into More Than One Lineage) by v-myb-Containing Viruses

Two viruses transform immature myeloid cells: the AMV strain expressing a 55-kDa v-*myb* protein and the E26 strain that transforms via a 135-kDa fusion protein ($p135^{gag\text{-}myb\text{-}ets}$) containing viral *gag* sequences as well as the two mutated transcription factors v-*myb* and v-*ets*. The E26 virus also transforms immature progenitors that can develop into more than one lineage.[10] The self-renewal of immature progenitor cells that can develop

[37] K. Briegel, K. C. Lim, C. Plank, H. Beug, J. D. Engel, and M. Zenke, *Genes Dev.* **7,** 1097 (1993).

[38] D. Barettino, T. H. Bugge, P. Bartunek, M. M. Vivanco Ruiz, V. Sonntag-Buck, H. Beug Zenke, and H. G. Stunnenberg, *EMBO J.* **12,** 1343 (1993).

[39] T. Graf, K. A. von, and H. Beug, *Exp. Cell Res.* **131,** 331 (1981).

into more than one lineage is also induced by the v-*ski* oncogene.[11,39a] In contrast to erythroid cells transformed by tyrosine kinase oncogenes, the immature myeloid and multipotential cells transformed by the latter viruses retain their strict dependence on avian hematopoietic growth factors (cytokines). Therefore, we will first describe some sources for avian cytokines as well as methods to produce them.

Production and Purification of Avian Cytokines

METHOD I. Two avian cytokines are available as cDNA clones; avian cMGF (chicken myelomonocytic growth factor[40]) and avian SCF (stem cell factor[26]), the ligand for the avian receptor tyrosine kinase c-*Kit*.[41] cMGF displays weak homology to the mammalian cytokines G-CSF and interleukin 6 (IL-6); structural considerations suggest that cMGF is the avian IL-6 homolog.[42] Avian SCF shows weak, but distinct homology to mammalian *Kit* ligand, but is unable to activate the mammalian c-*Kit* receptor tyrosine kinase.[42a]

The method of choice to mass produce both cMGF and SCF is to express these two factors in bacteria after modification of the cDNAs with an oligonucleotide encoding six histidines. This allows easy, one-step purification of the factors by affinity chromatography using metal-chelate columns.[43]

Solutions:

Buffer **A:** 6 *M* guanidine hydrochloride, 100 m*M* NaH_2PO_4, 100 m*M* Tris, 10 m*M* 2-mercaptoethanol (β-ME; add immediately before use), adjust pH to 8.0 with NaOH.

Buffers **B–E:** Make fresh every time: 8 *M* urea, 100 m*M* NaH_2PO_4, 10 m*M* Tris–HCl, 10 m*M* β-ME. For buffer **B,** adjust to pH 8.0; for buffers **C, D,** and **E,** adjust to pH 6.3, 5.7, and 4.5, respectively. Adjust pH accurately and just before use. Also prepare a 1 *M* stock of IPTG in doubly distilled H_2O and a 250 m*M* stock solution of L-cysteine in 0.5 *M* NaOH.

[39a] H. Beug, R. Dahl, P. Steinlein, S. Meyer, E. M. Deiner, and H. Beug, *Oncogene,* in press (1995).

[40] A. Leutz, K. Damm, E. Sterneck, E. Kowenz, S. Ness, R. Frank, H. Gausepohl, Y. C. Pan, J. Smart, and M. Hayman, *EMBO J.* **8,** 175 (1989).

[41] E. Sasaki, H. Okamura, T. Chikamune, Y. Kanai, M. Watanabe, M. Naito, and M. Sakurai, *Gene* **128,** 257 (1993).

[42] J. L. Boulay and W. E. Paul, *Curr. Biol.* **3,** 573 (1993).

[42a] O. Wessely, M. von Lindern, A. Levitzki, J. Jschenko, M. Hagman, and H. Beug, submitted.

[43] E. Hochuli, H. Döbeli, and A. Schacher, *J. Chromatogr.* **411,** 177 (1987).

1. Inoculate 1 liter of LB medium (+100 μg/ml ampicillin) with a 50-ml culture (inoculated from glycerol stock) of BL21 (DE3) bacteria containing the expression plasmids for cMGF or aSCF (pHis-SCF and pHis cMGF, P. Bartunek *et al.*, manuscript in preparation; the aSCF cDNA was a kind gift of F. Martin, AMGEN, and can only be made available with prior consent by AMGEN) and grow for 3–4 hr at 37°. Grow the bacteria up to OD_{600} of 0.7–0.9, and take a 10-ml aliquot (uninduced control), add IPTG to a final concentration of 1 m*M*, and grow for an additional 2–3 hr at 37°. Remove another 10-ml aliquot (induced control) and harvest the cells by 15 min centrifugation at 4500 rpm. This and all the next steps can be performed at room temperature.

2. Resuspend the bacterial pellet (ca. 4 g of wet mass), resuspend in 25 ml of buffer **A,** and stir for 1 hr on the magnetic stirrer. During this time prepare the agarose resin and buffers **B–E.** Take 8 ml of the Ni-NTA agarose slurry (QIAGEN, ca. 5 ml of the packed column) and wash once with 40 ml of buffer **A** in a 50-ml Falcon tube. Centrifuge the bacterial lysate at 10,000 rpm for 15 min. Transfer the supernatant to the tube with the Ni-NTA resin and incubate for 1 hr with continuous agitation. Spin at 4000 rpm and wash the resin again with buffer **A** (40 ml), then incubate 10 min with buffer **B** (urea pH 8.0) to change denaturant, spin, and wash with 45 ml of buffer **C** (pH 6.3) Spin, then resuspend the resin in 3 ml and load it into a column (e.g., Bio-Rad Poly-Prep).

3. Use 30–40 ml of buffer **D** (pH 5.7) to elute the His protein. Collect five fractions of 10 ml. Determine protein in all fractions and stored washing supernatants using Bradford's reagent and subsequently run the fractions together with whole lysates of positive and negative controls [−/+ isopropylthiogalactoside (IPTG)] on a 12% SDS–polyacrylamide gel.

4. Wash the column with 30 ml of buffer **E** (pH 4.5) to elute retained multimeric protein(s) from the column. Pool the fractions containing the desired His proteins, add L-cysteine to a final concentration of 10 m*M*, and incubate for 20 min to oxidize the reduced SH groups. Finally, dialyze the protein against 2 liters of 1 × PBS for 12 hr with one change of 1 × PBS. Measure the protein concentration by the Bradford method, freeze in aliquots at a concentration of 1 mg/ml or 100 μg/ml in liquid N_2, and store at −40° or −70°. Analyze biological activity by titration on normal SCF progenitors (His-SCF) and E26-transformed myeloblasts (His-cMGF) and analysis by thymidine incorporation or MTT assay (see above).

METHOD II. A rich alternative source for cMGF is the avian v-*myc*-transformed macrophage cell line HD 11.[30]

1. Grow HD 11 cells to confluency on two to four 245 × 245-mm dishes, using standard growth medium. Wash the cells twice with serum-free medium [500 ml modified DMEM, 5 μg/ml human recombinant insulin,

35 μg/ml iron-saturated conalbumin (ovotransferrin, Sigma) 20 μM ethanolamine, 10^{-4} M 2-mercaptoethanol, 10 mM HEPES, pH 7.3] and suitable antibiotics to minimize risk of contaminations. Resuspend the cells in this medium at 1–2 × 10^6/ml and incubate for 16–18 hr.

2. The factor yield is stimulated by bacterial lipopolysaccharide (LPS). Add to a final concentration of 5 μg/ml, from a 1-mg/ml stock solution.

3. Harvest the supernatant the following day and sterile filter. Concentrate 20- to 100-fold on a Amicon YM10 membrane. Remove virus by ultracentrifugation (1 hr, 100,000g, 4°). Dialyze exhaustively against Hanks' balanced salt solution lacking bicarbonate and phenol red, using a dialysis tubing with a size exclusion range of 5000–8000 Da.

4. Determine the activity of the conditioned media in a growth factor test[30] (see above).

METHOD III. By similar procedures, conditioned media containing various avian cytokines can also be produced from (i) the chicken embryo fibroblast cell line CEF 32,[14] (ii) chicken embryo fibroblasts infected with a c-*rel*-expressing retrovirus (P. Enrietto and H. Beug, unpublished), and (iii) the v-*rel*-transformed lymphoid cell line NPB4.[27,44] The CEF 32 cells produce a factor that is different from cMGF and act mainly on macrophages; it may be the avian CSF-1 counterpart. The c-*rel*-transformed CEF produces cMGF plus an uncharacterized cytokine(s) active on erythroid, monocytic, and granulocytic cells.

Finally, the v-*rel* lymphoblasts produce apparently one activity that acts on erythroid cells only (and cooperates with avian Epo present in anemic serum) as well as several myeloid cytokines. These erythroid and myeloid activities can be separated by a simple chromatographic procedure:

1. Dilute 1 liter of NPB4-conditioned medium with 2 liters of doubly distilled H_2O. Apply to a tandem array of two columns: the first containing ca. 40 ml DEAE Sephacel (Pharmacia) equilibrated with 0.05 M borate buffer, pH 8.0, and the second containing 40 ml CM Sephacel equilibrated with 0.05 M acetate buffer, pH 4.0.

2. When the whole amount of the diluted conditioned medium has been passed over both columns, separate columns. Wash each column with 10–20 vol of the respective equilibration buffer, then elute separately with 2–3 column volumes of the respective equilibration buffers containing 1 M NaCl. Concentrate eluted fractions by ultrafiltration using Amicon YM5 membranes, dialyze exhaustively against Hanks' balanced salt solution without sodium bicarbonate and phenol red, and store frozen in aliquots at

[44] H. Beug, H. Müller, S. Grieser, G. Doederlein, and T. Graf, *Virology* **115,** 295 (1981).

−40°. The erythroid factor(s) elutes preferentially from the DEAE column, and the myeloid ones from the CM column.

METHOD IV. Conditioned medium preparations containing cMGF and other uncharacterized factors can also be prepared from chicken spleen cells stimulated with concanavalin A.[45]

1. Remove spleens from several 6-week-old chickens and place in standard growth medium. Remove any connective tissue. Make an incision in the protective outer skin of the spleen and use a small round spatula to remove the soft inner part.
2. Using the spatula, press the tissue through a sterile stainless-steel sieve (100 μm) into a dish placed below the sieve and containing standard growth medium.
3. Suspend the cells well and pass the suspension through nylon mesh or a commercial cell strainer. Spin 5 min at 500–1000g (1000–1500 rpm).
4. Wash once more, then resuspend the cells in standard growth medium, count, adjust to 5×10^6/ml, and seed in 145-mm dishes containing 25 ml of medium. Add concanavalin A (Pharmacia) to a final concentration of 5 μg/ml (from sterile 1 mg/ml stock solution in PBS) and incubate the cells at 39° for 3 days.
5. Harvest the supernatant by centrifugation, add α-methylmannoside to a final concentration of 5 mM to inactivate the lectin, filter through a 0.45-μm membrane filter, and store frozen in aliquots at −20°. The conditioned medium must be titrated for activity on E26-transformed cells by the thymidine assay as described earlier.

Conditions for Transformation by E26 Virus of Myeloid, Erythroid, and Multipotent Cells

E26 virus infection proved to be difficult using virus-infected fibroblasts because this virus has not been available with a resistance marker and well-producing fibroblasts were essentially impossible to produce. For infection, 25×10^6 fresh BM cells are therefore cocultivated with 2×10^5 E26 virus myeloblasts from clones checked for high virus production and treated with mitomycin C as described for CEF. Infection can also be done with filtered supernatants from similar pretested clones of E26-transformed myeloblasts. To generate E26-transformed myeloblast clones, $2–5 \times 10^6$ infected BM cells are seeded into Methocel containing the additions of the standard growth medium plus 10 ng/ml cMGF (or conditioned media at concentrations giving a maximal response on E26 myeloblast test cells). Within 6–8 days, diffuse colonies of small round cells (much smaller than those in

[45] H. Beug, M. J. Hayman, and T. Graf, *EMBO J.* **1,** 1069 (1982).

contaminating macrophage colonies) will appear that can be expanded in standard growth medium plus 10 ng/ml of cMGF. Erythroid and multipotent colonies (that look much more compact than the myeloid cells[10,20]) can be obtained in Methocel containing the additions of complex growth medium supplemented with 3–5% of anemic chicken serum and 10 ng/ml of cMGF. These cells can be propagated in complex growth medium supplemented with 1 μg/ml of insulin; the addition of anemic serum and cMGF may be beneficial.

The conditions to differentiate myeloblasts transformed by ts mutants of the E26 virus are much less sophisticated than those described for the erythroid cells.[8,46] The best differentiation into macrophage-like cells or large, multinucleated, nondividing cells morphologically resembling osteoclasts is obtained in a modified standard growth medium (DMEM supplemented with 10% fetal calf serum, 100 μg/ml iron-saturated transferrin, 25 m*M* HEPES, and 10 ng/ml recombinant cMGF). The addition of 100 ng/ml SCF proved to be beneficial. To induce differentiation in erythroid cells transformed by the ts1.1 mutant of the E26 virus,[47] the very same conditions as used for kinase-oncogene-transformed erythroid cells can be used.

Lymphoid Cells

Transformation of Lymphoid Cells with v-rel-Containing Viruses

The turkey retrovirus REV-T and retroviral constructs containing the v-*rel* oncogene in a chicken retrovirus background (RCAS-v-*rel*[48]) give rise to transformed immature cells that share properties with pre-B–pre-T-lymphoblasts,[44] myeloid cells,[48] and perhaps multipotent cells.[49] Mass cultures of such transformed v-*rel* lymphoblasts can readily be obtained by cocultivating 25×10^6 fresh BM cells in standard or complex growth medium with 0.5 to 1×10^6 RCAS-v-*rel* expressing fibroblasts. Clones of v-*rel*-transformed cells can be obtained in Methocel containing complex growth medium additions. These colonies are very diffuse and consist of very small, irregular-shaped or elongated cells. They appear much later than those transformed by other retroviral oncogenes (10–14 days) so care must be taken that normal myeloid cells (macrophages) do not overgrow the culture. In particular, several concentrations of BM cells should always be tried. In

[46] H. Beug, P. A. Blundell, and T. Graf, *Genes Dev.* **1,** 277 (1987).

[47] N. Kraut, J. Frampton, K. McNagny, and T. Graf, *Genes Dev.* **8,** 33 (1994).

[48] L. E. Morrison, G. Boehmelt, H. Beug, and P. J. Enrietto, *Oncogene* **6,** 1657 (1991).

[49] G. Boehmelt, A. Walker, N. Kabrun, G. Mellitzer, B. H. M. Zenke, and P. J. Enrietto, *EMBO J.* **11,** 4641 (1992).

addition, the number of transformed colonies is always very small (1–10 colonies/5 × 10^6 BM cells seeded), thus multiple dishes have to be put up in order to get enough colonies. It is important that the v-*rel*-transformed colonies are allowed to get large enough (>0.5–2 mm) before picking, otherwise it may be very dificult to expand them in 96-well plates. This can be done in standard growth medium, but complex growth medium (to which 5% anemic serum can be added) is likely to be beneficial to grow up a larger percentage of the isolated clones.

Test Procedures for Media and Media Ingredients

Since essentially all commercially available components of tissue culture media have been optimized for mammalian immortalized cell lines, one cannot necessarily expect such products to be suitable for culture of primary, nonimmortalized avian cells. To ensure reproducibility of the techniques described so far, it is therefore required to subject any new batch of those ingredients listed below to comparative tests with respective batches that proved to support optimal growth and/or differentiation of the cells in question. The only real solution to the problem of how to find suitable ingredients when the just-mentioned techniques are set up for the first time is to purchase many different batches of the ingredient in question from different companies and compare them in the test systems described next.

Media Test

We have observed that different batches of powdered media may differ quite substantially in their ability to support the growth of avian cells. We therefore obtained test batches from powdered media (purchased in 1000-liter batches to minimize the frequency of testing) prior to ordering and tested them in proliferation assays on chicken embryo fibroblasts, ts-v-*sea*-transformed erythroblasts, the AEV-transformed cell line HD3,[50] and E26-transformed myeloblasts, comparing the new medium batch to an already tested one and using the same batches of pretested medium additions. In addition, the new medium batch should be used to make up Methocel-containing medium and be tested for the ability to support colony growth of ts-v-*sea*- and E26-transformed erythroblast and myeloblast colonies, respectively. Unless stated otherwise, the following procedures apply also for the test of other medium components.

[50] H. Beug, G. Doederlein, C. Freudenstein, and T. Graf, *J. Cell. Physiol., Suppl.* **1,** 195 (1982).

Adherent Cells (Fibroblasts)

Chicken embryo fibroblasts are seeded into 35-mm petri dishes at concentrations of 5, 2, and 1×10^5 cells per 2 ml of test medium. Duplicate dishes are set up for each cell concentration and each medium batch to be tested; four dishes are set up for the 1×10^5 cell sample. On consecutive days (starting at day 2 with the highest cell concentration), cells are washed twice with PBS, any remaining fluid is carefully removed, and 0.5 ml of trypsin solution is added to each dish. After 10 min, the cells are carefully suspended and an aliquot is counted in an electronic cell counter. When the spare dishes containing 1×10^5 cells have become confluent, they are trypsinized for $<$2–3 min, counted, and used to set up another series of dishes similar to the first one. In this fashion, the fibroblast growth rate can be monitored over 6–7 days.

Suspension Cells (AEV and Sea Erythroblasts, E26 Myeloblasts, etc.)

Seed cells at 1×10^6 cells/ml in test and reference media, allowing duplicate or triplicate dishes for each test batch and cell type. Keep dishes in moist chambers to minimize evaporation. Every day, remove 0.9 ml of supernatant (taking care not to remove any cells), add 1 ml of fresh medium, suspend, and remove sample for counting in an electronic counter. If the cell concentration exceeds 1.5 to 2×10^6/ml, remove cell suspension instead of supernatant and add fresh medium to readjust the cell concentration to $\approx 1 \times 10^6$/ml. (With some experience, the cell concentration can be judged from the appearance of the culture under the microscope, then cell suspension can be removed instead of supernatant from those dishes that are too dense.) Cell growth should be monitored in this fashion for 5–7 days.

Methocel Test

Make up Methocel media from the batches of powder to be tested (using a pretested powder batch as control) and add pretested ingredients for standard and complex growth media. Infect fresh BM cells by cocultivation with ts-v-*sea* CEF and E26 myeloblasts as described earlier and seed 0.2, 0.5, 1, and 3×10^6 infected BM cells per 4 ml of respective methocel medium. Set up duplicate dishes for each cell concentration and medium batch. After 6–9 days (when colonies have developed optimally) count colony numbers under an inverted microscope. In addition, 10–20 erythroblast and myeloblast colonies should be picked and grown to a few million cells to ensure their viability. In addition, seed 5000, 15,000, 50,000, and 150,000 HD3 cells per 3–4 ml of Methocel-containing complex growth medium additions plus 1 μg/ml insulin and count the number of clones 5–7 days later.

This extended test procedure is particularly important when switching to a new batch of Methocel powder is required. In this case, prepare small test batches of Methocel medium and purchase large amounts (1–2 kg) of those batches that were optimal in the tests. New batches of Methocel-containing medium made from an existing, pretested batch of powder also require testing, but the cloning of HD3 cells and the picking of colonies as well as the use of duplicates may be dispensible. To minimize testing, large batches of Methocel-containing medium should be prepared in this case. Since the most frequent cause for problems is too old or inappropriately frozen and thawed 2× concentrated growth medium, the above instructions for its storage and thawing should be closely followed.

FCS Test

New batches of fetal calf serum should be tested on CEF, HD3 cells, ts-v-*sea*-transformed fibroblasts, E26 myeloblasts, and for Methocel colony formation of ts-*sea*- and E26-transformed cells as described earlier. We have found that sera can be kept at −20° for about 2–3 years without loss of quality, therefore large batches should be purchased once tested. Since different serum batches may be optimal for the growth of erythroid cells in complex growth medium on the one hand and for growth of fibroblasts and/or myeloblasts in standard growth medium on the other, it can become necessary to keep different serum batches for different purposes.

Chicken Sera Test

Chicken sera should be tested on CEF and erythroid cells in standard and complex growth media, respectively. We have found it unnecessary to test chick sera in Methocel or on myeloblasts. The current batches available from Sigma have shown the most constant usefulness in our cultures.

BSA Test

New batches of BSA prepared using an already tested batch of ion exchanger need only be tested on ts v-*sea* cells and HD3 cells in complex growth medium. If the ion exchanger batch has to be changed, the BSA also has to be tested in Methocel and for its ability to support erythroid differentiation in the respective assays.

In Vitro Growth of Normal Hematopoietic Cells

Possibilities and Restrictions in Growing Normal Avian Hematopoietic Cells

Hematopoietic progenitors present in bone marrow are commonly believed to possess a fixed program of terminal differentiation. While executing this program, they undergo a limited number of cell divisions (5–10) and become increasingly mature at the same time. Most studies with human and murine bone marrow cell cultures therefore concentrated on the development of colony assays in semisolid media supplemented with various growth factors, with the end product being a colony consisting of terminally differentiated cells unable to divide further. Since bone marrow contains a mixture of mature cells and numerous types of immature progenitors of different lineages and of varying maturity, such clonal cultures in semisolid medium seem to be the only straightforward approach to culturing defined types of hematopoietic cells. As a rule, however, this approach generates amounts of cells too small for biochemical or molecular characterization.

Several exceptions to this rule have been found, allowing to mass produce certain types of normal hematopoietic cells. Such cultures can be of at least two types: (i) Progenitors may exist that can undergo self-renewal, i.e., proliferation without apparent differentiation, for their entire *in vitro* life span (which is between 30 and 50 generations in the chicken), if they are supplied with a suitable combination of growth factors and hormones. (ii) When cultivated with certain growth factors or their combinations, one particular cell type may preferably grow out from normal bone marrow, allowing its purification at an early stage and thus cultivation as a homogenous mass culture for a limited time period. Known examples for both types of normal progenitor mass cultures are listed below.

Methods for Cultivation of Normal Progenitors

Erythroid Cells

Methods have been developed for the *in vitro* culture of two different types of progenitor cells that are both committed to the erythroid lineage.[26,51] One of these progenitors (SCF/TGFα progenitors) undergoes

[51] B. Pain, F. Melet, S. Pain, J., T. Flickinger, M. Raines, S. Peyrol, C. Moscovici, M. G. Moscovici, H.-J. Kung, P. Jurdic, F. Lazarides, and J. Samarut, *Cell (Cambridge, Mass.)* **65,** 37 (1991).

sustained self-renewal for its entire life span, if supplied with TGFα, a ligand for the avian EGF-receptor homolog (c-*erbB*/TGFαR), estradiol, and unknown factors from chicken serum. Thus, large-scale mass cultures ($>10^9$ cells) of these cells are easily possible. The second progenitor (SCF progenitors[26]) proliferates in response to avian stem cell factor (SCF, see above) for about 8–10 days. Since these progenitors can be purified in mass amounts after 3–4 days of culture and then continue to exponentially proliferate as a predominantly immature cell for another 5–7 days, they can be grown to cell numbers $>2 \times 10^8$ as well.

Cultivation of SCF Progenitors

1. Prepare a large batch of chicken bone marrow (four chicks of 2–4 days of age, one to two chicks of 1–3 weeks of age). The cell numbers obtained should be between 2 and 4×10^8 from 1- to 2-day-old chicks to $1–2 \times 10^9$ from older ones.

2. Purify the immature BM cells by centrifugation through Percoll of 1.070 g/cm^3. Seed 20×10^6 of the interphase cells into a 145-mm culture dish containing 25 ml of complex growth medium supplemented with 5 ng/ml of TGFα (Promega), 100 ng/ml of recombinant avian SCF, and $5 \times 10^{-7} M$ estradiol (made up as a $10^{-3} M$ stock in ethanol). The addition of TGFα and estradiol is not essential to obtain these cells, but increases cell yields.

3. At day 1, remove half of the supernatant and add a respective amount of medium. Add a full complement of fresh growth factors from stock solutions.

4. At day 2, remove half of the supernatant as above. Then gently suspend the nonadherent cells by swirling and gentle pipetting, directing the fluid jet from the pipet to the rim of the dish. Transfer nonadherent cells to a new dish and add fresh medium and factors as above.

5. At day 3, repeat procedure of day 2 and then count cells. If there are significantly more cells than at the beginning of the experiment (e.g., 70×10^6 cells from $40–50 \times 10^6$ initially seeded), the nonadherent cells are centrifuged through Percoll 1.072 g/cm^3 to remove erythrocytes and maturing granulocytes and the remaining cells (which should now be $<90\%$ erythroid) are either reseeded at 1×10^6/ml in CFU-E medium plus SCF or frozen until further use. Otherwise grow for an additional day before Percoll purification. These cells can be grown for an additional 4–6 days as immature progenitors, but will accumulate increasing numbers of partially mature and apoptotic cells. They can be used for differentiation induction experiments (by switching to differentiation medium with anemic serum plus insulin after washing away the self-renewal factors) until days 5–6.

Cultivation of SCF/TGFα-Progenitors

The main published method[26,34] for the successful growth of these continuously self-renewing, c-*erb*B expressing progenitors is dependent on special batches of chicken serum. After describing this method, we will also suggest an alternative way of growing similar cells (that potentially retain a dependence on SCF), a method that is likely to work with a larger variety of chicken serum batches.

METHOD I. This method gives rise to SCF/TGFα progenitors expressing high levels of c-*Erb*B and able to grow in the absence of SCF.

1. Prepare BM cells from chicks not older than 7 days as noted earlier, and seed at 50×10^6 cells per 100-mm dish containing 10 ml of complex growth medium plus 5 ng/ml of TGFα and $5 \times 10^{-7}M$ estradiol. During the next days, perform either partial medium changes or (if the adherent cell layer becomes more than semiconfluent) suspend by vigorous pipetting and transfer to a new dish in addition to performing partial medium changes.

2. During the first 6–9 days, the cell number in the cultures decreases. Perform daily medium changes with the addition of fresh TGFα and estradiol and keep the nonadherent cells at a density of $1–2 \times 10^6$ cells/ml by reducing the culture volume. Successful cultures are characterized by the appearance of clumps of healthy, proliferating cells that are perfectly round and absolutely nongranulated (usually around days 5–7). Make sure there are some adherent cells present in the culture but prevent these from becoming confluent by transferring nonadherent and weakly adherent cells to new dishes by vigorous pipetting.

3. When the erythroid cells are visibly proliferating (usually around days 10 to 14) start eliminating contaminating myeloid cells by gently suspending the erythroid clumps (see above) and transferring them to new dishes until the myeloid, adherent cells have disappeared. When cultures have proliferated to cell numbers of $20–50 \times 10^6$, they can be frozen until further use. For successful thawing of these cells, it is particularly important to follow the instructions given earlier.

METHOD II. This method gives rise to similar cells as Method I, but they may express lower levels of c-*Erb*B and therefore require SCF for growth and may differ in some respects from the just-mentioned ones.

1. Grow SCF progenitors as described earlier, keeping them in SCF + TGFα + estradiol at all times.

2. Keep cells at a density between 1 and 2×10^6 cells/ml by partial medium changes or by passaging the cells. Depending on the chick, some cultures will grow with reduced speed for a few days whereas others will go on growing without a change in growth rate.

3. At days 8–12 (earlier in continuously growing cultures, later in those that stop growing), purify the cells by centrifugation (i) through Percoll 1.072 g/cm^3 and (ii) through Ficoll,[44] wash twice in complex growth medium, and seed at 2×10^6 cells/ml in complex growth medium supplemented with TGFα and estradiol as described earlier, but lacking SCF. Many cells will die, but numerous clumps of healthy cells will remain. These clumps have to be kept alive by daily partial medium changes plus the addition of TGFα and estradiol until they start to grow out (usually around days 10–14).

4. Grow up and freeze cells as described under Method I.

Myeloid Cells

The only normal myeloid cells that can so far be grown to some extent in culture are macrophages and their progenitors. Since procedures do not yet exist to prevent withdrawal from the cell cycle and terminal differentiation to occur in such cultures, the amounts of cells that can be obtained from a given number of BM cells are quite variable.

1. Prepare BM cells and seed 50×10^6 cells/ml per 100-mm dish. Depending on the culture, good results have been obtained using complex growth medium supplemented with 10 ng/ml cMGF, CEF-32-conditioned medium, or the former combined with SCF.

2. During the first 2–3 days, suspend the non- and weakly adherent cells by vigorous pipetting and transfer to a new dish. Then, transfer to a larger dish (145 mm) and allow the cells to adhere. Cultivate with medium changes every second day until the layer is confluent.

3. For passage, chicken macrophages can be trypsinized with high concentrations of trypsin, but this usually stops their further growth. It is therefore recommended to remove them with EDTA.[52] Wash the dish five times with prewarmed PBS. Then rinse two times with a prewarmed EDTA solution, add 20 ml of fresh EDTA solution, and incubate at 37° for 10–15 min in a bacterial incubator (i.e., not gassed with CO_2). When the cells can be dislocated by vigorous pipetting, add 10 ml of fetal calf serum to stop the action of EDTA, spin down, wash once in growth medium and seed at half the cell density obtained before passaging. Some cultures can be passaged three and more times this way, allowing to grow $50–100 \times 10^6$ macrophages before they finally cease proliferating.

4. Chicken macrophages can also be grown from the peripheral blood of adult chickens (T. Graf and H. Beug, unpublished). For this, spin down 10–20 ml of chicken blood taken in heparin or citrate at 1000 to 1500*g*

[52] B. Royer Pokora, H. Beug, M. Claviez, H. J. Winkhardt, R. R. Friis, and T. Graf, *Cell (Cambridge, Mass.)* **13,** 751 (1978).

(2000 rpm in clinical centrifuge) and wash two times in Ca^{2+}/Mg^{2+}-free PBS. Then take up in 100 ml of standard growth medium and centrifuge through Ficoll.[44] The interphase, consisting mainly of platelets, lymphocytes, and monocytes, is retrieved, washed once in complex growth medium, counted in an electronic counter at a setting counting only the large monocytic cells, and seeded at 0.5–1 $\times$ 10^6 cells/ml in complex growth medium containing a saturating dose of CEF-32-conditioned medium (cMGF can probably replace the CEF-32-conditioned medium). Further cultivation is as for chicken bone marrow macrophages.

Additional Tests of Medium Components Required for Cultivation of Normal Erythroid Progenitors

As mentioned earlier, the successful growth of normal chick erythroid progenitors is critically dependent on tested batches of chicken serum and BSA. Batches of these two complex growth medium supplements that do well on transformed erythroblasts might still not support the outgrowth of SCF/TGFα progenitors. The hints given below may be helpful to overcome these difficulties.

1. One should be aware that the type of chicken flock used may influence the result. Although several laboratories have succeeded in growing normal SCF/TGFα progenitors from other chicken flocks as the SPAFAS chickens used by our laboratory, others have reported difficulties with certain lines of chickens. These difficulties do not so much concern the inability to grow these progenitors, but the cells seemed to have much shorter *in vitro* life spans than those described by us.

2. Follow the procedure given for preparing detoxified BSA as closely as possible. Commercially available detoxified BSA or procedures omitting the ion exchanger treatment regularly lead to batches of BSA that may work on transformed erythroblasts, but do not support the growth of normal progenitors. Since this only becomes apparent after 5–7 days, the medium tests used to test new BSA batches should be run on normal SCF/TGFα progenitors for at least 7 days. Only those batches that support growth of the cells to a similar extent as the control batch for this time period may be suitable.

3. The only way to identify chicken serum batches suitable for SCF/TGFα progenitor outgrowth is to test them in an actual outgrowth experiment, preferably against a reference batch. So far, four consecutive batches of chicken serum provided by Sigma have proven to be suitable; we have not been successful with batches from other companies.

[5] Avian Neuroretina Cells in Oncogene Studies

By PATRICIA CRISANTI, BERNARD PESSAC, and GEORGES CALOTHY

There are two main reasons to use cells from the avian embryonic neuroretina (NR) as an experimental system to study the effects of oncogenes on regulation of cell growth and differentiation. First, NR cultures are exclusively composed of neuroectodermal cells: this tissue is not vascularized and does not contain cells of mesenchymal origin. Second, when these cells are maintained in culture, they rapidly stop dividing and express several markers of neural dfferentiation.

The NR is part of the central nervous system and is, therefore, of ectodermal origin. It is derived from the optic cup which is an evagination from the wall of the diencephalon. The optic cup consists of an external layer which gives rise to the pigment epithelium and an internal layer which develops into the NR itself (Fig. 1).[1,2]

A schematic representation of the vertebrate NR is shown in Fig. 2. The adult NR is composed of three types of neurons, photoreceptors, bipolar cells, and ganglion cells, and of two types of interneurons, horizontal and amacrine cells and one type of glial cells (Müller cells). These cells are organized in a layered structure together with two synaptic layers: the outer and the inner plexiform layers (Fig. 2).

Two main types of synaptic complexes are present in the NR: the nonribbon (conventional synapses that are found in all neural tissues) and the ribbon synapses that are specific of the NR (Fig. 3).[3]

Neuroretina Development and Differentiation

The avian NR development proceeds through essentially three phases: proliferation of neuroectodermal precursor cells, lamination of cell strata, and differentiation of postmitotic cells.[4] During early phases of development, immature cells are present across the entire primitive NR, and mitotic cells are visible in the external part adjacent to the future pigment epithelium. At a precise stage of development, cells become postmitotic and migrate toward the inner face of the NR. They then start to differentiate

[1] R. Adler and D. Farber, Eds. "The Retina, Part 1: A Model for Cell Biology Studies." Academic Press, Orlando, FL, 1986.

[2] N. J. Berrill and G. Karp, "Development." McGraw-Hill, New York, 1976.

[3] P. Crisanti-Combes, A., Privat, B. Pessac, and G. Calothy, *Cell Tissue Res.* **185,** 159 (1977).

[4] C. J. Barnstable, *Mol. Neurobiol.* **1,** 9 (1987).

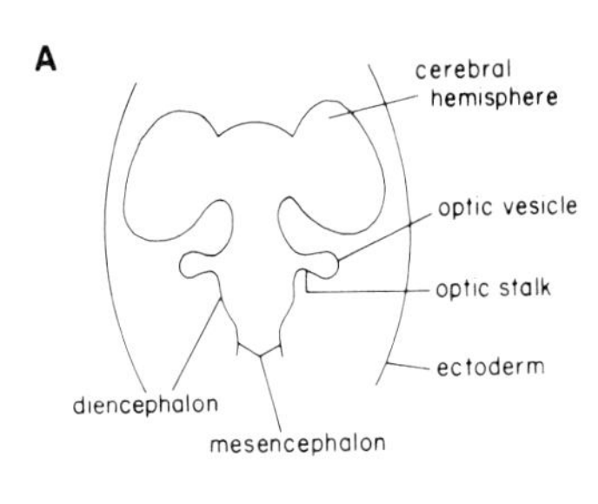

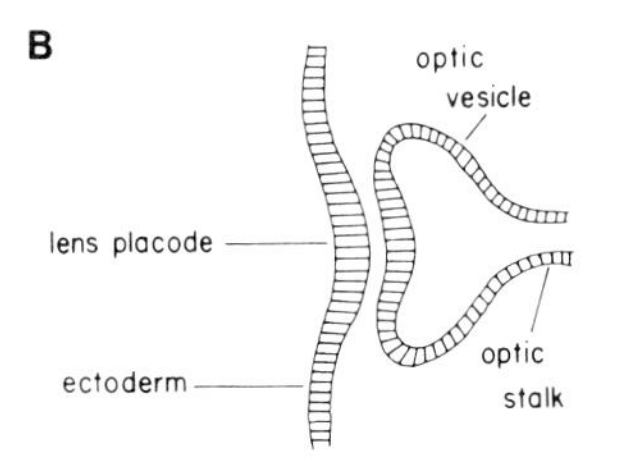

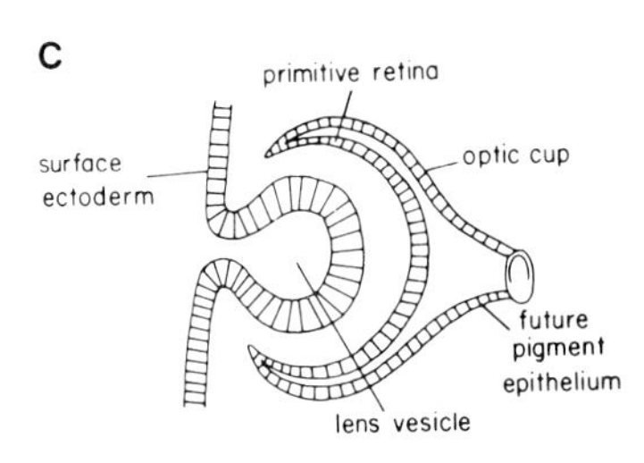

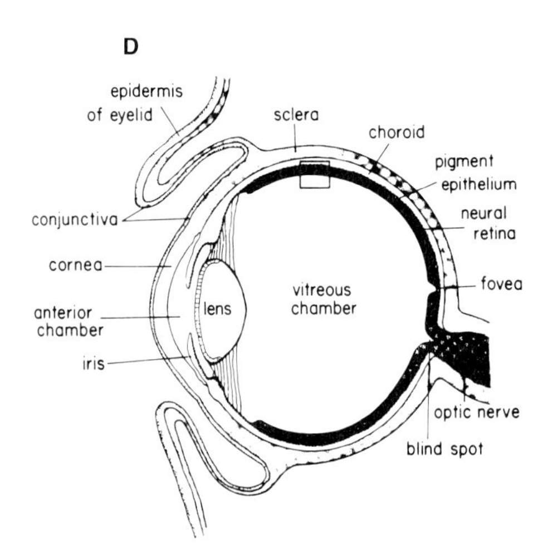

FIG. 1. Development of the eye. (A) Dorsal aspect of the embryo at the stage at which the optic vesicle makes contact with the overlying ectoderm. (B) Transverse section following induction of the lens placode by the tip of the optic vesicle. (C) Invagination of the optic cup and optic vesicle. (D) Horizontal section of the high vertebrate eye (from Adler and Farber[1]).

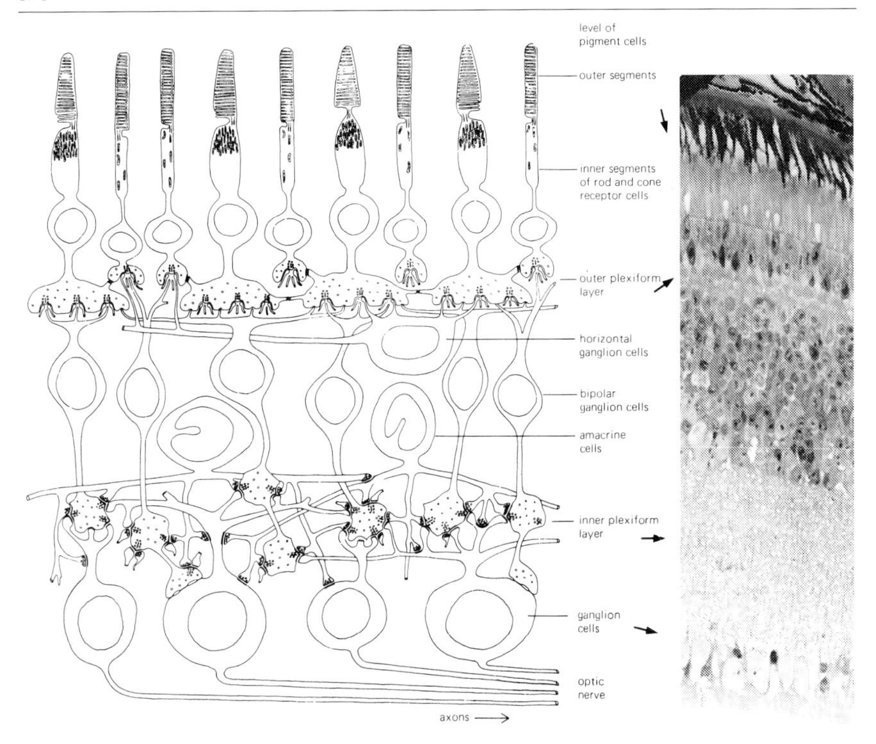

FIG. 2. Synaptic contacts between the main types of retinal cells. The upper layer consists of rods and cones. These sensory cells make synaptic contact with their neighboring cells and with bipolar and horizontal cells. These in turn make synaptic contacts with the inner ganglion cells whose axons constitute the optic nerve (from Berrill and Karf[2]).

when reaching positions that define their corresponding layers. Thymidine incorporation experiments, at various stages of chicken embryonic development, show that withdrawal from the cell cycle follows a posterior → anterior gradient: cells from the posterior pole, close to the optic nerve, are the first to withdraw from the cell cycle and to differentiate. Moreover, all cells located within a given region do not exit the cycle synchronously. Precursors of ganglion cells are the first to stop dividing between ED3 (Day 3 of embryonic development) and ED5, followed by precursors of cones and rods which in contrast will differentiate shortly before hatching. At later stages, most horizontal and amacrine cells become postmitotic. Precursors of bipolar and Müller cells are the last to exit from cell cycle. Finally, at ED10, the vast majority of chicken NR cells are postmitotic. Synaptogenesis begins between ED10 and ED11. Figure 4 describes representative steps of NR histogenesis in chicken embryos between ED6 and 1 day after hatching.

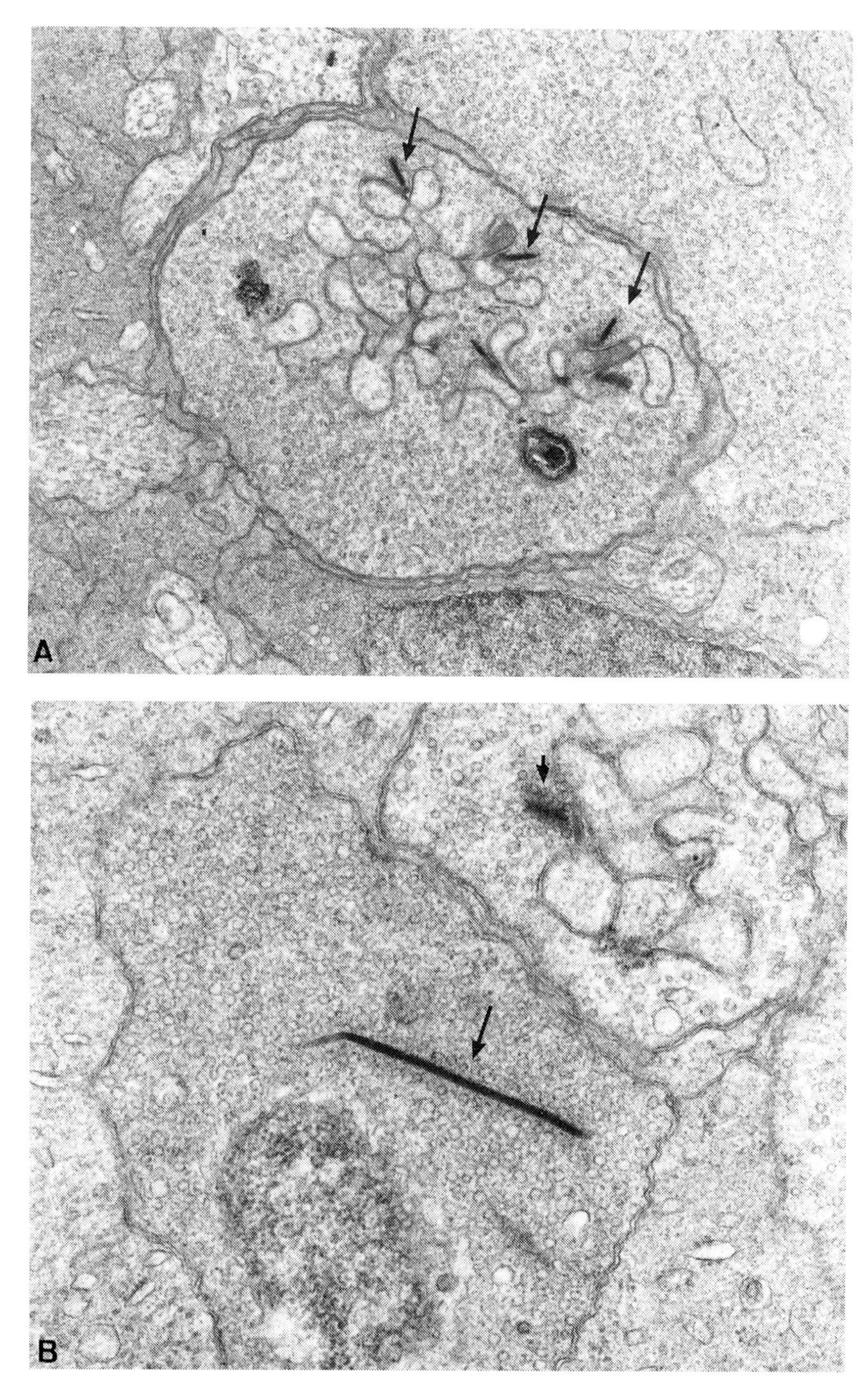

FIG. 3. (A) Typical diads and triads arrangement of synaptic ribbons associated with aligned synaptic vesicles. (B) A large presynaptic with a dense matrix containing asynaptic ribbons with vesicles. Identical synaptic junctions are observed in cultured NR cells.

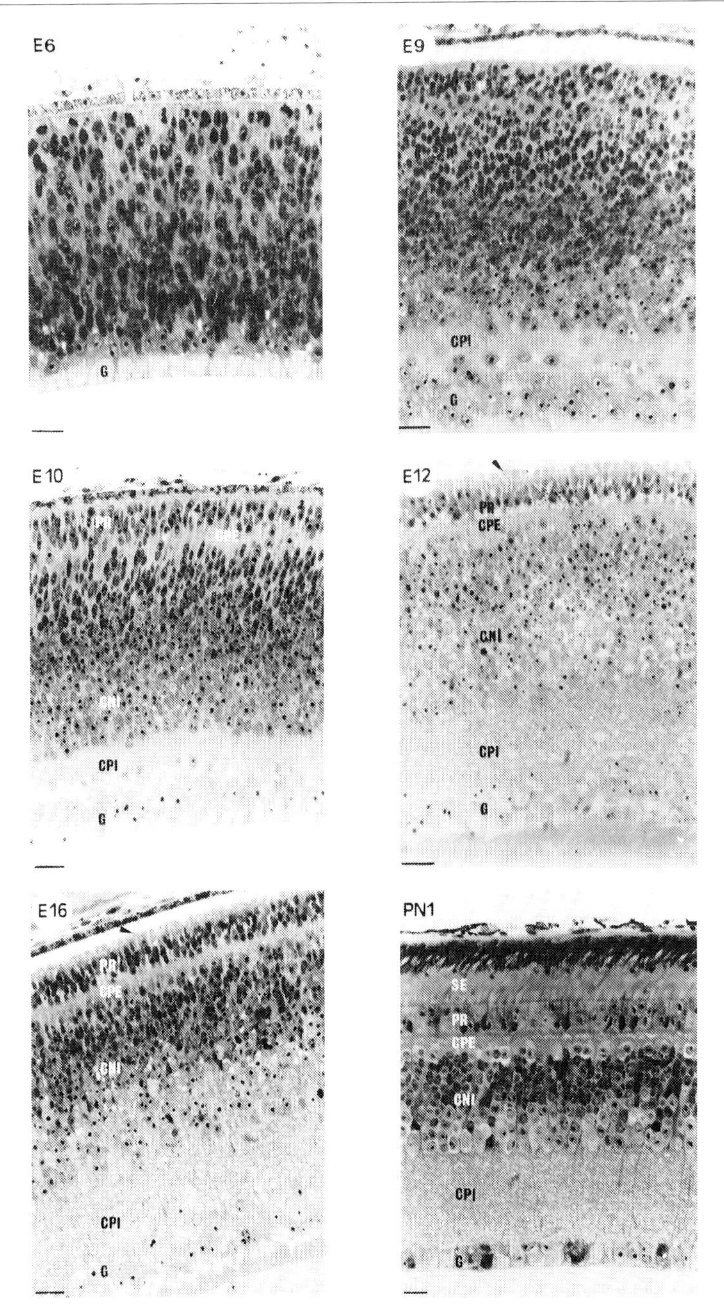

FIG. 4. Histogenesis of the quail neuroretina at different stages of development. See text.

Only a few reports have described the kinetics of cell cycle withdrawal in embryonic quail NR (QNR) cells.[5] Because the development of the quail proceeds at a faster rate than that of the chicken, both withdrawal from the cell cycle and lamination take place earlier (17 days vs 20 days in the chicken).

Investigation of the kinetics and topography of DNA arrest in the embryonic QNR, by bromodeoxyuridine (BrdU) incorporation in replicating DNA (see methods below), led to the following observations (Fig. 5):

At ED7 (Fig. 5A), the lamination process has just begun. Only the ganglion cells are organized in a distinct lamina which is not yet separated from the other cells by the plexiform layer. The remaining NR appears as a homogeneous population of elongated cells in the process of migration. At this stage, no labeled nuclei are observed in ganglion cells whereas a number of fluorescent nuclei are homogenously distributed over a wide zone of inorganized cells.

At ED9 (Fig. 5B), cells of the presumptive inner nuclear layer (INL) are still elongated and intermingled. Only the inner plexiform layer (IPL) starts to be individualized. The number of S phases is markedly reduced in comparison to ED7 and ED8. Labeled nuclei are localized in the presumptive INL. The posterior pole appears more differentiated and presents a reduced number of S phases as compared to the anterior pole.

At ED10 (Fig. 5C), the lamination process is not achieved but the IPL is visible, while the outer segments of photoreceptors are in formation. At this stage, no labeled nuclei are visible in either pole. Cells could not be labeled with BrdU thereafter.

Dissection

NR can be dissected from chicken (CNR) or quail (QNR) embryos as early as days 6–7 of incubation.[3] Heads are first immersed in phosphate-buffered saline (PBS) without Ca^{2+}/Mg^{2+} (PBS^-) and are parted in two along the median axis with fine forceps. Eyes are dissected out, the lenses are discarded, and the NR is carefully separated from the underlying pigment epithelium under a dissecting microscope. The retinas are spread in PBS^- and the few adhering pieces of pigment epithelium are cut away. Retinas are pooled, washed twice in PBS^-, incubated in PBS^- for 10 min at 37°, and then in a 0.25% (w/v) trypsin solution in PBS^- for 20 min. After centrifugation at room temperature, NR are washed three times in Eagle's

[5] A. A. Moscona and L. Degenstein, *Dev. Neurosci.* **4,** 211 (1981).

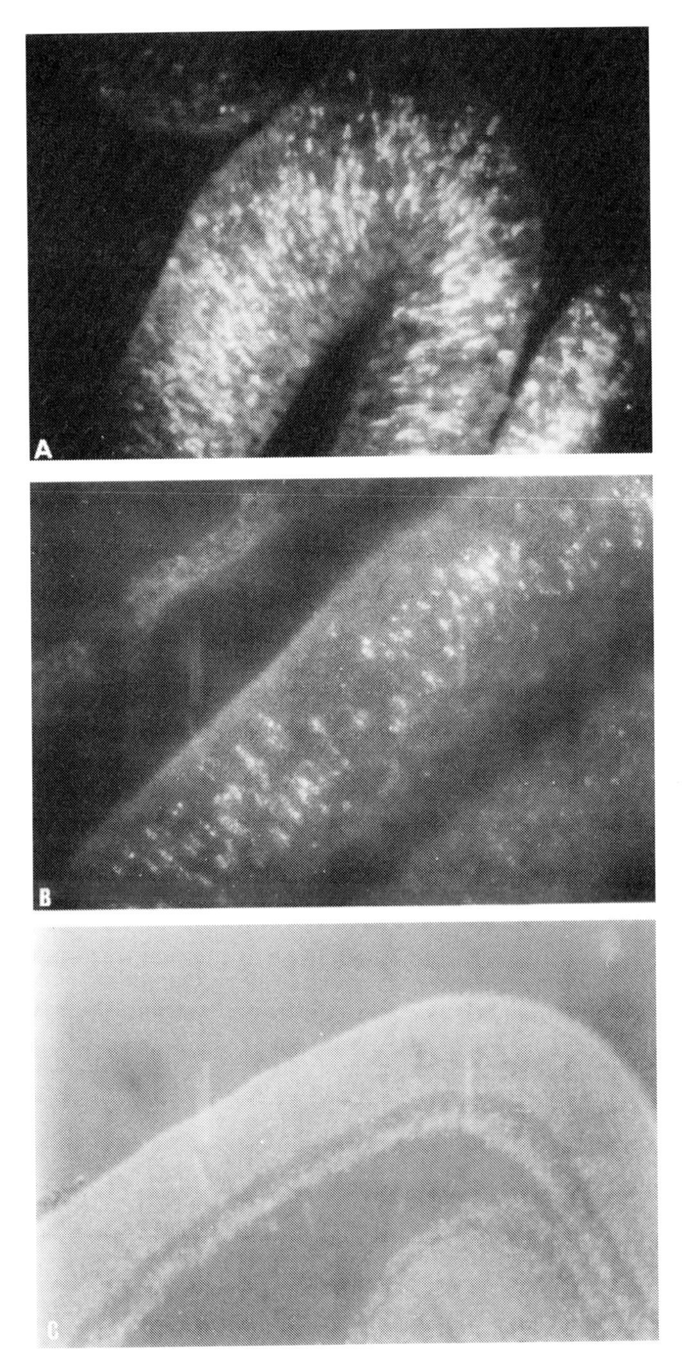

FIG. 5. Detection of S phases during various stages of development: (A) at ED7, (B) ED9, and (C) ED10. Incorporation of BrdU into replicating DNA was detected by indirect immunofluorescence. See text.

basal medium (BME), without dissociation, and finally, tissues are dissociated by gentle pipetting in BME supplemented with 5–10% fetal calf serum (FCS) and antibiotics (100 μg/ml kanamycin, 100 μg/ml streptomycin, and 100 units/ml penicillin). The cells are centrifuged at 200 *g*, resuspensed in culture medium, and counted in a hemocytometer. They appear as single cells with a few tiny clumps. One ED7 NR yields about 5×10^6 cells. They are usually seeded at a density of 3×10^5 cells/cm^2 and are incubated at 37° in a humidified atmosphere [5% (v/v) CO_2 in air]. Medium is renewed every second or third day.

Light and Electron Microscopy Studies of NR Cultures

Cultures are observed by light microscopy or processed for electron microscopy as follows: they are fixed in 2.5% glutaraldehyde in 0.1 *M* phosphate buffer for 1 hr and postfixed in 1% osmium tetroxide in the same buffer for 1 hr. Fixed cells are stained *en bloc* with uranyl acetate, dehydrated in ethanol–acetate, and flat-embedded in Araldite. Ultrathin sections are cut in a plane parallel to the substrate. NR dissected from embryos at the onset of the cultures are processed similarly for electron microscopy.

After 24 hr, NR cells from 6 to 7-day-old chicken embryos are composed of clusters of small, round cells (seven to eight times smaller than fibroblasts) that are undifferentiated neuroblasts. Two to 3 days later, these small clusters spread out and start to extend processes.

Round cells differentiate from very immature cell types, neuroepithelial like at the time of dissociation, into relatively mature neurons. This maturation process is visualized by the development of perikaryal organelles and transformation of the nucleus from a fusiform notched outline, with numerous chromatine clumps, to a regular, spherical shape with an evenly dispersed chromatin and a prominent nucleolus. In addition, processes grow out of these cells, progress with growth cones on the flattened cells, and form a reticular array between groups of round cells. However, it is difficult, if not impossible, to identify most of these neurons. Photoreceptors cells are identifiable because of the presence of ribbon synapses characteristic of cone receptors in the adult retina. Similarly, ganglion cells could be tentatively identified because of their size, larger than that of any other neuron in culture, and of their general morphology, especially because of the presence of an axon. Areas of neuropiles are distributed in the vicinity of neuronal clusters and, sometimes, in the center of such clusters. They more or less mimic the organization of the retina into cellular and plexiform layers. Synapses are found after 8 days *in vitro* and are still visible after 15 and 21 days in culture. They correspond to the different types found in the retina *in vivo,* although they most often lack the typical organization of postsynaptic elements. A second cell type becomes morphologically distinguishable and is composed of flat, transparent cells that may represent Müller cells for two reasons. First, they do not show neuronal characteristics. Second, they display some ultrastructural similarities to Müller cells of the *in vivo* retina, namely the presence of large arrays of filaments and microtubules, as well as numerous and extensive attachment plates associated with some of the fibrillar arrays (Fig. 6A).

In summary, NR cultures obtained from 7-day-old chicken embryos can be maintained as monolayers and are able to differentiate into neuronal and nonneuronal (glial) cells. This differentiation proceeds to the stage at which morphologically mature synapses are found. It, therefore, appears that synaptogenesis primarily depends on the genetic programs of the cells and that complex interactions, found in tridimensional structures, may not be required for the formation of synapses.

Transdifferentiation

From the fourth week onward, new phenotypes appear that might be due to transdifferentiation[6] of cells that were initially programmed into the neuronal or glial pathways or otherwise derived from uncommitted stem cells. One of the phenotypes is made of small pigmented cells, whereas the second consists of cells that progressively take on the appearance of lentoid bodies (LB). These structures are formed of densely packed refringent cells and of "bottle cells," both of which appear in culture in lens epithelial cells and follow a differentiation pathway similar to that of the lens (Fig. 6C). The LB are always associated with crystallins identical to those present in lens during development or in cultured lens epithelial cells. Crystallins constitute the bulk of lens proteins and, in birds, fall into three major classes (α, β, and δ) that are the products of three multigene families.[7,8] α- and δ-crystallins are present on day 28 in cultured QNR and progressively increase until day 42. The amounts of α- and δ-crystallins remain stable thereafter. In QNR cultures passaged once after 10 weeks, the levels of α- and δ-crystallins remain elevated.[9]

[6] T. S. Okada, Y. Itoh, K. Watanabe, and G. Eguchi, *Dev. Biol.* **45,** 318 (1975).

[7] D. I. De Pomerai, D. J. Pritchard, and R. M. Clayton, *Dev. Biol.* **60,** 416 (1977).

[8] R. M. Clayton, I. Thompson, and D. I. De Pomerai, *Nature* (*London*) **282,** 628 (1979).

[9] L. Simonneau, P. Crisanti, A. M. Lorinet, F. Alliot, Y. Courtois, G. Calothy, and B. Pessac, *Mol. Cell. Biol.* **6,** 3704 (1986).

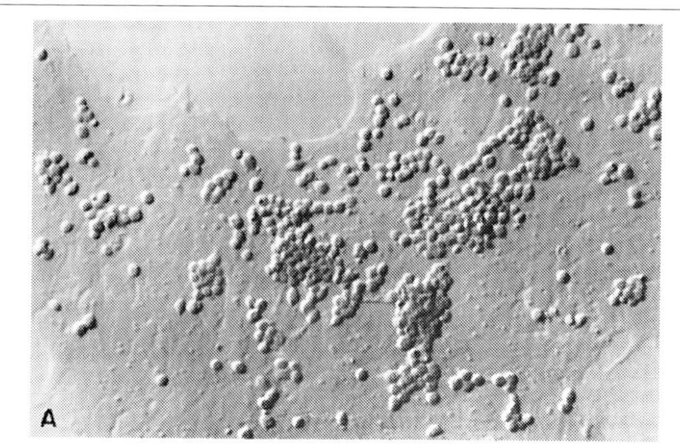
A

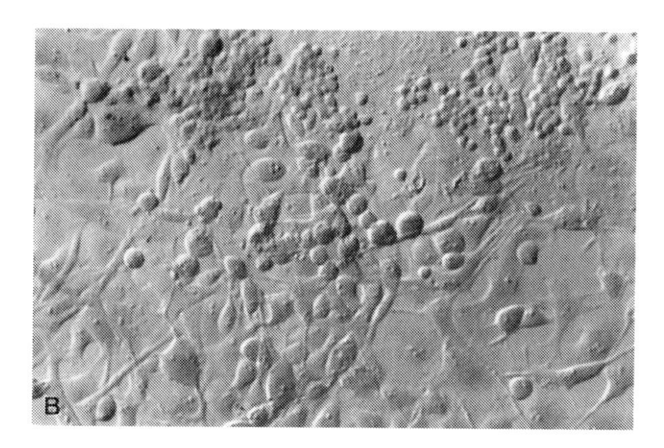
B

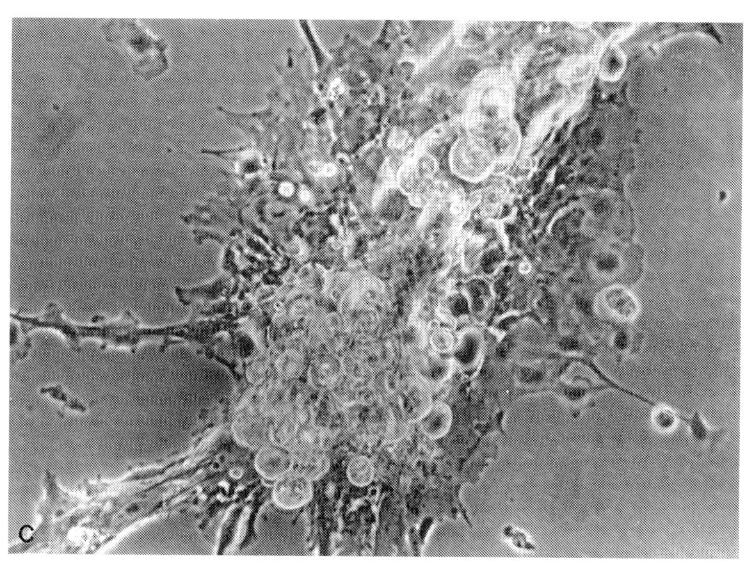
C

Withdrawal from Cell Cycle of Cultured NR

Cultures of CNR or QNR cells dissected at ED6–ED7 undergo two or three doublings at the level of cell population under the culture conditions described earlier (BME + 5–10% FCS). After 10 days in culture, less than 1% of NR cells are able to synthesize DNA, as shown by thymidine incorporation. Similar results on the kinetics of growth arrest in cultured NR cells are obtained by studying incorporation of BrdU in replicating DNA (see Methods below). At days 1 and 2, after plating, the great majority of cells are in the S phase, whereas at day 5 only a few cells are labeled (Figs. 7A and 7B).

NR cells can be maintained in culture for several weeks. Clusters of neuronal cells will disappear progressively; by day 21, most cells appear flattened. Eventually, all cells will degenerate.

Changing culture conditions may influence the growth pattern of NR cells. We have shown that these cells undergo about 10 divisions when cultured in Dulbecco's modified Eagle's medium (DMEM) supplemented with 10% FCS, 10% tryptose phosphate broth, and 2% heated chicken serum.[10] The mechanisms by which specific components in different media may act on NR cell division are unknown and are worth studying. It would also be interesting to investigate potential effects of growth factors on NR cell division.

Expression of Retrovial Oncogenes in NR Cells

Infection of CNR and QNR cultures with acutely transforming retroviruses, such as Rous sarcoma virus (RSV), results in transformation and sustained cell proliferation (Fig. 6B). An increase in the cell number becomes evident within 5–7 days. Dividing cells can be propagated during 20–30 generations, which represents a dramatic increase in their life span. However, RSV-infected CNR cells will eventually stop proliferating and undergo senescence. This process is accompanied by a progressive loss of the transformed phenotype.[11]

[10] C. Béchade and G. Calothy, *Oncogene* **6,** 2311 (1991).
[11] G. M. Seigel and M. F. D. Notter, *J. Virol.* **66,** 6242 (1992).

Fig. 6. Morphology of normal and RSV-transformed NR cells. (A) Cultured ED7 QNR cells 3 days after plating. Note the presence of clusters of small differentiating neurons atop flat glial (Müller) cells. (B) RSV-transformed QNR cells 7 days after infection. Note the presence of typical round cells. These cells are larger than normal neurons. (C) Cultured ED7 QNR cells 5 weeks after plating showing the presence of typical lentoid bodies.

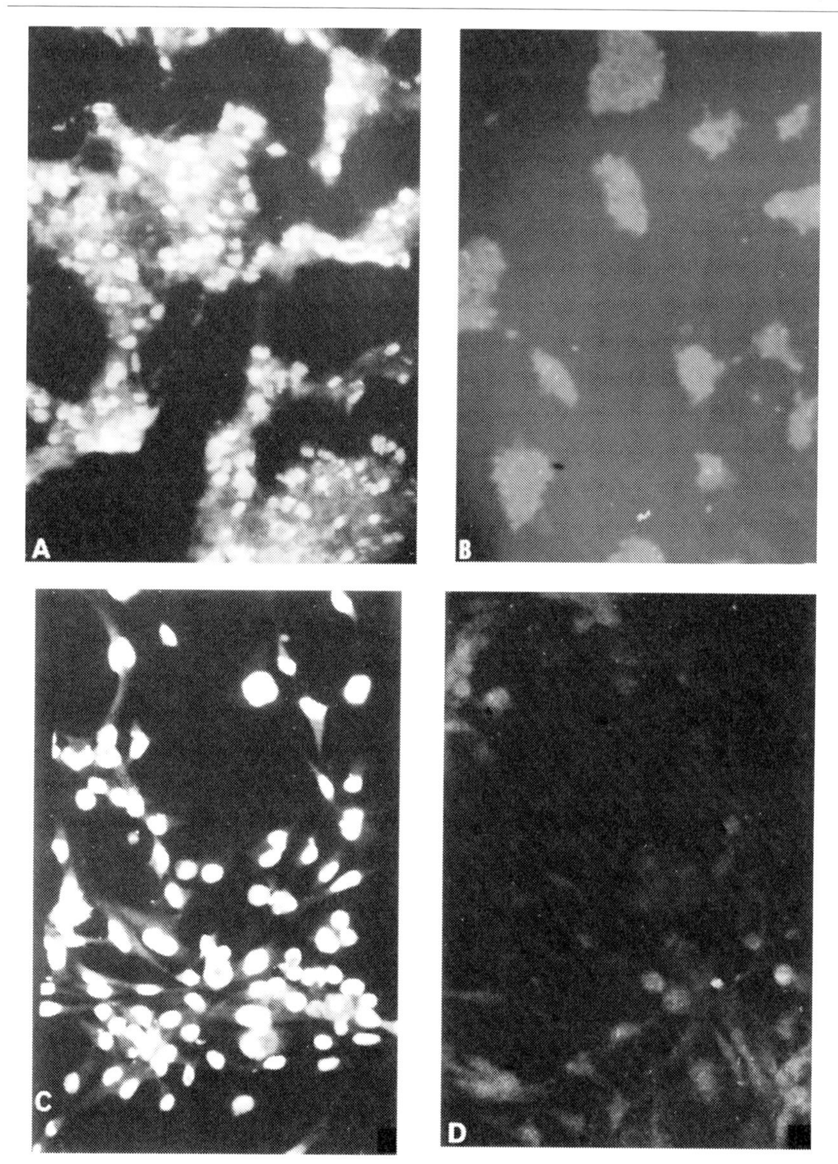

FIG. 7. Detection of S phases with anti-BrdU antibodies in cultured cells. (A) Cultured ED7 QNR cells 2 days after plating. (B) Same culture after 5 days. (C) Clonal K_2 cell line derived from *ts* NY68-infected QNR cells at 36°. (D) K_2 cells, made quiescent by transferring the cultures to 41.5° for 2 days.

In RSV-infected cells, proliferation requires the expression of a functional p60$^{v\text{-}src}$ as demonstrated by the existence of conditional mutants that are temperature sensitive (*ts*) for their mitogenic capacity.[12,13] Proliferation of cells infected with such *ts* mutants can be, at will, switched on and off on their transfer to the appropriate temperature (Figs. 7C and 7D). Therefore, these cells are most suitable for dissecting mechanisms that lead to cell division. However, NR cells expressing an active p60$^{v\text{-}src}$ also require the presence of growth factors present in the serum for their proliferation, in contrast to other cell types transformed by v-*src*. Such cells could be made quiescent by lowering the serum concentration to 0.1–0.5%, suggesting the possibility that two synergistic pathways are involved in the regulation of NR cell division.[14]

NR cell proliferation can be measured either at the level of the cell population by cell counting, [^{3}H]thymidine incorporation,[12] or flow cytometry[14] or at the single cell level by determination of mitotic indexes by autoradiography[11] or by BrdU labeling (see below).

Results of flow cytometry analysis indicate that NR cells infected with *ts* mutants of RSV are arrested in G_0/early G_1 following inactivation of p60$^{v\text{-}src}$ at a nonpermissive temperature. The S phase resumes synchronously within 12 hr following transfer to the permissive temperature. Similarly RSV-infected NR cells can be arrested in G_1/G_1, in the presence of 0.5% FCS at 37°.[14]

Methods

Infection with Retroviruses

Viral stocks should be filtered (0.22-μm filters) to remove potential dividing cells. The addition of Polybrene (2–4 μg/ml) may increase virus uptake. CNR and QNR can be infected as cell suspensions, e.g., at the time of seeding, or as monolayers. The susceptibility of cultured NR cells to retroviral infection decreases as a function of time, in correlation with their withdrawal from the cell cycle. Similarly, NR cells dissected after ED10 are less responsive to viral infection.[15] Infection of NR cells, dissected at ED6–8, with avian leukosis viruses (ALV) which do not induce NR cell proliferation yields virus titers that are comparable to those obtained in fibroblasts. Moreover, the virus production in proliferating NR cells infected

[12] G. Calothy, F. Poirier, G. Dambrine, P. Mignatti, P. Combes, and B. Pessac, *Cold Spring Harbor Symp. Quant. Biol.* **44,** 983 (1980).

[13] F. Poirier, G. Calothy, R. E. Karess, E. Erikson, and H. Hanafusa, *J. Virol.* **42,** 780 (1982).

[14] G. Gillet, D. Michel, P. Crisanti, M. Guérin, Y. Herault, B. Pessac, G. Calothy, G. Brun, and M. Volovitch, *Oncogene* **8,** 565 (1993).

[15] B. Pessac and G. Calothy, *Science* **185,** 709 (1974).

with acutely transforming viruses usually exceeds 10^7 infectious units (IU)/ml. NR cells can also serve as host cells for nonreplicative retroviral vectors expressing selection markers and/or specific genes.

DNA Transfer in NR Cells

NR cells (5×10^6) (equivalent to one ED7 retina) are placed in a 35-mm dish. The medium is changed the following day. Transfection is done 3 hr later as follows:

1. Prepare solutions A and B:

Solution A: 250 μl 2× HBS and 5 μl 100× Phosphate.
Solution B: 30 μl 2 *M* $CaCl_2$, 5–30 μg DNA (5 μg of DNA is used for transfection of recombinant plasmids containing a replicative viral genome, whereas up to 30 μg of DNA is required when transfecting recombinant DNA containing a nonreplicative genome), and up to 250 μl H_2O.

2. Add 250 μl of solution B dropwise to 250 μl of solution A. The resulting precipitate is immediately added to the plate, without removing the medium. Five to twenty-four hr later, plates are washed with PBS, and fresh medium is added.

Solutions

2× HBS: 10 g/liter HEPES and 16 g/liter NaCl, pH 7.1; sterilize by autoclaving or filtering.
100× Phosphate: 70 m*M* Na_2HPO_4 and 70 m*M* NaH_2PO_4; sterilize by autoclaving or filtering.
2 *M* $CaCl_2$, filter.
H_2O, sterilize by autoclaving or filtering.

Detection of S Phases with anti-BrdU Antibodies

ON CULTURED QNR CELLS INFECTED WITH A *ts* MUTANT OF RSV. Cells are plated in complete medium on 2-cm^2 dishes (2×10^4 cells/dish) and maintained at 36.5° for 5 days without medium renewal. At time $t = 0$, the culture medium is renewed and part of the cultures are shifted to 41.5°. At each interval, BrdU is added to the medium, at a final concentration of 10 μM, for 1 hr. Cells are incubated, fixed in 90% ethanol, and air dried. Slides are then immersed in 4 *N* HCl, neutralized in 0.1 *M* $Na_2B_4O_7$, pH 8.5, and treated with 50 μl of diluted mouse monoclonal anti-BrdU antibody (1/10 in PBS + 0.5% Tween 20) for 30 min. After PBS washing, slides are incubated for 30 min with 50 μl of 0.5% Tween 20 in PBS containing a 1/20 dilution of fluorescein isothiocyanate (FITC)-conjugated $F(ab')_2$

goat anti-mouse, washed again in PBS, incubated for 1 min in 0.04 μg/ml Evans blue (1/10,000) iodide, dried, and mounted for microscopic examination.

ON EYE SECTIONS. Eyes are removed daily from incubated ED7–ED10 embryos. The iris is cut out and the vitrous humor is discarded, allowing contact between the NR and incubation medium. Eyes are incubated for 4 hr in complete medium containing 10 μM BrdU, fixed by immersion in paraformaldehyde (4% in PBS), and treated with 30% sucrose overnight. Eyes are frozen and sectioned at 10 μm. Sections are dehydrated successively in 70, 95, and 100% ethanol and dried. BrdU incorporation is visualized as described earlier.

Results of BrdU incorporation *in ovo* and in cultured cells are shown in Figs. 5 and 6.

Establishment of Clonal Cell Lines Derived from Quail Retinas

We derived clonal lines from QNR cells transformed with a conditional mutant of RSV, *ts*NY68.[16] NR cells infected with this virus are transformed and proliferate at permissive temperatures (35–37°). They are morphologically normal and quiescent at nonpermissive temperatures (40–42°). Infected cells were first maintained at 39.5° for 4 weeks without subcultivation. At this intermediate temperature, QNR cells proliferate without being fully transformed. They were subsequently passaged three or four times over a period of 4 months after which the cells ceased to grow and underwent a "crisis" which lasted a few weeks. Cell growth then resumed and was maintained.

Single cell derived clones were obtained by diluting the cultures according to Poisson distribution into 96-well plates in BME + 20% FCS. Cell suspensions were mixed with irradiated homologous cells, serving as feeder layers. Established clonal lines remained *ts* for cell proliferation (see Figs. 6C and 6D).

Use of NR Cells in Genetic and Functional Analysis of Oncogenes

As mentioned earlier, the v-*src* gene is responsible for the mitogenic and transforming properties of RSV, as shown by the existence of v-*src* mutants that are *ts* for their capacity to induce NR cell division. However, these parameters are not always coordinately expressed in NR cells. We previously reported that NR cells infected with certain *ts* mutants of RSV

[16] B. Pessac, A. Girard, G. Romey, P. Crisanti, A. M. Lorinet, and G. Calothy, *Nature* (*London*) **302,** 616 (1983).

were induced to proliferate, at a nonpermissive temperature, in the absence of morphological transformation.[12,17]

Based on this observation, we used NR cell proliferation as a phenotypic marker to select for mutants expressing phenotypic dissociation of mitogenic and transforming properties at permissive temperatures (35–37°). Two such mutants (PA101 and PA104) were extensively characterized.[12,13,18–20] Interestingly, both viruses exhibit very low tyrosine kinase activity,[13,20] suggesting that there was no direct correlation between the levels of tyrosine kinase activity and the ability to induce cell division. Molecular and functional analysis of these mutants showed the importance of the kinase domain in controlling all aspects of transformation. They also established the importance of the amino-terminal protein of $p60^{v\text{-}src}$ in modulating transformation parameters and interacting with the kinase domain. It was proposed that the N-terminal portion of $p60^{v\text{-}src}$ could be involved in cell substrate recognition, a hypothesis widely confirmed since through the molecular and functional dissection of specific subdomains of $p60^{v\text{-}src}$ (SH_3 and SH_2 domains). Along these lines, we also showed that mutations that prevent myristoylation and binding of $p60^{v\text{-}src}$ to the membrane do not abolish its mitogenic property.[21] Taken together, these results are consistent with the possibility that interaction of $p60^{v\text{-}src}$ with a subset of cellular substrates leads to changes in growth regulation of NR cells.

In another set of experiments, NR cells were used to analyze the functions of avian retrovirus MH2 which carries both the *mil/raf* and the *myc* oncogene. Infection of NR cells with this virus results in morphological transformation and sustained proliferation. Isolation of NR cells of spontaneous MH2 mutants bearing a deletion of the v-*myc* gene showed that v-*mil/raf* was sufficient to induce NR cell proliferation.[22]

Induction of NR Cell Proliferation by Avian Leukosis Viruses: A Model for Studying Oncogene Transduction by Retroviruses

We reported that avian leukosis viruses, which do not carry an oncogene, also induce NR cell proliferation. However, cell division is observed in a

[17] G. Calothy and B. Pessac, *Virology* **71,** 336 (1976).

[18] B. J. Mayer, R. Jove, J. F. Krane, F. Poirier, G. Calothy, and H. Hanafusa, *J. Virol.* **60,** 858 (1986).

[19] R. Jove, B. J. Mayer, H. Iba, D. Laugier, F. Poirier, G. Calothy, T. Hanafusa, and H. Hanafusa, *J. Virol.* **60,** 840 (1986).

[20] R. Jove, E. A. Garber, H. Iba, and H. Hanafusa, *J. Virol.* **60,** 849 (1986).

[21] G. Calothy, D. Laugier, F. R. Cross, R. Jove, T. Hanafusa, and H. Hanafusa, *J. Virol.* **61,** 1678 (1987).

[22] C. Béchade, G. Calothy, B. Pessac, P. Martin, J. Coll, F. Denhez, S. Saule, J. Ghysdael, and D. Stéhélin, *Nature* (*London*) **316,** 559 (1985).

small minority of cells after a long delay (several weeks), which contrasts with the massive and rapid proliferation caused by oncogene-containing retroviruses. This suggests that ALV activate, presumably by insertional mutagenesis, genes that are responsible for NR cell multiplication. By serially passaging supernatants of proliferating cells on fresh NR cultures, we reproducibly isolated new acutely mitogenic retroviruses that are capable of inducing NR cell division within a short delay. We showed that all retroviruses, thus far isolated, transduced the catalytic domain of two related serine/threonine protein kinases: c-*mil*/c-*raf* and c-R*mil*/B-*raf*.[23,24]

By analyzing retroviruses isolated at different passages on NR cells, we were able to establish that a crucial step in oncogene transduction is the formation of hybrid transcripts containing viral and activated cellular sequences (e.g., *mil* and R*mil*). Such transcripts are presumably generated by occasional read-through transcription escaping termination signals in the 3′ LTR[25] and are responsible for NR proliferation.[26] They are subsequently packaged and acquire 3′ viral sequences by recombination with the viral genome[26] favored by partial sequence identity.[27,28]

Characterization of such read-through transcripts in correlation with the acquisition of growth capacity in NR cells infected with ALV may prove useful in identifying genes involved in the regulation of growth or proliferation of these cells.

Analysis of Retina-Specific Differentially Regulated Genes by *in Situ* Hybridization

NR cells infected with conditional mutants of RSV represent a useful model to study mechanisms regulating the growth of differentiated cells. Our experimental approach in this direction was to isolate, by differential screening, cDNAs corresponding to mRNAs specifically expressed in nondividing NR cells (postmitotic NR cells *in ovo* and cells made quiescent after shifting *ts* RSV-infected cultures to nonpermissive temperature) and down-regulated in proliferating cells. We used *in situ* hybridization to study

[23] M. Marx, P. Crisanti, A. Eychène, C. Béchade, D. Laugier, J. Ghysdael, B. Pessac, and G. Calothy, *J. Virol.* **62,** 4627 (1988).

[24] M. Marx, A. Eychène, D. Laugier, C. Béchade, P. Crisanti, P. Dezélée, B. Pessac, and G. Calothy, *EMBO J.* **7,** 3369 (1988).

[25] S. A. Herman and J. M. Coffin, *J. Virol.* **60,** 497 (1986).

[26] M. P. Felder, D. Laugier, A. Eychène, G. Calothy, and M. Marx, *J. Virol.* **67,** 6853 (1993).

[27] A. Eychène, C. Béchade, M. Marx, D. Laugier, P. Dezélée, and G. Calothy, *J. Virol.* **64,** 231 (1990).

[28] M. P. Felder, A. Eychène, J. V. Barnier, I. Calogeraki, G. Calothy, and M. Marx, *J. Virol.* **65,** 3633 (1991).

the transcription of two retina-specific genes, QN1[29] (Fig. 8) and QR1[30] (Fig. 9), during development of the NR. Both genes encode presumably secreted proteins that could play a role in the regulation of cell division and/or cell to cell interactions.

Methods

Eyes of embryonic or postnatal quails were fixed overnight in 4% paraformaldehyde in PBS and then treated overnight in 30% sucrose. Sections of 15 μm were freeze-cut and hybridized as described with RNAs probes or oligonucleotides probes.

RNA Probes

POLYMERASE. The DNA to be transcribed should be cloned into the polylinker site of an appropriate transcription vector which contains a promoter for SP6 T3 or T7 RNA polymerase.

LABELING REACTION: Transcription buffer 10× concentrated: 400 mmol/liter, Tris–HCl, pH 8 (20°), 60 mmol/liter $MgCl_2$, 100 mmol/liter disthiothreitol, 20 mmol/liter, spermidine, 100 mmol/liter NaCl, and 1 U/μl RNase inhibitor.

1. The linearized DNA to be transcribed should be purified by phenol/chloroform extraction and ethanol precipitation.
2. Add the following mixture to a microfuge tube on ice:
   1 μg of linearized DNA
   2 μl DIG NTP labeling mixture, 10× concentration
   2 μl 10× transcription buffer made up to 18 μl with sterile water and add
   2 μl SP6 (40 U) T7 or T3 RNA polymerase.
3. Centrifuge briefly and incubate for 2 hr at 37°. A longer incubation does not increase the yield of labeled RNA.
4. As the amount of the DIG-labeled RNA transcript is far in excess of the template DNA, it is not usually necessary to remove the template DNA by DNase treatment. If desired, the template DNA can be removed by the direct addition of 20 U DNase I, RNase-free and incubation for 15 min at 37°.
5. With or without prior DNase, add 2 μl 0.2 *M* EDTA solution, pH 8, to stop the reaction.

[29] L. Bidou, P. Crisanti, C. Blancher, and B. Pessac, *Mech. Dev.* **43,** 159 (1993).

[30] M. Guermah, P. Crisanti, D. Laugier, P. Dezélée, L. Bidou, B. Pessac, and G. Calothy, *Proc. Natl. Acad. Sci. U.S.A.* **88,** 4503 (1991).

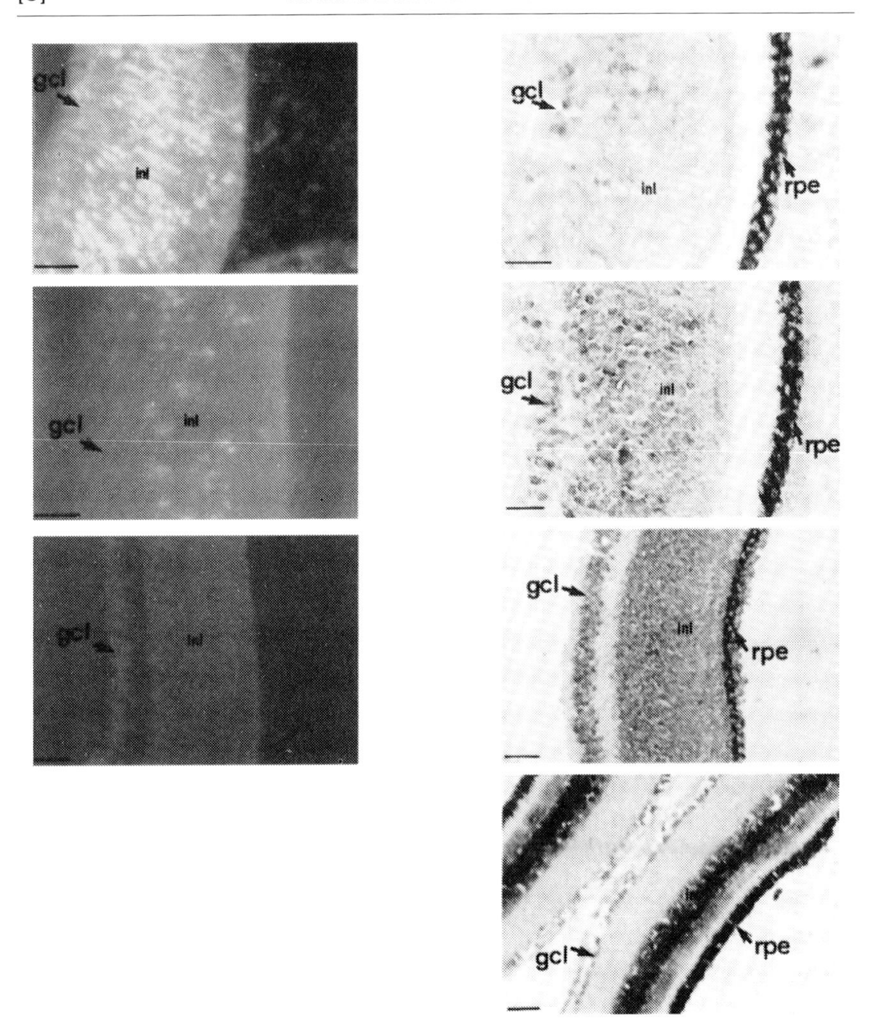

FIG. 8. Study of QN1 RNA expression during NR development by *in situ* hybridization. QN1 expression is compared to S phase detection (left) QN1 RNA is detected only in postmitotic neurons (right).

6. Precipitate the labeled RNA with 2.5 μl LiCl, 4 mol/liter and 75 μl prechilled (20°C) ethanol, mix well.

7. Leave for at least 30 min at −70° or for 2 hr at −20°.

8. Centrifuge (at 12,000 *g*), wash the pellet with 50 μl cold ethanol, 70% (v/v), dry under vacuum, and dissolve for 30 min at 37° in 100 μl

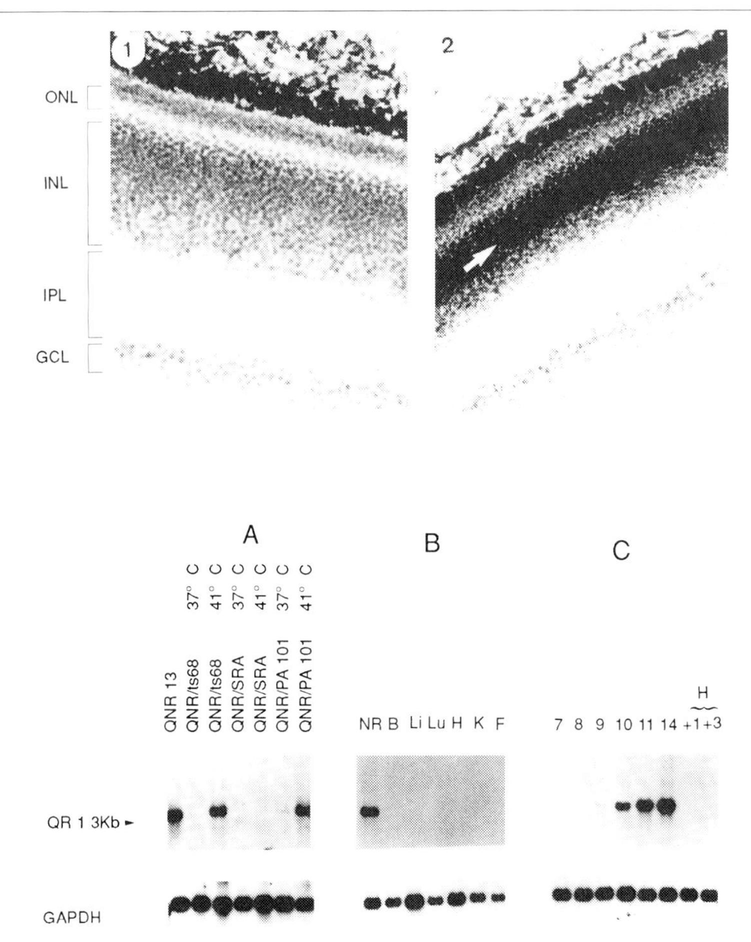

FIG. 9. Regulation of QR1 expression. (Top) Hybridization with sense (1) and antisense (2) RNA probes. QR1 is expressed in the middle of the inner nuclear layer. (Bottom) Northern blot analysis of QR1 transcription. (A) QR1 expression correlates with quiescence of NR cells infected with *ts* mutants of RSV. (B) QR1 is a retina-specific gene. (C) QR1 is regulated during NR development.

diethyl pryocarbonate-treated water; 20 U RNase inhibitor can be added to inhibit possible contaminating RNases.

9. The amount of newly synthesized labeled RNA depends on the amount, size (site of linearization), and purity of the template DNA.

10. The transcript can be analyzed by agarose gel electrophoresis and ethidium bromide staining; the yield of transcript can be estimated by comparing the ratio between DNA and RNA bands.

Oligonucleotide Probes

Oligonucleotides of 23- and 31-mer are tailed with DIG-dUTP as described in the Boehringer Mannheim kit. Hybridizations of retina sections were performed overnight at 37° with 10 μl of probes at a concentration of 10 to 20 ng/μl in 50% formamide, 5 × SSC, 0.02% SDS, DNA salmon sperm or 5% blocking reagent and 0.1% *N*-laurylsarcosine. After hybridization, sections were washed at room temperature in 2 × SSC for 30 min, in 1 × SSC for 30 min, and in 0.5 × SSC for 30 min; rapidly soaked in 100 m*M* Tris–HCl, 100 m*M* NaCl, pH 7.5 (buffer 1); and incubated in the 1% (w/v) blocking reagent (w:v) from Boehringer Mannheim (buffer 2) or with 50 μg/ml of DNA salmon sperm. Polyclonal sheep antidigoxigenin Fab fragments conjugated to alkaline phosphatase (750 U/mg) were diluted with buffer 1 and applied to sections for 30 min at room temperature. Slides were washed with buffer 1 at room temperature and equilibrated with 100 m*M* Tris–HCl, 100 m*M* NaCl, and 50 m*M* $MgCl_2$, pH 9.5 (buffer 3); finally, slides were incubated with freshly prepared NBT and X phosphate solution (45 μl of NBT at 70 mg/ml and 35 μl of X phosphate at 50 mg/ml) in 10 ml of buffer 3. This incubation was in a humidified light-tight box (in the dark) without shaking for 3 hr or more. Reactions were stopped with 10 m*M* Tris–HCl, 1 m*M* EDTA, pH 8.

Conclusion

NR cells proved useful in oncogene studies in several aspects. They allowed to establish model cell systems in which growth and differentiation are controlled by oncogene expression. They also provided a tool for genetic and functional analysis of oncogenes by showing that cell proliferation induced by oncogenes could be dissociated from the expression of other transformation markers. Finally, they provided a convenient model for *in vitro* studies on activation and transduction of protooncogenes by retroviruses.

Acknowledgments

We gratefully acknowledge support from the Association pour la Recherche sur le Cancer and the Association Française Retinitis Pigmentosa.

[6] Myogenic and Chondrogenic Cells

By DAVID BOETTIGER and CAMILLE DILULLO

I. Introduction

Muscle cells and cartilage cells can be cultured as relatively pure primary cell populations and have been used extensively for analysis of differentiation and for evaluation of the effects of oncogenes on differentiation.[1,2] Thus there is a considerable background of information for both normal as well as transformed cell types. The identification of the myogenic regulatory genes which control the expression of muscle-specific genes has provided a major impetus to the molecular analysis of muscle differentiation.[3–5] This chapter focuses on the production of primary cultures from embryonic muscle and cartilage with some reference to cell lines which may provide a useful alternative. The maintenance of differentiated cells in culture requires careful attention to culture conditions including batches of serum, specific media, components, and details of the procedure for isolation of the primary cells which, of necessity, varies from laboratory to laboratory. Thus, it is particularly important to pay attention to the methods for evaluating the cultures and to optimize the conditions for each laboratory using specific criteria.

II. Primary Skeletal Muscle Cell Culture

The muscle tissues used consist of cells at different stages of myogenic differentiation from replicating myoblasts to multinucleated myotubes. The age of the embryos used is chosen to maximize the yield of replicating mononucleated myoblasts. The method described is for chicken embryonic pectoralis muscle. A section on modifications for other tissue sources is given (see Section II, B).

A. General Procedure for Culture of Skeletal Muscle Cells from Avian Pectoralis

1. Prepare collagen-coated tissue culture dishes with or without Aclar (see Section VI, A, 4). Collagen must thoroughly dry before plating

[1] S. Alema and F. Tato, *Adv. Cancer Res.* **49,** 1 (1987).
[2] D. Boettiger, *Curr. Top. Microbiol. Immunol.* **147,** 31 (1989).
[3] R. L. Davis, H. Weintraub, and A. B. Lassar, *Cell (Cambridge, Mass.)* **51,** 987 (1987).
[4] W. E. Wright, D. Sassoon, and V. K. Lin, *Cell (Cambridge, Mass.)* **56,** 607 (1989).
[5] H. Weintraub, R. Davis, S. Tapscott, M. Thayer, M. Krause, R. Benezra, T. K. Blackwell, D. Turner, R. Rupp, and S. Hollenberg, *Science* **251,** 761 (1991).

cells or cells will not do well. Medium should also be prepared in advance.

2. Douse 11-day embryonated chicken eggs with 70% (v/v) ethanol in order to keep conditions as sterile as possible. Remove embryo from egg and put in a 100-mm petri dish with Hanks' balanced salt solution (HBSS; Life Science Technologies, Grand Island, NY). Place the embryo on its back and remove the skin from the chest area. On the right- and left-hand sides cut the breast muscle away from the sternum and clavicle. Place the excised muscle into HBSS in a separate glass dish.

3. Using a dissecting scope, clean the muscle tissue by removing blood vessels and connective tissue. These tissues are the major source of nonmuscle cells; therefore, careful dissection removing the overlying connective tissue is key to high-quality muscle cultures. We use Dumont No. 5 forceps (Fine Science Tools, Inc., Foster City, CA) for dissection. Put the cleaned muscle tissue into a second dish and mince it into very small pieces ($<1\ mm^3$).

4. Trypsinize minced muscle tissue in 10 ml of 0.25% trypsin in calcium- and magnesium-free balanced salt solution (CMF), mix completely, and incubate for 25 min at 37°. At the end of the incubation, stop the reaction by adding 1 ml of myo medium [minimum essential medium (MEM) with 10% horse serum, 10% chicken embryo extract, 1% glutamine, and antibiotics; see Section VI, B]. Collect the muscle cells in a 15-ml tube and spin at 1500 rpm for 5 min.

5. Discard the supernate and add 2 ml of fresh medium. Disperse the cells gently with a drawn out Pasteur pipette until no clumps are visible. Add medium to bring volume up to 10 ml. Filter the muscle cell suspension through two layers of lens paper in a 13-mm Swinnex filter holder (Millipore, Bedford, MA) attached to a 10-ml disposable syringe. Let cells filter through by gravity. Do not use the plunger as the pressure may damage the lens paper. Count cells and plate at a density of $2–7 \times 10^5$ cells/35-mm dish in myo medium (Eagle's MEM with 10% horse serum, 10% chicken embryo extract, 2 m*M* glutamine, and antibiotics, see Section VI, B). A lower density is suitable for morphological and immunofluorescent analysis whereas a higher density is better for biochemical experiments.

B. Skeletal Muscle Cultures from Other Species

1. Quail. Quail muscle cultures are prepared from 10-day embryos using the same protocol as for chick. The advantages of the quail system are that the cells can be cloned more easily and that cells continue to differentiate efficiently after subculture.[6]

2. Rat. Rat muscle cells are cultured from the limb muscles of newborn

[6] I. R. Konigsberg, this series, Vol. 63, p. 511.

rats. At the time of birth, substantial numbers of mononucleated cells still exist. By 10 days postnatal, few mononucleated cells remain, resulting in poor myogenic cultures. Use the hind leg thigh muscles of newborn rats and follow dissection and dissociation procedures given for chicken muscle. The recommended plating, density is about 3.5×10^6 cells/60-mm collagen-coated dishes in 1 : 4 Medium 199 to Dulbeccos modified Eagle's medium (DMEM) with 10% horse serum, 1% chick embryo extract, and antibiotics. Plating efficiency is about 50% and fusion rates of 40–60% can be achieved.[7]

3. Human. Human muscle cultures can be produced from muscle biopsy tissue. This tissue can produce 1×10^3 satellite cells per 0.1 mm^3 of tissue. (For a detailed procedure, see Blau and Webster[8].)

C. Myogenic Cell Lines

Although myogenic cell lines do not differentiate as fully as primary myoblasts, they do retain the ability to differentiate to varying degrees. The actual differentiation depends on the particular cell line, the subclone being used, and the culture conditions. It is generally wise to optimize these parameters before proceeding. Cell lines commonly used include murine cell lines C2C12[9] and MM14[10] and rat cell lines L6 and L8 (available from American Type Culture Collection, Rockville, MD). Human muscle cell lines which differentiate efficiently have not been developed. Some differentiation has been obtained with human rhabdomyosarcomas, but these are generally more limited models.

D. Evaluation of Skeletal Muscle Cultures

1. Morphological Analysis. With a little experience, one can assess the quality of a culture by examining cellular morphology under phase microscopy. The cells, as initially isolated from the embryo, consist of a large population of replicating myoblasts along with some fibroblast-like cells. The myoblasts tend to be small and either rounded or only moderately spread. They are difficult to distinguish from contaminating fibroblasts at this stage. After withdrawal from the cell cycle, myoblasts elongate and exhibit a bipolar morphology (Fig. 1A, solid arrows). This usually begins about day 1 and is prominent by day 2 of culture. The presence of these somewhat bifrigent, bipolar cells provides the first reliable indication of myogenesis and their number is a measure of the quality of the culture.

[7] D. Yaffe, *in* "Tissue Culture Methods and Applications" (P. F. Kruse and M. K. J. Patterson, eds.), p. 106. Academic Press, New York, 1973.

[8] H. M. Blau and C. Webster, *Proc. Natl. Acad. Sci. U.S.A.* **78,** 5623 (1981).

[9] D. Yaffe and O. Saxel, (*London*) *Nature* **270,** 725 (1977).

[10] T. A. Linkhart, C. H. Clegg, and S. D. Hauschka, *Dev. Biol.* **86,** 19 (1981).

These postmitotic, mononucleated myocytes further elongate (Fig. 1A, open arrows) and generally fuse to form multinucleated myotubes (Fig. 1B). The girth of the myotubes increases with time.

2. Quantitation of Differentiation. The quantitation of the number of nuclei present in myotubes is a good measure of the relative proportion of cells in the myogenic lineage in the original culture and their efficiency of differentiation. We have found that an easy means of visualizing muscle nuclei for counting is to stain the cells with acridine orange (see Section VI, C2). The stained cells are observed using a fluorescent microscope with FITC filters. In primary chicken muscle cultures, fusion efficiencies of 60–85% (nuclei in myotubes) can be achieved by days 3–4. Mononucleated cells continue to proliferate which results in a decrease of the proportion of nuclei in myotubes at later times. The continued proliferation of these cells tends to cause the detachment of the myotubes. One method to eliminate these replicating cells is to add cytosine arabinoside (10 μM) at day 3.

A more accurate method of nuclear quantitation is based on the identification of nuclei in cells that also express a muscle-specific protein. Cultures can be stained immunofluorescently with an antibody to a muscle-specific protein (we use MF20, a muscle myosin monoclonal antibody available from Developmental Studies Hybridoma Bank) and Hoechst 33258 (0.5 mg/ml, 10 min). The cells are scored using a fluorescent microscope with a filter cube (Chroma Technology, Brattleboro, VT) which provides for simultaneous viewing of the blue Hoechst 33258 and FITC.

3. Markers for Replicating Myoblasts. The replicating myoblasts initiate the synthesis of a cell surface antigen identified by the monoclonal antibody L4 which is not expressed by fibroblasts (the major contaminant of the cultures); however, it is not a unique muscle marker as it is also expressed in the chondrogenic lineage.[11] Nevertheless, immunofluoroescent staining of the primary muscle culture with this antibody provides the most sensitive current measurement of the proportion of potential muscle cells in the replicating cell population. Our cultures are routinely 85% $L4^+$. Note that the level of expression of L4 in the replicating progenitors is low and that its level continues to rise as myogenic differentiation proceeds to myotubes. The replicating cell population of myogenic cells also expresses mRNAs for MyoD (myogenic regulatory factors) (CMD),[12–14] myf5,[15] cardiac α-

[11] M. George-Weinstein, C. Decker, and A. F. Horwitz, *Dev. Biol.* **125,** 34 (1988).

[12] F. C. de la Brousse and C. P. J. Emerson, *Genes Dev.* **4,** 567 (1990).

[13] D. Sassoon, G. Lyons, W. E. Wright, V. Lin, A. B. Lassar, H. Weintraub, and M. Buckingham, *Nature (London)* **341,** 303 (1989).

[14] E. N. Olson, *Genes Dev.* **4,** 1454 (1990).

[15] T. Braun, E. Bober, G. Buschhausen-Denker, S. Kotz, K. Grzeschik, and H. H. Arnold, *EMBO J.* **8,** 2617 (1989).

actin,[16] and, at later stages but prior to withdrawal from the cell cycle, desmin.[17] These can be detected by Northern blots or polymerase chain reaction, but it may be necessary to do a quantitative comparison using a later muscle marker since these genes are also expressed, mostly at higher levels, in the mature myotube.

4. Markers for Differentiation. A variety of both mRNA probes and monoclonal antibodies have been developed for analysis of muscle-specific gene products. The synthesis of these products is initiated at the time of withdrawal from the cell cycle so they are not seen in the replicating cell population but can be seen in postmitotic, mononucleated myocytes. The most commonly used marker is the muscle-specific myosin heavy chain (there are several isoforms but many of the antibodies will react with more than one isoform). Other commonly used markers include the myosin light chains, muscle-specific actin, muscle tropomyosins, and acetylcholine receptor. cDNA probes and monoclonal antibodies have been made for many of these (see Developmental Studies Hybridoma Bank, Univ. of Iowa, Iowa City).

III. Cardiac Muscle Cultures

A. Preparation of Cardiac Muscle Cultures

1. Cardiac muscle cultures can be made from 6- to 8-day embryonic chicken hearts. Dissociated cardiac myocytes will be plated on uncoated sterile glass coverslips which should be prepared ahead of time.

2. Douse eggs with 70% ethanol to keep conditions as sterile as possible. Embryos are removed from their shells and placed in HBSS. For 7-day embryos, the chest cavity will be open and the heart visible. For older

[16] M. Shani, D. Zevin-Sonkin, O. Saxel, Y. Carmon, D. Katcoff, U. Nudel, and D. Yaffe, *Dev. Biol.* **86,** 483 (1981).

[17] C. S. Hill, S. Duran, Z. Lin, K. Weber, and H. Holtzer, *J. Cell. Biol.* **103,** 2185 (1986).

FIG. 1. (A) Phase micrograph of a skeletal muscle culture prepared from 10-day chicken embryonic pectoralis muscle at day 2 in culture. The closed arrow identifies a mononucleated myocyte beginning to extend fine bipolar processes. The open arrows identify more mature, postmitotic myocytes whose bipolar processes are elongating. The remaining cells are replicating myoblasts and "fibroblasts." (B) Phase micrograph of a skeletal muscle culture (as above) at day 8 in culture showing two large multinucleated myotubes formed by using of mononucleated myocytes which continue to increase in size. (C) Phase micrograph of substrate adherent vertebral chondrocytes which have been replated and grown as a monolayer for 3 days. A pair of chondrocytes is beginning to detach. The detached cell population continues to divide and may be reattached by digestion of their extracellular matrix.

A

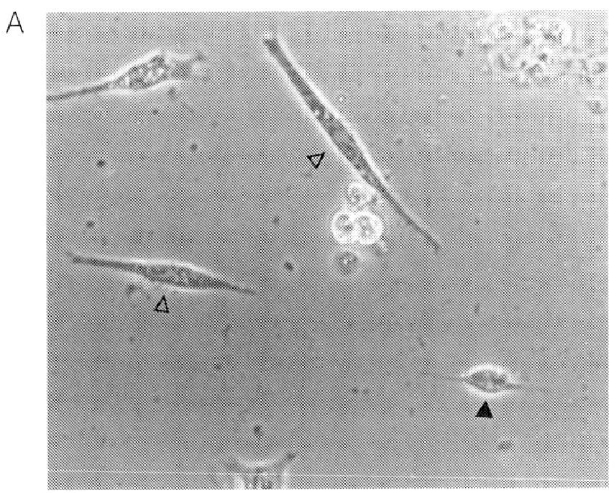

B

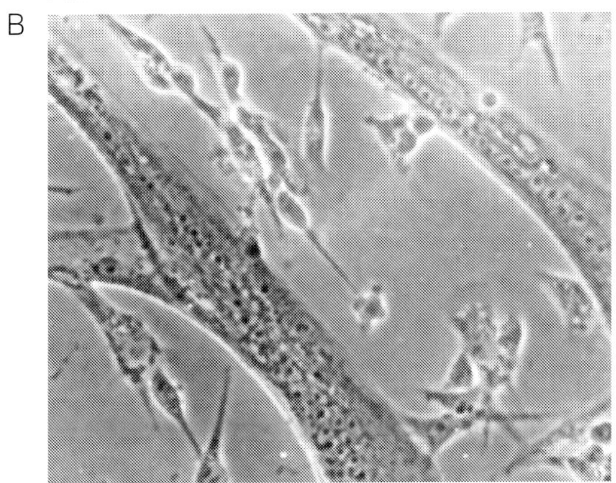

C

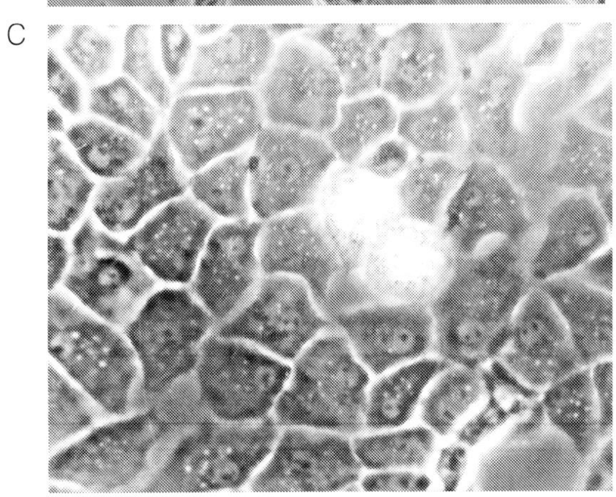

embryos the chest cavity must be carefully opened by cutting through the sternum to expose the heart. The heart can be removed with a forceps. The vessels will probably remain attached. Collect hearts in a small plastic dish with HBSS. It is best to work with plastic at this point because cardiac cells tend to stick to glass.

3. The heart must be cleaned of pericardia and attached vessels. The pericardia is a very thin sheet which adheres tightly to the heart. It is most easily identified on the ventricles. One should begin in this region and then proceed to remove it from the remainder of the heart. The vessels can generally be pulled right out. Firmly grasp the heart with a sturdy forceps in one hand, being careful not to damage it, and pull the vessels out by their proximal ends with a second forceps. The valve below the aorta should be removed as well as the valve on the other side of the heart. These are difficult to locate and may take some practice.

4. Put the cleaned heart into a 60-mm plastic dish without medium. It is advantageous to have additional hearts because it is easy to lose or damage the cells. Mince the heart into very small pieces. Add 1 ml of calcium- and magnesium-free HBSS (CMF) directly to the heart tissue. Transfer heart to a 15-ml plastic tube. (If using glass pipettes to transfer, rinse pipette with medium before using, otherwise cells will stick to the pipette.)

5. Add 8.8 ml of CMF and 0.2 ml of 2.5% trypsin. Invert tube approximately 20 times. Incubate tube at 37° for 10 min. Allow cells to settle (do not spin). Remove supernate from tube, and add 9.8 ml of fresh CMF and another 0.2 ml of 2.5% trypsin. Mix and incubate tube at 37° for another 15 min (20 min if there are a lot of cells). Shake tube every 5 min.

6. Prepare a 50-ml tube with 10 ml of cardiac plating medium. After the second trypsin incubation, collect pellet as well as supernate with remaining suspended cells and add to the 50-ml tube with plating media (MEM–L-glutamine with 5% fetal calf serum, 1% L-glutamine, and antibiotics). Remember, if using a glass pipette to transfer, rinse pipette first! Add an additional 1–2 ml of plating medium to the 15-ml tube and collect any remaining cell and add to the 50-ml tube. Centrifuge the 50-ml tube for 5 min at 1500 rpm. Discard the supernate and resuspend cells in fresh plating medium. Plate cells at a density of 1 heart per 35-mm dish. Note: Cardiac cells are fragile and too rigorous a mechanical dispersion will damage the cells.

7. Allow cells approximately 24 hr to attach. Remove plating medium and add cardiac growth media (same as plating medium but without L-glutamine). For best results, cells should be fed every day.

B. *Evaluation of Cardiac Muscle Cultures*

Cardiac cultures will be a mixture of fibroblasts and cardiac cells. The most definitive way to distinguish cardiac cells from fibroblasts in culture is by their spontaneous contractions. When initially plated, before the cells come out of suspension, cardiac cells can sometimes be seen to beat. Beating cannot be detected, however, as the cells settle on the dish. By day 3 when the cells are spread, but probably earlier, the cardiac cells will resume beating. At this stage, the cardiac cells may cluster in small islands and exhibit cobblestone-like pattern.

Spread cardiac cells are difficult to distinguish from fibroblasts. However, cardiac nuclei differ from the nuclei in surrounding fibroblasts. Cardiac nuclei are rounded and of a uniform shape whereas the nuclei in fibroblastic cells are more flattened and have a kidney shape. Differentiated cardiac cells with myofibrils can divide. Because this is a mixed culture, the best way to distinguish cardiac myocytes is by staining for muscle-specific proteins. Many antibodies for muscle-specific markers that will stain skeletal myocytes can also be used to stain cardiac myocytes (see Section II, A, 2).

IV. Chondrocyte Cultures

Chondrocyte cultures have been made from a variety of different species, but most commonly from chicken and rabbit. Cells that are cultured from vertebral or sternal cartilage are fairly mature chondrocytes and synthesize chondrocyte markers such as type II collagen and specific chondrocyte proteoglycans at the time of culture. *In vivo*, chondrocytes are surrounded by considerable extracellular matrix which is not completely removed in the preparation of the cells for culture. The presence of this matrix makes the cells difficult to infect with retroviral vectors (and probably also for DNA transfection, although we have not attempted this). As an alternative to the chondrocyte, one can use limb bud cells which are more susceptible to infection and will primarily differentiate into chondrocytes. Limb bud cultures are more heterogeneous than the vertebral chondrocyte cultures and also contain myogenic cells, fibroblasts, and osteogenic cells.

A. Preparation of Chicken Embryo Vertebral Chondrocytes

1. Douse 12-day embryonated chicken eggs with 70% ethanol. Remove embryos from the eggs with sterile forceps and place into a sterile 100-mm dish. With the embryo face down in a separate 100-mm dish, gently remove skin and gelatinous material with a fine (No. 5) forceps exposing the vertebral column. Insert one fine point of the forceps on each side of the base of the column and pull gently outward to cut off the column below ribs. Under a dissecting microscope, separate the solid ventral portion of the vertebral column from the unwanted dorsal side by gently cutting along each side with a fine curved (No. 7B) forceps. Hold down the column with a second pair of foreceps and clean away remaining muscle and soft tissue.

2. Transfer the cleaned ventral portions of the vertebral column to a drop of phosphate-buffered saline (PBS) on a glass dish and mince finely with two scalpels. Incubate fragments in a 15-ml conical centrifuge tube in 1 ml 0.25% trypsin + 1 ml of 0.25% collagenase in complete Hanks' balanced salts (with Ca^{2+} and Mg^{2+}) for 2 hr at 37°. Resuspended cells by pipetting every half hour to mix.

3. Spin down cells, resuspended in DMEM with 10% fetal calf serum, and filter through lens paper (Section II, A, 5). Dilute and plate at 1×10^6 cells/ml, 10 ml/100-mm dish in DMEM with 10% fetal calf serum and antibiotics.

4. Incubate for 5 days at 37° without refeeding. During this period the cells attach to the dish by day 1 and then many detach and remain as round floating cells in suspension. The cells do continue to divide in suspension. It is largely the synthesis of proteoglycans by the cells that is responsible for their detachment and the medium itself can become quite viscous due to this proteoglycan. The cells which remain attached are mainly fibroblasts.

5. The floating cells are collected, washed twice with PBS, treated with 0.25% trypsin for 5–10 min at 37°, centrifuged, resuspended in fresh medium, and replated at 1×10^6 cells/100-mm dish. The cells attach and show a cobblestone appearance (Fig. 1C). This trypsinization is the tricky part since this is the time at which oncogenes should be introduced. Harsher treatments such as longer trypsinization or use of pronase make for more efficient retrovirus infection but also can result in degradation of the differentiated phenotype, so a balance must be achieved. If the cultures are well differentiated they may synthesize sufficient proteoglycan to continue to detach in the subcultures. If this is a problem, they can be reattached by adding 0.1 ml of a 10-mg/ml solution of crude hyaluronidase (Sigma, St. Louis, MO) directly to the culture medium.

B. Preparation and Culture of Chick Embryo Limb Bud Cells

The limb buds provide a source enriched for chondrocyte progenitor cells. The culture system described here is chosen to promote the differentiation of this progenitor cell population into mature chondrocytes. The cells are plated in spot cultures to keep the cells in restricted areas of the dish as high density cultures differentiate much more efficiently. It should be noted that the maintenance of differentiation for these cells is more fragile than for vertebral chondrocyte cultures and the differentiation ability can be easily lost through improper handling, improper media components, or even elements present in inocula used for gene transfer (including retrovirus stocks grown in calf serum).

1. Limb buds are dissected from Hamburger and Hamilton stage 21–22 embryos (about 3.5 days of incubation) using a dissecting microscope.[18] The time of incubation of the eggs is important and is affected by temperature and humidity. Not all the eggs in a batch will be exactly the same age; therefore, visual inspection of the embryos is necessary to ensure they are at the appropriate developmental stage. Prior to stage 24, the width of the limb buds is greater than their length and this can be used as the criteria for selection of embryos at the appropriate stage. Once embryos pass this stage, chondrocytes mature and type II collagen is expressed so the cultures will be contaminated with mature chondrocytes which may complicate the experiment. The yield from four limb buds (one embryo) is about 1×10^6 cells.

2. Douse eggs with 70% ethanol and remove the embryos, placing them in a 100-mm dish with calcium- and magnesium-free Hanks' balanced salt solution. Hold the embryo with a forceps and use a scalpel or a pair of scissors to clip off the limb buds which protrude from the body of the embryo.

3. Dissected limb buds are transferred to a 15-ml conical centrifuge tube, washed with PBS, and resuspended in 0.25% trypsin/EDTA and incubated for 10 min at 37° with occasional pipetting to dissociate the cells. The resulting cell suspension is centrifuged, resuspended in DMEM with 10% fetal calf serum, and filtered through lens paper (Section II, A, 5).

4. Cells are diluted to $1.5–2.0 \times 10^7$/ml and plated in 20-μl spots on dry, untreated culture dishes ($3–4 \times 10^5$ cells/spot). Four to six spots are plated per 35-mm dish. Spots are incubated for 30 min to 1 hr at 37° to allow the cells to attach. The dishes are then flooded gently with DMEM containing 10% fetal calf serum and antibiotics.

Note. We have also employed an alternative to these high density cultures which selects for chondrocyte differentiation. In this variation the dishes are prepared with a 0.5% agar sublayer in (DMEM) with 10% fetal calf serum which is allowed to solidify before addition of the cells.

C. *Evaluation of Chondrocyte Cultures*

1. Morphological Evaluation. When vertebral chondrocytes are plated, many of the cells initially attach to the dish and appear as cobblestone-like colonies (Fig. 1C). Contaminating fibroblasts can usually be distinguished by their "fibroblastic" shape as opposed to the more "epitheloid" shape of the chondroblasts. Too many fibroblasts in the culture can affect chondrocyte differentiation adversely. Chondrocytes normally synthesize large amounts

[18] R. B. Birge, J. E. Fajardo, C. Reidman, S. E. Shoelson, Z. Songyang, L. C. Cantley, and H. Hanafusa, *Mol. Cell. Biol.* **13,** 4648 (1993).

of extracellular matrix consisting of type II collagen, proteoglycans, and glycosaminoglycans.[19] These products cause the cells to take on a more refractile appearance, to round up, and to detach from the dish. More detachment indicates that the cells have more successfully maintained the normal rate of synthesis of these products. Treatment with proteolytic enzymes, hyaluronidase, or condroitenases removes some of the extracellular matrix and allows the cells to reattach. The overall quality of these cultures is quite easy to determine with these simple morphological criteria. Because of the selection for floating cells, these cultures are usually >99% chondroblasts.

The limb bud spot cultures consist of small cells at high density and have few distinguishing characteristics at early times. As the cultures mature (2–7 days), nodules of cells appear. The individual cells in the nodules appear to be rounded. These cartilage nodules can be easily identified by staining with Alcian blue which localizes sulfated proteoglycans (see section VI, C, 1). By 4 days, the majority of cells in the spot should stain and show a nearly confluent array of dense clusters, often donut shaped. Myogenic cells are usually present at the periphery of the spots.

2. Chondroblast Markers. The easiest marker to use, currently, is type II collagen available as both monoclonal antibodies and cDNA probes. In contrast, the presence of fibronectin and type I collagen indicates the loss of a differentiated phenotype or the presence of contaminating cell types.

V. Introducing Oncogenes

A. *Introduction of Oncogenes Using Retroviral Vectors*

Two strategies have been developed which use retroviral vectors to introduce exogenous genes into primary differentiated cells.

1. Mass Cultures. Avian retroviruses have been isolated which carry a wide selection of oncogenes and a number of useful temperature-sensitive oncogene mutants.[20–22] When possible the use of temperature-sensitive mutants is recommended as the ability of the cells to express specific differentiated markers can and does occur due to culture conditions. Thus, the ability to revert to the differentiated phenotype following inactivation of the oncogene is a very important control. For genes that are not found in naturally occurring retroviruses, the RCAS and RCAN vectors based on

[19] G. Cossu, L. Warren, D. Boettiger, H. Holtzer, and M. Pacifici, *J. Biol. Chem.* **257,** 4463 (1982).

[20] H. Beug, A. Leutz, P. Kahn, and T. Graf, *Cell (Cambridge, Mass.)* **39,** 579 (1984).

[21] F. von Weizsaecker, H. Beug, and T. Graf, *EMBO J.* **5,** 1521 (1986).

[22] J. T. Parsons and M. J. Weber, *Curr. Top. Microbiol. Immunol.* **147,** 79 (1989).

the Rous sarcoma virus have been useful.[23] The most efficient way to infect many differentiated cell types is to resuspend the freshly trypinized cells in the virus inoculum at about 5 infectious units/cell and incubate the cells at 37° for 30 min before diluting and plating the cells. Note that the infection of myoblasts is less efficient than fibroblasts or cell lines; however, the infection of chondroblasts can even be several orders of magnitude less efficient. The trick with chrondroblasts is to remove as much of the extracellular matrix as possible. In practice, the removal of matrix is balanced by the need to maintain the differentiated phenotype which becomes progressively more fragile as more matrix is removed. In general, only a portion of the cell population is infected by the initial inoculum and one to three passages are usually required to ensure complete infection of the culture. Note that the virus to be used for infection should be grown in the same medium (serum) to be used for the differentiated cultures.

2. *Cloning.* A second approach that can be used is to clone the infected cells in soft agar suspension. For this procedure the cells are plated and infected as described earlier, although the high multiplicity of infection is less critical.

1. Prepare a 2.5% stock of agar (Difco Bacto-agar or Scott agar) and sterilize by autoclaving. Use the agar to make up a mixture of regular medium used for the differentiated cells with a final concentration of 0.5% agar. This will require the making of some double-strength base medium from powder and separate bicarbonate stock (as it precipitates from double-strength medium) to compensate for the lack of medium in the agar stock. Add enough agar medium to the plates to cover the bottom (3 ml/60-mm dish) and distribute 2-ml aliquots to glass tubes to be maintained at 45° in a heating block (one tube for each plate).

2. Prepare a suspension of infected cells. Remove one tube of agar quickly. Add 1.0 ml of cell suspension, mix, and pour onto a prepared base layer of agar. Allow agar to gel before incubation at 37°.

3. After 1–2 weeks (depending in part on the oncogene), compact macroscopic colonies are apparent in the agar. The colonies can be picked from agar under a dissecting or low power-inverted microscope using either hematocrit capillaries or pasture pipettes which have been heated and drawn to a fine tip. The size of the tip should be smaller than the colony diameter because an agar plug surrounds the picked cells and they have a reduced chance of growing out. The picked colonies are deposited into 24-well plates containing the appropriate medium for the cell type. In this method, it must be remembered that not all the colonies picked will express

[23] S. H. Hughes, J. J. Greenhouse, C. J. Petropoulos, and P. Sutrave, *J. Virol.* **61,** 3004 (1987).

differentiated markers. Thus this procedure is most practical when expression of the oncogene is conditional (such as temperature-sensitive mutants of pp60$^{v\text{-}src}$). It is important to test individual clones to ensure that they retain the ability to differentiate. Using this method, we have found that roughly one-third to one-half of the clones picked from ts-src-infected myoblasts retain the ability to differentiate following a shift to the nonpermissive temperature which inactivates the oncogene. There is also considerable variation in their efficiency of differentiation. Some clones will give 90% nuclei in myotubes (better than the primary mass cultures).

B. DNA Transfection into Cardiac and Skeletal Cultures

This section details the introduction of cDNA into primary muscle cultures.[24]

1. Cardiac or skeletal cultures should be prepared as described in the earlier sections. Skeletal muscle should be plated at a density of 4.5×10^5 per 35mm dish onto collagen-coated aclar (see Section VI, A, 4). Cardiac cells (from 8-day embryos) should be plated onto sterile glass coverslips using one heart per 35-mm dish.
2. Transfect approximately 24 hr after plating. Cultures should be about 60% confluent. Two to 3 hr before the transfection, change the medium to appropriate fresh medium (1.5 ml/dish).
3. A standard calcium phosphate precipitation method is used to transfect the cells.[25] Prepare the following two solutions ahead of time. Per 35-mm dish: Solution A: 12.4 μl 2*M* $CaCl_2$, 2 μg DNA; deionized water to make 110 μl. Solution B: 2x HEPES-buffered saline (280 m*M* NaCl; 10 m*M* KCl; 1.5 m*M* Na_2HPO_4; 12 m*M* dextrose, 50 m*M* HEPES, pH 7.5) to 110 μl. Thirty minutes before transfection, add Solution A drop by drop to Solution B, swirling gently. The solution should look cloudy but no large precipitate should appear. The final solution should stand at room temperature for 30 min.
4. At the end of 30 min, gently mix the solution with a pipette. Add 200 μl to each 35-mm dish (add drop by drop with the dish at a slight angle so that the solution is not placed directly onto the cells). Gently swirl the dish while adding solution so that it mixes with the medium. Return the dish to the incubator immediately.
5. Twelve hours later, wash the dish two times with MEM containing

[24] T. Schultheiss, J. Choi, Z. X. Lin, C. DiLullo, L. Cohen-Gould, D. A. Fischman, and H. Holtzer, *Proc. Natl. Acad. Sci. U.S.A.* **89,** 9282 (1992).

[25] J. Sambrook, E. F. Fritsch, and T. Maniatis, "Molecular Cloning: A Laboratory Manual," 2nd ed. Cold Spring Harbor, Lab., Cold Spring Harbor, NY, 1989.

no serum or with CMF. Be careful to get rid of most of the precipitate. Do not let a layer of precipitate form over the cells. Feed the cells with fresh myo medium. The time of exposure of the cells to Ca^{2+} can be critical. If it exceeds 12 hr, the cells may not develop well.

6. When myotubes are formed, the proportion of embryo extract in the myo medium can be reduced to 3%. Feed the cells every other day with this medium. Cytosine arabinoside (10 μM)[26] can be added to the culture at this stage to eliminate the replicating cells (primarily fibroblasts). Skeletal muscle development may be delayed by approximately 1 day in transfected cells.

VI. Subprotocols

A. *Coating with Extracellular Matrix*

The use of standard tissue culture dishes results in a relative selection for fibroblast-like cells and a reduction in the proportion of cells that differentiate. This is because the extracellular matrix (or more specifically fibronectin) is an essential "inducer" of muscle differentiation.[27] Although the muscle cells do synthesize fibronectin, additional fibronectin or collagen is necessary for efficient maintenance and differentiation. Dishes can be coated with fibronectin, type I collagen, or gelatin. The original procedures use rat tail collagen which is an acid extract (see Section VI, A, 1) whereas others use bovine collagen which is commercially available as a pepsin digest extract. For experiments involving myotubes and relating to muscle contraction, collagen is probably the substrate of choice. For early stages of differentiation (up to cell fusion), gelatin works as well, is cheaper, and coats more evenly. Fibronecting is preferable, particularly for low density cultures.

1. Coating Plates with Collagen. Collagen can be obtained commercially (Vitrogen 100 from Celtrix Laboratories, Palo Alto, CA). Use directly or dilute 1:3 with 0.01 *M* HCl, spread a thin coat on the dish with a rubber policeman, and allow to air dry.

Preparation of Rat Tail Collagen

1. Clip off tails in one piece. They can be stored in a plastic bag frozen at −20°. If dehydrated from freezing, rehydrate them in HBSS at 4° overnight or at 37° for 30 min.

[26] Z. X. Lin, J. Eshelman, C. Grund, D. A. Fischman, T. Masaki, W. W. Franke, and H. Holtzer, *J. Cell. Biol.* **108,** 1079 (1989).

[27] A. S. Menko and D. Boettiger, *Cell (Cambridge, Mass.)* **51,** 51 (1987).

2. Place the tails in a 100-mm culture dish containing 95% ethanol to sterilize.

3. Under sterile conditions, use bone clippers to remove a section of epidermis from the proximal end of the tail. Pull the epidermis away while keeping the tail intact to reveal the white collagen fibers. Remove these fibers working toward the distil end of the tail and place them into sterile deionized water. Remove any remaining blood vessels.

4. Transfer the collagen fibers to 1% glacial acetic acid (60 ml/tail). Stir at 4° for 1–2 days.

5. Centrifuge at 5000 rpm for 15 min to remove debris.

6. If collagen is too viscous, dilute with 1% acetic acid but not to more than 100 ml/tail. Aliquot to sterile tubes and store at −20°.

2. Coating Plates with Gelatin. Add 0.1% gelatin in distilled water to plates and incubate overnight. Remove solution and wash twice with PBS.

3. Coating Plates with Fibronectin. Fibronectin can be purchased commercially (Life Science Technologies, Collaborative Research and others) as a lyophilized powder. Make a 1-mg/ml solution and store at −70° in small aliquots to reduce freeze-thaw cycles. Dilute to 10 μg/ml in PBS, pH 7.3, and coat with enough to cover bottom of dish (about 1 ml for a 35-mm dish). Incubate for 2 hr (can be left overnight) in a humid atmosphere. The dishes should not be allowed to dry out. Rinse once with PBS before use.

4. Aclar Preparation. Aclar can be purchased from Allied Signal Inc. (Pottsville, PA). A sheet of Aclar should be washed in detergent and rinsed well. Cut the sheet into appropriate sizes for the dishes that are being used. Put the pieces into a glass container and cover with concentrated nitric acid. Shake for 30 min, stirring intermittently. Remove the acid (acid can be reused several times for this purpose). Add 95% ethanol and swirl for a short time. Remove ethanol. Add fresh ethanol and shake for 5–10 min. Dry on sterile towels. Store in a sterile container.

To prepare dishes with collagen-coated Aclar, put a drop of collagen on the dish with a Pasteur pipette and place a piece of Aclar on top (will attach Aclar to dish). Coat Aclar with collagen as described earlier.

B. Culture Medium

The culture medium is a critical factor and can make large differences in the maintenance of differentiation. Considerations for choosing each particular medium are given below.

1. Base Medium. A variety of media have been used as a base, including minimal essential medium (Eagle's), Dulbecco's modified essential medium, and Ham's F10 medium. These are convenient and relatively inexpensive. MCDB120 has become commercially available (Clonetics, San Diego CA)[28] and our experience shows that this is superior to the simpler medium and gives excellent results in the absence of serum.

2. Serum. Many laboratories (particularly in the culture of cells of mammalian origin) use two distinct media: one for growth and one for differentiation. Growth medium usually contains 10–20% fetal bovine serum which promotes maximal growth of the cells at the expense of differentiation. Differentiation medium has reduced serum (2–4%) and switches from fetal calf serum to horse serum. Sera vary considerably between different lots. It is best to test individual lots to select the combination which gives optimal results.

3. Embryo Extract. Embryo extract from chicken embryos is used in most media formulations. It is available from Life Sciences Technologies (Grand Island, NY) but can also be made. Eleven-day chicken embryos are removed from their shells and rinsed in HBSS. The embryos are forced through a 50-ml syringe, diluted 1:1 with BSS, and allowed to sit at room temperature for 1 hr. The mixture should be aliquoted and frozen at $-20°$. The yield is about 3.5 ml/embryo. Before use, thaw and spin the extract in a clinical centrifuge at top speed for 10 min. The extract should be tested at 1, 3, 5, 7, and 10% prior to use.

C. Staining

1. Alcian Blue. (1) Rinse culture with 3% acetic acid for 3 min. (2) Stain with Alcian blue for 30 min (Alcian blue 8GX, 1.0 g; $CaCl_2$, 0.5 g; distilled water to 100 ml). (3) Rinse in tap water. (4) Counterstain with methanil yellow (methanil yellow, 0.25 g; distilled water to 100 ml; glacial acetic acid, 2 drops). (5) Rinse in tap water.

2. Acridine Orange. (1) Rinse culture briefly with 1% acetic acid, then rinse with distilled water. (2) Stain in acridine orange for 3 min (0.1% in distilled water, store in dark). (3) Destain for 1 min in phosphate buffer (60 m*M*, pH 6.0). (4) Differentiate in 0.1 *M* $CaCl_2$ for 30 sec. (5) Wash with phosphate buffer. The nuclei appear yellow-green and the cytoplasm orange-red.

Acknowledgments

The authors acknowledge Drs. Howard Holtzer and Zhong Xiang Lin for helpful discussions. This research was supported by grant CA16502 from the National Cancer Institute.

[28] R. G. Ham, J. A. St. Clair, C. Webster, and H. M. Blau, *In Vitro Cell Dev. Biol.* **24,** 833 (1988).

[7] Cell Synchronization

By WILHELM KREK and JAMES A. DECAPRIO

Introduction

In order to understand how eukaryotic cells grow and divide, it is important to identify and characterize the molecular components controlling entry into and execution of the two major events of the cell cycle, S and M phase. Detailed functional analyses of the proteins involved in these processes require techniques that allow one to synchronize or arrest cell populations at selected phases of the cell cycle.

For example, serum starvation is an effective method to synchronize a variety of cell types in G_0/G_1. Serum-starved cells can remain healthy for a long period of time in this nonproliferating or quiescent state. The addition of high concentrations of serum to these cells induces them to leave the G_0 state in a synchronous manner, to progress through the G_1 phase, to replicate their DNA, and to divide. These cell systems are particularly useful for studying the molecular events contributing to the G_0 to S phase traversal.[1,2] In addition, there are several types of cells that are naturally arrested in the G_0/G_1 state, such as T and B lymphocytes. These cells can be readily isolated from human blood or spleen, respectively, and can be induced to enter the cell cycle synchronously with a variety of agents.[3]

Among the most widely used methods to synchronize cells in the G_1 to S phase interval are starvation in methionine-free media[4] or culturing cells in the presence of mimosine,[5] hydroxyurea,[6] aphidicolin,[7] or excess thymidine.[8] When the metabolic block is removed, cells progress in a synchronous fashion through the subsequent S and G_2 phase. Nocodazole is commonly used to arrest cells at metaphase.[9] Following removal of this agent, cells progress synchronously into G_1 and then further through the

[1] A. B. Pardee, *Science* **246,** 603 (1989).
[2] B. J. Rollins and C. D. Stiles, *Adv. Cancer Res.* **53,** 1 (1989).
[3] Y. Furukawa, J. A. DeCaprio, A. Freedman, Y. Kanakura, M. Nakamura, T. Y. Ernst, D. M. Livingston, and J. D. Griffin, *Proc. Natl. Acad. Sci. U.S.A.* **87,** 2770 (1990).
[4] J. A. DeCaprio, J. W. Ludlow, D. Lynch, Y. Furukawa, J. Griffin, H. Piwnica-Worms, C. Huang, and D. M. Livingston, *Cell (Cambridge, Mass.)* **58,** 1085 (1989).
[5] M. Lalande, *Exp. Cell Res.* **186,** 332 (1990).
[6] J. A. Hubermann, *Cell (Cambridge, Mass.)* **23,** 647 (1981).
[7] P. Reichard and A. Ehrenberg, *Science* **221,** 514 (1983).
[8] D. Bootsma, L. Budke, and O. Vos, *Exp. Cell Res.* **33** (1964).
[9] G. W. Zieve, D. Turnbull, J. M. Mullins, and J. R. McIntosh, *Exp. Cell Res.* **126,** 397 (1980).

S phase. Generally, these agents are very useful for cell synchronization studies since they arrest cells at a single specific point in the cell cycle and the cell cycle block is reversible, allowing for synchronous progression of cells through subsequent cell cycle stages. However, it is important to remember that the addition of any drug may perturb the normal biochemical processes of the cell.

Cell synchronization by centrifugal elutriation[10] does not cause perturbations in either the biochemistry of the cell or the subsequent progression of cells through the cell cycle. Since this technology exploits the difference in the volume of cells as a function of their position in the cell cycle, it allows the separation of an asynchronous cell population into fractions enriched in G_1, S, and G_2/M phase cells.

This chapter describes methods that permit the isolation of cell populations that are unique with respect to their position within the division cycle. In particular, we detail protocols for synchronization of continously dividing cells by either imposing a metabolic block or by centrifugal elutriation.

Cell Synchronization of Continously Dividing Cells

General Considerations

It is important to remember that the extent to which a particular cell line will arrest at a specific point in the cell cycle in response to the various treatments described in this chapter will vary. For example, only normal, untransformed, and some immortalized cell lines respond to serum deprivation with growth arrest, whereas transformed cells serum starve very poorly or not at all. Examples of cell lines commonly used for serum arrest/stimulation experiments are WI-38 human fetal lung fibroblasts, Hs68 human foreskin fibroblasts, and BALB/3T3 clone A31 mouse embryo fibroblasts.

The use of metabolic agents for synchronization of cells requires first the determination of the parameters governing optimal cell synchronization in a specific cell line. First, the doubling time of an exponentially growing culture has to be known. The doubling time is the minimum time of exposure to the drug required to synchronize an asynchronous population of cells. Second, the minimum concentration of the drug of choice to enrich for cells in a particular phase of the cell cycle must be determined. A satisfactory level of enrichment is reached when approximately 75% of a cell population is arrested in a particular phase of the cell cycle. Exposure to any drug for too long or at too high of a concentration can lead to cell death and

[10] R. J. Grabske, *Fractions* **1,** 1 (1978).

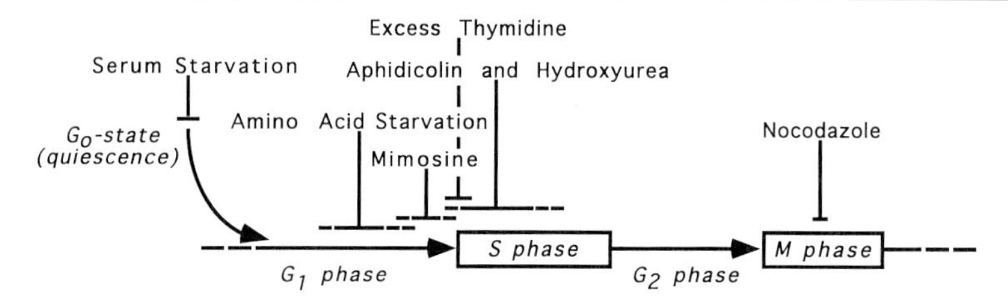

FIG. 1. Schematic illustration showing where in the cell cycle that cells arrest following application of the various synchronization methods described in detail in the text. Note that not all the various treatments outlined here are applicable for cell synchronization of every cell type. Also note that a given treatment will arrest cells, depending on its type, at slightly different points in a certain interval in a specific cell cycle phase (illustrated by horizontal bars).

irreversible cell damage. Third, it is critical to know if cells will recover and synchronously progress through an entire cell cycle after removal of the drug. In this regard, one should consider possible effects of temperature changes on the progression of cells after release from the inhibitor. Specifically, removal of cells from the incubator, addition of cold medium, or even medium kept at room temperature will slow down cell cycle progression.

Once these parameters are known, synchronous cell populations of many different cell types can be obtained using the protocols outlined here as guidelines. We emphasize that to apply any of these synchronization methods successfully, all cells in a population must be growing exponentially and the cell density at the time of the blockage has to allow active growth to be maintained following the release of the cells from the block.

In addition, the cell cycle progression of synchronized cell populations should always be monitored by processing individual cell samples either for flow cytometric (FACS) analysis[11] of DNA content or for [^{3}H]thymidine[12] or 5-bromodeoxyuridine (BrdU)[13] incorporation into DNA as a quantitative measure of new DNA synthesis.

The following discussion details protocols commonly used for synchronizing cells at specific stages of the cell cycle. Figure 1 summarizes where in the cell cycle cells arrest in response to the treatments described here.

Cell Cycle Arrest in G_0 Phase

Serum Starvation. As discussed earlier, certain cell types enter a state of quiescence (G_0) when grown in low serum containing media. The following

[11] Elsewhere in this volume.

[12] S. O. Freytag, *Mol. Cell. Biol.* **8,** 1614 (1988).

[13] F. Dolbeare, W. Kuo, W. Beisker, M. Vanderlaan, and J. W. Gray, *in* "Flow Cytometry" (Z. Darzynkiewicz and H. A. Crissman, eds.), p. 207. Academic Press, San Diego, CA, 1990.

protocol has been used to growth arrest NIH 3T3 (ATCC, Rockville, MD CRL 1658) and MRC-5 (ATCC CCL 171) cells.[14]

To initiate the growth arrest, wash exponentially growing cells at 50–60% density three times with serum-free media. Then add complete media containing 0.5% serum and antibiotics and incubate cells in a CO_2 incubator at 37°. After 48 hr remove the media, add fresh media containing 0.5% serum, and continue the incubation for another 24 hr. At that time about 95% of the cells show a 2n DNA content by FACS analysis. To release cells from growth arrest, feed cells directly with complete media containing 10% serum. To monitor cell synchrony, take individual cell samples at appropriate time points following the release of cells from serum starvation and process them for FACS analysis. The S phase is reached between 12 and 16 hr after the addition of serum.

Cell Cycle Arrest in Late G_1 Phase

Mimosine Synchronization. Mimosine, a plant amino acid, reversibly inhibits cell cycle progression in late G_1[5] by a mechanism which has not yet been precisely determined. Mimosine blocks the cell cycle of human lymphoblastoid cells at a point prior to a block mediated by aphidicolin, an inhibitor of S phase entry.[7] Additionally, mimosine-arrested ML-1 cells have been used to enrich for E2F/cyclin E-kinase complexes, which occur in late G_1.[15] We have used mimosine to synchronize human U-2 OS (osteogenic sarcoma) cells and have reproducibly obtained cell populations that have been 70–80% enriched in the late G_1 phase.[14]

In the protocol outlined here, the length of mimosine treatment for human U-2 OS cells (ATCC HTB 96) is based on a doubling time of about 18–20 hr, and the concentration of mimosine required to arrest these cells in G_1 is 0.2 mM. U-2 OS cells are maintained in Dulbecco's modified Eagle's medium (DMEM) supplemented with 10% fetal calf serum and antibiotics in a CO_2 incubator at 37°. Split cells 16–20 hr prior to the addition of mimosine in such a way that they reach approximately 60% confluency the next day. To initiate the mimosine block, remove the growth medium and replace it with fresh growth medium containing 0.2 m*M* mimosine. Mimosine (Sigma, St. Louis, MO) is stored at −20° as a sterile-filtered 10 m*M* stock in phosphate-buffered saline (PBS). Incubate cells in the presence of mimosine for 18 hr. This time is sufficient for cells that just passed the G_1/S boundary at the time of the mimosine addition to accumulate in G_1 of the following cell cycle. Then, remove the mimosine-containing medium and process cells either for the appropriate biochemical assays or release

[14] W. Krek and D. M. Livingston, unpublished (1993).

[15] E. Lees, B. Faha, V. Dulic, S. I. Reed, and E. Harlow, *Genes Dev.* **6,** 1874 (1992).

them from the block by rinsing the cell monolayer three times with prewarmed DMEM (37°) followed by the addition of fresh prewarmed growth medium. Take individual cell samples at appropriate times following release of cells from the block and process the samples for FACS analysis to monitor their cell cycle distribution.

An example of a typical FACS profile, describing the DNA distributions of U-2 OS cells as a function of time following synchronization by mimosine, is presented in Fig. 2. About 80% of the mimosine-treated cells are blocked in late G_1. The G_2/M compartment is almost completely depleted of cells, whereas a certain fraction of cells remains in S phase. Following release from the mimosine block, there is little dispersion of the G_1 population during its traversal through the S phase, which is about 6 hr postrelease. At about 12 hr postrelease, the initial G_1 population has entered the G_2/M phase as a relatively homogenous population.

Amino Acid Starvation. The growth of certain cell types in the absence of an essential amino acid such as methionine leads to a highly synchronized cell population arrested in early G_1.[4] For a given cell type it is important to evaluate first the sensitivity to methionine starvation. To test this, split cells into methionine-free media containing 10% dialyzed fetal calf serum and antibiotics and culture them in a CO_2 incubator at 37°. After approximately two doubling times, evaluate cells for viability by trypan blue exclusion and by FACS analysis to determine whether cells are enriched for the

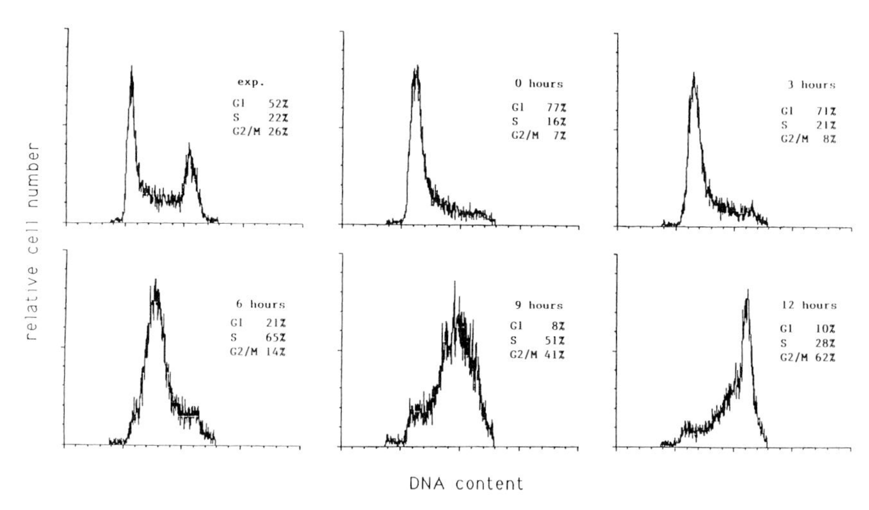

FIG. 2. A series of representative histograms describing the DNA content of exponentially growing human U-2 OS cells (exp.), synchronized by mimosine for 18 hr (0 hr), or at time points (3–12 hr) following release from the block for indicated times (3–12 hr). The relative percentages of G_1, S, and G_2/M phase cells are presented on top of each histogram.

G_1 phase. To release cells from methionine starvation, feed cells either directly with complete media containing 10% fetal calf serum or trypsinize cells first and then replate them into complete media containing 10% fetal calf serum. The latter method selects for cells capable of replating and enriches for healthy cells and may improve synchronization. The S phase is usually reached 3–6 hr after release.[4] Some cell types are not viable in methionine-free media. In such a case, isoleucine starvation has been successfully used.[16,17]

Cell Cycle Arrest in S Phase

Hydroxyurea or Aphidicolin Synchronization. Hydroxyurea and aphidicolin interfere with cell cycle progression by inhibiting DNA synthesis. The former acts on ribonucleotide reductase, which in turn leads to cell cycle arrest by limiting the amount of ribonucleotides available for DNA synthesis,[6] whereas the latter prevents the binding of deoxynucleotide triphosphates to DNA polymerase α.[7] Cell cycle arrest with either agent prevents the entry of cells into the S phase or blocks progression of cells that have already entered the S phase at the time of drug addition. Both inhibitors are in widespread use since they are applicable for cell cycle blockage on a variety of cell types. We have used these drugs successfully to synchronize cell lines of different origins, including human U-2 OS cells,[14] monkey CV-1P cells,[4] or chicken DU249 cells[18] and the quality of synchronization obtained was similar to that using mimosine.

Cell cycle synchronization by either hydroxyurea or aphidicolin is achieved by following the procedures described for the mimosine block/release experiment. Add hydroxyurea (Boehringer Mannheim) to a final concentration of 0.5 to 1 mM from a sterile-filtered 100 mM stock in H_2O stored at $-20°$. Use aphidocolin (Sigma, St. Louis, MO) at a concentration of 5 μg/ml. A 100× aphidicolin stock is prepared in dimethyl sulfoxide (DMSO) and stored at $-20°$.

Cells can also be synchronized at the G_1/S boundary by a procedure termed the double-thymidine block. A detailed description of synchronization of HeLa cells by this method has been reported by Stein and Stein.[19] For these experiments, an asynchronously dividing cell population is first synchronized at the G_1/S boundary in the presence of excess thymidine (2 mM final). Cells are then incubated for a few hours in the absence of

[16] L. R. Gurley, R. A. Walters, and R. A. Tobey, *J. Biol. Chem.* **250,** 3936 (1975).

[17] T. A. Woodford and A. B. Pardee, *J. Biol. Chem.* **261,** 4669 (1986).

[18] W. Krek and E. A. Nigg, *EMBO J.* **10,** 305 (1991).

[19] G. S. Stein and J. L. Stein, *in* "Cell Growth and Division" (R. Baserga, ed.), p. 133. IRL Press at Oxford University Press, Oxford and New York, 1989.

thymidine and are then treated again with 2 m*M* thymidine for a time interval corresponding to an entire cell cycle. Hydroxyurea or aphidicolin can replace the second thymidine block.

Cell Cycle Arrest at Metaphase

Nocodazole Synchronization. When adherent growing cells enter the M phase, they become rounded and detach from the culture plate. Thus, M phase cells can be purified by gentle shaking or by pipetting them off the dish. Although this mitotic selection method[20] yields highly synchronized cell populations, it is limited by the fact that mitotic cells represent only a very small fraction of an exponentially growing cell culture. Therefore, the application of nocodazole, a drug that specifically blocks cells at metaphase by destabilizing the microtubule structure,[9] leads to the recovery of a cell population that is highly enriched in the M phase of the cell cycle. Depending on the cell line, between 40 and 80% of the cells can arrest at metaphase following nocodazole treatment. Most cell lines recover completely from a nocodazole block and enter the subsequent G_1 and S phase in a highly synchronous fashion. In some cases, mitotic arrest and hence the accumulation of metaphase cells can be significantly increased when exponentially growing cells are first synchronized at the G_1/S boundary with excess thymidine or hydroxyurea and then released from this block in the presence of nocodazole. The nocodazole/block release synchronization of cells has been successfully used to study the timing of phosphorylation/dephosphorylation of the retinoblastoma gene product[4,21] and the cell cycle-dependent phosphorylation of the p34cdc2 kinase.[18,22]

As an example, we describe here a detailed nocodazole block/release protocol using the chicken hepatoma cell line DU249. Chicken DU249 cells are maintained in DMEM supplemented with 5% fetal calf serum and antibiotics in a CO_2 incubator at 37°. Seed DU249 cells 16–20 hr prior to the addition of the blocking agent. At 70% confluency, add thymidine (Sigma, St. Louis, MO) to a final concentration of 2 m*M* from a sterile-filtered 100 × stock prepared in PBS. Incubate cells in the presence of thymidine for 12 hr. Remove the thymidine containing medium, rinse the cell monolayers carefully three times with prewarmed DMEM, and add prewarmed growth medium containing nocodazole at 500 ng/ml. Nocodazole (Janssen Life Science Products, Piscataway, NJ) is stored at −20° as a 1000 × stock solution in DMSO. In this particular cell line, 12 hr after

[20] T. Terasima and L. J. Tolmach, *Nature (London)* **190,** 1210 (1961).

[21] J. W. Ludlow, C. L. Glendening, D. M. Livingston, and J. A. DeCaprio, *Mol. Cell. Biol.* **13,** 367 (1993).

[22] A. O. Morla, G. Draetta, D. Beach, and Y. J. Wang, *Cell (Cambridge, Mass.)* **58,** 193 (1989).

the addition of nocodazole, 80% of the cells are arrested in metaphase, as indicated by their round shape and condensed chromosomes.[23] Since metaphase cells are very loosely attached to the culture dish, collect them by mechanical shakeoff or by gentle pipetting. Centrifuge cells at 1000*g* for 3 min at room temperature (or preferably at 37°), resuspend them carefully in prewarmed DMEM (37°), and centrifuge again. Repeat this washing step once more to ensure efficient removal of the drug. Subsequently, resuspend cells in prewarmed growth medium and incubate them in a CO_2 incubator at 37°. At timed intervals, remove cells from the incubator and analyze their DNA content by FACS to monitor cell cycle progression. DU249 cells adhere to the dish and enter G_1 by 4 hr after replating.

Centrifugal Elutriation

General Considerations

An asynchronous population of cells can be separated into fractions enriched for G_1, S, and G_2/M phase by centrifugal elutriation.[10] For mammalian cells in culture, the early G_1 cells are the smallest whereas cells in late G_2/M phase cells are the largest. Centrifugal elutriation can separate a large number of cells into populations of uniform size. Soon after mitosis, a mammalian cell will divide into two identical cells that are each half the size of the original mitotic cell. These early G_1 cells will continue to grow as they advance through each phase of the cell cycle so that S phase cells are of intermediate size.

All mammalian cells in culture can be elutriated. Cells grown in suspension are easier to elutriate since they tend not to adhere to each other and they have a rounded, uniform shape. However, many types of adherent cells can also be elutriated, including NIH 3T3, HeLa, U-2 OS, Saos-2, and CV-1P cells. After an adherent cell is trypsinized and resuspended in media, it will assume a rounded shape and its cell size will correlate with its position in the cell cycle.

Centrifugal elutriation can be used to synchronize an exponentially growing population of cells in two different ways. A single fraction of the small, early G_1 cells can be collected and then reinoculated into culture.[24] Aliquots of cells can be removed at various times after reinoculation and progression through the cell cycle can be monitored. Alternatively, cells can be elutriated into fractions enriched for early G_1, S, or G_2/M phase.

[23] J. Nakagawa, G. T. Kitten, and E. A. Nigg, *J. Cell Sci.* **94,** 449 (1989).

[24] G. Draetta and D. Beach, *Cell (Cambridge, Mass.)* **54,** 17 (1988).

Preparing Cells for Elutriation

Between 5×10^8 and 2×10^9 exponentially growing cells are required for elutriation. If fewer than 2×10^8 cells are introduced into the elutriation chamber, then an equilibrium will not be established and the cells will simply wash out the chamber. Conversely, attempting to load too many cells will crowd the elutriation chamber and result in poor separation.

Cells should be elutriated in culture media rather than in buffered salt solutions. The nutrients may allow the cells to remain viable for the 1 to 2 hr required for elutriation. To prevent cell–cell adhesion, calcium-free media, e.g., S-MEM, Joklik's, or RPMI 1640, should be used. Serum (1%) is added to prevent the cells from adhering to the tubing and elutriation chamber. Antibiotics, such as penicillin/streptomycin or gentamycin, are added to prevent gross bacterial contamination.

Cells grown in suspension are collected by centrifugation or filtration and are resuspended in 7 ml of elutriation media. Adherent cells are first treated with trypsin, washed free of the trypsin, and then resuspended in elutriation media. The resuspended cells should be collected into a 10-ml syringe without a needle attached to prevent shearing.

Preparing Centrifuge for Elutriation

The basic setup includes a Beckman J-6B centrifuge, with a strobe light assembly, a JE-5.0 rotor with a standard 4-ml elutriation chamber, a pressure gauge, and a sensitive peristaltic pump such as the Cole-Parmer MasterFlex Digi-Staltic. This pump can be conveniently calibrated which is essential for reproducible elutriation. Ideally, a centrifuge is dedicated to elutriation to prevent excessive wear and tear on the rotor assembly resulting from assembly and disassembly of the chamber each time an elutriation is performed.

The apparatus is set up as specified in the instruction manual from Beckman (The JE-5.0 Elutriation System). However, it is not necessary to have an air purge system for introducing the cells into the system. A single, three-way stopcock is positioned between the elutriation media reservoir and the peristaltic pump. The syringe with the resuspended cells can be attached directly to the remaining port on the stopcock.

After assembling the apparatus, the system is primed with water supplied by an inflow line placed in a beaker of water. The peristaltic pump is turned on and is set to a speed of approximately 25 ml/min. After the water exists from the chamber to the outflow line, the centrifuge is turned on and the rotor is spun at 250 rpm/min. This low speed allows air bubbles to escape from the chamber. Once the bubbles have been removed from the chamber, the rotor speed is increased to 1200 rpm, while the pump speed is maintained at 25 ml/min.

Once the rotor has reached full speed, adjust the strobe light so that the elutriation chamber can be viewed through the window on the centrifuge lid. If any air remains in the system or if there are any leaks, then the water will stop returning from the chamber when the rotor speed reaches approximately 900 rpm. If this occurs, turn off the rotor, and check for leaks, specifically from the seal between the two halves of the elutriation chamber. This chamber may need to be tightened or the seal replaced. In the absence of any leak during the spin, the amount of liquid pumped into the system should equal the collected volume. After priming the system, the peristaltic pump should be calibrated according to the manufacturers directions.

A pressure gauge is very useful for determining if air is trapped in the system. During normal operation, the pressure gauge will read less than 4 psi. However, if air becomes trapped, then the pressure will rise very quickly to greater than 30 psi. Very high pressure may disconnect the tubing, resulting in loss of media and cells into the centrifuge chamber. If the pressure gauge rises suddenly, turn off the rotor and the pump and, when the rotor stops, check for leaks and air bubbles.

The system can be sterilized if the cells in culture are to be reinoculated after elutriation. First, pump several hundred milliliters of 70% ethanol through the system and then purge the alcohol with several hundred milliliters of sterile PBS. The system is then flushed with at least 30 ml of elutriation media. It is usually not necessary to collect the cells in a laminar flow hood to prevent contamination.

Performing Elutriation

Within the elutriation chamber, an equilibrium is established between the centrifugal force generated by the spinning rotor and the counterflow of the media established by the pump. The exact rotor speed and pump speed required to establish this equilibrium need to be determined for each cell line that is used. However, once the appropriate parameters for a particular cell line have been established they will remain remarkably reproducible from run to run. For larger cells, slower rotor speeds of 1200 rpm and pump speeds of 12 to 30 ml/min are recommended. Smaller cells may require rotor speeds from 1800 to 2200 rpm.

For the initial run, set the pump speed to 12 ml/min and the rotor speed to 1800 rpm. Turn the pump and centrifuge on, allow the rotor to reach full speed, and adjust the strobe light to obtain a clear view of the spinning elutriation chamber. Attach the syringe containing the cells to the three-way stopcock and open the valve to the syringe and the rotor while closing it to the media reservoir. Allow the pump to take up all but 0.5 ml of the cells from the syringe and then close the stopcock to the syringe and open it to the media reservoir. The cells remaining in the syringe can serve as the unelutriated control.

The cells should collect within the chamber and be visible when viewed through the window. At equilibrium, the larger cells (G_2/M) are maintained near the bottom of the chamber, whereas the smaller (G_1) cells are maintained higher in the chamber, closer to the line of elutriation. The cells should be evenly distributed from the top to the bottom of the elutriation chamber. Adjust the rotor speed to achieve an equal distribution of cells within the chamber. If the cells are all forced into the lower half of the chamber, decrease the rotor speed by 10%. On the other hand, if the cells are not retained within the chamber, the rotor speed should be increased. If any cells are expelled from the chamber before equilibrium is established, simply stop the rotor with the pump on. Collect the cells by centrifugation, resuspend them in 7 ml of elutriation media, and start again.

Successful elutriation requires complete collection of all the small, early G_1 cells before collection of the larger cells. This necessitates collecting large fractions of 100 ml each, while very slowly increasing the pump speed with each successive fraction. Each fraction is conveniently collected into two 50-ml polypropylene tubes. After collecting the first fraction, increase the pump speed by 1 ml/min and then collect the second 100-ml fraction. Usually, the pump speed is increased by 1 ml/min for each of the first five fractions. The pump speed should be increased to 2 ml/min then to 3 ml/min for collection of later fractions. This protocol ensures that small G_1 cells are completely removed from the chamber and will not contaminate the later fractions and that late fractions are highly enriched for large G_2/M cells. After 10–12 fractions are collected, any cells remaining in the chamber can be flushed out by stopping the rotor with the pump running.

The fractionated cells can be collected by centrifugation and washed with media or PBS. Ten percent of the cells from each fraction are fixed for FACS analysis of the DNA content and the remaining cells can be assayed as desired. The cells can be metabolically labeled with [^{35}S]methionine or ortho [^{32}P]phosphate, or directly lysed in the appropriate buffers.

Acknowledgments

We are grateful to Dr. D. M. Livingston (Dana-Farber Cancer Institute, Boston) and Dr. E. A. Nigg (Swiss Institute for Experimental Cancer Research, Epalinges), in which laboratories these experiments have been performed, for generous and continuous support. We also thank Drs. J. Griffin, E. Neumann, and S. Shirodkar for critical reading of this manuscript.

[8] Analysis of Cell Cycle Checkpoint Status in Mammalian Cells

By THEA TLSTY, AMY BRIOT, and BENJAMIN POULOSE

Ever since the 1970s investigators have gained considerable insight into the biochemical processes that drive the cell through the phases of proliferation.[1-5] These processes, governed by the cyclins (cyclin A, B1, C, D1–3, E, etc.) and their partner kinases (CDC2, CDK2, CDK4, etc.), can be modulated by external influences. Analyses in yeast and mammalian cells have demonstrated the presence of a molecular circuitry, which monitors the state of the genome at different phases of the cell cycle. Specific events, such as the completion of DNA synthesis or the alignment of chromosomes on the metaphase plate, must be completed before the cell proceeds to the next phase of the cycle.[6] These feedback controls, which allow the cell cycle to proceed or arrest, have been termed checkpoint controls.[7] Multiple checkpoints exist in each phase of the cell cycle. Initially, mutants in these processes were identified by the morphological variants of yeast that resulted from alterations in cell cycle progression.[8] More recently, however, flow cytometry has provided a sensitive method to detect changes in cell cycle progression even in the absence of morphological changes. The accumulation of cells in one phase of the cell cycle compared to another is evidence of the integrity of one checkpoint and the relaxation of another.

Flow cytometry allows the analysis of various parameters of single cells in suspension. This is accomplished as the cells are aligned, single file, in a fluid sheath and are passed in front of an exciting beam of light.[9] Cells that are passing through this beam can cause light to be scattered in all directions and/or absorb the exciting light and emit fluorescence. For fluorescence emission, cells are usually stained with a fluorescent dye.[10] Optical

[1] T. Hunter, *Cell (Cambridge, Mass.)* **75,** 839 (1993).
[2] C. J. Sherr, *Cell (Cambridge, Mass.)* **73,** 1059 (1993).
[3] A. W. Murray and M. W. Kirschner, *Sci. Am.* March, p. 56 (1991).
[4] S. van den Heuvel and E. Harlow, *Science* **262,** 2050 (1993).
[5] J. Pines, *Trends Biochem. Sci.* **18,** 195 (1993).
[6] A. Murray and T. Hunt, "The Cell Cycle: An Introduction." Freeman, New York, 1993.
[7] L. H. Hartwell and T. A. Weinert, *Science* **246,** 629 (1989).
[8] L. H. Hartwell, J. Culotti, and B. Reid, *Proc. Natl. Acad. Sci. U.S.A.* **66,** 352 (1970).
[9] H. B. Steen, *in* "Flow Cytometry and Sorting" (M. R. Melamed, T. Lindmo, and M. L. Mendelsohn, eds.), p. 11. Wiley-Liss, New York, 1990.
[10] A. S. Waggoner, *in* "Flow Cytometry and Sorting" (M. R. Melamed, T. Lindmo, and M. L. Mendelsohn, eds.), p. 209. Wiley-Liss, New York, 1990.

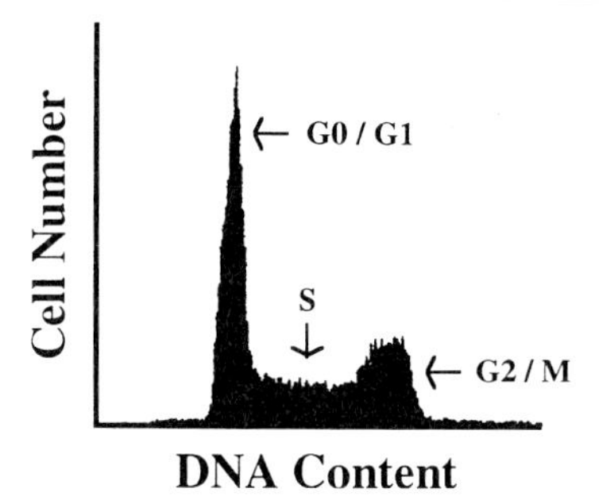

FIG. 1. Representative histogram of proliferating normal fibroblasts indicating phases of the cell cycle based on cellular DNA content. Cells were fixed and stained with propidium iodide and analyzed on the Becton-Dickinson FACScan flow cytometer.

sensors measure the amount of scattered light or fluorescence and assign these values to each individual cell. The flow rate of the sheath fluid, as well as several other adjustable variables, allow analysis of 5–5000 cells per second, making this a rapid and powerful method of analysis.[11] The simultaneous measurement of multiple parameters for an individual cell provides a quantitative value for a range of cellular attributes. Some attributes, such as relative cell size and cell viability, can be measured directly, but others, such as DNA content or enzyme activity, rely on the use of cytochemical or immunologic reagents to identify macromolecules.[12]

Presently, the most common approach to analyzing cell cycle progression is by analysis of the DNA content of the cell.[11–13] Several dyes bind in a stoichiometric manner to DNA and yield fluorescence proportional to the DNA content. Intercalators (propidium iodide, ethidium bromide) and molecules which position themselves in the major and minor grooves of DNA (mithramycin, chromomycin) are used extensively, but only on fixed cells. The Hoechst dyes an be used to stain the DNA of live cells and permit the sorting of viable cells on the basis of DNA content. An ultraviolet laser is needed for analysis of these cells. Variables such as differential chromatin condensation and binding preferences of different dyes must be carefully controlled if reproducible, quantitative data are to be obtained.[11]

Past measurements of DNA content have relied on the method of analysis shown in Fig. 1.[11] Propidium iodide (PI) stains DNA by intercalating between nucleotide bases and shows no preferences for AT-rich or GC-

[11] J. W. Gray, F. Dolbeare, and M. G. Pallavicini, *in* "Flow Cyometry and Sorting" (M. R. Melamed, T. Lindmo, and M. L. Mendelsohn, eds.), p. 445. Wiley-Liss, New York, 1990.

[12] K. A. Muirhead and P. K. Horan, *Adv. Cell Cult.* **3,** 57 (1984).

[13] Z. Darzynkiewicz and F. Traganos, *in* "Flow Cytometry and Sorting" (M. R. Melamed, T. Lindmo, and M. L. Mendelsohn, eds.), p. 469. Wiley-Liss, New York, 1990.

rich regions.[14] Cells, stained by PI, are accumulated and displayed as a histogram. Noncycling cells (G_0) and post mitotic cells (G_1) both have a diploid DNA content and are indistinguishable from each other. Cells with a DNA content between diploid and tetraploid are determined to be in their S phase of the cell cycle. Cells that have doubled their DNA content to the tetraploid amount (G_2) and those that have not yet divided (M) are found in the same peak. Instrument variability and staining variability contribute to the broadening of the histogram distributions. The information in the single parameter histogram is displayed using 256 channels in which cells are assigned to the appropriate channel based on the fluorescence intensity. A cumulative number of cells in each channel is displayed as a function of channel number (fluorescence intensity).

Although the relative number of cells in G_1, S, and G_2/M can be obtained from this type of histogram, the calculation of the individual peaks is somewhat ambiguous because of the variability of the peak width. Different mathematical models have been designed to resolve the peaks and calculate the individual populations.[15–17] As a significant improvement, a two parameter histogram provides much more information by allowing correlations among several variables.[13] In the analysis of cell cycle progression, the use of an antibody to bromodeoxyuridine (BrdU)-substituted DNA allows the distinction between G_1 and early S phase cells or late S phase and G_2 cells.[11] As demonstrated in Fig. 2, S phase cells incorporate BUdR, a thymidine analog, and on subsequent incubation with a BUdR antibody and flow analysis, the population of cells that contains newly synthesized DNA (i.e., BUdR-containing DNA) is displaced on the x axis providing unambiguous identification.

Analysis of Checkpoint Function

The arrest in different phases of the cell cycle that occurs on activation of a checkpoint response is the result of a complex series of biochemical reactions. These reactions, and the molecules that are involved in this signal transduction pathway, are being elucidated at the present time.[18,19] To study

[14] S. A. Latt and R. G. Langlois, *in* "Flow Cytometry and Sorting" (M. R. Melamed, T. Lindmo, and M. L. Mendelsohn, eds.), p. 249. Wiley-Liss, New York, 1990.

[15] J. W. Gray, P. N. Dean, and M. L. Mendelsohn, *in* "Flow Cytometry and Sorting" (M. R. Melamed, P. F. Mullaney, and M. L. Mendelsohn, eds.), p. 383. Wiley-Liss, New York, 1979.

[16] L. E. Sheck, K. A. Muirhead, and P. K. Horan, *Cytometry* **1,** 109 (1980).

[17] M. H. Fox, *Cytomery* **1,** 71 (1980).

[18] L. Livingstone, A. White, J. Sprouse, E. Livanos, and T. D. Tlsty, *Cell (Cambridge, Mass.)* **70,** 923 (1992).

[19] M. B. Kastan, O. Onyekwere, D. Sidransky, B. Vogelstein, and R. W. Craig, *Cancer Res.* **51,** 6304 (1991).

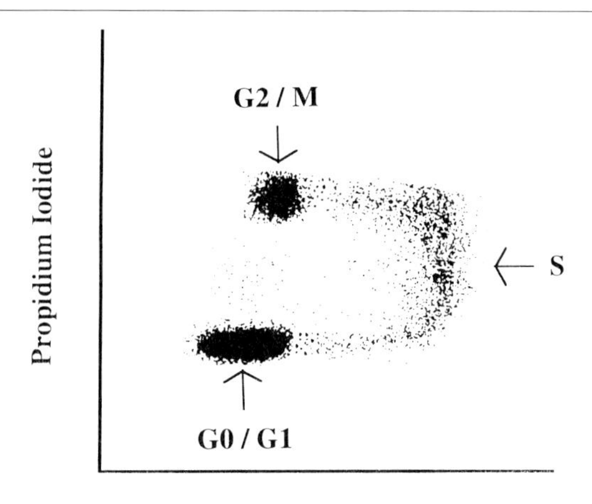

FIG. 2. A two-dimensional histogram showing total DNA content versus active DNA synthesis in normal diploid fibroblasts. The *x* axis depicts the degree of BUdR fluorescence; cells actively synthesizing DNA incorporate BUdR and are displaced to the right. The *y* axis portrays propidium iodide fluoresence; cells with 2N or 4N DNA content are represented as G_0/G_1 or G_2/M, respectively. Those cells showing intermediate levels of propidium iodide fluorescence and increased BUdR fluorescence indicate S phase.

the checkpoint response in cells, one needs to apply a stimulus and then monitor proliferation kinetics over a period of time. Figure 3 illustrates the activation of the G_1 and G_2 checkpoints in normal human fibroblasts that are exposed to a metabolic inhibitor. A similar response, albeit with more rapid kinetics, occurs when the cells are exposed to gamma radiation.[19] Stimuli for the activation of different checkpoints within the cell can be of endogenous or exogenous origin. The stimuli also determine which checkpoints are activated.

Figure 3F is an area graph which displays the cumulative data obtained after normal human fibroblasts are exposed to the metabolic inhibitor, *N*-(phosphonoacetyl)-L-aspartate (PALA), for an extended period of time. Hours after exposure to PALA, an exponentially growing population of normal, diploid cells will accumulate in the S phase of the cell cycle. This increased S fraction gradually decreases until, at day 2, there is a complete absence of cells in the S phase, indicating an arrest in cell cycle progression. The population is distributed between the G_1 and G_2 phases of the cell cycle for the duration of the selection. This temporal distribution of cells within each phase of the cell cycle is presented as a scatter plot (data directly from the flow cytometer) in Figs. 3A–3E and as a compilation over time in Fig. 3F. The white area on the graph in Fig. 3F represents the

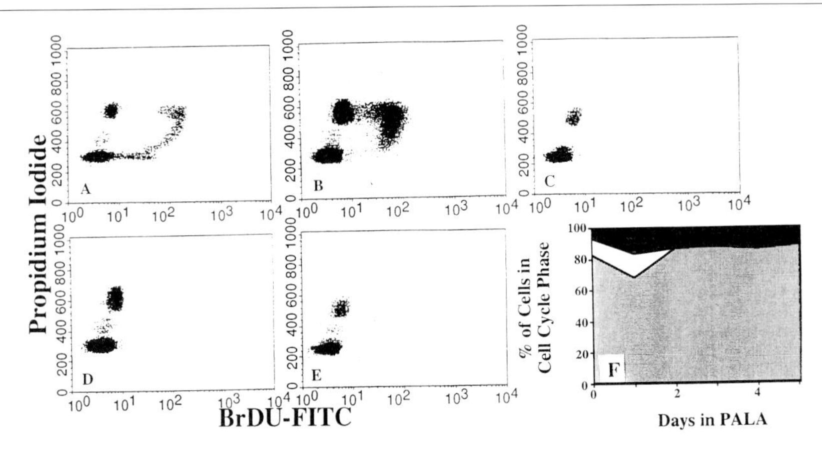

FIG. 3. Flow cytometric analysis of NHF cell population in 9 × LD_{50} PALA. Cells were labeled with BrdU, fixed, and stained with anti-BrdU-FITC and propidium iodide, then analyzed using a Becton-Dickinson FACScan instrument. The x axis represents cells undergoing DNA synthesis. Cells at the right have more BrdU incorporated and are therefore in the S phase of the cell cycle. The y axis represents DNA content as measured by propidium iodide staining. The upper population has a G_2 phase DNA content, and the lower population has a G_1 phase DNA content. (A) NHF with no PALA exposure and (B–E) NHF days 1–5 in PALA. Cell cycle data were compiled and presented in the form of an area graph, providing a characteristic cell cycle pattern that could be conveniently compared among the cells (F). Each graph indicates the percentage of cells in each phase of the cell cycle (ordinate) as a function of time (axis; 0, 1, 2, 3, 4, and 5 days in PALA). The G_2, S, and G_1 phases of the cell cycle are represented by the top, middle, and bottom areas of the graph, respectively. There was no significant portion of the parental NHF cells in the S phase after 2 days in PALA.

percentage of cells in the S phase fraction, the lower gray area represents those in G_1, and the upper dark area represents those in G_2. This visual representation of the relative fractions of the population in the G_1, S, or G_2/M phase is helpful for comparing the response of normal cells in drug with that of tumorigenic cells in drug. We find that most malignant cell lines and many premalignant cell populations demonstrate alterations in checkpoint control when analyzed by this method (Fig. 4). These cells fail to arrest at the various checkpoints that would normally be activated in primary diploid cells. Similar alterations are also seen in cells expressing DNA-transforming viral oncoproteins.[20]

Procedure

Staining of nucleic acid with fluorescent dyes to measure individual cellular quantities coupled with the incorporation of a nucleic acid analog

[20] A. White, E. Livanos, and T. D. Tlsty, *Genes Dev.* **8,** 666 (1994).

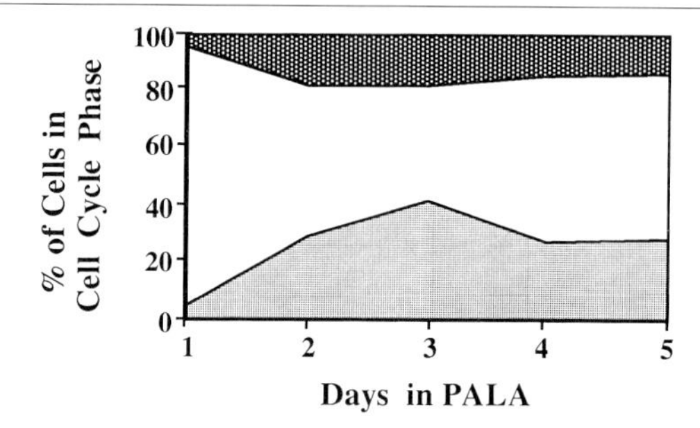

FIG. 4. HT1080 cells, a fibrosarcoma line, were exposed to 9 × LD_{50} PALA. They were labeled for 4 hr with BUdR at daily intervals, fixed, and stained with anti-BUdR-FITC and propidium iodide. Cells were analyzed and the data are displayed as in Fig. 3. The above representation by area graph indicates a prominent S phase demonstrating an abrogation of normal cell cycle checkpoints.

to detect cells that are actively synthesizing DNA provides the basis for the cell cycle checkpoint assay described in this procedure. The amount of DNA per cell, indirectly indicated by the amount of fluorescence, indicates whether the cell is in the G_1, S, or G_2/M phase of the cell cycle (Fig. 1). The combination of this basic analysis with the incorporation of precursor analogs allows the detection of cells that are synthesizing DNA. The spatial displacement of the S phase population facilitates the measurement of small changes in cells entering the S phase of the cell cycle (Fig. 2).

Either intact cells or cell nuclei can be stained. Staining intact cells is faster but the cells have a tendency to adhere to each other, forming clumps that are difficult to analyze.[21] Although removing the cytoplasm requires additional time, these preparations are free of cellular debris and the nuclei have less tendency to aggregate. The following procedure, which details nuclei staining, combines aspects of protocols described previously.[22,23]

Materials and Reagents

1 m*M* Bromodeoxyuridine—wear gloves and mask; this chemical is a known carcinogen. To make 1 m*M* stock solution in phosphate-buffered saline (PBS), add 3.07 mg to 10 ml PBS. Filter and aliquot to 1 ml. Store at −20°. Do not thaw and refreeze vials.

[21] C. A. Hoy, L. C. Seamer, and R. T. Schimke, *Cytometry* **10,** 718 (1989).

[22] M. B. Kastan, A. I. Radin, S. J. Kuerbitz, O. Onyekwere, C. A. Wolkow, C. I. Civin, K. D. Stone, T. Woo, Y. Ravindranath, and R. W. Craig, *Cancer Res.* **51,** 4279 (1991).

[23] R. A. White, N. H. A. Terry, M. L. Meistrich, and D. P. Calkins, *Cytomery* **11,** 314 (1990).

0.08% pepsin in 0.1 *N* HCl—wear gloves and mask to weigh powder. Add 4.13 ml of 12.1 *N* concentrated HCl to ~400 ml distilled H_2O, dissolve 0.4 g pepsin (EM Science) by stirring 30′ – 60′ at room temperature. Bring to 500 ml volume with distilled H_2O and filter. Store at 4°.

2 *N* HCl—wear gloves and use hood to measure concentrated HCl. Slowly add 82.6 ml of 12.1 *N* concentrated HCl to ~400 ml distilled H_2O. Add distilled H_2O and bring to 500 ml, filter, and store at room temperature.

0.1 *M* $NaBO_3$—wear gloves and mask; dissolve 38.3 g in 1 liter of distilled H_2O over low heat. Filter and store at room temperature.

IFA (0.5% Tween 20)—wear gloves when handling sodium azide: 10 m*M* HEPES, pH 7.4 (10 ml 1 *M* stock solution), 150 m*M* NaCl (30 ml 5 *M* stock solution), 4% fetal calf serum (40 ml FCS stock solution), 0.1% sodium azide (1 g stock), and 0.5% Tween 20 (5 ml stock). Omit Tween 20 when IFA only. Bring to 1 liter with distilled H_2O and filter. Store at room temperature.

Pulse Labeling of Cells with Bromodeoxyuridine and Fixation with Ethanol

1. Add 80 μl BUdR (1m*M*) to a p100 plate of cells containing 8 ml of media (final concentration 10 μM BUdR). Incubate at 37° for 1–4 hr (depending on experiment).

2. To trypsinize cells for removal from plate: Aspirate media. Add 5 ml HBSS/p100 plate, rinse thoroughly, and aspirate. Add 1 ml trypsin/p100 and incubate at 37° for 1–5 min. Check for dislodged cells, then add media with 10% serum to terminate the trypsinization reaction. Serum has been filtered to remove serum precipitate which would interfere with subsequent flow cytometry. Remove cells by repeated pipetting on a petri dish, then place suspension in a 15-ml tube and vortex to a single cell suspension. Be sure to obtain cell counts at this point. Spin for 5 min at 4° at 800 rpm in a tabletop Sorvall to pellet cells.

3. Aspirate supernatant. Vortex pellet well before adding solutions to assure that cells are well dispersed. The following volumes are for ~10^6 cells or less (if more, increase proportionately): While vortexing on high, add 1.5 ml *cold* 1 × PBS (Ca^{2+}/Mg^{2+}-free). Continue to vortex but turn down to the half-speed setting and add 3 ml *cold* 95% ethanol (directly from −20° freezer). Continue gentle vortexing for 30 sec, cover the tube with aluminum foil, and store at 4°. Leave overnight at 4° or longer (cells have been held for 1 month and longer with good results) to accumulate samples for preparation of nuclei.

Preparation of Nuclei for Flow Cytometry

1. Vortex fixed cells in a 15-ml tube for 15 sec. Spin for 5 min at 4° at 1200 rpm in a tabletop Sorvall. Aspirate supernatant and vortex pellet.
2. While vortexing gently, add 3 ml 0.08% pepsin (in 0.1 *N* HCl) per 15-ml tube. Incubate cells on a rocker at 37° for 20 min. Visualize by microscopy to assure that nuclei are free of cytoplasm. Spin for 5 min at 4° at 1400 rpm.
3. Aspirate supernatant and vortex pellet for 5–10 sec to get a single nuclear suspension. Add 1.5 ml 2 *N* HCl per 15-ml tube while vortexing. Incubate for 20 min at 37°. Shake twice during incubation. Add 3 ml 0.1 *M* sodium borate per 15-ml tube while vortexing; continue to vortex 10 sec longer. Spin for 5 min at 4° at 1400 rpm.
4. Aspirate supernatant, vortex pellet, and add 2 ml IFA/0.5% Tween 20 to each 15-ml tube while vortexing. Transfer to a 12 × 75-mm snap top tube (Falcon No. 2065) for flow work.
5. Spin for 5 min at 4° at 1400 rpm, aspirate, and vortex pellet.

Staining of Nuclei

1. Add 100 μl anti-BrdU fluorescein isothiocyanate (FITC) (obtained from Becton-Dickenson: used at a 1:5 dilution with IFA) to each tube and incubate on ice and in the dark for 30 min.
2. Wash cells once with IFA/0.5% Tween 20 (2 ml per tube) and spin for 5 min at 4° at 1400 rpm. Aspirate and vortex pellet. Resuspend in appropriate volume IFA (we have found that 500 μl works well with numbers of cells $<10^6$).
3. Add 5 μg/ml RNase A (final concentration) to each tube. Add 50 μg/ml propidium iodide (final concentration) and incubate at 37° for 15 min. Place tubes in the dark and on ice for at least another 15 min (or an extended period of time; see below.) Analyze on a flow cytometer.

Points of Practice

Stability of Fixed Cells

Duplicate samples of fixed cells were held for various amounts of time at 4° in the dark. Individual samples were stained and analyzed by flow cytometry. We found that fixed samples can be stored from days to several months and produce identical profiles upon further preparation. The accumulation of experimental samples after fixation provides a useful stopping point during an extended experimental protocol.

Stability of Stained Cells

We also determined the stability of the samples after they had been completely prepared and stained. Sometimes it is necessary to delay analysis due to operator availability or instrument availability. Cells were fixed, stained at an initial time point, and then analyzed on various days thereafter. The profile remains identical for at least 1 week under proper storage conditions.

Adaptation to Different Cell Types

The above procedure has been used for both fibroblasts and epithelial cells that are grown in a monolayer tissue culture. To analyze suspension cells such as lymphoblasts, use the same procedure but omit the trypsinization step. The incubation time with pepsin should be empirically determined for each cell type.

Quantity of Cells

With healthy cycling cells, a minimum of 2×10^5 cells per time point are needed for the protocol. Some cells are lost due to adherence to the tube during the manipulations. Siliconization of tubes could be helpful if a limited sample size is available. Ultimately, a minimum of 50,000 events is collected for analysis of cell cycle distribution.

Acknowledgments

We thank Mary Hixon and Todd Gray for critical reading of the manuscript. We thank Jack Vincent for assistance with flow cytometry.

[9] Microcell Fusion

By ANN MCNEILL KILLARY and R. E. K. FOURNIER

Introduction

The first documented fusions of somatic cells in culture in the early 1960s resulted in the production of viable hybrid cells and produced a technology that would revolutionize genetic analysis *in vitro*.[1] Parasexual

[1] G. Barski, S. Sorieul, and F. Cornefert, *J. Natl. Cancer Inst.* **26,** 1269 (1961).

approaches have been instrumental in the physical mapping of the human genome and in the identification and mapping of genes involved in growth regulation, differentiation, and neoplasia.[2,3]

Somatic cell hybrids formed by fusing cells of different species (interspecific hybrids) undergo unilateral chromosome segregation.[4] This results in the generation of independent clonal lines that contain different partial genomes of one of the parental cells. Gene mapping using interspecific hybrids relies on statistical correlations between retention of specific chromosomes of the segregating parental cell and expression of a particular phenotype or detection of specific DNA sequences in clonal isolates. The utility of interspecific hybrids for physical mapping has been well documented; however, mapping panels such as these require a collection of hybrid clones with various chromosome contents. The karyotypic complexity and instability of interspecific somatic cell hybrids has limited their use for the study of complex phenotypes or as starting materials for regional mapping and gene isolation.

Microcell fusion is a method of physical mapping which involves the transfer of single or small clusters of intact chromosomes from one cell to another.[5,6] It is perhaps the most attractive technique for interspecific and intraspecific gene mapping because it affords a way to introduce a single chromosome, rather than the entire genome, from a donor cell into a recipient cell background. The transferred chromosome can be stably retained in the recipient cell background under dominant selective pressure. This type of monochromosomal hybrid has several advantages over whole cell hybrid clones. Monochromosomal microcell hybrids necessarily contain only the introduced chromosome, whereas in segregating whole cell hybrids, other chromsomes and potentially translocations and chromosomal fragments could be present and are difficult to detect. Second, by applying a dominant selectable marker to the donor chromosomes, they can be maintained stably and at high frequency in the recipient cell background, resulting in a nearly homogeneous hybrid population for genetic analysis.

Microcell fusion is accomplished in five essential steps (diagramed in Fig. 1) which include: (1) micronucleation of donor cells; (2) enucleation of the micronucleate cell population; (3) isolation of microcells; (4) fusion of microcells to recipient cells; and (5) selection of viable microcell hybrid clones.

[2] R. Sager, *Adv. Cancer Res.* **44,** 43 (1985).
[3] H. Gourdeau and R. E. K. Fournier, *Annu. Rev. Cell Biol.* **6,** 69 (1990).
[4] M. C. Weiss and H. Green, *Proc. Natl. Acad. Sci. U.S.A.* **58,** 1104 (1967).
[5] R. E. K. Fournier and F. H. Ruddle, *Proc. Natl. Acad. Sci. U.S.A.* **74,** 319 (1977).
[6] C. A. McNeill and R. L. Brown, *Proc. Natl. Acad. Sci. U.S.A.* **77,** 5394 (1980).

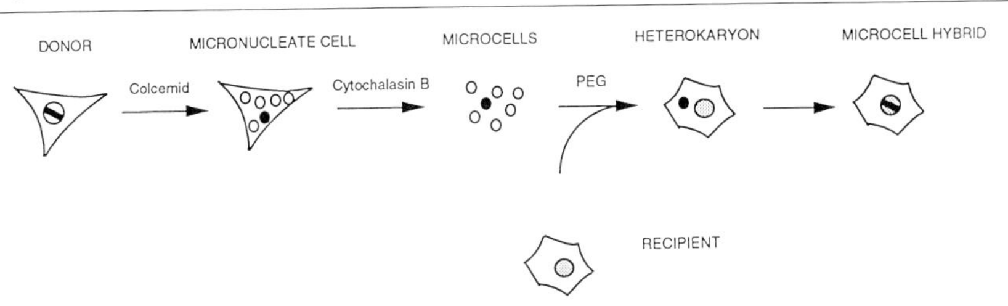

FIG. 1. Diagram of the technique of microcell fusion.

In outline, donor cells are exposed for a prolonged period to the mitotic inhibitor Colcemid. This aberrant mitotic arrest induces the formation of micronuclei containing single or small numbers of chromosomes; micronuclei are extruded from the cell by centrifugation and cytochalasin B treatment resulting in an assortment of cell particles including microcells, enucleated whole nuclei (karyoplasts), whole cells, and cytoplasmic vesicles; microcells are next purified and fused to recipient cells. Viable microcell hybrid clones can be isolated using a variety of selection schemes.

This chapter describes a detailed protocol for microcell fusion. We also discuss the utility of microcell hybrids for mapping the human and mouse genomes and for studying genetic mechanisms that underlie cell growth, differentiation, and neoplasia.

Micronucleation of Donor Cells

The microcell fusion technique involves the use of micronuclei formed by prolonged treatment of growing cells with the mitotic inhibitor Colcemid.[7–9] In the presence of this microtubule polymerization inhibitor, cells cannot form a functional mitotic spindle. During Colcemid-induced mitotic arrest, chromosomes appear scattered throughout the cell. With prolonged Colcemid arrest, many cells within the population attempt to divide; the nuclear membrane, rather than surrounding the entire chromosome complement, reforms around single or small clusters of chromosomes, forming micronuclei.[8,10] The following is a general protocol for the micronucleation of mouse L cells (A9). Conditions for other cell lines will also be discussed.

[7] A. Levan, *Hereditas* **40,** 1 (1954).
[8] E. Stubblefield, "Cytogenetics of Cells in Culture." Academic Press, New York, 1964.
[9] S. G. Phillips and D. M. Phillips, *J. Cell Biol.* **40,** 248 (1969).
[10] T. Ege, N. R. Ringertz, H. Hamberg, and E. Sidebottom, *in* "Preparation of Microcells" (D. M. Prescott, ed.), p. 339. Academic Press, New York, 1977.

Materials and Methods of Procedure

T75 and T150 tissue culture flasks [Corning (T75s, 25110-75; T150s, 25120-150)]
150-mm tissue culture plates (Falcon; 3025)
Hot wire cutter
Dulbecco's modified Eagle's medium (DMEM) (GIBCO-BRL Gaithersburg, MD; Catalog No. 430-2100EF)
Fetal bovine serum (Hyclone; Hyclone Defined Fetal Bovine Serum)
Colcemid (Demecolcine, Sigma; Catalog No. D-7385)
95% ethanol
Acridine orange (Sigma; A-6014)

Preparation of Solutions and Materials for Micronucleation

For micronucleation and subsequent enucleation, donors cells are plated onto thin plastic sheets or "bullets"[6,11,12] such that following Colcemid treatment, micronucleated cell populations on bullets can be placed directly inside 50-ml conical centrifuge tubes for enucleation.

1. Bullets are cut from the bottom and sides of Corning T75 flasks using a hot wire cutter (a nickel–chromium wire attached to a variable current transformer). They should be cut to resemble microscope slides of approximately 25 × 90 mm with one rounded end so as to fit upright in 50-ml centrifuge tubes. The number of bullets used per experiment depends on the percentage of micronucleation in the donor cell population. For A9 cells, 16 bullets are generally used per experiment. The bullets should be washed in glassware detergent, rinsed thoroughly in water, and placed in 95% ethanol 17–24 hr prior to cell plating. Before plating the cells, bullets are removed from ethanol using sterile forceps, placed in T150-mm tissue culture dishes (4 bullets per dish), and left in a tissue culture hood to dry (30–45 min). Donor cells are plated directly onto the bullets in complete medium containing 50 μg/ml gentamycin. Plates containing bullets are incubated for 48 hr in the presence of Colcemid to induce micronucleation.
2. Following enucleation, bullets are placed in a 3% solution of bleach overnight, washed with detergent, and rinsed in water. The bullets can be stored and reused until cracking occurs from excessive usage.
3. Solutions for micronucleation include (1) Dulbecco's modified Eagle's medium containing 10% fetal bovine serum; and (2) Colcemid dissolved in 0.9% NaCl, filter sterilized, and aliquoted into 1-mg/ml stocks in

[11] J. J. Lucas, E. Szekely, and J. R. Kates, *Cell (Cambridge, Mass.)* **7,** 115 (1976).
[12] R. E. K. Fournier, *Proc. Natl. Acad. Sci. U.S.A.* **78,** 6349 (1981).

15-ml conical tubes. Tubes can be stored at −20° for approximately 6 months if protected from light.

Experimental Protocol: Micronucleation by Prolonged Mitotic Arrest

1. For micronucleation of A9 cells or A9-based hybrids containing single human chromosomes, four T150 flasks of logarithmically growing cells at 80–90% confluence are needed. Each T150 is trypsinized and replated into four 150-mm tissue culture dishes, each containing four bullets in complete medium. The tissue culture dishes containing the cells are incubated at 37° for at least 3–4 hr to allow for cell attachment prior to Colcemid addition.

2. Colcemid (0.06 μg/ml) is added to each dish, and the cells are incubated at 37° for 48 hr to induce micronucleation.

3. Following the micronucleation step, the percentage of cells containing micronuclei is quantitated. For some cell lines like A9 it may be possible to determine frequencies simply by phase-contrast microscopy. One bullet may also be used, if necessary, for staining with acridine orange (see section on quantitation of microcells), which differentially stains the nucleus (green) versus the cytoplasm (red) using fluorescence microscopy. Micronucleation for A9 cells routinely ranges from 90 to 95% (Fig. 2a). Micronucleation of human foreskin fibroblasts (Fig. 2b) ranges from 40 to 60% and requires considerably higher Colcemid doses, as discussed below.

Parameters Affecting Micronucleation

Two parameters should be determined for each donor cell line to be used. The most critical variable in the procedure is the determination of the optimal Colcemid concentration. Optimal conditions should be determined for each cell line by performing a 48-hr concentration curve. Most rodent lines micronucleate in the range of 0.01–0.1 μg/ml Colcemid.[5] Human foreskin fibroblasts and some human cell lines can also be induced to micronucleate; however, the concentration of Colcemid required to induce micronucleation is in the range of 10–20 μg/ml.[6] Higher concentrations of the mitotic inhibitor result in increased cell toxicity and a decreased yield of micronucleated cells. Another important variable for micronucleation of human fibroblasts is the culture age.[6] Early passage cells micronucleate optimally; however, after population doubling 17–18, the percentage of micronucleate cells decreases dramatically.

The second parameter to be determined for each cell line is the duration of mitotic arrest. The incubation period in Colcemid that is required to induce micronucleation varies depending on the doubling time of the particular cell line. Slower growing cell lines require longer arrest times. In

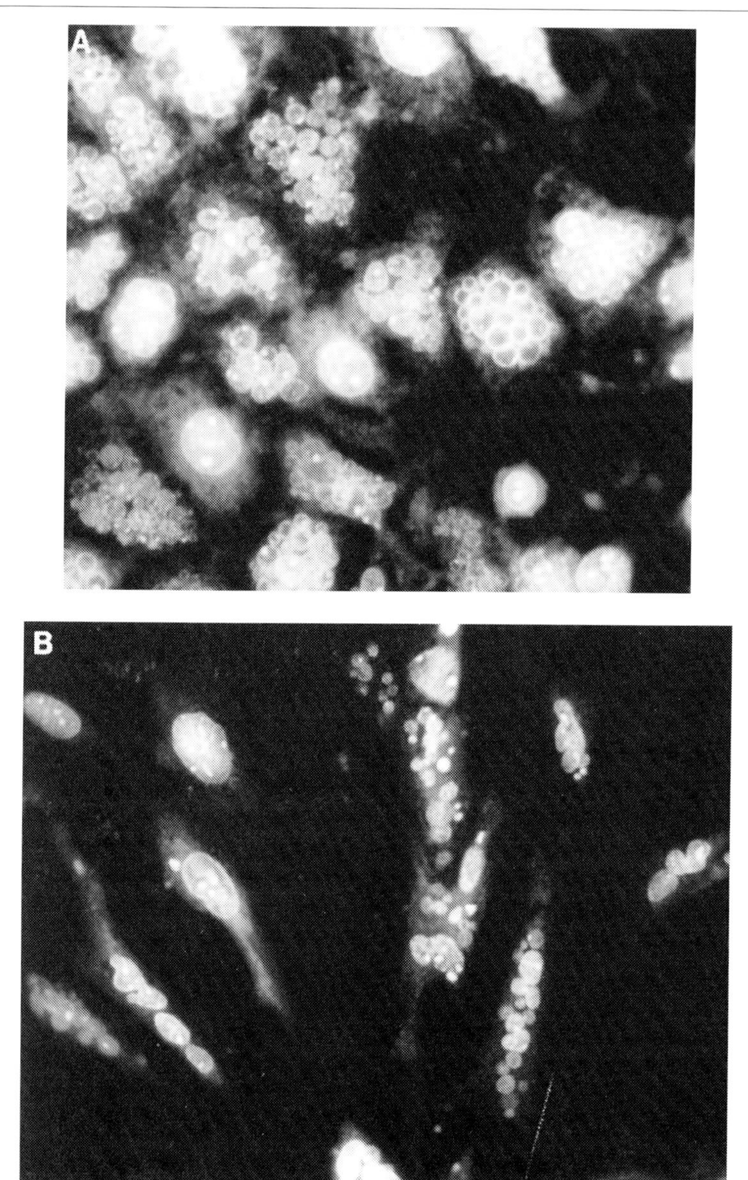

Fig. 2. Micronucleation by prolonged mitotic arrest. Acridine orange-stained micronuclei from (A) mouse A9 cells and (B) human foreskin fibroblasts.

addition, we have observed that the number of microcell hybrid clones containing intact donor chromosomes also depends on the length of time in the mitotic inhibitor. In general, the longer the incubation time in Colcemid, the greater the frequency of fragmented donor chromosomes. Conditions that result in the highest percentage of micronucleate cells with the shortest possible incubation in Colcemid should be established for each cell line.

Enucleation of Micronucleate Cell Populations

Once micronucleate cell populations are generated, they can be centrifuged in the presence of cytochalasin B to induce enucleation. Cytochalasin B is a fungal metabolite that interferes with microfilament attachment to the cell membrane.[13] Exposure of cells to high concentrations of this drug (10 μg/ml) results in nuclear extrusion in a proportion of cells. Large-scale enucleation of whole cells was first accomplished by incubation in cytochalasin B combined with centrifugation.[14–16] Centrifugation in the presence of this drug causes nuclei to protrude on thin cytoplasmic stalks and to pellet in the bottom of the centrifuge tube (Fig. 3). The cytoplasm remains attached to the surface on which the cells were plated.

Application of this technology to micronucleate populations results in the extrusion of micronuclei to form microcells.[17] Each microcell consists of a single micronucleus surrounded by a small amount of cytoplasm and an intact plasma membrane. Following centrifugation of micronucleate populations in the presence of cytochalasin B, the pellet consists of microcells of various sizes, enucleated whole nuclei (karyoplasts), whole cells that stripped off prior to enucleation, and cytoplasmic particles. A detailed protocol for enucleation of mouse A9 cells is described below, as are methods for quantitation of microcell yield and enucleation conditions for other cell lines.

Materials and Methods of Procedure

Cytochalasin B (Sigma; Catalog No. C6762)
Dimethyl sulfoxide (DMSO) (Sigma; Catalog No. D-5879)
50-ml polycarbonate centrifuge tubes (Nalgene; Catalog No. 1) with lids (Catalog No. 3111-0030)

[13] S. B. Carter, *Nature (London)* **213,** 261 (1967).
[14] D. M. Prescott, D. Myerson, and J. Wallace, *Exp. Cell Res.* **71,** 480 (1972).
[15] G. Poste and P. Reeve, *Exp. Cell Res.* **73,** 287 (1972).
[16] W. E. Wright and L. Hayflick, *Exp. Cell Res.* **74,** 187 (1972).
[17] T. Ege and N. R. Ringertz, *Exp. Cell Res.* **87,** 378 (1974).

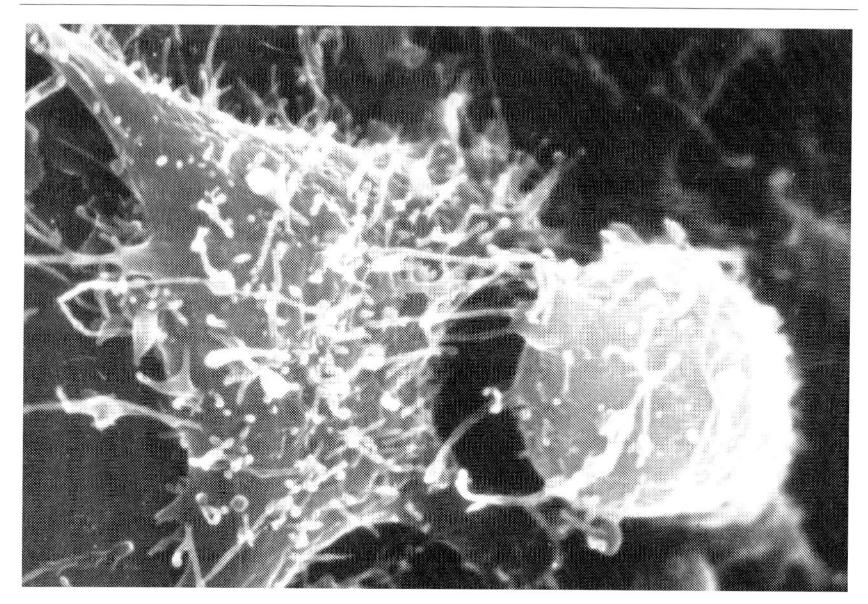

FIG. 3. Electron micrograph of an enucleating cell. (Photograph courtesy of Dr. Ronald L. Brown.)

Dulbecco's modified Eagle's medium (serum-free) (Gibco BRL; Catalog No. 430-2100EF)
Superspeed refrigerated centrifuge
Fixed angle rotor for 50-ml tubes [example, Sorvall RC-5B (SS-34 rotor)]

Preparation of Solutions for Enucleation

1. The cytochalasin B stock solution (1–2 mg/ml) is prepared by dissolving powder in DMSO. The solution is transferred to sterile tubes wrapped in foil to protect from light.

2. The enucleation medium consists of DMEM, pH 7.3–7.5, containing 10 μg/ml cytochalasin B.

Experimental Protocol: Enucleation

1. Wash and autoclave 50-ml Nalgene tubes and lids. To each of eight autoclaved tubes, add 38.5 ml serum-free DMEM (37°) containing 10 μg/ml cytochalasin B. Incubate at 37°.

2. Prewarm superspeed centrifuge and rotor to 28–33°.

3. Aseptically remove 16 bullets containing micronucleate cells and place 2 bullets back-to-back inside each 50-ml tube. Place sterile lids on the tubes and secure with tape.

4. Immediately place tubes containing bullets in the fixed angle rotor such that bullets are parallel to rotor radii.

5. Centrifuge at 27,000*g* (15,000 rpm) for 70 min at 28–33°.

6. Following centrifugation, carefully remove bullets from tubes and place in a 3% bleach solution. Save 1 bullet to monitor enucleation efficiency (see below). Enucleation should be >90–95%.

7. Decant medium from the tubes and resuspend pellets. Combine pellets into a single conical tube containing 10–12 ml serum-free DMEM. Resuspend the microcell preparation and save an aliquot for hemocytometer counts and quantification of percent microcells using acridine orange or acetoorcein.

Enucleation Using Concanavalin A-Coated Bullets

The protocol just outlined is for use with cell lines that adhere well to plastic. For cell lines that attach poorly to the substratum or that grow in suspension, treatment of bullets with concanavalin A (Con A) prior to cell plating will allow most cells to attach firmly for enucleation.

Materials and Methods of Procedure

Concanavalin A (Sigma; Catalog No. C2631)

1-Cyclohexyl-3-(2-morpholinoethyl)carbodiimide metho-*p*-toluene sulfonate [water-soluble carbodiimide (WSC) (Sigma; Catalog No. C1011)]

Solution Preparation for Enculeation with Con A Bullets

1. The Con A solution is prepared in 0.9% saline at 15 mg/ml and is incubated at 37° for 30 min. The solution is passed through Whatman No. 1 paper and is filter sterilized.

2. The WSC cross-linking agent is prepared in 0.9% saline at 75 mg/ml and is filter sterilized.

Experimental Protocol: Enucleation Using Con A Bullets

1. Remove bullets from ethanol and place four bullets in each 150-mm tissue culture dish. Allow to dry. Layer 0.6 ml of WSC solution on each bullet followed by 0.6 ml of Con A solution and spread mixture evenly across bullets. Incubate for 1–2 hr at room temperature.

2. Remove the ConA–WSC solution by aspiration, and rinse the bullets twice with sterile phosphate-buffered saline (PBS).

3. Micronucleate cells (induced by prolonged Colcemid arrest in T75 or T150 flasks) are harvested by trypsinization, counted, and resuspended in PBS at 10^6 cells/ml. The cell suspension (1.2 ml) is added to each bullet, and the cells are allowed to attach for 10–15 min at room temperature. Complete medium is added, and the cells are incubated for 30–60 min at 37° to allow for cell spreading.

4. Bullets containing micronucleate cells can be processed for enucleation as described in the preceding section.

Suspension Enucleation

For cell lines that fail to attach using Con A, it is possible to enucleate the cells in suspension. This strategy involves the centrifugation of micronuleate cells through either Ficoll[18] or Percoll[19] gradients containing cytochalasin B. Gradient separation is based on differences in buoyant densities of nucleoplasm versus cytoplasm in a centrifugal force. Using this protocol, large numbers of cells can be processed. For example, human lymphoblastoid cell lines, which micronucleate at low frequency, can nonetheless be used as donors if a sufficient number of cells can be processed.[19] However, enucleation is often incomplete in these gradients, and many partially enucleated cells will be lost because they partition away from isolated microcells. Nevertheless, gradient enucleation provides a means to enucleate cell lines that can not be enucleated on bullets.

Parameters Affecting Enucleation

There are three main parameters that affect the efficiency of enucleation: (1) the concentration of cytochalasin B, (2) the centrifugation speed, and (3) the centrifugation time. The concentration of cytochalasin B is fairly standard and ranges from 5 to 10 μg/ml. Cytochalasin B in serum-free medium is stable for approximately 6 months such that following each enucleation, media containing cytochalasin B can be filter sterilized, stored at 4°, and reused. Cytochalasin B-containing medium should be discarded if the enucleation efficiency decreases. Other variables that should be determined for individual cell lines include centrifugation speed and time. If incomplete enucleation occurs, adjusting either of these variables can increase the efficiency of enuclation. However, care must be taken to examine bullets following enucleation. Excess centrifugal force or excess time can

[18] M. H. Wigler and I. B. Weinstein, *Biochem. Biophys. Res. Commun.* **63,** 669 (1975).

[19] J. A. Sanford and E. Stubblefield, *Somatic Cell Mol. Genet.* **13,** 279 (1987).

result in complete stripping of cytoplasm from the bullet. Table I describes enucleation conditions for some commonly used cell lines.

Quantitation of Microcells

Following enucleation, the microcell preparation consists of microcells of various sizes, usually containing one to five interphase chromosomes, enucleated whole nuclei (karyoplasts), whole cells that have stripped off the bullet prior to enucleation, and cytoplasmic vesicles. Total particle counts can be obtained using a hemocytometer. To determine the percentage of microcells in the total particle count, quantitation of microcells is required using bright field or fluorescence microscopy. Two staining protocols are outlined; acetoorecein and bright-field microscopy or acridine orange staining and fluorescence microscopy.

Materials and Methods of Procedure

Orcein (Sigma; Catalog No. 07380)
Acetic acid
Acridine orange (Sigma; Catalog No. A-6014)

Staining Protocol: Acridine Orange

Acridine orange is a fluorochrome that stains the nucleus and the cytoplasm differentially. Nuclei and micronuclei stain green, whereas cytoplasmic particles stain red using fluorescence microscopy. This stain is excellent for the quantitation of micronucleation, enucleation efficiency, or microcell yield. However, acridine orange is a carcinogen, and should be handled with caution.

1. Prepare a stock solution of acridine orange at 2 mg/ml in sterile water. Store in foil-wrapped conical tubes at 4°.

TABLE I
ENUCLEATION CONDITIONS FOR CELL LINES

Cell line	Cytochalasin B concentration (μg/ml)	Centrifugation: Speed (*g*)	Centrifugation: Time (min)
Mouse A9 hybrids containing single human chromosome	10	27,000	70
Mouse embryo fibroblasts	10	27,000	30
Human foreskin fibroblasts	5	18,800	35

2. For staining, prepare a 10-μg/ml solution of acridine orange in water or normal saline. A foil-wrapped Coplin jar is ideal for processing bullets for staining.

3. To check for micronucleation or enucleation efficiency, remove bullets from medium and place in a Coplin jar containing a 1:1 mixture of 95% methanol and serum-free medium for 10 min. Transfer bullets to a second Coplin jar containing 95% methanol for 10 min and allow bullets to air dry. Place each bullet in acridine orange staining solution for 30–60 sec. Rinse twice with water and coverslip using 0.05 *M* citrate phosphate mounting buffer, pH 4.1. Bullets can be viewed by fluorescence microscopy using a barrier filter which excites in the 500-nm range.

4. To quantitate microcell yield, place a drop of microcell preparation before and after filtration (see below) on a microscope slide. To that drop, mix one drop of working solution of acridine orange and coverslip (30–60 sec). View the particles by fluorescence microscopy.

Staining Protocol: Acetoorcein

Acetoorcein will also differentially stain the nucleus from the cytoplasm. Nuclei will appear red and cytoplasm pink using this dye. Interpretation of acetoorcein-stained material is sometimes difficult because discrimination between shades of the same color is required rather than visualization using fluorochromes of contrasting colors. However, the method is simple to perform.

1. Prepare acetoorecein stain by making a 0.5% orcein solution in 50% acetic acid. Dissolve by refluxing for several hours.

2. For quantitation of microcell yield, mix one drop of microcell preparation with one drop of acetoorcein stain on a microscope slide. Overlay with a coverslip and stain for 2–5 min. View by bright-field microscopy. The staining is time dependent; after 5–10 min, all particles are deep red, and discriminating nuclei from cytoplasm is difficult.

Purification of Microcells

Purification of the microcell population following enucleation is an essential step in the protocol if monochromosomal hybrids are desired. The simplest procedure involves the use of Nuclepore filters[6,12] to selectively isolate the smallest microcells in the preparation. A technique for unit gravity sedimentation is also described.[5,12]

Materials and Methods of Procedure

Filtration

Swinnex filter units (Swinnex; 25 mm, Catalog No. SX0002500)
Silicone gaskets (Swinnex; 25 mm, Catalog No. SX0002501)
Nucleopore filters [VWR Scientific (3 μm, Catalog No. 28158-602; 5 μm, Catalog No. 28158-668)]
10-ml syringes

Unit Gravity Sedimentation

Bovine serum albumin (BSA)
Gradient mixing device (LKB; Catalog No. 8121)
Peristaltic pump (e.g., Cole Parmer; Catalog No. 7545)
Three-way valve

Purification by Filtration

1. Place 8-, 5-, or 3-μm Nuclepore filters in separate Swinnex filter units and autoclave.
2. Attach filter units to sterile 10-cm^3 syringes.
3. Resuspend the enucleation pellet in 10–15 ml serum-free medium.
4. Pour one-half of crude microcell preparation into a syringe containing a 5-μm filter. Allow solution to filter through by gravity as much as possible; apply slight pressure on syringe with plunger if necessary. Repeat with the remaining half of the microcell preparation using a new 5-μm filter set. Following this step, most of the whole cells and karyoplasts should be filtered out of the microcell preparation (Fig. 4). Preparations containing large numbers of intact cells should be passed through 8-μm filters prior to 5-μm filtration.
5. The remaining microcell solution should contain various sizes of microcells and cytoplasmic particles. To select for the smallest microcells, containing most probably single chromosomes, the solution can be filtered as in step 4 using two 3-μm filter units. Care should be taken not to use too much pressure during filtration through the 3-μm filters which might result in cracking of the filter. Aliquots should be taken after filtration for quantitation using acridine orange or acetoorcein.

Purification by Unit Gravity Sedimentation

Unit gravity sedimentation on density gradients has been successfully used to separate microcell preparations based on size.[5,12]

1. Prepare a linear gradient of 1–3% (BSA) using a gradient mixing device connected to a peristaltic pump.

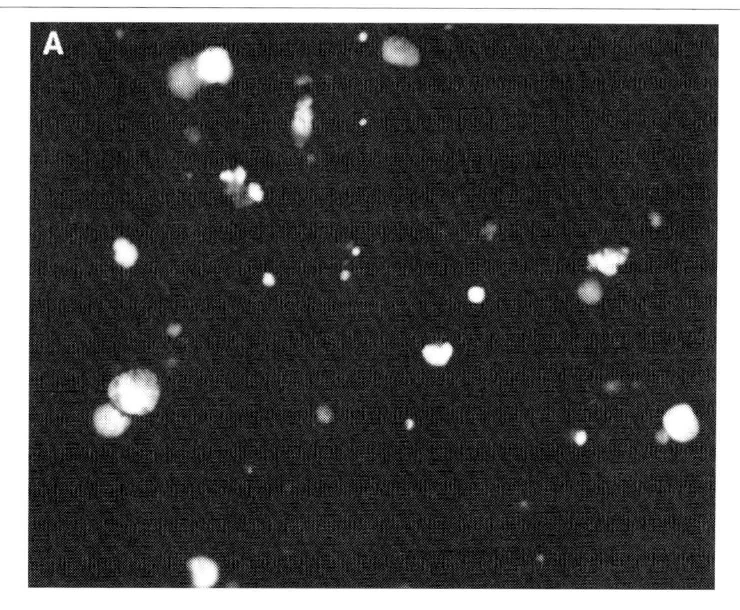

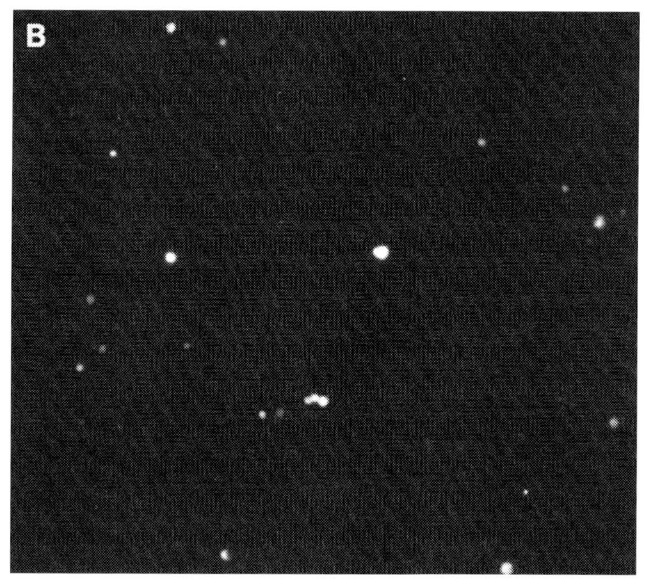

FIG. 4. Purification of microcells. Microcell preparation (A) unfiltered and (B) following filtration through 5- and 3-μm Nuclepore filters.

2. Connect tubing from the gradient mixing device through the peristaltic pump to a three-way valve. Connect one valve outlet to a 5-cm^3 syringe which serves as a bubble trap. The second valve should connect via tubing to another three-way valve. The upper outlet of this valve is connected to a 50-cm^3 syringe used as the gradient chamber. Place three glass beads over the inlet of the chamber. Connect a small piece of tubing to the other outlet which serves as the sampling tube. Autoclave the entire apparatus.

3. Add sterile PBS to the gradient mixing device and pump to fill the tubing just at the bottom of the gradient chamber. Pump excess PBS into the bubble trap.

4. Add 25 ml 3% BSA in PBS to one side of the mixing device and 25 ml 1% BSA in PBS (containing phenol red) to the other. Open the connection between the chambers and start the stirrer motor.

5. Resuspend the microcell preparation in 2 ml 0.5% BSA in PBS and add to the gradient chamber from the top. Start the pump and allow PBS in the lines to be pumped into the bubble trap. When the BSA solution (red) reaches the trap, turn the valve to allow solution to enter the gradient chamber. The BSA gradient should be pumped into the chamber at a rate of 2–3 ml/min.

6. Allow the microcell preparation to settle through the gradient for 3–3.5 hr, and collect fractions by dripping through the sample tube. Size fractionation efficiently separates whole cells and karyoplasts in the bottom 15–20 ml of the 50-ml gradient. The purified microcell preparation is recovered from the top 20–25 ml.

Parameters Affecting Purification Schemes

Purification of the microcell preparation is an important step in the protocol for microcell fusion if the experiment requires monochromosomal microcell hybrid clones. Without purification, fusion of the crude microcell preparation will result in a fraction of clones containing only the chromosome for which selection is applied, and the remaining clones will consist of whole cell hybrids and microcell hybrids containing multiple chromosomes. The relative proportion of whole cell hybrids to microcell hybrids recovered per experiment is dependent on the percentage of micronucleate cells in the population. For A9 hybrids which micronucleate at >95% efficiency, the numbers of whole cell hybrids recovered are generally small. The obvious advantage of purification of the donor microcell population is that the recovery of hybrid clones containing a single, intact transferred chromosome or a fragment of that chromosome is greatly enhanced. However, the total yield of microcell hybrid clones per experiment will decrease following purification.

Filtration through Nucleopore filters is the simplest technique for purification; passage through 5- and then 3-μm filters results in the elimination of whole cells and 99% of karyoplasts. The efficiency of this purification step is critically dependent on the proper use of the filter. Overloading of the filter with the microcell preparation could result in clogging of the filter and a subsequent decrease in the yield of purified microcells for fusion. Total particle counts of 1.5×10^6 particles/ml can be easily passed through Nucleopore filters.

Perhaps the most efficient method for purification of microcell preparations is the unit gravity sedimentation method. This technique virtually eliminates whole cells and karyoplasts from the crude microcell preparation. Separation on density gradients, however, is a time-consuming procedure and therefore may not be the method of choice unless absolute purification of the preparation is required.

Fusion of Purified Microcells to Recipient Cells

Microcells were first fused to recipient cells using inactivated Sendai virus.[5] A more efficient and simple technique for fusion involves the use of phytohemagglutinin P(PHA-P) to adhere microcells to recipient cells followed by fusion using polyethylene glycol (PEG).[6,12]

Two protocols are described for the fusion of microcells to recipient cells. The first protocol involves the agglutination of microcells to a monolayer of recipient cells using PHA-P followed by fusion using polyethylene glycol. The second method is a suspension protocol for use with cells which do not adhere to monolayer culture or for which PHA-P is toxic.

Materials and Methods of Procedure

Polyethylene glycol (Koch Chemical United, MW 1540; Catalog No. 00-14806)

Phytohemagglutinin-P (Difco; Catalog No. 3110-56-4)

Preparation of Solutions

1. PEG is prepared 50% (w/w) in serum-free DMEM. The mixture is first incubated at 37° until PEG has dissolved, and then it is filter sterilized. The pH should be between 7.5 and 8.0.
2. PHA-P is a lyophilized powder which is 50% salt by weight. PHA-P is prepared at 100 μg/ml in serum-free DMEM and filter sterilized.

PHA-P/PEG Monolayer Fusion

1. Prepare monolayer cultures of recipient cells in 25-cm^2 flasks. Two flasks are used routinely for fusion and one flask is reserved for a control. Cultures should be subcultured 4–24 hr prior to fusion to allow for firm attachment to the substratum. Each flask should be at 70–80% confluence at the time of fusion.

2. Centrifuge filtered microcell preparation at 2000*g* for 10 min (37°). Resuspend the resulting microcell pellet in 4 ml PHA-P solution (100 μg/ml). The suspension is redistributed by pipette to disperse clumps of agglutinating microcell particles. Medium is aspirated off recipient cell monolayers, and the microcell solution is immediately added to each of two fusion flasks (2 ml/flask) and incubated at 37° until microcells have agglutinated to the monolayer (usually 10–20 min).

3. Carefully aspirate medium off fusion monolayers. To each monolayer, add 0.5 ml PEG (50%, w/w) and gently rock flask for 1 min. Quickly remove PEG solution and rinse monolayer three times with 4 ml serum-free medium. Add complete nonselective medium and return flasks to the 37° incubator. Incubate control flask without donor microcells with PHA-P solution as described, rinse three times with medium without serum, and refeed in medium containing 10% FBS.

Protocol: Suspension Microcell Fusion

1. Count recipient cells and resuspend at 1 × 10^6 cells/ml in 5 ml of serunm-free medium.

2. Combine recipient cells with filtered microcell preparation in a 15-ml conical centrifuge tube. Centrifuge at 2000*g* for 15 min at 37°.

3. Resuspend pellet well. Carefully add 0.5 ml PEG (50%, w/w) drop-wise while gently dispersing pellet for 1 min. Immediately add 9.5 ml serum-free medium over 1 min with gentle swirling.

4. Prepare 20 T25s per fusion in complete nonselective medium. Repeat for the control flask. Aliquot 0.5 ml of fusion or control suspension per T25. Incubate overnight (17–24 hr) at 37°.

5. Refeed fusion and control flasks with complete selective medium. Hybrid colonies should be visible within 2–3 weeks.

Parameters Affecting Microcell Fusion

The two most important variables to be determined for each recipient cell line prior to microcell fusion are (1) the optimal concentration of fusogen and (2) the concentration of PHA-P which results in agglutination without cell toxicity. The first parameter can be determined easily using the recipient cell line alone to quantitate the numbers of binucleate heterokaryons postfusion with increasing concentrations of PEG. Optimal concentrations of PEG for many cell lines are in the range of 44–50%. Higher concentrations of the fusogen can result in large numbers of cells fused *en masse* with increased cell toxicity. PHA-P toxicity experiments should be performed in increasing concentrations of PHA-P (50–200 μg/ml) to determine the optimal concentration of PHA-P for agglutination of recipient cells with the least amount of cell toxicity. Agglutination can be monitored by phase microscopy, and the degree of toxicity can be measured by trypan blue staining.

In general, the monolayer protocol for fusion is much more efficient than the suspension protocol, with transfer frequencies in the range of 0.5–2.0×10^{-5}. The suspension protocol, however, works well for cell lines which grow in suspension, adhere poorly to the monolayer, or for which PHA-P is cytotoxic.

Microcell Hybrid Selection

A number of selective systems are available which allow for the transfer of a specific chromosome into a recipient cell background. One of the attractive uses for microcell fusion has been in complementation mapping. Using this strategy, cells harboring a recessive, conditional lethal mutation are recipients for microcell transfer of a wild-type chromosome encoding a gene which complements the mutant phenotype. Nucleotide salvage pathway enzymes hypoxanthine phosphoribosyltransferase (HPRT) and thymidine kinase (TK) via the HAT selective system[20] and adenine phosphoribosyltransferase (APRT) via the AAT selective system[21] are the best known complementation systems for somatic cell hybridization.

The most conceptually attractive microcell hybrid clone would be one in which a single chromosome is transferred into a recipient cell and maintained under dominant selective pressure. A number of cloned selectable markers are available, including pSV2*neo*, a plasmid vector containing the bacterial gene *neo* which confers resistance to the antibiotic G418,[22] and pSV2*gpt*, a similar vector containing the bacterial gene *gpt* which can be selected in an $HPRT^-$ background using HAT selection or in a wild-type background using mycophenolic acid.[23] Plasmid vectors such as these as well as retroviral vectors have been successfully introduced into donor cell

[20] J. W. Littlefield, *Science* **145,** 709 (1964).

[21] T. Kusano, C. Long, and H. Green, *Proc. Natl. Acad. Sci. U.S.A.* **68,** 82 (1971).

[22] P. J. Southern and P. Berg, *J. Mol. Appl. Genet.* **1,** 327 (1982).

[23] R. C. Mulligan and P. Berg, *Proc. Natl. Acad. Sci. U.S.A.* **78,** 2072 (1981).

lines prior to micronucleation, and these tagged chromosomes subsequently are transferred into a variety of recipient cell lines.[24–26]

Applications of Microcell Fusion

Microcell hybrids have proven to be powerful tools not only for the physical mapping of the human and mouse genomes, but also as defined reagents to study genes associated with differentiation, development, and disease.[3,27] Microcell fusion is the only available technology which results in the transfer of a single, intact chromosome from one cell to another. Microcell hybrids, by definition, are karyotypically simple, which allows for the unambiguous assignment of genetic loci without the complication of additional chromosomes or fragments and translocations of other chromosomal regions. Microcell hybrids have been used to identify genes involved in cellular senescence[28] and to isolate genes involved in the control of tissue-specific gene expression.[29,30] Transfer of specific human chromosomes via microcell fusion into malignant cell lines (containing deletions or showing allele losses on that chromosome) has been instrumental in the functional definition of novel tumor suppressor genes.[31–35]

Perhaps one of the most attractive uses for microcell hybrid clones is in deletion mapping. Microcell hybrids generated in a single experiment typically yield clones containing intact donor chromosomes as well as clones that contain fragments of the introduced chromosome. Fragment-containing microcell hybrids have been successfully used for detailed intrachro-

[24] P. J. Saxon, E. S. Srivatsan, G. V. Leipzig, J. H. Sameshima, and E. J. Stanbridge, *Mol. Cell. Biol.* **5,** 140 (1985).

[25] T. G. Lugo, B. Handelin, A. M. Killary, D. E. Housman, and R. E. K. Fournier, *Mol. Cell. Biol.* **7,** 2814 (1987).

[26] Y. Sanchez, M. Garcia-Heras, T. Giambernardi, M. E. Wolf, A. U. Banks, and A. M. Killary, in preparation.

[27] E. J. Stanbridge, *Annu. Rev. Genet.* **24,** 615 (1990).

[28] Y. Ning, J. L. Weber, A. M. Killary, D. H. Ledbetter, J. R. Smith, and O. M. Pereira-Smith, *Proc. Natl. Acad. Sci. U.S.A.* **88,** 5635 (1991).

[29] A. M. Killary and R. E. K. Fournier, *Cell* (*Cambridge, Mass.*) **38,** 523 (1984).

[30] K. W. Jones, M. H. Shapero, M. Chevrette, and R. E. K. Fournier, *Cell* (*Cambridge, Mass.*) **66,** 861 (1991).

[31] P. J. Saxon, E. S. Srivatsan, and E. J. Stanbridge, *EMBO J* **5,** 3461 (1986).

[32] B. E. Weissman, P. J. Saxon, S. R. Pasquale, G. R. Jones, A. G. Geiser, and E. J. Stanbridge, *Science* **236,** 175 (1987).

[33] J. M. Trent, E. J. Stanbridge, H. L. McBride, E. U. Meese, G. Casey, D. E. Araujo, C. M. Witkowski, and R. B. Nagle, *Science* **247,** 568 (1990).

[34] K. Tanaka, M. Oshimura, R. Kikuchi, M. Seki, T. Hayashi, and M. Miyaki, *Nature* (*London*) **349,** 340 (1991).

[35] M. Shimizu, J. Yokota, N. Mori, T. Shuin, M. Shinoda, M. Terada, and M. Oshimura, *Oncogene* **5,** 185 (1990).

mosomal mapping.[36–38] Deletion microcell hybrids have been used in the mapping and isolation in critical disease genes including neurofibromatosis I (NF1)[39] and hereditary nonpolyposis coli (HNPCC).[40]

Another use of fragment-containing microcell hybrids has been in the regional localization of novel tumor suppressor genes.[41–43] Defined *neo*-marked fragments of normal human chromosomes have been successfully transferred into human malignant cell lines. Using these kinds of strategies, two novel tumor suppressor loci were functionally identified on the short arm of human chromosome 3 and mapped to intervals of 2 Mb within 3p21–3p22 and 10–15 Mb within 3p12–3p14.[42,43]

Fragment-containing microcell hybrids are particularly useful for generation of large chromosomal fragments. More detailed physical mapping may be accomplished using radiation hybrid mapping.[44] Using this technique, microcells are gamma-irradiated and rescued by fusion to a recipient cell. Using radiation microcell hybrids, a high resolution physical map can be generated; the extent of chromosome fragmentation becomes a function of the dosage of radiation. Fragment-containing hybrids generated by either route are useful sources of molecular probes for defined chromosomal segments and have been instrumental in the positional cloning of several important human genes.[30,39,40]

[36] R. E. K. Fournier and J. A. Frelinger, *Mol. Cell. Biol.* **2,** 539 (1982).

[37] R. E. K. Fournier and R. G. Moran, *Somatic Cell Mol. Genet.* **9,** 69 (1983).

[38] R. J. Leach, M. J. Thayer, A. J. Schafer, and R. E. K. Fournier, *Genomics* **5,** 167 (1989).

[39] J. W. Fountain, M. R. Wallace, A. M. Berereton, P. O'Connell, R. L. White, D. C. Rich, D. H. Ledbetter, R. J. Leach, R. E. K. Fournier, A. G. Menon, J. F. Gusella, D. Barker, K. Stephens, and F. S. Collins, *Am. J. Hum. Genet.* **44,** 58 (1989).

[40] F. S. Leach, N. C. Nicolaides, N. Papadopoulos, B. Liu, J. Jen, R. Parsons, P. Peltomäki, P. Sistonen, L. A. Aaltonen, M. Nyström-Lahti, X.-Y. Guan,. J. Zhang, P. S. Meltzer, J.-W. Yu, F.-T. Kao, D. J. Chen, K. M. Cerosaletti, R. E. K. Fournier, S. Todd, T. Lewis, R. J. Leach, S. L. Naylor, J. Weissenbach, J.-P. Mecklin, H. Järvinen, G. M. Petersen, S. R. Hamilton, J. Green, J. Jass, P. Watson, H. T. Lynch, J. M. Trent, A. de la Chapelle, K. W. Kinzler, and B. Vogelstein, *Cell (Cambridge, Mass.)* **75,** 1215 (1993).

[41] S. F. Dowdy, C. L. Fasching, D. Araujo, K.-M. Lai, E. Livanos, B. E. Weissman, and E. J. Stanbridge, *Science* **254,** 293 (1991).

[42] A. M. Killary, M. E. Wolf, T. A. Giambernardi, and S. L. Naylor, *Proc. Natl. Acad. Sci. U.S.A.* **89,** 10877 (1992).

[43] Y. Sanchez, A. El-Naggar, S. Pathak, and A. M. Killary, *Proc. Natl. Acad. Sci. U.S.A.* **91,** 3383 (1994).

[44] S. J. Goss and H. Harris, *Nature (London)* **255,** 1445 (1975).

[10] Tumor Cell Culture

By MARY PAT MOYER

Introduction

There are innumerable ways to initiate primary cell cultures or to isolate cells from fresh tissue specimens (e.g.[1-4]) and many parameters that should be used as a guideline for cell culture selection (Table 1). This chapter describes generally successful methods for culture initiation of tumor cells from multiple human and animal source tissues. Some aspects of these methods and their applications, including the study of normal cells, have been published elsewhere.[5-9] The success rate of 80–90% for the initiation of viable primary cultures with the propensity to maintain the ability to divide and differentiate for at least two to three subcultures (which is 4–10 population doublings, depending on the culture) has been observed for the attempted culture of more than 10,000 source specimens from surgically resected and fetal donor tissues. A lower success rate (about 25%) has been generally observed with cadaver donor tissues or those in which there is a long delay from harvest to culture initiation. However, this is highly dependent on the organ site harvested since some organs (e.g., the alimentary tract and liver) are unsuitable for culture within a shorter time period than others. Thus, success is defined as the ability to propagate the initiated culture for at least 2 to 3 population doublings. It does not imply that continuous cell lines are easily derived. In fact, that is generally much more difficult as it requires selection of a subpopulation of cells that is readily maintained indefinitely *in vitro*. The overall success

[1] R. I. Freshney, "Culture of Animal Cells: A Manual of Basic Technique." Wiley-Liss, New York, 1994.

[2] R. I. Freshney, "Culture of Epithelial Cells." Wiley-Liss, New York, 1992.

[3] M. M. Webber, ed., "In Vitro Models for Cancer Research," Vol. 1. CRC Press, Boca Raton, FL, 1985.

[4] R. Pollack, "Readings in Mammalian Cell Culture," 2nd ed. Cold Spring Harbor Lab., Cold Spring Harbor, NY, 1981.

[5] M. P. Moyer, C. P. Page, and R. C. Moyer, *in* "In Vitro Models of Human Cancer" (M. Webber and L. Sekely, eds.), Vol. 1, Chapter 11. CRC Press, Boca Raton, FL, 1984.

[6] M. P. Moyer, *J. Tissue Cult. Methods* **8,** 63 (1983).

[7] M. P. Moyer, *J. Tissue Cult. Methods* **13,** 107 (1991).

[8] M. P. Moyer, P. S. Dixon, A. L. Culpepper, and J. B. Aust, *in* "Colon Cancer Cells" (M. P. Moyer and G. Poste, eds.), p. 85. Academic Press, San Diego, CA, 1990.

[9] T. J. Goodwin, W. F. Schroeder, D. A. Wolf, and M. P. Moyer, *Proc. Soc. Exp. Biol. Med.* **202,** 181 (1993).

TABLE I
PARAMETERS USED AS A GUIDELINE FOR THE SELECTION OF CELL CULTURES

Scientific question(s) being asked
Available, replaceable, or renewable resource
Ease and cost of propagation: reagents and labor
Experience of the technical staff and the investigator
Quality control for origins, propagation, and microbial contaminants
Reproducibility of phenotype with routine culture and cryopreservation
Culture longevity issues, e.g., immortalization status, population doubling times
Species of origin and need for human and/or animal subjects approval(s)
Need for follow-up pathology documentation or other source validation
Introduced genes and/or "spontaneous" genetic alterations in the cells
Viruses or viral gene(s) present in, or expressed by, the cells or host from which they were derived
Malignant and/or invasive potential of the cells
Organ site or specialized tissue source
Expression of differentiation-associated characteristics
For cell transfection studies: potential ability of cells to use regulatory elements (e.g., enhancers and promoters)

rate is much less than 50% for all tissue sources. The greatest chance for successful cell line establishment is from metastatic tumors or those with high metastatic potential, particularly tumors of blood cell origin, melanomas, and solid tumors that have the ability to grow in suspension as effusions or ascites.

Culture of Tumor Cells from Freshly Isolated Tissues

It is an incorrect assumption that tumors are easily propagated *in vitro* simply because they apparently are not being normally regulated *in vivo*.[10-12] The growth and maintenance of cells (both normal and malignant) are very complex processes of local and systemic stimuli tempered by inhibitory regulators.[13-17] Important concepts are that there are numerous

[10] I. L. Cameron and T. B. Pool, eds., "The Transformed Cell." Academic Press, New York, 1981.

[11] P. Skehan and S. J. Friedman, "Growth, Cancer and the Cell Cycle." Humana Press, Clifton, NJ, 1984.

[12] L. M. Franks and C. B. Wigley, eds., "Neoplastic Transformation in Differentiated Epithelial Cell Systems In Vitro." Academic Press, New York, 1979.

[13] G. Serrero and J. Hayashi, "Cellular Endocrinology: Hormonal Control of Embryonic and Cellular Differentiation." Alan R. Liss, New York, 1986.

[14] G. H. Sato, W. L. McKeehan, and D. W. Barnes, *in* "Molecular Mechanisms in the Regulation of Cell Behavior" (C. Waymouth, ed.), p. 65. Alan R. Liss, New York, 1987.

[15] B. Alberts, D. Bray, J. Lewis, M. Raff, K. Roberts, and J. D. Watson, "Molecular Biology of the Cell," 2nd ed. Garland Publishing, New York, 1989.

TABLE II
ADVANTAGES AND DIFFICULTIES IN CULTURING TUMOR CELLS FROM FRESH TISSUE SPECIMENS

Advantages
Replace use of *in vivo* models or subjects (i.e., animals and humans)
Multiple cell types can be propagated from various organ sites
Numbers of viable, replicating cells can be expanded in culture
Cells may display differentiation and functional characteristics
When possible, comparisons can be made to directly test normal cells versus tumor cells
More easily assessed biochemical, metabolic, and genetic studies
Potential for direct testing of drugs, including antitumor agents
Clinical specimens
Closer to *in situ* than continuous cell lines
Comparisons among individual patients are possible
Difficulties
Type(s) of cells represented in total population that can be propagated
Limited viability or cell division, especially from normal tissues and solid tumors
Subpopulation of cells may be selected by *in vitro* culture
Cultured cells may lose characteristics of differentiation that were present *in situ*
Investigators may lack experience and intuition about cell culture
Choice of optimal culture medium and/or supplements
Sensitivity of cells to chemical dissociating agents (e.g., enzymes, chelators)
Clinical specimens
Long delay between tissue harvest and processing
Inappropriate handling (e.g., nonaseptic, allowed to dry out)
Microbial contamination of tissue site (e.g., colon or oral cavity)
Lack of pathology documentation

chemical, physical, and cell–cell interactions and that tumors, especially solid tumors, are frequently heterogeneous. Although the propagation of cells outside of the body has improved over the past 40 years of cell culture, and there are many advantages to their use, there are still some disadvantages (Table II).

Although tumors may display features of autocrine growth factor regulation, this is not mediated by a single factor. That most tumor cells need numerous growth factors and nutrients for *in vitro* growth is supported by several common observations of our group and others: (i) Low seeding densities are often lethal to tumor cell cultures, particularly those derived from primary tumors. (ii) Metastatic tumors, especially those in pleural

[16] J. E. Darnell, H. Lodish, and D. Baltimore, "Molecular Cell Biology." Freeman, New York, 1990.

[17] W. H. Schwesinger and M. P. Moyer, *in* "The Physiologic Basis of Surgery" (J. P. O'Leary, ed.), p. 1. Williams & Wilkins, Baltimore, MD, 1993.

effusions or ascites, are generally easier to grow and can be propagated in less complex growth media. This may be because they have already undergone several *in vivo* selections, are producing more autocrine/paracrine factors that allow them to grow away from their site of origin, or that they no longer depend on the presence of matrix or mesenchymal cell-provided factors. (iii) There is a general inability to readily develop cell lines using standard limiting dilution or semisolid media cloning methods. If the cells were completely autocrine in their growth regulation, the factors should maintain growth and not be diluted out. Thus, they may be functioning more as paracrine factors acting on adjacent cells.

Another key problem is that a tumor specimen is not composed of tumor cells only. Other cells that are present include mesenchymal cell types (e.g., fibroblasts, endothelial cells, and smooth muscle cells), blood cells (e.g., macrophages, granulocytes, and lymphocytes, red cells), and normal cells from the tissue of origin. Adding to the complexity is that growth of primary cultures of tumor cells is very frequently augmented when mesenchymal cells are cocultured with them. These can be autologous cells from the tumor (as often occurs with the methods described herein during culture initiation), but some investigators use irradiated or mitomycin C-treated monolayers of fibroblasts to provide a "feeder cell" function to the tumor cells (e.g., see Freshney,[1] for details). The irradiation and mitomycin C treatments prevent new rounds of cell cycle progression through mitosis, but the cells remain attached to the substrate and provide matrix/adhesive molecules, as well as paracrine factors or other signals, to the tumor cells. Notably, feeder cells would provide similar growth components to normal cells that might be present in the specimen. But, if there are many fewer normal cells, it is more probable that tumor cells will comprise the outgrowth of cells from suspensions of enzyme-generated single cells or multicellular aggregates. Depending on the long-term use of the cells, e.g., for medical applications,[18] a potential concern that should be preevaluated is whether or not the feeder layer substrate contains endogenous viruses (a common occurrence in most rodent cell lines).

Materials and Methods

The following general guidelines used by the UTHSCSA Center for Human Cell Biotechnology should be followed for the processing of fresh tissue specimens for cell culture from humans or animals. This protocol was originally developed for human gastrointestinal (GI) tract speci-

[18] M. P. Moyer, *Med. Prog. Technol.* **15,** 83 (1989).

mens,[5–8] but has been successfully adapted for many types of normal and malignant cells.

A. Materials

Biosafety Level II laminar flow cabinet
95% ETOH to flame sterilize instruments (if they are not presterilized and packaged)
Clean scissors, forceps, scalpel handle with No. 10 blade
Sterile disposable petri dishes (70 and 100 mm) or other suitable vessels (for scraping and mincing of tissue)
Sterile pipettes (1, 5, and 10 ml volume)
Sterile, conical centrifuge tubes (15 or 50 ml depending on specimen size)
Nalgene or other freezer containers (for slow freezing during cryopreservation)

B. Media/Reagents

TRM (for transport and rinsing of tissues; InCell Corporation, Ltd., San Antonio, TX)
M3 derivative media (e.g., M3:2, M3:10, M3:20, and M3:M5) are from InCell Corporation, Ltd. These highly specialized media, supplemented with numerous growth factors and nutrients, were derived from previous base formulations.[5–8]
Standard formulation media (e.g., RPMI, KSFM; from GIBCO, Grand Island, NY)
Fetal bovine serum (FBS) (Hyclone Laboratories, Logan, UT)
M3 cryopreservation medium (InCell Corporation, Ltd.)
Omnifix (An-Con Genetics, Melville NY) and cassettes (for histologic preparation)
Red blood cell (RBC) lysing buffer (for bloody samples) prepared as follows: For 1L, weight out 8.29 g ammonium chloride, 1.00 g potassium bicarbonate, and 0.0371 g ethylenediaminetetraacetic acid (EDTA). qs to 1L with ddH_2O. Autoclave to sterilize at 250° for 30 min. To use: Add 10 times the volume of the cell pellet containing the RBCs. Leave at room temperature for 10 min, centrifuge, and discard supernatant. May need to be repeated once or more, as needed.
Dissociation reagents: Collagenase (culture grade; 1.5 mg/ml final concentration; Worthington; Sigma); collagenase-dispase (Boehringer Mannheim Biochemicals); trypsin:EDTA (TE) = 0.25% (1:250)

trypsin: 0.02 *M* EDTA (many vendors; GIBCO); and ECD Lite (Specialty Media, Inc., Lavallette, NJ).

Sterile spinner flask (for mesenchymal tissues, if large specimen)

C. General Guidelines, Planning, and Record Keeping

Safety notes: All work is done with aseptic technique in Biosafety Level II laminar flow hoods and with the assumption that any human tissues may contain potential pathogens. Gloves, lab coats, and masks are worn during tissue processing. Technical staff handling human tissue specimens are vaccinated against hepatitis (Heptavax).

1. Prior to processing, prepare all materials and reagents. The tissue specimen should be assigned a standard alphanumeric designation: a letter prefix (denoting the tissue type) followed by a sequential number (for the new specimen). Check past records for the tissue prefix and the current sequence number to be used. Examples of some prefixes are HCC (human colon cancer); NCM (normal colon mucosa); NCM-SM (normal colon mucosal-derived submucosal mesenchyme); HST (human stomach tumor); TESP (tumor of the esophagus); and NBLAST (neuroblastoma). Similar abbreviations are used for animal source tumors but they are preceded by a letter designating species (e.g., M, mouse; R, rat; H, hamster).

2. All tissue specimens should be divided and allocated serially as follows unless a large enough specimen is not available (e.g., a biopsy). In that case, cell culture and cryopreservation followed by histology and then quick-freezing are the preferred order of priority.
   a. A small piece (about 10 mm^3) is put into a cassette, fixed in Omnifix, then processed for routine histology.
   b. At least one piece (0.5 to 0.8 cm^2, but as many as two or three, depending on the size of the specimen), is placed into 1-ml cryovials and quick frozen at $-70°$, for subsequent biochemical analysis, as required.
   c. The remaining tissue is processed for culture and cryopreservation. Enough tissue is processed for at least two cultures (25-cm^2 flasks) that are seeded with cells to assess viability, cell subpopulations, and possible contamination.
   d. All remaining cells from the processed tissue not required for current research projects are cryopreserved for future use in M3 cryopreservation media.

3. Record keeping:
   a. Any special or detailed notes on processing should be kept in the individual staff member's notebook.

b. Specimen data should also be entered into one of the central computer databases for cell and tissue inventory.
c. A hard copy of the specimen data sheet (for an example see Fig. 1) should be included with existing hard copy of the computerized inventories kept in central notebooks.

Tissue type is designated in the notebook labeled "letter designations for normal and tumor cells in culture" and there will be no spaces in the tissue designations. If the tissue is a pair, then put a slash between the different tissue types. For example, NCM55/HCC88 is a paired sample of normal colon mucosa (NCM) and human colon cancer (HCC) **Other tissues (if any)**: List all tissues that are from the same patient.
If any information is unknown fill in the blank with the abbreviation **UNK** for unknown.

The **date of collection** is written as 00-00-00.

Patient Information:

Patient number will be written with no dashes, commas, slashes, or spaces.

Age is written as 00.

Sex is typed as M: male or F: female.

Date of birth is written as 00-00-00 (month-day-year)

Ethnic group is written as H for Hispanic (including Mexican American, Latin American and all other races usually included in the ethnic group), B for black, C for Caucasian and UNK for unknown. For any other races type the description.

Hospital should be typed as UH for University Hospital, VA for the Audie L. Murphy veterans hospital. If the tissue is from another hospital, type out the entire name.

Cassette and **quick freeze** are yes or no answers (typed Y or N).

Vials cryopreserved: indicate the total number of vials frozen the day of culture initiation only. Specify how many vials per tissue source type in the "notes" field.

In the **notes** field write anything that is not of standard protocol and specify exactly what was frozen away or prepared as reference source material. For example,6 vials for HCC and 5 vials for NCM. Also under notes is where pediatric donor sources (age 0 to 10) should be indicated.

Proc. tech refers to the processing technician. It must be filled in with your first initial and full last name.

Path report: Type N for no if you do not have a pathology report at the time you are processing the tissue, NA for not applicable (for example, some types of discard tissue or if the pathologist has indicated that a report was unnecessary for that patient.). Use **UR** for unretrievable (for any reason).

FIG. 1. Example of CHCB cell inventory database.

4. After initial tissue preparation, additional information that becomes available at a later date should also be provided as part of the normal record keeping:

a. For clinical specimens, a copy of the anatomic pathology report should be requested and then put with the corresponding specimen information in the central notebooks as soon as it arrives. If there is a pathology report for animal tissues, it should be included. Protocol approval numbers of the Institutional Review Board (IRB) for use of human subjects and the Laboratory Animal Use and Care Committee should be noted as part of that record.

b. The cultures should be characterized a week or two after plating and these data should be added to the computer database and central notebooks.

D. Detailed Protocol for Tissue Collection, Processing, and Culture Initiation of Epithelial, Neural, and Other Soft Tissue Tumors

1. Supply TRM (25 ml/tube) in 50-ml centrifuge tubes to the operating suite/pathology area for storage at 4° prior to tissue collection.

2. After the specimen is transported to the laboratory, remove all fat and necrotic tissue, and discard. This can be done in petri dishes utilizing TRM with care to always keep the tissue wetted and the pH close to neutral. *Note*: The time between specimen collection and processing should be kept to a minimum. Culture designations must be determined immediately to properly label cassettes, cryopreservation vials, and culture flasks.

3. If there is a large amount of blood the tissue can then be removed to a new petri dish containing fresh medium.

4. A small (1–2 mm^3) cross section of the specimen should now be removed sharply and placed in a histology cassette which is in turn placed in Omnifix. Label the cassette with an indelible marker.

5. Mince the tumor into the smallest pieces possible in a small volume of TRM. This can be done in a sterile petri dish using curved iris scissors and forceps that have been autoclaved or flame sterilized.

6. With a wide bore pipette, transfer and triturate the minced tumor pieces to a 15- or 50-ml test tube(s) for rinsing. The tissue to TRM volume ratio should be 1:10. Thus, the size and numbers of tubes needed will depend on the amount of tissue.

7. Dissociation procedures: These will vary depending on the cell type(s). It is optional if cultures are to be seeded as small (1–2 mm^3)

explants or if the tissue is soft and friable enough to simply be dispersed into aggregates and single cells by trituration.

a. Physical dissociation: As previously detailed,[6] a tissue press can be used to generate a tumor brei (from nonmesenchymal origin tumors and those which are not extensively fibrotic) while leaving most of the support tissue on the screens. The disadvantage of the method is aseptic assembly and handling of the tissue press, particularly if large numbers of samples are being processed. Another approach is to use a mesh wire screen to generate tumor brie.

b. Chemical dissociation: Several chemical dissociation methods work well, but all have the disadvantage of also releasing mesenchymal cells that may overgrow the culture if selective conditions are not used.

(i). Continuous exposure to collagenase, which is prepared in complete growth medium, is detailed in Freshney.[1] The collagenase is rinsed out by centrifugation after several hours or days of incubation.

(ii). Collagenase-dispase: A 1- to 2-hr treatment is followed by trituration and centrifugation of the dissociated cells. The tissue fragments that remain after processing can be separately seeded into culture vessels and grown as explants.

(iii). Trypsin–EDTA (TE): Standard TE is easily used to generate single cells from tumors as well as other tissues, but it has the disadvantage of being toxic to many tumors, especially those of epithelial origin. It is preferred for mesenchymal source tumors and is used according to procedures described below for dissociation of mesenchymal tissues. After cells are dissociated with this reagent, the trypsin action must be stopped with the addition of FBS and, preferably, by rinsing in complete medium prior to seeding.

(iv). ECD Lite is a gentle cell dissociation agent. Use to dissociate cell aggregates at room temperature for several minutes or to subculture. Rinse cells or cell pellet once with calcium- and magnesium-free phosphate buffer and rinse once (about 1 min) with ECD Lite. Follow by a 1- to 5-min incubation with the cells to dissociate. Time will depend on the type of cells and their relative adherence properties. Triturate to make single cell suspensions, then seed culture vessels.

8. At the end of the dissociation procedure, the cell suspension or tumor brei should again be transferred to centrifuge tubes. They can be in TRM or in the preferred culture medium (M3 : 2 for epithelial origin tumors; M3 : M5 or M3 : 10 for mesenchymal origin tumors; M3 : 20 or RPMI 1640 with 10 to 20% FBS for blood cell origin tumors).

9. The specimens should be in a test tube with TRM (or complete culture medium if they were dissociated) at this step. They are then centrifuged at 1200 rpm for 8–10 min in a swinging bucket centrifuge.

10. Carefully discard the supernatants into a bucket containing bleach. If the sample is bloody, mix the sample with 10 ml RBC lysing buffer. Centrifuge for 8–10 min at 1200 rpm.

11. Discard the supernatants as above and add 10–15 ml TRM or 10 times the volume of the cell pellet. Resuspend and disperse the cell pellet with a pipette or gentle vortexing.

12. Repeat seven more times for a total of eight rinses for potentially contaminated specimens (e.g., GI tract specimens, oral cavity specimens, and cadaver donors). Other tissue sites require only three to four rinses.

13. After the final rinse, discard the supernatants and remove the remaining volume of the cell pellet, except for about 0.5 ml, for cryopreservation. To the 0.5 ml remaining, add M3 : 10 at about 10–20 times the volume to give a cell suspension (count with a hemocytometer) of about $1–2 \times 10^6$ cells/ml, and transfer to appropriate sized culture flask(s). Place the flask in an incubator at 37° with a 5% CO_2 atmosphere.

14. The cells that were removed for cryopreservation should now be resuspended in M3 cryopreservation medium. (*Helpful hint* for tissue fragments: Use a disposable, wide orifice pipette to add freezing medium and mix well, then use a shortie pipette with large orifice for *uniform* distribution into freezing vials.) Aliquots (1.5 ml) of these cells should be transferred to 2-ml cryovials. These are then placed in freezing containers at −70° overnight and the following day are placed at −135°.

15. Observe the cultured cells within 24 hr to check for viability and contamination. Feed the cultures within 24–48 hr of initial setup to help maintain viable cell populations. Make note of cell appearance and growth on appropriate log sheets.

16. Subculture and maintain as follows:

a. Cultures seeded as single cells in a standard medium that selects for attachment often have a large number of nonviable cells which will not be adherent even after 24 hr postinitiation. These can be removed by aspirating off the culture medium, then adding fresh replacement medium to the culture. However, it is noteworthy that cells in suspension are not necessarily dead. In addition to cultures derived from blood cells, many types of tumors and secretory epithelial cells (e.g., GI cells), in contrast to mesenchymal cells and keratinocytes, are viable and able to divide in suspension. Thus cultures must be carefully observed daily for increasing numbers of cells and viability (i.e., intact, glowing refractive cells by phase microscopy). If there are large numbers of viable cells in the supernatant, they should be transferred to a fresh flask and the monolayer cultures refed.

b. Seeding cells as multicellular aggregates will yield both viable and nonviable cells in suspension as well as cells attached (singly, but more often as colonies) onto the monolayer. Secretory epithelial origin cells and many epithelial tumors seeded in this fashion frequently grow as large clusters or multicellular aggregates in suspension. These may form gland or "organoid"-like structures with secretory properties such as lumen formation and/or be loosely associated, often within newly synthesized biomatrix materials, such as collagen, fibronectin, and proteoglycans. Alternatively, the aggregates may be more tightly associated. Serial subculture from the suspension may yield populations that preferentially grow in suspension, as monolayers or both, but will be highly enriched in epithelial cells.

c. The mesenchymal cells (primarily fibroblasts, smooth muscle, and endothelial cells) present in the original tissue generally have a greater tendency than the epithelial cells to attach and spread onto the plastic substrate *in vitro*. This provides one method for selection of the nonmesenchymal cell types. Depending on the culture, but usually 2–5 days postseeding, the culture supernatants that are enriched with cellular aggregates or organoids can be removed and centrifuged. The cell pellets can be resuspended and put into new flasks in medium [usually at a 1:2 or 1:4 (v/v) subculture ratio and at least 25% of the conditioned medium from the original culture]. Subculture can be done without dissociation or they can be gently dissociated by trituration or with a nontrypsin dissociation agent (e.g., collagenase-dispase or ECD Lite). After further culture adaptation, epithelial cell populations can be specifically selected by differential subculture using this type of "washout" method and the appropriate cell culture medium. The best general medium in our hands is designated M3:2 for the selective propagation of epithelial and many neural origin cells. RPMI 1640 or M3 supplemented with 10–20% fetal bovine serum (FBS) allow propagation of blood cell origin tumors (e.g., lymphomas).

E. Protocol for Preparation of Mesenchymal Origin Tumors

1. For mesenchymal source tumors, such as sarcomas and gliomas, cut the tumor into small pieces (1–2 mm^3) and rinse three to four times with TRM in petri dishes or centrifuge tubes, depending on the amount of tissue.
2. Dissociation
   a. Large sample: Place pieces in STERILE Spinner Flask with a 1:4 (v:v) dilution of TE:TRM (i.e., 25 ml TE + 100 ml TRM). Spin at 4° overnight at moderate speed (setting of 3–4 on Bellco stir plate). Shake vigorously in the morning to release more cells. Centrifuge the supernatant containing the cell suspension and rinse the pellet at least four additional times with TRM. If necessary, use RBC lysing buffer as described earlier for epithelial tumors. Mince remaining fragments to very small pieces and cryopreserve (as described earlier) *or* initiate as explant cultures. *Optional*: Do a second (third or fourth) treatment of the tissue fragments to release more cells. *Note*: Do not use an Erlenmeyer containing a stir bar on the bottom as that can physically damage the tissue and cells.
   b. Small sample: Place pieces in a 15- or 50-ml conical centrifuge tube in the cold overnight with a 1:4 (v:v) dilution of TE:TRM (e.g., 2 ml TE + 8 ml TRM). Shake vigorously in the morning, centrifuge, and rinse at least four additional times with TRM.
3. Explant cultures: Seed some pieces of tissue into flasks as explant cultures in M3:M5 or M3:10 media (e.g., 7–10 ml per 75-cm^2 flask). *Note*: M3:10 selects for mostly fibroblasts, split with TE to select. M3:M5 selects for more of a mixed population of epithelial and fibroblast cells. Subculture with collagenase, collagenase/dispase, or ECD Lite to maintain the mixed population with selection toward epithelial cell growth. *Optional*: As culture outgrowth occurs, serially transfer tissue fragments to new flasks for repeated outgrowth or remove and cryopreserve them. Refeed monolayers with fresh medium and 10–25% conditioned medium (concentration depends on cell density; use higher percentage for lower cell densities).
4. Subculture and maintain as follows: Observe on a fairly frequent basis until cells begin to grow. Continue to monitor until a confluent monolayer is observed. Then subculture using standard monolayer maintenance protocol. In general, mesenchymal cells of many types and species are readily propagated in M3:10 or M3:M5 media and are best maintained using standard trypsinization procedures with TE for subculture at split ratios that usually begin at 1:2 to 1:4 but may be greater on subculture. These cells usually grow better with complete replacement of the growth

medium, although 10–25% conditioned medium may help the primary or earliest passages of the culture get started. These cells are easily damaged by scraping methods for subculture and are mitotically stimulated with surface proteolysis. Thus, cell scraping is not a subculture method of choice unless there is interest in selectively removing a colony of cells from the monolayer without all of the other cells in the vessel.

Section II

Molecular Clones

[11] Superfamilies of Protooncogenes: Homology Cloning and Characterization of Related Members

By TADASHI YAMAMOTO and YASUNORI KADOWAKI

Introduction

Oncogenes were first identified in the retroviral genome as genes responsible for tumor induction. The first molecularly well-characterized oncogene was the v-*src* gene of the Rous sarcoma virus which produces sarcomas in susceptible chickens. Other retroviral oncogenes, for example, v-*yes* and v-*fgr*, were found in the genome of the Y73 and Esh sarcoma viruses and the Gardner-Rasheed sarcoma virus, respectively. The products of these genes are associated with protein-tyrosine kinase activity. Structural analysis of v-*src*, v-*yes*, and v-*fgr* revealed that they are similar to each other, having a highly conserved sequence for the kinase domain. Analysis of the genomic structure of the cellular counterparts to the viral genes showed that the nucleotide sequence and the exon–intron organization of the kinase domain are conserved, suggesting that they originate from a single ancestral gene. In addition, accumulating evidence shows that there are at least six more cellular genes[1] highly similar to the c-*src* gene, suggesting existence of a family of *src*-like protooncogenes.

Similarly, cellular homologs of other oncogenes, including *erbB, ras, raf, jun,* and *erbA*, are able to form families of protooncogenes. Each family member is sometimes identified as an oncogene of an independent RNA tumor virus, but in many cases, family members are found by means of cross-hybridization. This chapter describes examples for molecular cloning of human protooncogenes of the *src, erbB,* and *erbA* families.

Principle

The widely employed strategy of homology cloning is based on filter hybridization, namely screening of a genomic or cDNA library trapped on filters with a specific DNA probe at relaxed conditions of hybridization. The hybridization of denatured DNA probes (excess amount) to filter-bound DNA occurs through two different manners of annealing: either with perfectly matched sequences (specific hybridization) or with mismatched

[1] M. Sudol, *in* "The Molecular Basis of Human Cancer" (B. Neel and R. Kumar, eds.), p. 203. Futura, New York, 1993.

sequences (cross-hybridization). The temperature dependency of specific hybridization differs from that of cross-hybridization: one can choose a condition that is suitable for homology cloning.[2] The former detects the gene or cDNA identical to the probe and the latter detects the genes nonidentical but homologous to the probe DNA. A perfect match hybrid is formed under stringent hybridization conditions, whereas both specific and mismatch hybridizations occur under relaxed conditions. The mismatch hybrids formed under relaxed conditions can be removed by washing filters with a stringent washing buffer.

Critical parameters for efficient filter hybridization are the factors that affect T_m (melting temperature) and include ionic strength, base composition, and temperature. For DNA : DNA hybrid formation, T_m can be estimated using the following formula: $T_m = 81.5° + 16.6(\log[Na^+]) + 0.41$ $(\%G + C) - 0.63$ (%formamide) $-$ (600/l), where l is the length of the hybrid in base pairs.

Other factors such as viscosity, pH, proportion of mismatched base pairs, and type of filters are also important and are well described in previously published experimental protocols.[3,4] To allow for cross-hybridization of a probe with distant homologs, a temperature 40 to 50° below T_m of the specific hybridization should be employed, as the T_m of a DNA : DNA hybrid decreases by 0.5 to 1.5° with a 1% degree of mismatch. This can be achieved by decreasing the temperature of the reaction, increasing the salt concentration, or decreasing the formamide concentration. A onefold increase in the monovalent cation (in the range of 0.01 to 0.40 *M* NaCl) results in a 16.6° increase of T_m whereas each 1% formamide reduces T_m by about 0.6°. The rate of hybrid formation between a DNA probe with the capacity to reanneal (namely double-stranded DNA) and filter-trapped DNA follows essentially second-order kinetics. However, since the concentrations of the available DNA probe at the initial step of hybridization and that of the trapped DNA are difficult to control, empirical data are valuable. The length of time (*t*: in hours) to achieve 50% hybridization between a given probe and filter-trapped DNA can be estimated by adapting the following formula[4]:

$$t = 2(1/x)(y/5)(z/10) = yz/25x,$$

[2] G. A. Beltz, K. A. Jacobs, T. H. Eickbush, P. T. Cherbas, and F. C. Kafatos, this series, Vol. 100, p. 266.

[3] F. M. Ausubel, R. Brent, R. E. Kingston, D. D. Moore, J. G. Seidmon, J. A. Smith, and K. Struhl, eds., "Current Protocols in Molecular Biology." Wiley (Interscience), New York, 1987.

[4] J. Sambrook, E. F. Fritsch, and T. Maniatis, eds., "Molecular Cloning: A Laboratory Manual," 2nd ed. Cold Spring Harbor Lab., Cold Spring Harbor, NY, 1989.

where x is the weight of probes in micrograms; y, is the length of probe DNA in kilobase pairs; and z is the reaction volume in milliliters. Three times t would be almost enough to complete hybridization.

Materials, Reagents, and Fundamental Methods

Genome DNA

Genomic DNAs for Southern blot hybridization and library construction are prepared from human embryo fibroblasts (HE2144 and TIG-1), human placenta, A431 human vulva carcinoma cells, and mouse B82 cells following a standard method as described.[3]

Complementary DNA

Total RNAs are extracted by the guanidine isothiocyanate–cesium chloride method[5] from human embryo fibroblast TIG-1[6] and human placenta. TIG-1 cells are cultured in Dulbecco's modified Eagle's medium with 10% fetal calf serum. Polyadenylated, poly(A)$^+$ RNAs are selected by oligo(dT) chromatography and are used as templates for cDNA synthesis.[7] Poly(A)$^+$ RNA may be size fractionated to enrich certain mRNA species.

DNA Libraries

For the experiments described here as examples of the homology cloning method, the following genomic and cDNA libraries are prepared. A human genomic library is constructed from *Alu*I–*Hae*III partial digests of human placenta DNA and the Charon 4A *Eco*RI arm by a previously described method.[3] A cDNA library from mRNAs of TIG-1 human embryo fibroblasts is constructed in λgt10. Phages are mixed with 0.3 ml indicator *Escherichia coli* (overnight culture suspended in one-half volume of 10 m*M* $MgSO_4$) and 7 ml of medium (0.5% yeast extract, 0.5% NaCl, and 0.2% $MgCl_2$) in 0.7% agarose, and the mixture is spread over the bottom agar (LB plate: 1.0% Bacto-tryptone, 0.5% yeast extract, 0.5% NaCl and 1.3% Bacto-agar; NZY plate: 1.0% NZ-amine Type A, 0.5% yeast extract, 0.5% NaCl, 0.2% $MgSO_4$, and 1.3% Bacto-agar in a 150-mm petri dish). Although *E. coli* suspended in $MgSO_4$ can last for more than 2 weeks, a freshly

[5] J. M. Chirgwin, A. E. Pysybyla, R. J. McDonald, and W. J. Rutter, *Biochemistry* **18,** 5294 (1979).

[6] M. Ohashi, S. Aizawa, H. Ooka, T. Ohsawa, K. Kaji, H. Kondo, T. Kobayashi, T. Nomura, M. Matsuo, Y. Mitui, S. Murota, K. Yamamoto, H. Itoh, H. Shimada, and T. Utakoji, *Exp. Gerontol.* **15,** 121 (1980).

[7] V. Gubler and B. J. Hoffman, *Gene* **25,** 477 (1986).

prepared suspension should be used for the first screening of the library to achieve efficient plaque formation. NZY plates are supposed to be more suitable than LB plates for λ phages but we have not seen much difference. We recommend preparing the LB or NZY plate at least 1 day in advance. Phages grown on bacterial lawn are transferred to nitrocellulose filters (BA85; Schleicher & Schuell) and phage DNAs denatured with freshly prepared 0.2 *N* NaOH/1.5 *M* NaCl are trapped on the filter as described.[3,4] The nitrocellulose filters are then baked in a vacuum oven at 80° to immobilize DNA on the filters. Nylon filters (Hybond-N: Amersham) can be utilized instead of nitrocellulose filters. With nylon filters, DNA can be covalently linked to the filters by ultraviolet cross-linking. DNA probes utilized for screening (Fig. 1) are prepared from cloned DNAs of chicken RNA tumor viruses: v-*yes*-specific pYS-2 insert DNA,[8] v-*erbB*-specific DNA fragment from AEV-H DNA,[9] and v-*erbA*-specific sequence from AEV-R DNA.[10] All DNA probes are labeled with [α-^{32}P]dCTP (Amersham) by nick translation to a specific activity of about 2×10^8 cpm/μg of DNA. Using random oligonucleotides primers in conjunction with the DNA polymerase reaction (Megaprime DNA labeling kit; Amersham), the DNA probe may be labeled to a specific activity of about 10^9 cpm/μg. Approximately 10^6–5×10^6 cpm of the radioactive DNA probe is used during hybridization.

Reagents

Hybridization buffer (relaxed): 0.6–0.75 *M* NaCl, 60–75 m*M* sodium citrate, 50 m*M* HEPES [*N*-2-hydroxyethylpiperazine-*N*′-2-ethane sulfonate], pH 7.0, 10× Denhardt's solution (1× is 0.02% (w/v) polyvinylpyrrolidone, 0.02% (w/v) Ficoll, and 0.02% (w/v) bovine serum albumin), 30 to 35% (v/v) formamide, and denatured salmon sperm DNA. Denatured salmon sperm DNA may be dispensable.

Hybridization buffer (stringent): Same as for "relaxed" buffer except that the formamide concentration is 50% (v/v) instead of 30 to 35%.

Rinse buffer: 0.3 *M* NaCl, 30 m*M* sodium citrate, and 0.1% sodium dodecyl sulfate.

Washing buffer (relaxed): 66 to 90 m*M* NaCl, 6.6 to 9 m*M* sodium citrate, and 0.1% sodium dodecyl sulfate.

Washing buffer (stringent): 15 m*M* NaCl, 1.5 m*M* sodium citrate, and 0.1% sodium dodecyl sulfate.

[8] K. Semba, Y. Yamanashi, M. Nishizawa, J. Sukegawa, M. Yoshida, M. Sasaki, T. Yamamoto, and K. Toyoshima, *Science* **227,** 1038 (1985).

[9] T. Yamamoto, T. Nishida, N. Miyajima, S. Kawai, T. Ooi, and K. Toyoshima, *Cell (Cambridge, Mass.)* **35,** 71 (1983).

[10] N. Miyajima, Y. Kadowaki, S. Fukushige, S. Shimizu, K. Semba, Y. Yamanashi, K. Matsubara, K. Toyoshima, and T. Yamamoto, *Nucleic Acids Res.* **16,** 11057 (1988).

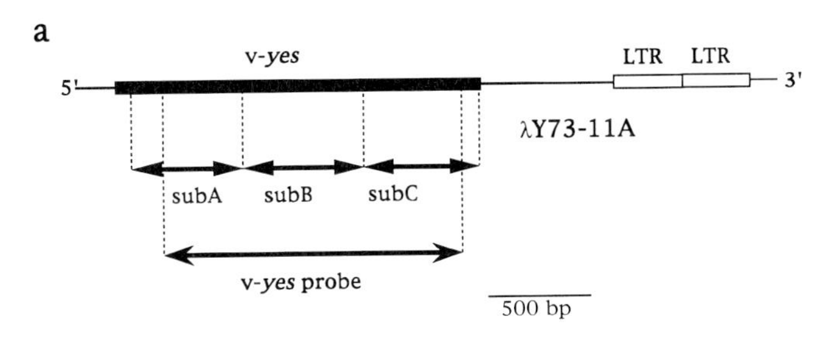

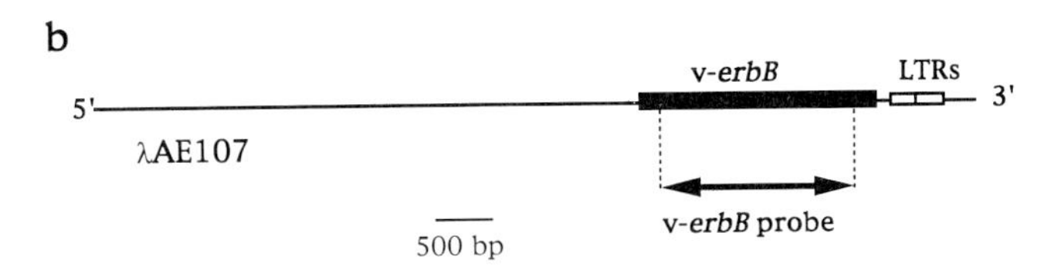

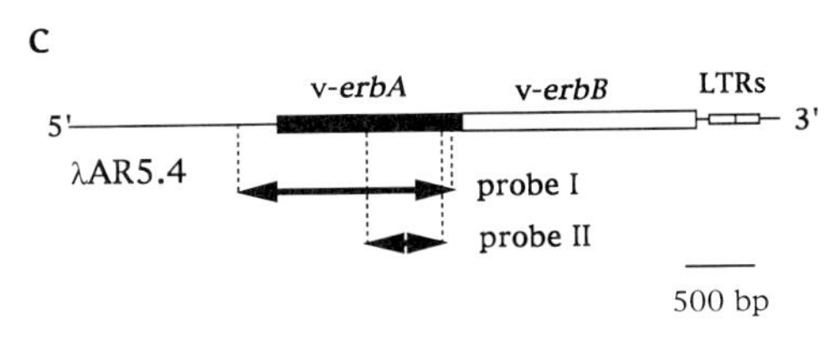

FIG. 1. Schematic illustration of the DNA fragment used for probes in hybridization analysis. (a) A *Sin*I fragment represents the v-*yes* probe. The *Sma*I–*Pst*, *Pst*–*Eco*RI, and *Eco*RI-*Pst* fragments were prepared as subdivided probes; subA, subB, and subC, respectively. (b) A *Sst*I–*Stu*I fragment represents the v-*erbB* probe. (c) Two *erbA* probes, probe I and probe II, were prepared: probe I was generated by *Ava*I–*Sst* digestion and probe II by *Pvu*II digestion.

Screening

Filters containing the genomic or cDNA libraries (about 5×10^5 plaques on 10 150-mm filters) are pretreated with hybridization buffer overnight and then incubate with the ^{32}P-labeled DNA probe in the new hybridization buffer (either relaxed or stringent) at 42° for 16 hr. After hybridizaton, the filters are rinsed with the rinse buffer three times at room temperature and then washed with the washing buffer (either relaxed or stringent) four times (15 min each) at 50°. The concentration of formamide used in each screening is indicated (see below). Ample amounts of washing buffer should be used; use at least 1 liter of the buffer for 10 filters (150-mm filter). In the examples of homology cloning shown below, the washing buffer contains 0.44 × SSC (66 m*M* NaCl and 6.6 m*M* sodium citrate), 0.5 × SSC (75 m*M* NaCl and 7.5 m*M* sodium citrate), or 0.6 × SSC (90 m*M* NaCl and 9 m*M* sodium citrate). The following are the specific conditions employed in each experiment: (1) for screening the genomic library with the v-*yes* probe use 30% formamide, 0.6 *M* NaCl for hybridization, and 66 m*M* NaCl for wash; (2) for screening the cDNA library with the v-*yes* probe use 35% formamide, 0.75 *M* NaCl for hybridization, and 0.75 *M* NaCl for wash; and (3) for screening the genomic library with the v-*erbB* probe and the cDNA library with the v-*erbA* probe use 30% formamide, 0.6 *M* NaCl for hybridization, and 90 m*M* NaCl for wash.

Nucleotide Sequencing

The cloned genomic DNAs are subjected to restriction mapping followed by Southern blot hybridization to identify restriction fragments that contain possible exon sequences. The DNA sequences of the putative exons or the cDNA inserts are determined essentially by the dideoxy chain-termination method[11] using pUC18 and pUC19 as cloning sequencing vectors. The plasmids, pUC118, pUC119, and pBluescript (Stratagene, La Jolla CA), may also be used.

Procedures: Examples of Homology Cloning

Identification of Multiple Human Homologs of v-yes and v-erbB

High molecular weight genomic DNAs (10 μg/lane) are digested with *Eco*RI and the digests are fractionated by electrophoresis on agarose gels (0.9 – 1%). The fragments are transferred onto nitrocellulose filters and hybridized with ^{32}P-labeled v-*yes*[8] (Fig. 2a) and v-*erbB*[9] (Fig. 2b) probes. Hybridization and filter washing are carried out in the relaxed condition. The presence of multiple hybridizing fragments with the v-*yes* probe suggests that either the mammalian c-*yes* gene has a large complexity, having small exons separated by long introns, or the presence of cellular genes similar to, but distinct from, the c-*yes* gene. Of the 10 *Eco*RI fragments (25.0-, 12.0-, 11.0-, 9.4-, 8.7-, 8.2-, 7.0-, 5.7-, 5.0-, and 1.9-kbp) of human origin, weakly hybridizing fragments likely represent cellular genes other than c-*yes*.

Figure 2b shows that the v-*erbB* probe hybridized with at least 12 *Eco*RI fragments including 13-, 8.1-, 6.4-, 5.9-, 5.4-, 4.3-, 3.7-, and 3.5-kbp DNAs

[11] F. Sanger, S. Nicklen, and A. R. Coulson, *Proc. Natl. Acad. Sci. U.S.A.* **74,** 5463 (1977).

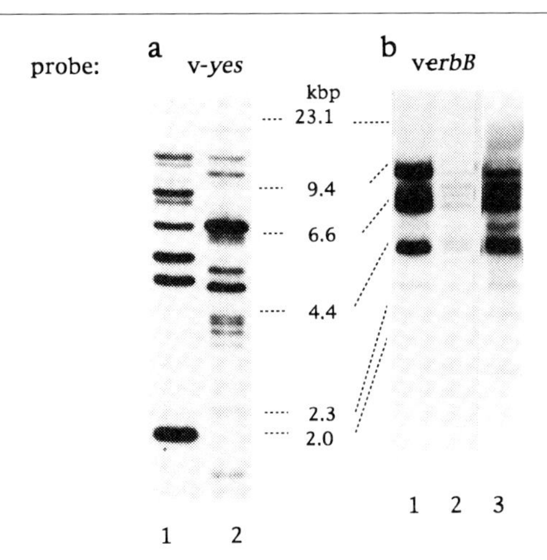

FIG. 2. Southern blot analysis of genomic DNA with v-*yes* and v-*erbB* probes. (a) Hybridization with the ^{32}P-labeled v-*yes* probe. *Eco*RI-digested human (lane 1) and mouse (lane 2) cellular DNAs (15 μg/lane) were fractionated by electrophoresis on a 0.9% agarose gel and then subjected to Southern blot hybridization using (b) hybridization with the ^{32}P-labeled v-*erbB* probe. DNAs prepared from A431 cells (lane 1) and placenta (lanes 2 and 3) were digested with *Eco*RI. The digests (10 μg DNA/lane) were fractionated by electrophoresis on a 1.0% agarose gel and were then subjected to Southern blot hybridization. Lane 3 is the autoradiogram obtained by longer exposure of lane 2. In both (a) and (b), hybridization was carried out under relaxed condition.

of human origin. Of these, four fragments (8.1-, 5.9-, 5.4-, and 3.5-kbp) gave strong signals with DNAs from A431 cells in which the EGF receptor gene is amplified at least 20-fold compared with placenta.[12] This observation is consistent with previous reports that these amplified sequences are from the EGF receptor gene. Several other restriction fragments may represent another v-*erbB*-related gene(s). The 13- and 6.4-kbp *Eco*RI fragments are indeed from the c-*erbB*-2 gene.

Cloning of v-yes Homologs

From the Genomic Library

The human genomic library described earlier was screened with the 1.5-kbp v-*yes* DNA under relaxed condition (30% formamide) and 26 independent genomic clones were isolated.[13] Dot-blot hybridization of the

[12] G. T. Merlino, Y.-H. Xu, S. Ishii, A. J. L. Clarke, K. Semba, K. Toyoshima, T. Yamamoto, and I. Pastan, *Science* **224,** 417 (1984).

[13] M. Nishizawa, K. Semba, M. Yoshida, T. Yamamoto, M. Sasaki, and K. Toyoshima, *Mol. Cell. Biol.* **6,** 511 (1986).

cloned DNA with the subdivided v-*yes* probes (subA, subB, and subC, Fig. 1a) shows that some clones (clones 003, 008, 015, 019, 020, 021, and 022) have sequences that correspond to the junction of subB and subC. However, restriction maps of the DNA inserts of these clones are different from each other, suggesting that they represent distinct genes. Further analysis of clones 003, 008, and 015 reveals that they correspond to the c-*src*, c-*yes*, and c-*fgr* genes, respectively. Other clones such as clones 029, 016, and 014 do not represent known cellular genes.

c-src Clone. Figure 3a shows a restriction map of the DNA insert of λ003. A Southern blot analysis shows that the 320-bp *Bam*HI–*Xba*I fragment corresponds most 3′ sequence of the subB probe. Nucleotide sequencing of the fragment identified a sequence that encodes 59 amino acids identical to the sequence encoded by exon 8 of human c-*src*.[14] Restriction mapping data show that clones 017, 019, 020, 021, and 022 are also derived from the c-*src* gene.

c-yes Clone. Figure 3b is a restriction map of the λ008 insert. That all the *Eco*RI fragments (4.2-, 0.6-, 6.4-, 3.7-kbp) hybridize to the v-*yes* probe suggests that each fragment contains an exon(s). A nucleotide sequence analysis of the 0.6-kbp *Eco*RI fragment reveals a sequence coded for 25 amino acids homologous to the corresponding sequence of the v-*yes*[15] (identity, 23/25), c-*src* (21/25), and v-*fgr*[13](18/25) proteins. This sequence corresponds to exon 9 of c-*src*.[16] Further nucleotide sequencing of the v-*yes*-hybridizing regions within the λ008 insert results in identification of sequences corresponding to four other exons (exons 7, 8, 10, and 11). In total, an amino acid sequence of 232 residues was encoded by the λ008 insert, highly homologous to the v-*yes* protein[15] (97% identity). Comparing the restriction map of each clone reveals that clones 018, 026, and 028 overlap with clone 008. The restriction map of λ013 does not overlap these clones but does contain the c-*yes* sequence corresponding to exon 4 of c-*src*.

c-fgr Clone. Restriction maps of the inserts of λ015 and λ010 show that these clones have an overlap of about 1.0-kbp (Fig. 3c). A Southern blot analysis of the cloned DNAs shows that the two *Eco*RI fragments (7.7- and 2.9-kbp) contain the v-*yes*-related sequence. The nucleotide sequence of the 7.7-kbp *Eco*RI fragment predicted the presence of seven exons that correspond to exons 4 to 10 of c-*src*. The total length of the exon sequences is 920-bp, coding for 306 amino acid residues. The exon sequence has a

[14] S. K. Anderson, C. P. Gibbs, R. A. Tanaka, H.-J. Kung, and D. J. Fujita, *Mol. Cell. Biol.* **5,** 1122 (1985).

[15] N. Kitamura, A. Kitamura, K. Toyoshima, Y. Hirayama, and M. Yoshida, *Nature* (*London*) **297,** 205 (1982).

[16] T. Takeya and H. Hanafusa, *Cell* (*Cambridge, Mass.*) **32,** 881 (1983).

higher homology to the v-*fgr* gene than to the v-*yes* and human c-*src* genes (Table I).

fyn Clone. The restriction maps of DNA inserts of clones 029, 016, and 014 overlap partly and are different from those of the genomic clones representing the c-*src*, c-*yes*, c-*yes*-2 (a pseudogene[17] of c-*yes*), and c-*fgr* genes (Fig. 3d). Therefore, these three clones may represent a novel v-*yes*-related gene, which is termed the *fyn* gene.[18] By analyzing the nucleotide sequence of the cloned insert that hybridizes with the v-*yes* probe, one putative exon flanked by the splicing consensus sequence has been identified. The 180-bp sequence of the exon is homologous with the corresponding sequence of the v-*yes* gene (79%) and corresponds to exon 4 of c-*src*.

From a cDNA Library: The lyn Clone

A cDNA library was made from poly(A)$^+$ RNA prepared from human placenta and hybridized with the v-*yes* probe under relaxed conditions. By screening 3×10^4 independent clones of this library, a unique cDNA clone has been obtained in addition to a c-*yes* cDNA clone. The number screened is small because the cDNA library was constructed using size-fractionated mRNA. Nucleotide sequence of the DNA insert of the former clone revealed that this clone contains a 2741-bp sequence that encodes a polypeptide related to the v-*yes* protein. A part of the sequence corresponds to that of exon 4 of c-*src*. The predicted amino acid sequence is highly homologous to that of *lck*[19] (75% in the kinase domain) and v-*yes* (69% in the kinase domain). The gene for this clone is thus termed *lyn (lck/yes*-related novel tyrosine kinase).[20]

Cloning of v-erbB Homolog

By screening the human genomic library constructed from placental DNA, six independent clones have been isolated.[21] A restriction map analysis showed that all the cloned inserts except the λ107 represent one gene, the epidermal growth factor (EGF) receptor gene (=c-*erbB-1*). As expected, four exon-containing *Eco*RI fragments in these clones are amplified

[17] K. Semba, M. Nishizawa, H. Sato, S. Fukushige, M. C. Yoshida, M. Sasaki, K. Matsubara, T. Yamamoto, and K. Toyoshima, *Jpn. J. Cancer Res.* **79,** 710 (1988).

[18] K. Semba, M. Nishizawa, N. Miyajima, M. C. Yoshida, J. Sukegawa, Y. Yamanashi, M. Sasaki, T. Yamamoto, and K. Toyoshima, *Proc. Natl. Acad. Sci. U.S.A.* **83,** 5459 (1986).

[19] J. D. Marth, R. Peet, E. G. Krebs, and R. M. Perlmutter, *Cell (Cambridge, Mass.)* **43,** 393 (1985).

[20] Y. Yamanashi, S. Fukushige, K. Semba, J. Sukegawa, N. Miyajima, K. Matsubara, T. Yamamoto, and K. Toyoshima, *Mol. Cell. Biol.* **7,** 237 (1987).

[21] K. Semba, N. Kamata, K. Toyoshima, and T. Yamamoto, *Proc. Natl. Acad. Sci. U.S.A.* **82,** 6497 (1985).

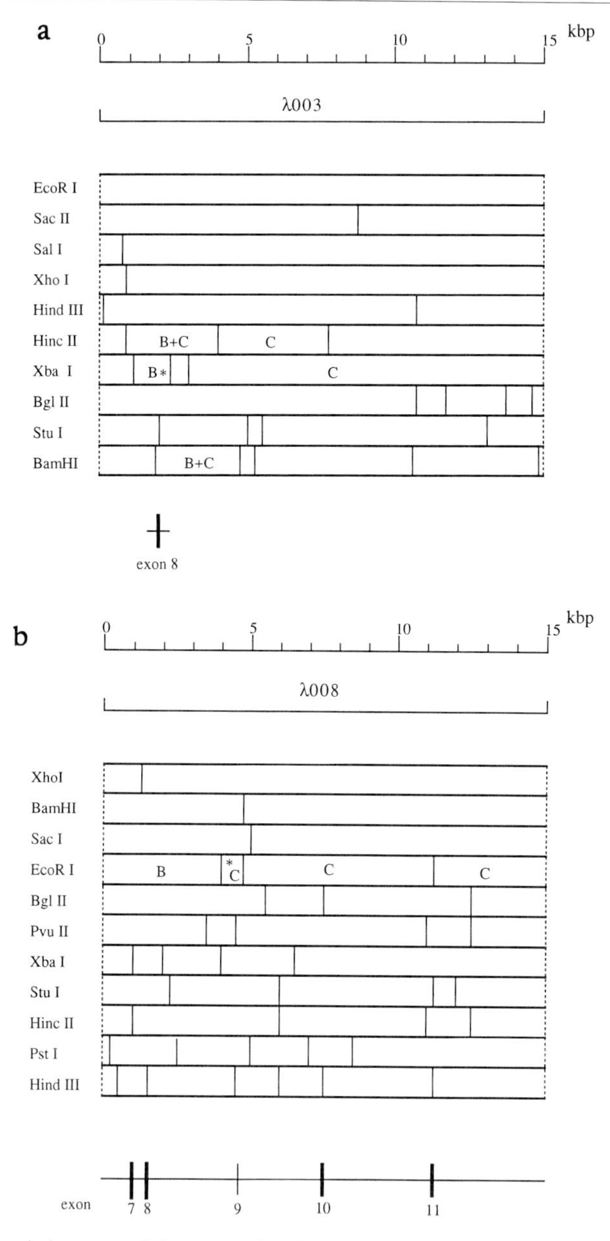

FIG. 3. Restriction map of the v-*yes* related genes. Restriction maps of the λ003 (a), λ008 (b), λ015 and λ010 (c), and λ029, λ016, and λ014 (d) clones that represent the human c-*src*, c-*yes*, c-*fgr*, and *fyn* genes, respectively. Restriction mapping was done by analysis of double digestion products. A, B, and C represent fragments that hybridized with subA, subB,

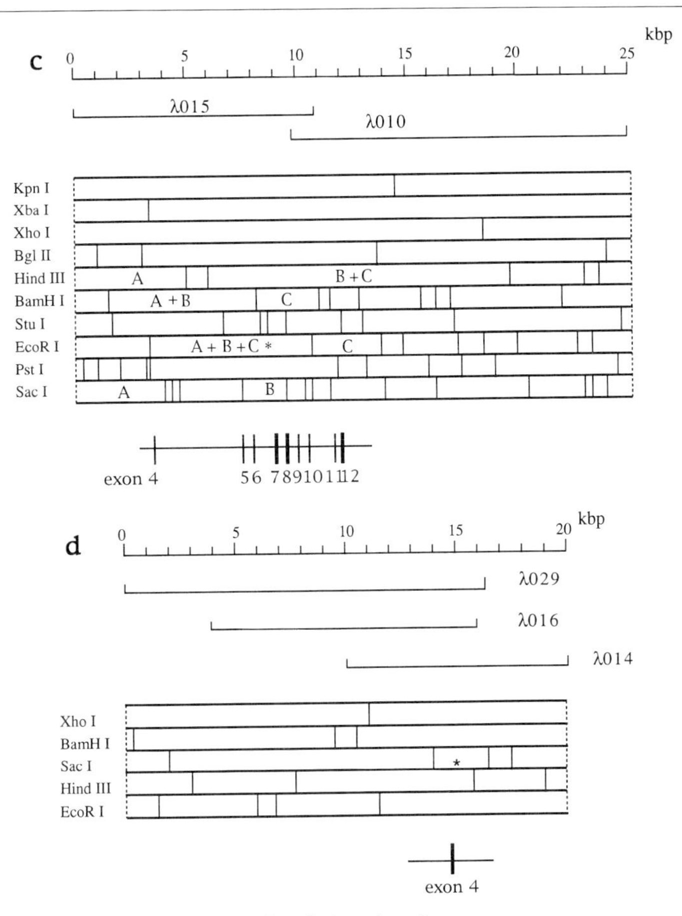

FIG. 3. (*continued*)

and subC probes, respectively. The nucleotide sequences of the fragments (marked with an asterisk) that carry sequences corresponding to or near the junction of subB and subC DNAs were first determined (a, b, and c). The nucleotide sequence of the *Sac*I fragments (marked with an asterisk) that hybridized with the v-*yes* probe was determined (d).

in A431 cells. Restriction mapping in conjunction with Southern hybridization of the λ107 DNA insert shows that the 440-bp *Kpn*I–*Xho*I fragment (*KX* DNA) contains a v-*erbB*-hybridizing sequence. The *KX* DNA reacts with the 13- and 6.4-kbp *Eco*RI fragments that are not amplified in A431 cells in Southern blot hybridization. These data suggest that the λ107 insert does not represent the EGF receptor gene but instead represents another v-*erbB*-related gene that we have termed c-*erbB*-2. Determination of the nucleotide sequence of the λ107 insert revealed seven putative exons

TABLE I
NUCLEOTIDE AND AMINO ACID SEQUENCE HOMOLOGY[a]

Oncogene	% Homology	
	Nucleotide	Amino acid
λ015/λ010 (c-*fgr*)	100	100
v-*fgr*	90.0	90.7
v-*src*	68.7	77.5
v-*yes*	71.9	73.0
c-*src*	72.9	73.1

[a] Nucleotide and amino acid sequence homologies between c-*fgr* and v-*fgr*, v-*src*, v-*yes*, and human c-*src* were determined. The 52-bp (17 amino acids) 5′ terminal sequences were excluded from the comparison of c-*fgr* and v-*fgr* since this portion of v-*fgr* codes for the actin-like sequence. The homology between c-*fgr* and c-*src* was calculated at the exons 6 through 10.

flanked by a consensus sequence of splicing junctions. The nucleotide sequences of all seven exons are highly homologous with the corresponding regions of the cDNA clone for the EGF receptor[22] (74%). The amino acid sequence deduced from an open reading frame found in the sequence of the c-*erbB-2* exons is highly homologous to that of the EGF receptor kinase domain (82%) and is distantly related to those of the Src-family kinases (i.e., 43% homology with v-Src). The c-*erbB*-2 gene has been shown to be conserved in vertebrates, corresponding to the rat *neu* gene.[23]

Cloning of erbA Homologs

RNA blot hybridization analysis with the ^{32}P-labeled v-*erbA* probe (probe I, Figs. 1 and 4C), which is prepared from the cloned AEV-R genomic DNA at low stringency, reveals at least two species of c-*erbA* mRNA (2.7- and 5.0-kb) in human embryo fibroblasts.[10] In order to obtain the human homolog of v-*erbA*, a cDNA library was constructed in the λgt10 vector using mRNAs from human embryo fibroblasts. On screening 3.5×10^5 independent clones of this library with probe I under conditions of low stringency, 21 positive clones are obtained. A Southern blot hybridization of DNA inserts of these clones shows that all inserts hybridize with

[22] A. Ullrich, L. Coussens, J. S. Hayflick, T. J. Dull, A. Gray, A. W. Tam, J. Lee, Y. Yaden, T. A. Libermann, J. Schlessinger, J. Downward, E. L. V. Mayes, N. Whittle, M. D. Waterfield, and P. H. Seeburg, *Nature (London)* **309,** 418 (1984).
[23] C. I. Bargmann, C.-C. Hung, and R. A. Weinberg, *Cell (Cambridge, Mass.)* **45,** 649 (1986).

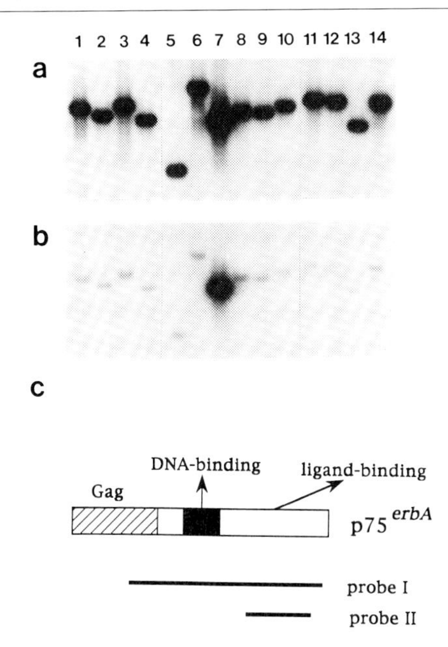

FIG. 4. Southern blot analysis of cDNA inserts of v-*erbA*-hybridizing clones. Fourteen out of 21 v-*erbA* positive clones were analyzed. Complementary DNA inserts (*Eco*RI digests) of recombinant phages in the λgt10 vector were subjected to Southern blot hybridization using probe I (a) or probe II (b) specific for v-*erbA* under relaxed condition. (c) The position of the DNA probes was oriented to the corresponding protein sequence. Probe II does not cover the sequence for the DNA-binding domain that is conserved among the family member.

probe I to a similar extent, except the 2.5-kbp insert of λA7 which hybridizes slightly stronger than the inserts of the other clones (Fig. 4a). Using probe II, which corresponds to the carboxy-terminal portion of the v-*erbA* protein (Fig. 4c), the DNA insert of λA7 hybridizes strongly whereas those of the other clones hybridize weakly (Fig. 4b). Restriction mapping and mutual hybridization of the DNA inserts reveal that the 21 clones represent at least four novel *erbA*-related genes termed *ear-1, ear-2, ear-3*, and *ear-7* (*ear: erbA*-related).[10] The above hybridization data suggest that the *ear-7* gene represented by the λA7 insert is most closely related to the v-*erbA* gene. Partial nucleotide sequencing of the cDNA inserts reveals that four *erbA*-related genes encode proteins that contain sequences of 66 to 68 amino acid residues characteristic of the DNA binding domain of the steroid/thyroid hormone receptor gene. Of the 21 clones, 4 are of *ear-2*, 7 clones are of *ear-3*, 1 clone is of *ear-1,* and 1 clone is of *ear-7*. The homology of the products of these genes to the v-*erbA* gene product is 57, 62, 60, and 94%, respectively, within the sequence of the DNA binding domain. The nucleotide sequence homology at the respective corresponding sequences is 70, 71, 70, and 86%. Homology of the remaining portion is much lower. The high homology between chicken c-*erbA* and *ear-7* suggests that the *ear-7* gene is the human counterpart of the chicken c-*erbA* gene. This has been proven to be correct.

The amino acid homologies in the DNA binding domain between the v-*erbA* gene product and other members of the steroid/thyroid hormone receptor, such as the estrogen receptor, glucocorticoid receptor, and retinoic acid receptor α, are 53, 46, and 58%, respectively. The nucleotide sequence homology at the respective corresponding regions is 53, 46, and 67%. Thus, it is possible that some of the other clones that have not been well characterized represent retinoic acid receptor α. However, a lower stringency may be required to obtain DNA clones for the estrogen receptor or the glucocorticoid receptor, or those clones may represent the other novel *erbA*-related genes.

Comments

Several examples of cross-hybridization experiments that were carried out in our laboratory and have allowed cloning of new members of the *src*, *erbB*, and *erbA* families have been described. Using the same strategy with slight differences in hybridization and washing conditions, others have also cloned these genes. For example, the c-*erbB-2* gene has been cloned by two other groups; their conditions (conditions A and B) of homology cloning with the ^{32}P-labeled v-*erbB* probe are[24,25]:

Condition A: Hybridization is conducted in a solution of 40% formamide, 0.75 *M* NaCl, 75 m*M* sodium citrate, 5 × Denhardt's solution, 20 m*M* sodium phosphate (pH 6.5), 100 μg/ml sonicated and denatured salmon sperm DNA, and 10% (w/v) dextran sulfate 500 at 42° in a polyethylene bag. Filters are washed first in 0.3 *M* NaCl and 30 m*M* sodium citrate at room temperature and then in 15 m*M* NaCl, 1.5 m*M* sodium citrate, and 0.1% sodium dodecyl phosphate at 42°.

Condition B: Hybridization is performed in 20% formamide, 0.75 *M* NaCl, 75 m*M* sodium citrate, 50 μg sonicated salmon sperm DNA, 50 m*M* sodium phosphate buffer (pH 6.8), 1 m*M* sodium pyrophosphate, and 100 m*M* ATP. After hybridization, filters are washed

[24] C. R. King, M. H. Kraus, and S. A. Aaronson, *Science* **229,** 974 (1985).

[25] L. Coussens, T. L. Yang-Feng, Y.-C. Liao, E. Chen, A. Gray, J. McGrath, P. H. Seeburg, T. A. Libermann, J. Schlessinger, U. Francke, A. Levinson, and A. Ullrich, *Science* **230,** 1132 (1985).

four times for 20 min in a solution of 30 mM NaCl, 3 mM sodium citrate, and 0.1% sodium dodecyl sulfate at 40°.

As exemplified by the above empirical data, the stringency of library screening can be regulated by three parameters: formamide concentration, salt concentration, and temperature. Another relaxed condition[26] that has been used for the cloning of a *src*-family member, *hck*, is to apply 0.45 M NaCl, 45 mM sodium citrate, and 40% formamide, at 41°. To elevate T_m during hybridization, every group utilizes a salt concentration that is higher than 0.4 M NaCl (above the range where a onefold increase in monovalent cation results in 16.6° increase of T_m). Empirically, we feel it is easy to regulate the stringency of hybridization by altering the formamide concentration and fixing the NaCl concentration at around 0.6 M. It is possible to alter our condition for homology cloning by lowering the formamide concentration if a distantly related family member is to be obtained. Finally, our experience leads us to conclude that dextran sulfate, ATP, and denatured salmon sperm DNA are not critically important and seem to be dispensable.

The homology cloning with low-stringency screening described here depends on the conservation of both the amino acid and nucleotide sequences among the family members. This strategy is still applicable.[27] However, other methods can be used to clone multigene family members. The use of degenerate primers designed from the conserved amino acid sequences in polymerase chain reaction amplification was found to provide another convenient way to clone related cDNAs. For example, application of this technique allows identification of protein-tyrosine kinase genes that are expressed specifically in the central nervous system.[28] In the case of the protein-tyrosine kinase gene, immunoscreening of an expression cDNA library with an antiphosphotyrosine antibody has also been adopted.[29]

Acknowledgments

We thank K. Semba, Y. Yamanashi, N. Miyajima, M. Nishizawa, and K. Toyoshima for contributions to the study described in this chapter. We also thank M. Sudol for critical reading of the manuscript.

[26] N. Quintrell, R. Lebo, H. Varmus, J. M. Bishop, M. J. Pettenati, M. M. Le Beau, O. Diaz, and J. D. Rowley, *Mol. Cell. Biol.* **7,** 2267 (1987).

[27] M. Sudol, H. Greulich, L. Newman, A. Sarkar, J. Sukegawa, and T. Yamamoto, *Oncogene* **8,** 823 (1993).

[28] C. Lai and G. Lemke, *Neuron* **6,** 691 (1991).

[29] F. G. Sajjadi, E. B. Pasquale, and S. Subramani, *New Biol.* **3,** 769 (1991).

[12] Polymerase Chain Reaction Cloning of Related Genes

By MARY ANN SELLS and JONATHAN CHERNOFF

When a cDNA clone becomes available for a protein of interest, it is often desirable to determine whether related genes exist. In general, there are two methods for finding such homologous genes. First, functional similarities may be exploited by complementation analysis. This method, although powerful, requires an experimental system in which the protein under study can be rendered functionally defective or absent and for which a selectable phenotype is available. If at least two members of a gene family are known, a second method, exploiting conserved structural elements of the encoded proteins, can be used to identify additional members within a species or related genes from other species. In the past, such analyses were most often undertaken either by screening expression libraries with antibodies directed against conserved epitopes or by low stringency hybridization of cDNA libraries, using DNA probes derived from conserved regions of the proteins under study. More recently a variation of this latter method, utilizing the speed and specificity of the polymerase chain reaction (PCR), has gained favor for the cloning of related genes within a single organism or between species.

PCR cloning, in which primer sets derived from conserved amino acids are used to amplify gene fragments from a cDNA pool, has several key advantages over more traditional cloning methods. The most important of these are greater speed and specificity. Under ideal circumstances PCR amplification and product verification can be done in a few days whereas traditional screening procedures can take weeks. More importantly, the PCR-based procedure is much more specific. This is due to the inherent nature of the PCR process, which requires specific priming from both the sense and antisense oligonucleotides for the formation of a product, which in turn is exponentially amplified during the remaining PCR cycles. Mispriming events, which can plague standard oligonucleotide hybridization schemes, are less often a problem in PCR, as they are unlikely to propagate during the amplification process. The catalog of genes which have been identified in this manner is very large and amply demonstrates the power and utility of this cloning procedure.

Principles Underlying PCR Cloning of Related Genes

General Considerations

In designing a PCR protocol, several factors must be considered. Of primary importance are the degree of conservation of the protein domains on which the oligonucleotide primers will be based and the degeneracy of these primers. The length of the expected PCR product can also be an important factor, as products longer than about 1000 bp are much more difficult to obtain than are shorter fragments. Of somewhat lesser concern is the source of cDNA for amplification, the choice of PCR buffer and polymerase, and the parameters of the PCR cycle.

Assuming that the protein family under consideration has more than one conserved stretch of amino acid sequence, one needs to consider the fitness of these sequences to serve as the basis for PCR primers. Although there are computer programs designed to make this evaluation, the calculations are usually simple enough to be done by hand. The optimum primers would be based on an invariant stretch six to eight amino acids, or low (i.e., <1024-fold) degeneracy, and would have about 50% GC (guanine : cytosine) content. Rarely can such primers be designed. In practice, oligonucleotides with degeneracies greater than 10^6 have been successfully used to isolate gene fragments,[1] as have primers whose GC content strays far from the ideal. Therefore, although the investigator has to balance the desire to make the largest, most specific oligonucleotide possible with the resultant inevitable increase in primer degeneracy, there are no infallible rules to guide the final selection. In a later section, we will demonstrate the use of these principles in choosing primers for the amplification of protein-tyrosine-phosphatase (EC 3.1.3.48) gene fragments from the fission yeast.

In examining possible sets of PCR primers, it is useful to try to ensure that the resulting DNA product will include one or more internal regions of conserved sequence. In the event of a low (i.e., subnanogram) yield, such sequences can be used as a probe to determine whether a product has been obtained, and, if so, can also serve as nested primers for reamplification of the product of the primary PCR reaction. In addition, such sequences can be used to verify the authenticity of the amplified DNA since it is not uncommon for the desired products and artifactual bands to have similar sizes.

Finally, if possible, more than one set of PCR primers should be designed, based on different conserved amino acid stretches of the protein under study. Such redundancy is particularly important when trying to PCR

[1] G. A. Gonzalez, K. K. Yamamoto, W. H. Fischer, D. Karr, P. Menzel, W. Biggs, 3rd, W. W. Vale, and M. R. Montminy, *Nature* (*London*) **337,** 749 (1989).

clone-related genes from an evolutionarily distant organism, where the conservation of amino acid sequence may be weak. Often, the best designed primer set from the point of view of complexity, melting temperature, etc., fails to yield a desired PCR product, while a nominally inferior primer set is successful, simply because the principle underlying the entire process—conservation of amino acid sequence—holds true for only one of the pairs of primers.

Choice of Polymerase

Although *Taq* polymerase has been used most often for PCR procedures, newer heat-stable polymerases offer certain advantages. Proofreading polymerases such as Vent (New England Biolabs) and *Pfu* (Stratagene) have been reported to have higher fidelity than *Taq* polymerase.[2,3] On the minus side, both of these high-fidelity enzymes result in a product with flush ends, which may not be as convenient for subsequent subcloning procedures (see Cloning into T Vector, below).

Sources of cDNA Template

If a cDNA library from an appropriate tissue source is available, purified library DNA can serve as a convenient template for PCR. If such material is not available, it is also feasible to use RNA as template. In this case, the PCR is carried out following a reverse transcription step. Several protocols have been published in which the entire process can be accomplished in a single reaction tube.[4] Total RNA is usually adequate for this purpose, but mRNA isolation may be required for rare transcripts. The amount of template has varied widely in published reports, but 1 μg of cDNA or total RNA per reaction is usually more than sufficient.

PCR Parameters

PCR amplification is a resilient procedure, tolerating many variations from ideal reaction conditions. In particular, a degree of mispriming will be tolerated, as long as the 3′-most two to three bases do not contain a

[2] P. Mattila, J. Korpela, T. Tenkanen, and K. Pitkanen, *Nucleic Acids Res.* **19,** 4967 (1991).

[3] K. S. Lundberg, D. D. Shoemaker, M. W. W. Adams, J. M. Short, J. A. Sorge, and E. J. Mathur, *Gene* **108,** 1 (1991).

[4] E. S. Kawasaki, *in* "PCR Protocols: A Guide to Methods and Applications" (M. A. Innis, D. H. Gelfand, J. J. Sninsky, and T. J. White, eds.), p. 21. Academic Press, San Diego, CA, 1990.

mismatch.[5,6] The most important variables are the annealing temperature and the length of the elongation step. When trying to amplify new genes with degenerate oligonucleotides, the exact annealing temperature is not known, so a trial of several different temperatures may need to be done. We usually choose 40° as the initial annealing temperature, and increase in 2–3° increments if nonspecific products are obtained in the pilot experiment. It should also be noted that some investigators have reported that a slow ramp rate between the annealing and elongation temperatures may be critical during early PCR cycles when degenerate oligonucleotides are used as primers.[7] Although not well understood, this effect presumably stems from the weak interaction between the template and imperfectly matched primers. While we keep this phenomenon in mind when designing PCR protocols, in practice most of our amplifications are performed in capillary tubes using an air-driven PCR machine (1605 Air Thermo Cycler, Idaho Technology Inc.), which has an extremely fast ramp rate (~4°/sec).

When using degenerate primers for PCR, it is tempting to increase the primer concentration to compensate for the degeneracy of the oligonucleotides, but this usually causes increased nonspecific priming. Our standard reaction contains 1 μM of each primer. Although in most cases only a minority of the total oligonucleotide molecules are sufficiently homologous to bind to the template, enough mispriming is tolerated that significant amounts of product can nonetheless be obtained.

The length of time allotted for elongation is determined by the expected product length, using 1000 nucleotides per minute as a conservative guide. Varying $MgCl_2$ levels from the usual 1.5–2.5 mM range is occasionally necessary, and should be tried if no specific product is visible. Additives such as 10% formamide or dimethyl sulfoxide can aid amplification, particularly if the GC content of the desired product is high. Thirty cycles of amplification are standard. Increasing the number of PCR cycles beyond 35 is rarely necessary and usually results in the generation of artifacts.

Finally, when performing a PCR with degenerate primers, one should, if possible, include a template which serves as a positive control. This control template usually consists of 10–50 ng of an existing cDNA clone encoding a member of the protein family under study, in which the amino acid regions from which the primers were designed are conserved. Such a

[5] S. Kwok, D. E. Kellogg, D. Spasic, L. Goda, C. Levenson, and J. J. Sninsky, *Nucleic Acids Res.* **18,** 999 (1990).

[6] R. Sommer and D. Tautz, *Nucleic Acids Res.* **17,** 6749 (1989).

[7] T. Compton, *in* "PCR Protocols: A Guide to Methods and Applications" (M. A. Innis, D. H. Gelfand, J. J. Sninsky, and T. J. White, eds.), p. 39. Academic Press, San Diego, CA, 1990.

control can be very useful in evaluating the utility of degenerate primers and in optimizing PCR parameters.

Analysis of Results

Following the PCR reaction, the products are most often analyzed on a high percentage, low melting point agarose gel or, for very small DNA fragments, on a polyacrylamide gel. Ethidium bromide staining can be used to identify the products of the reaction. If no specific ethidium bromide-stained band is seen, the gel may be blotted and subjected to Southern analysis using a radiolabeled oligonucleotide derived from an internal sequence expected to be conserved in the PCR product. If a specific product is detected, it can usually be reamplified by excising the relevant region from the agarose gel, melting the slice, and using 1 μl of the molten material as a template from reamplification. The reamplification may be carried out using either the original primer set or internally nested primers.

Although it is possible to directly sequence PCR products without cloning, this procedure has certain disadvantages, as it will yield the sequence of the predominant species present in the PCR product, and obscure other less abundant, potentially interesting molecules which may be present. We therefore opt to first clone the PCR product into a standard vector, then sequence the ends of several (~20) of the resulting clones. As all of the current generation of heat-stable DNA polymerases have a significant error rate, the sequence of any one clone cannot be considered to be completely accurate, and must be authenticated by comparison to the actual cDNA clone, when this is eventually isolated.

Troubleshooting

If the expected DNA product is not obtained following the PCR reaction, the failure can usually be attributed to one of a few common causes. First and foremost, it is possible that there is no template to amplify because there are no transcripts in the tissue or organism under study that encode proteins with domains similar to the probe. Second, such transcripts may exist, and are potentially amplifiable, but the oligonucleotides chosen for the PCR are too degenerate. Finally, the product may in fact have been obtained, but in amounts too small to be easily detected. Although no amount of troubleshooting can amplify a cDNA that does not exist, the choice of amino acid regions on which the primers are based, primer degeneracy, and product detection are subject to manipulation by the investigator. These factors should be addressed if no appropriate product is obtained in the initial PCR reactions.

As highly degenerate oligonucleotides will decrease the specificity of PCR amplification, it is desirable to decrease the complexity of the primers. This can be done in several ways: (a) a different amino acid region can be chosen as the basis of the PCR primer; (b) the length of the oligonucleotide can be decreased; (c) codon preference tables can be used to eliminate rarely used codons or to construct nondegenerate "guess-mers"; (d) inosine can be used at "wobble" positions; or (e) the primers can be specially constructed using linked di- or trinucleotide synthesis.[8] We favor either of the first two options. Rigorous studies of primer length and degeneracy versus the specificity of PCR amplification have not been carried out, but some reports indicate that long (20–30 nucleotides), highly degenerate (10^5-fold) primers are less efficient than short (14–20 nucleotides), moderately degenerate (8- to 128-fold) primers.[9]

If minute, non-ethidium bromide-stainable quantities of the desired PCR product are obtained, one may reamplify these products as described earlier. If reamplification is required to generate sufficient DNA for cloning and/or sequencing, the sequence of the final product should be viewed with caution, as the total number of polymerase errors increases with each amplification cycle.

Cloning PCR Products

Many procedures have been described for cloning PCR products into plasmid vectors. Typically, restriction sites (with one to three extra bases for "padding") are incorporated into the 5′ ends of primers used for amplification. Following amplification, the gel-purified PCR product is then digested with the relevant restriction enzyme(s) and ligated to the vector of interest. Although this method usually yields satisfactory results, certain restriction enzymes do not cut well near the termini of linear duplexes,[10] necessitating the addition of extra "padding" bases to the primers or concatimerization of the PCR product prior to digestion. As these factors add to the cost and time of analysis, alternative methods have been developed. We have found that the most convenient of these is "TA" cloning, in which the ability of *Taq* polymerase to add an untemplated adenosine residue to the 3′ ends of PCR products is exploited for cloning purposes.[11,12]

[8] A. M. H. van Huijsduijnen, G. Ayala, and J. F. DeLamarter, *Nucleic Acids Res.* **20,** 919 (1992).

[9] A. F. Wilks, R. R. Kurban, C. M. Hovens, and S. J. Ralph, *Gene* **85,** 67 (1989).

[10] F. M. Ausubel, R. Brent, R. E. Kingston, D. D. Moore, J. G. Seidman, J. A. Smith, and K. Struhl, eds., "Current Protocols in Molecular Biology." Wiley (Interscience), New York, 1987.

[11] D. Marchuk, M. Drumm, A. Saulino, and F. S. Collins, *Nucleic Acids Res.* **19,** 1154 (1991).

[12] T. A. Holton and M. W. Graham, *Nucleic Acids Res.* **19,** 1156 (1991).

Preparation of T Vector

Digest 1–10 μg of a suitable cloning vector (e.g., pBluescript, Stratagene) with a restriction enzyme that produces a blunt end (e.g., *Sma*I). Add *Taq* polymerase (1 unit/μg plasmid, 20 μl volume) in standard *Taq* buffer (10 m*M* Tris–HCl, pH 8.3, 50 m*M* KCl, 1.5 m*M* $MgCl_2$, 0.1% TritonX-100, 0.01% gelatin) containing 2 m*M* dTTP for 2 hr at 74°. In the absence of other nucleotides, *Taq* polymerase will add a single thymidine to the 3′ ends of the cut vector. Extract once with phenol/chloroform, ethanol precipitate, and bring up the DNA pellet in TE.

Cloning into T Vector

Run out products of PCR amplification on an agarose gel. Excise the band corresponding to the expected molecular weight, and purify by standard methods [we prefer glass-powder absorption or, for small products, β-agarase (Gelase, Epicentre Technology) digestion followed by ethanol precipitation]. The purified PCR product is then ligated to the T vector at 14° overnight, followed by transformation and standard blue–white selection. In our experience, nearly all bacteria colonies harbor authentic recombinants.

Application to Protein Tyrosine Phosphatases

In order to illustrate some of the principles and practices used for the PCR cloning of related genes, we will show the application of this method to identify protein tyrosine phosphatase (PTP) genes in the yeast *Schizosaccharomyces pombe*. As can be seen in Fig. 1, a comparison of the amino acid sequences of various PTP catalytic domains shows several short stretches of homology. Since the catalytic domain is only about 240 amino acids in length, any PCR primer set based on amino acids within this region would be expected to produce a small DNA fragment, well within the size limits for efficient PCR amplification. All of these conserved amino acid stretches (labeled A–F, Fig. 1)[13–17] appear at first inspection to be reasonable candidates for the design of PCR primers. Further examination of the resulting

[13] J. Chernoff, A. R. Schievella, C. A. Jost, R. L. Erikson, and B. G. Neel, *Proc. Natl. Acad. Sci. U.S.A.* **87,** 2735 (1990).

[14] D. E. Cool, N. K. Tonks, H. Charbonneau, K. A. Walsh, E. H. Fischer, and E. G. Krebs, *Proc. Natl. Acad. Sci. U.S.A.* **86,** 5257 (1989).

[15] S. J. Ralph, M. L. Thomas, C. D. Morton, and I. S. Trowbridge, *EMBO J.* **6,** 1251 (1987).

[16] M. Streuli, L. R. Hall, Y. Saga, S. F. Schlossman, and H. Saito, *J. Exp. Med.* **168,** 1523 (1988).

[17] M. Streuli, N. X. Krueger, A. Y. M. Tsai, and H. Saito, *Proc. Natl. Acad. Sci. U.S.A.* **86,** 8698 (1989).

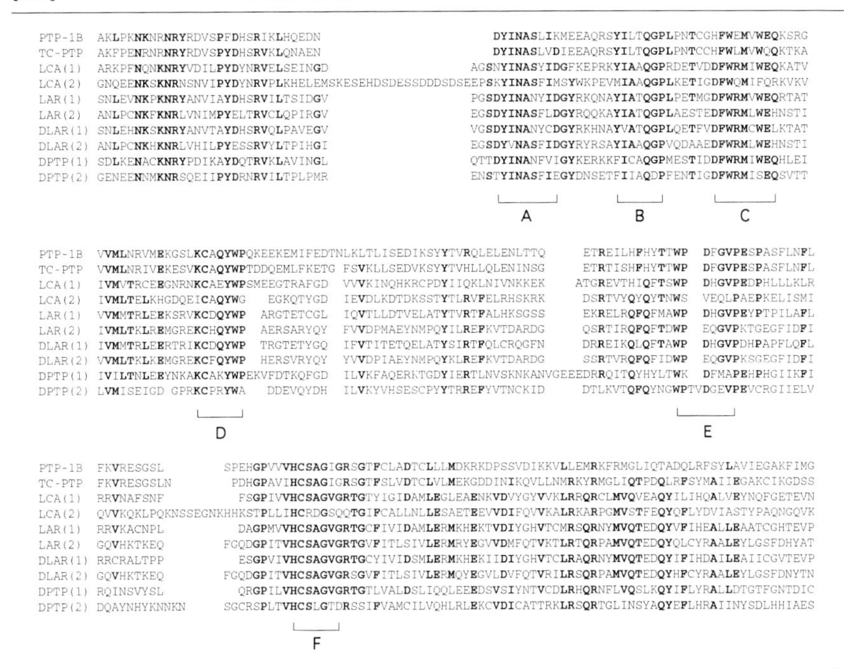

FIG. 1. Alignment of PTP sequences. The deduced amino acid sequences of human PTP1B,[13] T-cell PTP,[14] leukocyte common antigen (LCA) domains 1 and 2,[15] leukocyte-related antigen (LAR) domains 1 and 2,[16] *Drosophila* LAR (DLAR) domains 1 and 2,[17] and PTP (DPTP) domains 1 and 2[17] were aligned by eye. Gaps were inserted for optimal alignment and are represented by spaces. Amino acids in common to more than half of the clones are in boldface. Conserved regions that appear to meet the requirements for PCR primer design are designated A through F. Amino acids are in a single letter code.

primers shows that two candidates (oligonucleotides A and C) are highly degenerate due to the need to encode arginine or serine residues (Fig. 2). Because of their respective positions in the protein, oligonucleotides A–D were synthesized in the sense orientation, whereas E and F were synthesized as antisense primers.

Fifty nanograms of purified cDNA from a *S. pombe* library[18] was used as a template for PCR, using 1 μM sense primers A, B, C, or D in combination with antisense primers E or F. The PCR amplification was carried out in a 10-μl volume in a thin-walled glass capillary tube, in a 1605 Air Thermo Cycler (Idaho Technology Inc.) in a buffer containing 50 mM Tris–HCl, pH 8.3, 1.5 mM $MgCl_2$, 20 mM KCl, 200 μM dNTPs, 0.5 μg/ml bovine serum albumin (BSA), 0.4 U *Taq* polymerase, plus the primers and template. The

[18] Y. Yamawaki-Kataoka, T. Tamaoki, H.-R. Choe, H. Tanaka, and T. Kataoka, *Proc. Natl. Acad. Sci. U.S.A.* **86,** 5693 (1989).

PRIMER	ORIENTATION	NT SEQ	LENGTH	AA SEQ	DEGENERACY
A	S	C T T C AG CT G GA TA ATAAA GCN TN N TNGA T C C T TC TA C	26	SLI DYINA Y D NFV	294,912
B	S	T ACT G A TA ATCG N CNCA GGNCC C TTC A G	20	AA YI QGP LT	9,216
C	S	C C C CG C G A TT TGGG NATGNTNTGG A CANAA G T T AA G A	29	D R I QQ FW MVW K H E L EH	73,728
D	S	A T A G A T AA TG G N A TA TGGCC G C C C G C	20	DE KC YWP AQ	256
E	A	AA GGNACNCCN NTCNGGCCA TG	20	DF WP HGVP EQ	4,096
F	A	G CTA A CCNA NCCNGCN CA TG C GAG G	20	I HCSAG G V	8,192

FIG. 2. Design of PCR primers. Sequences of the degenerate oligonucleotides used to generate PTP fragments, beginning with the 5′ end, are listed from left to right. For each primer, the nucleotide sequence (NT SEQ), the size in nucleotides (LENGTH), the amino acids encoded (AA SEQ), and the fold degeneracy (*x*-fold) are shown.

BSA is included to prevent denaturation of the polymerase on the walls of the glass capillary tube, but is not required for PCRs in standard plastic vessels. The reaction tube was held at 94° for 30 sec to fully denature all double-standard DNA, followed by 30 cycles of amplification using temperatures of 94° for 1 sec, 45° for 1 sec, and 72° for 30 sec. Finally, a 3-min extension at 74° was carried out to allow all nascent nucleotide chains to be completed. The products were analyzed on a 1.5% agarose gel containing ethidium bromide. The results of this amplification are presented in Fig. 3. The PCR reactions performed with the most degenerate primers (A and C) did not yield a specific product (Fig. 3 lanes 1, 2, 5, and 6), whereas PCRs carried out with primers sets B/E and B/F, as well as D/E and D/F, gave rise to distinct products of the expected size (Fig. 3, lanes 3, 4, 7, and 8). In the case of primer set B/F, in addition to the expected product, another band of about 800 bp was also detected (Fig. 3, lane 4). Gel fragments containing these bands were excised, melted, and treated with β-agarase (Gelase, Epicentre Technology). Following isolation of the DNA, the fragments were cloned without further modification into pBluescript SK^+ (Stratagene), prepared as a "T" vector (see Preparation of T Vecor, above).

For each PCR product, plasmids from 30 bacterial colonies were analyzed for the presence of PTP-related inserts. In the case of the ~300-bp product (lane 8, product of PCR reaction containing primers D/F), 10 of the 30 plasmids encoded various nonspecific sequences. Most of these consisted of pUC8 vector sequences. Eighteen of the plasmids contained

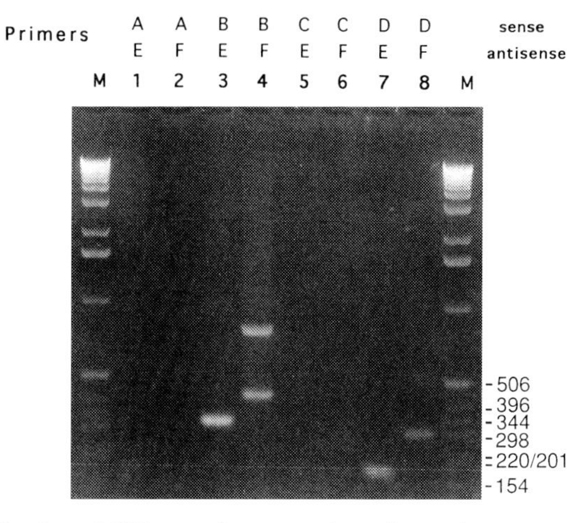

FIG. 3. Amplification of PTP gene fragments from *S. pombe*. DNA from an *S. pombe* cDNA library was used as a template in PCR reactions with various combinations of sense (primers A, B, C, D) and antisense (primers E, F) degenerate oligonucleotide primers, as described in the text. PCR products were electrophoresed in a 1.5% agarose gel and visualized by staining with ethidium bromide. Lanes marked M contain size markers (1-kb ladder, Gibco Life Sciences).

a fragment of DNA corresponding to *pyp2*, which was previously identified by PCR amplification using a very similar primer set and reaction conditions.[19] One of the remaining plasmids bore a *pyp1*[20] fragment, whereas the other corresponded to *pyp3*.[21] It is of some interest to note that on screening the cDNA library with the *pyp* probes we found that *pyp1* was represented at about a 10-fold excess over *pyp2* and *pyp3*; the opposite of the ratio detected by PCR. As the library screened was the same as that used for the PCR, it is apparent that the PCR reaction is heavily biased toward amplification of *pyp2*. Inspection of the actual sequences of the native *pyp* genes does not reveal a obvious explanation for this phenomenon, but we have occasionally noted similar biased amplification when using other degenerate primers for PCR cloning of other gene families. A similar analysis of the major PCR products in lanes 3, 4, and 7 (Fig. 3; resulting from amplification with primers B/E, B/F, and D/F, respectively) revealed that, with the exception of the ~800-bp band (lane 4, Fig. 3), these DNA

[19] S. Ottilie, J. Chernoff, G. Hannig, C. S. Hoffman, and R. L. Erikson, *Proc. Natl. Acad. Sci. U.S.A.* **88,** 3455 (1991).

[20] S. Ottilie, J. Chernoff, G. Hannig, C. S. Hoffman, and R. L. Erikson, *Mol. Cell. Biol.* **12,** 5571 (1992).

[21] J. B. A. Millar, G. Lanaers, and P. Russell, *EMBO J.* **11,** 4933 (1992).

	NT SEQ	MISMATCHES	AA SEQ
	`  C  T  T  C   AG CT G`		`     SLI`
Primer A	`GA TA ATAAA GCN TN  N TNGA`		`DYINA Y D`
	`  T  C  C  T   TC TA C`		`     NFV`
PYP1	GATTATATCAATGCTTCCTTT**A**TC**A**A	2/26	DYINASFI**K**
PYP2	GATTATATTAATGCTTCACAT**A**TAGA	1/26	DYINAS**H**ID
PYP3	GATTACATTAATGCTTCC**A**TAGTA**A**A	2/26	DYINAS**I**V**K**
	`  T  ACT G    A`		` AA`
Primer B	`TA ATCG N CNCA GGNCC`		`YI  QGP`
	`  C  TTC A    G`		` LT`
PYP1	TACATCGCT**TG**CCAAGGT**T**C	3/20	YIA**C**QG**S**
PYP2	TATATTGCC**TG**CCAGG**C**CCC	3/20	YIA**C**Q**A**P
PYP3	TTTATCGCTACTCAGGGTCC	0/20	**F**IATQGP
	`C C  C   CG            C G`		`D  R I QQ`
Primer C	` A TT TGGG NATGNTNTGG A CANAA`		` FW MVW  K`
	`G T  T   AA            G A`		`H  E L EH`
PYP1	GATTTCTGGCACATGGTATGGGA**CA**AT**GT**	4/29	DFW**H**MVW**DNV**
PYP2	GACTTTTGGGAAATGGTTTGGCA**TA**AC**TC**	4/29	DFWEMVW**H**N**S**
PYP3	G**T**TTTTTGGAAGATGGTCTGGCAA**TC**T**GT**	5/29	**V**FW**K**MVWQ**SV**
	`  A  T A G A  T`		` DE`
Primer D	`AA TG G N A TA TGGCC`		`KC  YWP`
	`  G  C C C G  C`		` AQ`
PYP1	A**T**GTGT**A**CTG**C**ATATTGGCC	2/20	**M**C**TA**YWP
PYP2	AAATGT**T**CTCAATATTGGCC	1/20	KC**S**QYWP
PYP3	AAATGTGAT**ATC**TATTGGCC	2/20	KCD**I**YWP
	`         AA`		` DF`
Primer E	`GGNACNCCN  NTCNGGCCA`		`WP HGVP`
	`         TG`		` EQ`
PYP1	GGA**GA**A**TT**A**C**AATCAG**A**CCA	5/20	W**S**D**CNS**P
PYP2	**TC**AAACCAC**GT**GT**G**C**AC**CCA	6/20	W**VHTWFD**
PYP3	GGT**G**CACCAAAGTCAGGCCA	1/20	WPDFG**A**P
	`    G        CTA  A`		`      I`
Primer F	`CCNA NCCNGCN   CA TG`		`HCSAG G`
	`    C        GAG  G`		`      V`
PYP1	CCAACACCGGCAGAGCAGTG	0/20	HCSAGIG
PYP2	CCTACGCCTGCTGAACAGTG	0/20	HCSAGVG
PYP3	CCA**CA**ACCCGCTGAACAATG	2/20	HCSAG**C**G

FIG. 4. Comparison of nucleotide and amino acids encoded by primers with actual *pyp* sequences. The sequences of the PCR primers and *pyp1*, *2*, and *3*, are aligned. Primers A–D are in the sense orientation whereas E and F are in antisense orientation. Sequences in boldface indicate mismatches between the primers and the authentic *pyp* sequences.

bands are composed of *pyp3* fragments only. A comparison of the PCR primer sequences with the actual nucleotide sequences of the three known *pyp* genes readily explains these results (Fig. 4) and illustrates why it is important to design more than one primer set for performing such amplifications. In retrospect, the design of primers A and C might have been improved by shortening them to ~20 nucleotides. This would have simultaneously decreased their complexity and 3' mismatches to *pyp* sequences.

Although our analysis in each case revealed the same three *pyp* gene fragments, it is possible that other *pyp* genes exist in *S. pombe*, but are not structurally similar enough to known PTPs for successful amplification by PCR with these primers. Indeed, the distantly related *S. pombe cdc25* gene, which encodes a dual-specificity (threonine–tyrosine) protein phosphatase,[22,23] was not identified in our screen. These results demonstrate some of the strengths and weaknesses of the PCR approach to cloning related genes.

Summary

PCR amplification is a powerful technique for cloning related genes. Compared to traditional structure-based methods, it is faster and more specific. When properly designed and executed, the isolation of related genes by PCR amplification can be accomplished in less than 1 week. The key element to success is the proper choice of PCR primers, which are designed according to the degree of conservation of amino acid sequence in the protein(s) of interest and the degeneracy of the resulting oligonucleotides. We demonstrate the utility of this approach for the isolation of PTP genes from distantly related organisms.

Acknowledgments

This work was supported by grants from the National Institutes of Health (NRSA CA 08563 and RO1 CA 58836-01). We thank Dr. T. Kataoka for the *schizosaccharomyces pombe* cDNA library.

[22] W. G. Dunphy and A. Kumagai, *Cell* (*Cambridge, Mass.*) **67,** 189 (1991).

[23] J. Gautier, M. J. Solomon, R. N. Booher, J. F. Bazan, and M. W. Kirschner, *Cell* (*Cambridge, Mass.*) **67,** 197 (1991).

[13] Isolation of Oncogenes by Expression cDNA Cloning

By TORU MIKI and STUART A. AARONSON

Introduction

By means of genomic DNA transfection–transformation assays utilizing NIH/3T3 mouse fibroblasts as recipient cells, various oncogenes have been isolated from both human and rodent tumors. However, this conventional oncogene isolation procedure has inherent difficulties. First, the procedure is very labor intensive, involving: (1) extraction of genomic DNA from tumor samples, (2) transfection of NIH/3T3 cells, (3) extraction of genomic DNA from transformants and multiple rounds of transfection to segregate unrelated human genomic sequences from the oncogenic sequence, (4) construction of genomic libraries from transformants, (5) screening libraries with human repetitive sequences, (6) identification of exon sequences, and (7) screening cDNA libraries with the exon probe. Because of the large size of some genes, it is difficult to isolate the complete sequence in an intact form. Even when possible, gene transfer steps involving very large DNA molecules may be inefficient. Another problem in gene transfer experiments is that introduced oncogenes may not be expressed if their promoters are not active in the recipient cell. These inherent difficulties may be overcome by means of expression cDNA cloning. In the cDNA form, most genes are sufficiently small to achieve efficient transfer and are certain to be expressed when a strong constitutive promoter in the vector is used.

Many mammalian expression cloning systems are based on transient high levels of expression driven by the simian virus 40 (SV40) promoter in COS cells.[1,2] However, selection of cells showing certain biologic phenotypes may only be practical by means of stable expression. In order for a single cDNA to register a stable phenotype, it is necessary to develop a system capable of efficient synthesis of full-length cDNAs, unidirectional insertion of cDNA fragments into expression vectors, and a high level of gene expression. Moreover, a method to rescue integrated cDNA clones from mammalian cells is also required. We have developed such an efficient cDNA cloning system using stable expression.[3,4] This chapter describes,

[1] H. Okayama and P. Berg, *Mol. Cell. Biol.* **3,** 280 (1983).

[2] A. Aruffo and B. Seed, *Proc. Natl. Acad. Sci. U.S.A.* **84,** 8573 (1987).

[3] T. Miki, T. Matsui, M. A. Heidaran, and S. A. Aaronson, *Gene* **83,** 137 (1989).

[4] T. Miki, T. P. Fleming, M. Crescenzi, C. J. Molloy, S. B. Blam, S. H. Reynolds, and S. A. Aaronson, *Proc. Natl. Acad. Sci. U.S.A.* **88,** 5167 (1991).

in detail, steps which are needed to isolate transforming cDNAs using this system.

Strategy to Isolate Oncogenes by Expression cDNA Cloning

The structure of our eukaryotic expression vector, λpCEV27, is schematically shown in Fig. 1. λpCEV27 is a λ-plasmid composite vector. The λ vector is useful in generating high complexity libraries by *in vitro* packaging. In this expression cloning system, the λ genome also plays a role as a shuttle vehicle for stable introduction of cDNAs into mammalian cells. The steps needed to isolate transforming cDNAs using the λpCEV27 system are shown in Fig. 2. Following integration, the cDNA is expressed stably. If plasmid libraries are used, it is often difficult to rescue the cDNA clones from the transfectant DNA because of rearrangements that occur at the site of integration. However, the plasmid part of the composite λpCEV27 vector can be cut out from the λ DNA by one of the infrequent cutters in the multiple excision sites. The digested DNA is then circularized and used

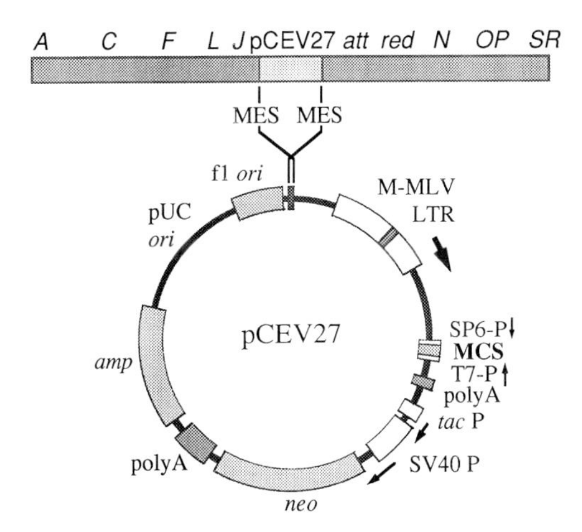

FIG. 1. The eukaryotic expression vector λpCEV27. Structure of the λpCEV27 vector is shown at the upper half with the location of λ genes. The plasmid portion is enlarged and is shown at the lower half as a circular map. The multiple excision site (MES) contains restriction sites for infrequent cutters: *Not*I, *Xho*I, and *Mlu*I. The multiple cloning site (MCS) contains restriction sites for *Bam*HI, *Sal*I, *Sfi*I(A), *Eco*RI, *Bgl*II, *Hin*dIII, *Sfi*I(B), *Sal*I, and *Bst*EII, in clockwise order. The two *Sfi*I sites are used to insert cDNA molecules by the automatic directional cloning (ADC) method[3] and the two *Sal*I sites are used to release the inserts. SP6-P and T7-P represent the phage promoters for SP6 and T7 RNA polymerases, respectively. The SV40 early and *tac* promoters were used to express the *neo* structural gene in mammalian cells (G418 resistance) and in *E. coli* (kanamycin resistance), respectively.

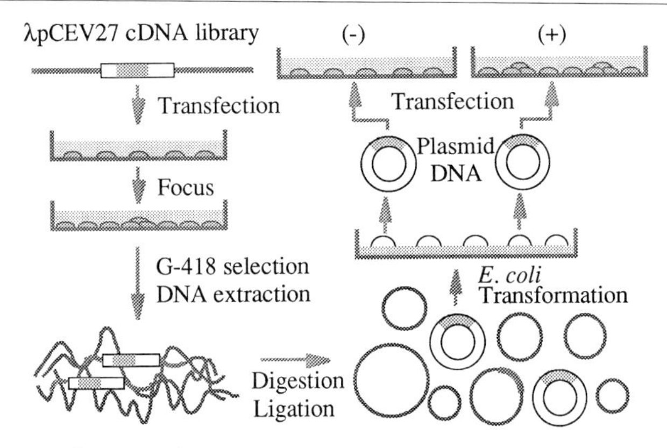

FIG. 2. Strategy for expression cDNA cloning of transforming cDNAs. NIH/3T3 cells are transfected with λpCEV27-cDNA library DNA and are scored at 14 to 17 days for transformed foci. Transformed cells are assayed for G418 resistance to confirm the presence of integrated vector sequences. Following expansion to mass culture, the genomic DNA is isolated and subjected to plasmid rescue by digestion with *Not*I, *Xho*I, or *Mlu*I, followed by ligation at a low DNA concentration and transformation to a suitable bacterial strain. Bacterial colonies resistant to both the ampicillin and kanamycin are isolated. Plasmid DNA extracted from each colony is tested by transfection analysis on NIH/3T3 cells to identify the transforming cDNA clone.

to transform competent bacterial cells to rescue the liberated plasmids. This approach has proven to be very efficient in the isolation of a rodent oncogene.[4]

In addition to isolation of oncogene cDNAs from tumors and tumor cell lines, we have devised a strategy to clone potential oncogenes from normal tissues and cell lines.[5–8] The introduction of an appropriate expression cDNA library prepared from the mRNA of one cell type into another results in the forced expression of genes which might normally be silent in recipient cells. Such ectopic expression could result in morphological transformation. A transformed phenotype may also be obtained by the increased expression of a normal gene product or by the expression of a structurally altered product such as a truncated protein (Fig. 3). We have isolated three distinct transforming cDNAs, designated *ect1*, *ect2*, and *ect3*,

[5] T. Miki, T. P. Fleming, D. P. Bottaro, J. S. Rubin, D. Ron, and S. A. Aaronson, *Science* **251,** 72 (1991).

[6] T. Miki, D. P. Bottaro, T. P. Fleming, C. Smith, W. H. Burgess, A. M.-L. Chan, and S. A. Aaronson, *Proc. Natl. Acad. Sci. U.S.A.* **89,** 246 (1992).

[7] T. Miki, C. L. Smith, J. E. Long, A. Eva, and T. P. Fleming, *Nature* (*London*) **362,** 462 (1993).

[8] A. M.-L. Chan, E. S. McGovern, G. Catalano, T. P. Fleming, and T. Miki, *Oncogene* **9,** 1057 (1994).

from an epithelial cell expression library. *ect1* encodes a receptor of keratinocyte growth factor (KGF).[5,6] Ectopic expression of the receptor in cells which express its ligand (KGF) created an autocrine transforming loop, leading to focus formation. *ect2* is activated by N-terminal truncation of the normal product.[7] *ect3* is the N-*ras* cDNA. In this case, overexpression of the normal N-*ras* product resulted in focus formation. Therefore, cDNA libraries from normal as well as tumor cell lines transferred into appropriate recipient cells may give rise to the induction of stable transformants.

Preparation of λpCEV27 DNA

A well-isolated single plaque of λpCEV27 may be used as a starting material. Since λpCEV27 contains a plasmid sequence, bacterial strains without plasmids, such as LE392, are preferable host cells. No suppressor is required for the growth of λpCEV27. Approximately 6×10^5 plaque-forming units (PFU) of phage lysate and 2 ml of LE392 cell suspension in SM buffer (50 m*M* Tris–HCl, pH7.5/10 m*M* NaCl/10 m*M* $MgCl_2$) are mixed in a plastic tube (e.g., Falcon 2070) and incubated at 37° for 20 min. Two liters of prewarmed NZY medium containing 7 m*M* $MgCl_2$ is inoculated with the λ-adsorbed cells and shaken at 37° for 6–12 hr vigorously, until the culture shows clear lysis. Following the addition of 2 ml of chloroform and shaking for an additional 10 min, the lysate is clarified by centrifugation at 8000 rpm for 30 min.

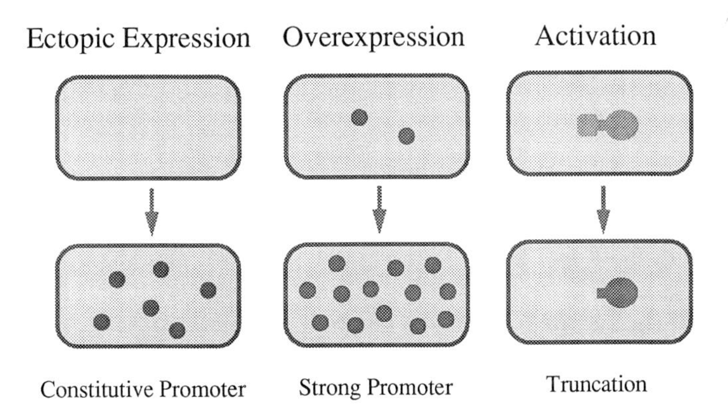

FIG. 3. Expression cloning of transforming cDNAs from normal cells. When expression cDNA libraries are constructed from normal cells and introduced into host cells, morphologically transformed foci might be generated by ectopic expression (left), overexpression (middle), or activation of the gene product by removal of a regulatory domain (right).

The supernatant (2 liters) is transferred to a sterile flask, followed by the addition of 1 μg/ml each of DNase I and RNase A and incubation at 37° for 1 hr with shaking. Phage particles are precipitated by the addition of 80 g of NaCl and 280 g of polyethylene glycol 8000 and overnight incubation at 4°. Pellets are collected by centrifugation at 8000 rpm for 30 min at 4° and are resuspended into 120 ml of SM buffer. The suspension is dispensed as 20-ml aliquots into plastic tubes, 20 ml of chloroform is added, and mixed well. After transfer of the suspension into tubes for the Sorval SS-34 rotor and centrifugation at 16,000 rpm for 20 min, the upper phase is collected. Phage particles are pelleted by centrifugation in a Beckman 50.2 Ti rotor at 25,000 rpm for 2 hr at 4°. Immediately after centrifugation, the supernatants are discarded and tubes are kept in an inverted position for 5 min. The pellets are dissolved in 0.5 ml/tube of SM buffer by incubation at 4° overnight.

Insoluble materials are removed by centrifugation in a microcentrifuge for 5 min. After the volume of the phage solution is measured, 1.31 vol of CsCl solution saturated at 0° in SM buffer is added. The sample is then transferred to nitrocellulose or ultraclear tubes for a SW50.1 rotor and overlayed with paraffin oil. Following centrifugation in a SW50.1 rotor at 23,000 rpm for 22 hr at 4°, phage bands are collected by a needle attached to a syringe. The phage solution is dialyzed in SM buffer at 4° overnight.

To extract phage DNA, proteinase K solution is added to a final concentration of 50 μg/ml and incubated at 37° for 15 min. Then, 1/10 vol of 100 m*M* Tris–HCl, pH 8.0/100 m*M* EDTA/1% sodium dodecyl sulfate (SDS) is added and incubated at 65° for 1 hr. The sample is then transferred to a plastic tube, diluted to 20 ml with TE (10 m*M* Tris–HCl, pH 8.0/1 m*M* EDTA), and extracted with phenol–chloroform (1 : 1) twice and chloroform once. In each extraction, phases are separated by centrifugation in a Sorvall SS-34 rotor at 16,000 rpm for 15 min at room temperature. A half volume of 7.5*M* ammonium acetate and the same volume of cold 2-propanol are added and mixed very gently several times to form a clear viscous precipitate. The precipitate is recovered by heat-sealed Pasteur pipettes and is washed in cold 70% ethanol twice. The DNA preparation is partially air-dried, dissolved in 1–2 ml of TE, and stored at 4°.

Construction of Expression cDNA Libraries

The composite λpCEV27 vector is designed to use the automatic directional cloning (ADC) system for cDNA synthesis. Although the multiple cloning sites (MCS) of λpCEV27 contain restriction sites for several other cDNA library construction methods, the ADC system is well suited for expression cDNA cloning. In this method, cDNA is synthesized from a linker–primer containing the *Sfi*I(B) site. Following second strand synthesis, an adaptor containing a *Sfi*I(A) half-site is attached to one end, and the other end is trimmed by *Sfi*I. This results in molecules containing different nonsymmetrical single-stranded extensions. Similarly, λpCEV27 DNA is digested with *Sfi*I, and the arms are purified by means of potassium acetate gradient centrifugation to remove the stuffer fragment.[3,9,10] The vector and insert fragments are then ligated and packaged to obtain phage particles. Since *Sfi*I fragments do not self-ligate because of the presence of asymmetrical single-stranded extensions, the cloning efficiency is very high and the direction of cDNA is strictly oriented. In addition, background (phages without cDNA inserts) is very low. Thus the ADC method is well suited to expression cDNA cloning.

cDNA library construction in λpCEV27 is easily performed using commercially available cDNA synthesis kits. Oligo(dT) primers in a kit can be replaced by the ADC linker–primers. Following second strand cDNA synthesis, an *Sfi*I adaptor is ligated to the ends of cDNA molecules, and the cDNA is trimmed by *Sfi*I. After the cDNA fragments are ligated in the λpCEV27 arms to generate an expression cDNA library, the library is amplified by a plate method to avoid possible recombination between different clones. Detailed protocols can be found elsewhere.[3,9,10] Phage particles are purified and DNA is extracted as described in the protocol for λpCEV27 DNA preparation. Next, the expression cDNA library DNA is used to transfect recipient cells.

Introduction of a cDNA Library to NIH/3T3 Cells

One day prior to transfection, NIH/3T3 cells are plated at a density of 1×10^5 cells/100-mm culture dish in Dulbecco's minimum essential medium (DMEM) containing 5% (v/v) calf serum. Fifty to 100 plates are usually prepared for a library transfection. To measure the transformation efficiency, an expression vector with a known oncogene is used as a positive control. As controls for the appearance of spontaneous foci, several plates are transfected with carrier DNA alone. These plates serve as a reference to discriminate transformed foci induced by the library DNA.

Library DNA (5 μg/plate) and high molecular weight calf thymus carrier DNA (40 μg/plate) are added to a plastic tube (e.g., Falcon 2070). Double distilled H_2O is added to the tube to a final total volume of 0.5 ml per

[9] T. Miki and S. A. Aaronson, *in* "Methods in Molecular Genetics" (K. W. Adolph, ed.), Vol. 1, Part A, p. 3. Academic Press, San Diego, CA, 1993.

[10] T. Miki and S. A. Aaronson, "Plasmids: A Practical Approach," p. 177. Oxford University Press, London, 1993.

plate. Then, 0.5 ml of 0.5 *M* $CaCl_2$ solution is added to the tube containing DNA, mixed, and incubated at room temperature for 15–20 min. HN buffer (50 m*M* HEPES/280 m*M* NaCl, pH 7.1) is mixed with 70 m*M* $NaPO_4$ at a ratio of 1.95 ml:45 μl to prepare HNP buffer. HNP buffer (1 ml per plate) is then added to a fresh plastic tube.

While vigorously bubbling air into the HNP buffer, the DNA–$CaCl_2$ solution is slowly added (drop by drop) to the tube. The tubes are incubated at room temperature for 15–20 min. It is important to form fine precipitates at this stage. The transfection mixture (2 ml) is added to the medium overlaying the culture and swirled gently. After incubation in a 37° incubator for 18–20 hr, the medium and precipitates are aspirated, and 5 ml of DMEM containing 10% calf serum is added. Following incubation for 2 hr, the medium is aspirated and 10 ml of DMEM containing 5% calf serum is added. Cultures are incubated at 37°. To measure transfection efficiency, G418 (Geneticin) is added to a concentration of 750 μg/ml and the resulting colonies are measured at around 10 to 14 days.

Focus Isolation and G418 Selection

Foci of transformed cells will appear at around 2 weeks after transfection. If negative control plates have many spontaneous foci, it may not be possible to discern transformed foci induced by the library cDNAs. If the background is relatively low, each individual focus can be isolated from plates of library-transfected cells by means of cloning cylinders. The location of foci can be marked on the bottom of plates. After the medium is removed, a cloning cylinder with silicone grease is placed over the area of the focus. Several drops of a warm trypsin solution are added to the cylinder followed by incubation at 37° for 5 min. Cells are suspended by means of a Pasteur pipette and transferred in duplicate to 12-well plates, one of which is selected for G418 resistance. If a focus is induced by a cDNA clone, the majority of the cells isolated will show G418 resistance. Thus, foci resistant to this antibiotic are considered to be induced by the cDNA library. Cells from positive foci are cultured in selective medium for preparation of DNA and plasmid rescue.

Small-Scale Genomic DNA Preparation and Plasmid Rescue

Genomic DNA is extracted from G418 resistant cells, and the plasmid portion of the integrated vector is cut out by one of the restriction enzymes, *Not*I, *Xho*I, and *Mlu*I. These enzymes are infrequent cutters for genomic DNA, since *Not*I recognizes 8 base pairs and other two enzymes contain the CpG sequence which is very rare in mammalian genomes. λpCEV27 contains these enzyme sites in the multiple excision sites (MESs), which are present at the λ/plasmid junctions. Digested DNA is circularized by ligation at a low concentration and is used to transform *Escherichia coli* competent cells. *E. coli* has several mechanisms for identifying foreign DNA and destroying it. Therefore, to achieve a higher efficiency of plasmid rescue, it is important to use mutant strains deficient in these mechanisms. *Eco*K restriction systems encoded by the *hsdRMS* genes can destroy DNAs, including mammalian genomic DNA, which have not been modified by the *Eco*K system. The products of *mcrA*, *mcrBC*, and *mrr* genes can also attack DNAs which have been methylated in specific sequences. Competent cells defective in these loci are available from several venders.

Two dishes (10 mm) of confluent monolayer cells are prepared from each G418-resistant focus. The medium is discarded, cells are washed with phosphate-buffered saline, and 0.5 ml of lysis solution (50 m*M* Tris–HCl, pH 8.0/10 m*M* EDTA pH 8.0/200 m*M* NaCl/0.5% SDS/0.5 mg/ml proteinase K) is added per dish. Cells are scraped and transferred to a microcentrifuge tube. Following incubation at 60° for 30 min, 0.5 ml of phenol–chloroform (1 : 1) is added and mixed vigorously for 30 min. Following centrifugation at top speed for 30 min at room temperature, the upper layer is transferred to a new tube containing 0.5 ml of cold 2-propanol. The tube is inverted gently several times and precipitates are collected by a sealed Pasteur pipette and washed in 75% ethanol. Ethanol is drained and the pellet is dissolved in 0.5 ml TE containing RNase A (2 μg/ml). The samples are mixed by inversion at 37° for 2 hr and then extracted with phenol–chloroform (1 : 1) once as above and then with chloroform once. The aqueous layer is mixed with 0.25 ml of 7.5 *M* ammonium acetate and 0.75 ml 2-propanol. Pellets are collected and washed as above, and then resuspended into 0.2 ml of TE (10 m*M* Tris–HCl, pH 8.0/1 m*M* EDTA, pH 8.0).

For plasmid rescue, 5 μg of genomic DNA is digested by *Not*I, *Xho*I, or *Mlu*I. The digested DNA is extracted by phenol/chloroform followed by chloroform. To the aqueous layer, 150 μl of 7.5 *M* ammonium acetate and 1 ml of cold ethanol are added and incubated on ice for 10 min. DNA is precipitated by centrifugation and washed with 70% ethanol. The pellet is briefly dried at 37° in a heat block and is resuspended in 100 μl of TE. Approximately 2 μg of the DNA is ligated in 400 μl under standard conditions and is then extracted with phenol–chloroform (1 : 1) followed by chloroform. DNA is precipitated as above and resuspended in 10 μl of TE.

Competent cells of a *mcrA*$^-$ *mcrB*$^-$ strain [e.g., DH5αMCR (BRL), DH10B (BRL), PLK-F′ (Stratagene), or SURE (Stratagene)] are thawed and 100-μl aliquots are dispensed into prechilled Falcon 2059 tubes using a prechilled pipette tip in the cold room. DNA (2 μl) is added to a tube. After incubation on ice for 30 min and at 42° for 45 sec, the tube is chilled on ice for 2 min. One milliliter of S.O.C. medium containing 1 m*M* IPTG is added and incubated at 225 rpm for 1 hr at 37° in a rotary shaker to allow expression of the *amp* gene and to induce expression of the *tac-neo* gene in pCEV27. Cells are spread on 150-mm plates of NZY agar containing 100 μg/ml ampicillin, 25 μg/ml kanamycin, and 1 m*M* IPTG. Plates are incubated at 37° overnight.

Analysis of Rescued Plasmids

Plasmid DNA is prepared by standard procedures and is used for both restriction analysis and NIH/3T3 cell transfection as described elsewhere.[3,10] *Sal*I digestion releases cDNA inserts from pCEV27 (approximately 6 kb). Usually, multiple distinct plasmids are rescued from a given transformant. To determine which clone is responsible for transformation, distinct plasmids are used to transfect NIH/3T3 cells. Thus, cDNA inserts from plasmids exhibiting high-titered transforming activity can be identified as transforming cDNAs.

If a transforming cDNA is not rescued from a focus, there may be several possibilities: (1) An insufficient number of plasmids have been analyzed. (2) The transforming plasmid cannot be released because the restriction sites in MESs have been methylated or lost during integration into the mammalian genomes. (3) The focus was generated spontaneously but cDNA clones were coincidentally introduced. In the first case, more plasmids may be rescued and analyzed. In the second case, secondary transfection of NIH/3T3 cells with genomic DNA of the primary transfectant may be useful in releasing the restriction sites from methylation. If a cDNA clone has lost at least one MES, random shearing and blunt end ligation of genomic DNA may rescue the plasmid. However, construction of a genomic library or PCR amplification of the insert will be necessary to obtain the cDNA clone if a larger portion of the plasmid was lost by rearrangement. The third possibility should be suspected if plasmid rescue is unsuccessful despite extensive attempts.

Identification and Sequencing of cDNA Clones

Foci induced by known oncogenes can be identified by immunoblotting when G418-resistant cells are obtained or by PCR analysis from primers which span more than one exon when genomic DNA is extracted from the transformant. If it is known that specific oncogenes are abundantly expressed in the cells which are the source of the library used for transfection, such analyses may be helpful. However, at the final stage of the expression cloning procedure, cDNA clones can be identified by rapid sequencing at the 5′ end. For this purpose, commercially available SP6 primers are useful. Database searches of the sequence determined by the primers will identify the gene. If no matches are found, it may be necessary to determine the internal sequence. This is because protooncogenes can be activated by rearrangement with unknown genes. If no matches are found, the entire sequence can be determined by several strategies, including the construction of overlapping clones containing unidirectional deletions. Since use of double-stranded DNA as a template for sequencing usually results in multiple compressions in the sequence ladders, a protocol to prepare single-stranded DNA from pCEV27 is described below.

F′ strains including MV1190: Δ(*lac-proAB*), *thi*, *supE*, Δ(*srl-recA*)*306*: :Tn*10* (*tet*^{r}), [F′ :*traD26*, *proAB*, *lacI*^{q} *Z*Δ*M15*] may be used as a host strain of pCEV27. Competent cells are prepared, transformed by a pCEV27–cDNA clone, and cells doubly resistant to ampicillin and kanamycin are selected. An overnight culture of the transformant in NZY broth containing ampicillin (100 μg/ml) and kanamycin (25 μg/ml) is prepared, and 5 ml of NZY broth containing ampicillin is inoculated with 0.1 ml of the overnight culture. The culture is then incubated at 37° for 2 hr with vigorous shaking. For a negative control, a similar culture of MV1190 in NZY broth without ampicillin is prepared. Ten microliters of M13KO7 stock (3×10^{11} PFU/ml) is added to each tube and shaken vigorously at 37° for 4 hr. Then, the culture is centrifuged in a bench-top centrifuge at 3000 rpm for 20 min and the supernatant is transferred into a fresh tube. Ten microliters of the supernatant and 2 μl of 6 × loading dye solution (0.06 *M* EDTA/0.6% SDS/0.06% bromphenol blue/0.06% xylene cyanole/ 15% Ficoll type 400) are mixed, heated at 68° for 5 min, and analyzed on a 0.7% agarose gel. At this step, clones for DNA extraction are selected. Each supernatant (1.5 ml) is transferred into a microcentrifuge tube and centrifuged at top speed for 5 min. Supernatants are transferred by decantation into a fresh tube containing 0.2 ml of 20% PEG/2.5 *M* NaCl and incubated at room temperature for 15 min. Tubes are centrifuged at top speed for 5 min, and the supernatants are discarded after removal by pipette. Pellets are dissolved into 0.1 ml of TE, and 1 μl of RNase A (1 mg/ml) is added, mixed, and incubated at 37° for 15 min. One microliter of proteinase K (20 mg/ml) is added to the tubes, mixed, and incubated at 65° for 20 min. Samples are extracted wtih phenol–chloroform twice and the phases are separated by centrifugation for 15 min following each extraction. Aqueous layers are transferred into fresh tubes containing 40 μl of 7.5 *M* ammonium acetate and are mixed by pipetting. Cold 2-propanol (120 μl) is added, mixed, and incubated on ice for 5 min. Tubes are centrifuged for

15 min at top speed, supernatants are discarded, and the pellets are washed with 75% ethanol. DNA is dried at 37° for 5 min in a heat block and dissolved into 10 μl of TE.

Concluding Remarks

The automatic directional cloning method[3] used in λpCEV27 and λpCEV-LAC systems can generate cDNA libraries very efficiently. The cDNA inserts are strictly enforced to orient to the direction of transcription from the promoter within the vector. Sizes (number of independent clones) of cDNA libraries made by the ADC method may be much higher and backgrounds (clones without inserts) are much lower than with libraries prepared by other methods. These features have enabled us to isolate several signal transducing molecules by expression cloning strategies. Since a small amount of cDNA is usually enough to generate relatively large libraries, it is possible to prepare prokaryotic and eukaryotic libraries at the same time. Using expression cloning strategies, cDNA clones can be isolated based on induction of a stable phenotype such as transformation or drug resistance. It may be possible to isolate genes which can not be cloned by conventional methods by incorporation of additional new strategies for these efficient expression systems.

[14] Replication-Competent and -Defective Retrovirus Vectors for Oncogenic Studies

By JACQUES SAMARUT, FRÉDÉRIC FLAMANT, and JEAN-MICHEL HEARD

Oncogenes were first discovered as gene components of acute transforming retroviruses. This discovery further revealed the function of retroviruses as natural vectors for cellular genes. These two properties have been widely exploited to use recombinant retroviruses as vectors for efficient gene transfer into animal cells *in vitro* and *in vivo*. The main interest in retrovirus vectors comes from their ability to stably integrate their genome into that of infected cells. Following its introduction into the host cell genome, the virus genome behaves as a new gene structure that can be genetically transmitted to daughter cells. For their accurate and efficient use, the constructed vectors should fulfill certain criteria. They should faithfully integrate foreign sequences into the genome of targeted cells, and these sequences should be properly expressed in these cells. Moreover, the vector by itself should be innocuous to the target cells so that the effect of the transferred gene of interest is not obscured.

These rules are sometimes purposely ignored to use retrovectors as unstable mutagenic agents that can rearrange foreign sequences, to reveal their oncogenic potential.[1,2] Most *in vivo* applications imply the infection of scarce target cells in heterogeneous tissues. Therefore the main technical effort is to achieve a high gene transfer efficiency.

The purpose of this chapter is to present the various vectors that are available for oncogenic studies in avian and mammalian species and to give some technical data on their preparation and use.

I. Basic Principles of Retrovirus Vectors

A. *Retrovirus Cycle*

Extensive data on retrovirus structures and functions can be found in Weiss *et al.*,[3] therefore only basic knowledge essential for understanding the construction and the use of retrovirus vectors will be reviewed in this chapter. Because the retroviral genetic information exists in both RNA and DNA forms, the structure of retrovirus genome carries specific control sequences useful for both DNA transcription and RNA processing. Since DNA provirus integration occurs only after the nuclear membrane breakdown, infection is strictly dependent on the occurrence of cell division soon after virus entry.

This chapter focuses on the viruses that are most commonly used as vectors: murine leukemia viruses (MLV) and avian leukosis viruses (ALV). Spleen necrosis virus vectors developed mainly in the laboratories of H. Temin and J. Dougherty have also been used but will not be described here as they were the subject of a recent review.[4]

B. *Comparative Structures of ALV and MLV*

Although the overall strategy for retroviral genome expression and function is the same for avian and mammalian retroviruses, the viruses differ in some part of their backbone structure, leading to slightly different functional mechanisms.

[1] T. Metz and T. Graf, *Cell (Cambridge, Mass.)* **66,** 95 (1991).

[2] A. Yoshimura, G. Longmore, and H. F. Lodish, *Nature (London)* **348,** 647 (1990).

[3] R. Weiss, N. Teich, H. Varmus, and J. Coffin, eds., "RNA Tumor Virus." Cold Spring Harbor Lab., Cold Spring Harbor, NY 1984.

[4] K. A. Boris-Lawrie and H. M. Temin, *Curr. Biol.* **3,** 102 (1993).

1. Splicing and Translation Initiation Signals

One major difference between ALV and MLV genomes is found in the noncoding leader sequence upstream of *gag* and in the structure of the *gag* gene itself (Fig. 1). In both types this leader sequence contains a splicing donor site (SD) that is essential for the production of subgenomic mRNA, working as a template for translation of retrovirus proteins. In ALV this DS is located 15 bp downstream from the translation initiation codon of the *gag* frame, whereas in murine retrovirus genomes it is located upstream from the *gag* initiation codon. As a consequence, avian retrovirus products encoded by subgenomic mRNAs are in most cases initiated at the *gag* initiation site and then carry the first six amino acids of *gag* at their amino terminus. The Rous sarcoma virus (RSV) uses a different translation strategy for the expression of the v-*src* oncogene. The v-*src* subgenomic RNA carries a stop codon 4 bases downstream from the splice acceptor site, which allows the *src* frame to be initiated at its own initiation codon located 63 bases further downstream. This specific structure has been exploited in ALV replication competent vectors (see below).

Another major element of the leader sequence is the packaging signal Ψ which promotes packaging of the viral genome RNA into the virus

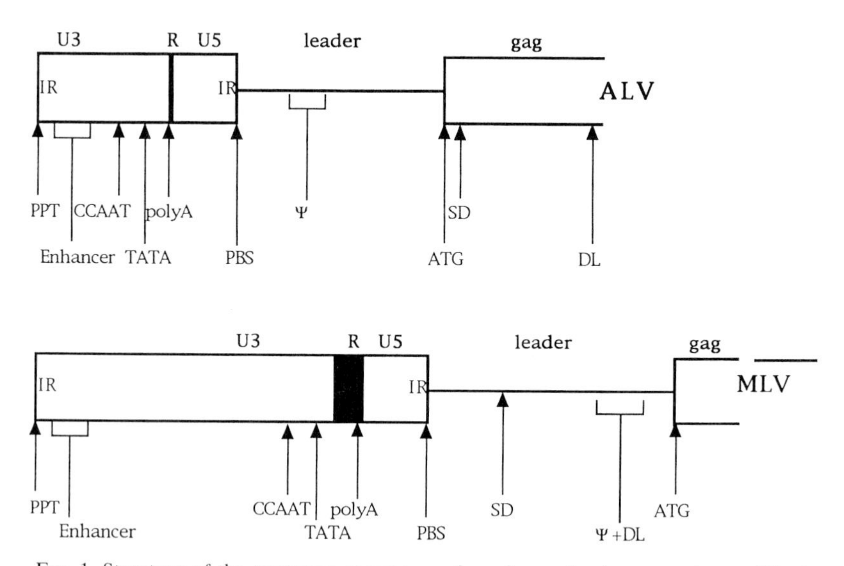

FIG. 1. Structure of the upstream sequences of murine and avian retroviruses. IR, short inverted repeat; PPT, polypurine tract; DL, signal for RNA dimerization; Ψ, signal for RNA packaging; PBS, tRNA primer binding site (ALV use $tRNA^{Trp}$, MLV use $tRNA^{Pro}$).

particle. In avian retroviral genomic RNA, this signal lies between nucleotides 215 and 238. In the Mo-MLV genome, sequences encompassing nucleotides 21 to 563 are sufficient for packaging, but extension to 1 kb into the *gag* gene provides more efficient packaging.[5]

2. *Long Terminal Repeats*

Long terminal repeats (LTRs) contain essential elements for the genome integration after reverse transcription and transcription of the integrated viral genome. This structure contains an enhancer and sequences directly involved in the initiation of transcription (CAAT box and TATA box) and in the termination of transcription and maturation of mRNAs, such as the polyadenylation signal. In both MLV and ALV, different LTRs possess different transcription efficiencies that are tissue specific. A careful choice should be made when designing a vector for preferential expression in a given tissue.

C. Basic Rules for Retrovirus Vector Design

The design of recombinant retroviral vectors is always somewhat empirical because retrovirus RNA transcription, splicing, packaging, stability, or protein translation are not yet fully understood and can be influenced by the inserted sequences. For these reasons, it is recommended to consider different vector configurations for a given purpose and to take into account that the simplest construct often gives the highest infection titers. Nevertheless, the following rules are considered as well established: (1) The maximal size of packaged retrovirus RNA molecules should not exceed 9 kb. Smaller vectors usually perform better. (2) Foreign introns introduced into the vector genome will be quickly lost after virus replication. If the retrovirus *gag* splice donor is present in the vector, the presence of cryptic splice acceptor sites in the inserted foreign sequences can generate truncated vector genomes. If introns have to be preserved in the insert, the gene should be inserted in an orientation inverse to LTR-driven transcription. Moreover, this gene should be associated to an autonomous promoter especially active in the target cells. (3) Polyadenylation signals should not be inserted into the vector (consensus AAUAAA): They trigger RNA cleavage and prevent full-length virus genome transcription, thereby inhibiting the production of infectious vector viruses. Such signals are present in the 3′-untranslated region of most cloned cDNAs.

[5] M. A. Bender, T. D. Palmer, R. E. Gelinas, and A. D. Miller, *J. Virol.* **61,** 1639 (1987).

TABLE I
AVIAN REPLICATION-COMPETENT VECTORS

Name	Insert	Note	Reference
RCAS	No	RSV LTR	6
RCOS	No	RAV(0) LTR	6
RCAN	No	No splice acceptor	6
RCASBP(A)	No	Subgroup A *env* gene	7
RCASBP(E)	No	Subgroup E *env* gene	7
RCASBP AP(A)	Alkaline phosphatase	Subgroup A *env* gene	7
RCASBP-AP(E)	Alkaline phosphatase	Subgroup E *env* gene	7
RCAS-CAT	CAT		8
RCAS-αCAT	CAT	Muscle-specific internal α-actin promoter	9
RCAMV	v-*myb*	AMV LTR active in hematopoietic cells	10
RCAS c-*jun*	Avian c-*jun*, murine c-*jun*, *junB*, *junD*		11
RCAS-*erbA*	Avian c-*erbA*, v-*erbA*		12
RCAS-*ras*	Harvey v-*ras*		13

II. Avian Retrovirus Vectors

A. *Replication-Competent Vectors*

Replication-competent (RC) vectors are constructed on the model of the natural RSV in which the *src* oncogene is substituted by the cDNA to be transferred. Such vectors have been widely used for reporter genes like CAT and alkaline phosphatase, as well as oncogene and protooncogene cDNAs (Table I).[6–13]

1. *Principle of Construction*

A very convenient device has been provided by the laboratory of S. Hughes.[6,8,13] All the RC vectors carry a unique *Cla*I site to insert foreign

[6] S. Hughes and Y. Gluzman, "Viral Vectors." Cold Spring Harbor Lab., Cold Spring Harbor, NY, 1988.
[7] D. M. Fekete and C. L. Cepko, *Proc. Natl. Acad. Sci. U.S.A.* **90,** 2350 (1993).
[8] C. J. Petropoulos, W. Payne, D. W. Salter, and S. H. Hughes, *J. Virol.* **66,** 3391 (1992).
[9] C. J. Petropoulos and S. H. Hughes, *J. Virol.* **66,** 3728 (1991).
[10] R. D. Press, A. Kim, D. L. Ewert, and E. P. Reddy, *J. Virol.* **66,** 5373 (1992).
[11] M. Castellazzi, G. Spyrou, N. Lavista, J. P. Dangy, F. Piu, M. Yaniv, and G. Brun, *Proc. Natl. Acad. Sci. U.S.A.* **88,** 8890 (1991).
[12] C. Desbois, B. Pain, C. Guilhot, M. Benchaibi, M. Ffrench, J. Ghysdael, J.-J. Madjar, and J. Samarut, *Oncogene* **6,** 2129 (1991).
[13] S. H. Hughes, J. J. Greenhouse, C. J. Petropoulos, and P. Sutrave, *J. Virol.* **61,** 3004 (1987).

cDNA. On the other hand, several adaptor plasmids simplify the conversion of cDNA into usable *Cla*I cassettes. A series of different adaptor plasmids provide combinations of signal elements necessary for proper expression of the cDNA-encoded products (Fig. 2). One version of the RC vector has been constructed with a short internal promoter to direct expression specifically to muscle cells.[9]

2. Swapping Genetic Elements

Avian cells exhibit various susceptibilities to infection by avian retroviruses depending on the subgroup of the virus envelope. RC vectors can be produced with any of the A, B, D, or E subgroup *env* genes which facilitates their use on various avian species or strains (Table I). The subgroup D

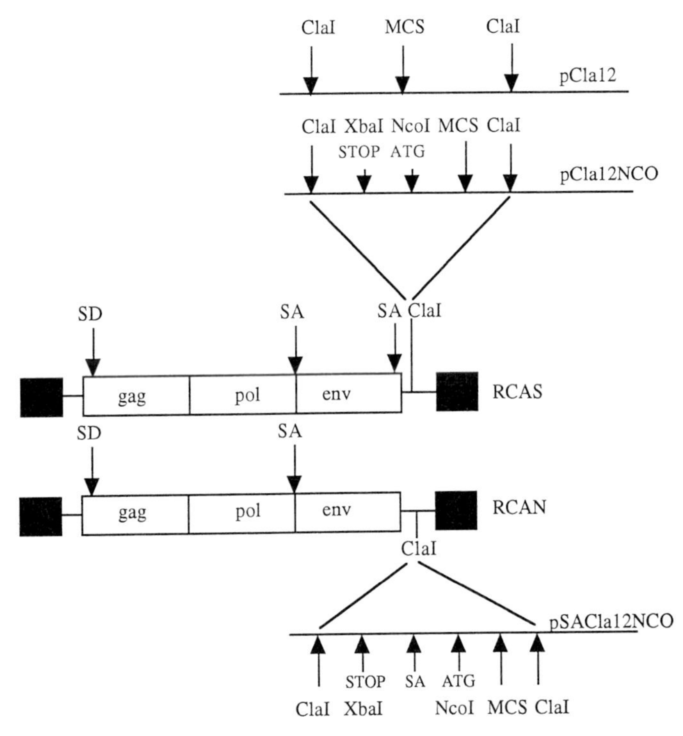

FIG. 2. Replication-competent avian retrovirus vectors and adaptor plasmids. SD, splice donor; SA, splice acceptor; MCS, multiple cloning site: 5′-*Eco*RI, *Sst*I, *Sma*I, *Bam*HI, *Xba*I, *Sal*I, *Pst*I, *Hin*dIII-3′. pCla12, pCla12NCO, and pSACla12NCO are the adaptors plasmids. The distance between the two *Cla*I sites is less than 200 nucleotides. RCAN vector differs from RCAS as it lacks the SA. Sequences inserted in RCAN should carry either a SA or a short transcription promoter.

virus can also be used to infect mammalian cells in culture. That several subgroup vectors are available is interesting in experimental oncogenesis, as it allows coinfection of the same cells with several different vectors to analyze oncogene cooperation.

The original vectors developed by Hughes were derived from the Schmidt Ruppin-A (SRA) strain of Rous sarcoma virus (RCAS). The Bryan high titer strain *gag-pol* sequences were later substituted for the homologous SRA sequences to generate the high titer RCAS-BP vectors. Different LTRs have also been substituted to the original RCAS LTRs. LTRs from the RAV(O) have been used to decrease the pathogenicity of the vectors (vectors RCOS and RCON) and AMV LTRs have been used to increase expression of the vector in hematopoietic cells (RCAM).

3. *Advantages and Drawbacks of RC Vectors*

One major advantage of replication competent vectors is that the genetic elements which control replication and spreading of the virus and the insert are physically linked. Consequently, these viruses can be produced at titers similar to natural RSV titer (around 10^6 infectious particles/ml). Unfortunately, these vectors exhibit a tendency to delete part of their genome, and vectors deleted of their insert can emerge and progressively predominate in the virus stock. For example, when a RCAS vector carrying the Neo^r gene was used to infect QT6 cells, 50% of the Neo^r-expressing cells carried a genome with large deletions.[14] Another drawback of these vectors is that the size of foreign sequences is limited to about 2 kb. As a consequence they cannot accommodate an additional selection gene in addition to the sequence to be transferred. This prevents any selection procedure in infected cells. Last, but not least, when they are injected *in vivo*, these vectors can elicit long-term (1–2 months) neoplasia (lymphoma, nephroblastoma, erythroblastosis) which might obscure investigation of the oncogenic properties of foreign sequences.

B. Replication-Defective (RD) Vectors

These vectors are deleted of most of viral genes and can then accommodate large inserts and several foreign sequences. Except for the first 200 nucleotides of *gag*, all the virus coding sequences can be replaced. A wide series of RD vectors with different inserts are available now (Table

[14] J. C. Olsen, C. Bova-Hill, D. P. Grandgenett, T. P. Quinn, J. P. Manfredi, and R. Swanström, *J. Virol.* **64**(11), 5475 (1990).

II)[15–28] and most of these vectors can be easily changed by substituting different oncogenic sequences.

1. Vectors with Single Gene

As they are short and simple to design, vectors with a single gene are preferred. The lack of a selection gene within the vector makes the screening for high producer cell clones more tedious. Foreign sequences can be inserted in frame with the remaining *gag* sequences and then a *gag*-fusion product is made whose functional properties are usually not altered. Thus *gag-LacZ* fusion products still exhibit efficient β-galactosidase activity, and corresponding vectors have been used for cell lineage studies.[15] This situation mimics that of several natural viral oncogenes.

2. Vectors with Two or More Genes

These vectors are usually designed to carry a dominant selection gene in addition to other genes of interest. Neo^r or $Hygro^r$ genes have been used in avian cells. However, strong selection can bias gene expression and can sometimes be detrimental to the expression of the nonselected gene.[25] In our hands these alterations are not predictable and seem to depend on the vector structure. To avoid an unbalanced expression of several genes, three strategies have been used.

[15] D. S. Galileo, G. E. Gray, G. C. Owens, J. Majors, and J. R. Sanes, *Proc. Natl. Acad. Sci. U.S.A.* **87,** 458 (1990).

[16] F. Flamant, D. Aubert, C. Legrand, F. L. Cosset, and J. Samarut, *J. Gen. Virol.* **74,** 39 (1993).

[17] F.-L. Cosset, C. Legras, J.-L. Thomas, R. M. Molina, Y. Chebloune, C. Faure, V. Nigon, and G. Verdier, *J. Virol.* **65,** 3388 (1991).

[18] C. F. Boerkoel, M. J. Federspiel, D. W. Salter, W. Payne, L. B. Crittenden, H. J. Kung, and S. H. Hughes, *Virology* **195,** 669 (1993).

[19] M. Fuerstenberg and B. Vennström, *Anal. Biochem.* **209,** 375 (1993).

[20] J. R. Sanes, S. Galileo, I. R. Ghattas, G. E. Gray, and J. E. Majors, *in* "Gene Transfer and Therapy in the Nervous System" (F. H. Gage and Y. Christen, eds.), p. 59. Springer-Verlag, Berlin, 1992.

[21] M. Benchaibi, F. Mallet, P. Thoraval, P. Savatier, J. H. Xiao, G. Verdier, and V. M. Nigon, *Virology* **169,** 15 (1989).

[22] M. Garcia and J. Samarut, *J. Virol.* **64,** 4684 (1990).

[23] C. Guilhot, M. Benchaibi, J. E. Flechon, and J. Samarut, *Oncogene* **8,** 619 (1993).

[24] M. Garcia and J. Samarut, *Proc. Natl. Acad. Sci. U.S.A.* **88,** 8837 (1993).

[25] N. La Vista, F. Piu, L. Loiseau, G. Brun, and M. Castellazi, *Oncol. Rep.* **1,** 185 (1994).

[26] N. Bachnou, V. Laudet, T. Jaffredo, B. Quatannens, S. Saule, and F. Dieterlen-Lievre, *Oncogene* **6,** 1041 (1991).

[27] C. Garrido, R. Ping Li, J. Samarut, D. Gospodarowicz, and S. Saule, *Virology* **192,** 578 (1993).

[28] D. Aubert, M. Garcia, M. Benchaibi, D. Poncet, Y. Chebloune, G. Verdier, V. Nigon, J. Samarut, and C. V. Mura, *J. Cell Biol.* **113,** 497 (1991).

TABLE II
AVIAN REPLICATION-DEFECTIVE VECTORS

Name	5′LTR	Structure	Note	Reference
pLZ10	RSV	gag-lacZ		15
OVA-D	RSV-D	gag-Neo, PRO, lacZ	SV40 promoter	16
OVA-Zcrip	RSV-D	gag-Neo, PRO, lacZ	Self-inactivating	16
NLB	RAV-2	SD gag-Neo, SA lacZ	RSV-D or RAV-1 LTR available	17
BBANβgal	RSV-BH	SD gag pol, SA v-src		18
SFCV-hisD	AEV	SD gag hisD, SA		19
Blacsarc2	AEV	SD lacZ, SA v-src		20
LZIS	RSV	lacZ, IRES, v-src	Translation reinitiation	20
XJ12	AEV	SD v-erbA, SA Neo-B		21
XJE26	AEV	SD gag-myb-ets, SA Neo-B		22
JN	AEV	SD gag-v-jun, SA Neo-A		22
JBN	AEV	SD gag-v-jun, SA v-erbA, SA Neo-A		23
ABN	AEV	SD v-erbA, SA v-erbB, SA Neo-A		23
HβR.N	AEV	SD gag-hRAR, SA Neo-A		24
HβR.BN	AEV	SD gag-hRAR, SA v-erbB, SA Neo-A		24
Rneo-JUN	AEV	SD gag-Neo, SA c-jun		25
MAHEVA	AEV	SD v-erbA, SA v-myc		26
JXF	AEV	SD gag-v-fos, SA Neo-A		27
TXN5′12S	AEV	SD gag-Neo, SA E1A 12S		23
TXN3′12S	AEV	SD SA E1A 12S, SA Neo-B		23
DAH5	AEV	polyA H5, PRO, PRO, Neo	Self-inactivating	28

a. Alternate Splicing. We have generalized the use of the Neo^r selection gene and have constructed convenient Neo^r cassettes that can be inserted in any replication defective vector. In the Neo-A cassette, the Neo^r gene is downstream from the splicing acceptor site derived from the RSV-*src* oncogene and produces the native Neo^r product. In the Neo-B cassette, the Neo^r gene is fused downstream from the splicing acceptor site of the AEV-v-*erbB* oncogene. In this case, the Neo^r product is initiated at the *gag* AUG and contains six Gag amino acids at its amino terminus which do not alter the enzymatic activity of the Neo^r product. The Neo-B cassette was used to generate vectors of the XJ series which have been widely used to derive multiple oncogene-carrying RD vectors (Table II). These cassettes were also used to construct vectors containing three different sequences, two of which are expressed from different subgenomic RNAs.[22,23] Selection genes can also be fused to the *gag* sequence but in this case, virus titers are lower.

Foreign sequences fused to *gag* can also be tested for their oncogenic potential. This strategy was used successfully to demonstrate the oncogenicity of altered retinoic acid receptors.[24]

b. Internal Promoter. RD vectors can accommodate an internal promoter, such as the simian virus 40 (SV40) early promoter.[16] In the case of Neo^r and LacZ the virus titer is five times lower than the one obtained with the alternate splicing strategy. Internal promoters should be used when LTR transcription activity is weak in target cells. Internal promoters with tissue-specific regulation might be usable but no experimental data have been provided yet.

c. Internal Entry Sites for Ribosomes. When two coding sequences are inserted in tandem into a vector the second one is poorly translated. Intercalation of the picornavirus virus entry site for translation reinitiation can restore the translation of the downstream coding sequence and then promote balanced expression of the two gene products in infected cells.[20]

3. Self-Inactivating (SIN) Vectors

These vectors carry a deletion in the 3′LTR to abolish LTR-driven transcription after one round of infection. This strategy is useful to prevent further dissemination of the vectors following accidental infection of the target cells with replication-competent retrovirus, and also to carry internal promoters whose function will not interfere with that of the LTR promoter. Such a vector has been used successfully to express histone H5 under a metallothionein inducible promoter in cultured avian fibroblasts.[28] SIN vectors are produced at a low titer presumably because of reduced packaging efficiency of the vector RNA genome.[16] Vector-producing cells as well as target cells infected with the vectors should be carefully screened for proper inactivation of the vector because recombination events can regenerate infectious secondary vectors.[16] Despite their low titer these vectors can be used to infect cells in culture.

C. Production and Titration of Avian Retrovirus Vectors

1. Production of RC Vectors

Viruses are recovered on transfection of the plasmid genomes into primary chicken embryo fibroblasts (CEFs) using calcium phosphate transfection as follows:

Cells (5×10^5) are seeded in a 60-mm dish in CEF medium (Dulbecco's modified Eagle's medium supplemented with 5% of fetal calf serum, 5% newborn calf serum, 1% chicken serum, and 5% tryptose phosphate broth (Gibco BRL) and left at 37° in a 10% CO_2 incubator.

The next day 250 μl of supercoiled DNA solution (5 μg in 0.2 *M* $CaCl_2$) is mixed with 250 μl of 2 × HBS [2 × HBS is 1.6% (w/v) NaCl, 0.074% KCl, 0.027% $Na_2HPO_4(2H_2O)$, 0.2% D-glucose, 1% HEPES, adjusted to pH 7.05 with NaOH] and left at room temperature for 30 min.

Cells are rinsed once with 1 × HBS (prepared by dilution of 2 × HBS) and 500 μl of the DNA solution is laid over the cells. After 20 min at room temperature, 2.5 ml of the CEF medium is added and the dishes are returned to the incubator for 4 hr.

The medium is then replaced by 500 μl of a glycerol solution [15% (v/v) in 1 × HBS] for 90 sec. Cells are then rinsed with CEF medium and returned to the incubator.

RC vectors are usually harvested 1 week later. Selected batches of fibroblasts producing virus at a high titer can be stored alive frozen in liquid nitrogen for future virus production. If the expression of the insert is detrimental to CEF growth, the virus is harvested 2 days after transfection. Virus passage is not recommended in every case.

The virus is collected from subconfluent cultures of producing cells after overnight incubation in serum-free medium. Cell debris is removed by centrifugation or filtration through a 0.45-μm pore size membrane. Polybrene (10 μg/ml) is added to the subgroup E virus only. The virus can be concentrated (10×) by ultracentrifugation (30 min at 35,000 rpm in SW41Ti) or by ultrafiltration through Centricon 30 (Amicon) before quick freezing and storage at −80° or in liquid nitrogen. The use of unfrozen supernatant on the day of harvest or the cocultivation of virus-producing cells and target cells provide a higher gene transfer efficiency.

2. Titration of RC Vectors

Titration of RC vectors can be performed on the quail fibroblast cell line QT6 for subgroup A and E viruses, or primary C/O CEFs for other subgroups according to the following protocol.[29,30]

Cells (10,000 cells per 10-mm-diameter well) are infected overnight with 10 μl of serial dilutions of the virus stock and the *gag*-p27 product is revealed *in situ* 48 hr later as follows:

cells are rinsed with phosphate-buffered saline (PBS) and are fixed for 10 min with a cold paraformaldehyde solution (4% in PBS, pH 7.0);

cells are rinsed several times with PBS, then incubated for 1 min in PBS containing 0.25% Triton $\times$-100, and then rinsed again with PBS. They are then incubated for 1 hr with a polyclonal rabbit anti-p27 antibody (Lifescience) at a 1/1000 dilution in PBS containing 4% bovine serum albumin (BSA);

cells are rinsed several times with PBS and then incubated for 30 min with a goat anti-rabbit IgG–alkaline phosphatase conjugate (Byosis) at a 1/500 dilution in PBS containing 4% BSA.

cells are extensively rinsed with PBS, then with AP buffer (0.1 *M* Tris, pH 9.5, 5 m*M* $MgCl_2$, 100 m*M* NaCl). Phosphatase activity is revealed in a solution containing 150 μg/ml of NBT (Promega), 300 μg/ml of BCIP (Promega), and 1 m*M* levamisole in AP buffer for 20 min at room temperature. Virus-infected cells appear as dark brown.

Expression of the insert itself can also be addressed either through immunoassay or *in situ* hybridization of its RNA. A functional titer can be assessed for vectors inducing oncogenic transformation of CEF (see below).

3. Production and Titration of RD Vectors

Defective ALV vectors can be produced in the same way as RC vectors by cotransfecting the vector plasmid with a plasmid carrying a helper genome (5 mol of vector plasmid/1 mol of helper plasmid) according to the protocol described earlier. In this case a mixture of helper and vector particles is produced. To avoid pathogenicity of the helper virus, it is safer to use packaging cells which produce helper-free vector viruses at high titers. Stably transfected helper cells which produce vectors containing a selection gene can be selected directly on transfection of the vector plasmid. For the vectors carrying a single non selectable gene, the vector plasmid is cotransfected with a selection plasmid (5/1 molar ratio). In both cases

[29] A. W. Stocker and M. L. Bissell, *J. Gen. Virol.* **68,** 2481 (1987).

[30] P. Savatier, C. Bagnis, P. Thoraval, D. Poncet, M. Belakebi, F. Mallet, C. Legras, F.-L. Cosset, J.-L. Thomas, Y. Chebloune, C. Faure, G. Verdier, J. Samarut, and V. Nigon, *J. Virol.* **63,** 513 (1989).

an extensive screening by supernatant titration of producer clones is recommended.

Isolde (subgroup A envelope) and Senta (subgroup E envelope) are the safest and the most widely used packaging cell lines for ALV vectors.[31,32] They have been derived from QT6 cells after the transfection of a *gag-pol-hygro* genome and a *env-phleo* genome. Preservation of the helper function requires regular selection of the cells with hygromycin (50 μg/ml, Boehringer-Mannheim) and phleomycin (30 μg/ml, Cayla Institute, Toulouse, France). Selection for Neo^r expression following vector plasmid transfection is ensured by G418 selection (400 μg/ml). A higher titer is achieved if drug selection is stopped several days before virus harvest. These two cell lines can be cross infected to increase the titer of the released helper-free vectors.[33]

Titration of RD vectors uses the same assay as RC vectors except that the detection of *gag*-p27 is not applicable in this case. The *gag*-p27 test is, however, usable to reveal any accidental rescue of the replication-competent helper virus. In this case, cells to be tested have to be passed two or three times to ensure dissemination and then better detection of any contaminating helper virus. For vectors carrying a selection gene, induction of drug resistance can be used for titration. Fibroblast cultures are infected with a serial dilution of the virus suspension (supplemented with 10 μg/ml of Polybrene for subgroup E), and G418 selection (0.2 mg/ml for primary CEF, 0.4 mg/ml for QT6) is applied 24 hr later and is maintained for about 8 days when resistant foci can be scored. For oncogenic vectors, oncogenic transformation of primary CEFs can be used similarly to RC vectors (see below).

D. Infection of Target Cells with RC and RD Vectors

The susceptibility of chicken cells to each virus subgroup is under genetic control. We use C/O chicken embryos and chicks available from Spafas Inc. (Norwich CT). Other avian strains can be used,[34] but the susceptibility of these strains to the various subgroups must be known. Subgroups A and E can infect quail cells. Because of interference with virus entry, cells that

[31] F.-L. Cosset, C. Legras, Y. Chebloune, P. Savatier, P. Thoraval, J.-L. Thomas, J. Samarut, V. Nigon, and G. Verdier, *J. Virol.* **64,** 1070 (1990).

[32] F.-L. Cosset, C. Ronfort, R. M. Molina, F. Flamant, A. Drynda, M. Benchaibi, S. Valsesia, V. Nigon, and G. Verdier, *J. Virol.* **66,** 5671 (1992).

[33] F.-L. Cosset, A. Girod, F. Flamant, A. Drynda, C. Ronfort, S. Valsesia, R. M. Molina, C. Faure, V. Nigon, and G. Verdier, *Virology* **193,** 385 (1993).

[34] J. L. Thomas, M. Afanassieff, F. L. Cosset, R. M. Molina, C. Ronfort, A. Drynda, C. Legras, Y. Chebloune, V. M. Nigon, and G. Verdier, *Int. J. Dev. Biol.* **36,** 215 (1992).

replicate a retrovirus of a given *env* subgroup cannot be further infected with particles from the same *env* subgroup.

1. Infection of Fibroblasts

Fibroblasts like CEFs and QT6 cells are cultured in CEF medium. For infection of these cells in a petri dish, the culture medium is removed from nonconfluent cultures and is replaced by a small amount of virus suspension sufficient to cover the cells. Cultures are then incubated overnight at 37°; fresh medium then is added. For detection of transformed foci, the culture medium is replaced by the same medium containing 0.35% Bacto-agar (Difco) and 0.5% dimethyl sulfoxide. An agar–medium overlay is added 3–5 days later and foci can usually be scored after a total of 10 days.[35]

2. Infection of Hemopoietic Cells

Chicken hemopoietic cells are rather difficult to infect *in vitro* because of the low frequency of potential target cells. Cells freshly collected from either bone marrow from late embryo or young chick, or embryonic yolk sac from 10- to 13-day embryos are incubated with the virus suspension during 60 min at 4° in a test tube. The cells are then directly seeded into semisolid culture containing G418 at 3 mg/ml for selection of erythrocytic cells and 1 mg/ml for selection of myeloid cells. Alternatively, cocultivation can be performed between a layer of virus-producing cells and hemopoietic cells. The coculture is maintained overnight at the culture temperature, and the hemopoietic cells are then collected and freed of contaminating fibroblasts by buoyant density fractionation. We use centrifugation at 1000*g* during 20 min at room temperature through a cushion of NycoPrep (Nycomed AS, Norway) or LSM (Organon Tecknica Corp.).[36] Virus-producing cells can be treated with mitomycin (10 μg/ml) for 2 hr before cocultivation, but the treatment is not necessary when these cells can be easily separated from the hemopoietic cells after cocultivation or do not impair growth of the infected target cells.

3. In Vivo Infection

Virus suspension (500 μl) can be injected *in vivo* into the wing vein of 2- to 3-week-old birds or into the femoral vein of newly hatched chicks. A higher efficiency in tumor and leukemia development is obtained with *in ovo* injections. The virus suspension (100 μl) is injected into a chorioallan-

[35] O. Gandrillon, P. Jurdic, M. Benchaibi, J. H. Xiao, J. Ghysdael, and J. Samarut, *Cell (Cambridge, Mass.)* **49,** 687 (1987).

[36] O. Gandrillon, P. Jurdic, B. Pain, C. Desbois, J. J. Madjar, M. G. Moscovici, C. Moscovici, and J. Samarut, *Cell (Cambridge, Mass.)* **58,** 115 (1989).

toic vein of 9- to 11-day-old chicken embryos or 6- to 7-day-old quail embryos. In most cases, neoplasia is evident soon after hatching. To ensure wide dissemination into most tissues, the virus can be injected in a small volume (10 μl) into either the subgerminal cavity of blastodermal embryos[34,37] or the coelom or heart of 3- to 5-day-old embryos.[26] We have obtained leukemias after *in ovo* injection of helper-free virus produced in a packaging cell line. However, a higher frequency of leukemia is reached with virus stocks containing a helper virus such as RAV-1 or RAV-2. In this case, vector-induced leukemias must develop soon after hatching (1–2 weeks) so that they cannot intermingle with the long-term neoplasias induced by the helper virus.

III. Murine Retrovirus Vectors

Most of the currently used murine retrovirus vectors are derived from the Moloney murine leukemia virus (Mo-MLV) provirus, but other strains such as the Harvey sarcoma virus,[38] the Friend MLV,[39] or the Friend spleen focus forming virus[40] have been used as well. These vectors are usually produced as helper-free stocks. They have proven to be efficient vectors for transferring genes into many different cell types in culture or for direct *in vivo* gene transfer. Human gene therapy protocols involving the use of murine retrovirus vectors are currently underway, including several trials in cancer patients.

A. Choosing Retroviral Vector Construct

Murine retrovirus vectors must possess *cis*-acting structures necessary for the completion of the retrovirus life cycle (Fig. 3A). All other elements, including the *gag-pol* and *env* genes, are dispensable and have been removed from provirus manipulated as part of a bacterial plasmid to generate MLV vectors. Convenient cloning sites are available in most vectors for inserting foreign DNA sequences. Although up to 8 kb of foreign sequences could theoretically be inserted, higher titers are obtained with an insert of less than 5 kb. This is large enough to accommodate sequences identified as oncogenes or of therapeutic relevance in cancers.

[37] S. T. Reddy, A. W. Stoker, and M. J. Bissell, *Proc. Natl. Acad. Sci. U.S.A.* **88,** 10505 (1991).
[38] T. J. Velu, L. Beguinot, W. C. Vass, M. C. Willingham, G. T. Merlino, I. Pastan, and D. R. Lowy, *Science* **238,** 1408 (1987).
[39] R. A. Feldman, D. R. Lowy, W. C. Vass, and T. J. Velu, *J. Virol.* **63,** 5469 (1989).
[40] J. P. Li, R. K. Bestwick, C. Spiro, and D. Kabat, *J. Virol.* **61,** 2782 (1987).

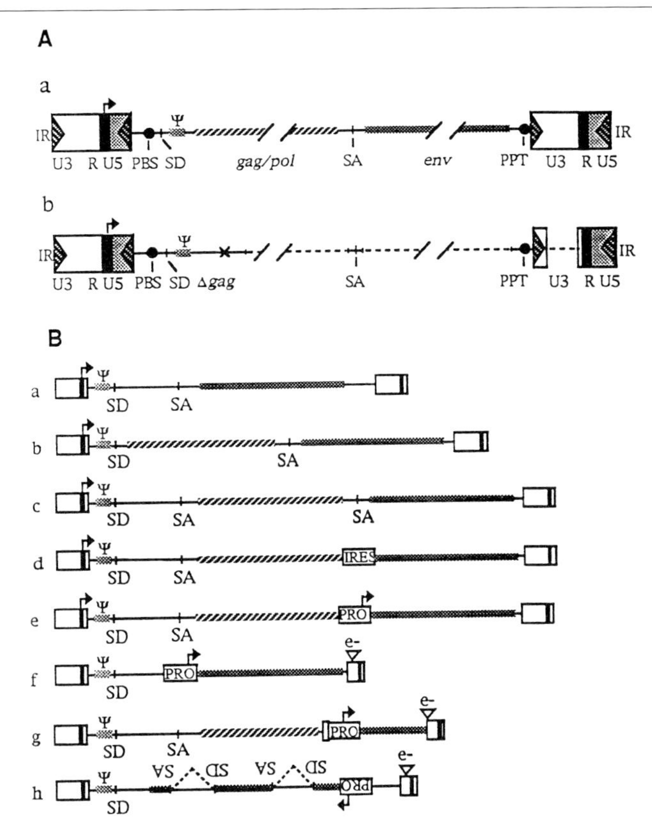

FIG. 3. Examples of murine defective retrovirus vectors. (A) Backbone structure of normal vector (a) and SIN vector deleted of the U3 enhancer in the LTR (b). (B) The structure of different MLV vector genomes. Boxes at the extremities represent the LTRs. Striped boxes and stippled boxes show inserts of foreign DNA material. PRO, internal promoter; e^-, enhancer deletion in the U3 domain.

1. Vectors Containing Foreign cDNA Controlled by the Retroviral LTR

cDNAs can be positioned at the *gag* initiation codon immediately downstream from the Ψ signal. However, packaging efficiency increases when 1 kb of the *gag* sequence is retained.[5] Synthesis of the *gag* product is abolished in both cases. In this construct the *env* splicing acceptor (SA) site is conserved and the cDNA is inserted at the *env* initiation codon as in the MFG vector (Fig. 3B,a). This vector often generates high infectious titers and provides a high level of expression in various cell types, including cancer cells.[41] Expression of the inserted cDNA seems even higher in vectors with a mutation (called B2) within the PBS sequence.[42]

To construct vectors containing two genes, one solution is to place one gene in a "*gag* configuration" and the other one in the "*env* configuration," as shown in Fig. 3B,b. To ensure balanced expression of both RNAs carrying the respective sequences, an additional SA site can be inserted upstream from the cDNA in *gag* (Fig. 3B,c). Vectors expressing a single bicistronic RNA (Fig. 3B,d) have been designed with the picornavirus IRES inserted between the two coding sequences.[5]

2. Vectors Containing Foreign cDNAs Controlled by Exogenous Promoters

These vectors are useful in the following situations: (1) expression in cell types where the MLV LTR is silent or poorly active, such as mouse embryonic stem cells[43] and mammalian primary fibroblasts implanted *in vivo*[44]; (2) when there is a need for a preferential or inducible[45] expression in designated cell types or tissues; or (3) as an alternative solution for expressing two cDNAs controlled by the LTR and the internal promoter, respectively. The widely used LXSN vector (Fig. 3B,e),[46] in which the cDNA of interest is inserted in the "*gag* configuration" and a Neo^r gene is expressed from an internal SV40 early promoter, usually generates high vector titers. pZEN vectors, which have been largely used for experimental oncogenesis, have a similar structure.[47]

Vectors containing an internal promoter are preferentially SIN vectors[48] carrying a deletion of the U3 enhancer domain of the LTRs. Advantages of SIN vectors containing an internal promoter are (1) to decrease interference between the two transcription units which often result in suppression of the expression[49]; and (2) to generate mRNAs with a nonviral structure that cannot be packaged into virions in case of accidental superinfection of the target cells with a replication-competent MLV. Figure 3B,f shows a SIN vector carrying one single gene. In some SIN vectors the cDNA of interest

[41] G. Dranoff, E. Jafee, A. Lazenby, P. Golumbek, H. Levitsky, K. Brose, V. Jackson, H. Hamada, D. Pardoll, and R. C. Mulligan, *Proc. Natl. Acad. Sci. U.S.A.* **90,** 3539 (1993).

[42] E. Barklis, R. C. Mulligan, and R. Jaenisch, *Cell* (*Cambridge, Mass.*) **47,** 391 (1986).

[43] B. C. Guild, M. H. Finer, D. E. Housman, and R. C. Mulligan, *J. Virol.* **62,** 3795 (1988).

[44] R. Scharfmann, J. H. Axelrod, and I. M. Verma, *Proc. Natl. Acad. Sci. U.S.A.* **88,** 4626 (1991).

[45] M. L. McGeady, P. M. Arthur, and M. Seidman, *J. Virol.* **64,** 3527 (1990).

[46] A. D. Miller, D. G. Miller, J. Victor-Garcia, and C. M. Lynch, this series, Vol. 217, p. 581.

[47] D. D. L. Botwell, S. Cory, G. R. Johnson, and T. J. Gonda, *J. Virol.* **62,** 2464 (1988).

[48] S. Yu, T. von Ruden, P. W. Kantoff, C. Garber, M. Seiberg, U. Rüther, W. F. Anderson, E. F. Wagner, and E. Gilboa, *Proc. Natl. Acad. Sci. U.S.A.* **83,** 3194 (1986).

[49] M. Emerman and H. M. Temin, *Mol. Cell. Biol.* **6,** 792 (1986).

fused to a foreign promoter is inserted in place of the 3′LTR U3 region, and a selective marker gene is inserted in the "*env* configuration"[50] (Fig. 3Bg). In this case the gene in U3 is duplicated at both extremities of the integrated provirus. These vectors provide a high level of expression of the carried cDNA but their titers may be low.

3. *Vectors Carrying Genes with Intron Sequences*

When intron sequences are necessary for tissue-specific expression, as for the β-globin gene,[51] genomic sequences instead of cDNA must be accommodated into retrovirus vectors. This can be performed by inserting the sequences in the opposite direction relative to the retroviral transcription (Fig. 3B,h). Such vectors should be preferentially self-inactivating to prevent promoter interference. Nevertheless, so far they have been characterized by frequent rearrangements, are produced at low infectious titers, and usually promote poor tissue-specific expression.

B. *Production of Stocks of Helper-Free Retrovirus Vectors*

MLV packaging cell lines are used for the production of helper-free retrovirus vector particles. Packaging cell lines ensuring a total absence of replication-competent particles in vector stocks are currently available.[52] However, accidental contamination of vector-producing cells with a replication-competent virus is still possible and must be prevented by confining cultures to restricted and controlled areas. It is recommended that the occurrence of replication-competent viruses in packaging cells is regularly checked with appropriate highly sensitive assays (see below). Packaging cell lines have been isolated that produce Mo-MLV capsids coated with envelope glycoproteins of different origins such as (1) the ecotropic envelope to infect mouse cells; (2) the amphotropic or gibbon ape leukemia virus[53] envelope to infect mammalian cells; and (3) the vesicular stomatitis virus envelope to infect both mammalian and nonmammalian cells.[54]

[50] P. A. Hantzopoulos, B. A. Sullenger, G. Ungers, and E. Gilboa, *Proc. Natl. Acad. Sci. U.S.A.* **86,** 3519 (1989).

[51] E. A. Dzierzak, T. Papayannopoulou, and R. C. Mulligan, *Nature* (*London*) **331,** 35 (1988).

[52] O. Danos, *in* "Construction of Retroviral Packaging Cell Lines" (M. K. L. Collins, ed.), p. 8. Humana Press, Clifton, NJ, 1991.

[53] A. D. Miller, J. V. Garcia, N. von Shur, C. M. Lynch, C. Wilson, and M. Eiden, *J. Virol.* **65,** 2220 (1991).

[54] J. C. Burns, T. Friedmann, W. Driever, M. Burrascano, and J. K. Yee, *Proc. Natl. Acad. Sci. U.S.A.* **90,** 8033 (1993).

1. *Production of Vectors after Transient Plasmid Expression*

The first step in preparing a vector stock is the transfection of the proviral construction into packaging cells. Any transfection procedure can be used for this purpose. Vector particles are transiently released in culture medium within 48 to 96 hr following transfection. Infectious titers are usually low, except if the transfection process is highly efficient.[55] Quality controls are not easy to perform on this material. Nevertheless, producing retrovirus vectors in transient assays is useful (1) to rapidly verify that the vector particles are capable of infecting a given target cell (see below); (2) to produce vector particles which carry a gene expressing a product that is toxic to the packaging cells; and (3) to efficiently introduce vector genomes into a second packaging cell line through infection. In this latter case the first and second packaging cells must express envelope glycoproteins of different subgroups to avoid inhibition of infection due to interference. Using this procedure, cells producing high virus titers can be isolated.[56] However, iterative reciprocal superinfections of packaging cells in a "Ping-Pong" system, which allows the production of very high virus titers, often lead to the generation of replication-competent particles.[57] These last procedures are of course not applicable to the production of SIN vectors.

2. *Production of Vectors after Stable Plasmid Expression*

Transfection into packaging cells is most frequently used to isolate stable cell clones that are further screened for the production of infectious particles. We use the following procedure. Transfected cells are selected on the basis of their resistance to G418 (1 mg/ml), which is conferred either by the vector genome or by cotransfecting a Neo^r expression plasmid. Around 50 cell clones are picked after trypsin treatment in cloning rings and are individually grown in 24-well plates. When cells reach confluence, the supernatant from each clone is assayed for infectious vector particles, using a rapid assay (see below) that avoids amplifying and/or freezing numerous cell lines. Three to six selected clones are further submitted to quality controls (see below). NIH/3T3 fibroblasts are convenient target cells from this assay, but any cell time susceptible to MLV infection can be used.

[55] W. S. Pear, G. P. Nolan, M. L. Scott, and D. Baltimore, *Proc. Natl. Acad. Sci. U.S.A.* **90,** 8392 (1993).
[56] R. K. Bestwick, S. L. Kozack, and D. Kabat, *Proc. Natl. Acad. Sci. U.S.A.* **85,** 5404 (1988).
[57] D. M. Bodine, K. T. McDonagh, P. A. Brandt, B. Ney, B. Agricola, E. Byrne, and A. W. Nienhuis, *Proc. Natl. Acad. Sci. U.S.A.* **87,** 3738 (1990).

3. *Method for Rapid Screening for High-Producer Cell Clones*

Supernatants (0.5 ml) from candidate vector-producing clones grown in 24-well plates are collected, supplemented with 8 μg/ml polybrene, filtered through a 0.45-μm filter, and incubated for 2 hr at 37° with target cells also grown in 24-well plates. After the addition of 3 vol of fresh culture medium, cells are grown to confluence and are analyzed for the transferred vector using any of the following techniques:

1. Vector-specific PCR can be used for any constructed vector. DNA is rapidly extracted from infected cells and primers located in the region of the Mo-MLV upstream of the 3′ LTR (5′-GACCACTGATATCCTGT-CTTTAAC-3′), and in the carried foreign DNA are used to amplify a vector-specific fragment. A signal is generated only from infected target cells.
2. Testing the biological activity of the vector is a convenient means of identifying high producers when the vector carries and expresses a marker gene such as LacZ or a gene inducing drug resistance.
3. Looking for the expression of the protein product generated by the transferred foreign sequence in infected cells, provided that a rapid enzymatic, immunological, or functional assay is available.

Even poorly sensitive assays are useful since only the clones conferring high biological activity in target cells are of interest.

4. *Analysis of Selected Vector-Producer Cell Clones*

Once candidate high-producer clones have been recognized, a precise quantification of the number of vector genomes transferred to target cells must be assessed by Southern blot analysis of the infected cells. In addition, this assay allows detection of undesirable clones that transfer highly re-arranged vector copies.[58] Vector-producing packaging cell clones are grown in petri dishes. The medium is changed and 10 ml of culture supernatant is collected 24 hr later and is processed as described earlier to infect subconfluent target cells (usually NIH/3T3) grown in 10-mm petri dishes. The infected cells are grown to confluence and are used to prepare high molecular weight DNA. This DNA is digested to generate an internal vector fragment and is hybridized with a specific vector probe. Amounts of plasmid DNA corresponding to, respectively, 1 and 0.1 copies of vector genome per cell are added to negative control cell DNA and are run in parallel on the same gel. The most accurate quantification of the number of vector genomes per infected cell is obtained by analyzing blots with a phosphorimager. The detection of a DNA sequence endogenous to the cells can often

[58] H. Xu and D. Littman, *Cell (Cambridge, Mass.)* **74,** 633 (1993).

be used as an internal standard for estimating the amount of loaded DNA. Clones transferring an average of one or more vector genomes per target cell are considered high producers. For most vectors, the transfer of one vector copy per target cell is approximately equivalent to a concentration of 10^6 infectious particles per ml of virus stock.

Direct titration of infectious particles is feasible only for vectors encoding either a selection marker like Neo^r or a product readily detectable in the cells like β-galactosidase, or for vectors inducing oncogenic transformation of NIH/3T3 cells. Such measurements depend on the level of expression needed for a phenotypic change in target cells and is not as accurate as the direct quantification of the presence of a vector genome by Southern blotting.

A higher titer is obtained when the virus is collected from confluent cultures over 24 hr. The vector virus concentration can be increased up to 100× by ultracentrifugation (30 min at 35,000 rpm in SW41Ti) when only small volumes are required.[55] Tangential ultrafiltration in a Sartocon minimodule with a polysulfone membrane (porosity 100,000, Sartorius) allows the treatment of large or very large volumes, but the concentration of the virus cannot exceed 25-fold.[59] Yields are usually up to 80%. Crude or concentrated vector preparations can be stored for a few days at 4° without loss of vector titers. For a period longer than a week, storage at −80° or in liquid nitrogen is necessary.

5. *Detection of Contaminating Replication-Competent Retrovirus Particles*

This is performed by infecting target cells that are further passed and grown for at least 2 weeks. Their culture supernatant is used to infect an indicator cell line containing a defective retrovirus that can be rescued by a helper virus. Various indicator cell lines are available expressing either an oncogenic transforming retrovirus (S^+L^-) or a retrovirus vector containing a marker such as the Neo^r or lacZ gene.[46,52,60]

The supernatant of these infected indicator cells is then tested for its ability to transfer the indicator genome to naive target cells. The target and indicator cell lines should be chosen according to the tropism of the potential helper virus to be detected.

C. Infection of Target Cells

In most situations, MLV vectors are used for *in vitro* gene transfer. Infections must always be performed in the presence of 8 μg/ml Polybrene.

[59] R. W. Paul, D. Morris, B. Hess, J. Dunn, and R. W. Overell, *Hum. Gene Ther.* **4,** 609 (1993).
[60] R. H. Bassin, N. Tutle, and P. J. Fishinger, *Nature* (*London*) **229,** 564 (1971).

TABLE III
ONCOGENIC SEQUENCES INTRODUCED INTO MAMMALIAN CELLS USING MURINE RETROVIRUS VECTORS[a]

Oncogenic sequence	Vector type[b]	Vector structure	Reference
v-*abl*	Abelson virus		61
v-*abl*	a		62
v-*fms*	FeSV-MD virus		63
v-*src*	?		64
v-*erb*	e	(Neo^r, TK-v-*erb*)	65
Py. mT/v-*src*	e	(Neo^r, TK-v-*onc*)	66
GM-CSF	a		67
IL-3	c	(Neo^r, IL-3)	68
IL-6	e	(IL-6, PGK-Neo^r)	69
bcr/abl	e	(bcr/abl, TK-Neo^r)	70
bcr/abl	e	(bcr/abl, TK-Neo^r)	71
bcr/abl	a		72
Anti-sens bcr/abl	e	(Neo^r, SV-anti-bcr/abl)	73
E2A-Pbx1	a		74
PML-RARα	e	(PML-RARα, SV-Neo^r)	75
Hox-2.4	e	(Neo^r, SRα-Hox2.4)	76

[a] Py-mT, polyoma virus middle T antigen, IL-6, interleukin 6.
[b] See Fig. 3B.

However, the most efficient procedure to infect cultured cells is to cocultivate them for 48–72 hr with vector-producing cells in the presence of Polybrene. Growth inhibition of vector-producing cells by irradiation (single dose of 3000 rads) or by treatment with mitomycin (10 μg/ml) immediately before cocultivation does not decrease virus infection but avoids further contamination of the population of target cells.

Examples of murine retrovirus vectors used to introduce oncogenic sequences into mammalian cells are given in Table III.[61–76]

[61] P. L. Green, D. A. Kaehler, and R. Risser, *Proc. Natl. Acad. Sci. U.S.A.* **84,** 5932 (1987).
[62] M. L. Scott, R. A. Van Etten, G. Q. Daley, and D. Baltimore, *Proc. Natl. Acad. Sci. U.S.A.* **88,** 6506 (1991).
[63] J. M. Heard, M. F. Roussel, C. W. Rettenmeier, and C. J. Sherr, *Cell (Cambridge, Mass.)* **51,** 663 (1987).
[64] G. Keller and E. F. Wagner, *Genes Dev.* **3,** 827 (1989).
[65] T. von Rüden, S. Kandels, T. Radaszkiewicz, A. Ullrich, and E. Wagner, *Blood* **79,** 3145 (1992).
[66] A. Aguzzi, P. Kleihues, K. Heckl, and O. D. Wiestler, *Oncogene* **6,** 113 (1991).
[67] G. R. Johnson, T. J. Gonda, D. Metcalf, I. K. Hariharan, and S. Cory, *EMBO J.* **8,** 441 (1989).
[68] C. E. Dunbar, T. M. Browder, J. S. Abrams, and A. W. Nienhuis, *Science* **245,** 1493 (1989).
[69] R. G. Hawley, A. Z. Fong, B. F. Burns, and T. S. Hawley, *J. Exp. Med.* **176,** 1149 (1992).
[70] G. Q. Daley, R. A. Van Etten, and D. Baltimore, *Science* **247,** 824 (1990).

IV. Conclusion and Future Prospects

Retrovirus vectors have been especially useful for demonstrating the oncogenicity of human rearranged genes in animal experimental models. More recently they have been used for the marking of tumor cells *in vivo*. This experimental approach is useful in analyzing the behavior of tumor cells when these cells are replaced in the environment of normal tissues.[77] The use of this methodology for tracing human tumor-infiltrating lymphocytes[78] or residual cancer cells after chemotherapy[79] brings more promising prospects. Retroviruses also provide vehicles from which new therapeutic interventions in cancer cells are made feasible, such as the introduction of the suicide gene[80] or cytokine cDNA supposed to increase antitumor immunity.[42]

Acknowledgments

We thank M. Castellazzi, O. Danos, M. Garcia, O. Gandrillon, C. Guilhot, P. Jurdic, and B. Pain for suggestions and all those who provided us with their data.

[71] J. McLaughlin, E. Chianese, and O. N. Witte, *Proc. Natl. Acad. Sci. U.SA.* **84,** 6558 (1987).
[72] A. G. Elefanty, I. K. Hariharan, and S. Cory, *EMBO J.* **9,** 1069 (1990).
[73] P. Martiat, P. Lewalle, A. S. Taj, M. Philippe, Y. Larondelle, J. L. Vaerman, C. Wildmann, J. M. Goldman, and J. L. Michaux, *Blood* **81,** 502 (1993).
[74] M. P. Kamps and D. Baltimore, *Mol. Cell. Biol.* **13,** 351 (1993).
[75] F. Grignani, P. F. Ferrucci, U. Testa, G. Talamo, M. Fagioli, M. Alcalay, A. Mencarelli, F. Grignani, C. Peschle, I. Nicoletti, and P. G. Pelicci, *Cell (Cambridge, Mass.)* **74,** 423 (1993).
[76] A. C. Perkins and S. Cory, *EMBO J* **12,** 3835 (1993).
[77] C. Bagnis, F.-L. Cosset, J. Samarut, G. Moscovici, and C. Moscovici, *Oncogene* **8,** 737 (1993).
[78] S. A. Rosenberg, P. Aebersold, K. Cornetta, A. Kasid, R. A. Morgan, R. Moen, E. M. Karson, M. T. Lotze, J. C. Yang, S. L. Topalian, M. J. Marino, K. Culver, A. D. Miller, R. M. Blaese, and W. F. Anderson, *N. Engl. J. Med.* **323,** 570 (1990).
[79] M. K. Brenner, D. R. Rill, R. C. Moen, R. A. Krance, J. Mirro, W. F. Anderson, and J. N. Ihle, *Lancet* **341,** 85 (1993).
[80] K. W. Culver, S. Wallbridge, H. Ishii, E. H. Oldfield, and R. M. Blaese, *Science* **256,** 1550 (1992).

[15] Identification of Protein–Protein Interactions by λgt11 Expression Cloning

By ELIZABETH M. BLACKWOOD and ROBERT N. EISENMAN

Protein–protein interactions govern a wide variety of fundamental biological processes, ranging from the control of enzymatic activity through subunit association and enzyme–substrate recognition, to the specific interactions of protein ligands with their receptors. Intermolecular associations are also central to the manner in which oncogene-encoded proteins regulate cell proliferation. For example, the signal transduction pathways within which tyrosine and serine/threonine kinases [such as the epithelial growth factor (EGF) and colony-stimulating factor-1 receptors, Src and Raf] and G proteins (such as the Ras family) function appear to be largely mediated by highly specific interactions between pathway components. Nuclear oncoproteins, many of which act as transcription factors, also function through homotypic or heterotypic interactions. Consequently, the DNA binding activities of the oncogenic transcription factors Fos, Jun, ErbA, and Myc are dependent not only on DNA contact surfaces, but also on domains that permit specific dimer formation. Other nuclear oncoproteins are also critically regulated by protein–protein interaction. For example, the transcription factor NF-κB is regulated through association with I-κB leading to cytoplasmic sequestration. The Myb protein binds to DNA as a monomer but depends on interaction with other factors for positive and negative modulation of its transcriptional activity. Furthermore, the mechanism by which these oncoproteins influence gene transcription is likely to occur through interaction with specific components of the basal transcription machinery.

The realization that highly specific protein–protein interactions mediate cell behavior at almost every level has led to attempts to develop general methods with which to identify molecules that physically associate with a given protein. These approaches fall roughly into two classes: biochemical and genetic. The yeast "two-hybrid" interaction cloning system, discussed in [16] in this volume, is an example of the latter. Although it is a highly sensitive assay and relatively straightforward to set up, it employs transcriptional activity as a readout and therefore requires special modification for use with "target" proteins that are themselves transcriptionally active. In addition, the limited repertoire of yeast-specific post-translational modifications may not confer biological activity to either of the hybrid polypeptides. Finally, it is possible that the interaction observed is mediated indirectly through a yeast component.

When correct processing and post-translational modification are a prerequisite for molecular interaction, a modified genetic strategy that employs transfection of a cDNA expression library into mammalian cells can be used. Such a technique has been used quite successfully to identify several cell surface receptors on the basis of their ligand binding activities.[1-3] The limitations of this approach lie in its dependence on a sensitive biological assay and the eventual recovery of the transfected DNA.

A frequently used biochemical approach to identify specific molecular interactions has been to employ an immobilized target protein to adsorb unknown binding partners from cell extracts. Low-affinity interactions that might be stabilized by the high concentrations of protein (as in the two-hybrid system) are likely to be missed by using this method. Furthermore, any polypeptide that specifically associates with the target protein must be purified in quantities sufficient for amino acid sequence analysis, and then cloned, before the nature of its interaction can be established with any certainty.

Another method that has had considerable success is "expression cloning" (Fig. 1). In this procedure, bacteriophage that have been engineered to express a cDNA library are allowed to produce plaques on a bacterial lawn. The lawn is transferred to nitrocellulose filters and is probed with a labeled target protein. After washing to decrease nonspecific binding, positive plaques are identified by autoradiography. Following phage purification through successive rounds of screening, sequence analysis of the cDNA insert frequently leads to direct identification of the protein and its interaction domain. This approach does not depend necessarily on any intrinsic activity of the protein, aside from its ability to interact with the target protein, and, because it is carried out in prokaryotic extracts, is less likely to involve a host protein intermediate (unless the target itself is prokaryotic).

One advantage of the expression cloning method is that the actual screening is carried out *ex vivo*. This permits a greater degree of control over the preparation and labeling of the probe than in the two-hybrid system. For example, in a search for proteins that recognize a phosphotyrosine-containing peptide, expression cloning would be the method of choice. Whereas many peptide probes can be phosphorylated *in vitro* using a

[1] A. Aruffo and B. Seed, *Proc. Natl. Acad. Sci. U.S.A.* **84,** 8573 (1987).

[2] D. P. Gearing, J. A. Kling, N. M. Gough, and N. A. Nicola, *EMBO J.* **8,** 3667 (1989).

[3] X.-F. Wang, H. Y. Lin, E. Ng-Eaton, J. Downward, H. F. Lodish, and R. A. Weinberg, *Cell (Cambridge, Mass.)* **67,** 797 (1991).

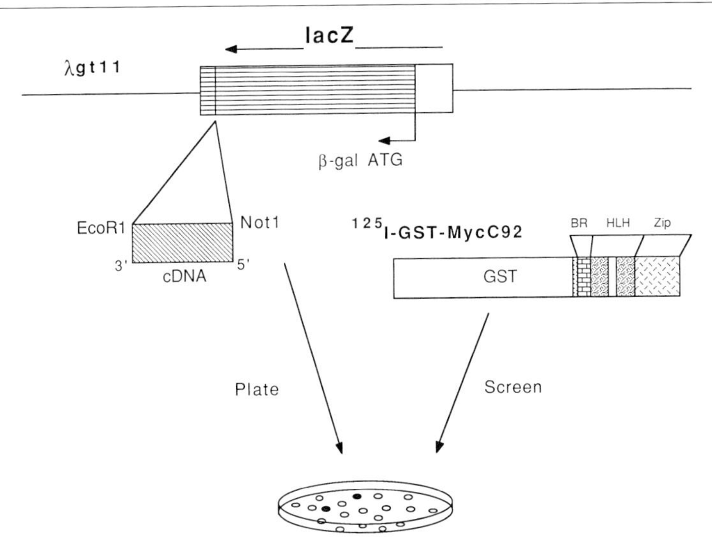

FIG. 1. Diagram of the cloning strategy used to identify the heterodimeric partner(s) for c-Myc. Phage from a λgt11 expression library produce nearly full-length β-galactosidase (*lacZ*) proteins fused with the open reading frames of a directionally inserted cDNA. A radiolabeled Myc fusion protein (GST-MycC92), which contains both helix–loop–helix and leucine zipper dimerization motifs, was used to identify reactive plaques (Max) on the basis of specific protein–protein interaction (BR, basic region; HLH, helix–loop–helix; Zip, leucine zipper; GST, glutathione *S*-transferase). [Adapted with permission from Blackwood and Eisenman, *Science* **251,** 1211 (1991). Copyright 1991 American Association for the Advancement of Science.]

tyrosine kinase, the two-hybrid system must rely on *in vivo* tyrosine phosphorylation which is not known to occur efficiently in yeast. Expression cloning also allows the use of a wide variety of probes, including RNA and DNA as well as peptides and proteins. Although originally employed for the detection of proteins reactive with monoclonal and polyclonal antibodies,[4] expression cloning has also been successful in detecting proteins that specifically interact with defined DNA[5,6] and RNA[7] sequences. Specific protein–protein interactions identified by using this strategy include those between Fos and BP1[8]; Myc, Max, and Mad[9,10]; and between the phosphory-

[4] R. A. Young and R. W. Davis, *Proc. Natl. Acad. Sci. U.S.A.* **80,** 1194 (1983).

[5] C. R. Vinson, K. L. LaMarco, P. F. Johnson, W. H. Landschulz, and S. L. McKnight, *Genes Dev.* **2,** 801 (1988).

[6] H. Singh, R. G. Clerc, and J. H. LeBowitz, *BioTechniques* **7,** 252 (1989).

[7] Z. Qian and J. Wilusz, *Anal. Biochem.* **212,** 547 (1993).

[8] P. F. Macgregor, C. Abate, and T. Curran, *Oncogene* **5,** 451 (1990).

[9] E. M. Blackwood and R. N. Eisenman, *Science* **251,** 1211 (1991).

[10] D. E. Ayer, L. Kretzner, and R. N. Eisenman, *Cell (Cambridge, Mass.)* **72,** 211 (1993).

lated EGF receptor and GRB-1.[11] The following sections describe in detail methods used for the detection of specific heterotypic interactions using a protein probe, based largely on our experiences with the identification of Max and Mad.

Protein Probe

In consideration of the chemical and structural determinants that mediate protein–protein interaction, the protein probe should be designed to contain a region that shares structural homology with other known dimerization motifs [e.g., leucine zippers, helix–loop–helix (HLH) domains] or that has been implicated functionally in bimolecular interactions (e.g., phosphopeptides, transcriptional activation domains, and binding "pockets"). In the case of Myc, we chose the C-terminal 92 amino acids that contain contiguous HLH and leucine zipper (LZ) dimerization motifs. This region was expressed as a fusion protein with glutathione *S*-transferase (GST-MycC92).

The engineering of a glutathione *S*-transferase fusion protein afforded several advantages. The fusion protein can be purified readily by affinity chromatography under nondenaturing conditions, and subsequently, the enzyme moiety provides a nonessential domain for labeling purposes. Though fusion proteins containing more extensive regions of c-*myc* were poorly expressed as well as insoluble, the C-terminal 92 residues are quite hydrophilic and retained solubility.

Since GST-MycC92 contained a basic region as well as dimerization domains, the functional properties of this fusion protein could be characterized based on its limited homodimerization and DNA binding capabilities. One potential drawback to the use of bacterially expressed HLH or LZ proteins, however, is their ability to self-associate, either as physiological homodimers or at elevated concentrations. Hence, the efficiency of subunit exchange at low temperature or the difference between solution and solid phase kinetics should be considered. As Myc homodimerizes rather weakly, irreversibly stable dimers were less of a concern.

The method of choice for the detection of specific protein–protein interactions is to radiolabel the probe with either ^{125}I or ^{32}P (see Table I for abbreviated protocols). As a nonradioactive alternative, the probe might be directly conjugated to enzymes such as HRP or alkaline phosphatase; however, these large protein moieties could sterically hinder subsequent molecular interactions. Although biotin is a much smaller molecule, and as

[11] E. Y. Skolnik, B. Margolis, M. Mohammadi, E. Lowenstein, R. Fischer, A. Drepps, A. Ullrich, and J. Schlessinger, *Cell (Cambridge, Mass.)* **65,** 83 (1991).

TABLE I
LABELING OF PROTEIN PROBE

A. Iodination on tyrosine residues
1. Prepare fusion protein in the absence of reducing agents.
2. Mix 50 μg of protein [in 950 μl Tris or phosphate-buffered saline (PBS)[a]] with 5 mCi $Na^{125}I$ (Amersham IMS30) and 3 iodobeads (immobilized chloramine-T; Pierce).
3. Incubate for 15 min at room temperature with occasional shaking.
4. Quench reaction with 50 μl of 5 mg/ml sodium metabisulfite.
5. Desalt immediately on a PD-10 column (Pharmacia; Sephadex G-25) preequilibrated with PBS/0.1% NP-40 (v/v)/0.5% powdered milk (w/v). The presence of carrier protein reduces loss of the labeled probe to the column resin.
6. Store probe at 4° with an antimicrobial agent. Depending on the functional stability, the labeled probe can be used for 6–8 weeks.

B. Phosphorylation
1. Prepare 5 μg of protein in 50 m*M* Tris, pH 7.5, 100 m*M* NaCl, 5 m*M* $MgCl_2$ (or other required divalent cation), 5 μM ATP, and 1.0% aprotinin (v/v).
2. Add 1μCi/μl [γ-^{32}P]ATP (3000 Ci/mmol, Dupont, New England Nuclear) and the recommended activity units of kinase.
3. Incubate for 30 min at 30° with occasional mixing.
4. Remove unincorporated [γ-^{32}P]ATP by desalting on a Sephadex G-50 column (or its equivalent) equilibrated in PBS/0.1% NP-40/10% glycerol/10 m*M* EDTA/10 m*M* β-glycerophosphate. The collected void volume may be supplemented wtih carrier protein.
5. Store the labeled probe at 4° with antimicrobial agents [e.g., 0.01% NaN_3 (w/v)].

[a] PBS, 137 m*M* NaCl/2.7 m*M* KCl/4.3 m*M* Na_2HPO_4/1.4 m*M* KH_2PO_4, pH 7.3.

such less likely to sterically interfere, biotinylation often generates excessive background problems in the expression screen.

Depending on the amino acid composition of the putative oligomerization interface, primary amines or tyrosine residues can be chemically iodinated with Bolton–Hunter reagent or chloramine-T, respectively. Iodobeads (immobilized chloramine-T) were used in the iodination of GST-MycC92. Theoretically, the 17 tyrosines found in GST should incorporate most of the radioisotope into the "nonfunctional" half of the fusion protein. In practice, however, the single tyrosine residue within helix II of Myc became disproportionally labeled. Although high specific activities could be achieved (estimated 1.4 [^{125}I]/GST-MycC92 molecule; 72 μCi/μg), preparations with 20-fold lower specific activities worked equally well in the screen and yielded a proportionate ratio of signal-to-noise.

Labeling of the probe with protein kinases can rely on naturally occurring phosphorylation sites within the polypeptide or protein. For instance, multiple casein kinase II phosphorylation sites within the C terminus of Max were exploited in the expression screen that identified Mad as a novel heterodimeric partner. If the protein probe lacks phosphorylation sites or if phosphorylation is suspected to negatively regulate specific association, the pGEX expression vector can be engineered to create a unique kinase recognition sequence.[12,13]

Expression Library

Central to the success of any expression screen are the biological considerations given to the choice of a suitable expression library (i.e., species-specific, spatial, temporal, or developmental patterns of function or expression). For example, in the screen for Myc dimerization partners, the λgt11 expression library was constructed from the RNA of a baboon lymphoid cell line (594S). Since c-*myc* is involved in many forms of B-cell neoplasia, we expected to find a Myc-associated factor represented in such a B-cell library. Myc expression also correlates closely with cell proliferation, which gives RNAs derived from a cultured cell line a distinct advantage over those from tissues of unknown mitotic state. The 594S λgt11 library had previously been used to isolate overlapping clones of CD44, both by hybridization and antibody screening. Access to such a characterized library afforded a real advantage.

Since dimerization interfaces need not be large, nor the protein expressed in its entirety, cDNAs should be randomly primed, size selected (350 bp minimum), and cloned into a λgt11-based expression vector. Directional cloning reduces the number of potential reading frames from 6 to 3, and thus effectively doubles the representation of relevant clones. To illustrate how size selection and the choice of cDNA primer can influence the generation of functional fusion proteins, we identified only two interactive Max inserts from the 10^6 plaques screened. In retrospect, this is not surprising given that the dimerization motifs are in the extreme amino terminus of the molecule and that an in-frame stop codon located two codons upstream of the initiating AUG of Max limits the numer of functional fusion points with *lacZ* to under 40 (i.e., those constructs that retain the contiguous HLH and LZ motifs later demonstrated to be necessary and sufficient for dimerization).

Two types of λ bacteriophage vectors are available for the construction of expression libraries, and each offers different advantages. With λgt11 expression vectors, cDNA inserts are cloned into the carboxy terminus of

[12] M. A. Blanar and W. J. Rutter, *Science* **256,** 1014 (1992).

[13] W. G. Kaelin, Jr., W. Krek, W. R. Sellers, J. A. DeCaprio, F. Ajchenbaum, C. S. Fuchs, T. Chittenden, Y. Li, P. J. Farnham, M. A. Blanar, D. M. Livingston, and E. K. Flemington, *Cell* (*Cambridge, Mass.*) **70,** 351 (1992).

the β-galactosidase protein (116 kDa); this fusion confers uniform production and stability to most products. In contrast, Lambda ZAP vectors (Stratagene) have multiple cloning sites within the amino terminus of the *lacZ* coding region (fusion proteins contain 36 N-terminal residues of β-galactosidase), which leads to greater variation in the production and stability of the expressed proteins. Lambda ZAP vectors, however, can accept larger cDNA inserts (up to 10 kb), and subsequent manipulations, such as sequencing and subcloning, are simplified by *in vivo* excision of the insert-containing pBluescript phagemid. It is also important to note that not all bacterial strains are suitable for propogating both types of bacteriophage vectors. To illustrate, bacterial host strains that contain the plasmid pMC9 (e.g., Y1088, Y1089, and Y1090) could undergo undesired homologous recombination with an introduced Lambda ZAP vector. The laboratory manuals edited by Ausubel *et al.*[14] and Sambrook *et al.*[15] are excellent sources of information on the construction, propagation, and use of bacteriophage vectors.

Screen Conditions

The λ expression library can be plated on bacterial host strains that produce high levels of the *lacI* repressor from a plasmid or episome (e.g., Y1088 or XL-1 Blue, respectively). In consequence, potentially toxic *lacZ*–cDNA fusion proteins are not expressed until the plaques are well established and are subsequently induced with isopropylthiogalactoside (IPTG). This yields a more uniform placque size, avoids loss by overgrowth of more robust plaques, and ensures that all recombinants are represented during the single round of library amplification. Phage can be plated on bacterial strains such as LE392 and KM392 to give somewhat larger plaques; IPTG induction is not required with such host strains due to the constitutive expression of the β-Gal fusion protein. Plating densities should not exceed 33,000 plaque-forming units (PFU)/150-mm agar plate.

Lifts and probings of the library are made much easier by the use of Hybond C Extra, a supported nitrocellulose filter that is considerably less friable than traditional filters (Amersham). If the library is to be induced, filters are impregnated with a solution of IPTG, air dried, and applied to the lawns shortly after the plaques are visibly established. The filters are left on overnight to enhance binding of proteins produced during plaque

[14] F. M. Ausubel, R. Brent, R. E. Kingston, D. D. Moore, J. G. Seidman, J. A. Smith, and K. Struhl, eds., "Current Protocols in Molecular Biology." J Wiley (Interscience), New York, 1987.

[15] J. Sambrook, E. F. Fritsch, and T. Maniatis, eds., "Molecular Cloning: A Laboratory Manual," Vol. 2. Cold Spring Harbor Lab., Cold Spring Harbor, NY, 1989.

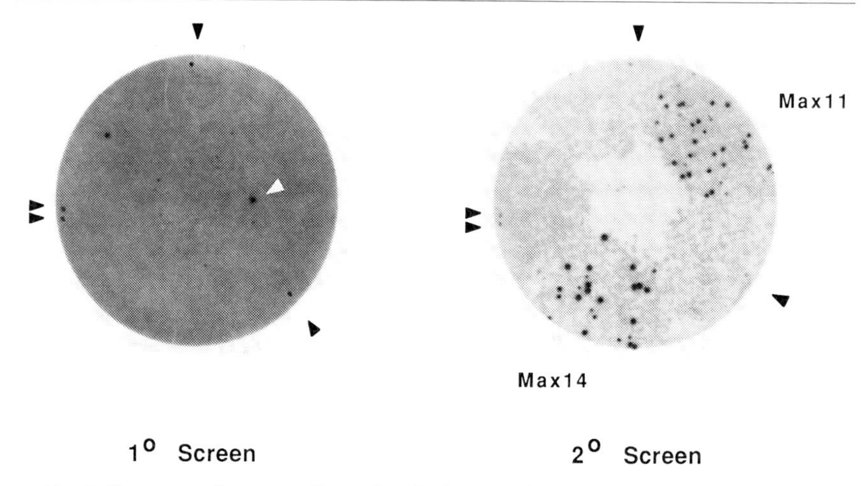

FIG. 2. Representative autoradiographs of primary and secondary expression screens using iodinated GST-MycC92. In the primary (1°) screen, the λgt11 expression library was plated at 40,000 PFU/150-mm plate (plaque size <1.0 mm). For secondary (2°) analysis, putative positives were replated as individual puddles on a 150-mm dish (1–2 mm plaques). Exposure times were 8 and 3 hr, respectively. The asymmetric India ink markings used for alignment are indicated with bold arrowheads, whereas the primary Max14 plaque is denoted with a white arrowhead. Other "positive" plaques seen on the 1° filter were eliminated in the secondary screen. [Adapted with permission from Blackwood and Eisenman, *Science* **251,** 1211 (1991). Copyright 1991 American Association for the Advancement of Science.]

growth at the agarose/nitrocellulose interface. Each plaque produces an estimated 100 pg of fusion protein (1 fmol). Higher expression levels can be achieved by lysogenizing the library, followed by screening of the bacterial colonies.[16] Although "duplicate" lifts can be made from the same agar plates, the heterogeneity in protein production limits any direct comparison. Secondary screening with its characteristic constellation of positive plaques is more reliable at validating reactivity (see Fig. 2).

Once the filters are removed from the lawns, denaturation and renaturation may be used to correct aberrant folding or to dissociate insoluble aggregates of fusion protein produced as a consequence of overexpression. In our experience with Myc, cycling through guanidine hydrochloride increased the background, but it did not enhance or diminish the Myc : Max interaction. In some cases, the probability of irreversibly denaturing a protein may outweigh the potential for regeneration of a functionally interactive molecule.

The absence of a positive control can be an advantage in that it forces

[16] R. Mutzel, A. Baeuerle, S. Jung, and H. Dammann, *Gene* **96,** 205 (1990).

the researcher to empirically evaluate screening conditions. In this instance, we chose to first optimize conditions on parental λgt11 filters. Besides denaturation and renaturation, blocking agents, buffers (salt, detergent, reducing agents), incubation times and temperatures, probe concentrations, and washes were varied until conditions were found that minimized the background on an overnight exposure. By focusing on background reduction, we hoped to enhance signal-to-noise ratios to a level that still permitted detection of true positives. Though Table II outlines a representative protocol, the variables we considered are discussed in the next few paragraphs.

For blocking residual protein binding sites on the filters, proteins such as powdered milk or bovine serum albumin worked better than nonionic detergents alone [Nonidet P-40 (NP-40), Triton X-100, or Tween 20]. A minimal incubation time for blocking seemed to be 1 hr. Significant differences between commercial sources of nonfat powdered milk have been noted; thus if one supermarket brand gives an excessive background, try another.

Although a variety of buffers may be suitable for this technique, we chose a HEPES-based buffer for its pK_a of 7.5, zwitterionic character, and lack of interaction with divalent cations. A pH range of 7.2 to 7.4 was considered physiological. Salt concentrations varied from 0 to 500 m*M* NaCl; a little salt was preferable to none, whereas increasing concentrations appeared to have no dramatic effect on background reduction. Therefore, 50 m*M* NaCl was used in the screen to minimize salt disruption of specific complexes. The inclusion of divalent cations (Mn^{2+} or Mg^{2+}) had no effect on background, although their presence may be required with other protein probes. Nonionic detergent and dithiothreitol (DTT) were the two most critical components in the buffer. The detergent maintained probe solubility, whereas DTT prevented nonspecific formation of disulfide bonds (MycC92 contains one cysteine, whereas GST has four). Milk protein served as a nonspecific competitor. Depending on the application, other additives may be considered: glycerol to enhance protein stability, sucrose to restore tonicity to low ionic strength buffers, protease or phosphatase inhibitors, etc.

The predicted affinity for each bimolecular interaction should be considered in the choice of probe concentration. For comparison, the dissociation constants (Kd) for several known protein–protein interactions span a large range: trypsin–substrate, 1.25×10^{-4} *M*; antigen–antibody, 10^{-6} to 10^{-8} *M*; E12 or MyoD homodimers, 10^{-6} *M*; Fos–Jun heterodimers, 10^{-7} *M*; λ repressor dimers, 2×10^{-8} *M*; PDGF–PDGF receptor, $<10^{-10}$ *M*; avidin–biotin, 10^{-15} *M*. Titrations of the probe in the microgram to nanogram per milliliter range (nanomolar to picomolar) should allow the choice of a concentration that gives a reasonable background signal. If the amount of probe required is a limitation, volume excluders (inert polymers such as

TABLE II
EXPRESSION LIBRARY SCREENING

A. Plating the library

1. Grow the appropriate strain of *Escherichia coli* (e.g., Y1088) to saturation in LB containing 0.2% maltose (w/v) and 10 m*M* $MgCl_2$.
2. Mix 15 ml of bacterial culture with 10^6 PFU of a titered λgt11 library. Aliquot 0.5 ml each into 30 tubes and incubate at room temperature for 15 min to promote phage adsorption. Warm to 37° (10 min).
3. Add 7 ml of molten top agarose (42°) to each tube, mix, and dispense to a prewarmed 150-mm LB agar plate. Try to minimize the formation of air bubbles. Allow top layer to solidify, then transfer plates to 37°.
4. As plaques become visible ("pinpricks" at ~6–8 hr), induce β-galactosidase fusion protein expression by overlaying the lawns with IPTG-impregnated nitrocellulose filters (saturate filters in 10 m*M* isopropyl β-D-thiogalactopyranoside, then air dry). Avoid adjusting the filters once in contact with the lawn. Incubate overnight at 37–42°.
5. To maintain adhesion between the top agarose and the agar underlayer during filter removal, cool the plates to room temperature. Mark each plate (agar and nitrocellulose) asymmetrically, then carefully peel off the filter. Rinse the filters with PBS to remove bacterial debris.

B. Probing the expression library

6. (Optional) For denaturation/renaturation of filter-bound proteins, immerse the filters in 200 ml HEPES buffer (20 m*M* HEPES, pH 7.2, 50 m*M* NaCl, 1 m*M* DTT) containing 6 *M* GuHCl. Agitate gently at room temperature for 10 min. Decant and repeat. Renature by serially diluting the GuHCl solution 1 : 1 with HEPES buffer every 5 min.
7. Block excess protein binding sites with 5% powdered milk (w/v) in HND buffer [20 m*M* HEPES, pH 7.2, 50 m*M* NaCl, 0.1% NP-40 (v/v), and 5 mM DTT] for a minimum of 1 hr at 4°. Note: All blocking, probing, and washing procedures are performed on a rotating platform.
8. Incubate the filters with radiolabeled protein for 4 hr at 4° [HND buffer containing 1% powdered milk (w/v)]. Use 7–10 ml buffer/filter to adequately submerge the stack. While the appropriate concentration of the probe is best determined empirically, a good starting point is in the nanomolar range.
9. Wash the filters seven times with cold PBS/0.2% Triton X-100 or until the wash buffer is minimally radioactive. Ensure thorough washing by transferring the filters individually from one tray to the next.
10. Lightly blot the filters on 3MM paper and wrap in plastic film. Expose for autoradiography for 2–18 hr.

C. Positive plaque verification

11. Pick positive plaques into 1.0 ml SM (i.e., remove a core of top agarose from the region). Add 25 μl $CHCl_3$. Allow phage to diffuse >1 hr.
12. Plate 1–2 μl phage stock/50 μl saturated bacterial culture/1.0 ml top agarose onto a 150-mm LB agar plate (five to seven puddles per plate). Repeat expression screen.
13. Plaque purify the positive clones, then excise the phagemid by superinfection with filamentous helper phage (Lambda ZAP) or purify phage DNA and subclone the insert (λgt11 and derivatives; Qiagen Lambda mini kit).

polyethylene glycol, dextran, polyvinyl alcohol, or Ficoll) may be used to enhance macromolecular crowding.[17]

When probe incubation times and temperatures were evaluated, it was noted that the background was significantly higher at room temperature than at 4°. Longer incubation times coordinately increased both signal and noise; the optimal time seemed to be 2 to 4 hr. To remove unbound probe and reduce nonspecific binding, filters are subsequently washed multiple times with buffers containing nonionic detergents. Filters should be transferred individually between buffer baths to ensure a thorough washing, a process that can take anywhere from 45 min to 1 hr. Cold buffers are not required to maintain Myc: Max association, though chilled washes may be needed with lower affinity probes. Prevention of a high background proved to be the best strategy since washing filters at elevated salt concentrations (0.5–2.0 *M* NaCl) failed to reduce the nonspecific signal.

Washed filters are exposed to radiographic film for several hours or overnight. Reactive plaques are easily identified as spots of uniform density but of slightly irregular shape (i.e., not punctate). Figure 2 demonstrates the type of signal-to-noise seen on filters from the GST-MycC92 dimerization screen.

Authentication

Negative controls should be performed during the process of purifying a reactive plaque. For example, Max14 plaques were probed with [^{125}I]GST to demonstrate that Myc sequences were required for the observed association.[9] Autophosphorylating kinases, when not removed following labeling of the probe, could give a false positive under low stringency conditions. Competition with excess unlabeled protein or steric interference with antibodies can also substantiate specificity.

Once the plaque has been purified to homogeneity, lysogens can be established and the extracts analyzed by SDS–PAGE and Western blot. The apparent molecular weight of the β-galactosidase fusion protein permits an estimation of the size of the coding region present in the cDNA insert. In addition, Far-Western analysis of the blot, using the conditions developed in the expression screen, can confirm that bacterial or phage components present in a plaque are not required for probe interaction. Coprecipitation of the radiolabeled probe from a lysogen extract (using β-Gal antisera) can establish that stable association also occurs in solution.

Sequence analysis of the reactive cDNA insert may reveal a polypeptide sequence with functional implication: phosphopeptides recognizing SH2

[17] A. P. Minton, *Mol. Cell. Biochem.* **55,** 119 (1983).

domains, activation domains interacting with TAFs or known coactivators, etc. When sequence analysis uncovers a novel protein bearing little structural homology with known motifs, affinity chromatography of mutant proteins can be useful in defining those protein surfaces that are necessary and sufficient to mediate protein–protein interaction. One potential caveat to affinity chromatography, however, is that the resin, which contains high concentrations of target protein, may act as a mixed-mode ion-exchange matrix and thus nonspecifically retain proteins based on their general biochemical properties.

The presense of adjacent HLH and Zip domains within the Max inserts clearly suggested a relevance to the biology of Myc. However, it was a concern that at high concentrations, or in the absense of a true physiological partner, promiscuous interactions might occur with structurally similar proteins. Thus, the "gold" standard for authenticating a specific protein–protein interaction is the identification of both components in a heteromeric complex *in vivo*. In the case of Myc, Max could clearly be coprecipitated in such a complex from untransfected cells (i.e., at physiological concentrations.[18] In support of this physical association are biological observations that Max is both a requirement for, and potential antagonist of, Myc function.[19,20] Although functional interactions need not be direct, the pairing of biological assays with biochemistry provides a rational approach for substantiating interaction specificity.

Authentic protein–protein interactions, by definition, will be those wherein two molecules establish a direct physical association, at physiological concentrations and conditions, to mediate a biological function.

Acknowledgments

We acknowledge those researchers whose λgt11 expression screens we were unable to cite, but whose work provided great impetus for the general use of this technique. In addition, we thank T. Palmer, L. Kerrigan, R. Rivera, R. Kamakaka, J. Kadonaga, and D. Ayer for comments on the manuscript.

[18] E. M. Blackwood, B. Lüscher, and R. N. Eisenman, *Genes Dev.* **6,** 71 (1992).

[19] B. Amati, M. W. Brooks, N. Levy, T. D. Littlewood, G. I. Evan, and H. Land, *Cell (Cambridge, Mass.)* **72,** 233 (1993).

[20] L. Kretzner, E. M. Blackwood, and R. N. Eisenman, *Nature (London)* **359,** 426 (1992).

[16] Analyzing Protein–Protein Interactions Using Two-Hybrid System

By PAUL L. BARTEL and STANLEY FIELDS

Introduction

Protein–protein interactions play a critical role in most biological processes. For example, the identification of interactions between viral oncoproteins and cellular tumor suppressor proteins, between components of signal transduction pathways, and between proteins involved in the regulation of the cell cycle has greatly increased our understanding of cellular function. In addition, studies defining domains of proteins [e.g., SH2 and SH3 domains, retinoblastoma (Rb) pocket] that are responsible for specific interactions have contributed significantly to unraveling the mechanisms of tumorigenesis.

The two-hybrid system is a yeast-based genetic assay for detecting protein–protein interactions *in vivo*. It can be used to establish interactions between two known proteins or to search genomic or cDNA libraries for proteins that interact with a target protein. For this latter application, the gene encoding the protein that interacts with a target protein is immediately available on a plasmid, which is not the case for many biochemical methods to detect interacting proteins. The two-hybrid system has also been used to define the protein domains that mediate an interaction and to identify specific residues that are involved in a protein–protein interaction. We will briefly discuss the basis for this method and then present the protocols that are necessary to use this system.

Principle of Method

The two-hybrid system[1,2] exploits the two domain nature of many site-specific eukaryotic transcription factors to detect interactions between two different hybrid proteins. These factors consist of a site-specific DNA-binding domain which is distinct from a domain that is responsible for transcriptional activation.[3,4] Two key results in the development of this

[1] S. Fields and O.-K. Song, *Nature* (*London*) **340,** 245 (1989).

[2] C.-T. Chien, P. L. Bartel, R. Sternglanz, and S. Fields, *Proc. Natl. Acad. Sci. U.S.A.* **88,** 9578 (1991).

[3] L. Keegan, G. Gill, and M. Ptashne, *Science* **231,** 699 (1986).

[4] I. A. Hope and K. Struhl, *Cell* (*Cambridge, Mass.*) **46,** 885 (1986).

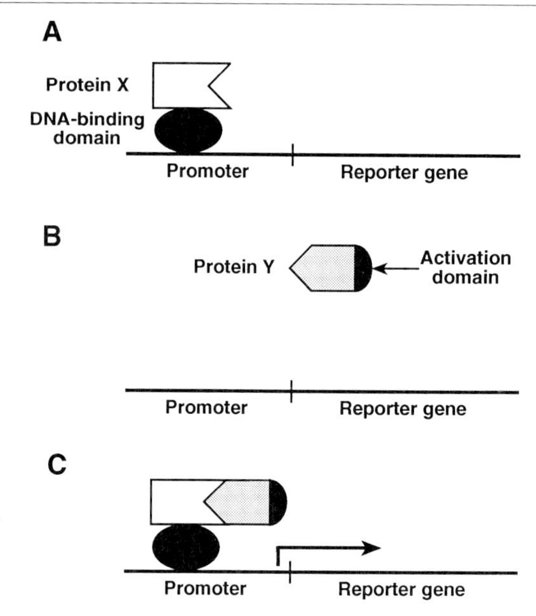

FIG. 1. The two-hybrid system. (A) A hybrid consisting of a DNA-binding domain fused to a protein X is unable to activate transcription of a reporter gene because it lacks a transcriptional activation domain. (B) A hybrid consisting of an activation domain fused to a protein Y fails to localize to the reporter gene. (C) If both hybrids are expressed in the same cell and proteins X and Y interact, the activation domain is anchored to the binding site and the reporter gene is expressed.

system were the demonstration that a hybrid protein containing domains from two different transcription factors can activate transcription[5] and the demonstration that the two domains need not be covalently attached to each other to function.[6,7] In the two-hybrid system, two fusion proteins must be generated. One is a fusion between the DNA-binding domain of a transcription factor and a test protein "X." The other is a fusion between the activation domain of a transcription factor and a test protein "Y." Plasmids encoding these fusions are introduced together into a yeast strain that contains one or more reporter genes with upstream binding sites for the DNA-binding domain present in the first hybrid. As illustrated in Fig. 1, the expression of either the hybrid of the DNA-binding domain with protein X or the hybrid of the activation domain with protein Y fails to

[5] R. Brent and M. Ptashne, *Cell (Cambridge, Mass.)* **43,** 729 (1985).
[6] J. Ma and M. Ptashne, *Cell (Cambridge, Mass.)* **55,** 443 (1988).
[7] J. McKnight, T. Kristie, and B. Roizman, *Proc. Natl. Acad. Sci. U.S.A.* **84,** 7061 (1987).

activate transcription of a reporter gene. However, if protein X and protein Y interact, the transcriptional activator is anchored to the binding site and leads to the expression of the reporter gene.

The two-hybrid system has been used to detect interactions among proteins produced by prokaryotic organisms such as *Escherichia coli* and by a wide range of eukaryotes including yeast, plants, and mammals. In addition, the protein interactions successfully detected by this system include those found normally in a variety of subcellular locations, including the nucleus, cytoplasm, and mitochondria, associated with membranes and extracellular. This system has been applied to the study of a variety of cellular processes, such as cell cycle progression, signal transduction, and oncogenesis. More specifically, interactions of oncoproteins in this system include ras and raf,[8–10] p53 and SV40 large T antigen,[11] Rb and large T antigen,[12] myc and max,[13] and p53 with itself.[11]

Materials and Methods

In theory, site-specific DNA-binding domains and transcriptional activation domains from many different transcription factors should work in the two-hybrid system. In practice, however, the most commonly used DNA-binding domains are from the yeast *Saccharomyces cerevisiae* transcription factor Gal4p and the *E. coli* repressor lexA. The most commonly used activation domains are from Gal4p and the herpes simplex virus VP16 protein. Most of our experience has been with the Gal4p DNA-binding and activation domains and this bias will be reflected in the remainder of this chapter. Reporter genes can encode any function detectable in *S. cerevisiae*, but generally encode the yeast His3p or Leu2p or *E. coli* β-galactosidase.

Vectors

Several laboratories have constructed vectors for use with the two-hybrid system. These plasmids have a number of features in common. The bacterial *ori* and *bla* sequences are included for maintenance and selection

[8] L. V. Aelst, M. Barr, S. Marcus, A. Polverino, and M. Wigler, *Proc. Natl. Acad. Sci. U.S.A.* **90,** 6213 (1993).

[9] A. B. Vojtek, S. M. Hollenberg, and J. A. Cooper, *Cell (Cambridge, Mass.)* **74,** 205 (1993).

[10] X.-F. Zhang, J. Settleman, J. M. Kyriakis, E. Takeuchi-Suzuki, S. J. Elledge, M. S. Marshall, J. T. Bruder, U. R. Rapp, and J. Avruch, *Nature (London)* **364,** 308 (1993).

[11] K. Iwabuchi, B. Li, P. L. Bartel, and S. Fields, *Oncogene* **8,** 1693 (1993).

[12] T. Durfee, K. Becherer, R.-L. Chen, S. H. Yeh, Y. Yang, A. E. Kilburn, W. H. Lee, and S. J. Elledge, *Genes Dev.* **7,** 555 (1993).

[13] A. S. Zervos, J. Gyuris, and R. Brent, *Cell (Cambridge, Mass.)* **72,** 223 (1993).

in *E. coli*, and yeast 2μ or CEN and ARS sequences are included to maintain the plasmids in yeast at high or low copy number, respectively. A yeast nutritional marker gene such as *HIS3, TRP1, LEU2,* or *URA3* is present to allow for selection of the plasmid in an appropriate yeast strain. The gene encoding either the DNA-binding or activation domain hybrid is under the regulation of a yeast promoter, usually derived from the *ADH1* gene, and is followed by unique restriction sites to allow for the creation of in-frame fusions. The Gal4p DNA-binding domain contains a nuclear localization sequence, and the large T antigen nuclear localization sequence has been engineered into many of the Gal4p activation domain vectors. These sequences localize the hybrid proteins to the yeast nucleus for activation of the reporter gene. Some of the vectors that have been developed for the two-hybrid system are listed in Table I,[14-21] and maps of commonly used DNA-binding domain vectors are provided in Fig. 2 and of activation domain vectors in Fig. 3.

Yeast Reporter Strains

Reporter strains that are compatible with either a Gal4p-based system or a lexA-based system are available. Strains used for the Gal4p system are deleted for the *GAL4* and *GAL80* genes. Yeast reporter strains are auxotrophs for certain amino acids such that the nutritional markers present on the two-hybrid system vectors can complement these deficiencies. In this way, transformants of the reporter strain can be selected on media lacking the appropriate amino acids. Reporter strains also carry one or more reporter gene constructs. Most strains contain a *lacZ* reporter gene that is under the control of either Gal4p- or lexA-binding sites. This allows for the screening of colonies that express interacting hybrid proteins because they produce β-galactosidase. In addition, strains may carry a nutritional reporter gene, such as *HIS3* or *LEU2*, that is under the control of Gal4p- or lexA-binding sites. These strains will be mutant for the chromosomal

[14] J. Ma and M. Ptashne, *Cell (Cambridge, Mass.)* **51,** 113 (1987).

[15] P. L. Bartel, C.-T. Chien, R. Sternglanz, and S. Fields, *in* "Cellular Interactions in Development: A Practical Approach" (D. A. Hartley, ed.), p. 153. Oxford University Press, Oxford, 1993.

[16] J. W. Harper, G. R. Adami, N. Wei, K. Keyomarsi, and S. J. Elledge, *Cell (Cambridge, Mass.)* **51,** 805 (1993).

[17] P. M. Chevray and D. Nathans, *Proc. Natl. Acad. Sci. U.S.A.* **89,** 5789 (1992).

[18] E. Golemis and R. Brent, personal communication (1993).

[19] G. J. Hannon, D. Demetrick, and D. Beach, *Genes Dev.* **7,** 2378 (1993).

[20] J. Luban, K. L. Bossolt, E. K. Franke, G. V. Kalpana, and S. P. Goff, *Cell (Cambridge, Mass.)* **73,** 1067 (1993).

[21] S. Dalton and R. Treisman, *Cell (Cambridge, Mass.)* **68,** 597 (1992).

TABLE I
VECTORS FOR TWO-HYBRID SYSTEM

Plasmid	Domain	Restriction sites	Marker	Refs.
		DNA-binding domain vectors		
pMA424	GAL4 bd	*Eco*RI, *Sma*I, *Bam*HI, *Sal*I, *Pst*I	*HIS3*	14
pGBT9	GAL4 bd	*Eco*RI, *Sma*I, *Bam*HI, *Sal*I, *Pst*I	*TRP1*	15
pAS1	GAL4 bd	*Nde*I, *Nco*I, *Sfi*I, *Sma*I, *Bam*HI	*TRP1*	12
pAS2	GAL4 bd	*Nde*I, *Nco*I, *Sfi*I, *Sma*I, *Bam*HI	*TRP1*, *CYH2*	16
pCP62	GAL4 bd	*Sal*I, *Pst*I, *Sma*I, *Spe*I, *Xba*I, *Not*I, *Sac*II	*LEU2*	17
pBTM116	lexA	*Eco*RI, *Sma*I, *Bam*HI, *Sal*I, *Pst*I	*TRP1*	15
lex(1-202)PL	lexA	*Eco*RI, *Sma*I, *Bam*HI, *Sal*I, *Pst*I	*HIS3*	13, 18
		Activation domain vectors		
pGAD.F	GAL4 ad	*Bam*HI	*LEU2*	2
pGAD424	GAL4 ad	*Eco*RI, *Sma*I, *Bam*HI, *Sal*I, *Pst*I, *Bgl*II	*LEU2*	15
pGAD.GH	GAL4 ad	*Spe*I, *Bam*HI, *Sma*I, *Eco*RI, *Sal*I, *Xho*I	*LEU2*	19
pACT	GAL4 ad	*Eco*RI, *Bam*HI, *Xho*I, *Bgl*II	*LEU2*	12
pGADNOT	GAL4 ad	*Bam*HI, *Not*I, *Sal*I	*LEU2*	20
pPC86	GAL4 ad	*Sal*I, *Sma*I, *Eco*RI, *Bgl*II, *Spe*I, *Not*I	*TRP1*	17
pJG4-5	B42	*Eco*RI, *Xho*I	*TRP1*	13, 18
pVP16	VP16	*Bam*HI, *Not*I	*LEU2*	9
pSD.10	VP16	*Eco*RI, *Bst*XI, *Xho*I	*URA3*	21

copy of the nutritional gene that is used as a reporter gene. When these reporter strains are transformed with plasmids and plated onto media that lack the appropriate nutrient, only those transformants that express interacting hybrid proteins, and therefore activate the nutritional reporter gene, will survive. This system for selecting transformants that express interacting hybrid proteins is especially useful when searching an activation domain library for interacting proteins. Two different *HIS3* reporter constructs are available for use with the Gal4p system. The plasmid pBM1499[22] contains the *GAL1* upstream activating sequence (UAS), the *HIS3* minimal promoter, and the *HIS3* coding sequence. Because the *HIS3* promoter results in a low level of His3p expression in the absence of any two-hybrid interaction, the competitive inhibitor 3-amino-1,2,4-triazole (3-AT) must be included in the media to allow for histidine selection. The other vector is pGH1[19,23] which contains the *GAL1* UAS, the *GAL1* minimal promoter, and the *HIS3* coding sequence. This reporter gene does not require the

[22] J. Flick and M. Johnston, *Mol. Cell Biol.* **10,** 4757 (1990).
[23] P. Bartel and S. Fields, unpublished (1993).

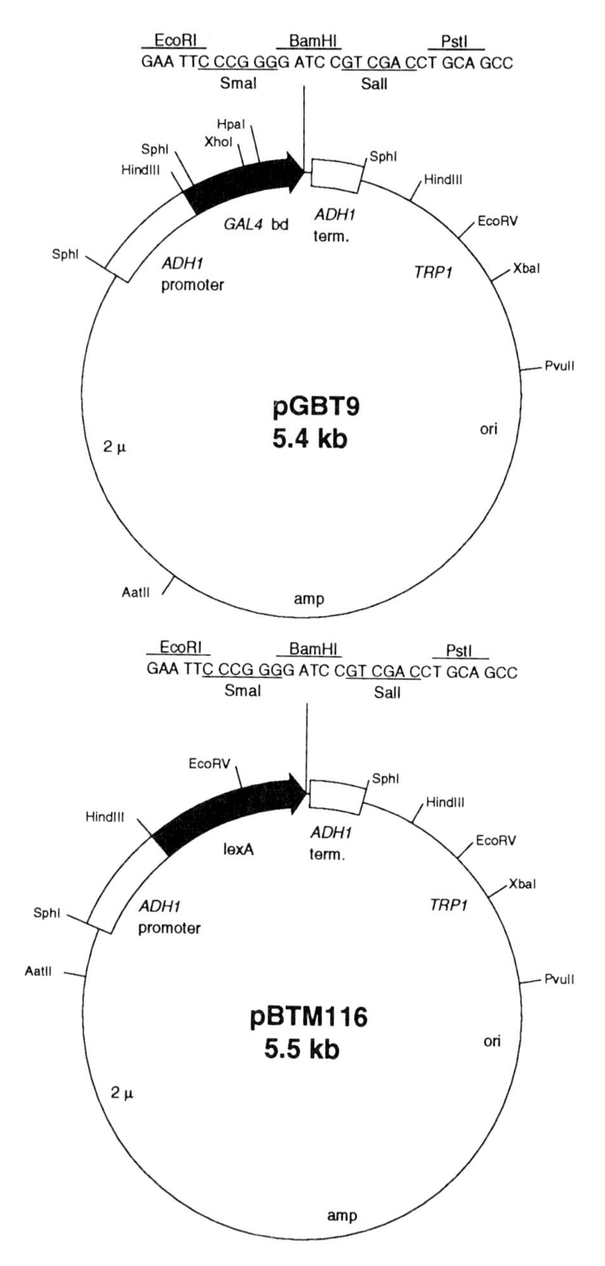

FIG. 2. Maps of DNA-binding domain vectors for the two-hybrid system. Plasmids pGBT9,[15] pBTM116,[15] pAS2,[16] and lex(1-202)PL[13] are shown. Important features include the *ADH1* promoter which drives transcription of either the *GAL4* DNA-binding domain (*GAL4 bd*) or the *E. coli* lexA (lexA) gene and the *ADH1* terminator (*ADH1* term.) Also included are a yeast origin of replication (2μ), an *E. coli* origin of replication (ori), an *E. coli* selectable marker for ampicillin resistance (amp), the yeast selectable markers *TRP1* or *HIS3*, and the

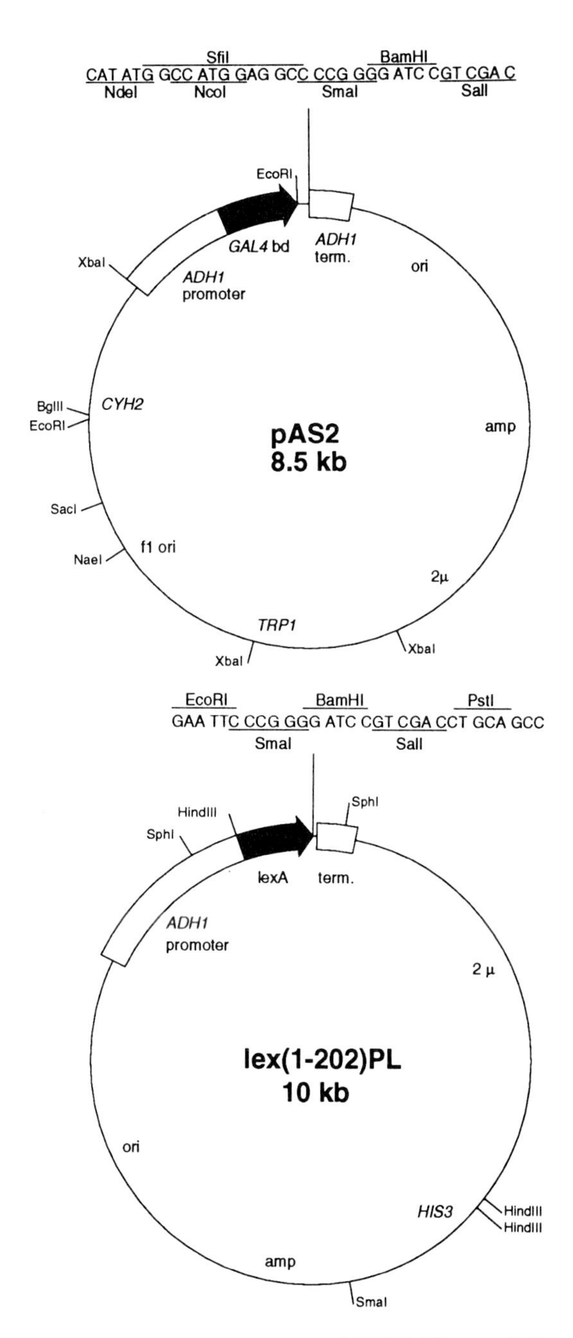

gene conferring cycloheximide sensitivity in yeast (*CYH2*). The restriction sites diagramed directly above each map are suitable for the insertion of heterologous genes. The nucleotide sequences are arranged in codons continuing in the same reading frame as the upstream DNA-binding domain.

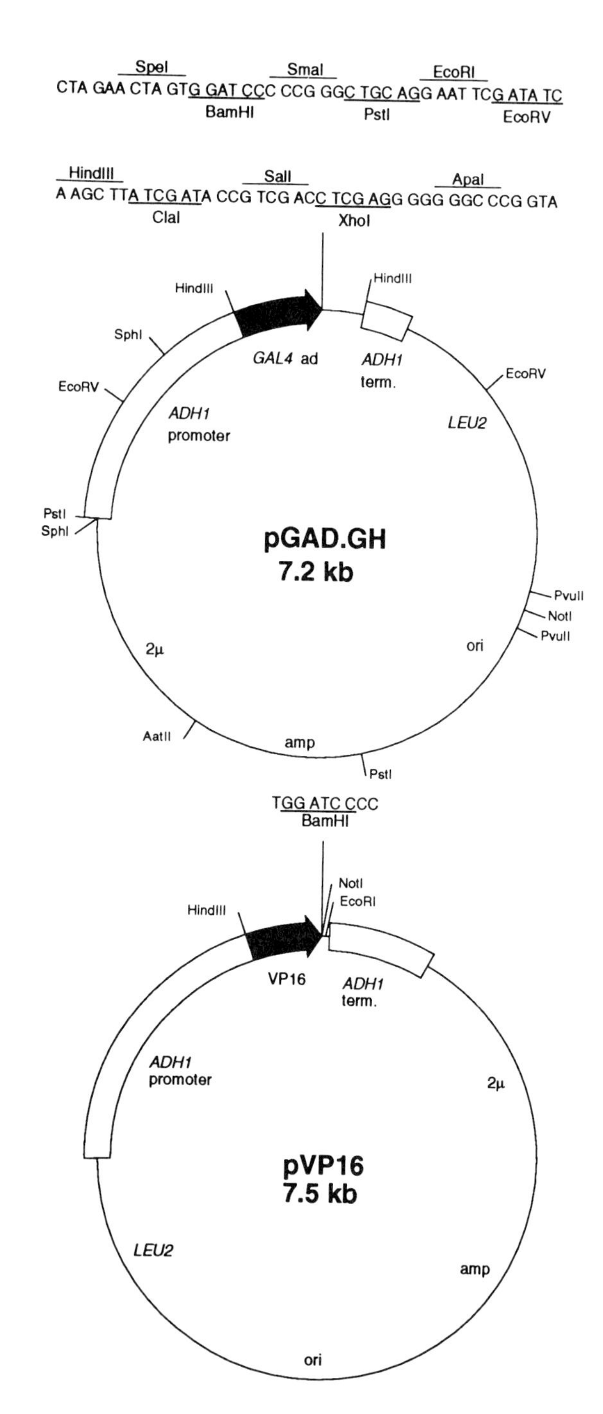

FIG. 3. Maps of activation domain vectors for the two-hybrid system. Plasmids pGAD.GH,[19] pACT,[12] pVP16,[9] and pJG4-5[13] are shown. Important features include the *ADH1* promoter

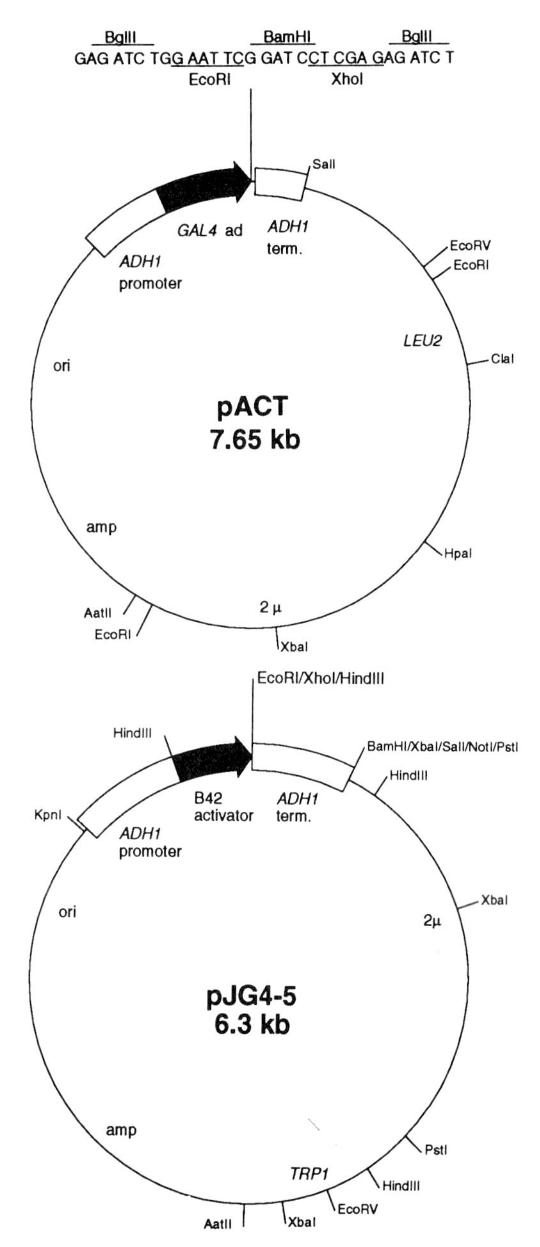

which drives transcription of the *GAL4* activation domain (*GAL4 ad*), the B42-activating sequence from *E. coli* (B42 activator), or the herpesvirus VP16 gene (VP16) and the *ADH1* terminator (*ADH1* term.) Also included are a yeast origin of replication (2μ), an *E. coli* origin of replication (ori), an *E. coli* selectable marker for ampicillin resistance (amp), and the yeast selectable markers *LEU2* or *TRP1*. The restriction sites diagramed directly above each map are suitable for the insertion of heterologous genes. The nucleotide sequences are arranged in codons continuing in the same reading frame as the upstream activation domain.

presence of 3-AT. A description of some reporter strains is provided in Table II.[24-26]

Construction of Hybrid Genes

Construction of hybrid genes requires that in-frame fusions of the genes of interest be made to the DNA encoding the DNA-binding domain or activation domain. This can be performed using standard molecular biology techniques.[27,28] Sometimes a fusion gene can be generated easily if compatible restriction sites exist in the test gene and the vector. If not, the gene fragment can be generated by polymerase chain reaction (PCR) with useful restriction sites incorporated into the primers. Alternatively, a restriction site can be changed into a different site or put into a different reading frame by incorporation of a short adapter oligonucleotide into a plasmid carrying one of the test genes.

Constructions can be verified by restriction or DNA sequence analysis. Useful sequencing primers for Gal4p-based vectors are 5′-TCA TCG GAA GAG AGT AG-3′, which corresponds to amino acid residues 132–137 of the Gal4p DNA-binding domain and 5′-TAC CAC TAC AAT GGA TG-3′, which corresponds to amino acid residues 858–863 of the Gal4p activation domain. For sequencing lexA fusions, the primer 5′-CTT CGT CAG CAG AGC TTC-3′, which corresponds to amino acid residues 181–186 of lexA, can be used. The Gal4p activation domain primer is also useful for sequencing inserts of plasmids isolated from two-hybrid library searches.

Transformation of Yeast

Plasmids can be introduced into yeast by electroporation, transformation of spheroplasts, or transformation of chemically treated cells. We use a modification of the lithium acetate method developed by Ito[29] and improved by Schiestl and Gietz[30] and Hill[31] for transformation of reporter strains with two-hybrid system plasmids. Yeast cells are grown in rich media, treated with lithium acetate, and incubated in the presence of plasmid DNA.

[24] G. Gill and M. Ptashne, *Cell (Cambridge, Mass.)* **51,** 121 (1987).

[25] P. Bartel, C.-T. Chien, R. Sternglanz, and S. Fields, *BioTechniques* **14,** 920 (1993).

[26] G. Hannon, H. Feilotter, and D. Beach, personal communication (1993).

[27] J. Sambrook, E. F. Fritsch, and T. Maniatis, eds., "Molecular Cloning: A Laboratory Manual," 2nd ed. Cold Spring Harbor Lab., Cold Spring Harbor, NY, 1989.

[28] F. M. Ausubel, R. Brent, R. E. Kingston, D. D. Moore, J. G. Seidman, J. A. Smith, and K. Struhl, eds., "Current Protocols in Molecular Biology." Greene Publishing Associates and Wiley (Interscience), New York, 1990.

[29] H. Ito, Y. Fukuda, K. Murata, and A. Kimura, *J. Bacteriol.* **153,** 163 (1983).

[30] R. H. Schiestl, P. Manivasakam, R. A. Woods, and R. D. Gietz, *Methods* **5,** 79 (1993).

[31] J. E. Hill, A. M. Myers, T. J. Koerner, and A. Tzagoloff, *Yeast* **2,** 163 (1986).

TABLE II
REPORTER STRAINS FOR TWO-HYBRID SYSTEM

Strain	Genotype	Reporter genes	Markers	Refs.
GGY1::171	*MATα*, *ura3*, *leu2*, *his3*, *tyr*, *Δgal4*, *Δgal80*, *URA3::GAL1-lacZ*	*GAL1-lacZ*	*his3*, *leu2*	24
PCY2	*MATα*, *Δgal4*, *Δgal80*, *URA3::GAL1-lacZ*, *lys2-80*, *his3-200*, *trp1-63*, *leu2*, *ade2-101*	*GAL1-lacZ*	*his3*, *trp1*, *leu2*	17
Y526	*MAT**a**, *ura3-52*, *his3-200*, *ade2-101*, *lys2-801*, *trp1-901*, *leu2-3*, *112*, *can*^{r}, *gal4-542*, *gal80-538*, *URA3::GAL1-lacZ*	*GAL1-lacZ*	*his3*, *leu2*, *trp1*	25
Y527	*MAT**a**, *ura3-52*, *his3-200*, *ade2-101*, *lys2-801*, *trp1-901*, *leu2-3*, *112*, *can*^{r}, *gal4-542*, *gal80-538*, *URA3::*Gal4 binding sites-*CYC1-lacZ*	*GAL-CYC1-lacZ*	*his3*, *leu2*, *trp1*	25
Y153	*MAT**a**, *leu2-3*, *112*, *ura3-52*, *trp1-901*, *his3-200*, *ade2-101*, *gal4Δ*, *gal80Δ*, *URA3::GAL1-lacZ*, *LYS2::GAL1-HIS3*	*GAL1-lacZ* *GAL1-HIS3*	*trp1*, *leu2*	12
Y190	*MAT**a**, *leu2-3*, *112*, *ura3-52*, *trp1-901*, *his3-200*, *ade2-101*, *gal4Δ*, *gal80Δ*, *URA3::GAL1-lacZ*, *LYS2::GAL1-HIS3*, *cyh*^{r}	*GAL1-lacZ* *GAL1-HIS3*	*trp1*, *leu2*	16
Hf7c	*MAT**a**, *ura3-52*, *his3-200*, *ade2-101*, *lys2-801*, *trp1-901*, *leu2-3*, *112*, *can*^{r}, *gal4-542*, *gal80-538*, *URA3::*Gal4 binding sites-*CYC1-lacZ*, *LYS2::GAL1-HIS3*	*GAL-CYC1-lacZ* *GAL1-HIS3*	*his3*, *leu2*, *trp1*	26
CYT10-5d	*MAT**a**, *ade2*, *trp1-901*, *leu2-3*, *112*, *his3-200*, *Δgal4*, *Δgal80*, *URA3::lexAop-lacZ*	*lexAop-lacZ*	*his3*, *leu2*, *trp1*	15
L40	*MAT**a**, *trp1*, *leu2*, *his3*, *LYS2::lexAop-lacZ*	*lexAop-lacZ*	*his3*, *trp1*, *leu2*	9
EGY48	*MAT**a**, *trp1*, *ura3*, *his3*, *LEU2::pLex-Aop6-LEU2*	*lexAop-LEU2*	*trp1*, *leu2*, *ura3*	13

The cells are then plated onto a minimal media that lacks the appropriate nutrients for plasmid selection and, if appropriate, reporter gene expression. The recipes for standard yeast media can be obtained from other sources[28,32] and are not included here.

[32] C. Guthrie and G. R. Fink, "Guide to Yeast Genetics and Molecular Biology." Academic Press, San Diego, CA, 1991.

Protocol

1. Inoculate 50 ml YEPD with yeast cells from a freshly grown culture. Incubate at 30° with shaking overnight.

2. Dilute the cells into 300 ml YEPD to give a final OD_{600} of 0.2. Incubate at 30° with shaking for 3 to 4 more hours. The culture should have an OD_{600} of about 0.5.

3. Pellet the cells by centrifugation (4000*g*, 5 min, room temperature)

4. Resuspend the cells in 10 ml sterile H_2O.

5. Pellet the cells by centrifugation (6000*g*, 5 min, room temperature)

6. Resuspend the cells in 1.5 ml TE/lithium acetate [10 m*M* Tris–Cl, 1 m*M* EDTA, 100 m*M* $C_2H_3O_2Li$ made fresh from stocks of 100 m*M* Tris–Cl (pH 7.5), 10 m*M* EDTA, and 1 *M* $C_2H_3O_2Li$ (adjusted to pH 7.5 with dilute acetic acid)].

7. Prepare the plasmid DNAs in a 1.5-ml microcentrifuge tube. Add 1–5μg of each plasmid and 100 μg of single-stranded carrier DNA to each tube. The carrier DNA is prepared by dissolving DNA (salmon sperm or calf thymus) in H_2O, sonicating to reduce its viscosity, extracting with phenol/chloroform [50% phenol (equilibrated to pH 8.0), 50% v/v chloroform] and then chloroform, and precipitating with ethanol. The carrier DNA is resuspended in TE [10 m*M* Tris–Cl (pH 8.0), 1 m*M* EDTA] at a concentration of 10 mg/ml and is denatured in a boiling water bath for 30 min, then is chilled on ice. Aliquots can be stored frozen until they are needed.

8. Add 200 μl cell suspension to each tube of plasmid DNA.

9. Add 600 μl polyethylene glycol (PEG)/TE/lithium acetate [40% PEG, 10 m*M* Tris–Cl, 1 m*M* EDTA, 100 m*M* $C_2H_3O_2Li$ made fresh from stocks of 50% PEG 3340, 100 m*M* Tris–Cl (pH 7.5), 10 m*M* EDTA, and 1 *M* $C_2H_3O_2Li$ (adjusted to pH 7.5 with dilute acetic acid)] to each tube. Mix gently.

10. Incubate at 30° for 30 min.

11. Add 90 μl fresh dimethyl sulfoxide (DMSO). Heat shock at 42° for 15 min.

12. Pellet the cells by centrifuging for 10 sec at high speed in a microcentrifuge.

13. Resuspend the pellet in 1 ml TE. Plate 100 μl of the cell suspension onto the appropriate selective media.

Transformation of a reporter strain with a single plasmid yields 10^4 to 10^5 transformants/μg plasmid DNA. If two plasmids are introduced simultaneously, the transformation frequency may drop by a factor of 10. Instead of introducing two plasmids simultaneously, some workers prefer to first transform with the plasmid expressing the hybrid of the DNA-binding domain with the target protein. The transformed strain is then grown in selective media, and transformed with the activation domain plasmid. This method works well unless expression of the DNA-binding domain hybrid is detrimental to the reporter strain, in which case expression of this hybrid is often lost before the second transformation. For these hybrids, simultaneous transformation with the activation domain vector is necessary.

Assaying for Reporter Gene Expression

After the transformants have grown into colonies, which usually takes 2 to 3 days in the absence of any selection for reporter gene expression, they must be tested for expression of a reporter gene. If the reporter gene encodes a protein, such as His3p or Leu2p, required for the synthesis of an amino acid, the colonies may grow slowly and take up to 10 days until they are visible. In either case, the most common reporter for the two-hybrid system is the *lacZ* gene, because β-galactosidase assays are simple, rapid, and easily quantified.

The filter assay presented below is a sensitive method to test whether colonies on a plate express β-galactosidase.[33] It is a useful method to assay transformants carrying defined hybrid proteins, usually on the third day after transformation. It is also used as a secondary screen of the rare positives that arise in a two-hybrid library search.

Protocol

1. Prepare Z buffer [16.1 g/liter $Na_2HPO_4 \cdot 7H_2O$, 5.5 g/liter $NaH_2PO_4 \cdot H_2O$, 0.75 g/liter KCl, 0.246 g/liter $MgSO_4 \cdot 7H_2O$, 2.7 ml/liter 2-mercaptoethanol (add this fresh)] with 5-bromo-4-chloro-3-indolyl-β-D-galactopyranoside (X-Gal). To each 100 ml of Z buffer, add 1.67 ml of X-Gal stock (20 mg/ml X-Gal in *N,N*-dimethylformamide).

2. Add an aliquot of the Z buffer containing X-Gal to a clean petri dish. Use 1.8 ml for a 100-mm dish and 5 ml for a 150-mm dish. Layer a clean paper filter disc onto the buffer, taking care to avoid air bubbles. We use 75- or 125-mm-diameter VWR grade 413 filters, although other paper filters will also work.

3. Layer a sterile paper or nitrocellulose filter onto the plate of transformants, taking care to avoid air bubbles. Allow the filter to wet completely.

4. Remove the filter from the plate and place it in a pool of liquid nitrogen such that the colonies face up. Let the filter freeze for 5–10 sec.

[33] L. Breeden and K. Nasmyth, *Cold Spring Harbor Symp. Quant. Biol.* **50,** 643 (1985).

5. Remove the filter from the liquid nitrogen and allow it to thaw completely at room temperature.

6. Layer the thawed filter, with the colonies facing up, onto the previously prepared filter that is soaked with Z buffer containing X-Gal. Take care to avoid air bubbles.

7. Incubate the filters at 30°. Strong positives may appear blue in as short as 30 min whereas weaker positives may take overnight or longer to turn blue.

The level of β-galactosidase expression in yeast transformants can be quantified using one of two different colorimetric assays. The first uses the inexpensive substrate *O*-nitrophenyl-β-D-galactopyranoside (ONPG),[34] and the second uses chlorophenol red-β-D-galactopyranoside (CPRG) (see, for example, Iwabuchi *et al.*[11]), which results in greater sensitivity. Prepare cultures for these assays by picking colonies to liquid selective media and incubating them at 30° on a roller or shaker.

Protocol

1. Grows cells in selective media to an OD_{600} of 0.5 to 1.
2. Pellet 0.1 to 1.5 ml of cells (depending on the expected enzyme activity) in a microcentrifuge tube. Resuspend the pellet in 0.8 ml Z buffer (see protocol for filter assay).
3. Add 50 μl $CHCl_3$ and 50 μl 0.1% (w/v) sodium dodecyl sulfate (SDS). Vortex the sample for 30 sec.
4. Add 0.16 ml of an ONPG stock solution [4 mg/ml in 100 m*M* phosphate buffer (pH 7.0)]. Mix well.
5. Incubate at 30° for a few minutes to 2 hr (depending on how rapidly the color changes). Quench the reaction by adding 0.4 ml of 1 *M* Na_2CO_3.
6. Spin the sample at high speed in a microcentrifuge for 10 min.
7. Measure the OD_{420} of the supernatant. Use a reaction mixture without added cells as a blank.
8. The β-galactosidase activity can be calculated as

$$\text{Units of } \beta\text{-galactosidase activity} = \frac{OD_{420}}{Vt(OD_{600})}$$

where V is the volume of culture used in milliliters and t is the reaction time in minutes.

[34] J. H. Miller, "Experiments in Molecular Genetics." Cold Spring Harbor Lab., Cold Spring Harbor, NY, 1972.

Many interactions in the two-hybrid system result in a low level of β-galactosidase expression not detected with ONPG as a substrate. For these cases, use CPRG as a substrate.

Protocol

1. Grow cells in selective media to an OD_{600} of 0.5 to 1.
2. Pellet 0.1 to 5 ml of cells (depending on the expected enzyme activity) and resuspend in HEPES buffer (prepare 100 mM N-2-hydroxyethylpiperazine-N'-2-ethanesulfonic acid, 154 mM NaCl, 1% bovine serum albumin, adjust this solution to pH 7.2 with HCl, and add DL-aspartic acid, hemimagnesium salt to 2 mM, and Tween 20 to 0.05%) to a final volume of 765 μl.
3. Add 55 μl $CHCl_3$ and 55 μl of 0.1% (w/v) SDS to each tube. Vortex for 1 min.
4. Add 125 μl of a 40 mM stock (prepared in H_2O) of CPRG.
5. Mix the contents of the tube well and incubate at 37°.
6. After a 20-min to 2-hr incubation (depending on how rapidly the color changes), stop the reaction with the addition of 10 μl of 100 mM $ZnCl_2$ to each tube. Mix the contents of the tube thoroughly.
7. Spin the tube at high speed in a microcentrifuge for 10 min.
8. Measure the OD_{574} of the supernatant. Use a reaction mixture without added cells as a blank.
9. The β-galactosidase activity can be calculated as

$$\text{Units of } \beta\text{-galactosidase activity} = \frac{1000 \times OD_{574}}{Vt(OD_{600})},$$

where V is the volume of culture used in milliliters and t is the reaction time in minutes.

Screening Libraries for Interacting Proteins

One of the most useful applications of the two-hybrid system is to detect, from a large number of hybrids encoded by a genomic or cDNA library, proteins that interact with a protein of interest. In this application, a protein of interest expressed as a fusion protein containing a DNA-binding domain and total genomic or cDNA expressed as fusion proteins with an activation domain are cotransformed into a reporter strain. The rare transformants that express the reporter gene(s) are identified, and the activation domain plasmids from these positives are analyzed.

The initial step in a two-hybrid search is to analyze the hybrid consisting of the target protein with the DNA-binding domain. Assay the DNA-binding domain hybrid alone for transcription of the reporter gene; there must be no activity for this hybrid to be used in a two-hybrid search. If it does activate transcription, a search can be performed with a portion of the target protein that does not result in transcriptional activation (e.g., Iwabuchi *et al.*[11]). In addition, test whether the DNA-binding domain hybrid is expressed; if an antibody is available, check for production of the DNA-binding domain hybrid by Western analysis. If possible, check whether the target protein retains its activity in yeast. One critical assay is to determine whether the target protein can interact with a known protein partner in a two-hybrid assay if such a partner is available. If the target protein is known to complement a defect in yeast, the DNA-binding domain hybrid can be assayed for this property.

The activation domain library is prepared by ligating the desired genomic DNA or cDNA to an activation domain vector. Table III provides a list of activation libraries that have been described in the literature. If the desired library is not available, it can be constructed using protocols found elsewhere.[27,28]

Two-hybrid searches can be conducted as screens, by assaying for transformants that express the *lacZ* reporter gene, but are most often conducted using strains that allow for the selection of transformants that carry interacting hybrid proteins. These strains are cotransformed with the DNA-binding domain hybrid and the activation domain library, plating on media that select for the presence of both vectors and the expression of the reporter gene. Only those transformants that can activate the reporter gene will grow on the plate. In addition, the surviving transformants can be screened for those that express the *lacZ* reporter gene. By using a strain that carries two reporter genes with dissimilar promoters, many false-positives can be eliminated quickly (see "Identification of False Positive Proteins").

TABLE III
ACTIVATION DOMAIN LIBRARIES

Vector	DNA source	Ref.
pPC86	Rat olfactory epithelium cDNA	41
pACT	Epstein–Barr virus-transformed human peripheral lymphocyte cDNA	12
pJG4-5	HeLa cell cDNA	13
pGADNOT	HL-60 (human leukocyte) cell cDNA	20
pVP16	Mouse embryo (9.5 and 10.5 day) cDNA	9
pGAD.F	*Saccharomyces cerevisiae* genomic DNA	2
pPC86	Mouse embryo (14.5 day) cDNA	17
pSD.10	HeLa cell cDNA	21
pGAD.GH	HeLa cell cDNA	19

Introduction of the DNA-binding domain hybrid and the activation domain library into the reporter strain can be performed as a scaled-up version of the transformation procedure[29–31] described previously, and outlined below.

Protocol

1. Grow the reporter strain (either with or without the plasmid encoding the DNA-binding domain hybrid) in 300 ml YEPD to an OD_{600} of about 0.5.
2. Pellet the cells by centrifugation (4000*g*, 5 min, room temperature).
3. Resuspend the cells in 10 ml sterile H_2O.
4. Pellet the cells by centrifugation (6000*g*, 5 min, room temperature).
5. Resuspend the cells in 1.5 ml TE/lithium acetate.
6. Prepare the plasmid DNAs in a 15-ml culture tube. Add 100 μg of each plasmid (or only the activation domain plasmid if the strain already carries the plasmid encoding the DNA-binding domain hybrid) and 2 mg of single-stranded carrier DNA to the tube.
7. Add 2 ml cell suspension to the tube of plasmid DNA.
8. Add 6 ml PEG/TE/lithium acetate to the tube. Mix gently.
9. Incubate at 30° for 30 min.
10. Add 0.8 ml fresh DMSO. Heat shock at 42° for 15 min.
11. Pellet the cells by centrifuging for 2 min at half speed in a tabletop centrifuge.
12. Resuspend the pellet in 10 ml TE. Plate 300 μl of the cell suspension onto 150-mm plates containing the appropriate selective media.

If the transformation mix is plated onto media that select both for the two-hybrid plasmids and for reporter gene expression, be sure to also plate an aliquot onto media that select only for the presence of the DNA-binding and activation domain plasmids. This procedure will allow the total number of transformants to be calculated. The plates should be incubated at 30° until colonies appear, at which point expression of β-galactosidase is assayed. Transformants positive both for growth and β-galactosidase activity are picked either directly from the filter or from the corresponding colony on the original plate, and transferred to fresh media.

Recovery of Plasmids from Yeast

After positive colonies have been purified, the activation domain plasmids should be recovered for further analysis. Since preparations of plasmid DNA from yeast are of poor quality and yield and are not useful for restriction or sequence analysis, these plasmids are introduced into *E. coli*.[35] A rapid procedure for preparing yeast plasmid DNA for introduction into *E. coli* is presented below.

Protocol

1. Using a toothpick, transfer cells from a colony to 50 μl lysis solution [2% Triton X-100, 1% SDS, 100 m*M* NaCl, 10 m*M* Tris–Cl (pH 8.0), and 1 m*M* EDTA].
2. Add 50 μl phenol/chloroform [50% phenol (equilibrated to pH 8.0), 50% chloroform] and ca. 0.1 g acid-washed glass beads (212–300 μm).
3. Vortex for 2 min.
4. Spin at high speed in a microcentrifuge for 5 min.
5. Transfer the supernatant to a clean tube and ethanol precipitate the DNA. Wash with 70% (v/v) ethanol and dry the pellet.
6. Resuspend the pellet in 10 μl H_2O. Use 1 μl to transform electrocompetent *E. coli*.

We usually obtain 100–1000 *E. coli* transformants when using this procedure. A similar but more rapid procedure has also been described.[36] If the activation domain plasmid carries the *LEU2* gene, it is helpful to introduce the yeast DNA into a bacterial strain [such as MH4 (*araD139* Δ(*lac*)*X74 galU galK hsr*$^-$ *hsm*$^+$ *strA leuB*$^-$)][37] that carries a *leuB* mutation because the *LEU2* gene complements this mutation and allows the strain to grow on minimal media lacking leucine. Plasmid DNA can be prepared from the *E. coli* transformants by standard methods.[27,28] Before being used for other analyses, recovered plasmids should be introduced into the reporter strain along with the plasmid expressing the target protein hybrid to confirm that the desired plasmid has been isolated. Often, multiple activation domain plasmids are introduced into individual cells during the transformation procedure, although only one of them is responsible for reporter gene activation. Maintaining selection for reporter gene (e.g., *HIS3*) activity on the positive yeast colonies while purifying them can increase the proportion of the plasmids being the desired ones, making these plasmids easier to isolate.

Identification of False-Positive Proteins

A positive signal may be obtained from a library search even though the target protein and the protein encoded by the activation domain plasmid

[35] C. S. Hoffman and F. Winston, *Gene* **57,** 267 (1987).
[36] P. Kaiser and B. Auer, *BioTechniques* **14,** 552 (1993).
[37] M. N. Hall, L. Hereford, and I. Herskowitz, *Cell* (*Cambridge, Mass.*) **36,** 1057 (1984).

do not interact with one another.[25] Although some of these false-positive proteins activate transcription in the absence of a DNA-binding domain hybrid, many require the presence of a DNA-binding domain hybrid. False positives are eliminated by verifying that reporter gene expression is specific for the presence of the target protein hybrid and is not generated in the presence of other nonrelated DNA-binding domain hybrids. In addition, strains that carry two reporter genes with dissimilar promoters that both contain binding sites for the same DNA-binding domain can help eliminate many false positives.[25] Select only the putative positive clones that activate transcription of both reporter genes in the presence of the target protein hybrid. Strains such as Y153[12] and Hf7c[26] contain such reporter genes. Positives that meet the criterion of specificity are best confirmed using a biochemical assay for interaction.

An additional genetic selection was developed to rapidly eliminate the DNA-binding domain vector from the reporter strain, leaving only the library plasmid present in the positive transformant.[16] The DNA-binding domain vector pAS2 (see Table I, Fig. 2) carries the *CYH2* gene which confers cycloheximide sensitivity to a resistant yeast strain such as Y190 (see Table II). Growth of positive transformants on cycloheximide-containing media selects for the loss of this plasmid. Plasmids encoding various DNA-binding domain hybrids can be introduced into the resulting strain by a mating procedure to test whether or not the positive signal is due to a specific interaction with the target protein.

Troubleshooting

Sometimes interactions that occur normally *in vivo* are not detected when tested in the two-hybrid system. For example, if high level expression of one of the fusion proteins is toxic to the reporter strain, transformants will be unable to grow. Sometimes truncation of the toxic protein will alleviate its detrimental effect, yet still allow the interaction to occur or, alternatively, the hybrid could be expressed using a conditional promoter. Other reasons for not obtaining a signal include hybrid proteins that are not stably expressed in yeast, DNA-binding or activation domains that occlude the site of interaction, failure of the hybrid proteins to fold properly in yeast, or failure of the hybrid proteins to enter the nucleus. Different hybrid constructions might help, but in some cases there may be no way to detect an interaction using this system.

A more common problem occurs when the protein fused to the DNA-binding domain activates transcription by itself. This activation is more likely to occur if the protein is a transcription factor but other proteins that are not involved in transcription may also activate transcription. In a test involving two defined proteins, the two proteins can be switched to the opposite vectors such that the activating protein is fused to the activation domain instead of the DNA-binding domain. It may also be possible to delete the activation domain of the target protein. Less frequently an activation domain hybrid will activate transcription in the absence of a DNA-binding domain hybrid.

Mapping Domains Involved in Protein–Protein Interactions

The two-hybrid system can be used to define interacting domains by assaying protein fragments for their ability to bind to a target protein or by screening large numbers of mutants to identify residues involved in a protein–protein interaction.[38] DNA fragments encoding portions of an interacting protein can be generated by restriction enzyme digestion, *Bal*31 deletion, or by PCR and these can be cloned into the appropriate two-hybrid vector. The ability of fragments encoded by each clone to interact with another hybrid can then be quickly assayed. For identifying specific residues that are involved in an interaction, a library of hybrid proteins containing point mutations is screened for mutants that have lost their ability to interact with the protein partner, as indicated below.

Protocol

1. Using PCR under suboptimal conditions,[39] generate mutant fragments of the gene encoding all or part of one member of an interacting pair.
2. Clone these fragments into the appropriate vector to create a library of mutant hybrid proteins. Pool the bacterial colonies and prepare plasmid DNA.
3. Transform this library and the plasmid expressing a hybrid of the protein partner into a yeast reporter strain. Select for transformants that receive both plasmids.
4. Perform a filter assay for β-galactosidase expression and identify colonies that are white or lighter blue than the unmutagenized control. These may represent mutants that have lost all or some of their ability to bind the protein partner.
5. Test these colonies with reduced β-galactosidase activity for expression of the mutant hybrid protein by Western blot analysis.

[38] B. Li and S. Fields, *FASEB J.* **7,** 957 (1993).

[39] D. W. Leung, E. Chen, and D. V. Goeddel, *Technique* **1,** 11 (1989).

6. For hybrid proteins that are full-length and stably expressed, isolate the plasmid encoding the mutant protein and determine the DNA sequence in the region that has been mutagenized.

Isolation of DNA-Binding Proteins

In addition to identifying protein–protein interactions, many of the reagents that are used to perform two-hybrid system library searches can be used for the identification of proteins that bind to specific DNA sequences.[40,41] The target DNA sequence of the putative binding protein is placed upstream of a reporter gene which is then integrated into yeast to create a new reporter strain. An activation domain library is introduced into the reporter strain and the transformants are then assayed for expression of the reporter gene. Hybrid proteins that bind to the test sequence are able to activate transcription of the reporter gene. The sequence of the DNA-binding protein is then obtained from the activation domain plasmid using methods discussed previously.

Vectors for constructing yeast reporter genes include pRS315HIS,[41] p601,[42] and pTH1,[23] which carry *HIS3* reporter genes. It is helpful to create, in parallel, a second strain carrying the same reporter gene with a mutated target sequence so that putative positive clones can be tested immediately to determine whether their binding is specific to the target DNA sequence. Included below is a brief protocol for screening activation domain libraries for site-specific DNA-binding proteins.

Protocol

1. Construct a plasmid containing the target DNA sequence upstream of a reporter gene such as *HIS3* or *lacZ* (or make both constructions).
2. Integrate the reporter gene(s) construct in an appropriately marked yeast strain. Verify that the vector has been integrated correctly by Southern blot analysis. Alternatively, transform a replicating plasmid to create the reporter strain and maintain selection for the reporter plasmid.
3. Check to ensure that the reporter strain does not express the reporter gene(s) constitutively.
4. Introduce the appropriate activation domain library into the reporter strain. Select for transformants that express the reporter gene(s).

[40] J. J. Li and I. Herskowitz, *Science* **262,** 1870 (1993).
[41] M. M. Wang and R. R. Reed, *Nature* (*London*) **364,** 121 (1993).
[42] C. Alexandre, D. A. Grueneberg, and M. Z. Gilman, *Methods* **5,** 147 (1993).

5. Isolate DNA from the yeast and electroporate into *E. coli* to isolate the activation domain plasmid.

6. Determine the sequence of the insert in the plasmid.

Other Applications

Other potential applications of the two-hybrid system include using it to screen for compounds that disrupt a protein–protein interaction, screening for peptides that bind to a target protein, and generating a "protein linkage map" that describes all the detectable protein–protein interactions in a cell. These applications are discussed briefly below.

Compounds could be screened for their ability to inhibit a protein–protein interaction by establishing a reporter system in which a readily detected protein, such as luciferase, is produced as a result of a two-hybrid interaction. This system could be yeast based or it could be a mammalian cell line.[43] The cells are then treated with each member of a library of compounds, and assays are performed to detect compounds that reduce the amount of reporter transcription. As a control, a parallel line carrying two hybrid proteins, which are unrelated to the target proteins and result in similar reporter gene expression, is also treated with each compound. This control eliminates from further study compounds that bind to the DNA-binding or activation domain, that are generally deleterious for transcription, or that are toxic to the cells.

The two-hybrid system is capable of detecting the interaction of a protein with a small peptide. An example of this interaction is that between the Rb protein and a peptide derived from the Simian virus 40 large T antigen.[44,45] With Rb fused to the Gal4p DNA-binding domain and the T antigen peptide fused to the Gal4p activation domain, only a Leu–X–Cys–X–Glu sequence present in the T antigen peptide is required for reporter gene expression. Screening an activation domain library that consists of synthetic oligonucleotides encoding 16 amino acid residue-long peptides resulted in the identification of several peptides containing the Leu–X–Cys–X–Glu motif.[45]

The two-hybrid system might also be used to identify large numbers of interacting proteins by preparing libraries in both the DNA-binding and activation domain vectors and screening them against one another. The plasmids encoding the DNA-binding and activation domain library fusions

[43] C. V. Dang, J. Barrett, M. Villa-Garcia, L. M. S. Resar, G. J. Kato, and E. R. Fearon, *Mol. Cell. Biol.* **11,** 954 (1991).

[44] T. Durfee, B. Gorovits, C. Hensey, and W.-H. Lee, submitted for publication.

[45] M. Yang, Z. Wu, and S. Fields, *Nucleic Acids Res.*, in press.

could be isolated and a short sequence of each insert would be obtained. Each pair of sequences from a positive serves as a tag for a protein–protein interaction. This screening should result in the identification of networks of interacting proteins to create a so-called protein linkage map. This information would identify new protein–protein interactions and would help place novel proteins into some cellular pathway.

Acknowledgments

We thank Gail Mandel for comments on the manuscript and for sharing unpublished data, Erica Golemis and Stan Hollenberg for plasmid maps, Bin Li and Kuniyoshi Iwabuchi for the CPRG assay protocol, and Judy Nimmo for assistance with the manuscript.

[17] Retrovirus Gene Traps

By GEOFFREY G. HICKS, ER-GANG SHI, JIN CHEN, MICHAEL ROSHON, DOUG WILLIAMSON, CHRISTINA SCHERER, and H. EARL RULEY

Introduction

The study of mammalian cells and animals has been hampered by the lack of efficient genetic systems as are commonly available for lower organisms. Despite improved molecular methods, genes responsible for organismal phenotypes are still difficult to clone. Moreover, relatively few mutant alleles are available for study, particularly those that are recessive or affect embryonic development.

To help identify mammalian genes responsible for recessive phenotypes, several types of gene trap retroviruses have been developed for use as insertional mutagens. The viruses permit direct selection of cell clones in which expressed genes have been disrupted as a result of virus integration. The U3 gene traps contain coding sequences for a selectable marker inserted into the U3 region of the long terminal repeat (LTR). Provirus integration positions the 5′ copy only 30 nucleotides from the flanking cellular DNA (Fig. 1A), and selection for U3 gene expression generates cell clones in which the proviruses have inserted in or near exons of transcriptionally active genes.[1–4] The splicing-activated gene traps (Fig. 1B) contain a selectable

[1] H. von Melchner and H. E. Ruley, *J. Virol.* **63,** 3227 (1989).
[2] H. von Melchner, J. DeGregori, H. Rayburn, C. Friedel, S. Reddy, and H. E. Ruley, *Genes Dev.* **6,** 919 (1992).
[3] S. Reddy, J. DeGregori, H. von Melchner, and H. E. Ruley, *J. Virol.* **65,** 1507 (1991).
[4] W. Chang, C. Hubbard, C. Friedel, and H. E. Ruley, *Virology* **193,** 737 (1993).

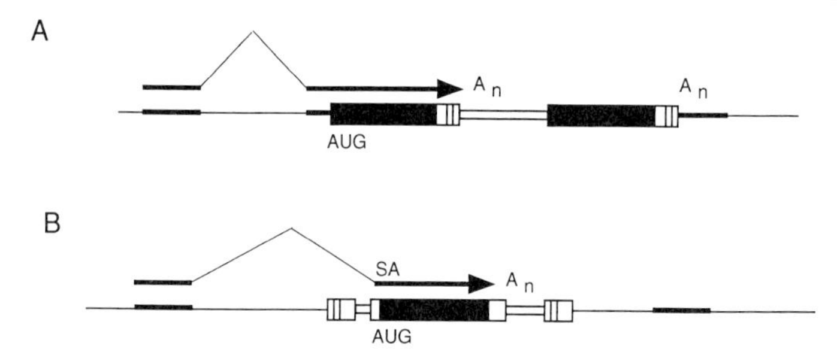

FIG. 1. Gene trap proviruses. The U3 gene trap vectors (A) contain coding sequences for a selectable marker (shown in black) inserted into the U3 region of the LTR. Integration places the leftward (5′) copy within 30 nucleotides of the flanking cellular DNA, and selection for U3 gene expression generates cell clones in which the provirus has inserted in or near expressed exons. The U3 gene is expressed from fusion transcripts (arrow) that extend into the virus from the flanking cellular DNA. Initiation codon of the U3 gene (AUG) and polyadenylation signals (A_n) are also indicated. The splicing-activated gene trap vectors (B) contain coding sequences for a selectable marker (shown in black) inserted into the body of the provirus, just downstream of a splice acceptor site (SA). Selection for virus gene expression generates cell clones in which the viral gene is expressed from cellular transcripts (arrow) that splice into the virus.

marker positioned next to a splice acceptor site. Consequently, the provirus inserts a functional exon that can be expressed when the virus integrates into introns of transcribed genes.[5,6]

Gene trap selection typically targets genes transcribed by RNA polymerase II. Thus, the disrupted genes are spliced[2,6–8] and are expressed in cells prior to virus integration.[2,9] Over 10% of the targets involve known genes,[2,10] and over 40% are linked to recessive phenotypes when introduced into the germline.[2,6,8] Pol I and Pol III promoters typically do not activate viral gene expression, probably because 5′ caps are required for efficient RNA processing and translation.[11,12]

[5] D. G. Brenner, S. Lin-Chao, and S. N. Cohen, *Proc. Natl. Acad. Sci. U.S.A.* **86,** 5517 (1989).
[6] G. Friedrich and P. Soriano, *Genes Dev.* **5,** 1513 (1991).
[7] J. V. DeGregori, A. Russ, H. von Melchner, H. Rayburn, P. Priyaranjan, N. Jenkins, N. Copeland, and H. E. Ruley, *Genes Dev.* **8,** 265 (1994).
[8] W. C. Skarnes, B. A. Auerbach, and A. Joyner, *Genes Dev.* **6,** 903 (1992).
[9] H. von Melchner and H. E. Ruley, *Proc. Natl. Acad. Sci. U.S.A.* **87,** 3733 (1990).
[10] G. G. Hicks, E.-G. Shi, D. Williamson, D. Roshon, J. Chen, and H. E. Ruley, unpublished results (1994).
[11] A. S. Banerjee, *Microbiol. Rev.* **44,** 175 (1980).
[12] M. Kozak, *J. Biol. Chem.* **266,** 19867 (1991).

Insertional Mutagenesis in Cultured Cells

In principle, gene trap mutagenesis may be used to isolate any gene for which a loss of function mutation results in a selectable phenotype. These include cellular genes involved in tumor suppression, programmed cell death (apoptosis), signal transduction, susceptibility of cells to killing by lytic viruses, and expression of proteins on the cell surface. A library of clones is selected which collectively contains proviruses in most expressed genes. Clones expressing phenotypes that result from loss of function mutations are subsequently selected and specific genes disrupted by the integrated proviruses are characterized.

Disruption of single-copy cellular genes by retroviruses typically requires between 5×10^6 and 10^8 integration events; however, gene trap selection enriches for insertional mutations by 100- to 1000-fold.[4] Since the typical mammalian cell expresses only about 10,000–20,000 genes,[13] a library of 10^5 gene trap clones is sufficient to disrupt all readily targeted genes. Such libraries are probably small enough to allow loss of diploid gene functions, due to preexisting hemizygosity or loss of heterozygosity (e.g., by gene conversion, nondisjunction, or mutation). Recovery of clones with null mutations may be enhanced by increasing the selective pressure[14] or by using hypodiploid cells.[4]

Methods for preparing retrovirus stocks and mutagenizing cells with gene trap vectors have been described elsewhere.[15] The host range of the gene trap retrovirus must allow efficient infection of the target cells. Ecotropic viruses are suitable for mouse and rat cells, whereas amphotrophic viruses allow a wider variety of cell types to be infected.[16] Host range can also be extended by using pseudotype viruses that incorporate the VSV (vesicular stomatitis virus) G protein[17] or by expressing the ecotropic retrovirus receptor in the target cells.[4,18]

Cells are infected at a relatively low multiplicity of infection so that most mutagenized clones will contain only a single provirus. This greatly simplifies efforts to determine if mutant phenotypes are virus induced.

[13] B. Lewin, *Cell (Cambridge, Mass.)* **4,** 77 (1975).

[14] R. M. Mortensen, D. A. Conner, S. Chao, A. A. Geisterfer-Lowrance, and J. G. Seidman, *Mol. Cell. Biol.* **12,** 2391 (1992).

[15] J. Chen, J. DeGregori, G. Hicks, M. Roshon, C. Scherer, E.-G. Shi, and H. E. Ruley, *Methods Mol. Genet.* **4,** 123 (1994).

[16] A. D. Miller, *Curr. Top. Microbiol. Immunol.* **158,** 1 (1991).

[17] J. C. Burns, T. Friedmann, W. Driever, M. Burrascano, and J. K. Yee, *Proc. Natl. Acad. Sci. U.S.A.* **90,** 8033 (1993).

[18] L. M. Albritton, L. Tseng, D. Scadden, and J. M. Cunningham, *Cell (Cambridge, Mass.)* **57,** 659 (1989).

Ideally, multiple mutants will contain viruses inserted into the same locus.[19] Alternatively, the unoccupied alleles may be missing, indicating loss of heterozygosity.[4] Finally, genes disrupted by the virus should be tested for the ability to complement the mutation.

Identification of Transcriptionally Regulated Genes

Gene trap retroviruses have employed a variety of selectable markers to disrupt expressed cellular genes. Those which permit selection both for and against viral gene expression may be used to identify cellular genes that are regulated by hormones or other factors.[3–5] Collections of clones are isolated following selection for or against U3 expression in the presence or absence of factor. Growth conditions are reversed while selecting for changes in viral gene expression. The resulting clones include those in which the provirus has inserted downstream of transcriptionally regulated promoters.

The process has several potential advantages over cDNA subtraction as a means to isolate differentially regulated genes. First, gene trap selection is sensitive, detecting even weaky expressed genes. This is important since most genes are expressed at levels (<0.01% of total mRNA) too low for efficient recovery after cDNA subtraction. Second, gene representation is not as biased for highly expressed genes. Third, gene traps frequently insert near 5′ exons that may be missing from cDNA clones. Fourth, differential gene expression is likely to reflect transcriptional control rather than changes in RNA stability or processing. Finally, cells recovered after gene trap selection contain a reporter gene expressed from the natural promoter of the gene and may be used to study transcriptional regulation.

Insertional Mutagenesis in Mice

Gene traps provide an effective means to analyze gene functions in mice. Genes disrupted in embryonic stem (ES) cells are introduced into the germline, and gene functions are studied in animals homozygous for the mutations. Several features of the process are particularly useful. First, the disrupted genes are relatively easy to isolate from sequences adjacent to the integrated provirus.[7] Second, genes expressed in ES cells are typically expressed in early embryos and are not expressed simply because the cells are in culture[3,6,20] Third, mutations affecting genes required for embryonic development are readily identified by the failure to recover offspring that

[19] S. C. Hubbard, L. Walls, H. E. Ruley, and E. A. Muchmore, *J. Biol. Chem.* **269,** 3717 (1994).
[20] A. Gossler, A. L. Joyner, J. Rossant, and W. C. Skarnes, *Science* **244,** 463 (1989).

are homozygous for the provirus. This has been observed with over 40% of the disrupted genes analyzed.[2,6] Finally, most genes, including weakly expressed genes, can be targeted by gene trap mutagenesis.[4,7]

Excellent results have also been obtained by introducing gene traps by DNA-mediated gene transfer.[6,20] The vectors can be constructed more rapidly and clones isolated after electroporation frequently contain only a single insert. However, retrovirus vectors are easier to use, especially for large-scale mutagenesis, and the structure of the recombination products is more predictable.

Because many insertions disrupt widely expressed or "housekeeping" genes, it is frequently desirable to screen ES cell clones for mutations that involve specific genes or genes that are developmentally regulated. One strategy uses a reporter gene that expresses both neomycin phosphotransferase and β-galactosidase (*lacZ*) activities.[6] Mutant clones are isolated by Geneticin (G418) selection, and developmentally regulated fusion genes are identified by staining differentiated and undifferentiated cells with X-Gal. Expression of *lacZ* fusion genes can be monitored *in vivo*, either before or after germline transmission,[6,20] and *in vitro*.[21]

Mutations that disrupt specific genes can be identified by sequencing DNA adjacent to the integrated provirus. Since U3 gene traps integrate in or near transcribed exons, it is only necessary to sequence regions within 300 nucleotides of the provirus. The flanking sequences are readily isolated by inverse polymerase chain reaction.[2] Alternatively, segments of genes disrupted by both U3 trap and splice trap vectors can be amplified by 5′ RACE (rapid amplification of cDNA ends).[8]

The flanking sequences also represent promoter-tagged sites (PTSs) and can be used for the same purposes (e.g., gene identification) as expressed sequence tags (ESTs).[22] In addition, PTSs complement information provided by ESTs since gene trapping is probably less biased for highly expressed genes than is cDNA cloning. In addition, cDNA probes are unable to distinguish between pseudogenes and functional genes, whereas PTSs are linked to expressed genes.

To simplify the process of analyzing genes disrupted by gene trap mutagenesis, a shuttle vector has been developed (Fig. 2) which contains a neomycin resistance gene in the U3 region, an ampicillin resistance gene, the pBR322 plasmid origin of replication, and the *lac* operator (*lacO*) sequence in the body of the provirus. Depending on the restriction enzyme

[21] S. Reddy, H. Rayburn, H. von Melchner, and H. E. Ruley, *Proc. Natl. Acad. Sci. U.S.A.* **89,** 6721 (1992).

[22] M. D. Adams, M. Dubnick, A. R. Kerlavage, R. Moreno, J. M. Kelley, T. R. Utterback, J. W. Nagle, C. Fields, and J. C. Venter, *Nature* (*London*) **13,** 632 (1992).

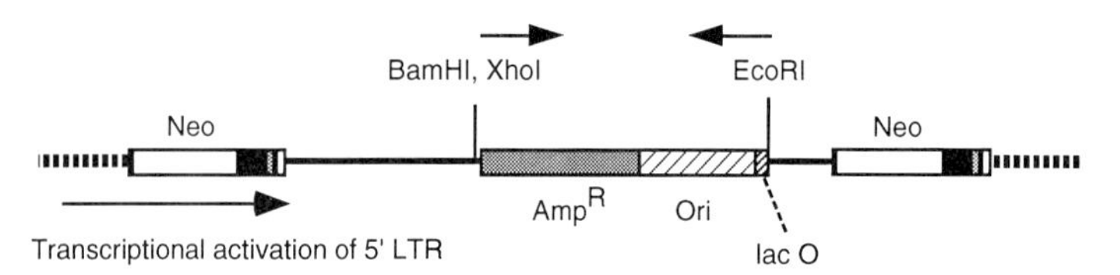

FIG. 2. U3NeoSV1 Shuttle Vector. U3NeoSV1 is a gene trap retrovirus that allows flanking cellular sequences (dashed line) to be cloned directly by plasmid rescue. Different restriction enzymes allow sequences 5′ (*Eco*RI) or 3′ (*Bam*HI or *Xho*I) of the provirus to be cloned as indicated by arrows. The positions of the aminoglycoside phosphotransferase gene (Neo), an ampicillin-resistance gene (Amp^R), the pBR322 plasmid origin of replication (Ori), and the *lac* operon operator (*lacO*) sequence are also shown.

used, cellular sequences 5′ or 3′ of the provirus can be cloned by plasmid rescue. *lacO* permits partial purification of flanking sequence fragments by binding to the *lac* repressor protein.[23,24]

Computer searches of the sequence databases reveal instances in which the provirus has disrupted known genes or cDNAs. Thus far, approximately 10% of flanking sequences have identified mutations in known genes or cDNAs.[10] Eventually, most flanking regions are expected to match known genes since the nucleic acid sequence databases double every 18 months.[25] In addition, large-scale sequencing of random cDNAs[22] will have a significant impact on the identification of targeted genes. Decisions to pass mutations into the germline will depend on the predicted structures of the protein products.

The capacity to induce, characterize, and maintain mutations in ES cells circumvents many limitations associated with conventional mammalian genetics. Large numbers of genes can be mutagenized even without detailed information on the genomic structure of the genes, and for a fraction of the cost of targeting genes by homologous recombination. This is important since genes required for biological function (and disease) frequently cannot be predicted in advance. It is also likely that many genes will be required for previously unknown biological processes. For example, nearly 80% of the genes identified from the sequence of yeast chromosome III displayed no significant similarity with previously sequenced genes.[26] In short, gene

[23] V. K. Pathak and H. M. Temin, *Proc. Natl. Acad. Sci. U.S.A.* **87,** 6019 (1990).

[24] J. A. Gossen, W. J. F. de Leeuw, A. C. Molijin, and J. Vijg, *BioTechniques* **14,** 624 (1993).

[25] D. Benson, D. J. Lipman, and J. Ostell, *Nucleic Acids Res.* **21,** 2963 (1993).

[26] S. G. Oliver, Q. J. M. van der Aart, M. L. Agostoni-Carbone, M. Aigle, L. Alberghina, D. Alexandraki, *et al.*, *Nature* (*London*) **357,** 38 (1992).

trap mutagenesis will facilitate a functional analysis of the mouse genome and will provide mouse models of human genetic diseases.

Detailed Methods

Routine Maintenance of ES Cells

ES cells are grown on irradiated mouse embryo fibroblast (MEF) feeder layers, prepared from 13- to 14-day gestation embryos. A liberal amount of 70% ethanol is applied to the abdomen, and the uterine horns are removed through an incision in the abdominal wall, transferred to a 10-cm bacteriological petri dish, and rinsed with sterile phosphate-buffered saline (PBS). Each embryo is dissected from the uterus and placenta using two pairs of watchmaker forceps and rinsed in fresh PBS. The heads and dark-colored tissues (e.g., liver) are removed and the embryos are rinsed in fresh PBS. The embryos are transferred to a fresh dish and minced with two scalpels or fine scissors. A solution of 25 ml trypsin/EDTA [GIBCO; 0.25% (w/v) trypsin, 1 m*M* ethylenediaminetetraacetic acid, pH 8.0, in PBS] is added and the tissues are disrupted further by pipetting up and down through a 5-ml pipette. The tissue suspension is transferred to a 100-ml bottle and incubated at 37° for 10 to 15 min with vigorous stirring. Fifty milliliters of culture media [Dulbecco's modified Eagle's medium (DME), 10% (v/v) fetal bovine serum (FBS)] is added, and large clumps are allowed to settle for 2 min. The decanted cells are centrifuged at 1000 rpm for 5 min at room temperature, resuspended (1 ml for each embryo) in ice-cold freezing medium [DME, 50% FBS, 12% dimethyl sulfoxide, v/v (DMSO)], and frozen slowly (−1°/min).

To expand the number of MEF cells, one vial of frozen cells is thawed and plated onto a 15-cm culture dish and expanded to 27 dishes by splitting the cells 1:3 each time the cells become confluent for a total of three passages. After expansion, each 15-cm culture dish is washed with PBS and 5 ml trypsin/EDTA is added. The cells are incubated at 37° for 2 min, serum is added to final concentration of 10%, and the cells are centrifugated at 1000 rpm for 5 min at room temperature. Cell pellets are resuspended in an equal volume of feeder cell medium and an aliquot is counted using a hemocytometer. The cells are exposed to 3000 rad γ-irradiation, in a 50-ml conical tube, pelleted by centrifugation, resuspended at 2×10^6 cells/ml in freezing medium, and frozen in 1-ml aliquots.

Feeder cells (2×10^6) are seeded onto a 10-cm tissue culture dish precoated with 0.1% gelatin. Feeders can be plated either before or along with ES cells. If plated before, thawed MEFs are plated directly and are allowed to adhere for 4 hr. The medium is changed to ES medium and ES cells are added. If plated together with ES cells, the thawed MEFs should be centrifuged first to remove DMSO which inhibits the growth of ES cells.

Several general procedures should be observed to maintain ES cells in an undifferentiated state. (1) Since overgrown ES colonies tend to differentiate, the cells should be passaged frequently (at least twice weekly). (2) Established ES cells divide rapidly and the medium [DME supplemented with 15% heat-inactivated (55°, 30 min) fetal bovine serum; 100 m*M* nonessential amino acids, 0.1 m*M* 2-mercaptoethanol, and 1000 U/ml leukemia inhibitory factor (Esgro, GIBCO)] should be changed every day. Serum lots are pretested for ES cell culture. The plating efficiency of ES cells should be at least 30% and the cells should grow quickly (doubling every 12–14 hr) and maintain an undifferentiated morphology. (3) Cell aggregates tend to differentiate and, consequently, the cells are well trypsinized when passaged. For this, the cells are washed once with PBS, trypsinized in 2 ml trypsin/EDTA solution [0.25% (w/v) trypsin, 1 m*M* EDTA, pH 8.0, in PBS], and incubated at 37° for 3 to 4 min. The cell suspension is vigorously pipetted up and down to disrupt cell aggregates, incubated at 37° for another 3 min, and then checked visually under the microscope to ensure that a single cell suspension is generated. Trypsinization is inhibited by the addition of 5 ml of ES cell medium and the cell suspension is split at a ratio of 1 : 10 on fresh gelatinized plates with feeders.

Selective Disruption of Genes by U3neo Shuttle Vector

Methods for preparing retrovirus producer cells and virus stocks have been described elsewhere.[15] ES cells are infected at an MOI (multiplicity of infection) of 0.1, as titered on NIH 3T3 cells (the actual titer on ES cells may be somewhat lower) by adding 2 ml of diluted virus stocks to 10^5 ES cells (plated 12 hr before infection on a 15-cm dish) in the presence of 4 μg/ml Polybrene. The cells are incubated for 1 hr at 37° with occasional rocking, at which time 18 ml of fresh ES cell medium is added. The following day, cells are placed in ES medium containing 300 μg/ml active G418 and maintained for 7 days. The medium is changed every day. Culture dishes are marked to indicate the positions of undifferentiated Neo^R colonies, and the colonies are gently washed with PBS. Marked colonies are drawn into a disposable tip of a micropipettor set at 10 μl under 40× magnification. We use a Nikon TMS inverted microscope placed in the hood. The cells are placed in a microtiter well containing 100 μl trypsin/EDTA solution for 5 min and are pipetted up and down to obtain a single-cell suspension. The ES medium (100 μl) is added to each well and the cells are pipetted again and are transferred to a 24-well plate with feeder cells. A multichannel pipettor facilitates processing large numbers of clones. Finally, cells are expanded into two 60-mm dishes. Cells from one dish are used for DNA isolation while the remainder are cryopreserved in liquid nitrogen.

It is important to pick only undifferentiated colonies. Differentiated colonies are flat and contain larger cells with an epithelial morphology. Undifferentiated colonies are thicker and consist of small cells surrounded by a sharp refractile boundary under phase contrast. Cells within the colonies are densely packed obscuring individual cells.

Plasmid Rescue

Cellular DNA is extracted from dense monolayers of ES cells as follows. Cells on one 60-cm culture dish are washed once with 3 ml PBS and are lysed for 15–30 min at 37° in 2 ml of tail buffer [100 m*M* Tris–HCl, pH 8.5; 5 m*M* EDTA; 0.2% sodium dodecyl sulfate; 200 m*M* NaCl; 10 μg/ml RNase A]. An additional 2 ml of tail buffer with 200 μg/ml proteinase K is then added by gentle pipetting using a large bore transfer pipette, and the lysate is transferred to a 15-ml conical tube and incubated overnight at 58°. A single extraction with an equal volume of phenol–chloroform solution (phenol : chloroform : isoamyl alcohol at 25 : 24 : 1; equilibrated with 100 mM Tris–HCl, pH 8.0; 0.1 m*M* EDTA), greatly improves ligation efficiency but is not necessary for Southern blot analysis. Genomic DNA is precipitated by the addition of 2.5 vol of ethanol (10 ml) and gentle rocking. DNA is spooled onto a flame-sealed Pasteur pipette, washed twice with 70% ethanol, briefly air dried, and dissolved in 400 μl 10 m*M* Tris–HCl, pH 8.0, 0.1 m*M* EDTA. For complete and uniform resolution, samples are heated to 68° for 30 min, gently agitated, and left at room temperature for 1–2 days. The DNA is then stored at 4°. This protocol yields 15–30 μg of high molecular weight DNA.

DNA (10–20 μg) from each ES cell line is digested with 50 U *Eco*RI (NEB high concentration) for 2–3 hr in a volume of 250 μl. The digests are heat-inactivated (68°, 20 min), allowed to cool to room temperature, and purified through a Wizard DNA Clean column as specified by the manufacturer (Promega). The eluate (75 μl) is ligated at a concentration of 5 μg/ml. At this concentration a 6-kb *Eco*RI fragment is twice as likely to ligate with itself than with another fragment.[27] Each reaction (0.5 ml) contains 2.5 μg of *Eco*RI-digested DNA, 50 μl 10× ligation buffer [50 m*M* Tris, pH 7.6, 10 m*M* $MgCl_2$, 1 m*M* dithiothreitol], 1.0 m*M* ATP, and 4.0 Wiess U of T4 DNA ligase (GIBCO/NEB). Samples are heated to 68° for 10 min, rapidly cooled on ice, ligase reagents are added at 0°, and the reactions are incubated overnight at 16°. Samples are then heat inactivated

[27] A. Dugaiczyk, H. W. Boyer, and H. M. Goodman, *J. Mol. Biol.* **96,** 171 (1975).

for 20 min at 68°, purified over the Wizard columns, and precipitated as before.

One microgram of ligated DNA (2 μl) is carefully transferred to the inside wall of a prechilled 0.1-cm electroporation cuvette. Electrocompetent DH10B *Escherichia coli* cells (GIBCO) are quickly thawed and immediately placed on ice. Cells (25 μl) are added to the droplet of DNA so that the samples mix completely while minimizing manipulation of the cells. Electroporation is performed at 200 ohm, 25 μF, and 1.8 kV. Time constants of 4.3 to 4.8 typically give 2×10^9 to 2×10^{10} colonies/μg with supercoiled plasmid controls. SOC (800 μl) is added to electroporated cells within 2 sec. The bacteria are transferred to a 6-ml tube and incubated for 1 hr at 37° with shaking; 400 μl is plated onto a 150-mm LB-Amp (50 μg/ml) dish and colonies are counted after 16 hr at 37°. The average efficiency of plasmid rescue is 100 colonies/μg of genomic DNA.

Identification of Genes Disrupted by Gene Trap Mutagenesis

The flanking sequences are compared to a nonredundant DNA sequence database by using the BlastN[28] or FastDB (Intellegenetics, Menlo Park, CA) programs. For most genes in the database, only cDNA sequences are known, and consequently only exon sequences in the flanking region will be able to match known genes. Matches with Poisson probability scores of 10^{-9} or lower are significant. Typically, such matches are several orders of magnitude more significant than the next nearest score; the provirus is in the same transcriptional orientation as the target gene and is located at the 5′ end of the gene. Probabilities higher than 10^{-7} are usually insignificant. Such scores are likely to occur by chance alone, and matches with unrelated genes have similar scores. Moreover, the provirus is frequently in the wrong transcriptional orientation with respect to the target sequence or is located toward the 3′ end of the gene. Intermediate scores may or may not be significant. An inconclusive score may result when only a small segment of exon is present. In such cases, a short region with a high degree of similarity will end abruptly at a splice donor or acceptor site, depending on whether the virus has integrated into an exon or an intron.

To illustrate, searches were conducted (BlastN) using 300 nucleotides of a genomic sequence that contained various lengths of exon sequence. This simulated integration into an exon at various distances downstream from a splice acceptor site. As summarized in Table I, the correct target gene was identified when the query contained exon segments 40 nucleotides

[28] S. F. Altschul, W. Gish, W. Miller, E. W. Meyers, and D. J. Lipman, *J. Mol. Biol.* **215,** 403 (1990).

TABLE I
DATABASE SEARCHES WITH EXON-CONTAINING QUERY SEQUENCES[a]

Exon length	Errors	Highest score with exon	Poisson probability	Relative rank	Next highest score	Poisson probability
50	0	276	1.7e−13	1	134	8.1e−07
	2	258	5.6e−12	1	134	8.1e−07
	5	235	5.0e−10	1	134	8.1e−07
40	0	226	2.9e−09	1	134	8.1e−07
	2	208	9.6e−08	1	134	8.1e−07
	5	<127	<0.48	<50	134	8.1e−07
30	0	176	4.9e−05	2	154	0.0036
20	0	<127	<0.49	<50	—	—
23 Exon	0	115	0.028	1	80	1.0
18 Exon	0	90	.97	1	76	1.0

[a] Sequences containing different lengths of exon 2 of the mouse *REX1* gene were used to search a nonredundant nucleic acid database by the BlastN program. The query included adjacent intron sequences so that the total length was 300 nucleotides, except for searches using 23 and 18 nucleotides of exon only. This simulates integration into the third exon as previously observed,[2] but at various distances downstream from the splice acceptor site. In some cases two or five sequence changes (Errors) were introduced to assess the effects of sequencing errors or interspecies variation. The highest BlastN score of matches involving the cognate *REX1* cDNA, the Poisson probability of that match, the ranking of that match relative to all matches, and the score and probability of the next match retrieved in the search are listed. Scores of matches not involving *REX1* were higher than random because the intron contains a partial B1 repetitive element. Searches were performed at the NCBI using the BLAST E-mail network service.

or more in size if the sequences were identical or by segments of 50 and 40 nucleotides in size containing five and two sequence changes, respectively. However, correct matches were not found when the query contained 30 nucleotides of exon or 40 nucleotides of exon with five mismatches. This is because random matches are likely to occur with sequences in the longer intron segment. Improved results were obtained by using short segments which only contained exon sequences. The cognate gene was correctly identified with both 18 and 23 nucleotide segments, but only in the latter case was the correct match statistically better than the best random match.

The BlastN program is optimized to identify short regions of sequence identity, whereas programs such as FastDB are better suited for finding imperfect matches. Even so, it is not possible to match short exon sequences (e.g., when the provirus has inserted near the promoter or a splice acceptor site). It is useful to rescue and analyze 3′ flanking regions whenever the 5′ sequence contains a splice acceptor sequence $(T/C)_{11}N(T/C)AG$. Since the average size of 5′ and internal exons is 80 and 150 nucleotide, respectively,[29] searches involving combinations of both 3′ and 5′ flanking sequences will usually identify the cognate genes, if they are present in the database.

Construction and Analysis of Germ Line Chimeras

The process of transmitting fusion genes into the germline involves: (i) injecting ES cells into preimplantation blastocysts, (ii) transferring the blastocysts into the uterus of outbred foster mothers, and (iii) breeding the resulting chimeric mice to obtain offspring that have inherited genes from the ES cells.[30]

The ES cells are derived from a mouse whose coat color differs from that of the strain from which the blastocysts were isolated. Consequently, chimeras and offspring that have inherited genes from ES cells are readily identified on the basis of coat color. Most laboratories use male ES cells since the efficiency of germ cell chimericism is greater than when female ES cells are used. Moreover, male chimeras generate more offpspring than female chimeras. Finally, the transfer of male ES cells into a female blastocyst can cause the resulting chimera to develop as a phenotypic male. Consequently, the majority of germ cells are ES cell derived, resulting in a high frequency of germline transmission.

We routinely use D3 ES cells[31] or a D3 subclone, J1, provided by Dr. Janet Rossant, Mount Sinai Hospital, Toronto. The cells are derived from the 129 strain and are homozygous for the Agouti coat color marker. Typically, 15–20 ES cells are injected into C57BL6 blastocysts (3.5 days) and implanted into the uterus of pseudopregnant ICR recipients.[30] After chimeric males are mated with C57BL6 mice, 3–100% of the offspring from 50% of the male chimeras inherit the Agouti coat color marker. Female chimeras will occasionally transmit genes from male ES cells, but the efficiency is low (ca. 5% of female chimeras).

Since the ES cells contain a single provirus, 50% of the Agouti offspring are typically heterozygous for the U3 fusion gene. To identify these mice, DNA is extracted from tail segments[32] and is analyzed by Southern blot hybridization. DNA is isolated from 1.0-cm tail segments as described earlier except the samples are not extracted with phenol and the DNA is

[29] J. D. Hawkins, *Nucleic Acids Res.* **16,** 9893 (1988).

[30] B. Hogan, F. Constantini, and E. Lacy, "Manipulating the Mouse Embryo: A Laboratory Manual." Cold Spring Harbor Lab., Cold Spring Harbor, NY, 1986.

[31] T. C. Doetschman, H. Eistetter, M. Katz, W. Schmidt, and R. Kemler, *J. Embryol. Exp. Morphol.* **87,** 27 (1985).

[32] P. W. Laird, A. Zijderveld, K. Linders, M. A. Rudnicki, R. Jaenisch, and A. Berns, *Nucleic Acids Res.* **19,** 4293 (1991).

spooled from 2-propanol. The yield is approximately 100 μg/cm. Mice that carry the provirus are crossed to produce offspring of which 25% are homozygous for the transgene. The genotype of the animals is assessed by Southern blot analysis using flanking cellular sequences as probes.

Acknowledgments

This work was supported by Public Health Service Grants R01HG400684, RO1GM48688, and RO1CA40602 to H. E. R. and by postdoctoral fellowships from the American Cancer Society to J. C. and the National Cancer Institute of Canada to G. H. M. R. and D. W. are supported by a Medical Scientist Training Grant and C. S. is supported by the Department of Biology, Massachusetts Institute of Technology.

[18] Fingerprinting of DNA and RNA by Arbitrarily Primed Polymerase Chain Reaction: Applications in Cancer Research

By MANUEL PERUCHO, JOHN WELSH, MIGUEL ANGEL PEINADO, YURIJ IONOV, and MICHAEL MCCLELLAND

Introduction

Arbitrarily primed polymerase chain reaction (PCR) or AP-PCR is a modification of the polymerase chain reaction that generates an information-rich and unbiased fingerprint of genomic DNA.[1,2] The method relies on the selective amplification of genomic sequences that, by chance, are flanked by adequate matches to a primer whose nucleotide sequence is arbitrarily chosen. Different template genomic DNA sequences display electrophoretic banding patterns in their AP-PCR fingerprints with a number of differences that is inversely proportional to the extent of similarity of the two genomes. Such differences can be exploited in ways largely analogous to the uses of restriction fragment length polymorphisms (RFLP), including the detection of mutations in phylogeny and ontogeny, with applications in genetic mapping, taxonomy, and cancer (reviewed in McClelland and Welsh[3]).

[1] J. Welsh and M. McClelland, *Nucleic Acids Res.* **18,** 7213 (1990).

[2] J. G. Williams, A. R. Kubelik, K. J. Livak, J. A. Rafalski, and S. V. Tingey, *Nucleic Acids Res.* **18,** 6531 (1990).

[3] M. McClelland and J. Welsh, *PCR Methods Appl.* **4,** 559 (1994).

Moreover, the assay is performed under competitive conditions (from 25 to 50 different sequences are coamplified with each arbitrary primer), which provides the requirements for quantitative amplification. Therefore, the arbitrarily primed PCR permits the quantitative detection of differences in the relative amounts of target sequences. This offers immediate applications in cancer research as an alternative approach to estimate the quantitative genomic changes underlying the aneuploidy of the cancer cell.

A simple modification of arbitrarily primed PCR enables the detection of differentially expressed genes by RNA fingerprinting.[4,5] The approach relies on the quantitative properties of the AP-PCR, whereby given two or more RNA populations, quantitative differences in the fingerprints result when corresponding templates are represented in different amounts.

In this chapter, we focus first on the application of arbitrarily primed PCR to the detection and characterization of genetic alterations that accompany malignancy. Such somatic genetic alterations include allelic losses or gains in anonymous sequences[6] and deletions or insertions of one or a few nucleotides in simple repeated sequences or microsatellites.[7]

In the second part of this chapter, we describe briefly a new approach based on arbitrarily primed PCR for the detection of differentially expressed genes. The strategy uses arbitrary priming in conjunction with reverse transcription and PCR to detect differences in the abundance of messenger RNAs isolated from cells in a number of situations, for instance, tissues that have differentiated along different developmental pathways, cultured cells that have been subjected to different growth conditions, or cells before and after transformation to the malignant state.

A. DNA Fingerprinting by Arbitrarily Primed PCR

In contrast with the standard PCR reaction where the two primers that flank the desired sequence are annealed to the template DNA at relatively high stringency so that only the desired sequence is amplified, arbitrarily primed PCR is based on the ability of PCR to generate a reproducible

[4] J. Welsh, K. Chada, S. S. Dalal, D. Ralph, R. Cheng, and M. McClelland, *Nucleic Acids Res.* **20,** 4965 (1992).

[5] P. Liang and A. Pardee, *Science* **257,** 967 (1992).

[6] M. A. Peinado, S. Malkhosyan, A. Velazquez, and M. Perucho, *Proc. Natl. Acad. Sci. U.S.A.* **89,** 10065 (1992).

[7] Y. Ionov, M. Peinado, S. Malkhosyan, D. Shibata, and M. Perucho, *Nature (London)* **363,** 558 (1993).

array of products when the annealing step is performed at low stringency. A single oligonucleotide of arbitrary sequence is used to initiate DNA synthesis from sites along the template with which it matches only imperfectly. A few of these sites exist, stochastically, on opposite strands of the template, within several hundred nucleotides of each other. In the low stringency cycles, the sequence of the primer flanks a handful of anonymous sequences. After these low stringency cycles, when longer primers are used (i.e., ~20-mers) the annealing temperature is raised and the reaction is allowed to continue under standard, high stringency PCR conditions. The products of this reaction are then resolved by gel electrophoresis. For a determined set of experimental conditions, the DNA band pattern is remarkably reproducible and this fingerprint is exquisitely dependent on the sequence of the arbitrary primer. Figures 1 and 2 show typical AP-PCR fingerprints of human colorectal tumors with two distinct arbitrary primers.

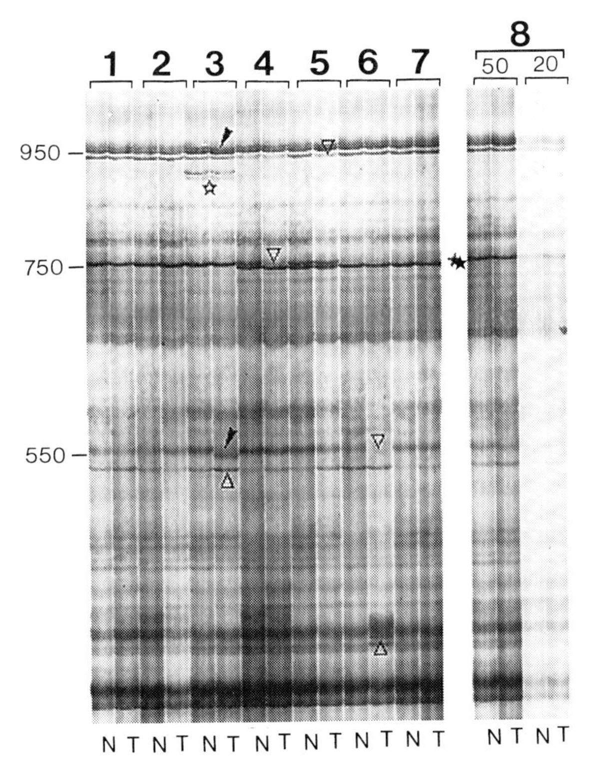

FIG. 1. AP-PCR fingerprints of colorectal tumors. The symbols are discussed in the text.

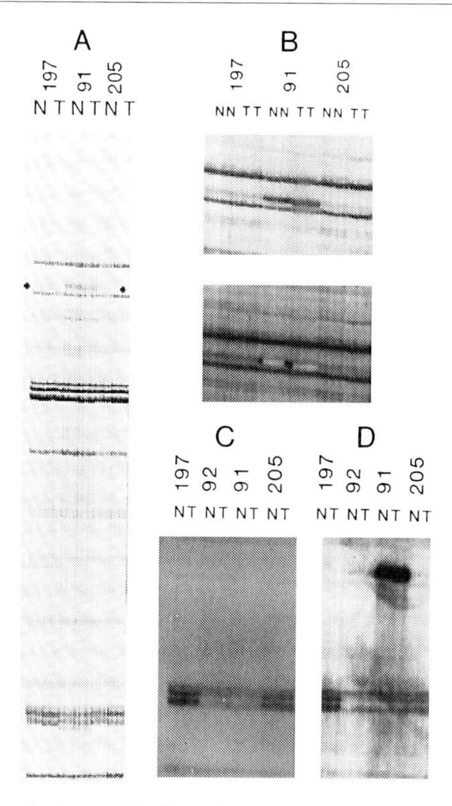

FIG. 2. Isolation and characterization of AP-PCR bands. (A) Autoradiogram of the [α-^{35}S]dATP-labeled fingerprints of genomic DNA from tumors and normal tissues (N and T, respectively) of cases indicated at the top obtained with the arbitrary primer K3US.[7] The polymorphic band of case 91 flanked by dots (A) was excised from the gel (B), labeled with [α-^{32}P]dCTP, and used as a probe in a Southern blot experiment of the AP-PCR bands transferred to a nylon membrane. (C and D) Membrane before and after hybridization, respectively.

Arbitrarily Primed PCR

Equipment and Reagents

Thermal cycler (Perkin-Elmer 480 Model)
Sequencing gel electrophoresis apparatus (40 cm × 30 cm × 0.4 mm)
Gel drier, X-ray film, exposure cassettes, and autoradiogram markers
Arbitrary primer, usually from 10 to 25 nucleotides long, concentration: 100 pmol/ml
5 U/μl *Taq* polymerase or 10 U/μl *Taq* polymerase Stoffel fragment (Perkin-Elmer Cetus)

TBE buffer (0.089 *M* Tris–borate, 0.025 *M* disodium EDTA, pH 8.3)
6% (w/v) polyacrylamide/8 *M* urea gel/1 × TBE buffer
10 × PCR buffer: 100 m*M* Tris, pH 8.0, 500 m*M* KCl, 25 m*M* $MgCl_2$, 0.01% (w/v) gelatin, 100 m*M* $MgCl_2$ for *Taq* polymerase, and 100 m*M* Tris, pH 8.0, 100 m*M* KCl, and 25 m*M* $MgCl_2$ for *Taq* polymerase Stoffel fragment
10 × dNTPs (1.25 m*M* each dNTP)
[α-^{35}S]dATP > 1000 Ci/mmol (5 μCi/reaction tube) or [α-^{32}P]dCTP > 3000 Ci/mmol
Denaturing loading buffer (95% (v/v) formamide, 0.1% (w/v) bromphenol blue, 0.1% (w/v) xylene cyanol, 10 m*M* Na_2EDTA)

Procedure A

Step 1. Preparation of DNA

Total cellular DNA is prepared by conventional methods.[6,8] We use a simplified procedure consisting of homogenization of freshly frozen tissue with a Polytron tissumizer (Tekmar) in a isotonic buffer (50 m*M* Tris–HCl, 0.15 *M* NaCl, and 10 m*M* EDTA, pH 8.5) containing 200 μg/ml of proteinase K (Beckman) and 0.2% sodium dodecyl sulfate (SDS), followed by incubation at 65° from 2 to 4 hr and at 37° overnight. After digestion, the solution is extracted with phenol/chloroform in a high salt buffer (same as above with 0.4 *M* NaCl) and spooling the DNA after ethanol precipitation. The DNA is air dried and dissolved in TE (1 m*M* Tris–HCl, pH 7.5, 1 m*M* EDTA). Digestion with RNAse A and further proteinase K/SDS treatment, phenol/chloroform extraction, and ethanol precipitation may be also implemented to increase the quality of the DNA preparation, especially when using large amounts of tissue.

The DNA concentration is checked spectrophotometrically and equal aliquots are electrophoresed on a 0.5% agarose gel and ethidium bromide is stained to estimate the quality (extents of degradation and contaminating RNA) of the DNA. DNA from normal and tumor tissue of the same individual is diluted to 20 and 10 ng/ml in TE. The measurement of DNA concentration is often not very accurate, so adjustments may be needed if the intensities of most of the arbitrarily primed PCR products are not uniform between DNA samples. This is one of the reasons why it is wise to fingerprint in duplicate using template DNA at concentrations differing by two- or fourfold. In this example, 25 and 50 ng of template DNA are used in the duplicate reactions.

[8] M. Perucho, M. Goldfarb, K. Shimizu, C. Lama, J. Fogh, and M. Wigler, *Cell* (*Cambridge, Mass.*) **27,** 467 (1981).

Step 2. Arbitrarily Primed PCR

A reaction mixture for 20 tubes is prepared to compare 10 genomes; add components sufficient for 21 reactions in this order:

Distilled water	16.05 μl	337 μl
10 × PCR buffer	2.5	52.5
100 m*M* $MgCl_2$*	0.5	10.5
10 × dNTP mix	2.5	52.5
Arbitrary primer	0.5	10.5
5 μCi [α-^{35}S]dATP	0.5	10.5
1 unit *Taq* polymerase (or 2 units *Taq* polymerase Stoffel fragment)	0.2	4.2
Total volume per tube	22.5 μl	400 μl

* Considering the content of the PCR buffer, the final $MgCl_2$ concentration is 5 m*M*.

Distribute 22.5 μl to each reaction tube. Add 50 ng of genomic DNA (2.5 μl of 20 ng/μl) to one set of tubes and 20 ng (2 μl of 10 ng/μl) to the other set of tubes. Add one drop of mineral oil to each tube. Perform the reaction in a thermal cycler for 5 low stringency cycles (95°, 1 min; 50°, 1 min; 72°, 2 min), then for 30 high stringency cycles (95°, 30 sec; 65°, 30 sec; 72°, 2 min). For 10-mers use 35° annealing throughout.

Dilute 10 μl of the reaction mix with 10 μl of formamide-dye buffer and incubate at 90–95° for 3 min. Immediately chill the solution on ice and load 2 μl on an 8 *M* urea/6% polyacrylamide sequencing gel and electrophorese for 6 hr at 55 W. Dry the gel under vacuum at 80° and directly expose to a X-ray film (Kodak AR-5 X-Omat) at room temperature, without an intensifier screen. Stick three or more luminescent labels to the dried gel in order to localize the bands in case isolation of some of the bands is desired.

Figure 1 shows the arbitrarily primed PCR analysis of eight matched colorectal normal-tumor (labeled at bottom N and T, respectively) tissue DNA pairs using the *Kpn*X arbitrary primer (primer sequence: 5′ CTTGCGCGCATACGCACAAC 3′[6,9]) under the conditions described earlier. The relative independence of the DNA fingerprints on DNA concentration is shown at the right, with the results obtained for 50 and 20 ng of genomic DNA from case 8. DNA fragments of sizes ranging from 100 to near 2000 nucleotides are reproducibly amplified, although the figure only shows fragments of sizes between 300 and 1000 nucleotides.

[9] J. Welsh, C. Petersen, and M. McClelland, *Nucleic Acids Res.* **19,** 303 (1991).

Differences in the banding patterns are apparent. Some of these differences represent polymorphisms in the human population because they are present in both normal and tumor tissues from only some individuals (for instance, the band indicated with an empty star in case 3). Other differences are due to length polymorphisms, for instance, the bands indicated with a solid star. Some individuals are homozygous (or hemizygous) for a larger allele (cases 1, 2, 3, and 8), some for a smaller allele (cases 6 and 7) and some others are heterozygous, amplifying both alleles (cases 4 and 5).

Other tumor-specific changes include (a) new bands in the tumor tissue DNA, but not in the matching normal (no such example is shown in this figure); (b) changes in the mobility of bands (shown by solid arrowheads in case 3); (c) decreases in the intensity of bands (empty triangles pointing down in cases 4, 5, and 6); and (d) increases in the intensity of bands (empty triangles pointing up, cases 3 and 6). These quantitative changes in the intensities of the bands represent allelic losses and gains as confirmed by cloning and Southern blot hybridization analyses of the genomic sequences without *in vitro* PCR amplification and by RFLP linkage analysis.[6,10] The mobility changes (b) represent somatic deletions of a few nucleotides in simple repeated sequences such as the poly(A) tails of *Alu* repeats.[6,7]

Isolation and Characterization of AP-PCR DNA Bands

Once a fingerprint has been generated, it is often necessary to purify and clone bands that display polymorphisms or other genetic alterations for further characterization or for use as biomarkers. As an example of the procedures used to isolate and characterize DNA sequences amplified by AP-PCR, we describe in some detail the analysis of the band representing a length polymorphism showing allelic loss in case 4 of Fig. 1. This band was characterized as a chromosome 17p single-copy genomic sequence of alternating purines/pyrimidines. The length polymorphisms resulted from deletion/insertion of the repeated hexanucleotide ATATGT.[6]

In many cases, hybridization analysis may be necessary to confirm that the cloned band corresponds to the band visualized in the arbitrarily primed PCR fingerprint. A dried arbitrarily primed PCR gel can be transferred to a blotting membrane. If the cloned band is used as a probe in a Southern hybridization of this filter, successful cloning can be easily determined if the probe hybridizes to the band of expected size and/or that shows the expected mobility or intensity changes according to the AP-PCR fingerprint. Figure 2 shows the typical results of such an experiment.

[10] S. Malkhosyan, J. Yasuda, T. Sekiya, J. Yokota and M. Perucho, in preparation.

Equipment and Reagents

PCR and electrophoresis material as in the previous protocol
Nitrocellulose or nylon blotting membrane (Bio-Rad Zeta-Probe, Amersham Hybond)
3MM Whatman paper
UV crosslinker (Stratalinker, Stratagene)
3 *M* NaCl, 0.5 *M* Tris–HCl, pH 7.4
2 × SSPE (0.036 *M* NaCl, 0.02 *M* sodium phosphate, pH 7.7, 2 m*M* EDTA)
Other material and reagents for ligation of PCR products to plasmid vectors and transformation of *Escherichia coli* cells as described.[11]

Procedure B

Step 1. Gel Purification and Cloning of Reamplified Bands

Align autoradiogram markers on the gel with their exposed images. Use a needle to mark in the dried gel the exact position of the band, then excise with a scalpel or razor blade (reexposure of the gel will confirm the accuracy in the excision of the band, see Fig. 2B). Place the excised portion of the gel in 50–100 μl of water and incubate at 60° for 10–20 min to elute the DNA. Reamplify 1 μl of the eluted DNA with the same arbitrary primer and under the same conditions as in Procedure A, step 2, except that no extra $MgCl_2$ is added (final concentration of $MgCl_2$ is 1.5 m*M* instead of 5.0 m*M*).

Amplify for 30 high stringency cycles (95°, 30 sec; 65°, 30 sec; 72°, 2 min). For 10-mers use 35° annealing throughout. Analyze the PCR product by electrophoresis next to an AP-PCR fingerprint to verify its size and purity. If other bands are coamplified, the desired band can again be cut from the gel and reamplified by PCR as before.

Cloning of the reamplified AP-PCR bands can be done by standard protocols for isolation of PCR products. The cloning can be facilitated if restriction sites are incorporated to the arbitrary oligonucleotide primers to produce DNA fragments with staggered ends compatible with cloning sites in commercially available plasmid or phagemid vectors. In this case, both the vector and reamplified band DNA can be digested with the appropriate restriction enzyme(s), ligate, and transform using standard protocols. Methods for cloning in the absence of restriction sites in the arbitrary primer have been published elsewhere.[9,11]

[11] J. Welsh, M. Perucho, M. Peinado, D. Ralph, and M. McClelland, *in* "PCR II: A Practical Approach" (G. Taylor, M. McPherson, and D. Hames, eds.). IRL Press, Oxford and New York, 1994 (in press).

Step 2. Southern Hybridization of Arbitrarily Primed PCR Gels

Cut out the part of the dried gel that is going to be transferred to the nitrocellulose or nylon membrane. Cut at least 10 sheets of Whatman 3MM paper and the blotting membrane to the same size as the gel segment. Dip 4 sheets of paper and the dried gel in 3 *M* NaCl, 0.5 *M* Tris–HCl and lay them on a glass plate with the gel on top. Soak the blotting membrane with water and lay it on top of the gel. Place 1 wet sheet of 3MM paper and 5 or more sheets of dry 3MM paper on top of the membrane. Cover with another glass plate and a weight (~0.5 kg). Leave for 2 hr to overnight. Wash the blotting membrane with water and 2 × SSPE and irradiate with a UV crosslinker or place face down on clear plastic wrap on a transilluminator for 15 min to crosslink the DNA. Dry the membrane and expose to X-ray film. This gives a pattern of the transferred radioactive bands to the filter before hybridization to the radioactive probe.

The radioactive probe is prepared using 20–100 ng from the purified PCR product or from a plasmid obtained by a miniprep. Unincorporated nucleotides are removed by elution of the probe using resin columns such as NucTrap probe purification columns (Stratagene). Hybridization is performed using standard methods.[12] The experiment is especially informative when using arbitrarily primed PCR gels with ^{35}S-labeled DNA fragments. The efficiency of the transfer to blotting membranes can be easily monitored by following the radioactivity during transfer (see Fig. 2C). Using this procedure, 10–50% of the radioactive material is transferred to the blotting membrane. The transfer efficiency is not the same for all the bands depending on size: bands larger than about 1000 bp will transfer with lower efficiency. Hybridization to the ^{32}P-labeled probe yields distinguishable signals over the background ^{35}S radioactive bands (Figs. 2D and 2C, respectively). The presence of these background bands may facilitate the identification of the new hybridizing bands.

Step 3. DNA Sequencing of Isolated AP-PCR Bands

Once the bands have been isolated and characterized, their nucleotide sequence can be readily determined by standard protocols. We have used a cycle sequencing method[13] that uses a double-stranded PCR product as a template and does not require sample purification.[14] The substrate is generated by PCR using two primers that anneal to the vector sequence.

[12] J. Sambrook, E. F. Fitsch, and T. Maniatis, eds., "Molecular Cloning: A Laboratory Manual," 2nd ed. Cold Spring Harbor Lab., Cold Spring Harbor, NY, 1989.

[13] G. Ruano and K. K. Kidd, *Proc. Natl. Acad. Sci. U.S.A.* **88,** 2815 (1991).

[14] M. A. Peinado, M. Fernandez-Renart, G. Capella, L. Wilson, and M. Perucho, *Int. J. Oncol.* **2,** 123 (1993).

Alternatively, the cloned product can be rescued as a phagemid for single-stranded DNA sequencing. A detailed procedure for the sequencing of AP-PCR bands has been described.[11]

Cycle sequencing can be also done without cloning the AP-PCR bands, if the AP-PCR fingerprint is generated by two arbitrary primers.[7,15] In this case, many of the amplified bands are products of the annealing of both primers. These bands can be reamplified directly from the fingerprint gels (Procedure B, step 1) under cycling sequencing conditions, directly yielding sequence information of the AP-PCR band.

Step 4. Chromosomal Localization of AP-PCR Bands

The rapid chromosomal localization of DNA fragments amplified by arbitrarily primed PCR can be achieved in two different manners. In the first method, the AP-PCR band of interest is first cloned and sequenced as described earlier and the nucleotide sequence information is used for the synthesis of specific PCR primers. Standard PCR amplification of this DNA fragment using panels of somatic cell hybrids containing defined human chromosomes permits the identification of the human chromosome where these sequences are located. Single-copy sequences produce a unique band. The identity of this band with the previously isolated band from the AP-PCR gel must be confirmed by sequencing and/or RFLP analysis.

In the second method, arbitrarily primed PCR fingerprinting is applied directly to rodent/human somatic cell hybrid DNA, and the resulting fingerprints are compared to those of rodent (hamster or mouse) and human total genomic DNA. The chromosomal origins of many of the human bands specific for the particular arbitrary primer can be simultaneously determined in a single experiment because these bands are amplified only from one of the cell hybrids containing a single human chromosome.[16]

The two methods are complementary since it is convenient or necessary to confirm the chromosomal origins for the bands of interest obtained by the second, shortcut method by the first, straight method. Moreover, the second method cannot determine the chromosomal derivation for all the AP-PCR fingerprint bands because some of them exhibit identical mobility in the two species of the somatic hybrid (i.e., human/mouse or human/hamster). The assigning of the chromosomal localization of the AP-PCR band of 750 nucleotides of Fig. 1 to chromosome 17 was previously described.[6,11] The second method will be described somewhere else.

[15] J. Welsh and M. McClelland, *Nucleic Acids Res.* **19,** 5275 (1991).

[16] A. Velazquez and M. Perucho, unpublished observations.

Summary

The application of arbitrarily primed PCR allows for an unbiased examination of genetic alterations of the cancer cell genome and has led to a better understanding of the genomic instability of cancer cells. The genomic fingerprints generated by arbitrarily primed PCR disclosed the presence of ubiquitous somatic mutations in the cancer cell genome. These are the molecular symptoms of a defect in mismatch repair that unveiled a new mutator molecular pathway for oncogenesis.[7]

In addition, it is possible to detect losses or gains in the number of copies of a target genomic sequence by the differences in the intensity of the corresponding band in the arbitrarily primed PCR pattern. DNA fingerprinting by arbitrarily primed PCR provides therefore a complementary and alternative molecular approach to the cytogenetics of solid tumors. This is based predominantly on a plethora of propitious properties of the procedure: (1) the amplified bands usually originate from single-copy sequences instead of from repetitive elements; (2) there is no apparent bias for the chromosomal origins of the amplified bands and, therefore, fingerprints representative of the full chromosomal complement can be obtained by the use of a few arbitrary primers; and (3) the amplification is semiquantitative in that the intensity of an amplified band is proportional to the concentration of its corresponding template sequence.

Therefore, the degree of aneuploidy of a tumor cell genome is reflected in differences in the intensities of arbitrarily primed PCR bands compared to those from the normal diploid genome from the same individual. The possibility of detecting moderate gains of genetic material, such as those corresponding to triploidy and tetraploidy,[10] represents a significant technical development because such genomic changes cannot be readily detected by conventional allelotyping by RFLP or by typing of microsatellites.

There are obvious direct applications of this strategy, for instance, for cancer prognosis. The relative differences in the extent of genomic damage undergone by the cancer cell can be estimated by comparing the number of quantitative changes in the AP-PCR fingerprints of tumor DNA relative to normal tissue DNA. These arbitrary values, which reflect the different degrees of aneuploidy of the tumors, may have a predictive prognostic value.

B. RNA Fingerprinting by Arbitrarily Primed PCR

We and others have developed a novel approach for the detection and cloning of differentially expressed genes. This method is based on a modification of arbitrarily primed PCR.[1,17] Here, we describe one variation

[17] M. McClelland, D. Ralph, R. Cheng and J. Welsh, *Nucleic Acids Res.* **22,** 4419 (1994).

of the method and rationale behind the adaptation of arbitrarily primed PCR to the fingerprinting of RNA, thereby facilitating the detection of RNA transcripts which are differentially expressed. A similar method has been developed by Liang and Pardee[5] and is also presented in Liang *et al.* [20] in this volume.

In RAP-PCR, first-strand synthesis is initiated from an arbitrarily chosen primer at those sites in the RNA that best match the primer. DNA synthesis from these priming sites is performed by reverse transcription. Second-strand synthesis is achieved by adding *Taq* polymerase and the appropriate buffer to the reaction mixture. Once again, priming occurs at the sites where the primer finds the best matches. Poorer matches at one end can be compensated for by very good matches at the other. The consequence of these two enzymatic steps is the construction of a collection of molecules that are flanked at their 3′ and 5′ ends by the exact sequence (and complement) of the arbitrary primer. These serve as templates for high stringency PCR amplification, which can be performed in the presence of radioactive label such that the products can be displayed on a sequencing-type polyacrylamide gel, as in AP-PCR fingerprinting of genomic DNA.[4,18] Because priming is arbitrary in the first step the lack of poly(A) tails in most bacterial RNAs is also not a problem for this method.[19]

Although these fingerprints contain anywhere from 10 to 40 bands, typical RNA populations for eukaryotic cells have complexities in the tens of thousands of molecules. Therefore, searching for a particular differentially expressed gene would be futile. Instead, the method is more appropriate for problems where many differentially expressed genes are anticipated. Many developmental and pathological phenomena are accompanied by many dozens or even hundreds of alterations in gene expression. Figure 3 shows the result of a RAP-PCR fingerprinting experiment of rat fibroblasts transformed *in vitro* by human c-K-*ras* oncogenes.

RNA Fingerprinting by Arbitrarily Primed PCR

Equipment and Reagents

96-well format thermocycler

An arbitrarily chosen primer; we suggest trying the M13 sequencing primer or reverse sequencing primer

Multichannel micropipet

[18] K. K. Wong, S. C.-H. Mok, J. Welsh, M. McClelland, S.-W. Tsao, and R. S. Berkowitz, *Int. J. Oncol.* **3,** 13 (1993).

[19] K. K. Wong and M. McClelland, *Proc. Natl. Acad. Sci. U.S.A.* **91,** 639 (1994).

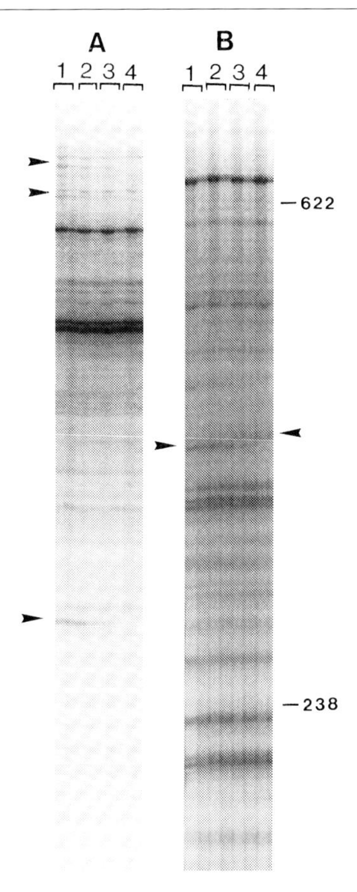

FIG. 3. RNA fingerprinting by arbitrarily primed PCR (RAP-PCR). RNA fingerprints obtained from total cellular RNA (40 ng, left and 10 ng, right, respectively) from pools of neomycin-resistant colonies of Rat 2 fibroblasts untreated (1) or transfected with a *neo* plasmid vector with a wild-type c-K-*ras* protooncogene (2), the gene with a codon 13 weak mutation (3), or a codon 12 strong mutation (4). The fingerprints were obtained with two different arbitrary primers: 5′ CCATATGCGCATATGAGA 3′ (A) and 5′ CACACGCACACG-GAAGAA 3′ (B). Arrows indicate differentially expressed RNAs. The numbers on the right-hand side are molecular weight markers (in nucleotides).

2 × DNase 1 treatment mixture (20 m*M* Tris–HCl, pH 8.0, 20 m*M* $MgCl_2$, 40 U/ml RNase-free DNase I)

First-strand reaction mixture (10 m*M* Tris–HCl, pH 8.3, 50 m*M* KCl, 4 m*M* $MgCl_2$, 20 m*M* DTT, 0.2 m*M* each dNTP, 1 μM primer, and 5 U MuLVRT)

Second-strand reaction mixture (10 mM Tris–HCl, pH 8.3, 25 mM KCl, 2 mM $MgCl_2$, 1 μM primer [as in first-strand reaction mix), 0.2 μCi/μl [α-^{32}P]dCTP and 1 U *Taq* polymerase (AmpliTaq, Cetus)]

DNase I (Boehringer Mannheim Biochemicals)

Denaturing loading buffer and other components (see section on arbitrarily primed PCR above)

Procedure C

Step 1. RNA Purification

The guanidinium thiocyanate–cesium gradient method of RNA purification of Chirgwin *et al.*[20] has been used successfully, but other methods may also work. Most purification methods yield RNA that is not entirely free of contaminating genomic DNA. This can be a serious problem because the genome is more than 10 times as complex as the RNA population, resulting in better matches with the primer. We therefore routinely treat the RNA with RNase-free DNase I prior to fingerprinting. The RNA concentration is checked spectrophotometrically and equal aliquots are electrophoresed on a 1% agarose gel and ethidium bromide stained to compare large and small ribosomal RNAs, qualitatively.

Step 2. RNA Fingerprinting by RAP

1. Dissolve the final pellet of the RNA preparation in 100 μl of water.
2. Add 100 μl of 2 $\times$ DNase 1 treatment mixture and incubate at 37° for 30 min. Phenol extract and ethanol precipitate.
3. Prepare treated RNA at two concentrations of about 20 and 10 ng/μl by dilution in water.
4. Add 10 μl of the first-strand reaction mixture to 10 μl RNA at each concentration. Allow the reaction to proceed at 37° for 30 min to 2 hr.
5. Add 20 μl of the second-strand reaction mixture to each first strand synthesis reaction. Cycle through one low stringency step (94°, 5 min; 40°, 5 min; 72°, 5 min) followed by 40 high stringency steps (94°, 1 min; 40°, 1 min; 72°, 2 min).
6. Add 2 μl of each reaction to 10 μl of denaturing loading buffer and electrophorese on a 4 or 6% polyacrylamide sequencing-type gel containing 40 to 50% (w/v) urea in 0.5 $\times$ Tris–borate–EDTA buffer.
7. Wrap the gel in plastic film or dry the gel and autoradiograph.

Primers are chosen with several criteria in mind. First, the primers should not have stable secondary structure. Second, the sequence should

[20] J. Chirgwin, A. Prezybyla, R. MacDonald, and W. J. Rutter, *Biochemistry* **18,** 5294 (1979).

be chosen such that the 3′ end is not complementary to any other sequence in the primer. In particular, palindromes should be avoided. Third, primers of 10 to 20 nucleotides in length can be used. Longer primers can contain more sequence information designed to aid in subsequent steps in the experiment, such as cloning and sequencing. Primers 10 bases in length can be obtained in kits from several companies such as Genesys (Woodlands, TX) and we have successfully employed them.[17]

The RNA fingerprint experiment by RAP-PCR (Fig. 3) shows the fluctuations in gene expression after transfection of cultured rat fibroblasts with mutated human c-K-*ras* oncogenes possessing different oncogenic mutations at codons 12 and 13. These point mutations (aspartic acid substitutions at these positions) exhibit different transforming potency in *in vitro* transformation assays of immortalized rat fibroblasts and *in vivo* tumorigenicity assays with these transfected cells in syngenic animals.[14,21] Quantitative changes in some RAP-PCR bands are apparent, representing transcripts both up-regulated (arrow pointing left) and down-regulated (arrows pointing right) by activated *ras*.

Fingerprints generated by this simple version of RNA arbitrarily primed PCR may represent differences in abundant RNAs more faithfully than differences in low abundance messages. The overrepresentation of abundant messages can be a serious problem when the genes of interest give rise to rare messages. However, Northern blot analysis has indicated that RNA from differentially expressed genes revealed by RAP-PCR vary over about two orders of magnitude. Nonetheless, it is known from hybridization kinetics that RNA abundance can vary over many orders of magnitude.

We have developed a partial solution to this problem, termed "nested RAP."[22] The strategy is very similar to standard nested PCR methods except that we do not know a priori the internal sequences of the amplified products. In this method, the RNA fingerprinting protocol is applied to the RNA, except that only 10 cycles of PCR amplification are performed instead of the 40 cycles in step 5 (Procedure C above). A small aliquot of the first reaction is then further amplified using a second nested primer having one, two, or three additional arbitrarily chosen nucleotides at the 3′ end of the first primer sequence.

This strategy partially normalizes the sampling that occurs during RNA fingerprinting for the relative abundance of the transcripts. If two messages have equally good matches and equally good amplification efficiency but differ by 100-fold in abundance, then the products derived from them will differ by 100-fold after RAP. Thus RAP fingerprinting produces a

[21] S. Malkhosyan, G. Capella, N. Stone and M. Perucho, in preparation.

[22] D. Ralph, J. Welsh, and M. McClelland, *Proc. Natl. Acad. Sci. U.S.A.* **90,** 10710 (1993).

background of products that are not visible on the gel that includes products derived from low abundance messages. A secondary round of amplification using a primer identical to the first except for an additional nucleotide at the 3′ end of the primer can be expected to selectively amplify those molecules in the background that, by chance, share this additional nucleotide. The high sequence complexity, low abundance class is more likely to contribute products with this extra matching nucleotide.

Summary

RNA fingerprinting by RAP-PCR is a powerful tool for the temporal and spatial analysis of differential gene expression. Many biological situations exist where differential gene expression results in distinguishable phenotypes, including, for example, tissue and cell types, responses to hormones, growth factors, stress, and the heterologous expression of certain genes. There are several methods for detecting differential gene expression and cloning differentially expressed genes that do not rely on a biological assay of phenotype. Most of these methods fall into two general categories: subtractive hybridization and differential screening. RAP-PCR offers numerous advantages over these methods, including its simplicity and its ability to compare the fluctuations in gene expression between multiple samples simultaneously using minute amounts of RNA. In addition, RAP-PCR can yield information on the overall patterns of gene expression between different cell types or between different physiological conditions of the same cell type. Comparison of the RAP-PCR fingerprints from these different experimental groups permits one to draw inferences regarding the overall cellular states of gene expression and the interrelation between gene transcripts belonging to the same or different regulatory pathways. Hypotheses regarding signal transduction pathways can be obtained using this information. RAP-PCR offers applications in cancer research in the detection of tumor-specific alterations in gene expression, providing a bountiful source of tumor markers. The pleiotropic impact of oncogene activation, tumor suppressor gene inactivation, and mutator mutations, in gene regulation, can be readily assessed by RAP-PCR in model systems both *in vitro* and *in vivo*.

Acknowledgments

This work was supported in part by National Institute of Health Grants CA 33021, CA 63585 and CA 38579 to M. P., CA 68822, AI34829 and HG00456 to M. M., and AI322644 to J. W.

[19] Representational Difference Analysis in Detection of Genetic Lesions in Cancer

By NIKOLAI LISITSYN and MICHAEL WIGLER

I. Introduction

Several approaches have proven useful for the analysis of the genetic lesions in cancer cells. For example, transfection studies led to the discovery of several dominant oncogenes, the most important of which were the *RAS* oncogenes[1]; positional cloning of chromosomal abnormalities, such as translocation, have led to the discovery of activated *MYC, BCL2, ABL* oncogenes,[2] among others; and positional information combined with brute force screening of regional specific probes led to other important discoveries such as the loss of the tumor suppressor genes such as retinoblastoma gene[3] and Wilms' tumor gene.[4] This chapter presents a global approach to the analysis of the sequence differences between the genomes of the cancer and normal cells that uses a method we call representational difference analysis (RDA).[5] In principle, RDA can be used to derive probes for genomic losses, rearrangements, amplifications, point mutations, and pathogenic organisms found within the cancer cell. RDA is a powerful, versatile, but complex procedure. To perform it properly requires not only diligence at the bench but also a clear headed understanding of how to use it. This chapter addresses both of these aspects. We shall discuss the application of RDA and provide detailed protocols.

Overview

RDA belongs to the general class of DNA subtractive methodologies. These methodologies all have in common that one DNA population (the "driver") is hybridized in excess against a second population (the "tester") to remove common hybridizing sequences, thereby enriching "target" sequences unique to the tester. Pure subtractive methodologies have limited

[1] M. Barbacid, *Annu. Rev. Biochem.* **56,** 779 (1987).
[2] T. H. Rabbitts, *Cell (Cambridge, Mass.)* **67,** 641 (1991).
[3] S. H. Friend, R. Bernards, S. Rogelj, R. A. Weinberg, J. M. Rapaport, D. M. Albert, and T. P. Dryja, *Nature (London)* **323,** 643 (1986).
[4] K. M. Call, T. Glaser, C. Y. Ito, A. J. Buckler, J. Pelletier, D. A. Haber, E. A. Rose, A. Kral, H. Yeger, W. H. Lewis, C. Jones, and D. E. Housman, *Cell (Cambridge, Mass.)* **60,** 509 (1990).
[5] N. A. Lisitsyn, N. M. Lisitsyn, and M. Wigler, *Science* **259,** 946 (1993).

usefulness in the analysis of complex genomes. This is because the enrichment required to purify target sequences is very high (on the order of 10^5-fold) and because the sequence complexity of DNA from higher organisms is too great for single copy sequences to hybridize sufficiently to completion. RDA combines subtractive enrichment with two further elements: kinetic enrichment and representation.

By a "representation" we mean any means of reproducibly producing a subpopulation of DNA fragments derived from a given DNA population such that the sequence complexity of the subpopulation is lower than the sequence complexity of the initial DNA. We perform our subsequent difference analysis on representations of driver and tester. The reduced complexity of the representation enables us to achieve more complete reannealing of driver. Representations that reduce complexity at least 10-fold over the complexity of the mammalian genome are generally required for the success of these subsequent steps.

Examples of representations include the set of all *Bam*HI fragments between 3.0 and 3.5 kbp in length or the set of all *Eco*RI fragments with a buoyant density in CsCl between 1.45 and 1.48 g/ml. For all of our applications of RDA we have used as our representation the set of all DNA fragments that are amplified after 20 cycles of polymerase chain reaction (PCR) following cleavage with a given restriction endonuclease and ligation to a defined oligonucleotide adaptor. In effect, this representation is composed of the subpopulation of small restriction fragments because small fragments are much more efficiently amplified than large fragments. Since our subsequent difference analysis is also PCR based, this form of representation is the most natural choice. We call such a representation an "amplicon," and speak of a "*Bgl*II amplicon," for example, as the fragments that amplify after cleavage with *Bgl*II followed by ligation to an oligonucleotide PCR adaptor.

The use of a representation reduces sequence complexity, but also has two other important consequences that must be understood. First, the representation of two nearly identical genomes will differ not only when there are absolute differences between the sequence content in the two genomes, but also sometimes when rearrangements or point mutations alter restriction endonuclease fragment lengths. Because of this, RDA will detect a portion of restriction fragment length polymorphisms between two humans (those that come in large and small fragment sized alleles). Thus care must be used to match cancer cell DNA with the normal DNA from the same individual from which it derived. Some of these issues are discussed in the paragraphs below. Second, because the representation does not have the complexity of the whole genome, not all of the potential differences between two genomes will be found. From the examples given below the reader can grasp some of the quantitative factors that must be considered.

Kinetic enrichment is based on the second-order kinetics of DNA reannealing: the rate of formation of double-stranded DNA is higher for DNA species of higher concentration. Thus, the more abundant DNA species in a mixture of fragments can be further partitioned from less abundant species by reannealing for low C_0t values (the product of initial concentration times time), usually below the C_0t required for half of the abundant species to reanneal, and then collecting only double-stranded molecules. The molar ratio of abundant to less abundant sequences in the product will then be on the order of the square of the initial ratio of the concentrations, so that, upon reiteration, kinetic enrichment leads to remarkable degrees of purification. In RDA, kinetic enrichment and subtractive enrichment are combined in a single step that we call hybridization/amplification. The method is so powerful that even a single difference between two DNA populations can be found.[5] However, when there are many differences, only a few, on the order of a dozen, can be readily detected.[5]

Prior to the hybridization/amplification step, only tester molecules are fitted with defined oligonucleotides by their 5′ ends. After reannealing tester and driver, the mixture of molecules is treated with DNA polymerase. This adds the complement of the defined oligonucleotides to the 3′ ends of only tester reannealed to itself, rendering only such molecules amplifiable by PCR when the defined oligonucleotide is used as primer. This serves three purposes. First, molecules of tester that reanneal to the excess of driver are "subtracted out" since the heteroduplex they form with driver will not be amplified. Second, abundant sequences of tester will reanneal faster than less abundant ones (kinetic enrichment) so that sequences which are enriched become even more enriched. Third, PCR amplification increases the yield so that the hybridization/amplification step can be reiterated or the product can be cloned and analyzed. Driver is always added in sufficient excess to tester to effectively remove repetitive DNA from tester; tester concentrations are kept sufficiently low to achieve kinetic enrichment. The kinetic component of RDA also enables the method to be used for isolation of probes to amplified regions, as discussed below.

In summary, the RDA procedure consists of the preparation of amplicons followed by multiple cycles of hybridization/amplification. Because RDA is a complex procedure, we recommend the inclusion of controls. Care and thought must be given to the source of DNA for analysis, and this also will be discussed.

II. Applying RDA to Cancer

RDA can be applied to the detection of genetic lesions in cancer in two different ways: using tumor DNA as driver and normal DNA from the same patient as tester, or in the reverse way. In either case, driver must be relatively free (at least 95%) of tester sequences. This effectively rules out the use of tumor DNA directly derived from cancer biopsies as driver since tumor tissue samples inevitably contain normal elements from blood, capillaries, and connective tissue. One can isolate pure tumor DNA from tumor cell lines or from biopsies using either cells or nuclei sorted by flow cytometry. For example, nuclei can be sorted into aneuploid and diploid populations[6] sufficient for use with RDA. Tester DNA can include significant quantities of driver, up to 90%,[7] but should not contain Epstein–Barr or other viruses used for immortalization of human cells. For most applications, DNAs isolated from immortalized B and T cells make good drivers, despite the rearrangements they undergo in immunoglobulin or T cell receptor genes.

When tumor DNA is used as driver, RDA yields probes that detect loss of heterozygosity (LOH), homozygous deletions, and hemizygous loss (loss of Y chromosome or deletion of other sequences present on one chromosome only in normal cells). The percentage of probes of each type generally depends on the total length of each loss, which may comprise up to 10% of a genome for LOH, up to 1% for loss of Y chromosome (probes of this type can be eliminated by using female normal/tumor DNA pairs), and up to 0.3% for homozygous deletions. Note that in the case of loss of heterozygosity, only the absence of polymorphic fragments in the tumor is detectable because one copy of the lost region is still present in tumor DNA. Since not more than 3% of fragments in human genome are polymorphic, this leads to a strong bias toward the cloning homo- and hemizygous losses. Probes detecting homo- and hemizygous losses can be distinguished in many cases based on their hybridization to Southern blots containing large panels of normal human DNA digests because probes to hemizygous losses will sometimes be missing on both normal chromosomes of some individual. In our experience,[8] based on RDA of colon cancer and renal cell carcinoma in female DNA pairs, probes detecting LOH comprise around 70% of difference product, and probes detecting homozygous and/or hemizygous losses contribute 30%. Some of the latter have been shown to be hemizygous.

When tumor DNA is used as tester, RDA might detect small restriction fragments acquired by tumor as a consequence of mutations, genomic rearrangements (deletions, translocations, inversions, etc.), and presence of DNA pathogens. It is also possible to clone highly amplified sequences[8] since

[6] B. J. Reid, R. C. Haggitt, C. E. Rubin, and P. S. Rabinovitch, *Gastroenterology* **93,** 1 (1987).

[7] N. A. Lisitsyn, J. A. Segré, K. Kusumi, N. M. Lisitsyn, J. H. Nadeau, W. N. Frankel, M. H. Wigler, and E. S. Lander, *Nat. Genet.* **6,** 57 (1994).

[8] N. A. Lisitsyn, N. M. Lisitsina, G. Dalbagni, P. Barker, C. A. Sanchez, J. Gnarra, W. M. Linehan, B. J. Reid, and M. H. Wigler, *Proc. Natl. Acad. Sci. U.S.A.* **92,** 151 (1995).

these cannot be completely subtracted by driver (normal DNA) and become enriched due to kinetic factors. Genomic rearrangements always create a bridging fragment that is not present in the normal DNA. The probability that this fragment will be present in the amplicon is around 10% (depending on the complexity of the representation). Thus it is necessary to use several different representations to have a high probability of finding a particular bridging fragment. The same is true for the search for an unknown DNA pathogen, if its genome is small (10 kb or less). One can calculate that for a six cutter, roughly only 1 in 5000 point mutations will create a new amplifiable fragment in a tumor. This is a large enough effect to be a nuisance.

III. Methodology

A. *Design of Oligonucleotides: Adaptors and Primers for PCR*

RDA depends heavily on the use of PCR. Pairs of single-stranded adaptor oligonucleotides, one long and one short, are used to alter the ends of DNA fragments so that they can be amplified. The long oligonucleotide of the pair is also used as primer. In all cases the adaptors are designed so that they can be removed by restriction endonuclease cleavage after amplification of the DNA template. Multiple rounds of PCR amplifications are used: one for preparing amplicons and two to four for the hybridization/amplification stages. We never use the same adaptors in two consecutive rounds of PCR in order to get the highest possible enrichment, which otherwise will be limited by the efficiency of restriction endonuclease digestion. In fact, we use three sets of primer pairs: one for the representation stage, and two used alternatingly in the hybridization/amplification stage so that the adaptors from the representation stage cannot interfere with the hybridization/amplification stage.

For the hybridization/amplification stage it is essential that the oligonucleotide adaptors be ligated to only the 5′ end of the double-stranded DNA fragments. We therefore use unphosphorylated adaptors. We also use one long and one short adaptor strand so that at elevated temperatures the short oligonucleotide will dissociate from duplex leaving a 3′-recessed end of the double-stranded DNA that can be filled in with a heat-stable DNA polymerase. We use this method also to modify the ends of digested genomic DNA as the first step in the representation stage in order to avoid the problem of oligonucleotide concatemerization.

We present three pairs of adaptor/primer oligonucleotides for *Bam*HI, *Bgl*II, and *Hin*dIII restriction endonucleases (see Table I). The optimal annealing temperature of all 24-mer primers in PCR amplification is 72°, except the J *Bgl*24 primer used in *Bgl*II and *Hin*dIII representations which has a 70°

TABLE I
SEQUENCES OF OLIGONUCLEOTIDES USED FOR RDA[a]

Primer pair set	Name	Sequence
1	R *Bgl*24	5′-AGCACTCTCCAGCCTCTCACCGCA-3′
	R *Bgl*12	5′-GATCTGCGGTGA-3′
2	J *Bgl*24	5′-ACCGACGTCGACTATCCATGAACA-3′
	J *Bgl*12	5′-GATCTGTTCATG-3′
3	N *Bgl*24	5′-AGGCAACTGTGCTATCCGAGGGAA-3′
	N *Bgl*12	5′-GATCTTCCCTCG-3′
1	R *Bam*24	5′-AGCACTCTCCAGCCTCTCACCGAG-3′
	R *Bam*12	5′-GATCCTCGGTGA-3′
2	J *Bam*24	5′-ACCGACGTCGACTATCCATGAACG-3′
	J *Bam*12	5′-GATCCGTTCATG-3′
3	N *Bam*24	5′-AGGCAACTGTGCTATCCGAGGGAG-3′
	N *Bam*12	5′-GATCCTCCCTCG-3′
1	R *Hin*d24	Same as R *Bgl*24 (see above)
	R *Hin*d12	5′-AGCTTGCGGTGA-3′
2	J *Hin*d24	Same as J *Bgl*24 (see above)
	J *Hin*d12	5′-AGCTTGTTCATG-3′
3	N *Hin*d24	5′-AGGCAGCTGTGGTATCGAGGGAGA-3′
	N *Hin*d12	5′-AGCTTCTCCCTC-3′

[a] Primer pair set 1 (R series) is used for representation stage and sets 2 (J series) and 3 (N series) are used for odd and even hybridization/amplification steps, respectively.

annealing temperature. One can use various other restriction endonucleases for RDA, modifying the sequences of the sets of primer pairs indicated in Table I. The changes must recreate the site for the chosen restriction endonuclease after ligation of the adaptor to DNA fragments. If designed adaptor sequences significantly differ from the prototypes indicated in Table I, one should use existing programs (e.g., OLIGO, National Biosciences) in order to check for the presence of alternative secondary structures (hairpins, primer dimers, and undesirable duplexes, which might be formed by short and long oligonucleotides), and keep calculated hybridization temperatures of oligonucleotides close to the prototypes. Introduction of additional restriction sites in the sequences of long primers is undesirable.

B. Representation: Preparation of Amplicons

1. Restriction of DNA

a. Digest separately 1–2 μg of driver and tester DNA in a volume of 200 μl with a restriction endonuclease chosen for the representation. Use 10 U of restriction endonuclease per μg of DNA. Check the digest by 1% agarose gel electrophoresis.

b. Add 10 μg of tRNA (Sigma, type X-SA, Cat. No. R 8759) after digestion and extract with equal volumes of phenol and phenol/chloroform (1 : 1).

c. Add 1/10 volume of 3 *M* sodium acetate, mix by pipetting, and add 3 vol of ethanol. Mix again by inverting the tube several times, incubate at −70° for 10 min, recover DNA by centrifugation (Eppendorf microcentrifuge, 13,000*g*, 10 min), wash the pellet with 1 ml of 70% ethanol, dry *in vacuo*, and resuspend at a concentration of 0.1 mg/ml.

Note: It is important always to treat driver and tester DNAs in the same way, except for where indicated.

All solutions on all steps of the procedure should be prepared using Milli-Q purified (Milli-Q Water System, Millipore) and sterilized water.

To avoid cross-contamination from the beginning of the RDA procedure, all pipettings should be made using aerosol barrier pipette tips.

DNA taken for difference analysis should be sufficiently pure to achieve complete digestion since partial digestion may cause multilple problems. For preparation of DNA from blood or cell cultures, usage of the Cell Culture DNA kit (Qiagen Inc.) proved to be reliable. Partial degradation of DNA (mean fragment length higher than 5 kb) does not interfere with RDA.

If one plans to use pools of different DNAs, it is necessary to prepare amplicons for each DNA sample independently.[7]

DNA concentrations at each step should be carefully determined by comparing ethidium bromide staining with appropriate dilutions of the standards after gel electrophoresis. If the concentration of DNA is less than 10 μg/ml on some of the steps described below, add tRNA as carrier at 20 μg/ml.

If the quantity of DNA is a limiting factor, one can take 100 ng of DNA for digestion. In this case add 10 μg of tRNA to the DNA sample before restriction endonuclease cleavage. Use ligation conditions described below with the modification of the PCR step indicated in the note to Section III,B,4.

One can use various restriction endonucleases for preparation of amplicons, designing new sets of primers, as indicated in Section III,A. To estimate the applicability of each particular restriction endonuclease, one should use the following results obtained on human and mouse genomes as guidelines:

(i) The complexity of human *Hin*dIII amplicon is close to the limit of RDA. Optimal hybridization and PCR conditions in this case slightly differ from those which are used for *Bam*HI or *Bgl*II representations (see below).

(ii) For restriction endonucleases producing mean fragment lengths significantly larger than *Bam*HI (e.g., *Sal*I), the preparation of amplicons becomes irreproducible.

(iii) All restriction endonucleases with mean fragment lengths in between these two extremities (e.g., *Eco*RI, *Pst*I) could in principle be used for difference analysis of human DNAs, if the sequence of the cleavage site is unambiguous. Our protocols are designed for restriction endonucleases generating 3′ or 5′ protruding ends only, although with minor modifications the method should work with blunt ends. For restriction endonucleases yielding mean fragment lengths smaller than *Hin*dIII, one can use additional simplification by cutting part of the driver and tester amplicons after gel electrophoresis and subsequent reamplification of chosen fractions. Consult the literature for the mean restriction lengths of restriction endonucleases.[9]

2. Purification of Oligonucleotides

a. Attach Sep-Pak cartridge (Waters, Millipore) to a 5-ml syringe and wash with 10 ml of acetonitrile and 10 ml of water.

b. Load 20 optical units (OD_{260}) of short or long oligonucleotide from primer pair sets 1–3 (deprotected by ammonium hydroxide at 55° and lyophilized in a SpeedVac) in 2 ml of water, wash with 10 ml of water, and elute with 60% methanol, collecting seven fractions in Eppendorf tubes (3 drops per tube).

c. Measure the concentration of 200-fold dilutions of the fractions at $\lambda = 260$ nm, combine fractions containing oligonucleotides (usually fractions 2, 3, and 4; total volume approximately 500 μl), and concentrate by lyophilization to a volume of 200–300 μl.

d. Add 1/10 vol of 3 *M* sodium acetate and 4 vol of ethanol, recover DNA by centrifugation as described in Section III,B,1,c, wash the pellet with 1 ml of 100% ethanol, dry *in vacuo,* and resuspend at 62 pmol/ml (12 OD_{260}/ml for 24-mers and 6 OD_{260}/ml for 12-mers).

Note: The purity of oligonucleotides is important during the ligation step since their concentration in the ligation mixture is high. For PCR amplification steps (see below), ethanol precipitation only (without reversed-phase chromatography) produces a sufficiently pure 24-mer primer.

3. Ligation of Adaptors

a. Mix 10 μl (1 μg) of digested driver or tester DNA, 7.5 μl each of the 12- and 24-mer (primer pair set 1, purified on Sep-Pak cartridge), 2 μl of water, 3 μl of 10× ligase buffer (New England Biolabs).

[9] D. T. Bishop, J. A. Williamson, and M. H. Skolnick, *Am. J. Hum. Genet.* **35,** 795 (1983).

b. To anneal the oligonucleotides, place the tube in a heating block (Thermoline Dri-Bath, holes filled with glycerol) at 50–55° and then place the block in a cold room for approximately 1 hr, until the temperature drops to 10–15°.

c. Place the tubes on ice for 3 min, add 1 μl of T4 DNA ligase (400 U/μl, New England Biolabs), mix by pipetting, and incubate overnight at 12–16°.

4. PCR Ampification

a. Prepare 2 tubes for preparation of tester amplicon and 12 tubes for preparation of driver amplicon, each containing 352 μl of standard PCR mixture composed of 80 μl of 5× PCR buffer [335 m*M* Tris–HCl, pH 8.9, at 25°, 20 m*M* $MgCl_2$, 80 m*M* $(NH_4)_2SO_4$, 50 m*M* 2-mercaptoethanol, 0.5 mg/ml of bovine serum albumin]; 32 μl of dNTPs solution (dATP, dGTP, dCTP, dTTP, 4 m*M* each; stock solutions supplied by Pharmacia); 240 μl of water.

b. Dilute each ligate from step III,B,3 with 970 μl of TE buffer (10 m*M* Tris–HCl, pH 8.0, 1 m*M* EDTA), containing tRNA (20 μg/ml).

c. Add 40 μl of the dilution of the ligate (40 ng of DNA) and 8 μl of 24-mer oligonucleotide (primer pair set 1, Sep-Pak purification is unnecessary) to each tube from above (Section III,A) and place the tubes in the thermocycler (Perkin-Elmer Cetus, Model 480) prewarmed at 72°. At this temperature the small oligonucleotide dissociates.

d. To fill in 3′-recessed ends of the ligated fragments, add 3 μl (15 U) of AmpliTaq DNA polymerase (Perkin Elmer Cetus) in each tube, mix by pipetting, overlay with 110 μl of mineral oil, and incubate for 5 min.

e. Amplify for 20 cycles (1 min at 95° and 3 min at 72°) with the last cycle followed by extension at 72° for 10 min.

f. To check the quantity and quality of amplicons, run 10 μl of the PCR product on a 2% agarose gel, loading appropriate dilutions of amplicons from previous experiments (if not available, use *Sau* 3A I digest of total human DNA). Note the presence of intensive bands of abundant repeats characteristic for each representation.[5]

Note: Take special precautions in PCR work; always use aerosol barrier pipette tips. Change all solutions often.

One should not lower the quantity of AmpliTaq DNA polymerase in the preparative PCR reaction, thus diminishing the ratio of the quantity of enzyme molecules to the quantity of the ends of DNA fragments. During the final PCR cycles it will lead to the formation of single-stranded DNA fragments and significantly lower the yield of the amplicon.

If DNA quantity is a limiting factor, 100 ng of DNA digest can be taken for ligation in the described conditions (Section III,B,3). After ligation add

10 μg of tRNA, 90 μl of TE buffer, 30 μl of 10 *M* ammonium acetate, and 380 μl of ethanol. Chill the tubes at −70° for 10 min, incubate at 37° for 2 min, recover the pellet by centrifugation as described, wash twice with 70% ethanol, and dry *in vacuo* (this stage removes ATP present at high concentration in the ligation buffer). Dissolve each pellet in 100 μl of TE buffer, prepare 2 tubes with a standard PCR mixture for preparation of driver amplicon and 2 tubes for preparation of tester amplicon. Add 40 μl of DNA ligate in each tube and make PCR amplification for 20 cycles as described. Remove mineral oil with a pipette. To obtain a sufficient quantity of driver amplicon DNA, prepare 12 tubes with a standard PCR mixture, add 40 μl of the obtained driver amplicon in each tube, and make a second PCR for 5 cycles as above (Section III,B,4,d). Make the second PCR amplification of tester amplicon in the same way (2 tubes).

5. *Removing Adaptors from Amplicons*

a. Remove mineral oil with a pipette and combine the contents of 2 PCR tubes in one Eppendorf tube for tester and 12 into six for driver. Extract with 600 μl of phenol and phenol/chloroform (1 : 1).

b. Add 1/10 vol of 3 *M* sodium acetate and 800 μl of isopropanol, incubate for 15 min in a wet ice bath, and recover the DNA by centrifugation, as described. Resuspend driver and tester amplicons in TE buffer at a concentration of 0.2–0.4 mg/ml (expecting 15 μg of DNA amplicon from one PCR tube) and check DNA by gel electrophoresis. Save 20 μg of driver and tester amplicons if you plan to prepare amplicon blots.

c. In order to cleave the adaptors, digest driver DNA (150 μg) and tester DNA (10 μg) with the initially chosen restriction endonuclease (10 U/μg) in the volumes 800 and 200 μl, respectively. Add 10 μg of tRNA to tester amplicon DNA digest, phenol, and phenol/chloroform extract and recover both DNAs by 2-propanol precipitation, as in Section III,B,5,b.

d. Resuspend driver and tester amplicon DNA digests in 125 μl of TE buffer. Use at least two vortex shakings, 30 sec each, to dissolve driver DNA pellet since the final DNA concentration in the driver tube is high (one should expect a concentration of about 1 mg/ml for driver and 0.1 mg/ml for tester amplicon DNA digest). Dilute an aliquot (2 μl) of driver amplicon DNA digest 10 times by adding 18 μl of water. Load approximately 0.2, 0.4, and 0.6 μg of driver and tester amplicon DNA digests and the same quantities of appropriate concentration standards on a 2% agarose gel. Load undigested amplicon DNA to check digestion.

e. Estimate concentrations of amplicon DNA digests by comparison of the intensities of the lanes. Adjust driver DNA concentration to 0.5 μg/μl and tester DNA concentration to 50 ng/μl with TE buffer. Check digestion of the amplicon by comparison of the mobilities of intensive bands of repeated sequences in the amplicon DNA before and after digestion (complete cleavage of adaptors from both ends should cause a visible shift in mobility of these bands, corresponding to 48 bp in total length).

Note: Prior to use of amplicons in the hybridization/amplification step, one should remove adaptors from both driver and tester amplicons. This step prevents driver amplicon fragments from forming end-annealed complexes during hybridization and is necessary in order to change adaptors on tester amplicon.

6. Change of Adaptors on Tester Amplicon

a. Load 5 μg of tester amplicon DNA digest in 100 μl of TE buffer on a preparative 2% NuSieve agarose gel (11 cm wide, 0.6 cm thick; low melting point, FMC Bioproducts). Make a short run (bromphenol blue dye marker should move 5 cm).

b. Cut an agarose slice (0.2–0.4 g) containing fragments 150–1500 bp in length and put it in a 5-ml polypropylene tube. Add 0.4 ml of 0.5 *M* MOPS buffer, pH 7.0, 0.4 ml of 5 *M* sodium chloride, and 3 ml of water.

c. Vortex, melt at 72° in a heating block for 10 min, vortex again, and melt for an additional 5 min.

d. Pass warm solution (40–60°) through Qiagen-tip20 (Qiagen Inc.), elute DNA as recommended by the supplier, add 10 μg of tRNA per each sample, and recover DNA by precipitation with 2-propanol as in Section III,B,5,b. Dissolve the DNA pellet in 30 μl of TE buffer, check DNA by gel electrophoresis, and adjust concentration to 0.1 mg/ml.

e. Ligate 1 μg of purified tester DNA amplicon DNA digest to primer pair set 2 (Sep-Pak purified), as in Section III,B,3. Adjust the DNA concentration to 10 μg/ml (25 μg/ml for *Hin*dIII representation) by adding TE buffer containing tRNA (20 μg/ml).

Note: Change of adaptors is necessary to enable only self-reannealed tester amplicon molecules to be amplified after hybridization.

We use a gel box dedicated only to the preparative electrophoresis of tester amplicon DNA digest to avoid the possibility of contamination by products from a previous RDA.

C. Subtractive/Kinetic Enrichment

1. First Round of Hybridization

a. Mix 80 μl of driver amplicon DNA digest (0.5 mg/ml) and 40 μl of diluted tester DNA ligate from previous step (0.4 μg for representations made with most six cutters, 1 μg for *Hin*dIII representation) in a PCR tube, extract once with an equal volume of phenol/chloroform.

b. Add 30 μl of 10 *M* ammonium acetate, mix by pipetting, add 380 μl of ethanol, chill at −70° for 10 min, incubate at 37° for 2 min, and recover the DNA by centrifugation. Wash twice with 70% ethanol and dry *in vacuo*.

c. Resuspend the pellet in 4 μl of EE × 3 buffer [30 m*M* EPPS buffer (Sigma), pH 8.0, at 20°, 3 m*M* EDTA] by vortexing for 2 min, collect the sample on the bottom of a tube by brief centrifugation, and overlay with 35 μl of mineral oil.

d. Denature DNA for 3–4 min at 98° in a heating block, carefully add 1 μl of 5 *M* sodium chloride to the DNA drop, and incubate at 67° for 20 hr.

Note: Purity and correct concentration of driver amplicon DNA digest are crucial at this stage since the DNA concentration is close to the limit of solubility. If the drop of DNA on the bottom of a PCR tube is too viscous and does not form a meniscus, add 1 μl of EE × 3 buffer and 1.5 μl of sodium chloride solution.

2. *Selective Amplification*

a. Remove oil, add 8 μl of tRNA solution (5 mg/ml) to reannealed DNA, mix thoroughly by pipetting, add 390 μl of TE buffer, and mix again.

b. To fill in the adaptor ends, make two tubes with 352 μl of standard PCR mixture (see Section B,4). Add 40 μl of DNA dilution in each tube, place in the Thermocycler at 72°, add 3 μl (15 U) of AmpliTaq DNA polymerase, mix by pipetting, and incubate for 5 min.

c. To amplify self-reannealed tester amplicon molecules, add 10 μl of 24-mer primer (primer pair set 2), mix, overlay with mineral oil, and perform 10 cycles of PCR as in Section III,B,4,e. For J *Bgl*24 primer (*Bgl*II or *Hin*dIII representations), make amplification under slightly different conditions: 1 min at 95° and 3 min at 70° (see Section III,A). This yields the first round difference product.

d. Remove mineral oil by pipetting, combine contents of two PCR tubes in one Eppendorf tube, and add 10 μg of tRNA. Phenol and phenol/chloroform extract and recover DNA by 2-propanol precipitation as in Section III,B,5,b. Dissolve the pellet in 40 μl of TE buffer.

e. Take 20 μl of the first round difference product, add 20 μl of 2× mung bean nuclease buffer (New England Biolabs), mix by pipetting, add 2 μl of mung bean nuclease (10 U/μl, New England Biolabs), mix again, and incubate at 30° for 30 min.

f. Add 160 μl of 50 m*M* Tris–HCl, pH 8.9, and inactivate the mung bean nuclease by a 5-min incubation at 98°. Prepare two tubes with a standard PCR mixture (352 μl, Section III,B,4), add 10 μl of 24-mer primer (primer pair set 2), add 40 μl of nuclease-treated difference product in each tube, and make PCR for 20 more cycles as in Section III,C,2,c. Use

70° as annealing/polymerization temperature for J *Bgl*24 primer and 72° for J *Bam*24 primer (see Section III,A).

g. Remove mineral oil and run 10 μl of the PCR product on a 2% agarose gel, estimate the quantity of DNA (usually 0.1–0.3 μg), and, if necessary to improve the yield, make one to three supplementary cycles adding 3 μl (15 U) of fresh AmpliTaq DNA polymerase, under PCR conditions used in Section III,C,2,c and f.

3. Change of Adaptors for Subsequent Hybridization/Amplification

a. Combine the contents of two PCR tubes in one Eppendorf tube, extract with phenol and phenol/chloroform as in Section III,B,1,b and recover DNA by isopropanol precipitation. Dissolve the pellet in 100 μl of TE buffer, check DNA by 2% agarose gel electrophoresis, and adjust DNA concentration to 0.1 mg/ml.

b. Digest 5 μg of the first difference product with the chosen restriction endonuclease (10 U/μg) in a volume of 100 μl, add 10 μg of tRNA, extract with phenol and phenol/chloroform, and recover DNA by ethanol precipitation as in Section III,B,1,c. Dissolve the pellet in 100 μl of TE buffer and adjust the DNA concentration to 20 ng/μl.

c. Take 5 μl (100 ng) of the digested difference product and directly ligate to primer pair set 3 in a volume 30 μl as described (Section III,B,3). Add 50 μl of TE buffer (10 μl for *Hin*dIII representation) containing tRNA (20 μg/ml) to the difference product DNA ligate.

4. Subsequent Hydridization/Amplification Steps

a. For the second hybridization, mix 40 μl of the first round difference product ligated to primer pair set 3 (50 ng for most restriction endonucleases, 100 ng for *Hin*dIII representation) and 40 μg in 80 μl of driver amplicon DNA digest (adaptors cleaved off). Proceed through hybridization and selective amplification steps as above (Sections III,C,1 and C,2). Change adaptor on the second round difference product with primer pair set 2, precisely as described in Section III,C,3. Adjust the concentration of DNA ligate to 2.5 pg/μl (10 pg/μl for *Hin*dIII representation) by making two consecutive dilutions with TE buffer containing tRNA (20 μg/ml).

b. For the third hybridization, mix 40 μl of difference product 2 ligated to primer pair set 2 (100 pg for most restriction endonucleases, 400 pg for *Hin*dIII representation) with 40 μg in 80 μl of driver amplicon DNA digest. Proceed through hybridization and selective amplification steps as above (Sections III,C,1 and C,2).

c. For *Hin*dIII representation a fourth hybridization is needed. Change the adaptor on the third round difference product with primer pair set 3

(Section III,C,3) and adjust the concentration of DNA ligate to 0.125 pg/μl by making three consecutive dilutions with TE buffer containing tRNA (20 μg/ml). Mix 5 pg in 40 μl of the third round difference product ligated to primer pair set 3 with 40 μg in 80 μl of driver amplicon DNA digest and proceed through hybridization and selective amplification steps as above (Sections III,C,1 and C,2) with the final amplification after mung bean nuclease treatment for an additional 27 cycles (Section III,C,2,f).

D. Controls

It is extremely useful to include positive controls in the first experiments, taking the same DNA for preparation of driver and tester amplicons and adding appropriate target into the tester DNA, e.g., 16 pg of λ phage DNA per μg of human DNA for *Bgl*II and *Hin*dIII representations prior to preparing amplicons.[5]

[20] Analysis of Altered Gene Expression by Differential Display

By Peng Liang, David Bauer, Lidia Averboukh, Peter Warthoe, Markus Rohrwild, Heiko Muller, Michael Strauss, and Arthur B. Pardee

Introduction

Temporal and spatial expression of the 100,000 different genes in a mammalian cell[1] must be highly regulated and tightly controlled since their specificity determines life processes from the fertilized egg to death of the organism. Only a small fraction of genes, perhaps 10–15%, are expressed in any individual cell at any given time.[1] It is the choice of which genes are expressed that determines all life processes: development and differentiation, homeostasis, response to insults, cell cycle regulation, aging, and even programmed cell death. The course of normal development as well as the pathological changes that arise in diseases such as cancer, whether caused by a single gene mutation or complex multigene effects, are driven by changes in gene expression.

A pressing problem is to find effective ways to identify and isolate those genes that are differentially expressed in various cell types or under altered

[1] B. Alberts, D. Bray, J. Lewis, M. Raff, K. Roberts, and J. D. Watson, "Molecular Biology of the Cell," 2nd ed. Garland Publishing, New York and London, 1989.

conditions. The activities of genes are reflected in the kinds and quantities of their mRNA and protein species. A fingerprinting technique for mRNAs would be very useful, for instance, by two-dimensional electrophoresis, similar to two-dimensional protein separation.[2]

Desirable criteria of such a method are that (i) most of the possible 15,000 mRNAs in a cell, including rare ones, can be visualized; (ii) reproducibility should be high; (iii) it should permit side-by-side comparisons of mRNAs from different sources; (iv) the identified spots should be usable for identifying and isolating the corresponding genes, mRNAs, or cDNAs, which would make it a much more powerful technique than protein gels where frustration often follows because of an inability to obtain enough of the identified proteins for molecular characterization[3]; and (v) it should be fast and easy.

Current methods to distinguish mRNAs in comparative studies rely largely on the differential or subtractive hybridization techniques. Several important genes implicated in tumorigenesis have been isolated by these methods.[4–6] Although subtraction is quite sensitive and can detect fairly rare mRNAs, this method recovers genes incompletely and selects for genes in only one direction at a time during a two-way comparison between a pair of cells. The process is laborious and time-consuming, and results are not seen until the end of the process.

A novel method that we named differential display[7–9] or DDRT-PCR, according to polymerase chain reaction (PCR) terminology, was developed.[10] This method provides a sensitive and flexible approach to the identification of genes that are differentially expressed at the mRNA level. The method is directed toward the identification of differentially expressed genes, detecting individual mRNA species that are changed in different sets of eukaryotic cells, and then permitting recovery and cloning of their DNA.

Differential display can be used for many purposes. One is to provide a picture of mRNA composition of cells, by displaying subsets of mRNAs

[2] R. Bravo, *Semin. Cancer Biol.* **1,** 37 (1990).

[3] R. G. Croy and A. B. Pardee, *Proc. Natl. Acad. Sci. U.S.A.* **80,** 4699 (1983).

[4] S. W. Lee, C. Tomasetto, and R. Sager, *Proc. Natl. Acad. Sci. U.S.A.* **88,** 2825 (1991).

[5] P. S. Steeg, G. Bevilacqua, L. Kopper, U. P. Thorgeirsson, J. E. Talmadge, L. A. Liotta, and M. E. Sobel, *J. Natl. Cancer Inst.* **80,** 200 (1988).

[6] P. Basset, J. P. Bellocq, C. Wolf, I. Stoll, P. Hutin, J. M. Limacher, O. P. Podhajcer, M. P. Chenard, M. C. Rio, and P. Chambon, *Nature (London)* **348,** 699 (1990).

[7] P. Liang and A. B. Pardee, *Science* **257,** 967 (1992).

[8] P. Liang, L. Averboukh, K. Keyomarsi, R. Sager, and A. B. Pardee, *Cancer Res.* **52,** 6966 (1992).

[9] P. Liang, L. Averboukh, and A. B. Pardee, *Nucleic Acids Res.* **21,** 3269 (1993).

[10] D. Bauer, H. Müller, J. Reich, H. Riedel, V. Ahrenkiel, P. Waethoe, and M. Strauss, *Nucleic Acids Res.* **21,** 4272 (1993).

as short cDNA bands. This display is useful in the same way as are two-dimensional protein gels, for example, in observing alterations in gene expression. Second, these cDNAs can be quickly sequenced; thereby a sequence unique for each mRNA can readily be obtained and compared with sequences in data banks. Third, individual bands can readily be cloned and then used as probes for northern or Southern blotting and to isolate genes from genomic or cDNA libraries. A comprehensive set of these genes obtained from a display are useful for genetic applications and for preparing antibodies via corresponding amino acid sequences.

Differential display meets many of the criteria just listed, which include important advantages over the alternative subtractive hybridization. Only a few micrograms of total RNA is required, compared to 50× or more for subtractive hybridization. It is much quicker; 2 months are required to isolate clones from cells by subtractive hybridization, but with differential display band patterns are obtained in 2 days and clones in 1 week. There are far fewer technical difficulties with differential display. Most mRNAs should be represented in the pattern as only one band. Redundancy and underrepresentation of rare mRNAs occurring in subtractive hybridization are therefore avoided. Reproducibility and sensitivity are obtained.[9] Multilane displays comparing sets of cells select for both positive and negative differences unique to a process.[9] Most importantly, since the assay can be checked at each step, it is no longer necessary, as with other methods, to wait until the end of the procedure to determine whether it worked properly. These advantages should make this a method of choice for detecting and isolating differentially expressed genes.

The general strategy for differential display is outlined in Fig. 1. It depends on a combination of three techniques: (i) reverse transcription from anchored primers; (ii) choice of arbitrary primers for setting lengths of cDNAs to be amplifed by the PCR, each corresponding to part of a mRNA (tags); and (iii) sequencing gels for high resolution. The objective is to obtain a tag of a few hundred bases, which is sufficiently long to uniquely identify a mRNA and yet short enough to be separated from others by size. Pairs of primers are selected so that each will amplify DNA from about 100 mRNAs because this number is optimal for display on one lane of the gel. As described earlier, for the initial reverse transcription one uses a 14-mer that consists of a stretch of 12 Ts plus 2 or more 3′ bases, which anchors the primer to the poly(A) tail of many mRNAs. To amplify cDNA tags by PCR one adds a definite, arbitrary 10-mer. Arbitrary primers have been used before to amplify DNA polymorphisms,[11,12] and

[11] J. G. K. Williams, A. R. Kubelik, K. J. Livak, J. A. Rafalski, and S. V. Tingey, *Nucleic Acids Res.* **18,** 6531 (1990).

[12] J. Welsh and M. McClelland, *Nucleic Acids Res.* **18,** 7213 (1990).

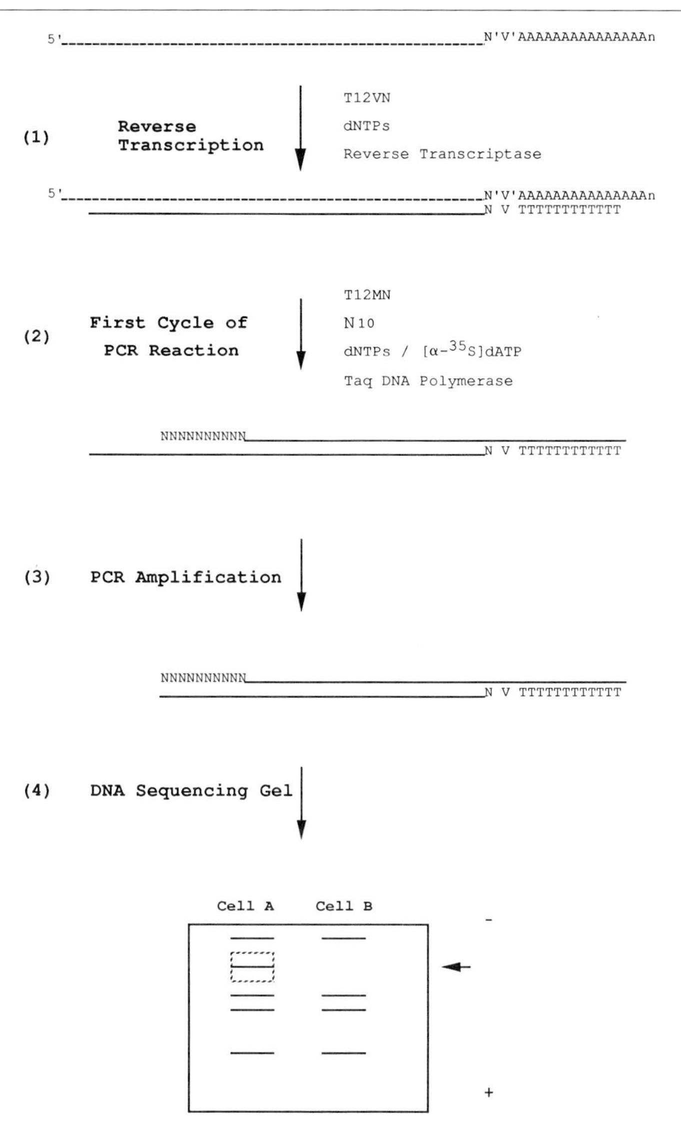

FIG. 1. Differential display. Dotted lines indicate mRNA and solid lines indicate DNA.

more recently to obtain mRNA tags for differentially expressed genes.[13] The majority of mRNAs should be represented if the 12 possible 3′ primers and a set of 5′ primers are used in all combinations for independent amplifi-

[13] J. Welsh, K. Chada, S. S. Dalal, R. Cheng, D. Ralph, and M. McClelland, *Nucleic Acids Res.* **20,** 4965 (1992).

cations and displays. Successful applications of the method are emerging, using either a limited number of primer combinations[14,15] or more impressively using 312 combinations of primer pairs made of all 12 possible anchored oligo(dT) primers and 26 arbitrary decamers and nondenaturing gel.[10]

Experimental Method

RNA Isolation

Although both total RNA or poly(A) RNA can be used for differential display, the frequent high background smear problem associated with the use of poly(A) RNA discourage its use as an RNA source (see below in discussion). Total RNA is the template of choice for differential display also because of its easy isolation and integrity verification by agarose gel. Total cellular RNA isolation can be carried out using either the standard CsCl gradient[16] or a simplified method developed by Chromczynski and Sacchi.[17] The isolated RNA should be stored in diethyl pyrocarbonate (DEPC)-treated H_2O or as an ethanol precipitate at $-80°$, before either Northern blot analysis or DNase I treatment is applied to remove contaminant chromosomal DNA prior to differential display.

DNase I Treatment of Total RNA

Removal of all contaminating chromosomal DNA from the RNA sample is essential before carrying out differential display. Incubate 10 to 100 μg of total cellular RNA with 10 units of human placental ribonuclease inhibitor (BRL, Gaithersburg, MD), 10 units of DNase I (BRL) in 10 mM Tris–Cl, pH 8.3, 50 mM KCl, 1.5 mM $MgCl_2$ for 30 min at 37°. Inactivate DNase I by adding an equal volume of phenol : chloroform (3 : 1) to the sample. Mix by vortexing and leave the sample on ice for 10 in. Centrifuge the sample, at full speed, for 5 min at 4° in a Eppendorf centrifuge. Save the supernatant, and ethanol precipitate the RNA by adding 3 vol of ethanol in the presence of 0.3 M sodium acetate and incubate at $-80°$ for 30 min. Pellet the RNA by centrifuging at 4° for 10 min. Rinse the RNA pellet with 0.5 ml of 70% ethanol (made with DEPC–H_2O) and redissolve the

[14] R. Sager, A. Anisowicz, M. Neveu, P. Liang, and G. Sotiropoulou, *FASEB J.* **7,** 964 (1993).
[15] L. Zhang and D. Medina, *Mol. Carcinog.* **8,** 123 (1993).
[16] F. Ausubel, R. Brent, R. E. Kingston, D. D. Moore, J. G. Seidman, J. A. Smith, and K. Struhl, eds., "Current Protocols in Molecular Biology." Greene Publishing Associates and Wiley (Interscience), New York, 1988.
[17] P. Chomczynski and N. Sacchi, *Anal. Biochem.* **162,** 156 (1987).

RNA in 20 μl of DEPC-treated H_2O. Measure the RNA concentration at OD_{260} with a spectrophotometer by diluting 1 μl of the RNA sample in 1 ml of H_2O. Check the integrity of the RNA samples before and after cleaning with DNase I by running 1–3 μg of each RNA on a 7% formaldehyde agarose gel. Store the RNA sample at a concentration higher than 1 μg/μl at $-80°$ before use for differential display.

Reverse Transcription of mRNA

Freshly dilute a cleaned total RNA sample to a concentration of 0.1 μg/μl in DEPC-H_2O, set on ice, and immediately freeze the remaining undiluted RNA sample to avoid degradation. Set up 4 reverse transcription reactions for each RNA sample in four 0.5-ml PCR tubes; each contains 1 of the 4 different $T_{12}VN$ primers ($T_{12}VG$, $T_{12}VA$, $T_{12}VT$, $T_{12}VC$, where V may be G, A, or C).[9] For a 20-μl reaction, add in order 9.4 μl of H_2O, 4 μl of 5X reaction buffer (250 m*M* Tris–HCl, pH 8.3, 375 m*M* KCl, 15 m*M* $MgCl_2$, 50 m*M* DTT), 1.6 μl of 250 μM each dNTPs, 2 μl of 10 μM $T_{12}MN$ primer, and 2 μl of 0.1 μg/μl DNA-free total RNA. Alternatively, 12 possible $T_{12}VN$ primers can be used separately in each reverse transcription reaction.[7,10] Program the thermocycler as follows: 65° for 5 min, 37° for 60 min, 95° for 5 min, and soak at 4°. Start the cycle, and after incubating for 10 min at 37° add to each tube 1 μl (200 units) of Moloney murine leukemia virus reverse transcriptase. At the end of the reverse transcription reaction, briefly spin tubes with 1.5-ml adapter tubes to collect condensation. Set tubes on ice for PCR or store at $-20°$ for later use.

PCR Amplification

Thaw [α-^{35}S]dATP (1200 Ci/mmol) or [α-^{33}P]dATP in advance and set on ice. Label PCR tubes to mark the reverse-transcribed cDNA and primer set used. For each primer set combination, add at room temperature in order: 8.2 μl H_2O, 2 μl of 10X PCR buffer (100 m*M* Tris–Cl, pH 8.4, 500 m*M* KCl, 15 m*M* $MgCl_2$, 0.01% gelatin), 1.6 μl of 25 μM dNTP, 2 μl of 2 μM arbitrary primer, 2 μl of 10 μM $T_{12}VN$ primer, 2 μl of corresponding $T_{12}VN$ reverse-transcribed cDNA, 1 μl of [α-^{35}S]dATP or [α-^{33}P]dATP, and 0.2 μl (1 unit) of AmpliTaq (Perkin-Elmer). Mix well by pipetting up and down and overlay 25 μl of mineral oil on top of the reaction mixture. Carry out PCR reactions as follow: 94° for 30 sec, 40° for 2 min, 72° for 30 sec for 40 cycles, then 72° for 5 min and soak at 4°.

It is advantageous to prepare batches of core mixes based on the above formula. For example, aliquot reverse-transcribed cDNA and arbitrary primers individually, but make 10X of the PCR core mix for each of the four $T_{12}VN$ primers to be used in combination with five different arbitrary primers for a pair of RNA samples being compared.

Denaturing Polyacrylamide Gel Electrophoresis

Use 3.5 μl of sample plus 2 μl of loading dye (95% formamide, 0.09% w/v bromphenol blue, and 0.09% w/v xylene cyanole FF) and incubate at 80° for 2 min before loading onto a 6% DNA sequencing gel. Electrophorese for about 3 hr at 60 W constant power until the xylene cyanole dye runs to within 10 cm from the bottom. After electrophoresis, transfer the gel directly onto a piece of Whatman 3MM paper. Cover the gel with a sheet of Saran wrap and dry under vacuum at 80° for 1 hr. Take off Saran wrap and expose the X-ray film at room temperature for 24 to 48 hr. Orient the film and dried gel with radioactive ink or needle punches.

Nondenaturing Polyacrylamide Gel Electrophoresis

Mix 3.5 μl of sample with 2 μl dye solution (20% glycerol, 0.1% bromphenol blue, and 0.1% xylene cyanole FF) and load onto a 6% polyacrylamide gel (3 $\times$ 40 cm) without urea. Run gel at 60 W until the xylene cyanole dye reaches to 5 cm from the bottom. Transfer, dry, and expose the gel to the X-ray film as above.

Reamplification of cDNA Probe

After developing the film, align the autoradiogram with the gel. Locate bands of interest, either by marking with a clean pencil underneath the film or cutting through the film. Soak the gel slice along with the 3MM paper in 100 μl of distilled H_2O for 10 min. Boil the tube with a tightly closed cap (e.g., with Parafilm) for 15 min. If the fragment is recovered from a nondenaturing gel, the ethanol precipitation step can be omitted. Spin the tube for 2 min to collect condensation and pellet the gel and paper debris. Transfer the supernatant to a new microfuge tube. Add 10 μl or 3 *M* sodium acetate, 5 μl of glycogen (10 mg/ml), and 450 μl of ethanol. Let sit for 30 min on dry ice or in a $-80°$ freezer. Centrifuge for 10 min at 4° to pellet DNA. After centrifugation the white glycogen powder should be visible. Discard the supernatant and rinse the pellet with 200 μl ice-cold 85% ethanol. Do not disturb the pellet. Spin briefly and remove the residual ethanol. Dissolve the pellet with 10 μl of distilled H_2O and use 4 μl for reamplification. Save the rest at $-20°$ for future use.

Reamplification should be done using the same primer set and PCR conditions except the dNTP concentrations are 20 μM (use 250 μM dNTP stock) instead of 2–4 μM and no isotopes are added. A 40-μl reaction is recommended. Thirty microliters of the reamplified PCR sample is run on a 1.5% agarose gel with xylene cyanole as a loading dye, and stained with ethidium bromide. Save the remaining PCR samples at −20° for cloning or future reamplification. About 90% of probes should be seen after the first round of PCR. If a cDNA probe fails to be amplified in the first round, 4 μl of the PCR sample from the first round can be used as a template for a second round of PCR using identical conditions. Check to see if the size of a reamplified PCR product is consistent with its size on the DNA sequencing gel. The reamplified cDNA probe is cut from the agarose gel under a UV lamp and extracted using the Qiaex kit (Qiagen) or purified with a Millipore spin column. The extracted cDNA probe is eluted in 20 μl of H_2O and saved for a probe for Northern blot analysis. If more cDNA probe needs to be reamplified, 4 μl of the original eluted cDNA from the sequencing gel or nondenaturing gel can be used again as a template.

Northern Blot Confirmation

One can use either a cloned or reamplified cDNA probe directly for the Northern blot analysis to verify the differential expression of the gene. Generally, 20 μg of each total RNA sample is run on a 7% formaldehyde agarose gel (1.1%) and transferred to a nylon membrane essentially as described.[18] Label 9 μl of Qiaex extracted probe with [α-^{32}P]dCTP by the random-prime method,[19] except 0.5 μM of the corresponding T_{12}MN primer is included as well as random hexamers. Remove the unincorporated isotope using a Sephadex G-50 spin column. Prehybridize and hybridize the membrane and follow the standard procedure.[16] Wash the membrane twice for 15 min at room temperature with 1X SSC, 0.1% SDS, and once at 55° for 20 min with 0.25X SSC, 01.% SDS. Dry the membrane briefly between two sheets of 3 MM paper and expose to a X-ray film with an intensifying screen at −80°.

Cloning and Sequencing of cDNA Probes

Reamplified cDNA probes were cloned into the PRII vector using the TA cloning system from Invitrogen (San Diego, CA) or the pCR-TRAP positive-selection cloning system from GenHunter Corporation (Brookline, MA). Plasmid DNA sequencing of cloned cDNA probes with either T7 or SP6 primer was carried out using the Sequenase kit from United States Biochemicals Co (Cleveland, OH) or fmole kit (Promega).

[18] K. Keyomarsi, L. Sandoval, V. Band, and A. B. Pardee, *Cancer Res.* **51,** 3602 (1991).

[19] A. P. Feinberg and B. Vogelstein, *Anal. Biochem.* **137,** 266 (1984).

Cloning of Full-Length cDNA

After a CDNA probe is confirmed by Northern blot to give an expected differential display pattern, the cloned cDNA probe can be used to screen a cDNA library for a full-length clone by the standard procedure.[16] If the reamplified cDNA probe before being cloned detects a single mRNA species on the Northern blot, it can be used directly as a probe to screen a cDNA library, bypassing an intermediate cloning step. To recover the 5′ upstream sequence of the gene if a cDNA library is unavailable, one can also use the 5′ RACE method[20] based on sequence information for the cloned cDNA probe generated by differential display.[21]

Automatic Analysis of Differential Display Pattern

If cloning of differentially expressed genes is not the first priority and a diagnostic pattern is required instead, it is advantageous to use a nonradioactive detection method in combination with automatic analysis of the resulting gel pattern.[10] For this purpose, an automatic DNA sequencing machine can be used with denaturing or nondenaturing gel electrophoresis. Arbitrary primers are labeled with fluorescent dyes FAM, JOE, ROX, and TAMRA (Applied Biosystems) according to the manufacturer's instruction. Since the fluorescence intensity of FAM and JOE is four- to sixfold higher than that of ROX and TAMRA, it is difficult to use all four dyes for the comparison of four different RNA samples in a single lane. In addition, different dyes have different effects on the hybridization property of the primer labeled. At the moment, only FAM or JOE is used to label the arbitrary primer so samples should be compared side by side as described in the standard procedure above.

For each reaction, add at room temperature 10μl H_2O, 2 μl of 10X PCR buffer (as above), 2 μl of 1 mM dNTP, 2μl of 25 μM arbitrary primer (dye labeled and HPLC purified), 2 μl of 5 μM anchored oligo(dT) primer, 2 μl of respective reverse-transcribed cDNA, and 0.2 μl (1 unit) of AMpliTaq (Perkin-Elmer). PCR reactions are carried out as described above. It is also advantageous to prepare core mixes as above. To avoid high background and to achieve maximal sensitivity, the amount of sample should be adjusted to achieve relative band intensities between 200 and 1000 arbitrary units in the case of the ABI machine. The electrophoretic runs are recorded and band patterns are compared according to the software instructions of the manufacturer.

[20] M. A. Frohman, M. K. Dush, and G. R. Martin, *Proc. Natl. Acad. Sci. U.S.A.* **85,** 8998 (1988).
[21] P. Liang and W. Zhu, unpublished result 1995.

Experimental Results and Discussion

Differential Display of mRNAs from Normal vs Tumor Cells Using Degenerate Anchored Oligo(dT) Primers

Theoretical consideration suggests that 12 possible anchored oligo(dT) primers (T_{12}VN where V may be dG, dA, or dC and N may be any of the 4 deoxynucleotides) used in all combinations with at least 20 different arbitrary 10-mers are necessary to display 10,000 of the mRNA species that are present in a mammalian cell.[7,10] The observation that the terminal 3′ base of the anchored oligo(dT) primer provides most of the selectivity enables one to cut down the anchored oligo(dT) primer from 12 to 4; each is degenerate (except T) at the penultimate 3′ base.[9]

Using the four degenerate T_{12}VN primers in combination with five arbitrary 10-mers, RNA from human breast cancer cells and normal mammary epithelial cells[22] were differentially displayed (Fig. 2). The result showed that the patterns of the display were dependent on both the T_{12}VN primers that differ in the last 3′ base and on the arbitrary 10-mers, as predicted by theory. Each primer set displayed 50–100 bands. A visual survey of the mRNA differential display between the normal and breast cancer cells revealed a few percent of bands that were differentially expressed, of which perhaps half were due to changes of gene expression. Thus, on the order of 100 genes are differentially expressed in cancer, consistent with an estimate from protein blots.[2]

Differential Display of mRNA from Normal and Regenerating Liver: Comparison of Denaturing and Nondenaturing Gels

mRNA expression patterns from normal and regenerating mouse or pig liver were compared using 12 anchored T_{11}VN primers and 26 arbitrary upstream primers.[10] Three hundred and twelve individual PCR incubations were carried out for every time point (normal liver, 12, 24, and 36 hr after heptectomy) resulting in a total of 1248 reactions. Half of the reaction mixtures were run on denaturing gels and the other half on nondenaturing gels, in each case 26 gels with 48 lanes. Whereas the bands in denaturing gels were sharp, they were slightly fuzzy in nondenaturing gels. However, doublets and triplets of bands were often seen on a denaturing gel at the lower molecular weight range (<200 bases). After sequencing, these bands turned out to be the same cDNA species, a result of separation of the complementary strands on the one hand and the presence or absence of an additional A added by the *Taq* polymerase on the other hand. This

[22] V. Band and R. Sager, *Proc. Natl. Acad. Sci. U.S.A.* **86,** 1249 (1989).

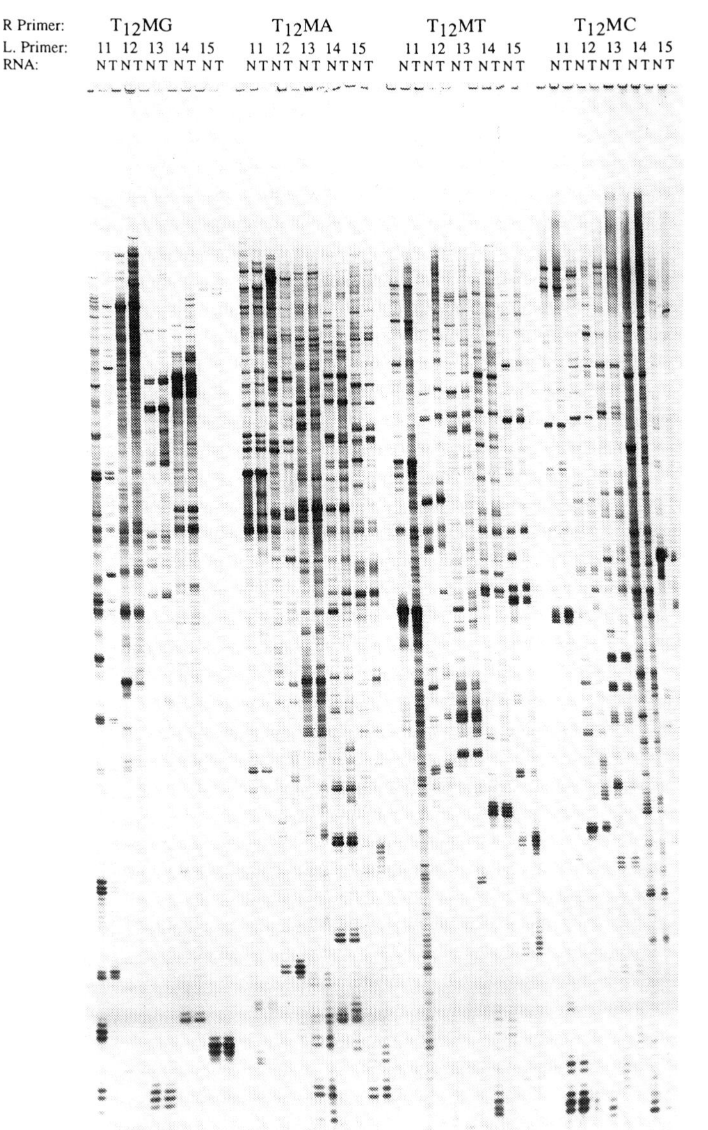

FIG. 2. Differential display of mRNA from normal mammary epithelial cell 76 N and a breast cancer cell line 21MT-2, using RNAmap kit C (GenHunter Corporation, Brookline, MA) with 20 combinations of primer sets made of $T_{12}MG$, $T_{12}MA$, $T_{12}MC$, $T_{12}MT$, and five arbitrary 10-mers AP-11 to 15. The five AP primers (L. Primer) used were AP-11 (CAGACCGTTC), AP-12 (TGCTGACCTG), AP-13 (AGTTAGGCAC), AP-14 (AATGGGCTGA), and AP-15 (AGGGCCTGTT). N, normal cell; T, breast cancer cell.

artificial complexity of the pattern is reduced in the nondenaturing gels. Figure 3 shows a comparison of both systems with selected primers and only three time points of pig liver RNA. Another advantage of the nondenaturing gel is the easier processing of the fragments, while ethanol precipitation can be omitted.

Some 70 bands out of 38,000 total bands in the nondenaturing gel run with mouse liver RNA-derived fragments were found to be different between normal and regenerating liver. About 50% of them turned out to correspond indeed to differentially expressed genes by Northern blot. Thirty percent of these bands contain only one species (as determined by cloning and sequencing of 6 clones of each), 40% contain two species, and the remaining 30% contain more than two species where generally only one species within the fragment turned out to be differentially expressed. It is interesting to note that in cases where the differential display suggested a plus/minus situation (absence of the band in one lane, presence in the other lanes, or vice versa), Northern blotting often revealed much less dramatic differences. An induction of four- to sixfold was quite common.[10]

Almost identical conclusions could be drawn from a similar complete comparison of RNA from cells which are either deficient for the tumor suppressor Rb or transfected with the Rb gene.[10]

Degeneracy of Arbitrary Primers

Theoretically, the arbitrary primers should ideally be 6- to 7-mers in order to amplify the optimal number of mRNAs to be separated on a sequencing gel; however, primers shorter than 8 bases did not display any mRNAs in combination with the anchored olig(dT) primers.[7] This was explained by the possibility that primers shorter than 9 bases may have too low melting temperatures to be used under the PCR conditions where the *Taq* DNA polymerase is active. Increasing the length of arbitrary primer to 10 gave a manageable number of amplified cDNA bands. This finding is contradictory to the theoretical prediction unless the 10-mers hybridize in a degenerate manner. Enough evidence has now been obtained to show that degeneracy is part of the explanation.[9] Often 2–3 base mismatches are found between the arbitrary primer sequence and the original cDNA template obtained from a cDNA library (Table I). The majority of these mismatches are located at the 5′ end of the arbitrary primer, not unexpectedly.

To further test if an arbitrary decamer can tolerate mismatches, mouse thymidine kinase (TK) cDNA was used as a template and amplified using $T_{11}CA$ as the anchored oligo(dT) primer and upstream primers with perfect matches and one to three mismatches in the 5′ ends (Fig. 4). With limited

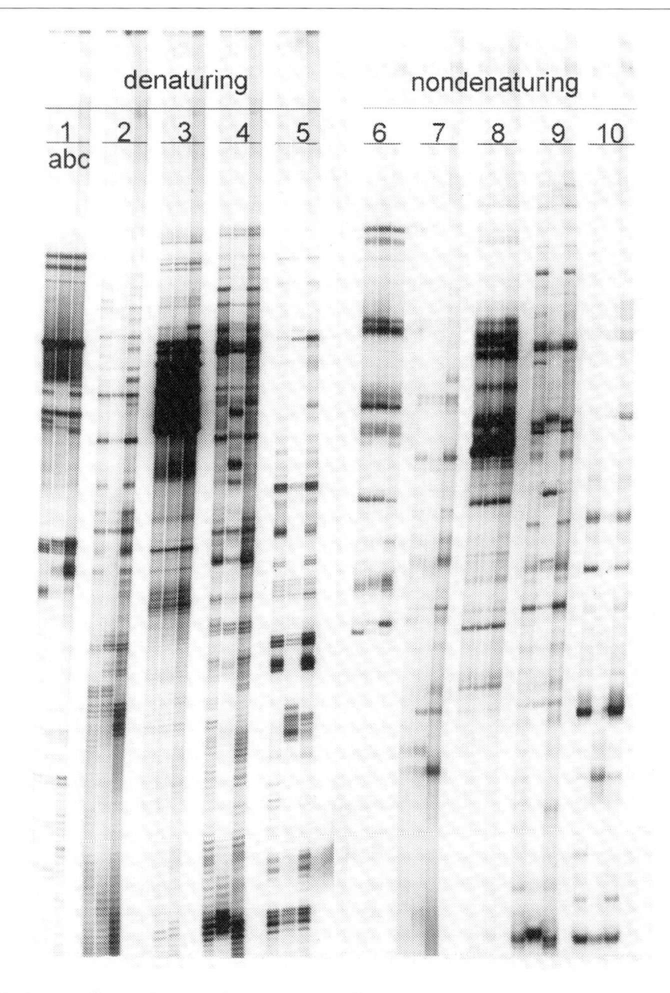

FIG. 3. Comparison of two gel systems for differential display of fragments derived from pig liver. RNA was obtained from normal and regenerating pig liver, reverse transcribed using anchored primer $T_{11}CGT$, and cDNA was amplified using the same anchored primer and five arbitrary primers 13–17 (10). Half of the PCR product was run on a denaturing gel (lanes 6a–c to 5a–c) and the other half on a nondenaturing gel (lanes 6a–c to 10a–c). The three lanes (a–c) run with each pair of primers represent resting liver (a), 12 hr (b), and 24 hr (c) after hepatectomy.

primers tested, primers with two mismatches were still able to amplify the TK cDNA while other bands were amplified as well because the primer specificity changed. A primer (M3) with three mismatches (two mismatches at G-C base pairs and one mismatch at an A-T base pair) failed to amplify

TABLE I
DEGENERACY OF ARBITRARY PRIMERS

dCNA clone	Arbitrary primer[a]	Mismatches (bp)	Ref.
N1	tTcGATTGCC	2	21
S1	ctGaTCCATG	3	8
FW	caGTAACGCC	2	21
F	tCtGTGCTGG	2	21
J	GACCGctTGT	2	21
Int-6	cagGCCCTTC	3	14
PC6	gGTAcTCCAC	2	21
AAT	gatcTAAGGC	4	10
Alb	cATCATGGTC	1	10
b-act	GAtcGCATTG	2	10
FO8	ctGGTCATAG	2	10
No. of mismatch	9553111000		

[a] The lowercase letters indicate bases where mismatch occurred between arbitrary primers and the original cDNA sequences isolated from a cDNA library.

the TK cDNA whereas many cDNAs were displayed with primers of three to four mismatches (Table II). Therefore the number of mismatches may depend on several factors including the GC content and distribution of individual primer and the PCR conditions. It is predicted that more mismatches at the positions of A-T base pairs may be tolerated than at the G-C base pairs for an arbitrary decamer with a given G-C content. Thus this findings provides a molecular basis for differential display methodology and will be helpful in future primer designs.

Comparison of Poly(A) RNA and Total RNA as Templates

One important aspect of the differential display method as described earlier is that either poly(A)RNA or total cellular RNA can be used as templates. As has been described,[9] the resulting cDNA patterns displayed by a given primer set were nearly identical. This finding indicated that $T_{12}MN$ primers preferentially select the mRNA over the abundant ribosomal RNA and transfer RNA during the reverse transcription reaction. Total RNA was routinely used in most of our experiments since it is generally much easier to purify than poly(A) RNA. The use of total RNA instead of poly(A) RNA also avoids the frequent oligo(dT) primer contamination which gives a problem of high background smear.[9] This was due to the oligo(dT) priming without anchoring to the beginning of the poly(A) tails of a mRNA.

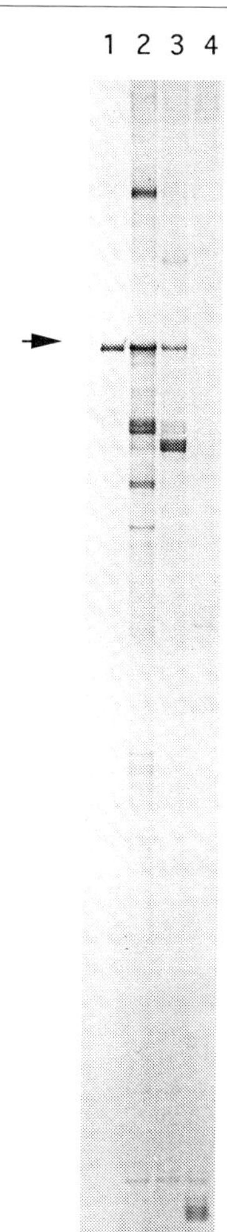

FIG. 4. Degeneracy of the arbitrary primer. Amplification of a 300-bp TK fragment (indicated by the arrowhead) using the $T_{12}CA$ downstream primer in combination with Ltk3, M1, M2, and M3 upstream primers (lanes 1 to 4, respectively; see Table II) and a 1156-bp mouse TK cDNA template from pAMTK plasmid.

TABLE II
MISMATCHED PRIMERS DERIVED FROM LTK3[a]

Primer	Sequence	Mismatches
Ltk3	5′-CTTGATTGCC-3′	0
M1	5′-**T**TTGATTGCC-3′	1
M2	5′-**T**T**C**GATTGCC-3′	2
M3	5′-**T**T**CC**ATTGCC-3′	3

[a] Showing 100% identity to position 867 to 876 of a 1156-bp mouse thymidine kinase cDNA.

Estimation of Number of mRNA Species in a Cell by Differential Display

Our result that each primer set reveals about 75 (50 to 100) bands for mammalian cells permits a calculation of the total number of mRNAs that can be identified by this method. Each $T_{12}MN$ should select 1/4 of the mRNAs (assuming little degeneracy) and each arbitrary primer represents 1/16,000 sequences if it actually hybridizes as a 7-mer. A given mRNA 3′ end sequence contains about 500 sites for recognition on the gel. Thus each arbitrary primer will recognize 500/16,000 = 0.03 of the mRNAs. Therefore the total number of mRNA recognized are about $4 \times 75/0.03 = 10{,}000$. This independent rough estimate is close to an earlier determination of 10,000 to 20,000 mRNAs in a mammalian cell.[1]

Estimation of Number of Primers Needed to Display Most of mRNAs in Cell

A calculation can also be made to determine how many PCR reactions must be done to detect a given fraction of the total mRNAs (Table III) using 4 $T_{12}VN$ primers or 12 T12VN primers in combinations with a number of arbitrary primers (n). Assume that an arbitrary primer detects 0.03 of

TABLE III
NUMBER OF ARBITRARY PRIMERS NEEDED IN COMBINATION WITH ALL FOUR 3′ PRIMERS[a]

Number of arbitrary primer (n)	Reactions	Fraction of mRNA detected [$P = 1 - (0.97)^n$]
25	100	0.53
30	120	0.60
40	160	0.70
75	300	0.90

[a] To detect a given fraction of mRNA by differential display.

the total mRNAs if it primes as a 7-mer within a distance of 500 bases from the poly(A) tail($500/4^7$), so 0.97 of the mRNAs will not be seen. With n arbitrary primers, $(0.97)^n$ of total mRNAs will not be detected. Therefore, with 25 arbitrary primers and the 4 T_{12}MN primers, about half of the mRNAs will be seen (Table II). mRNAs will be increasingly seen more than once as the number of lanes is increased. For example, based on this calculation, about 300 PCR reactions made up with 75 arbitrary primers and the 4 T_{12}VN primers will be required to detect 90% of the possible mRNAs in a mammalian cell. Alternatively, a similar result may be obtained with 25 arbitrary primers and 12 T_{12}VN primers assuming little redundancy.[7,10]

Problems and Limitations

Although differential display has many potential advantages over other conventional methods in isolating differentially expressed genes, some inherent problems and limitations should also be pointed out so they can be properly dealt with as the method becomes more mature. First is the noise level which has to be distinguished from the true differences. The second has to do with whether rare messages are represented by the method. Third, as with subtractive and differential hybridization, differential display will be unlikely to reveal differences involving mutational changes such as in p53 and *ras* genes which have point mutations and other nontranscriptional alterations. Fourth, the use of anchored oligo(dT) primers limits the region where mRNA sequence tags are obtained. Moreover, it has to be mentioned that certain fraction of bands are generated by the arbitrary primer alone from palindromic sequences within a mRNA molecule. The resulting fragments may be derived from different segments of a given mRNA. Some possible remedies are discussed below, and more are waiting to be formulated.

Strategy to Minimize Isolating False-Positive Probes

Although differential display gives highly reproducible patterns of amplified mRNAs, which are the basis on which differentially expressed genes can be identified and isolated, it should be pointed out the noise level (a few percent of nonreproducible bands) can be quite comparable to the number of real differences. Since one is looking for differences, a way of distinguishing the true differences in mRNA expression from the noise level obviously would significantly save time and energy in the process of further screening and Northern confirmation. Therefore, it is recommended that if one is to compare a pair of mRNAs, those PCR reactions which generate candidate differential expressed cDNA bands ought to be repeated, ideally starting from the reverse transcription step in order to determine which differences can be confirmed. The other approach takes advantage of side by side comparisons of more than two related RNA samples, which is possible with differential display; differentially expressed genes that are unique to a process, such as cell transformation, can be more surely isolated. Multiple displays therefore provide an internal control to minimize choosing false-positive or individual cell-specific genes that may not be relevant to the system under study.

Acknowledgments

We thank Dr. R. Sager for providing the normal and tumor mammary epithelial cells. The authors also acknowledge F. Wang and W. Zhu for their excellent technical assistance.

[21] Approaches to the Identification and Molecular Cloning of Chromosome Breakpoints

By STEVEN A. SCHICHMAN and CARLO M. CROCE

Introduction

Chromosomal rearrangements represent a special class of genetic abnormalities which are sometimes associated with malignant disease. Because these rearrangements often involve the breakage and rejoining of large pieces of chromosomal material, these abnormalities can usually be visualized under the light microscope by cytogenetic techniques. Translocations and inversions represent the most common types of chromosomal rearrangements. Reciprocal translocations, which involve the reciprocal exchange of genetic material between two different chromosomes, are frequently seen in cancers of blood cells, leukemia, and lymphoma.[1] Recurrent reciprocal translocations have also been identified in other types of tumors.[2] Chromosomal translocations have served as guideposts to the location of genes involved in malignancy.[3] In some cases, the gene has been found precisely at the site of the chromosomal breakpoint. In other cases, a candidate gene has been discovered at a distance—up to several hundred

[1] J. D. Rowley, *Semin. Hematol.* **27,** 122 (1990).
[2] C. Sreekantaiah, M. Landanyi, E. Rodriguez, and R. S. K. Chaganti, *Am. J. Pathol.* **144,** 1121 (1994).
[3] E. Solomon, J. Borrow, and A. D. Goddard, *Science* **254,** 1153 (1991).

kilobases—from the breakpoint. In either circumstance, the identification and cloning of chromosomal breakpoints have provided important tools for the discovery of genes involved in cancer.[4]

Chromosomal rearrangements can lead to malignant transformation through the genetic mechanisms of gene activation and gene fusion.[5] The transcriptional activation of a gene can occur when it is placed near the regulatory elements of an immunoglobulin (Ig) or T-cell receptor (TCR) gene locus by chromosomal translocation or inversion. The t(8;14) reciprocal translocation found in Burkitt's lymphoma provides one example of gene activation. This chromosome rearrangement places the c-*myc* gene, normally found at chromosome band 8q24, under the control of regulatory elements from the immunoglobulin heavy chain locus on chromosome band 14q32.[6] The resulting transcriptional activation of the c-*myc* gene, which is involved in the regulation of cell growth, plays a critical role in the development of Burkitt's lymphoma.[7]

Gene fusion refers to the physical juxtaposition of segments from two different genes. This type of rearrangement results in a chimeric structure consisting of the head of one gene and the tail of another gene. Fusion of genes can occur as a consequence of translocation or inversion when the chromosome breakpoints are located within the loci of two different genes. The t(9;22) chromosomal translocation found in chronic myelogenous leukemia (CML) involves a well-known example of gene fusion.[8] In this chromosomal rearrangement, the *c-abl* gene, normally located on chromosome band 9q34, is fused with the *bcr* gene, located on chromosome band 22q11.[9] The resulting hybrid *bcr/abl* gene gives rise to a chimeric protein which contributes to the expansion of the neoplastic myeloid clone.[10]

The precise identification and molecular characterization of a chromosome breakpoint begin with the cytogenetic analysis of tumors and tumor cell lines. Because various chromosomal aberrations are common in malignant cells, it is important to focus on recurring chromosomal rearrangements which are consistently associated with a particular type of malignancy. Cytogenetic analysis is used to identify the chromosomes involved in a rearrangement and to localize the breakpoint to a distinct band on each

[4] P. C. Nowell, *FASEB J.* **8,** 408 (1994).
[5] C. M. Croce, *Cell (Cambridge, Mass.)* **49,** 155 (1987).
[6] R. Dalla Favera, M. Bregni, J. Erikson D. Patterson, R. C. Gallo, and C. M. Croce, *Proc. Natl. Acad. Sci. U.S.A.* **79,** 7824 (1982).
[7] P. C. Nowell and C. M. Croce, *Am. J. Clin. Pathol.* **94,** 229 (1990).
[8] P. C. Nowell and D. Hungerford, *Science* **132,** 1497 (1960).
[9] J. Groffen, J. R. Stephenson, N. Heistercamp, A. de Klein, C. R. Bartram, and G. Grosveld, *Cell (Cambridge, Mass.)* **36,** 93 (1984).
[10] E. Shitvelman, B. Lifshitz, R. P. Gale, and E. Canaani, *Nature (London)* **315,** 550 (1985).

involved chromosome. For example, in many cases of acute myelogenous leukemia (AML), a recurring reciprocal translocation is seen that involves chromosome bands 8q22 and 21q22 [t(8;21)(q22;q22)].[11,12] The cytogenetic localization of a chromosomal breakpoint serves as the starting point for its molecular identification.

Chromosome bands defined by cytogenetic staining typically average 10 megabases of DNA. The molecular cloning of a chromosome breakpoint is therefore an exercise in positional cloning using the physical location of a chromosome band, or mapped loci within the band, as initial markers. The techniques used in positional cloning have advanced rapidly during the last few years. These technical advances, which have enabled the recent molecular identification of many chromosomal breakpoints, form the basis for a general approach to breakpoint cloning. From an historical perspective, however, some breakpoints were identified during the 1980s by their association with (1) immunoglobulin or T-cell receptor gene loci, or (2) known oncogenes. These fortunate associations with known markers provided shortcuts to the tedious process of positional cloning. This chapter presents a conceptual approach to the molecular identification of chromosome breakpoints using examples from the literature. First, the juxtaposition of Ig and TCR gene loci with chromosome breakpoints will be discussed. Second, the association of known genes with breakpoints will be summarized. Third, the identification of chromosome breakpoints by fluorescence *in situ* hybridization (FISH) using yeast artificial chromosomes (YACs) or cosmids will be described. Finally, a general flow scheme for the identification of chromosomal breakpoints will be presented.

Chromosomal Breakpoints Associated with Immunoglobulin or T-Cell Receptor Gene Loci

Mistakes in the normal rearrangement of Ig or TCR genes during B and T lymphocyte development are thought to give rise to many of the recurring chromosomal rearrangements found in lymphoma and leukemia.[13] In these cases, chromosomal translocation or inversion events are thought to be mediated, at least in part, by the VDJ recombinase enzyme system which is active during the early stages of T- and B-cell differentiation. The consequence of chromosomal rearrangements resulting from errors in the normal process of Ig or TCR gene rearrangement is the transcriptional

[11] J. D. Rowley, *Nature (London)* **243,** 290 (1973).

[12] H. Miyoshi, K. Shimizu, T. Kozu, N. Maseki, Y. Kaneko, and M. Ohki, *Proc. Natl. Acad. Sci. U.S.A.* **88,** 10431 (1991).

[13] L. R. Finger, R. C. Harvey, R. C. A. Moore, L. C. Showe, and C. M. Croce, *Science* **234,** 982 (1986).

activation of protooncogenes which are juxtaposed to the Ig or TCR gene loci.

Chromosomal rearrangements in many B-cell malignancies involve the immunoglobulin heavy chain locus (IgH) located on chromosome 14 band q32.[14] These include most follicular lymphomas, mantle cell lymphomas, some large cell lymphomas, and Burkitt's lymphoma.[15] Some variant types of Burkitt's lymphoma involve either the immunoglobulin light chain (IgL) κ locus at 2p12 or the IgL λ locus at 22q11.[16] Acute and chronic T-cell leukemias frequently have chromosomal rearrangements involving either the TCR β locus at 7q35 or the TCR α/δ locus at 14q11.[17–19] Translocations involving the TCR γ gene locus at 7p15 have not yet been identified. A partial list of chromosomal rearrangements in hematologic malignancies involving Ig or TCR gene loci is given in Table I.[20–35]

[14] F. G. Haluska, Y. Tsujimoto, and C. M. Croce, *Trends Genet.* **3,** 11 (1987).

[15] M. M. Le Beau, *in* "Neoplastic Hematopathology" (D. M. Knowles, ed.), p. 299. Williams & Wilkins, Baltimore, MD, 1992.

[16] B. S. Emanuel, J. R. Selden, R. S. K. Chaganti, S. Jhanwar, P. C. Nowell, and C. M. Croce, *Proc. Natl. Acad. Sci. U.S.A.* **81,** 2444 (1984).

[17] M. Isobe, J. Erikson, B. S. Emanuel, P. C. Nowell, and C. M. Croce, *Science* **228,** 580 (1985).

[18] C. M. Croce, M. Isobe, A. Palumbo, J. Puck, J. Ming, D. Tweardy, J. Erikson, M. Davis, and G. Rovera, *Science* **227,** 1044 (1985).

[19] M. Isobe, G. Russo, F. G. Haluska, and C. M. Croce, *Proc. Natl. Acad. Sci. U.S.A.* **85,** 3933 (1988).

[20] C. M. Croce, W. Thierfelder, J. Erikson, K. Nishikura, J. Finan, G. Lenoir, and P. C. Nowell, *Proc. Natl. Acad. Sci. U.S.A.* **80,** 6922 (1983).

[21] Y. Tsujimoto, J. Yunis, L. Onorato-Showe, J. Erikson, P. C. Nowell, and C. M. Croce, *Science* **224,** 1403 (1984).

[22] D. A. Withers, R. C. Harvey, J. B. Faust, O. Melnyk, K. Carey, and T. C. Meeker, *Mol. Cell. Biol.* **11,** 4846 (1991).

[23] Y. Tsujimoto, L. R. Finger, J. Yunis, P. C. Nowell, and C. M. Croce, *Science* **226,** 1097 (1984).

[24] Y. Tsujimoto and C. M. Croce, *Proc. Natl. Acad. Sci. U.S.A.* **83,** 5214 (1986).

[25] H. Ohno, G. Takimoto, and T. W. McKeithan, *Cell (Cambridge, Mass.)* **60,** 991 (1990).

[26] B. H. Ye, F. Lista, F. Lo Coco, D. M. Knowles, K. Offit, R. S. K. Chaganti, and R. Dalla-Favera, *Science* **262,** 747 (1993).

[27] J. D. Mellentin, S. D. Smith, and M. L. Cleary, *Cell (Cambridge, Mass.)* **58,** 77 (1989).

[28] R. Baer, *Semin. Cancer Biol.* **4,** 341 (1993).

[29] L. R. Finger, J. Kagan, G. Christopher, J. Kurtzberg, M. S. Hershfield, P. C. Nowell, and C. M. Croce, *Proc. Natl. Acad. Sci. U.S.A.* **86,** 5039 (1989).

[30] J. Erikson, L. Finger, L. Sun, A. ar-Rushdi, K. Nishikura, J. Minowada, J. Finan, B. S. Emanuel, P. C. Nowell, and C. M. Croce, *Science* **232,** 884 (1986).

[31] J. Kagan, L. R. Finger, J. Letofsky, J. Finan, P. C. Nowell, and C. M. Croce, *Proc. Natl. Acad. Sci. U.S.A.* **86,** 4161 (1989).

[32] M. Hatano, C. W. M. Roberts, M. Minden, W. M. Crist, and S. J. Korsmeyer, *Science* **253,** 79 (1991).

[33] E. A. McGuire, R. D. Hockett, K. M. Pollock, M. F. Bartholdi, S. J. O'Brien, and S. J. Korsmeyer, *Mol. Cell. Biol.* **9,** 2124 (1989).

TABLE I

CHROMOSOMAL REARRANGEMENTS IN HEMATOLOGIC MALIGNANCY INVOLVING IMMUNOGLOBULIN OR T-CELL RECEPTOR GENE LOCI

Rearrangement	Disease[a]	Rearranged gene at breakpoint	Activated gene near breakpoint	Ref.
B-cell malignancies				
t(8;14)(q24;q32)	BL, B-ALL	IgH (14q32)	c-MYC (8q24)	6
t(2;8)(p12;q24)	BL,B-ALL	IgL-κ (2p12)	c-MYC (8q24)	16
t(8;22)(q24;q11)	BL,B-ALL	IgL-λ (22q11)	c-MYC (8q24)	20
t(11;14)(q13;q32)	MCL	IgH (14q32)	BCL1/PRAD1/cyclin D1 (11q13)	21 22
t(14;18)(q32;q21)	FL	IgH (14q32)	BCL2 (18q21)	23 24
t(14;19)(q32;q13)	B-CLL	IgH (14q32)	BCL3 (19p13.1)	25
t(3;14)(q27;q32)	DLCL	IgH (14q32)	BCL6/Laz3 (3q27)	26
T-cell malignancies				
t(7;19)(q35;p13)	T-ALL	TCR-β (7q35)	LYL1(19p13)	27
t(7;9)(q35;q34)	T-ALL	TCR-β (7q35)	TAL2 (9q34)	28
t(1;14)(p32;q11)	T-ALL	TCR-α (14q11)	TAL1/SCL/TCL5 (1p32)	29 28
t(8;14)(q24;q11)	T-ALL	TCR-α (14q11)	c-MYC (8q24)	30 13
t(10;14)(q24;q11)	T-ALL	TCR-α (14q11)	HOX11 (10q24)	31 32
t(11;14)(p15;q11)	T-ALL	TCR-δ (14q11)	RBTN1/Ttg1 (11p15)	33 34
inv(14)(q11;q32)	T-CLL/T-PLL	TCR-Cα (14q11)	TCL1 (14q32.1)	35

[a] BL, Burkitt's lymphoma; MCL, mantle cell lymphoma; FL, follicular lymphoma; CLL, chronic lymphocytic leukemia (B or T cell); DLCL, diffuse large cell lymphoma; ALL, acute lymphoblastic leukemia (B or T cell); PLL, prolymphocytic leukemia.

If cytogenetic analysis of a hematologic malignancy shows chromosomal translocation or inversion involving chromosome bands 14q32, 7q35, or 14q11 (or less commonly 2p12 or 22q11), then there is a good chance that the rearrangement involves the corresponding Ig or TCR gene locus. Because these chromosomal rearrangements are thought to result from mistakes in the normal process of Ig or TCR gene rearrangement, the breakpoint is usually located adjacent to a segment of the Ig or TCR gene. Commonly, the chromosome breakpoint is next to a joining segment of an Ig or TCR gene, suggesting that the translocation or inversion event occurred during attempted V-(D)-J rearrangement.[36]

[34] T. Boehm, L. Foroni, Y. Kaneko, M. F. Perutz, and T. H. Rabbits, *Proc. Natl. Acad. Sci. U.S.A.* **88,** 4367 (1991).

[35] L. Virgilio, M. G. Narducci, M. Isobe, L. Billips, M. Cooper, C. M. Croce, and G. Russo, *Proc. Natl. Acad. Sci. U.S.A.* **91,** 12530–12534 (1994).

[36] Y. Tsujimoto, J. Gorham, J. Cossman, E. Jaffe, and C. M. Croce, *Science* **229,** 1390 (1985).

Clonal rearrangement of Ig or TCR genes may be demonstrated in lymphomas and lymphoid leukemias by Southern analysis.[37] Probes from the joining regions or constant regions of Ig or TCR genes are most frequently used for this purpose.[38] These probes have been used to clone chromosomal breakpoints associated with Ig or TCR genes. For example, a probe from the joining region of the immunoglobulin heavy chain gene (J_H) was used to clone the breakpoint of the t(11;14)(q13;q32),[21] associated with mantle cell lymphomas, and the breakpoint of the t(14;18)(q32;q21),[23] associated with follicular lymphomas. In these two examples, the J_H probe was used to screen bacteriophage λ libraries made from the genomic DNA of cell lines carrying the translocations. Single-copy sequences from recombinant λ clones were then mapped to human chromosomes using panels of rodent–human hybrid cells. For each of the two different translocations, a small restriction fragment spanning the chromosomal breakpoint was identified and eventually sequenced. The same general approach can be used to clone any chromosomal breakpoint associated with Ig or TCR genes if a suitable probe is available.

Although the chromosomal breakpoints associated with Ig or TCR genes can usually be cloned in a straightforward manner, the protooncogenes which are transcriptionally activated by the chromosomal rearrangement may be difficult to identify. In some cases, a candidate protooncogene has been found at or near the breakpoint. For example, the major breakpoint region for the t(14;18)(q32;q21) is located within the 3′ untranslated region of the *bcl-2* gene on chromosome band 18q21, leading to the rapid identification of this gene.[24,39,40] In other cases, however, the candidate gene has been found far away from the breakpoint. The *bcl-1* gene (PRAD1 or cyclin D1) on chromosome band 11q13 is located more than 100 kb away from the major breakpoint region of the t(11;14)(q13;q32).[22] The *tcl-1* gene on chromosome band 14q32.1 [associated with T-cell malignancies having inv(14)(q11;q32) or t(14;14)(q11;q32)] was identified within a 160-kb region flanked by two clusters of chromosomal breakpoints.[35] In cases where the activated protooncogene is distant from the chromosome breakpoint, techniques such as walking, exon trapping, and CpG island localization are required to identify a candidate gene involved in malignant transformation. Further analysis must then be performed to demonstrate

[37] J. Sklar, *in* "Neoplastic Hematopathology" (D. M. Knowles, ed.), p. 215. Williams & Wilkins, Baltimore, MD, 1992.

[38] C. A. Felix, D. G. Poplack, G. H. Reaman, S. M. Steinbery, D. E. Cole, B. J. Taylor, C. G. Begley, and I. R. Kirsch, *J. Clin. Oncol.* **8,** 431 (1990).

[39] Y. Tsujimoto, J. Cossman, E. Jaffe, and C. M. Croce, *Science* **228,** 1440 (1985).

[40] M. L. Cleary, S. D. Smith, and J. Sklar, *Cell (Cambridge, Mass.)* **47,** 19 (1986).

abnormal expression of the candidate gene in malignant cells containing the translocation.

Use of Rodent–Human Somatic Cell Hybrids

In the past, rodent–human hybrid cell lines have served as a valuable tool to aid in the identification of breakpoints associated with chromosomal rearrangements. The establishment of these cell lines—by fusion of tumor cells from a patient with immortalized rodent cells—provides a means of propagating abnormal chromosomes from human tumors.[41] This technique is quite valuable when a human tumor cell line cannot be established. In the case of reciprocal translocations, rodent–human hybrid cell lines are especially useful for isolating derivative human chromosomes away from their normal counterparts. For example, in the case of the t(11;14)(q13;q32) in B-cell lymphoma, individual hybrid cell lines have been established which (1) contain the human der(11) chromosome in the absence of the der(14) chromosome and the normal human chromosome 11, or (2) contain the human der(14) chromosome in the absence of the der(11) chromosome and the normal human chromosome 14.[42] These tools enable one to determine by Southern analysis or polymerase chain reaction (PCR) whether a particular probe maps centromeric or telomeric to the breakpoint in question. Such mapping strategies are often a first step in positional approaches to cloning a breakpoint. In addition, somatic cell hybrids containing derivative chromosomes can be used to readily determine if a candidate gene is split by a breakpoint. These studies are done by Southern analysis or PCR using probes from the 5′ and 3′ portions of the candidate gene. Although panels of somatic cell hybrids continue to be a valuable resource for mapping genes, the creation of hybrid cell lines for the isolation of derivative chromosomes involved in translocations has been largely superceded by newer techniques in human genome research using yeast artificial chromosomes and fluorescence *in situ* hybridization (see below).

Chromosomal Breakpoints Associated with Gene Fusions

Breakpoints Involving Previously Mapped Genes

In a few fortunate cases, chromosomal translocations giving rise to gene fusions have been shown to involve a previously characterized gene that

[41] C. M. Croce, M. Shander, J. Martinis, L. Cicurel, G. G. D'Ancona, T. W. Dolby, and H. Koprowski, *Proc. Natl. Acad. Sci. U.S.A.* **76,** 3416 (1979).
[42] Y. Akao, Y. Tsujimoto, J. Finan, P. C. Nowell, and C. M. Croce, *Cancer Res.* **50,** 4856 (1990).

had been mapped to one of the chromosome bands implicated in the rearrangement. Examples include the *c-abl* oncogene, located at 9q34, which is involved in the t(9;22)(q34;q11) associated with CML and some cases of ALL[9]; the *E2A* gene, located at 19p13, which is involved in the t(1;19)(q23;p13) in pre-B cell ALL[43,44]; and the retinoic acid receptor-α gene (*RARA*), located at 17q21, which is involved in the t(15;17)(q21;q11-22) associated with acute promyelocytic leukemia.[45]

Any characterized gene which has been mapped to a chromosome band involved in a translocation is usually considered a candidate gene for participation in the chromosomal rearrangement. At the beginning of each breakpoint cloning project, each candidate gene is tested for rearrangement by Southern analysis in tumor cells or cell lines carrying the translocation. These studies are done using pulsed-field and standard gel electrophoresis. Even if all initial candidate genes are negative, new candidate genes usually become available during the course of a positional cloning effort (see below). In some cases, testing the new candidate genes has been successful. Examples include the involvement of the platelet-derived growth factor receptor β gene in a case of chronic myelomonocytic leukemia with a t(5;12)(q33;p13)[46] and the involvement of the PAX3 gene in the pediatric solid tumor alveolar rhabdomyosarcoma with a t(2;13)(q35;q14).[47]

Positional Cloning through the Utilization of FISH with YACs and/or Cosmids

Fluorescence *in situ* hybridization using yeast artificial chromosomes or cosmids is a powerful tool for the identification of chromosome breakpoints associated with recurring chromosomal translocations.[48] This strategy involves the use of FISH to identify a YAC or cosmid (isolated from a human genomic library) which is split between the two derivative chromosomes of a reciprocal translocation (Fig. 1). A similar strategy can be used for recurring chromosome inversions. This approach exploits the unique fea-

[43] J. Nourse, J. D. Mellentin, N. Galili, J. Wilkinson, E. Stanbridge, S. D. Smith, and M. L. Cleary, *Cell (Cambridge, Mass.)* **60,** 535 (1990).

[44] M. P. Kamps, C. Murre, X.-H. Sun, and D. Baltimore, *Cell (Cambridge, Mass.)* **60,** 547 (1990).

[45] H. de-Thé, C. Chomienne, M. Lanotte, L. Degos, and A. Dejean, *Nature (London)* **347,** 558 (1990).

[46] T. R. Golub, G. F. Barker, M. Lovett, and D. G. Gilliland, *Cell (Cambridge, Mass.)* **77,** 307 (1994).

[47] F. G. Barr, N. Galili, J. Holick, J. A. Biegel, G. Rovera, and B. S. Emanuel, *Nat. Genet.* **3,** 113 (1993).

[48] M. Le Beau, *Blood* **81,** 1979 (1993).

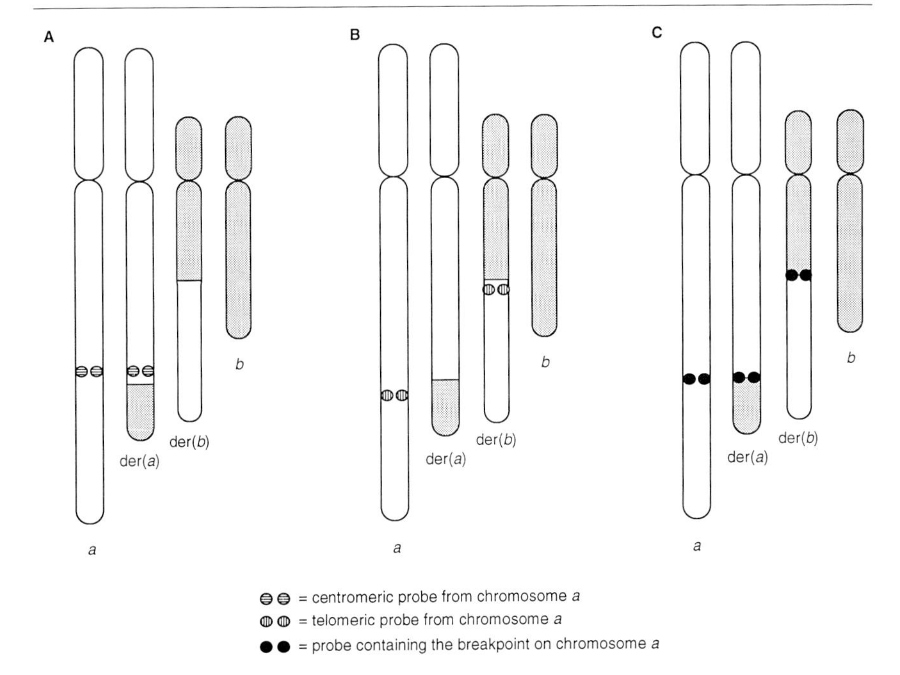

FIG. 1. Example of FISH analysis of metaphase chromosomes from a cell line or fresh tumor sample carrying a hypothetical t(*a*;*b*) reciprocal translocation. Probes are derived from chromosome *a*. (A) Pattern using probe centromeric to breakpoint, (B) pattern using probe telomeric to breakpoint, and (C) pattern using probe crossing the breakpoint on chromosome *a*.

ture of chromosome rearrangements—the physical separation of chromosomal regions which are normally contiguous.

A positional cloning strategy is required to find a YAC or cosmid which spans a chromosomal breakpoint.[49] The complexity of the positional strategy is, in part, dependent on the number and density of marker loci which have been previously mapped to the chromosome bands of interest. In the case of the t(4;11)(q21;q23) in acute leukemia, the 11q23 breakpoint was identified by FISH using a YAC containing a marker locus (the *CD3G* gene) which had previously been mapped to 11q23.[50] The 11q23 breakpoint was found to involve a new gene (called *ALL-1, MLL,* or *HRX*) which is

[49] J. M. Lu-Kuo, D. Le Paslier, J. Weissenbach, I. Chumakov, D. Cohen, and D. C. Ward, *Hum. Mol. Genet.* **3,** 99 (1994).

[50] J. D. Rowley, M. O. Diaz, R. Espinosa, III, Y. D. Patel, E. van Melle, S. Ziemin, P. Taillon-Miller, P. Lichter, G. A. Evans, J. H. Kersey, D. C. Ward, P. H. Domer, and M. M. Le Beau, *Proc. Natl. Acad. Sci. U.S.A.* **87,** 9358 (1990).

located about 100 kb telomeric to the *CD3G* gene.[51–53] This example represents a fortunate circumstance in which a mapped locus was relatively close to a translocation breakpoint.

In the case of the t(2;5)(p23;q35) in anaplastic large cell lymphoma, the 5q35 breakpoint was identified by a positional strategy based on dual-color FISH ordering of cosmid clones relative to the breakpoint.[54] In this example, cosmids were isolated from a human chromosome 5-specific cosmid library using single-copy probes derived from microdissection of the chromosome 5q35 region. After ordering groups of cosmids located centromeric and telomeric to the 5q35 breakpoint, the breakpoint was ultimately identified by bidirectional chromosome walks from two cosmids that closely flanked the breakpoint region. The complexity of this strategy is in part due to the paucity of loci which had been mapped to the 5q35 region at the time the project was undertaken.

After a YAC or cosmid which contains a chromosome breakpoint has been identified, the next step involves the molecular cloning of the breakpoint and the identification of the genes which have been fused by the chromosomal rearrangement. If a YAC serves as the starting point for cloning the breakpoint, the YAC is usually broken down into smaller pieces by the construction of cosmid or bacteriophage λ clones from YAC DNA.[51–53] Single-copy probes are then derived from these cosmid or λ clones. Using a sufficient number of these single-copy probes, it is usually possible to identify a restriction fragment in normal cells that is rearranged in cells containing the translocation. This type of mapping is initially done with rare-cutting restriction enzymes using pulsed-field gel electrophoresis. By continuing the mapping on a finer scale, the region containing the breakpoint may be narrowed down to a small restriction fragment. If a cosmid serves as the starting point for cloning the breakpoint, this process is simplified considerably. In principle, it is possible to test YAC-derived cosmid clones by FISH in order to identify a cosmid which crosses the chromosomal breakpoint.

In order to identify the genes which are fused by the chromosomal rearrangement, single-copy probes derived from the cosmid or phage clones which cross the breakpoint are used in Northern analysis of mRNA from

[51] Y. Gu, T. Nakamura, H. Alder, R. Prasad, O. Canaani, G. Cimino, C. M. Croce, and E. Canaani, *Cell (Cambridge, Mass.)* **71,** 701 (1992).

[52] D. C. Tkachuk, S. Kohler, and M. L. Cleary, *Cell (Cambridge, Mass.)* **71,** 691 (1992).

[53] S. Zieman-van der Poel, N. R. McCabe, H. J. Gill, R. Espinosa, III, Y. Patel, A. Harden, P. Rubinelli, S. D. Smith, M. M. Le Beau, J. D. Rowley, and M. O. Diaz, *Proc. Natl. Acad. Sci. U.S.A.* **88,** 10735 (1991).

[54] S. W. Morris, M. N. Kirstein, M. B. Valentine, K. G. Dittmer, D. N. Shapiro, D. L. Saltman, and A. T. Look, *Science* **263,** 1281 (1994).

panels of normal cells and from cells carrying the translocation.[54] If a probe is found which identifies in normal cells a transcript whose size is consistently altered in cells carrying the translocation, then the probe can be used to clone the two genes which are fused by the translocation. First, the probe is used to screen a cDNA library made from normal cells in order to clone the normal counterpart of one gene involved in the fusion. Next, the probe is used to screen a cDNA library made from cells carrying the translocation in order to clone the fusion cDNA. A new probe, corresponding to the second gene involved in the fusion, is then derived from the fusion cDNA. Finally, this second probe is used to screen a cDNA library made from normal cells in order to clone the normal counterpart of the second gene involved in the fusion. Prior to library screening, it is important to confirm that the second probe maps to the correct region on the partner chromosome involved in the translocation.

General Flow Scheme for Chromosome Breakpoint Identification

Using the approaches presented in this chapter, it is possible to develop a general flow scheme for the molecular cloning of chromosome breakpoints (Fig. 2). Two branching pathways, corresponding to the genetic mechanisms of gene activation and gene fusion, form the initial entrance into the flow scheme. For hematologic malignancies with cytogenetic evidence of Ig or TCR locus involvement, probes from these loci are used to determine whether the breakpoint involves a segment of an Ig or TCR gene.[21,23,31,55] For solid tumors and hematologic malignancies without evidence of Ig or TCR locus involvement, the first step involves the compilation of a list of loci which have been mapped to the chromosome bands involved in the rearrangement. Available probes from these loci consist of PCR and non-PCR-based DNA tools, including oligonucleotide primers, recombinant plasmids, cDNAs, bacteriophage λ clones, cosmids, and YACs.[55,56] Single-copy probes, such as a cDNA from a characterized gene, can be tested directly for rearrangement by the translocation using Southern analysis. Cosmids and YACs can be tested by FISH to determine whether they are split by the breakpoint.

If none of the available probes from marker loci directly demonstrates molecular rearrangement in a cell line or tumor sample containing the chromosomal rearrangement, a positional cloning strategy is undertaken.

[55] A. J. Cuticchia, P. L. Pearson, and H. P. Klinger, eds., "Chromosome Coordinating Meeting (1992)," Vol. 1. Karger, Basel, 1993.

[56] Human Genome Data Base, Laboratory for Applied Research in Academic Information, William H. Welch Medical Library, 1830 E. Monument Street, Baltimore, Maryland 21205-2100.

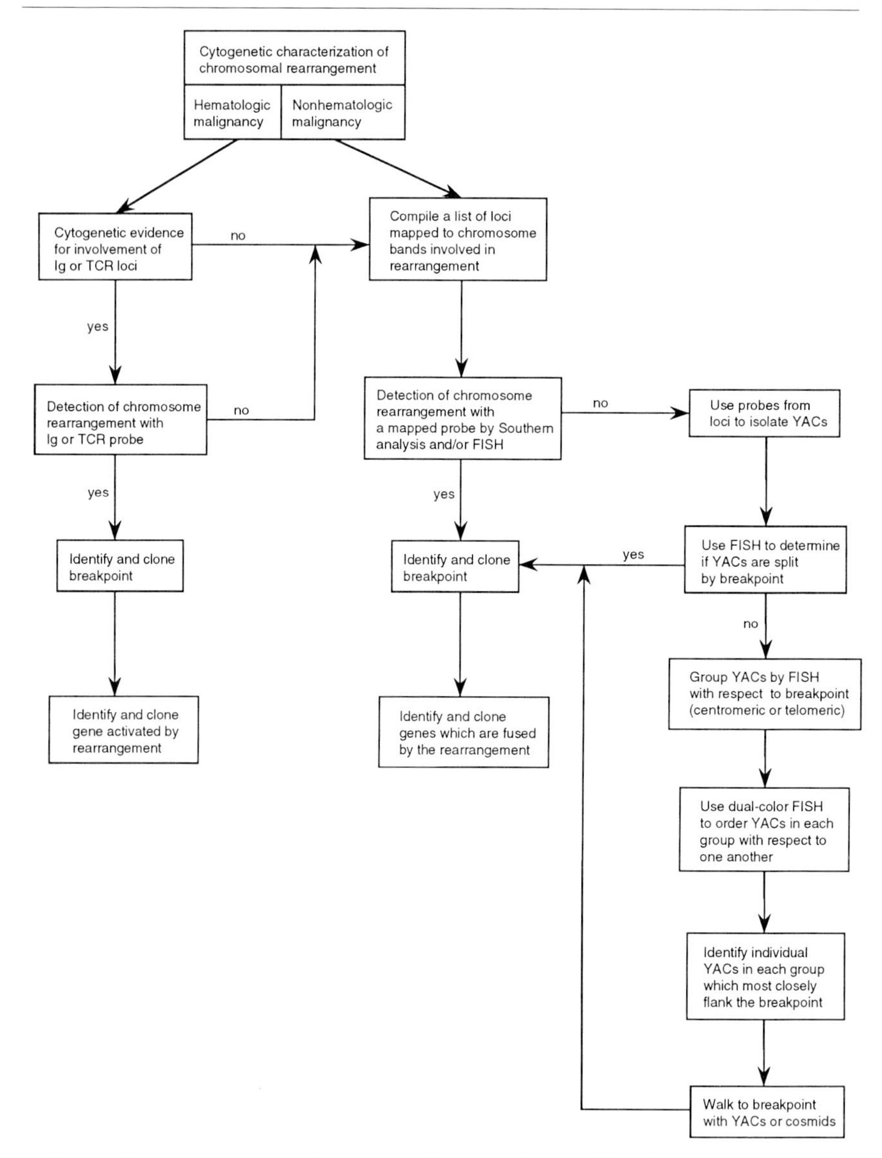

FIG. 2. Flow chart representing a general scheme for the identification and molecular cloning of breakpoints associated with chromosomal rearrangements.

Probes from marker loci are used to isolate YACs from a human YAC library, such as the CEPH library.[49] Screening of YAC libraries is most easily done with PCR primers utilizing commercially available pools of YACs.[57,58] Once a set of nonchimeric YACs corresponding to the marker loci have been isolated, FISH with each YAC is performed on cells which contain the translocation. This analysis will determine the relative position of each YAC to the breakpoint (i.e., centromeric, telomeric, or crossing). If one fails to find a YAC which crosses the breakpoint, the next step involves the ordering of YACs relative to one another using dual-color FISH. Pairwise ordering of YACs in the centromeric and telomeric groups will determine which YAC in each group is closest to the breakpoint. These two YACs, which most closely flank the breakpoint, serve as the starting points for a bidirectional walk with YACs or cosmids to identify a clone which crosses the breakpoint.[49,54]

Conclusions

The cloning of chromosomal breakpoints involved in recurring chromosomal rearrangements has led to the discovery of many new genes involved in hematologic malignancy and solid tumors. Although a number of breakpoints have been cloned by association with Ig or TCR genes or by association with previously identified genes, the general approach to the molecular identification of a chromosomal breakpoint is an exercise in positional cloning. Technical advances involving human YAC libraries and FISH have provided tools suitable for the identification of any chromosomal breakpoint, but the process of positional cloning remains slow and tedious at this time. In the future, progress in human genome research (including the physical mapping of chromosomes and the large-scale identification of human genes) will greatly simplify the process of breakpoint cloning and the identification of genes which are activated or fused by chromosomal rearrangements. When YAC contigs are successfully constructed for human chromosomes, the initial identification of a breakpoint will simply consist of testing commercially available YACs, which have been mapped to the chromosome region of interest, in order to find a YAC which is split by the breakpoint using FISH. Automated procedures for FISH using pools of YACs will greatly speed this process. Using these approaches with sophis-

[57] H. M. Albertson, H. Abderrahim, H. M. Cann, J. Dausset, D. Le Paslier, and D. Cohen, *Proc. Natl. Acad. Sci. U.S.A.* **87,** 4256 (1990).

[58] C. Bellanne-Chantelot, B. Lacroix, P. Ougen, A. Billault, S. Beaufils, S. Bertrand, I. Georges, F. Gilbert, I. Gros, G. Lucotte, L. Susini, J.-J. Codani, P. Gesnouin, S. Pook, G. Vaysseix, J. Lu-Kuo, T. Ried, D. Ward, I. Chumakov, D. Le Paslier, E. Barillot, and D. Cohen, *Cell (Cambridge, Mass.)* **70,** 1059 (1992).

ticated genome tools, it should be possible in the near future to identify and clone any chromosomal breakpoint in a few weeks time.

Acknowledgments

We thank Florencia Bullrich for critical reading of the manuscript and for graphic design of Fig. 1. This work was supported by the Falk Medical Research Trust and grants from the National Cancer Institute.

[22] Detection of Chromosomal Aberrations by Means of Molecular Cytogenetics: Painting of Chromosomes and Chromosomal Subregions and Comparative Genomic Hybridization

By PETER LICHTER, MARTIN BENTZ, and STEFAN JOOS

Introduction

In situ hybridization is a powerful tool in oncology allowing one to analyze the constitution of genomes in a very direct manner. Using preparations of selected nucleic acid sequences as hybridization probes in cellular preparations, this method provides a link between the fields of molecular genetics and classical cytogenetics. Based on the early work by Gall and Pardue in 1969,[1] routine isotopic *in situ* hybridization protocols were established, and in 1981 the feasibility of localizing single-copy sequences, cloned from individual genes, by isotopic *in situ* hybridization was demonstrated for the first time.[2–4] In the 1980s, protocols for nonisotopic *in situ* hybridization techniques were developed, and among these fluorescence *in situ* hybridization (FISH) became the most popular method.[5–17] There are a number of

[1] J. G. Gall and M. L. Pardue, *Proc. Natl. Acad. Sci. U.S.A.* **63,** 378 (1969).

[2] D. S. Gerhard, E. S. Kawasaki, F. C. Bancroft, and P. Szabo, *Proc. Natl. Acad. Sci. U.S.A.* **78,** 3755 (1981).

[3] M. E. Harper, A. Ullrich, and G. F. Saunders, *Proc. Natl. Acad. Sci. U.S.A.* **78,** 4458 (1981).

[4] S. Malcolm, P. Barton, C. Murphy, and M. A. Ferguson-Smith, *Ann. Hum. Genet.* **45,** 135 (1981).

[5] L. Manuelidis, P. R. Langer-Safer, and D. C. Ward, *J. Cell Biol.* **95,** 619 (1982).

[6] J. E. Landegent, N. Jansen in de Wal, R. A. Baan, J. H. J. Hoeijmakers, and M. van der Ploeg, *Exp. Cell Res.* **153,** 61 (1984).

[7] D. G. Albertson, *EMBO J.* **4,** 2493 (1985).

[8] J. B. Lawrence and R. H. Singer, *Nucleic Acids Res.* **13,** 1777 (1985).

[9] L. Manuelidis, *Focus* (*Bethesda, MD*) **7,** 4 (1985).

reasons for this, notably: (i) an increase in speed and (ii) an improved signal resolution compared to conventional isotopic *in situ* hybridization procedures using tritium-labeled probes; furthermore, (iii) the development of *in situ* suppression hybridization protocols[18–20] allowing to use virtually any genomic DNA as a probe greatly enhances the potential of the method in many applications, such as gene mapping, clinical diagnostics, and cell biology; and most of all (iv) nonisotopic *in situ* hybridization provides a means for simultaneous detection of multiple target regions in a single specimen. This can be performed best when different probes labeled with different techniques are combined during the hybridization and are simultaneously detected by different fluorochromes in multiple colors. The option for multicolor detection is one of the main reasons why fluorescence *in situ* hybridization has been preferred by many investigators. However, protocols have also been developed that allow multicolor detection by enzymatic procedures resulting in permanent signals suitable for the long-term storage of specimen as is often required in the pathology laboratory.[21] The potential of almost all applications of *in situ* hybridization is greatly enhanced by the multicolor detection of simultaneously hybridized probes. It is particularly useful when structural chromosome aberrations involving different chromosomal regions are to be diagnosed or when several numerical aberrations should be detected in parallel. Three fluorochrome colors can be easily distinguished, and by combinations of three different labeling/detection procedures, i.e., by labeling single probes with one, two, or three colors, an even higher number of targets becomes distinguish-

---

[10] A. H. N. Hopman, J. Wiegant, G. I. Tesser, and P. van Duijn, *Nucleic Acids Res.* **14,** 6471 (1986).

[11] D. Pinkel, T. Straume, and J. W. Gray, *Proc. Natl. Acad. Sci. U.S.A.* **83,** 2934 (1986).

[12] A. K. Raap, J. G. J. Marijnen, J. Vrolijk, and M. van der Ploeg, *Cytometry* **7,** 235 (1986).

[13] P. F. Ambros and H. I. Karlic, *Chromosoma* **77,** 251 (1987).

[14] J. A. Garson, J. A. van den Berghe, and J. T. Kemshead, *Nucleic Acids Res.* **15,** 4761 (1987).

[15] B. Bhatt, J. Burns, D. Flannery, and J. O. McGee, *Nucleic Acids Res.* **16,** 3951 (1988).

[16] J. B. Lawrence, C. A. Villnave, and R. H. Singer, *Cell (Cambridge, Mass.)* **52,** 51 (1988).

[17] P. Lichter, T. Cremer, C. C. Tang, P. C. Watkins, L. Manuelidis, and D. C. Ward, *Proc. Natl. Acad. Sci. U.S.A.* **85,** 9664 (1988).

[18] J. E. Landegent, N. Jansen in de Wal, R. W. Dirks, F. Baas, and M. van der Ploeg, *Hum. Genet.* **77,** 366 (1987).

[19] P. Lichter, T. Cremer, J. Borden, L. Manuelidis, and D. C. Ward, *Hum. Genet.* **80,** 224 (1988).

[20] D. Pinkel, J. Landegent, C. Collins, J. Fuscoe, R. Segraves, J. Lucas, and J. W. Gray, *Proc. Natl. Acad. Sci. U.S.A.* **85,** 9138 (1988).

[21] E. J. Speel, M. Kamps, J. Bonnet, F. C. Ramaekers, and A. H. Hopman, *Histochemistry* **100,** 357 (1993).

able.[22–25] Alternatively, hybridized target regions can be discriminated on the basis of different ratios of two combined fluorochromes.[23,25–27]

Nonisotopic *in situ* hybridization using specific DNA probes allows the analysis of numerical and structural chromosome aberrations not only in metaphase, but also directly in the interphase nucleus,[28,29] an approach which has been termed interphase cytogenetics.[29] This is of great significance in cases where metaphase chromosome spreads cannot be prepared in sufficient quality or quantity, or not at all as in noncycling cells. Thus, in principle, it overcomes a number of handicaps known to restrict the application of conventional cytogenetics in the analysis of tumors.

Analysis of numerical chromosome aberrations is in most cases performed by probes recognizing tandemly repeated, chromosome-specific DNA elements, such as alphoid DNA sequences, which are present in the heterochromatin of most human chromosomes. Since the target region is considerably large, expanding up to many hundred kilobases in length, it is easy to detect after *in situ* hybridization with a single repeated element or even part of it as a probe. Typically, such probes are cloned in plasmid vectors; however, the repeat sequences can also be amplified from genomic DNA sources via primers targeting chromosome-specific subsequences. [30–32] The repeats often exhibit some degree of homology to clustered repeats on other chromosomes, and therefore in many cases the hybridization stringency must be increased as compared to single-copy probes. For some human chromosomes, e.g., chromosomes 13 and 21, no chromosome-specific repeat sequence has yet been found. In these cases, complex probes such as genomic fragments (without clustered repeats)

[22] P. M. Nederlof, S. van der Flier, J. Wiegant, A. K. Raap, H. J. Tanke, J. S. Ploem, and M. van der Ploeg, *Cytometry* **11,** 126 (1990).

[23] J. G. Dauwerse, J. Wiegant, A. K. Raap, M. H. Breuning, and G. J. B. van Ommen, *Hum. Mol. Genet.* **1,** 593 (1992).

[24] T. Ried, A. Baldini, T. C. Rand, and D. C. Ward, *Proc. Natl. Acad. Sci. U.S.A.* **89,** 1388 (1992).

[25] J. Wiegant, C. C. Wiesmeijer, J. M. N. Hoovers, E. Schuuring, A. d'Azzo, J. Vrolijk, H. J. Tanke, and A. K. Raap, *Cytogenet. Cell Genet.* **63,** 73 (1993).

[26] P. M. Nederlof, S. van der Flier, J. Vrolijk, H. J. Tanke, and A. K. Raap, *Cytometry* **13,** 839 (1992).

[27] S. Du Manoir, M. R. Speicher, S. Joos, E. Schröck, S. Popp, H. Döhner, G. Kovacs, M. Robert-Nicoud, P. Lichter, and T. Cremer, *Hum. Genet.* **90,** 590 (1993).

[28] A. M. Joseph, J. R. Gosden, and A. C. Chandley, *Hum. Genet.* **66,** 234 (1984).

[29] T. Cremer, J. Landegent, A Brückner, H. P. Scholl, M. Schardin, H. D. Hager, P. Devilee, P. Pearson, and M. van der Ploeg, *Hum. Genet.* **74,** 346 (1986).

[30] J. E. Koch, S. Kølvraa, K. B. Petersen, N. Gregersen, and L. Bolund, *Chromosoma* **98,** 259 (1989).

[31] H.-U. G. Weier, R. Segraves, D. Pinkel, and J. W. Gray, *J. Histochem. Cytochem.* **38,** 421 (1990).

[32] I. Dunham, C. Lengauer, T. Cremer, and T. Featherstone, *Hum. Genet.* **88,** 457 (1991).

cloned in cosmids, P1, yeast artificial chromosomes (YACs), or similar vectors are applied (see below). In contrast to the repeat probes, these genomic fragments can be selected from a large set of possible probes. Therefore the probe can be optimized in such a way that a standardized hybridization stringency should be applied. This is of particular advantage when several probes have to be combined for a multicolor *in situ* hybridization experiment.

For the detection of structural chromosome aberrations, clustered repeat sequences are of very limited value. Current protocols that allow to delineate virtually any chromosomal region rely on the use of cloned genomic DNA fragments. The vast majority of cloned genomic fragments contain, in addition to unique sequences, interspersed repetitive sequences (IRS), such as small interspersed repetitive elements, e.g., Alu elements, and large interspersed repetitive elements, e.g., L1 elements. Because of their ubiquitous presence throughout the genome, the IRS generate hybridization signals distributed over the whole chromosome complement. Since the most abundant repetitive DNA is the Alu repeat family, which is preferentially located within R bands,[33] usually a R banding-like staining pattern results. In order to suppress the portion of the signal caused by IRS within a given probe, the labeled probe fragments are denatured together with an excess of unlabeled competitor DNA. Subsequent preannealing results in the formation of double-stranded IRS elements in the probe that can no longer hybridize to their chromosomal DNA targets. Furthermore, the remaining unlabeled single-stranded IRS elements will compete for the hybridization to chromosomal IRS targets during the following *in situ* hybridization. In this way, a highly specific signal is produced by the target site-specific single-copy or low copy repeat sequences.[18–20]

In order to stain entire chromosomes, the DNA fragments from sorted human chromosomes are used as a pooled probe,[17,19,20,34] a procedure also termed "chromosome painting.[20] The libraries commonly used are either cloned in λ phage[35] or pBS plasmid vectors.[36,37] The latter are generally easier to use since the portion of vector sequences is smaller, resulting in

[33] L. Manuelidis and D. C. Ward, *Chromosoma* **91,** 28 (1984).

[34] T. Cremer, P. Lichter, J. Borden, D. C. Ward, and L. Manuelidis, *Hum. Genet.* **80,** 235 (1988).

[35] M. A. van Dilla, L. L. Deaven, K. L. Albright, N. A. Allen, M. R. Aubuchon, M. F. Bartholdi, N. C. Brown, E. W. Campbell, A. V. Carrano, L. M. Clark, L. S. Cram, B. D. Crawford, J. C. Fuscoe, J. W. Gray, C. E. Hildebrand, P. J. Jackson, J. H. Jett, J. L. Longmire, C. R. Lozes, M. L. Luedemann, J. C. Martin, J. S. McNinch, L. J. Meincke, M. L. Mendelsohn, J. Meyne, R. K. Moyzis, A. C. Munk, J. Perlman, R. C. Peters, A J. Silva, and B. J. Trask, *Bio Technology* **4,** 537 (1986).

[36] J. C. Fuscoe, C. C. Collins, D. Pinkel, and J. W. Gray, *Genomics* **5,** 100 (1989).

[37] C. Collins, W. L. Kuo, R. Segraves, J. Fuscoe, D. Pinkel, and J. Gray, *Genomics* **11,** 997 (1991).

less background problems arising from high amounts of labeled vector. Painting probes generated by polymerase chain reaction (PCR) amplification of flow-sorted chromosomes provide another source of *in situ* hybridization probes with little inherent background.[38–42]

Although chromosome painting probes can be very useful for resolving certain structural chromosome aberrations such as gross rearrangements,[17,20,34,43] they are not suitable for answering many other questions in molecular cytogenetic diagnostics since they (i) do not allow the detection of small deletions, (ii) they are not of sufficient complexity to delineate small rearranged chromosome pieces, and (iii) they produce interphase signals that are very difficult or even impossible to assess quantitatively. Therefore, regional probes generating strong signals in metaphase and in interphase cells are being employed allowing one to detect certain chromosomal breaks with high specificity.

Probe sets for the specific staining of subchromosomal regions can be generated in several ways. These include the pooling of single clones from a particular region[17] and regional-specific DNA libraries established from microdissected chromosomal material[44] as well as genomic DNA of somatic cell hybrids containing, e.g., pieces of human chromosomes.[45,46] Alternatively, species-specific PCR amplification products from such hybrids constitute probe sets of high quality.[47,48] When using single probes for the diagnosis of structural aberrations, the detection efficiency needs to be considered. In general, the latter correlates well with the size of the targeted DNA. Cosmid probes containing 25–40 kb of a cloned insert DNA often delineate

[38] N. P. Carter, M. Ferguson-Smith, M. T. Perryman, H. Telenius, A. H. Pelmear, M. A. Leversha, M. T. Glancy, S. L. Wood, K. Cook, and H. M. Dyson, *J. Med. Genet.* **29,** 299 (1992).

[39] K. S. Chang, R. C. Vyas, L. L. Deaven, J.M. Trujillo, S. A. Stass, and W. N. Hittelman, *Genomics* **12,** 307 (1992).

[40] R. F. Suijkerbuijk, D. Matthopoulos, L. Kearney, S. Monard, S. Dhut, F. Cotter, J. Herbergs, A. G. van Kessel, and B. D. Young, *Genomics* **13,** 355 (1992).

[41] H. Telenius, N. P. Carter, C. E. Bebb, M. Nordenskjöld, B. A. J. Ponder, and A. Tunnacliffe, *Genomics* **13,** 718 (1992).

[42] M. Vooijs, L.-C. Yu, D. Tkachuk, D. PInkel, D. Johnson, and J. W. Gray, *Am. J. Hum. Genet.* **52,** 586 (1993).

[43] A. Jauch, C. Daumer, P. Lichter, J. Murken, T. Schroeder-Kurth, and T. Cremer, *Hum. Genet.* **85,** 145 (1990).

[44] C. Lengauer, A. Eckelt, A. Weith, N. Endlich, N. Ponelies, P. Lichter, K. O. Greulich, and T. Cremer, *Cytogenet. Cell Genet.* **56,** 27 (1991).

[45] A. L. Boyle, P. Lichter, and D. C. Ward, *Genomics* **7,** 127 (1990).

[46] T. Kievits, P. Devilee, J. Wiegant, M. C. Wapenaar, C. J. Cornelisse, G. J. B. van Ommen, and P. L. Pearson, *Cytometry* **11,** 105 (1990).

[47] C. Lengauer, H. Riethman, and T. Cremer, *Hum. Genet.* **86,** 1 (1990).

[48] P. Lichter, S. A. Ledbetter, D. H. Ledbetter, and D. C. Ward, *Proc. Natl. Acad. Sci. U.S.A.* **87,** 6634 (1990).

both homolog signals in more than 90% of all metaphase spreads. Larger fragments cloned in P1 or YAC vectors further improve the efficiency, which is particularly important for the diagnosis of chromosomal aberrations in interphase nuclei. The first chromosome aberrations detected by regional probes were trisomies,[49] translocations,[50–52] inversions,[53] and deletions.[51,54,55] The most common strategies to detect chromosomal aberrations in metaphase and interphase cells by FISH using regional probes are illustrated in Fig. 1.

For further information about *in situ* hybridization techniques in the analysis of chromosomes, in particular with respect to diagnostic applications, the reader is refered to review articles.[56–63]

Delineation of Specific Chromosomal Target Regions

Specimen Preparation

A large variety of specimen preparation techniques is compatible with fluorescence *in situ* hybridization. The analysis of constitutional diseases

[49] P. Lichter, A. Jauch, T. Cremer, and D. C. Ward, *in* "Molecular Genetics of Chromosome 21 and Down Syndrome" (D. Patterson, ed.), p. 69. Alan R. Liss, New York, 1990.

[50] E. P. J. Arnoldus, J. Wiegant, I. A. Noordemeer, J. W. Wessels, G. C. Beverstock, G. C. Grosveld, M. van der Ploeg, and A. K. Raap, *Cytogenet. Cell Genet.* **54,** 108 (1990).

[51] J. D. Rowley, M. O. Diaz, R. Espinosa, III, Y. C. Patel, E. van Melle, S. Ziemin, P. Taillon-Miller, P. Lichter, G. A. Evans, J. H. Kersey, D. C. Ward, P. H. Domer, and M. M. Le Beau, *Proc. Natl. Acad. Sci. U.S.A.* **87,** 9358 (1990).

[52] D. Tkachuk, C. Westbrook, M. Andreef, T. Donlon, M. Cleary, K. Suranarayan, M. Homge, A. Redner, J. Gray, and D. Pinkel, *Science* **250,** 559 (1990).

[53] J. G. Dauwerse, T. Kievits, G. C. Beverstock, D. van der Keur, E. Smit, H. W. Wessels, A. Hagemeijer, P. L. Pearson, G.-J. B. van Ommen, and M. H. Breuning, *Cytogenet. Cell Genet.* **53,** 126, (1990).

[54] S. E. Lux, W. T. Tse, J. C. Menninger, K. M. John, P. Harris, O. Shalev, R. R. Chilcote, S. L. Marchesi, P. C. Watkins, V. Bennett, S. McIntosh, F. S. Collins, U. Francke, D. C. Ward, and B. G. Forget, *Nature (London)* **345,** 736 (1990).

[55] T. Ried, V. Mahler, P. Vogt, L. Blonden, G. J. B. van Ommen, T. Cremer, and M. Cremer, *Hum. Genet.* **85,** 581 (1990).

[56] P. Lichter and D. C. Ward, *Nature (London)* **345,** 93 (1990).

[57] A. K. Raap, R. W. Dirks, N. M. Jiwa, P. M. Nederlof, and M. van der Ploeg, *in* "Modern Patholody of AIDS and Other Retroviral Infections" (P. Racz, A. T. Haase, and J. C. Gluckman, eds.) p. 17. Karger, Basel, 1990.

[58] P. Lichter, A. L. Boyle, T. Cremer, and D. C. Ward, *Genet. Anal. Tech. Appl.* **8,** 24 (1991).

[59] D. C. Tkachuk, D. Pinkel, W.-L. Kuo, H.-U. Weier, and J. W. Gray, *Genet. Anal. Tech. Appl.* **8,** 67 (1991).

[60] B. Trask, *Trends Genet.* **7,** 149 (1991).

[61] M. M. Le Beau, *Blood* **81,** 1979 (1993).

[62] M. Bentz, H. Döhner, G. Cabot, and P. Lichter, *Leukemia* **8,** 1447 (1994).

[63] J. Joos, T. M. Fink, A. Rätsch, and P. Lichter, *J. Biotechnol.* **35,** 135 (1994).

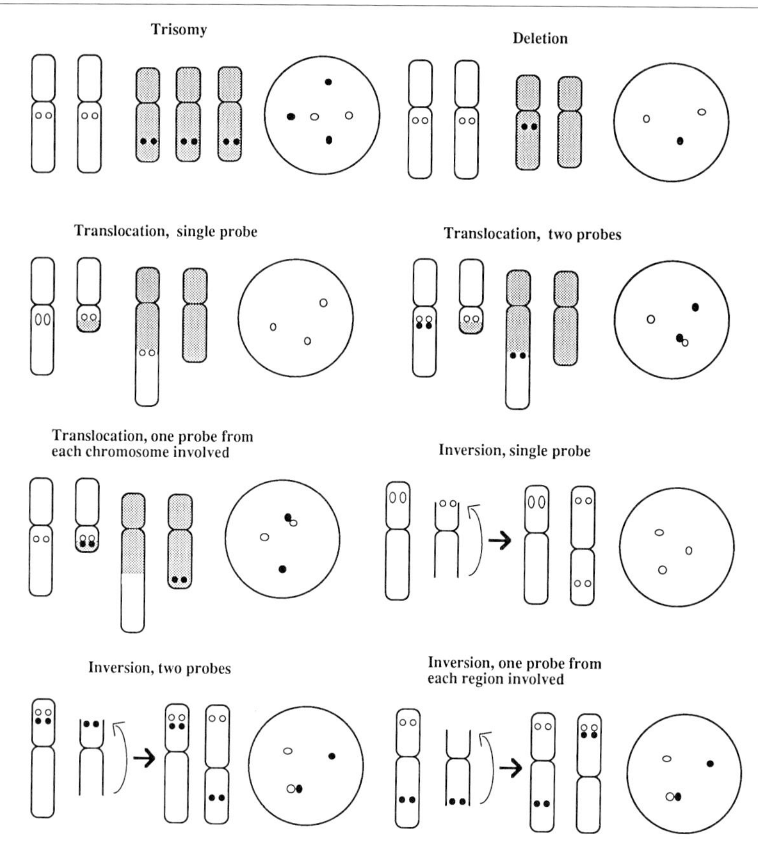

FIG. 1. Schematic illustrations of the detection of numerical and structural chromosomal aberrations on metaphase chromosomes (partial karyotypes on the left of each panel) and in interphase nuclei (on the right of each panel) using focal hybridization signals. The particular aberration as well as the number of hybridization probes used for the diagnosis of that aberration are indicated in the figure. Different chromosomes within one karyotype are distinguished by white and gray-colored drawings. Signals generated by a local probe are shown as a white circle; and for two probe/two color detection the signals are distinguished by white and black circles. Copy number changes of chromosomal material (e.g., trisomy or deletion) can be detected by the number of signals per nucleus generated by a probe from the corresponding region. Generally, a second cohybridization probe from a different chromosomal region is included as an internal control (see top row). Chromosomal breaks occurring in translocations and inversions can be detected by (i) a single probe spanning the breakpoint region, (ii) two probes flanking the breakpoint region (diagnosis by separation of one pair of different signals colocalized in normal cells), or (iii) two probes flanking the regions that are fused by the translocation or inversion event (diagnosis by colocalization of two different signals separated in normal cells) (see rows 2 to 4).

and hematological malignancies is routinely performed on methanol/acetic acid-fixed metaphase spreads and nuclei or on postfixed blood and bone marrow smears. The interphase analysis of solid tumors needs special treatment of the tissue to allow penetration of probe and detection reagents.[64,65] Alternatively, cells can be isolated from tissue, fixed, and hybridized on slides, allowing a more sensitive detection. However, in this case, morphological information regarding the hybridized cell might get lost.[65]

Preparation of Chromosomes and Nuclei from Peripheral Blood Cells. A 0.5- to 1-ml aliquot from heparinized peripheral blood is cultured in 10 ml RPMI 1640 medium in the presence of 10% fetal calf serum and 1.5% (v/v) phytohemagglutinin at 37° (optionally in an atmosphere of 5% CO_2) for 71 hr. Settled cells are resuspended once per day. Colcemid (stock: 10 μg/ml) is added to a final concentration of 0.1 μg/ml, in order to arrest cells in metaphase, and incubated for 10–60 min at 37° (different cell sources have different optima of the incubation time). Cells are pelleted by centrifugation, resuspended in 5 ml of hypotonic solution 0.075 *M* KCl and incubated for 15 min at 37°. After spinning the cells down, removing the supernatant, and resuspending the pellet, freshly prepared ice-cold fixative methanol/acetic acid (3/1) (v/v) is slowly added in drops to a final volume of 5 ml while agitating the tube. During the following procedure the cells are kept at 4° or on ice. Cells are collected by centrifugation, the supernatant is removed, and fresh fixative is added as before. This is repeated 6 to 10 times with an incubation on ice for 30–60 min in the second fixative solution, and a final resuspension in 0.5–1 ml fixative volume. Aliquots (20–50 μl) are dropped from a distance (of 10–50 cm) on microscopic glass slides which were precleaned in ethanol and water, dried, and moistened by breathing on their surface. During the dropping and the subsequent drying, slides should be kept in a moist environment. After air drying, specimens are dehydrated in a series of cold 70, 90, and 100% ethanol for 5 min each followed again by air drying. When used immediately, specimens are baked at 60° for 1 hr; for use within the following 2 weeks, slides are stored at room temperature; and for long-term storage, slides are sealed in bags and stored at −80° for months or years. When thawing the slides again, water condensation should be prevented or dehydration should be repeated. Although cell pellets can be stored in fixative at −20° for longer time periods before dropping, the quality of chromosome spreads prepared weeks or months later is usually inferior.

[64] A. H. N. Hopman, P. J. Poddighe, W. A. G. B. Smeets, O. Moesker, J. L. M. Beck, G. P. Vooijs, and F. C. S. Ramaekers, *Am. J. Pathol.* **135,** 1105 (1989).

[65] A. H. N. Hopman, P. Poddighe, O. Moesker, and F. C. S. Ramaekers, *in* "Diagnostic Molecular Pathology: A Practical Approach" (H. McGee, ed.), Vol. 1, p. 142. IRL Press, Oxford and New York, 1992.

Preparation of Chromosomes and Nuclei from Cell Cultures. In general, the same procedure is applied for cell cultures; however, a mitogen stimulus is not needed. Cell cultures (in log phase) should reach approximately 70% confluency before adding colcemid. The incubation time of colcemid is dependent on the growth rate of the culture and has to be optimized for each cell type/culture (usually >1 hr). The collection of mitotic cells from adherently growing cells is performed either by shaking off or by 10–15 min digestion with a trypsin–EDTA solution [0.25% in phosphate-buffered saline (PBS)] prewarmed to 37°. Trypsin is inhibited by adding culture medium 5–10 times the volume of the trypsin–EDTA solution. Mitotic cells are pelleted by centrifugation.

Pretreatment of Chromosomes and Nuclei by Pepsin Digestion. The penetration of probe and detection reagents can sometimes be improved by proteolytic digestion. However, since overdigestion can also be a problem, e.g., resulting in the loss of nuclei and chromosomes, this treatment should only be done when necessary [see also below for comparative genomic hybridization (CGH)]. Enzyme concentration and incubation time should be optimized for each preparation.

Slides with fixed chromosomes and nuclei are equilibrated for 5 min at 37° in PBS before incubating in 1–2 mg pepsin (porcine stomach mucosa, 500–3000 units/mg) in 100 ml 0.01 *M* HCl for 10 min at 37°. Following a wash in PBS for 5 min at room temperature, post fixation is performed by freshly prepared 0.5% paraformaldehyde (paraformaldehyde dissolved in PBS, 5 m*M* $MgCl_2$ is heated to 80°, the pH is adjusted to 7 using 1 *M* NaOH, and the solution is filtrated) for 5 min at room temperature. After three washes in PBS for 5 min each, slides are dehydrated in a series of 70, 90, and 100% ethanol for 5 min each and air dried.

Pretreatment of Blood Smears. In hematological laboratories, blood smears are routinely treated with Wright's stain and stored at room temperature for months or years. Such specimens can be subjected to molecular cytogenetic analysis.[66,67] Smears are first incubated in 10 m*M* Tris–HCl, pH 7.5, 1 m*M* EDTA for 5 min at 37°. However, when the smears are older than 2 weeks, this incubation is extended overnight. Fixation of blood smears is performed in freshly prepared methanol/acetic acid (19/1) (v/v) for 10 min at room temperature, followed by a second fixation in methanol/acetic acid (3/1) (v/v) for 10 min, and a series of 70, 90, and 100% ethanol, 5 min each, at room temperature before air drying. An optional pepsin digestion could be performed as described earlier.

[66] M. Bentz, M. Schröder, M. Herz, S. Stilgenbauer, P. Lichter, and H. Döhner, *Leukemia* **7,** 752 (1993).

[67] K. van Lom, A. Hagemeijer, E. M. E. Smit, and B. Löwenberg, *Blood* **82,** 884 (1993).

Pretreatment of Bone Marrow Smears. For the preparation of bone marrow smears, a proteolytic digestion step is included. Smears are incubated in 10 m*M* Tris–HCl, pH 7.5, 1 m*M* EDTA for 5 min at room temperature, followed by an incubation in pepsin solution (50–200 μg/ml 0.01 *M* HCl) for 8 min at room temperature. Following three washes in PBS and three washes in water, fixation is performed by incubation in methanol/acetic acid (3/1) (v/v) for 45 sec and in 70% acetic acid for 45 sec. Finally, slides are air dried, dehydrated by a series of 70, 90, and 100% ethanol, and air dried again.

Pretreatment of Paraffin-Embedded Tissue Sections from Solid Tumor Material. In many applications, tissue sections are mounted on glass slides prior to the treatment protocol. Alternatively, the treatment (as well as the hybridization procedure) of the sections can be performed in suspension, followed by the mounting. The adhesion of mounted tissue sections is significantly facilitated by poly-L-lysine coating of the glass slides. For coating, glass slides are incubated for 5 min in a poly-L-lysine solution [1 mg/ml poly-L-lysine (MW 70,000–150,000) in water] followed by five washes in H_2O. Activation of the coated slides is performed in 2.5% glutaraldehyde dissolved in PBS for 1 hr at room temperature followed by washes in PBS and water. Paraffin sections (about 4–6 μm thick) are mounted onto these slides, air dried, and baked at 56° overnight. For the extraction of paraffin, slides are incubated three times in 100% xylene for 10 min and twice in absolute methanol for 5 min each. Slides are then immersed in 1% H_2O_2 (dissolved in absolute methanol) for 30 min, washed twice in 100% methanol for 5 min each, and air dried. Subsequent immersion of the slides in 1 *M* sodium thiocyanate at 80° for 10 min followed by two washes in H_2O (5 min each) facilitates the later penetration of the probe and detection reagents. This step is followed by incubation in a solution of pepsin [from porcine stomach mucosa (2500–3000 units/mg), 4 mg pepsin/ml in 0.2 *M* HCl] for 5 to 15 min at 37°. Following brief washes (five times in water and five times in PBS), sections are dehydrated in a series of 70, 90, and 100% ethanol for 3 min each and air dried.

Pretreatment of Frozen Tissue Sections from Solid Tumor Material. Frozen tissue sections (about 50 μm thick) are mounted on poly-L-lysine-coated slides (see above) and air dried. Fixation in methanol/acetone (1/1) (v/v) is allowed for 20 min at −20°. After washing twice in PBS, 0.5% Tween 20 for 5 min each, slides are incubated for approximately 10 min in 200 μl of 50–400 μg pepsin/ml in 0.01 *M* HCl at 37°, followed by five washes in water and five washes in PBS. For fixation, slides are incubated in 1% paraformaldehyde, freshly dissolved in 0.1 *M* sodium phosphate buffer (pH 7.2–7.4), 5 m*M* $MgCl_2$, for 10 min at room temperature. Following washes as before, specimens are dehydrated in 70, 90, and 100% ethanol for 5 min each before air drying.

TABLE I
TYPICAL NUCLEOTIDE CONCENTRATIONS DURING LABELING REACTION

Labeling by			
Biotin	Digoxigenin (Dig)	Dinitrophenol (DNP)	Fluorochromes[a]
50 μM dATP	50 μM dATP	50 μM dATP	50 μM dATP
50 μM dGTP	50 μM dGTP	50 μM dGTP	50 μM dGTP
50 μM dCTP	50 μM dCTP	50 μM dCTP	50 μM dCTP
40 μM biotin-11-dUTP	12.5 μM Dig-11-dUTP	40 μM DNP-11-dUTP	40 μM fluoro-11-dUTP
10 μM dTTP	37.5 μM dTTP	10 μM dTTP	10 μM dTTP

[a] Nucleotides coupled typically with coumarin, fluorescein, or rhodamine.

Preparation of Single Cell Suspensions from Solid Tumor Material. Pieces of solid tumor tissue (0.5–1 g, if available) are cut mechanically in a cold petri dish as much as possible. This material is suspended in 70% ethanol (−20°) and is further homogenized in a douncer. The suspension of tumor cells should be adjusted to a concentration of approximately 5×10^5 cells/ml; 10 μl of it is applied on poly-L-lysine-coated slides (see above) and air dried for 15 min. Digestion in 200 μl 50–400 μg pepsin/ml in 0.01 *M* HCl is performed for 15 min, followed by five washes in water and five washes in PBS. For fixation, slides are incubated in 1% paraformaldehyde, freshly dissolved in 0.1 *M* sodium phosphate buffer (pH 7.2–7.4), 5 m*M* $McCl_2$, for 10 min at room temperature. After repeated washes as before, specimens are dehydrated in 70, 90, and 100% ethanol for 5 min each followed by air drying.

Probe Labeling by Nick Translation

A nucleic acid probe is labeled with a reporter molecule, which is either a fluorochrome allowing for direct detection of the probe or a hapten and can be detected by means of immunohistochemistry. Nick translation is performed in the presence of reporter-modified nucleotides as outlined in Table I. Direct detection of fluorochromated probes is particularly useful when large regions are targeted, e.g., by using repetitive DNA probes or whole chromosome painting probes.[68] Although the detection sensitivity is somewhat reduced as compared to indirect detection procedures, background problems seem also considerably reduced.

One of the most important requirements for a good *in situ* hybridization probe is the actual size of the probe molecules obtained after the labeling

[68] J. Wiegant, T. Ried, P. Nederlof, M. van der Ploeg, H. J. Tanke, and A. K. Raap, *Nucleic Acids Res.* **19,** 3237 (1991).

reaction. The probe molecules should be between 100 and 500 nucleotides in length which is obtained by adjusting the DNase concentration in a nick translation reaction.

The DNA template (isolation of insert fragments from vector sequences is not necessary) is subjected to a nick translation at a concentration of 20 μg/ml (or less) in 50 mM Tris–HCl, pH 8.0, 5 mM $MgCl_2$, 50 μg/ml BSA, 10 mM 2-mercaptoethanol, 50 μM of each type of nucleotide (see Table I), approximately 100 ng/ml DNase I (the volume of the diluted stock that is to be used in the reaction must be tested for each new batch of DNase I), and 0.2 units/μl *Escherichia coli* DNA polymerase I. After incubation for 2 hr at 15°, the reaction is inhibited by incubation on ice until the actual size of the probe molecules is determined. This is achieved by agarose gel electrophoresis of a 10-μl aliquot of the reaction which was heat denatured prior to loading on the gel. Denaturation is important since aggregates of multiple fragments lead to misinterpretation of the fragment sizes. The probe molecules, visible as a smear after ethidium bromide staining, should be 100–500 nucleotides in length; if the probe is smaller, it is discarded, if it is larger, more DNase I is added. The enzyme is inactivated by heating at 68° for 15 min in the presence of 15 mM EDTA and 0.1% sodium dodecyl sulfate (SDS). Unincorporated nucleotides are separated by gel filtration using a Sephadex G-50 spin column equilibrated in 10 mM Tris, 1 mM EDTA, and 0.1% SDS (the SDS is crucial to avoid sticking of the reporter-modified probe in the column).

Amplification of Human DNA Cloned in YACs by PCR. DNA fragments cloned in YACs are mapped by fluorescence *in situ* hybridization using either total yeast clone DNA or YAC DNA isolated from pulsed-field gels. However, the hybridization efficiency is often less than satisfactory. To overcome these problems, Alu-PCR protocols allowing to generate human-specific DNA probes were optimized.[69] Alu-PCR is based on oligonucleotide primers that anneal specifically to human-specific subsequences within Alu elements. Although Alu-PCR-amplified YAC clones often result in strong fluorescent FISH signals, the quality of a given probe relies on the distribution of Alu sequences within the probe, and the target sequence is not contiguous.

For the PCR experiment, 100 ng of genomic yeast DNA is used, which was either purified from the yeast culture or isolated from agarose plugs prepared for pulsed-field gel electrophoresis [plugs are equilibrated in 1 M Tris–HCl (pH 8.0) and subsequently in 10 mM Tris–HCl, pH 8.4, 50 mM KCl, 1.5 mM $MgCl_2$, 0.001% gelatin at room temperature, then melted for 5 min at 70° before an aliquot is taken for PCR]. PCR is carried out in a

[69] C. Lengauer, E. D. Green, and T. Cremer, *Genomics* **13,** 826 (1992).

100-μl volume containing the template DNA, 100 m*M* Tris–HCl, pH 8.4, 50 m*M* KCl, 1.5 m*M* $MgCl_2$, 0.001% gelatin, 250 m*M* of each dNTP, 250 n*M* Alu-Primer CL1 (5′-TCC CAA AGT GCT GGG ATT ACA G-3′), 250 n*M* Alu-Primer CL2 (5′-CTG CAC TCC AGC CTG GG-3′), and 25 units/ml native *Taq* DNA polymerase. The PCR conditions are as follows: initial denaturation is at 96° for 5 min, followed by 30 cycles of amplification with denaturation at 96° for 1 min, annealing at 37° for 30 sec, and extension at 72° for 6 min, whereas the last extension step is at 72° for 4 min. Aliquots of 10 μl are fractionated by agarose gel electrophoresis in order to analyze the PCR products. Following ethanol precipitation, the products are resuspended in H_2O and subjected to labeling by nick translation.

Filter-Based Assay to Test the Hapten Labeling. Aliquots of 1 μl of different concentrations of test DNA [0, 1, 2, 5, 10, and 20 pg/μl DNA diluted in 0.1 mg/ml sheared salmon sperm DNA, 6 × standard saline citrate (SSC)] are spotted in parallel with 1-μl aliquots of the same concentrations of standard DNA on a small piece of nitrocellulose filter (e.g., 3 × 4 cm). The filter is baked for 30–60 min at 80°, washed in AP 7.5 buffer (containing 0.1 *M* Tris–HCl, pH 7.5, 0.1 *M* NaCl, and 2 m*M* $MgCl_2$) for 1 min at room temperature and sealed in a plastic bag together with 5–10 ml blocking buffer (3% BSA or 5% dry milk in AP 7.5 buffer) to be incubated for 30–60 min at 37°. After the solution is substituted by 1μg/ml streptavidin-conjugated alkaline phosphatase (in case of a biotinylated probe) or sheep anti-digoxigenin antibody-conjugated alkaline phosphatase (in case of a digoxigenin-labeled probe), respectively, in AP 7.5 buffer and incubated for 30 min at 37°, the filter is washed twice in a dish with AP 7.5 buffer for 5 min each, followed by a wash in freshly mixed AP 9.5 buffer (containing 0.1 *M* Tris–HCl, pH 9.5, 0.1 *M* NaCl, and 50 m*M* $MgCl_2$) for 10 min. Then it is sealed with 7.5 ml of freshly prepared substrate buffer (330 μg/ml nitroblue tetrazolium and 165 μg/ml 5-bromo-4-chloro-3-indolyl phosphate in AP 9.5 buffer) and incubated in the dark at room temperature for 15–60 min. After a final wash in 10 m*M* Tris–HCl, pH 7.5, m*M* EDTA, the filter is air dried. The probe should yield signals comparable to the corresponding dilutions of the standard DNA, and the spotted 1 pg of the labeled probe should be visible.

In Situ Hybridization

In Situ Hybridization with the Whole Probe Contributing to the Signal. When all the sequences of a probe should contribute to the hybridization signal, such as chromosome-specific repetitive DNA probes or unique cDNA probes, a standardized protocol is applied. The amount of probe used depends on the type of probe and target sequences. Recommendations are listed in Table II regarding a 10-μl hybridization cocktail volume. Aliquots of 1–4 μl from the labeled probes (20 ng/μl) are combined with 2 μl of sheared salmon sperm genomic DNA (1 μg/μl) as a carrier, lyophilized, and resuspended in 5 μl deionized formamide (since most reporter-labeled nucleic acids dissolve better in formamide than in water) before adding 5 μl of 4 × SSC, 100 m*M* sodium phosphate buffer, pH 7.0, 20% dextran sulfate. When chromosome-specific repetitive DNAs are hybridized, dextran sulfate can be omitted and the stringency of the hybridization is increased by adjusting to a final concentration of 60–70% formamide, 2 × SSC, 50 m*M* sodium phosphate buffer, pH 7.0. Denaturation of the probe is achieved by incubation at 75° for 5 min followed by chilling on ice for approximately 5 min just before application to the denatured specimen (see below).

In Situ Hybridization with Only Part of the Probe Contributing to the Signal. Chromosomal *in situ* suppression (CISS) hybridization performed with probes containing interspersed repetitive DNA sequences requires higher concentrations of nucleic acids. Therefore, the labeled probe DNA (for the recommended amount, depending on the complexity of the probe or target sequences, see Table II) is combined with 3–5 μg human Cot 1 DNA (1 μg/μl, on the average 400–500 bp long) and 5–7 μg sheared salmon sperm DNA (1 μg/μl), and the mixutre is adjusted to 0.15 *M* sodium acetate, pH 5, before precipitation with 2 vol of cold ethanol. Pelleting in the

TABLE II

SUGGESTED AMOUNT OF LABELED PROBE FOR 10 μl HYBRIDIZATION VOLUMES

Satellite or alphoid DNA repeat sequence	2–20 ng
Single fragments cloned in cosmids, plasmids, or phages	20–60 ng
Band-specific probe pool (>100 kb)	100–200 ng
Fragment cloned in YAC, total yeast DNA	1 μg
Fragment cloned in YAC, amplified by Alu-PCR	100–150 ng
Fragment cloned in YAC, gel-purified	50–100 ng
Painting probes	
DNA from sorted human chromosomes cloned in pBS plasmids (vector to insert ratio is 2:1)	
A–C group chromosomes	300–400 ng
D–G group chromosomes	150–250 ng
DNA from sorted human chromosomes PCR amplified	
A–C group chromosomes	150–300 ng
D–G group chromosomes	100–150 ng
Reverse painting probes	
Total genomic DNA from somatic cell hybrids	1 μg
Alu-PCR products from somatic cell hybrids	100–300 ng
Total genomic DNA (e.g., for CGH)	200 ng–1 μg

centrifuge, wash of the pellet by 70% ethanol, recentrifugation, and lyophilization are followed by resuspension in 5 μl deionized formamide before adding 5 μl of 4 × SSC, 100 m*M* sodium phosphate buffer, pH 7.0, 20% dextran sulfate. After denaturation at 75°C for 5 min, preannealing of the repetitive sequences within the cocktail is allowed for 5–30 min before being applied to the denatured specimen (see below).

If the probes are enriched for Alu sequences, such as many Alu-PCR products, higher amounts of Cot 1 DNA must be used. For example, when using Alu-PCR-amlplified YAC clones as probes, ≥15 μg of Cot 1 DNA is added.

Denaturation of Metaphase and Interphase Chromosomes. Glass slides containing specimens are heated to 50–60° in order to prevent dropping of the temperature when the slides are put into the denaturation solution. Slides are transferred to a prewarmed solution of 70% deionized formamide, 2 × SSC, 50 m*M* sodium phosphate buffer, pH 7.0, and incubated at 70° for exactly 2 min. When chromosomes are overdenatured, they usually look fuzzy when counterstained with propidium iodide or 4,6-diamino-2-phenyl-indole (DAPI). However, for interphase analysis, some degree of overdenaturation might be helpful. Bright counterstaining of chromosomes and nuclei is often an indication for insufficient chromosomal denaturation. For tissue sections, denaturation has to be more stringent; for example, nuclei in paraffin-embedded tissue sections, pretreated as described earlier, are denatured at 85° for 4–5 min.[65,70] Following the denaturation step, specimens are dehydrated in a series of ice-cold 70, 90, and 100% ethanol for 3 min each before air drying and application of denatured probe cocktails (see above).

In Situ Hybridization. The 10-μl hybridization mixture with the denatured probe is directly applied to the denatured chromosome preparation. For CISS hybridization, prewarming of the denatured specimen to 37–42° is preferred. A coverslip (18 × 18 mm) is put on the droplet and sealed with rubber cement. Slides are incubated in a wet chamber at 37° usually overnight. However, repetitive probes do not need more than 1 hr of incubation. For comparative genomic hybridization, see below. Simultaneous denaturation of the probe and specimen in an 80° incubator is often performed, but can also yield in a higher variability with respect to the quality of hybridization signals.

Posthybridization Washes. After washing off the coverslip, specimens are agitated in prewarmed 50% formamide, 2 × SSC at 42° three times for 5 min each. Since the next step determines the stringency of the

[70] H. Scherthan and T. Cremer, *in* "Methods in Molecular Genetics, Chromosome and Gene Analysis" (K. W. Adolph, ed.), Vol. 2. p. 223. Academic Press, 1994.

wash, the salt condition can be adjusted to the experimental needs; washing is performed in prewarmed 1–0.1 × SSC at 60° three times for 5 min each.

Probe Detection

Direct Detection. When using directly labeled probes, counterstaining of chromosomes and nuclei is followed by incubation in either 200 ng/ml propidium iodide, 2 × SSC or 50 ng/ml DAPI, 2 × SSC for 15 min at room temperature. After a final wash in 2 × SSC, preparations are mounted in 30 μl of an antifade embedding medium [2.33% (w/v) 1,4-diazobicyclo (2.2.2.) octane (DAPCO), 20 m*M* Tris–HCl, pH 8.0, 90% (v/v) glycerol, or Vectashield from Vector Laboratories], reducing the fading of the fluorochromes during the following microscopic inspection.

Indirect Detection. For most diagnostic applications, detection of hybridized probe is performed using a protocol with one layer of detection reagent. However, other avenues with more layers of detection reagents resulting in stronger signals or signal amplifications are also applied. Multilayer detection protocols impose a higher risk for increased background staining. The avenues used most frequently for fluorescence detection are listed in Table III. The simultaneous detection of differently labeled probes can be performed by incubating detection reagents for different avenues (see Table III) in the same layer. However, the combination of detection reagents must be carefully considered and tested in order to avoid cross-reactions between them.

Following posthybridization washes, blocking is carried out by incubation with 200 μl 3–5% nonfat dry milk, 4 × SSC under a large coverslip (22 × 40 mm) at 37° for 30 min in a moist chamber. The first layer of detection reagent is applied in a 200-μl volume of 4 × SSC, 1% BSA, 0.1% Tween 20. Avidin-conjugated FITC is included with a final concentration of 5 μg/ml (other detection reagents are used in concentrations according to the recommendation of the supplier). Incubation under a coverslip for 20–30 min at 37° in a moist chamber is followed by three washes in 4 × SSC and 0.1% Tween 20 at 42°, 5 min each. More layers of solutions with detection reagents (see Table III) can be added, each layer followed by the same washing procedure. Counterstaining of chromosomes and nuclei is followed by incubation in either 200 ng/ml propidium iodide, 2 × SSC or 50 ng/ml DAPI, 2 × SSC for 15 min at room temperature. The appropriate counterstain, whose spectral range should not overlap with the fluorochromes used for probe detection, can be selected according to Table IV. Final mounting in antifade embedding medium is performed as described earlier.

TABLE III
TYPICAL AVENUES FOR DETECTION BY INDIRECT FLUORESCENCE

Probe	Layer 1	Layer 2	Layer 3
Fluorochrome conjugated (direct)	—	—	—
Fluorochrome conjugated	Anti-fluorochrome antibody	Fluorochrome-conjugated secondary antibody	—
Biotinylated	Avidin/streptavidin-conjugated fluorochrome	—	—
Biotinylated	Avidin/streptavidin-conjugated fluorochrome	Biotinylated anti-avidin antibody	Avidin/streptavidin conjugated fluorochrome
Biotinylated	Fluorochrome-conjugated anti-biotin antibody	—	—
Digoxigenin labeled	Fluorochrome-conjugated anti-digoxigenin antibody	—	—
Digoxigenin labeled	Fluorochrome-conjugated sheep anti-digoxigenin antibody	Fluorochrome-conjugated anti-sheep antibody	—
Digoxigenin labeled	Sheep anti-digoxigenin antibody	Fluorochrome-conjugated anti-sheep antibody	—
Digoxigenin labeled	Mouse anti-digoxigenin antibody	Digoxigenin-labeled anti-mouse antibody	Fluorochrome-conjugated anti-digoxigenin antibody
Dinitrophenol labeled	Rat anti-dinitrophenol antibody	Fluorochrome conjugated anti-rat antibody	—

Microscopic Evaluation

Most diagnostic applications of *in situ* hybridization are designed to yield intense signals that can be easily evaluated with standard equipment. FISH signals in diagnostic settings are usually analyzed using research quality epifluorescence microscopes. For evaluation by visual inspection, it is of advantage to use optical filter sets, which allow simultaneous visualization of the fluorescent dyes used in the experiment, such as multiple bandpass filters. However, for some applications it is necessary to use a

TABLE IV
FLUORESCENT DYES USED IN FISH[a]

Color	Conjugated fluorochromes	Chromosomal counterstain
Blue	Coumarins (e.g., AMCA)	DAPI, Hoechst 33258
Green	FITC, Bodipy, Spectrum Green	Quinacrine
Red	Rhodamines, Texas Red, Spectrum Orange, Cy3	Propidium iodide
Far red	Cy5	YOYO-3

[a] Because of the spectral overlap, dyes listed in the same line are not combined.

filter set that is very selective for a given fluorochrome (see, e.g., below, CGH). To exploit the full potential of multicolor fluorescence *in situ* hybridization in clinical and tumor cytogenetics, quantitative digital fluorescence microscopy is needed. Digitized images are not only easier to handle but can also be subject to powerful image processing. Furthermore, approaches based on quantitative measurement of signal intensities such as CGH (see below) rely on digital image analysis. Of the many camera systems available for acquiring digitized images from fluorescent signals, CCD (charged coupled device) cameras are the most popular ones based on their sensitivity. Configuration and quality of the CCD cameras required for the various applications in molecular cytogenetics are subjects of specialized literature.

Interphase Analysis

Ideal probes for interphase analysis should result in strong focal signals that are easy to evaluate following the schemes illustrated in Fig. 1. Chromosome-specific repeat sequences label heterochromatin which usually stays highly condensed throughout the cell cycle resulting in focal signals. However, in addition to their limited utility in detecting structural aberrations, other restrictions have been described: In some tumor cases the centromeric heterochromatin is highly decondensed, hampering the enumeration of hybridization signals considerably.[71] Furthermore, polymorphism in the copy number of these repeats or somatic pairing of centromeric regions, as has been described for certain cell types,[72] could result in misinterpretation of FISH signals. Therefore, other types of probes might be selected

[71] H. Döhner, S. Pohl, M. Bulgay-Mörschel, S. Stilgenbauer, M. Bentz, and P. Lichter, *Leukemia* **7,** 516 (1993).

[72] E. P. J. Arnoldus, A. C. B Peters, G. T. A. M. Bots, A. K. Raap, and M. van der Ploeg, *Hum. Genet.* **83,** 231 (1989).

for an interphase analysis. However, complex probes, as for example fragments of several hundred kilobases cloned in a YAC, detect cellular target regions which could be decondensed due to chromatin packaging phenomenons. Therefore, prior to a diagnostic study the utility of such probes must be tested for the cell type of interest.

Nucleic acid probes used in interphase cytogenetics should yield in high efficiencies (>90%) with respect to the delineation of the targeted chromosome region. This is particularly important when copy number changes, such as deletions, are to be assessed (see Fig. 2A). Furthermore,

---

FIG. 2. Examples of different approaches to detect chromosomal aberrations by fluorescence *in situ* hybridization. (A) Detection of a structural aberration in interphase nuclei. In this example, a probe delineating the RB-1 (retinoblastoma susceptibility) gene (green signals) is used to identify possible deletions in nuclei from patients with B-cell chronic lymphocytic leukemia.[74] Slides are only evaluated when the control probe (red signals) reveals two signals in >90% of the nuclei. Note that there are clearly two cell populations present containing one and two RB-1 copies, respectively. For the assessment of the proportion of cells with one signal that allows the diagnosis of a deletion, see text. (B) Visualization of chromosomal imbalances in a glioblastoma by reverse painting.[78] Gray level image of a metaphase spread of normal human chromosomes after hybridization with tumor DNA. The strong signals allow to identify a coamplification of sequences in two chromosomal regions (7p13 and 12q13–15). Furthermore, a weak painting of the chromosomes X and 10 indicates an underrepresentation of these chromosomes. The relevant chromosomes are indicated by numbers. (C–F) Comparative genomic hybridization of a prostate carcinoma case. (C) Image of a normal metaphase spread hybridized with genomic DNA from prostate tumor cells detected via FITC (green). (D) Hybridization pattern of the cohybridized, differentially labeled control DNA detected via rhodamine (red). Note the differences in the staining intensity along the chromosomes within C as well as when comparing the same regions in C and D. (E) Mixed-color image obtained by electronical overlaying of the images shown in C and D; overrepresentation of DNA sequences in the tumor cells is indicated by a more yellowish/greenish color, whereas underrepresentation is visible by a more reddish color of the corresponding region. (F) Copy number karyotype of the same metaphase spread shown in C–E. Illustration of chromosomal gains and losses defined by ratio values using a three-color look-up table (one homolog of chromosome 17 is not fully presented). Green, red, and white colors indicate gain, loss, and balanced state of chromosomal material, respectively. Note that the arrangement of the chromosomes in this karyogram reveals a fairly high consistency with respect to the ratio values on both homologs of a given chromosome. Images were acquired using a cooled black and white CCD camera; A, C–F are pseudocolored; A, E, and F represent electronically overlayed images. (G) Intensity ratio values of hybridization signals from tumor and control DNA[79] for all C-group chromosomes (indicated by numbers) from the prostate case shown in C–F. This profile results from averaging the ratio profiles of 10 metaphase spreads including the spread shown above. The lines next to each chromosome ideogram define a central value (center line) indicating a balanced state of the chromosomal region, as well as diagnostic thresholds for gains (right line) and losses (left line) occurring in at least 50% of the cells from which test DNA is prepared. Details are described elsewhere.[79] The gray-shaded area indicates the heterochromatic region 9q12 which is highly polymorphic, and therefore the corresponding ratio values are not considered in the analysis. Compare the profiles to the copy number karyotype in F.

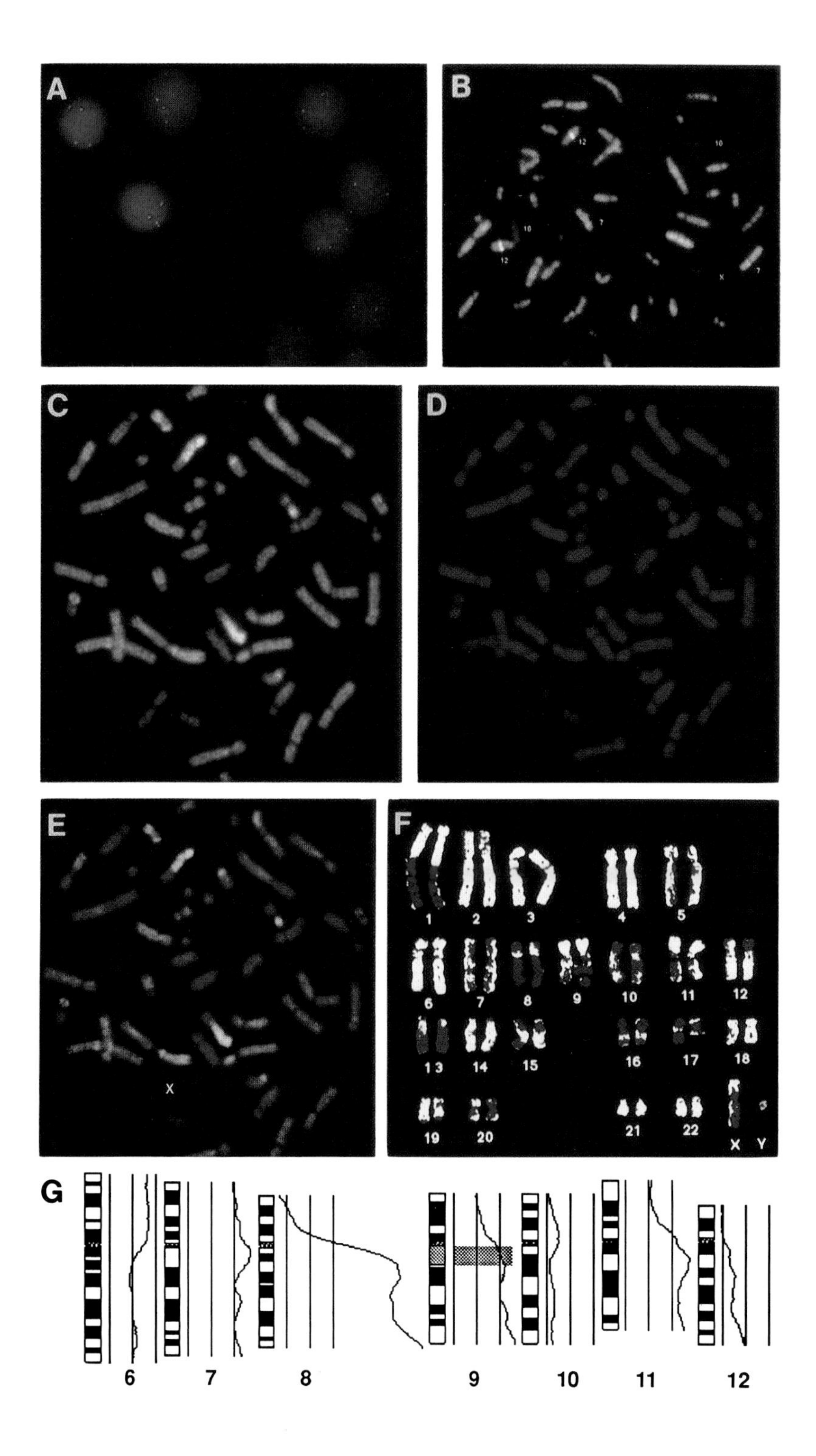
A
B
12
10
10
7
12
X
7
C
D
E
X
F
1
2
3
4
5
6
7
8
9
10
11
12
13
14
15
16
17
18
19
20
21
22
X
Y
G
6
7
8
9
10
11
12

multicolor probe sets should consist of probes with even higher efficiencies, e.g., three probes with an efficiency of 90% would be simultaneously visible in only 73% of the nuclei. In order to obtain an independent measure of the hybridization efficiency, it is recommended to cohybridize with a probe of similar complexity localized in a chromosomal region which is likely not affected by a chromosomal aberration in the analyzed tumor type (see Fig. 2A).

In most analyses the investigated cell population consists of different cell clones with variable numbers of cells. Therefore, it is important to define a diagnostic cutoff level for each experimental setup. For this purpose, the hybridization signals in control cells of several probands are evaluated in order to determine the mean values, e.g., one (or three) fluorescence spots. Adding three times the standard deviation to this mean value has been proven to be a very useful cutoff level.[71,73–75] It is important to note that the cutoff level has to be determined for each probe in each laboratory separately and cannot be adopted from other studies. Although it is relatively easy to prepare control cells for studies of hematological diseases, it is often difficult to obtain control specimens for the analysis of solid tumors. In such cases, alternative strategies to define diagnostic cutoff levels based on the signal distribution of cohybridized probes in the tumor sample might be applied.[76]

Diagnostic cutoff levels with single FISH probes are usually in the range of 1–15%. For all probes, the incidence of one signal in normal cells, e.g., based on incomplete hybridization or overlap of signals viewed two-dimensionally, is higher than the occurrence of three fluorescence spots. Therefore, screening for deletions often deals with higher diagnostic thresholds. However, the thresholds can be lowered by multicolor probe sets since the cutoff level is considerably lower when several probes are evaluated and combinations of signals are scored.[75]

Reverse Painting and Comparative Genomic Hybridization

Diagnosis of chromosomal aberrations by visualization of single chromosomal target regions is based on the preknowledge of chromosomal

[73] J. Anastasi, M. M. Le Beau, J. W. Vardiman, and C. A. Westbrook, *Am. J. Pathol.* **136,** 131 (1990).

[74] S. Stilgenbauer, H. Döhner, M. Bulgay-Mörschel, S. Weitz, M. Bentz, and P. Lichter, *Blood* **81,** 2118 (1993).

[75] M. Bentz, G. Cabot, M. Moos, M. R. Speicher, A. Ganser, P. Lichter, and H. Döhner, *Blood* **83,** 1922 (1994).

[76] H. Matsuyama, Y. Pan, L. Skoog, B. Tribukait, K. Naito, P. Ekman, P. Lichter, and U. S. R. Bergerheim, *Oncogene* **9,** 3071 (1994).

regions that might be affected in a certain disease. Furthermore, the analysis is restricted to a limited number of regions. Although this number can be increased by using multicolor probe sets, there are practical limitations with regard to the detection of large numbers of signals in cell nuclei. A new approach has been introduced, termed comparative genomic hybridization,[77] allowing a comprehensive analysis of chromosomal gains and losses. CGH permits the detection of unknown genetic imbalances in any cell type, even in previously unassessable specimens, as only preparation of genomic DNA is required. Thus, it is particularly useful for the analysis of tumors in which cytogenetic analysis is difficult to perform.

The important principle of CGH is the reverse painting of normal chromosomes by hybridization with whole genomic DNA as a probe (see Fig. 3). The genomic probe DNA to be analyzed (test DNA) labels more or less all of the chromosomes of a metaphase spread. However, when certain chromosomes or chromosomal segments are over- or underrepresented within the probe, the corresponding chromosomal regions are stained stronger or weaker, respectively[27,77,78] (see Fig. 2). Cohybridization of differently labeled normal DNA (control DNA) provides an internal standard, which allows to differentiate between variations in signal intensities that are due to differences in DNA sequence copy numbers and variations due to experimental parameters. The latter might be influenced by base composition (e.g., GC- versus AT-rich regions), chromatin packaging, and target accessability.

However, there is a second explanation why cohybridization with control DNA is required for CGH: after *in situ* hybridization the DNA probe covers only part of the denatured chromosomal target sequences. While sequences underrepresented in the probe will cover only a fraction of this part, overrepresented sequences are assumed to result in a more extended coverage. Test DNA alone should, in principle, after long hybridization lead to a (almost) complete coverage of all targets, not allowing detection of diagnostic signal variations. However, the inclusion of control DNA provides probe sequences specific for the same chromosomal targets at a constant ratio with regard to the test DNA. Since the hybridization of both DNAs follows the rules of stochastics, this implemented ratio remains constant even if the coverage of the chromosomal target region is close to saturation. Thus, for a robust protocol allowing to detect chromosomal

[77] A. Kallioniemi, O.-P. Kallioniemi, D., Sudar, D. Rutovitz, J. W. Gray, F. Waldman, and D. Pinkel, *Science* **258,** 818 (1992).

[78] S. Joos, H. Scherthan, M. R. Speicher, J. Schlegel, T. Cremer, and P. Lichter, *Hum. Genet.* **90,** 584 (1993).

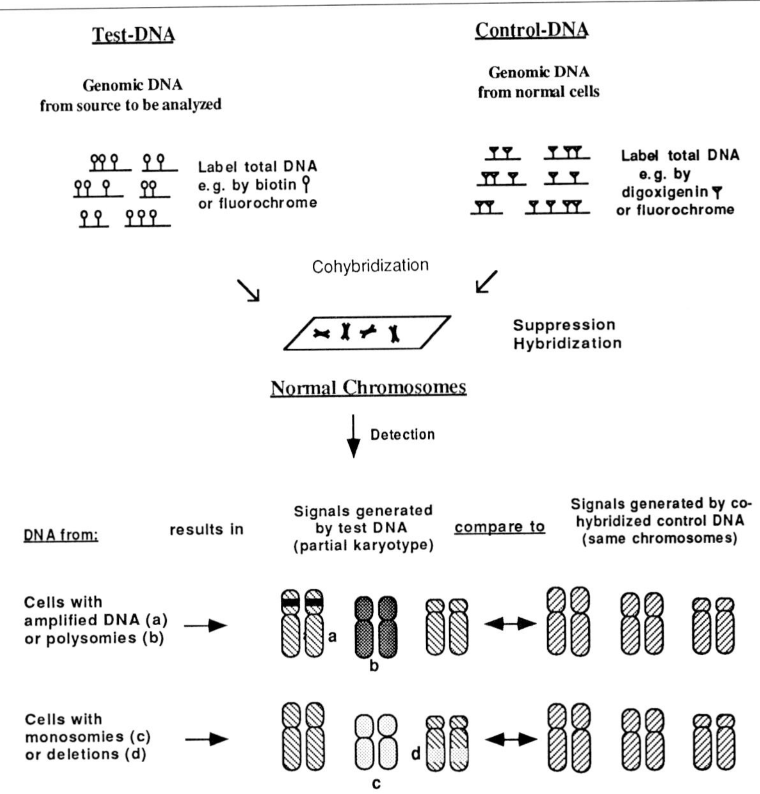

FIG. 3. Schematic illustration of the approach of comparative genomic hybridization.

imbalances, cohybridization with normal DNA seems to be an important factor.

Chromosomal gains and losses are assessed on the basis of differences in signal intensities generated by test and control DNA in different colors along the chromosomes. Although many of such differences can be detected by visual inspection, the quantitative assessment of the ratio of both fluorochromes by digital image analysis allows a much higher accuracy and resolution of CGH. The theoretical ratio of the signals from the test and control genomes is 1 (following normalization of fluorescence intensities) in case the chromosomal region is balanced; 0.5 when there is a deletion, monosomy, or hemizygosity; and 1.5 when there are partial or complete trisomies, etc. For higher degree amplifications, much higher ratios are anticipated. Generally, accepted criteria for CGH evaluation by digital image analysis are still a matter of discussion; for the current state of the art the reader is referred to specialized literature.[79–81] CGH evaluation can be done on different levels of sophistication and accuracy. For example, the screening for amplified sequences can be performed by reverse painting with total genomic test DNA and evaluation by visual inspection[78] (see Fig. 2). The sensitivity of detecting an amplicon of 100 to 200 kb present in 24 copies is similar to the detection efficiency of cosmid probes hybridized to metaphase chromosome spread.[78] However, for an accurate CGH analysis of low copy number changes, especially when present in only a portion of the analyzed cell population, quantitative measurements of fluorescence intensities are required. For this task, it is recommended to utilize sensitive cameras such as cooled CCD cameras on a high quality research fluorescence microscope and to standardize the image acquisition procedure as much as possible. Software application for CGH evaluation should be carefully selected on the basis of the criteria discussed in more detail elsewhere.[79,82,83] The appropriate diagnostic cutoff level of the ratio values is an issue of discussion, although definitions of underrepresented chromosomal regions by ratios <0.75 and overrepresented regions by >1.25 have been proven to be very useful (see Fig. 2).

Comparative Genomic Hybridization

Although the protocol for CGH is very similar to the FISH protocol described earlier, a considerable amount of time and energy is needed to establish CGH even in laboratories experienced in molecular cytogenetics. The success seems to be dependent on a summation of numerous steps in the protocol that have to be specially taken care of, rather than of one or two specific parameters only. In the following, a number of these steps are briefly addressed in order to provide guidelines for the establishment of CGH in a laboratory.

Specimen Preparation. One of the most critical parameters for successful CGH is the quality of metaphase spreads. As indicated earlier, pretreatment of the specimen by proteolytic digestion can improve the accessability of

[79] S. Du Manoir, E. Schröck, M. Bentz, M. R. Speicher, S. Joos, T. Ried, P. Lichter, and T. Cremer, *Cytometry* **19,** 27 (1995).

[80] C. Lundsteen, J. Maahr, B. Christensen, T. Bryndorf, M. Bentz, P. Lichter, and T. Gerdes, *Cytometry* **19,** 42 (1995).

[81] J. Piper, D. Rutovitz, D. Sudar, A. Kallioniemi, O.-P. Kallioniemi, F. M. Waldman, J. W. Gray, and D. Pinkel, *Cytometry* **19,** 10 (1995).

[82] S. Du Manoir, O.-P. Kallioniemi, P. Lichter, J. Piper, *et al., Cytometry* **19, 4** (1994).

[83] P. Lichter, M. Bentz, S. du Manoir, and S. Joos, *in* "Human Chromosomes" (R. Verma and S. Babu, eds.). McGraw-Hill, New York, 1995.

probes to specimen. However, such a treatment may also result in a higher granularity of the signal generated by the genomic DNA hampering the CGH analysis. Therefore, it is recommended for CGH to prepare new metaphase spreads instead of trying to improve suboptimal preparations by proteolytic digestion. Multiple fixations should be included in the chromosome preparation procedure.

Probe Labeling. DNA probes for CGH can be labeled either directly or indirectly. Although directly labeled probes generate smooth fluorescence along the chromosomes, which is preferred for the analysis, indirect detection procedures also result in good quality CGH experiments. Commonly used fluorochromes for test and control DNA are FITC and rhodamine, respectively, or vice versa.

The length of the probe molecules after labeling is a very critical factor for good quality *in situ* hybridizations (see above). In CGH it is important to adjust not only the length of the fragments, as outlined earlier, but also to achieve a similar length for test and control DNA.

In Situ Hybridization. The large amounts of probe DNA provide high frequencies of labeled interspersed repetitive sequences to be suppressed. On the other hand, to accurately assess chromosomal imbalances by variations in the fluorescence intensity ratios, suppression must be as complete as possible (this is in contrast to FISH with single probes, whose localization is still clearly visible as long as they yield signals stronger than the background staining). Therefore, an excess of unlabeled Cot1 DNA is used and/or the preannealing time (after probe denaturation) is elongated. With a preannealing time of 15 min, 1 μg of test DNA and 1 μg of control DNA are combined with 40 to 70 μg of Cot1 DNA, adjusted to 0.15 *M* sodium acetate, and coprecipitated by adding 2 vol of cold ethanol.

The large amounts of DNA also require an extensive agitation (30 min on a shaker) during dissolvation of the precipitated test, control, and competitor DNA in formamide.

The quality of CGH experiments is also improved by extension of the *in situ* hybridization time from overnight to 2 or 3 days.

Equipment. CGH analysis by quantitative measurements put high demands on the equipment necessary for the evaluation, namely the microscope, the camera system, and the computer software program. The relevant instrumentation parameters specific for CGH are described elsewhere.[82]

Universal PCR

CGH greatly stimulated studies analyzing genetic alterations in tumor tissues. However, since tumor material is often too small to yield DNA amounts sufficient for CGH, amplification of the DNA by universal PCR can be applied.[41,84] The possibility of performing CGH from very few cells by combining it with degenerated oligonucleotide-primer (DOP)-PCR,[41] a universal PCR amplification technique, has been demonstrated.[83,85]

Amplification of Small Amounts of DNA for CGH by DOP-PCR. Total genomic DNA (0.1–1 ng) is subjected to DOP-PCR in 2 m*M* $MgCl_2$, 50 m*M* KCl, 10 m*M* Tris–HCl (pH 8.4), 0.1 mg/ml gelatin, 200 μM of each dNTP, 2 μM oligonucleotide 5′CCG ACT CGA GNN NNN NAT GTG G 3′ (N = A, C, G, or T), and 25 units/ml *Taq* polymerase. After initial denaturation at 94° for 10 min, 5 cycles of low stringency amplification are performed with denaturation at 94° for 1 min, annealing at 30° for 1.5 min, transition from 30 to 72° for 3 min, and extension at 72° for 3 min. Exponential amplification of these products is achieved by subsequent 35 cycles with denaturation at 94° for 1 min, annealing at 62° for 1 min, and extension at 72° for 3 min (with an addition of 1 sec/cycle). The PCR products are analyzed by agarose gel electrophoresis and should be visible as a smear of fragments between 200 and 2000 bp long.

Conclusions

Visualization of chromosomal aberrations by *in situ* hybridization is a powerful tool for the analysis of the genetic constitution of tumor cells. Whereas the new approach of comparative genomic hybridization is so far mainly a research tool allowing to analyze chromosomal imbalances in a very comprehensive manner, delineation of specific chromosomal target regions in interphase nuclei has already become a standard method for tumor cytogeneticists and is increasingly applied in clinical settings. New developments in histochemical detection procedures as well as in camera systems are likely to facilitate the use of shorter probes in oncological diagnostics in the future. Detection of single hybridized oligomeres would allow to visualize and identify different alleles in cellular preparations. This would be of great benefit to tumor investigators since it would permit to detect loss of heterozygosity on the level of single cells and to differentiate, for example, cells with and without such losses within the same tissue preparation.

[84] S. K. Bohlander, R. Espinosa, M. Le Beau, J. D. Rowley, and M. O. Diaz, *Genomics* **13,** 1322 (1992).

[85] M. R. Speicher, S. du Manoir, E. Schröck, H. Holtgreve-Grez, B. Schoell, C. Lengauer, T. Cremer, and T. Ried, *Hum. Mol. Genet.* **2,** 1907 (1993).

Acknowledgments

We thank all the members of the laboratory as well as Thomas Cremer (Heidelberg) for continuous discussion, Stefan Stilgenbauer (Heidelberg) for providing photographic material, and Ulf Bergerheim (Stockholm) for providing prostate tumor DNA. This work was supported by the European Community (PL930055).

Section III

Gene Function

[23] Antisense Techniques

By CHERYL ROBINSON-BENION and JEFFREY T. HOLT

The ability to manipulate gene expression within mammalian cells has provided powerful experimental approaches for studies of gene function and gene regulation. Methods which inhibit gene expression are particularly important because they permit studies probing the normal function of a specific gene product within a cell. Antibody injection and antisense studies have shown that cellular protooncogenes, particularly transcription factor protooncogenes, have important roles regulating proliferation and differentiation.[1-9] The antisense methodologies produce inhibition of specific gene products by exploiting hybridization of complementary nucleic acids, resulting in decreased mRNA stability, or through a block in mRNA processing, transport, or translation.[10-12] Antisense inhibition can be produced either by the transfection of plasmids or by the addition of small single-stranded oligonucleotides. Many elegant modifications of these strategies have been developed, including inducible antisense vectors,[1,3] antisense retroviral vectors, and a variety of oligonucleotide modifications to facilitate delivery and enhance efficacy.[13,14]

Although antisense methods can provide useful and convincing data when appropriate care and necessary controls are included, there are a

[1] J. T. Holt, T. V. Gopal, A. D. Moulton, and A. W. Nienhuis, *Proc. Natl. Acad. Sci. U.S.A.* **83,** 4793 (1986).

[2] K. Yokoyama and F. Imamoto, *Proc. Natl. Acad. Sci. U.S.A.* **84,** 7363 (1987).

[3] K. Nishikura and J. M. Murray, *Mol. Cell. Biol.* **7,** 639 (1987).

[4] K. Heikkila, G. Schwab, E. Wickstrom, S. L. Loke, D. H. Pluznik, R. Watt, and L. M. Neckers, *Nature (London)* **328,** 445 (1987).

[5] J. T. Holt, R. L. Redner, and A. W. Nienhuis, *Mol. Cell. Biol.* **8,** 963 (1988).

[6] L. Kerr, J. T. Holt, and L. M. Matrisian, *Science* **242,** 1424 (1988).

[7] J. A. Pietenpol, J. T. Holt, R. W. Stein, and H. L. Moses, *Proc. Natl. Acad. Sci. U.S.A.* **87,** 3758 (1990).

[8] K. T. Riabowol, R. J. Vosatka, E. B. Ziff, N. J. Lamb, and J. R. Feramisco, *Mol. Cell. Biol.* **8,** 1670 (1988).

[9] K. Kovary and R. Bravo, *Mol. Cell. Biol.* **11,** 4466 (1991).

[10] R. Y. Walder and J. A. Walder, *Proc. Natl. Acad. Sci. U.S.A.* **85,** 5011 (1988).

[11] S. K. Kim and B. J. Wold, *Cell (Cambridge, Mass.)* **42,** 129 (1985).

[12] C. A. Scherczinger and D. A. Knecht, *Antisense Res. Dev.* **3,** 191 (1993).

[13] L. Perrouault, U. Asseline, C. Rivalle, N. T. Thuong, E. Bisagni, C. Givannangeli, T. Le Doan, and C. Helene, *Nature (London)* **344,** 358 (1990).

[14] M. Matsukura, G. Zon, K. Shinozuka, M. Robert-Guroff, T. Shimada, C. A. Stein, H. Mitsuya, F. Wong-Staal, J. S. Cohen, and S. Broder, *Proc. Natl. Acad. Sci. U.S.A.* **86,** 4244 (1987).

number of artifacts which must be avoided when using these strategies. Transfection of antisense plasmids generally requires the isolation of stable transformants which can produce artifactual results due to biased selection pressures and/or direct effects of inducing agents on experimental systems. The addition of antisense oligonucleotides can occasionally generate artifactual results due to toxicity and not as a direct consequence of hybridization and consequent antisense inhibition.[15] The focus of this chapter is on strategies and controls which can limit artifactual results: in particular, the use of reversal or rescue methods to hybrid arrest the antisense effect.

Principle of Method

Plasmid-based antisense methods employ expression vectors which generate RNAs containing sequences complementary to kcy rcgions of specific genes. These approaches mimic the naturally occurring antisense RNAs which serve important regulatory functions in bacteria.[16] Genetically engineered antisense expression vectors generally contain several elements: (1) prokaryotic and eukaryotic selectable markers allowing vector construction and identification of transfectants; (2) a gene promoter which controls the expression of the antisense RNA; (3) an antisense sequence which is complementary to a bindable region within the specific target gene sequence (5′ untranslated and/or translation initiation regions are often employed as target sites); and (4) RNA stabilizing sequences to assure stability of antisense RNAs. The design of antisense expression vectors requires combinations of these elements which are most suited for inhibition of the target gene in a specific cellular context. Plasmid-based antisense approaches can also be used for gene replacement. Optimal gene replacement protocols would include both inhibition of the endogenous gene and overexpression of the preferred (or mutant) gene. We have used an antisense rescue method to test whether antisense-resistant genes (designed by deletion of antisense RNA target sequences) can replace the function of endogenous genes.[17] This gene replacement method for inhibition of endogenous genes and replacement with preferred genes has implications for gene therapy of certain dominant hereditary hematologic disorders and for the correction or "repair" of gene mutations in cancer.

Oligonucleotide-based antisense methods employ synthetic single-stranded oligonucleotides which may range from simple deoxyribonucleotides to more complex molecules containing base modifications and/or covalent modifications which enhance delivery, uptake, or antisense effect.

[15] C. A. Stein and Y.-C. Cheng, *Science* **261,** 1065 (1993).
[16] K. M. Takayama and M. Inouye, *Crit. Rev. Biochem. Mol. Biochem.* **25,** 155 (1990).
[17] J. T. Holt, *Mol. Cell. Biol.* **13,** 3821 (1993).

Both plasmid- and oligonucleotide-based antisense methods have been employed to inhibit cellular protooncogenes, demonstrating important roles for specific protooncogenes in the regulation of cellular proliferation, differentiation, and in specific cellular processes such as transcription.

Materials and Reagents

Antisense expression vectors can be obtained from the laboratories where they were developed or can be constructed by combining the key elements described earlier. Many of these expression vectors include neomycin as a selectable marker so transfectants are isolated employing Dulbecco's modified Eagle's medium with 10% calf serum and the appropriate concentration of the antibiotic G418 (Bethesda Research Laboratories, Bethesda, MD). One milligram per milliliter of G418 is generally employed for selection of stable transformants in cultured fibroblasts, but lower concentrations must be employed in other cell lines which are more sensitive to G418.

Successful oligonucleotide antisense experiments require pure oligonucleotide preparations and nuclease-free culture conditions. Antisense oligonucleotides may be synthesized by solid state methods[18] and purified by chromatography or gel purification (or obtained from commercial sources). Following purification, we lyophilize oligonucleotides two to five times in sterile distilled water to remove volatile components remaining from synthesis or purification steps. Both unmodified and modified oligonucleotides may be suspended in 10 m*M* HEPES (*N*-2-hydroxyethylpiperazine-*N*′-2-ethanesulfonic acid)-buffered saline at pH 7.4. It is important to test the pH of these oligonucleotide solutions because the addition of even small amounts of solutions with nonphysiologic pH can have profoundly toxic effects on cells. This is a particular problem following gel purification of oligonucleotides unless much effort is devoted to the removal of residual acrylamide. Nuclease-free culture conditions are necessary for studies involving unmodified deoxyribonucleotides which are rapidly degraded by nucleases that are frequently present in serum and occasionally present in media. Methods to assure nuclease-free culture conditions are described subsequently in this chapter.

Methods

Design and Construction of Antisense Expression Vectors

Figure 1 shows the structure of a prototype antisense expression vector which we have employed for inhibition of c-*fos* expression in mammalian

[18] M. J. Gait, ed., "Oligonucleotide Synthesis." IRL Press, Oxford, 1984.

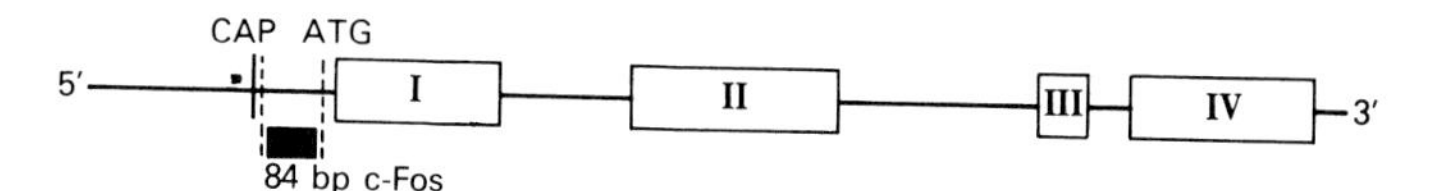

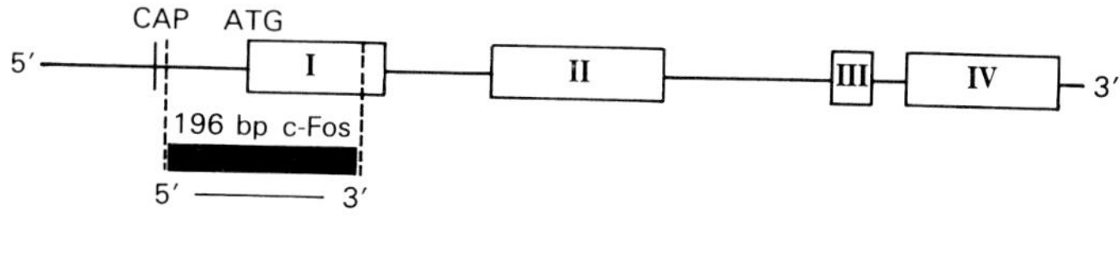

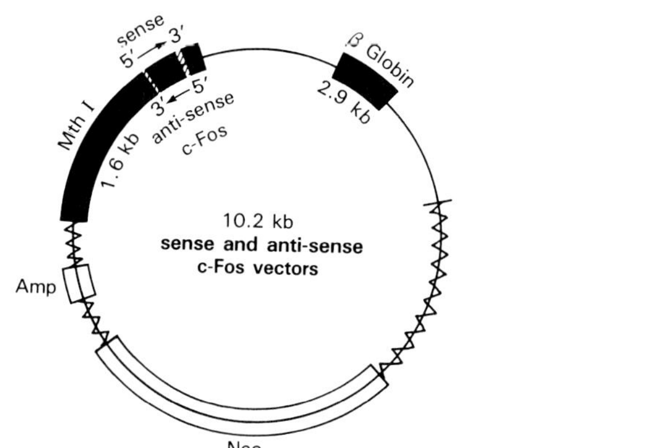

FIG. 1. Structure of antisense plasmids directed against the c-*fos* gene. The upper line drawing (mouse c-*fos*) shows the gene structure of the mouse c-*fos* gene. The CAP and ATG translation initiation sites are shown. The four exons are illustrated as clear boxes marked with roman numerals. The target site for antisense inhibition is the 84-bp sequence shown as a black box between the CAT and the ATG which is marked 84 bp c-*fos*. The lower line drawing (human c-*fos*) shows the human gene structure along with a 196-bp antisense target site. The circular plasmid drawing shows that sense (5′ to 3′) or antisense (3′ to 5′) sequences are cloned between the mouse metallothionein I promoter (Mth I) and β-globin splice and exon sequences (shown as black box). Note that the 84-bp antisense will inhibit endogenous c-*fos* genes but not transfected genes containing deletion of this region, but that the 196-bp antisense will inhibit both endogenous and transfected c-*fos* genes which contain exon I.

cells. This particular vector allows cloning of antisense and control "sense" gene fragments between the metal-inducible metallothionein promoter and β-globin sequences which are added to provide stability to the antisense sequences and has been employed in studies demonstrating the role of the c-*fos* gene in autoinduction of the transforming growth factor β gene.[19] We have also constructed vectors which differ only in the promoter region because the promoter region in these vectors is readily excised by 5′ *Eco*RI and 3′ *Hin*dIII sites: one series of vectors is regulated by the steroid-inducible mouse mammary tumor virus (MMTV) promoter,[1] and another series of vectors is regulated by the constitutive Rous sarcoma virus long terminal repeat (RSV-LTR). Antisense retroviral vectors have also been constructed which are compatible with our plasmid expression vectors: the promoter cartridge is flanked by unique 5′ *Xho*I and 3′ *Hin*dIII sites, and the antisense cartridge is flanked by unique 5′ *Hin*dIII and 3′ *Bam*HI sites. Because the target sequence for the 84-bp antisense vector resides entirely within the 5′-untranslated region, this antisense vector inhibits the endogenous c-*fos* gene but does not inhibit transfected c-*fos* genes in which the 5′-untranslated antisense target site has been deleted.[17] Antisense-resistant expression vectors were constructed by cloning the *Hae*II–*Eco*RI fragment of human c-*fos* cDNA into a Moloney LTR-regulated expression vector removing sequences complementary to the 84-bp antisense c-*fos* construct.[17] In order to test the antisense rescue approach, c-*fos* C-terminal deletion mutants were constructed by linker insertion of an in-frame termination codon. All of these mutant c-*fos* plasmids were subcloned into expression vectors for expression in cultured fibroblasts (Moloney-LTR promoter) or F9 embryonal carcinoma cells (Rous sarcoma virus LTR promoter). These studies demonstrated which domains of the c-*fos* gene contribute to cellular proliferation.[17]

Production of Stable Transformants Expressing Antisense RNA

Stable transformants expressing both the antisense-resistant vector and the anti-*fos* RNA construct were obtained by cotransfection of both plasmids using the following protocol which generally results in 20–50 transformants per 100-mm^2 tissue culture dish.[20]

Day 1. Plate 1.2 million cultured mouse fibroblasts (NIH 3T3, BALB c/3T3) per 100-mm^2 tissue culture plates in 10 ml of Dulbecco's minimal Eagle's medium supplemented with 10% calf serum and allow cells to attach overnight.

Day 2. Transfect each plate by calcium–DNA coprecipitation with 20 μg of cesium-banded plasmid DNA. For cotransfection studies, use 2 μg of the antisense vector containing the NEO selectable marker and 18 μg of the antisense-resistant expression vector which lacks this selectable marker.

[19] S.-J. Kim, F. Denhez, K.-Y. Kim, J. T. Holt, M. B. Sporn, and A. B. Roberts, *J. Biol. Chem.* **264,** 19373 (1989).

[20] C. Robinson-Benion, N. Kamata, and J. T. Holt, *Antisense Res. Dev.* **1,** 21 (1991).

Because the plasmid without the selectable marker is present in 10-fold excess, this protocol generally results in expression of both plasmids in the majority of stable transformants.[17] A DNA–calcium solution is prepared by mixing in the following order: 20 μg of plasmid DNA dissolved in TE (10 m*M* Tris, pH 7.5, 1 m*M* EDTA), 50 μl of 2.5 *M* $CaCl_2$, and TE to make 500 μl total. This solution is added slowly to 500 μl of 40 m*M* HEPES-buffered saline (pH 7.2) and is then mixed by bubble aeration to form an opalescent solution. After a 30-min incubation at 25°, the entire 1-ml coprecipitate is added to the plate. After incubating the DNA coprecipitate with the cells for 4 hr, the media is aspirated, and the cells are treated with 5% glycerol in 20 m*M* HEPES buffer for 3 min. After this glycerol shock, the cells are washed again with phosphate-buffered saline (PBS) and then refed with 10 ml of fresh medium containing serum.

Day 3. Aspirate the media and replace with 10 ml of Dulbecco's minimal Eagle's medium supplemented with 10% calf serum and 1 mg/ml of the antibiotic G418.

Day 5. Aspirate the medium (which by this time contains dead cells and cellular debris from G418-sensitive cells) and refeed with 10 ml of Dulbecco's minimal Eagle's medium supplemented with 10% calf serum and 1 mg/ml of the antibiotic G418.

Day 7. Aspirate the medium and refeed with 10 ml of Dulbecco's minimal Eagle's medium supplemented with 10% calf serum and 1 mg/ml of the antibiotic G418.

Days 9–12. Once individual clones are clearly apparent they can be isolated with cloning rings and expanded in individual wells of microtiter plates. Multiple clones should be studied for each combination of antisense vector and antisense-resistant rescue plasmid.

Characterization of Cotransfectant Clones

Cotransfectant clones are analyzed to quantitate the extent of antisense inhibition of the endogenous target gene, the level of expression of the antisense-resistant transfected gene, and the effect of target gene inhibition on cellular growth or other phenotypic events. Because inhibition of protooncogenes can alter cell growth properties, we herein describe assays to quantitiate cell growth and analyze antisense inhibition in the presence and absence of antisense RNA.

Serum-induced DNA synthesis is measured by the timed addition of serum to quiescent stable transformants. Fibroblasts from cloned stable transformants are plated at a concentration of 200,000 cells per ml and are grown to confluence on glass slide chambers (Nunc Incorporated, Naperville, IL) in complete media containing 10% serum. Twenty-four hours after achieving confluence, the serum-containing medium is replaced with medium containing 0.5% serum. Following an additional 48 hr in serum-depleted medium, the cells are preincubated with 1 μM dexamethasone to induce the MMTV promoter (or ethanol vehicle for controls) for 1 hr prior to stimulation with fresh medium containing 10% serum. [^{3}H]Thymidine is added to a concentration of 1 μCi/ml 12 hr after serum stimulation, and then labeling occurs for 12 hr. The cells are then washed with PBS, fixed in cold methanol, and processed for autoradiography as described. The labeling index is determined by the mean of 200 cell counts on three independent experiments. Cell viability can be monitored by the addition of trypan blue. In order to verify that the growth inhibition is due to antisense inhibition, it is important to show that transfectants obtained with control "sense" plasmids do not exhibit inducer-dependent growth inhibition. In addition, rescue with a wild-type antisense-resistant gene should overcome the antisense inhibition, providing additional assurance that the growth inhibition is actually due to specific inhibition of the target gene.

Antisense inhibition can be quantitated by measuring levels of either target gene mRNA or protein. Nuclease protection assays provide an excellent method for quantitation of antisense effect because the technique is both sensitive and quantitative. Cells from the stable transformant clones are placed in DMEM with 0.5% calf serum for 48 hr prior to steroid treatment and/or serum restimulation. RNA is isolated from cells by the guanidinium thiocyanate method and the total RNA from 2×10^6 cells is hybridized with 3×10^6 cpm of the labeled RNA probe for 16 hr at 45°. Samples are then treated with 2 μg/ml of RNase A and 4 U/ml of RNase T1 for 30 min at 25° and are deproteinized by sequential treatment with proteinase K and phenol–chloroform followed by ethanol precipitation, and electrophoresis on an 8% denaturing polyacrylamide gel.

Comparison of Results of Plasmid-Mediated and Oligonucleotide-Mediated Antisense Approaches

Clearly, antisense data demonstrating a required role for a certain gene are more convincing if it can be verified by two different antisense methods. Because both plasmid- and oligonucleotide-based methods have technical limitations, it may not always be possible to demonstrate effective antisense inhibition employing both of these strategies. For example, plasmid-mediated antisense vectors which inhibit essential genes may prevent the isolation of stable transformants (unless inducible promoters are employed) whereas antisense oligonucleotides may be unable to inhibit the expression of genes which produce stable proteins because the antisense effect is of shorter duration than the protein half-life. Our laboratory has demonstrated that antisense inhibition of the c-*fos* gene is necessary for normal growth of cultured mouse fibroblasts by both plasmid-mediated antisense and oligonucleotide-mediated antisense approaches.[1,17] These studies have been confirmed by other laboratories employing both antisense and antibody injection methods,[3,8,9] although one study showed less inhibition employing a c-*fos*-specific antibody.[9] The recent generation of c-*fos* null mice[21,22] has demonstrated that mice can grow to maturity without the c-*fos* gene, suggesting that the c-*fos* gene is not involved in growth or that some redundancy is present (e.g., FosB might replace c-*fos* function).

Antisense Inhibition with Oligonucleotides

Major considerations for design of antisense oligonucleotide include: (1) site of target sequence within gene; (2) length of antisense oligonucleotide; and (3) secondary structure of antisense oligonucleotide. Oligonucleotides should be designed that are directed against regions of the target mRNA which are available for hybridization: a region that is unbound by protein and free of secondary structure. Sequences within the 5′-untranslated region or translation initiation region are frequently employed for design of antisense oligonucleotides. The length of oligonucleotides employed for antisense inhibition varies between 12 and 20 nucleotides. Our study analyzing the effectiveness of mismatched oligonucleotides suggests that a melting temperature (T_m) of 50–55° is optimal for specific inhibition of target genes.[5] T_m for short oligonucleotides is best determined by a formula in which G and C residues equal 4° and A and T residues equal 2°. Thus optimal antisense oligonucleotides generally range from 14 to 19 nucleotides. Although questions of target gene site and oligonucleotide length can be approached on a rational basis, determining a reliable secondary structure for oligonucleotides is less predictable. Our approach has been to screen sequences visually for obvious stem structures or hairpins without undue reliance on computer predictive algorithms.

Determination of Oligonucleotide Stability

Unmodified deoxyribonucleotides can be employed for antisense studies in cultured cells if culture media and serum are screened for the absence of nuclease activity. Oligonucleotides are 5′ end labeled with polynucleotide

[21] S. Field, R. S. Johnson, R. Mortenson, V. E. Papaioannou, B. M. Spiegelman, and M. E. Greenberg, *Proc. Natl. Acad. Sci. U.S.A.* **89,** 9306 (1992).

[22] R. S. Johnson, B. M. Spiegelman, and V. Papaioannou, *Cell (Cambridge, Mass.)* **71,** 577 (1992).

kinase and [^{32}P]ATP and are then added to culture media and serum (plus unlabeled oligonucleotide to achieve a final oligonucleotide concentration of 5 μM: for a 15-mer this is approximately 20 μg/ml). Aliquots were removed periodically and resolved on denaturing 20% acrylamide gels as described.[5] Figure 2 shows the typical appearance of a time course study employing a serum with low nuclease activity. Because serum lots vary dramatically in nuclease activity and many serum lots will completely degrade deoxyribonucleotides within an hour, it is important to screen several lots of serum in order to identify those serum lots which have low nuclease activity. Heating the serum to 65° destroys complement activity and diminishes phosphatase activity, but unfortunately has little effect on a serum with high nuclease activity. Screening for nuclease activity is probably unnecessary for experiments employing phosphorothiorate oligonucleotides, although we generally test the stability of modified oligonucleotides to show that they are indeed nuclease resistant.

Detection of Intracellular Duplexes

In order to detect intracellular duplex, 20 μg of oligonucleotide is 5′ end labeled oligonucleotides with 20 mCi of [^{32}P]ATP to achieve a specific activity of 50 million cpm/μg. After incubation for 4 hr, unincorporated oligonucleotides are removed by washing the cells three times with HEPES-buffered saline prewarmed to 37° (to prevent the melting of duplexes). The cells are then lysed in 100 μl of Nonidet P-40 lysis buffer (10 mM Tris, pH 7.5, 10 mM NaCl, 3 mM $MgCl_2$, 0.05% Nonidet P-40) containing 0.5% sodium dodecyl sulfate, 100 μg of proteinase K per ml, and a 10,000-fold excess of unlabeled oligomer (as the carrier). The amount of carrier added should be based on the amount of cell-associated radioactivity remaining on washed cells, detected by scintillation counting of thrice-washed pellets. Following phenol/chloroform extraction and ethanol precipitation, a S1 nuclease protection assay was performed at 37°, and the products were analyzed on a 20% denaturing acrylamide gel (containing 42 g of urea of 100 ml). To demonstrate that the duplex is intracellular and not an artifact of RNA isolation, an "add-back" control may be performed in which the measured amount of cell-associated radioactivity is added with carrier (excess unlabeled oligonucleotide) to a lysate of cells that were previously unexposed to oligonucleotide.[5]

Reversal of Antisense Oligonucleotide Effects by Hybridization Competition

To confirm that antisense oligonucleotide effects are due entirely to target gene inhibition, reversal experiments may be performed employing

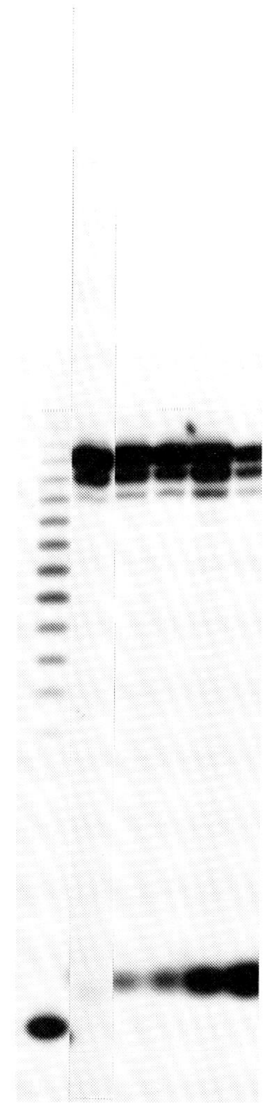

FIG. 2. Autoradiograph demonstrating the stability of 5′-end-labeled antisense c-*myc* oligonucleotide (deoxyribonucleotide of sequence 5′ AAC GTT GAG GGG CAT 3′) following incubation with HL-60 cells in media supplemented with 10% serum. Gel-purified oligonucleotides were incubated with logarithmically growing cells for the time indicated. Lane 1 shows a labeled molecular weight marker showing oligonucleotides ranging from 4 to 22 nucleotides. The broad dark band at the bottom is unincorporated ATP. Lanes 2–6 show a time course of oligonucleotide incubates with cells and media for the following periods of time: lane 2, gel-purified oligonucleotide prior to incubation with cells and media; lane 3, 1 hr; lane 4, 2 hr; lane 5, 4 hr; and lane 6, 8 hr. The low molecular weight bands represent a 5′-labeled single nucleotide, presumably produced by 5′ exonucleolytic cleavage.

an excess of anticomplementary ("sense") oligonucleotide: in essence this approach is to antisense the antisense effect. Figure 3 shows an example of this type of experiment employing sense c-*myc* and the complementary antisense c-*myc* oligonucleotide mixed in either a 1:1 ratio or a 10:1 ratio. The sense oligonucleotide can reverse the effects of the antisense oligonucleotide under three different experimental conditions: (1) if sense and antisense oligonucleotides are preannealed prior to addition to media; (2) if the cells are preloaded with sense oligonucleotide prior to addition of antisense oligonucleotide; and (3) if sense and antisense oligonucleotides are added simultaneously. This experiment has been published previously,[5] providing additional assurance that antisense inhibition of the c-*myc* gene produces phenotypic effects due to hybridization and consequent inhibition of the c-*myc* target gene.

Concluding Remarks

Both plasmid-based and oligonucleotide-based strategies inhibit the expression of specific protooncogenes and analyze the effects of gene inhibition on cell growth and differentiation. The use of sense control plasmids and the antisense rescue plasmids provides evidence that the observed

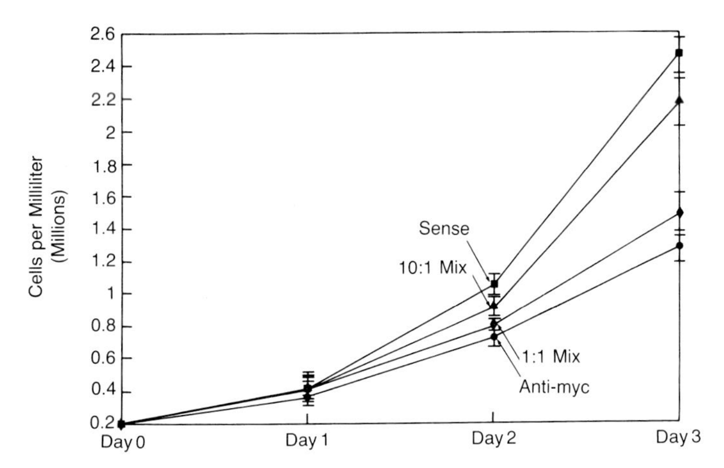

FIG. 3. Growth curves of HL-60 cells incubated with 4 μM antisense c-*myc* oligonucleotide. Points and bars show the mean, plus and minus standard error. Arrows point to lines marking growth rates of cells incubated with 4 μM sense oligonucleotide (closed squares), a 10:1 ratio of 40 μM sense and 4 μM antisense oligonucleotide added simultaneously (closed triangles), a 1:1 ratio of 4 μM sense and 4 μM antisense oligonucleotide added simultaneously (closed diamonds), and 4 μM antisense oligonucleotide (closed circles).

effects of anti-*fos* induction are a consequence of antisense-mediated gene inhibition and not merely a nonspecific effect of inducer or a toxic effect mediated by double-stranded RNA. Rescued clones provide a specificity control of a different kind: for example, in our antisense c-*fos* studies,[17] the endogenous c-*fos* gene is inhibited as demonstrated by immunoprecipitation studies, but expression of the antisense rescue gene restores the *fos* function. This provides a useful specificity control for the antisense RNA and suggests that the observed results are due to inhibition of endogenous gene expression and not merely toxicity due to unexpected effects of double-stranded RNA, etc.

Reversal experiments can provide a control for specificity of oligonucleotide-based studies in a manner similar to the antisense rescue method. If an effect is due to hybridization of antisense oligonucleotide sequences to specific gene target sequences, then addition of an excess of sense oligonucleotide should reverse the observed effects on cell growth or gene expression. However, a toxic effect due to impurities in oligonucleotide preparations or protein binding by oligonucleotide would presumably not be reversed by an excess of anticomplementary oligonucleotide.

Oligonucleotide preparations can have toxic effects on cells, thus antisense oligonucleotide experiments should employ multiple controls to assure that the observed results are not merely due to toxicity. Although it is of utmost importance to show that the target gene is inhibited by analyzing levels of mRNA and/or protein, decreased target gene expression may not assure sequence-specific inhibition since an inhibition of cell growth often correlates with decreased levels of protooncogene expression. For example, oligonucleotide-based inhibition of the c-*myc* gene in HL-60 cells need not be a consequence of c-*myc* gene inhibition since a variety of growth inhibitors and differentiation decrease c-*myc* expression. For these reasons we have employed two distinct approaches to demonstrate antisense effect in these c-*myc* experiments[5]: (1) synthesis and testing of oligonucleotides with sufficient sequence dyshomology to prohibit duplex formation, and (2) addition of an anticomplementary oligonucleotide to reverse the antisense oligonucleotide effects by hybridization competition.

Future directions in antisense development will include the search for nontoxic-modified oligonucleotides and methods to deliver oligonucleotides more effectively to cells, and ultimately to deliver oligonucleotides to specific tissues in animals. A number of novel and creative approaches are already being tested in a variety of laboratories (although beyond the scope of this chapter). As new antisense approaches are being tested and evaluated, it will be important to continually demonstrate that observed oligonucleotide effects are due to specific inhibition of hybridization. This can be assured by demonstrating that target genes are specifically inhibited and by judicious use of control oligonucleotides and reversal experiments.

Acknowledgments

We thank W. J. Pledger for helpful comments on the manuscript. This work was supported by Grant CA51735.

[24] Use of *lac* Activator Proteins for Regulated Expression of Oncogenes

By MARK A. LABOW

Introduction

Inducible expression systems have a variety of uses in the study of oncogenes. Inducible expression systems can be used to establish conditionally transformed cell lines by regulating expression of an oncogene. Such cell lines can be used to study various aspects of the transformation process including the temporal requirements for morphological transformation or reversion to the untransformed state. Controlled expression of an oncogene can be used to study the physiologic changes that occur over time in a cell line as a consequence of expression of a specific oncogene. Reversibly transformed cell lines can be used to determine if a particular oncogene is required for the initial transformation event and/or maintenance of the transformed phenotype. These cell lines can also be used to identify downstream genes that are activated or repressed by the regulated oncogene using subtractive or differential cDNA screening techniques.

An inducible expression system must have a variety of characteristics in order to be useful. First, the expression system must tightly regulate expression of the target gene such that the physiological effects of expression are also regulated. Second, the stimulus used to control expression must not have significant physiologic effects on cell growth or gene expression. Finally, the expression system should have features that allow it to be readily adapted to any cell type. Inducible artificial mammalian transcriptional transactivators based on the *lac* repressor (referred to as the *lac* activator proteins or LAPs) have been described which meet these criteria.[1,2] This

[1] M. A. Labow, S. B. Baim, T. Shenk, and A. J. Levine, *Mol. Cell. Biol.* **10,** 3343 (1990).
[2] S. B. Baim, M. A. Labow, A. J. Levine, and T. Shenk, *Proc. Natl. Acad. Sci. U.S.A.* **88,** 5072 (1991).

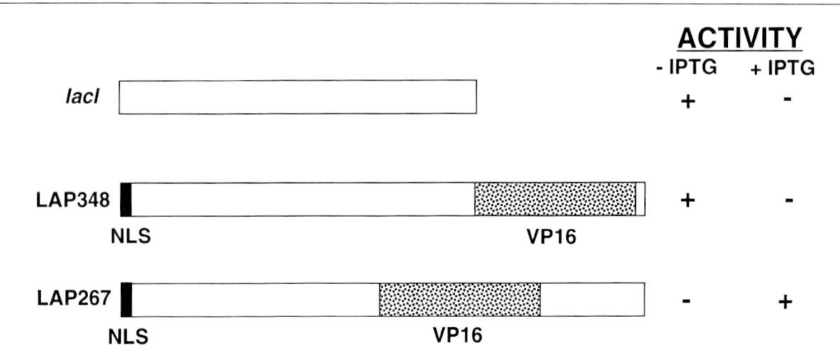

FIG. 1. Structure and regulatory properties of the *lac* activator protein (LAP) genes. The structure of the original *lacI* gene and LAP constructs are shown. *lac* repressor sequences are shown as open boxes. The coding regions for a nuclear localization signal and for the VP16 activation domain are shown as a solid box and stippled box, respectively, as indicated in the figure. The activity of the protein in the presence or absence of IPTG is indicated on the right. Activity for the *lacI* gene product represents its ability to bind DNA and act as a repressor, whereas an activity of the LAP gene products represents its ability to act as an transcriptional activator.

chapter outlines the LAP system, the currently available vectors, and their use in developing conditionally transformed cell lines using LAP-inducible oncogenes.

LAP Constructs and Their Properties

The LAP system is based on the isopropyl β-*d*-thiogalactoside (IPTG)-regulated *lac* repressor. Fusion of the *lac* repressor DNA-binding protein with a mammalian transcriptional activation domain results in the generation of an IPTG-regulated transcription factor. Several IPTG-regulated LAPs have been described that activate transcription of mammalian promoters. This activation is dependent on the presence and number of *lac* operator sequences within the promoter. The LAP constructs consist of the entire coding region of the *lac* repressor with two modifications. First, the coding region for a nuclear localization signal (NLS) derived from the SV40 (simian virus 40) tumor (T) antigen gene has been added at the very 5′ end of the *lac*I gene. Second, the coding region for the C-terminal transcriptional activation domain of the herpes simplex virus Type-1 VP16 gene[3,4] has been inserted in-frame within the *lac* repressor coding region. Two LAP genes are shown in Fig. 1. The two genes differ as to the location

[3] P. E. Pellet, J. L. C. McKnight, F. J. Jenkins, and B. Roizman, *Proc. Natl. Acad. Sci. U.S.A.* **82,** 5870 (1985).

[4] S. J. Triezenberg, R. C. Kingsbury, and S. L. McKnight, *Genes Dev.* **2,** 718 (1988).

of the VP16 coding region. The LAP267 gene contains the VP16 sequences inserted after the coding region for amino acid 267 of *lac* repressor and the LAP348 gene contains the insertion after the coding region for amino acid 348 of *lac* repressor.

Both LAP genes produce stable proteins that are efficiently transported to the nucleus of mammalian cells. However, the two LAPs differ dramatically in their activity and in their response to IPTG as summarized in Fig. 1. LAP348 behaves in a manner predicted from the regulatory properties of the native repressor; IPTG (in concentrations of 5 m*M* or less) added to the culture media of cells inhibits the transcriptional activation activity of LAP348, presumably due to the ability of IPTG to block the DNA binding activity of *lac* repressor. LAP267, however, behaves in the reverse. LAP267 is largely inactive at 37° or 39.5°. However, in the presence of 5 m*M* IPTG, LAP267 efficiently activates transcription of LAP-inducible reporters at these temperatures. LAP267 will also activate transcription at 32° in the absence of IPTG, although less efficiently than at higher temperatures in the presence of IPTG. Thus, insertion of the VP16 sequences in LAP267 produces a mutation that allows the activity of the *lac* repressor DNA binding domain to be rescued by either temperature or IPTG in a manner similar to that previously reported for i^{r} or i^{rc} mutants of *lac* repressor.[5,6] Thus IPTG can be used to turn off LAP348 or to turn on LAP267-mediated transcription.

Because the use of IPTG is critical to the use of the LAP system, several aspects of the use of IPTG in mammalian cell culture should be kept in mind. The first is that although high concentrations of IPTG may be toxic to mammalian cells, very little or no toxicity is seen in concentrations less than 10 m*M*.[7,8] This makes it unlikely that the concentrations used for regulating the LAP constructs (5 m*M* or less) will have adverse effects on mammalian cells. The second point is that uptake of IPTG appears fairly rapid in mammalian cells, indicating that IPTG will rapidly affect LAP activity.[2,8,9] The final point is that, although only IPTG has been used to regulate LAP activity, a variety of other inducers have been tested for *lac* repressor that may have increased utility in mammalian cells or in transgenic animals.[9]

[5] J. H. Miller, *in* "The Operon" (J. H. Miller and W. S. Reznikoff, III, eds.), p. 31. Cold Spring Harbor Lab., Cold Spring Harbor, NY, 1978.

[6] G. L. Myers and J. R. Sadler, *J. Mol. Biol.* **58,** 1 (1971).

[7] J. Figge, C. Wright, C. J. Collins, T. M. Roberts, and D. M. Livingston, *Cell (Cambridge, Mass.)* **52,** 713 (1988).

[8] G. G. Miao and T. Curran, *Mol. Cell. Biol.* **14,** 4295 (1994).

[9] D. L. Wyborski and J. M. Short, *Nucleic Acids Res.* **19,** 4647 (1991).

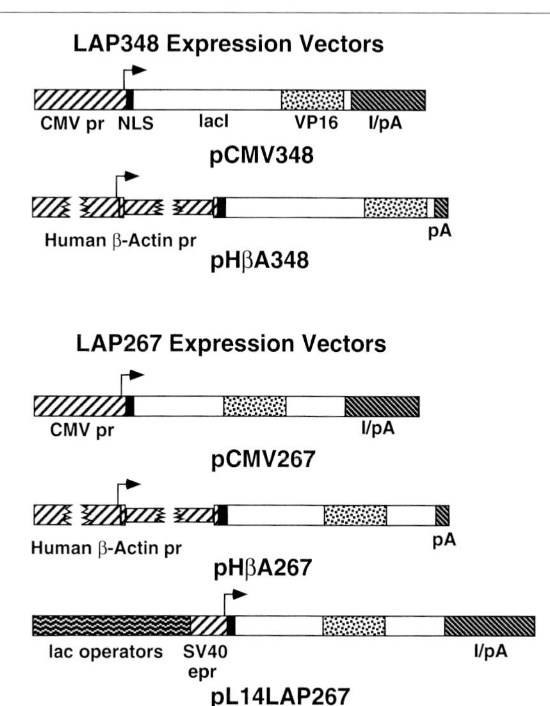

FIG. 2. Structure of LAP expression vectors. The LAP constructs are drawn as described for Fig. 1. The human CMV promoter (pr), human β-actin promoter, and SV40 early promoter (epr) are indicated and drawn with hatched boxes. The start site of transcription for each vector is indicated by an arrow. The human β-actin promoter vectors also include the first intron of the actin gene as represented by the narrower hatched box. Introns (I) and poly(A) addition sites (pA) present in the vectors are indicated as dark hatched boxes.

LAP Expression Vectors

A selection of LAP expression vectors are shown in Fig. 2. LAP348 and LAP267 have been expressed through a variety of promoters, although the most useful vectors have utilized either the cytomegalovirus (CMV) promoter[10] or the human β-actin promoter[11] (HβA). Expression of the LAPs with the CMV promoter results in high levels of expression ideal for use in transient assays. This level of expression, however, can result in high levels of nonspecific inhibition of transcription of other genes and is

[10] M. Boshart, F. Weber, G. Jahn, K. Dirsch-Hasler, B. Fleckenstein, and W. Schaffner, *Cell (Cambridge, Mass.)* **41,** 521 (1985).

[11] P. Gunning, J. Leavitt, G. Muscat, S.-Y. Ng, and L. Kedes, *Proc. Natl. Acad. Sci. U.S.A.* **84,** 4831 (1987).

detrimental to the production of stable LAP-expressing cell lines. LAP expression vectors utilizing promoters significantly weaker than the CMV promoter should be used for stable expression of either LAP348 or LAP267. The HβA promoter vectors have been particularly useful in this regard. In addition, a LAP267 gene under control of the pL14 promoter (pL14267) has been developed. This vector should produce very low levels of LAP267 in the absence of IPTG but should produce increasing levels on induction through positive autoregulation. This method of regulating LAP267 synthesis may be useful for decreasing toxic effects of high levels of LAP267 expression.

LAP-Inducible Promoters and Cloning Vectors

The LAP proteins have the ability to activate a number of mammalian promoters when the promoters are linked to *lac* operators. The most work has been carried out with operator sequences linked to a minimal SV40 promoter containing SP1 binding sites, a TATA box, and the SV40 origin of replication as shown in Fig. 3. The most efficient configuration of operators

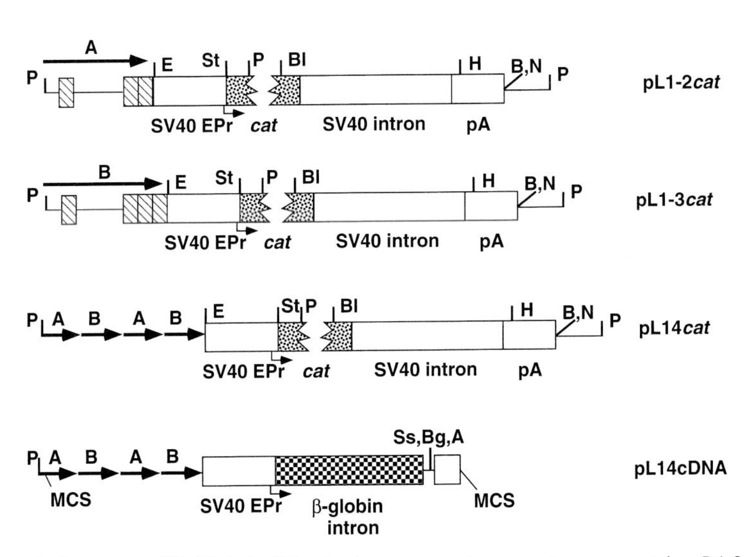

FIG. 3. Structure of LAP-inducible cloning vectors. *lac* operators present in pL1-2*cat* or pL1-3*cat* are indicated by hatched boxes. The repeated sets of operators from pL1-2*cat* and pL1-3*cat* present in pL14 constructs are designated by arrows A and B, respectively. Restriction sites are indicated as follows: *Pvu*II (P), *Eco*RV (E), *Stu*I (S), *Bal*I (BI), *Hpa*I (H), *Bam*HI (Bm), *Not*I (N), *Sse*8387 I (Ss), *Bgl*II (Bg), and *Asc*I (A). Multiple cloning sites (MCS) are indicated and include in order: *Sac*I, *Bst*XI, *Sac*II, *Not*I, *Xba*I, *Spe*I, *Bam*HI, *Sma*I, *Pst*I, and *Eco*RI sites.

appears to be pairs or groups of operators separated by approximately 100 bp. Linking spaced sets of operators increases the response of the promoter to the LAPs. All the vectors described in this figure contain upstream sets of operators. The operator sets in pL1-2*cat* and pL1-3*cat* were combined and multimerized to create pL14*cat*. The *cat* vectors can be used for insertion of any gene by removing the *cat* cassette with *Stu*I and either *Bal*I or *Hpa*I. The use of *Bal*I will leave a small segment of the 3′ end of the *cat* gene but will leave the SV40 intron and poly(A) addition site intact. Removal of the *cat* gene altogether with *Hpa*I will leave a functional poly(A) addition site. The pL14cDNA construct and a related vector carrying only a single set of operators, pL1-2cDNA (not shown), have been produced for use in directional cloning of cDNA libraries. These vectors are also useful for inserting specific sequences. Both vectors contain a 5′ β-globin intron followed by a short multiple cloning site and a SV40 poly(A) addition site. The bacterial vector contains a repeated polylinker on either side of the expression cassette as described in the legend to Fig. 3. Expression levels from the cDNA constructs containing a 5′ intron appear to be approximately twofold higher than that of comparable constructs containing the 3′ SV40 intron.

A special note should be made concerning cloning and subcloning with vectors containing reiterated *lac* operators. The repeated sets of operators are unstable in many bacterial strains. The use of recombination deficient hosts for propagation of these plasmids is essential. In addition, *lacI^q* strains of bacteria should not be used as the presence of the high levels of the *lac* repressor protein appears to greatly decrease the proportion and yield of plasmid containing largely intact arrays of operators. However, even when all precautions are taken, plasmid preparations containing the promoters of vectors such as pL14*cat* and pL21*cat* usually represent heterogeneous populations and must be screened carefully to determine the number of operators left in individual preparations of vector DNA. Restriction digestion with *Pvu*II and *Eco*RV are particularly useful for determining the number of operators. Plasmids appear to lose operators largely due to deletion of the repeated spaced arrays. Thus, when analyzing plasmid preparations, operator fragments often appear to be present in sizes corresponding to the deletion of one or more of these sets. In general, the number of operators left in a given plasmid can be estimated by the size of the operator containing fragments (each operator set is approximately 340 bp in length) which is an indication of the number of spaced sets remaining.

IPTG will regulate the activity of the different operator containing promoters in response to either LAP267 or LAP348. The most efficient levels of regulation require the use of different operator sets. The use of multiple sets of spaced operators dramatically increases the level of expression driven by LAP348 even in the presence of IPTG. Thus, when using LAP348 to control a gene whose expression needs to be turned off, an inducible promoter containing only a single set of operators should be used, such as that contained in the pL1-2*cat* or pL1-3*cat* constructs. Conversely, because the activity of LAP267 is somewhat lower than that of LAP348, promoters with large numbers of operator sets (such as those contained in the pL14 or pL21 constructs) are required for maximal induction in the presence of IPTG. The larger number of operators does not appear to increase the level of transactivation by LAP267 in the absence of IPTG at 39.5°.

Although this chapter focuses primarily on the use of the LAP vectors for controlled expression of target genes, this system may also be applicable for overexpression experiments. Expression levels from plasmid vectors containing large numbers of operators (e.g., pL21*cat*) are extremely high in the presence of LAPs, particularly when LAP348 is used. Using the *cat* gene as an indicator, LAP348 driven expression from pL21*cat* was approximately 30 times higher than that from a SV40 expression vector containing its endogenous enhancers when compared in a variety of cell lines.

Selection of LAP-Expressing Cell Lines

In order to facilitate the use of LAP348 in cell lines, several genetic selection schemes have been developed which greatly enrich for LAP-expressing cells. The selection for LAP activity is based on the observation that the LAP-inducible promoters described earlier exhibit very low levels of expression in the absence of LAP. When cells are transfected with selectable markers under control of the LAP-inducible promoters and a LAP expression vector, a large increase in the number of selectable colonies is observed compared to transfections without the LAP construct. The resultant cell lines, for the most part, contain both constructs and produce functional levels of the LAP. A variety of cell types and selection schemes have been used as shown in Table I. Two types of selectable markers have been used.[1] Drug resistance markers including the neomycin resistance gene[12] (*neo*) and the histidinol resistance gene[13] (*his*) can be used to select for LAP-expressing cell lines. LAP-inducible drug resistance gene constructs are outlined in Fig. 4. This strategy has several advantages. First, if cell lines are to be produced with a variety of LAP-inducible constructs, it is probably best to first derive a single $LAP348^+$ clone to use as a control

[12] P. J. Southern and P. Berg, *J. Mol. Appl. Genet.* **1,** 327 (1982).

[13] O. Danos and R. C. Mulligan, *Proc. Natl. Acad. Sci. U.S.A.* **85,** 6460 (1988).

TABLE I
GENETIC SELECTION SCHEMES FOR PRODUCING LAP348⁺ CELL LINES

Selection vector	Cell type used
Drug resistance markers	
pL1-2*neo*[a]	HeLa
	Ltk⁻
	3T3
	REFs
	D3 ES cells
pL14*his*[b]	Ltk⁻
	3T3
Transforming genes[c]	
pL2T-antigen	REF
pL1-3 p53	REF

[a] Amounts of G418 used in selections varied with the cell line. Optimal concentrations for each cell line were previously determined.

[b] Histidinol selection utilized histidine-free media along with recommended concentrations of histidinol.

[c] In all cases cell lines transformed with oncogenes were selected by isolating transformed foci from the lawn of nontransformed rat embryo firbroblasts (REFs)

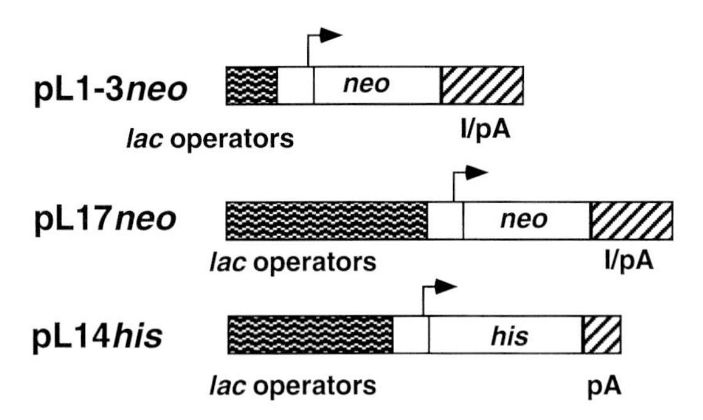

FIG. 4. LAP-inducible drug resistance marker genes. Operators are indicated by boxes containing wavy lines. The SV40 early promoter is indicated by the unmarked open boxes. The start site of transcription of each vector is indicated by an arrow. The presence of a 3′ SV40 intron (I) or poly(A) addition site (pA) is indicated.

in all experiments. This line, once characterized for LAP activity, can be used for introduction of the test gene by cotransfection with a second selectable marker. In addition, the drug resistance markers can be utilized in any type of cell.

An additional advantage to positive selection for LAP activity is that the drug-resistant markers can be used to coselect test genes that do not produce a sufficient morphological change or alteration in growth potential of cells to select for transformants directly. In this case the test gene can be introduced by a triple transfection with the LAP-inducible selectable marker and LAP348 or can be introduced in a second transfection with a second marker into a LAP$^+$ cell line. When carrying out triple transfections, the use of pL17*neo* as the selectable marker should be considered. The reason for this is that the large number of operators in pL17*neo* will allow it to be activated to some extent by LAP348 even in the presence of optimal amounts of IPTG. Thus, during experiments where the test gene (under control of a weaker promoter such as that in pL1-2) is turned off with IPTG, the cells should still remain *neo* resistant. This strategy has been used successfully in the production of cells conditionally transformed with *fos*.[8]

In addition to drug resistance genes, transforming oncogenes have been used as selectable markers. The oncogenes can be used as selectable agents only when their expression imparts a selectable morphological change in the cell or increased growth potential. This strategy has been used successfully to transform cells with LAP-inducible SV40 T-antigen[1] and p53[14] constructs.

Selection of Stable Cell Lines with LAP348

As already introduced, LAP-inducible drug resistance markers can be used to isolate LAP$^+$ cell lines at high frequency. These transformations are carried out using transfection techniques known to work in a given cell line. To date, our laboratory has produced LAP-expressing cell lines using either calcium phosphate transfection or electroporation methods. Whichever transformation method is used, several points should be considered. The first is that two types of transfection should be carried out. Cells should be transformed with the LAP-inducible selectable marker alone and in combination with LAP348. The number of drug-resistant colonies resulting from the transformation will indicate the relative success of the coselection. A large increase in the number of colonies on cotransfection with LAP348 is a good indicator that the majority of colonies will express LAP348. A

[14] G. P. Zambetti, D. Olson, M. Labow, and A. J. Levine, *Proc. Natl. Acad. Sci. U.S.A.* **89,** 3952 (1992).

large number of colonies in transfections without the activator may suggest that the selectable marker exhibits a relatively high level of LAP-independent expression in that cell line or that very low levels of expression are required for drug resistance. LAP348-expressing cell lines have been derived, however, even when LAP348 cotransfection causes a decrease in the number of colonies after cotransfection.[1] The amount of particular DNAs used in these transfections should also be considered. Typically, transfections have included 1 μg of LAP348 and 2.5 to 5 μg of the LAP-inducible selectable marker. If triple transfections are being done in order to introduce a test LAP-inducible gene which has no selection, the test construct should be included in higher molar amounts than the LAP348 and drug resistance genes. Test genes introduced by coselection should be linearized within the vector sequences. The LAP constructs and LAP-inducible selectable marker DNAs do not need to be linearized prior to transfection, although linearization may increase the number of colonies derived from a particular transfection.

LAP^+ cells can be identified in two ways. LAP348 expression can be examined using immunofluorescence as previously described. One procedure previously used successfully is listed below. This procedure was optimized for use with a rabbit polyclonal sera[1] (the antibody was developed in the laboratory of Dr. Thomas Shenk, Department of Molecular Biology, Princeton University, Princeton, NJ).

Immunofluorescence Analysis of LAP Expression

1. Cells grown on coverslips or 35-mm dishes are washed three times with phosphate-buffered saline (PBS) then fixed with PBS containing 4% (v/v) formaldehyde for 10 min at room temperature.
2. Cells are then permeabilized with PBS containing 0.1% Triton X-100 for 2 min at room temperature and then washed five times with PBS.
3. Cells are incubated at room temperature with a 1:100 dilution of rabbit polyclonal sera (in PBS also containing 5% goat serum) directed against *lac* repressor.
4. Cells are washed five times with PBS containing 0.5% bovine serum albumin and 0.1% Tween 20.
5. Cells are incubated with PBS containing 25 μg/ml fluorescein-conjugated goat anti-rabbit antibody, 0.5% bovine serum albumin, and 0.1% Tween 20 followed by washing five additional times in PBS containing 0.1% Tween 20. The cells are then analyzed by fluorescence microscopy. LAP-expressing cells should show distinct nuclear fluorescence.

Alternatively, LAP348-expressing cell lines can also be identified by testing the ability of cell lines to support expression of a LAP-inducible *cat* gene after transient transfection. The procedure shown below is for use with cells grown in monolayers.

Identification of LAP348 Expressing Cells by Comparative CAT Assays

1. Cells (after initial expansion of colonies) are split into duplicate wells of a six-well dish and are incubated for 6-18 hr under normal culture conditions. It is best to plate sufficient cells such that the well is 25–50% confluent at the time of transfection (approximately 5×10^5 to 1×10^6 cells per 60-mm dish depending on the growth rate and plating efficiency of the cell line). A cell line known not to express LAP348 should also be included as a control for the basal level of expression of pL14*cat* observed in the absence of LAP.

2. Cells are transfected with either pL14*cat* or pSV2*cat*. Because of the high level of expression from LAP-inducible promoters, the transfection technique does not need to be particularly efficient. For most cell lines, transfection of 2–5 μg of test plasmid using calcium phosphate transfection techniques should be sufficient. We always transfect the test gene in the presence 20–25 μg of high molecular weight carrier DNA (either calf thymus or salmon sperm double-stranded DNA) per ml of transfection precipitate. Calcium phosphate precipitates are made up in batch to try and eliminate variation due to precipitate consistency. For calcium phosphate transfections, cells are incubated overnight with the precipitate, washed with fresh medium, refed, and incubated an additional 24–36 hr.

3. Cells are washed twice with phosphate-buffered saline and then harvested by scraping in 1 ml of STE [40 m*M* Tris–HCl (pH7.4), 0.15 *M* NaCl, and 1 m*M* EDTA], pelleted by centrifugation, and resuspended in 100 μl 0.25 *M* Tris–HCl (pH 7.9). Cell extracts are prepared and CAT assays are carried out essentially as previously described.[15] We generally use 75 μl of the extracts for the first CAT assays and incubate the reactions for 2 hr at 37°. LAP348-expressing cells should express the pL14*cat* gene at significantly higher levels than that by pSV2*cat*. LAP-negative cell lines should show the opposite pattern.

Direct Selection of Cell Lines Transformed with LAP-Inducible Oncogenes

The second strategy for selecting LAP-expressing cell lines is to use oncogenes that produce selectable alteration in the growth of the recipient cell, e.g., the ability to produce identifiable foci. In this case, oncogenes

[15] C. M. Gorman, L. F. Moffat, and B. H. Howard, *Mol. Cell. Biol.* **2,** 1044 (1982).

under control of a LAP-inducible promoter are directly introduced into the test cell line (such as primary REFs or 3T3 cells) by cotransfection with LAP348. The most important aspect to this selection scheme is that the phenotype of cell lines isolated by coselection with LAP348 is often readily controlled by IPTG. REFs transformed with either LAP-inducible p53, T-antigen, or *fos* constructs were rapidly reverted to the untransformed state by growth in the presence of low amounts of IPTG. The morphological changes in such cell lines were correlated with a large inhibition of synthesis of the regulated oncogene. In these cases, removal of IPTG resulted in reexpression of the oncogene and resumption of the transformed state.

Typically, transformation experiments carried out with LAP-inducible oncogenes are carried out in a manner similar to that described earlier for drug resistance markers. Cell lines isolated in this manner can be directly tested for the ability of IPTG to control the selected phenotype. Foci are isolated and, after initial amplification, are split into two dishes with and without 5 m*M* IPTG in the media. The cultures are observed for 5–10 days, and cell lines which show distinct differences in growth or morphology are frozen and used for further analysis.

After isolation of cell lines that appear to be regulated by IPTG, two types of experiments should be carried out. Expression of the oncogene in the transformed cell lines should be analyzed as a function of the concentration of IPTG in the media. LAP348 activity in transformed cells can be regulated with concentrations of IPTG between 1 and 10 m*M*; 5 m*M* IPTG is typically used in initial experiments. This concentration of IPTG has not shown any toxicity in the cell lines tested so far and also appears to give near maximal inhibition of LAP348 in stable cell lines. However, amounts of IPTG as low as 100 μM can be effective.[8] The optimal amount of IPTG needed to affect expression in a given cell line must be determined empirically. In analyzing the regulation of the tested oncogene in LAP348-expressing cells, the half-life of the protein or mRNA should be taken into account in designing these experiments. If the oncogene has a particularly long half-life (as is the case for the SV40 T-antigen), then it would be difficult to determine the level of regulation by Western blotting. In general, the best way to measure the level of regulation is to examine levels of the oncogene via immunoprecipitation after a short period of metabolic labeling at various times after the addition of IPTG. Measuring mRNA levels is also a good indicator of the level of regulation of the LAP-inducible gene since the LAPs regulate expression at the level of transcription.

Finally, the growth and morphological characteristics of the cell lines should be examined in detail as a function of IPTG concentration and time in optimal amounts of IPTG. The half-life of the oncogene protein or mRNA will be of particular importance in this analysis. If the oncogene protein has a very long half-life, the morphological response of that cell will be correspondingly slower. For example, REF cell lines transformed with SV40 T-antigen (unpublished observations) took significantly longer to revert to a more untransformed phenotype than REF cell lines transformed with a mutant p53 construct.

Use of LAP267

Although LAP348 has proven to be effective in the establishment of reversibly transformed cell lines, it is probably difficult to use LAP348 in stable cell lines when the test gene is toxic or inhibitory to cell growth. In this case it would be advantageous to use a system where the test gene was normally off but could be rapidly turned on using some stimulus. LAP267 may be of particular use in these settings because of its very low level of activity in the absence of IPTG. Although it should be possible to use direct selection techniques for LAP267, in general, LAP267-expressing cell lines have been produced by cotransfection with a constitutively expressed drug resistance marker. A second gene can then be introduced into the LAP267 cells using a second selectable marker. It should be noted that LAP267 expression may be toxic to cells since it has been very difficult to produce cell lines expressing this protein, although mouse Ltk$^-$ cells, human 293 cells, and 3T3 cells have been used successfully. The description of the LAP267 vectors is included for completeness. To date, the best results have been obtained with a variant of mouse Ltk$^-$ cells which have been transformed with pHβA267. LAP267-expressing cells can be used to express a test gene by cotransection and coselection with the HSV*tk* gene. These transfections are carried out at 39° to minimize the activity of LAP267 since it appears activate transcription at a low level at 37°.

[25] Lineage Analysis Using Retrovirus Vectors

By CONSTANCE CEPKO, ELIZABETH F. RYDER, CHRISTOPHER P. AUSTIN, CHRISTOPHER WALSH, and DONNA M. FEKETE

I. Introduction

The complexity and inaccessibility of many types of vertebrate embryos have made lineage analysis through direct approaches, such as time-lapse microscopy and injection of tracers, extremely difficult or impossible. A genetic and clonal solution to lineage mapping in some species is by infection with retrovirus vectors. This chapter summarizes the basis for this technique and details the strategies and current methods in use in our laboratory.

Transduction of Genes via Retrovirus Vectors

A retrovirus vector is an infectious virus that transduces a nonviral gene into mitotic cells *in vivo* or *in vitro*.[1] These vectors utilize the same efficient and precise integration machinery of naturally occurring retroviruses to produce a single copy of the viral genome stably integrated into the host chromosome. Those that are useful for lineage analysis have been modified so that they are replication incompetent and thus cannot spread from one infected cell to another. They are, however, faithfully passed on to all daughters once viral integration has taken place, making them ideal for lineage analysis.

Retroviruses use RNA as their genome, which is packaged into a membrane-bound protein capsid. Immediately following entry into a cell, they produce a DNA copy of their genome via reverse transcriptase, a protein included in the capsid. The DNA copy is integrated into the host cell genome and is thereafter referred to as a "provirus." Complete synthesis of an integration-competent DNA viral genome requires S phase, and integration occurs during M phase. Thus, only mitotic cells will serve successfully as hosts for retroviral integration. Since integration occurs during the M phase and since there is only one DNA copy of the viral genome produced per infectious virion, only one daughter of the first mitosis following infection will inherit the viral genome.[2] However, all subsequent daughters of that cell will inherit the viral genome.

Most vectors began as proviruses that were cloned from cells infected with a naturally occurring retrovirus. Although extensive deletions of proviruses were made, vectors retain the cis-acting viral sequences necessary for the viral life cycle. These include the Ψ packaging sequence (necessary for recognition of the viral RNA for encapsidation into the viral particle), reverse transcription signals, integration signals, viral promoter, enhancer, and polyadenylation sequences. A cDNA can thus be expressed in the vector using the transcription regulatory sequences provided by the virus. Since replication-incompetent retrovirus vectors usually do not encode the structural genes whose products comprise the viral particle, these proteins must be supplied through complementation. The structural proteins, products of the *gag*, *pol*, and *env* genes, are typically supplied by "packaging"

[1] R. Weiss, N. Teich, H. Varmus, and J. Coffin, "RNA Tumor Viruses." Cold Spring Harbor Lab., Cold Spring Harbor, NY, 1984, 1985.

[2] T. Roe, T. C. Reynolds, G. Yu, and P. O. Brown, *EMBO J.* **12**(5), 2099 (1993).

cell lines. These lines are usually stable lines that contain the *gag*, *pol*, and *env* genes as a result of the introduction of these genes by transfection. However, these lines do not contain the packaging sequence, Ψ, on the viral RNA that encodes the structural proteins. Thus, the packaging lines make viral particles that do not contain RNA encoding gag, pol, or env proteins.

Retrovirus vector particles are essentially identical to naturally occurring retrovirus particles. They enter the host cell via interaction of a viral envelope glycoprotein (a product of the viral env gene) with a host cell receptor. The murine viruses have several classes of env glycoprotein which interact with different host cell receptors. The most useful class for lineage analysis of rodents is the ecotropic class. The ecotropic env glycoprotein allows entry only into rat and mouse cells via the ecotropic receptor on these species. It does not allow infection of humans, and thus is considered relatively safe for gene transfer experiments. Until 1988, the packaging line most commonly in use was the Ψ2 line.[3] It encodes the ecotropic env gene and, in our experience, it makes the highest titers of vectors, relative to other packaging lines (for unknown reasons). However, it can also lead to the production of helper virus (discussed below). Second generation ecotopic packaging lines, ΨCRE[4] and GP+E-86,[5] and third generation packaging lines, ΩE[6] and Bosc 23,[7] have not been reported to lead to production of helper virus to date. Any of the ecotropic packaging lines can be used to produce vector stocks for lineage analysis. Regardless of the packaging line used, however, all stocks should be assayed for the presence of helper virus.

Avian viruses also have multiple envelope types, called subgroups A–E. Two packaging lines, Q2bn[8] and Isolde,[9] are available for producing avian retroviruses and both make subgroup A envelope. Subgroup A envelope allows infection of most commercially available chicken strains. However, one thing to keep in mind is that commercially available chickens are not monitored or bred for infection of subgroup A viruses. If you have trouble with efficiency of infection, you might want to use another source of embryos in case the A envelope receptor is limiting.

[3] R. Mann, R. C. Mulligan, and D. Baltimore, *Cell* (*Cambridge, Mass.*) **33,** 153 (1983).
[4] O. Danos, and R. C. Mulligan, *Proc. Natl. Acad. Sci. U.S.A.* **85,** (1988).
[5] D. Markowitz, S. Goff, and A. Bank, *J. Virol.* **62,** 1120 (1988).
[6] J. P. Morgenstern, and H. Land, *Nucleic Acids Res.* **18,** 3587 (1990).
[7] W. S. Pear, G. P. Nolan, M. L. Scott, and D. Baltimore, *Proc. Natl. Acad. Sci. U.S.A.* **90,** 8392 (1993).
[8] A. W. Stoker and M. J. Bissell, *J. Virol.* **62,** 1008 (1988).
[9] F.-L. Cosset, C. Legras, Y. Chebloune, P. Savatier, P. Thoraval, J. L. Thomas, J. Samarut, V. M. Nigon, and G. Verdier, *J. Virol.* **64,** 1070 (1990).

It has been known for some time that nonretroviral enveloped viruses can contribute their envelope glycoprotein to retroviruses and thus endow a retrovirus with a different host range. The vesicular stomatitis virus (VSV) envelope protein, G protein, has been expressed in a murine retrovirus vector such that a broad host range virion carrying the neo gene was produced.[10] This virus was shown to infect and express in zebrafish, chum salmon, hamster, and canine cells *in vitro*. The vector unfortunately expressed poorly in the fish lines and thus newer generation vectors need to be designed to overcome this limitation. Such vectors are currently being developed.

Production of Virus Stocks for Lineage Analysis

Replication-incompetent vectors that encode a histochemical reporter gene, such as *Escherichia coli lacZ*, are the most useful for lineage studies as they allow analysis of individual cells in tissue sections or whole mounts. Stocks of such vectors are most easily obtained from packaging lines stably transduced with the vector genome. One can also produce stocks transiently by transfection of particular packaging cells, such as the Bosc 23 or Q2bn, or by cotransfection using the vector expressing VSV-G when one wishes to obtain the broadest host range. (A detailed description of protocols for making vector stocks, for titering and concentrating virus stocks, and checking for helper virus contamination has been published[11] and will not be given here.) When one can use stable lines, as in the case when infecting murine or rat cells, it is best to obtain lines that make high titered stocks of lineage vectors from the laboratories that have created stable lines. We have placed Ψ2 and ΨCRE producers of BAG,[12] a *lacZ* virus that we have used for lineage analysis, on deposit at the ATCC in Rockville, Maryland. They can be obtained by anyone and are listed as ATCC CRL Nos. 1858 (ΨCRE BAG) and 9560 (Ψ2 BAG). Similarly, Ψ2 producers of DAP,[13] a vector encoding human placental alkaline phosphatase (PLAP, described further below), is available as CRL No. 1949. Both of these vectors transcribe the reporter gene from the viral LTR promoter and are generally useful for expression of the reporter gene in most tissues. We compared

[10] J. C. Burns, T. Friedmann, W. Driever, M. Burrascano, and J.-K. Yee, *Proc. Natl. Acad. Sci. U.S.A.* **90,** 8033 (1993).

[11] C. L. Cepko, F. M. Ausubel, R. Brent, R. E. Kingston, D. D. Moore, J. G. Seidman, J. A. Smith, and K. Struhl, eds., "Current Protocols in Molecular Biology," Chapters 9.10–9.13. Greene Publishing Associates, New York, 1992.

[12] J. Price, D. Turner, and C. L. Cepko, *Proc. Natl. Acad. Sci. U.S.A.* **84,** (1987).

[13] S. C. Fields-Berry, A. L. Halliday, and C. L. Cepko, *Proc. Natl. Acad. Sci. U.S.A.* **89,** 693 (1992).

the expression of *lacZ* driven by several different promoters[14,15] and found that the LTR was generally the most reliable and non-cell type specific. This is an important consideration as it is desirable to see all of the cells descended from an infected progenitor and thus a constitutive promoter is the most useful for lineage studies. However, even with constitutive promoters, it has been noted that some infected cells do not express detectable β-Gal protein, even among clones of fibroblasts infected *in vitro*. Thus, it is important to restrict conclusions about lineage relationships to cells that are marked and not to make assumptions about their relationships to cells that are unmarked.

For lineage applications it is usually necessary to concentrate the virus in order to achieve sufficient titer. This is typically due to a limitation in the volume that can be injected at any one site. Viruses can be concentrated fairly easily by a relatively short centrifugation step. Virions also can be precipitated using polyethylene glycol or ammonium sulfate, and the resulting precipitate is collected by centrifugation. Finally, the viral supernatant can be concentrated by centrifugation through a filter that allows only small molecules, to pass (e.g., Centricon filters). Regardless of the protocols that are used, it must be kept in mind that retroviral particles are fragile, with short half-lives even under optimum conditions. One exception to this are the viruses bearing VSV-G, which are very stable. In order to prepare the highest titered stock for multiple experiments, we usually concentrate several hundred milliliters of producer cell supernatant. These concentrated stocks are titered and tested for helper virus contamination, and can be stored indefinitely at $-80°$ in small (20- to 50-aliquots).

Replication-Competent Helper Virus

The replication-competent virus is sometimes referred to as the helper virus as it can complement ("help") a replication-incompetent virus and thus allow it to spread from cell to cell. It can be present in an animal through exogenous infection (e.g., from a viremic animal), expression of an endogenous retroviral genome, or through recombination events between two viral RNAs encapsidated in retroviral virions produced by packaging lines. The presence of a helper virus is an issue of concern when using replication-incompetent viruses for lineage analysis as it can lead to horizontal spread of the marker virus, creating false lineage relationships. The most likely source of helper virus is the viral stock used for lineage analysis. The genome(s) that supplies the gag, pol, and env genes in packag-

[14] D. L. Turner, E. Y. Snyder, and C. L. Cepko, *Neuron* **4,** 833 (1990).

[15] C. L. Cepko, C. P. Austin, C. Walsh, E. F. Ryder, A. Halliday, and S. Fields-Berry, *Cold Spring Harbor Symp. Quant. Biol.* **55,** 265 (1990).

ing lines does not encode the Ψ sequence, but can still become packaged, although at a low frequency. If it is coencapsidated with a vector genome, recombination in the next cycle of reverse transcription can occur. If the recombination allows the Ψ^- genome to acquire the Ψ sequence from the vector genome, a recombinant that is capable of autonomous replication results. This recombinant can spread through the entire culture (although slowly due to envelope interference). Once this occurs, it is best to discard the producer clone as there is no convenient way to eliminate the helper virus. As would be expected, recombination giving rise to helper virus occurs with greater frequency in stocks with a high titer and with vectors that have retained more of the wild-type sequences (i.e., the more homology between the vector and packaging genomes, the more opportunity there is for recombination). Note that a murine helper genome itself will not encode a histochemical marker gene as apparently there is not room, or flexibility, within murine viruses that allows them to be both replication competent and capable of expressing another gene like *lacZ*. The way that spread would occur is by a cell being infected with both the *lacZ* virus and a helper virus. Such a doubly infected cell would then produce both viruses. In the case of avian viruses, a replication-competent helper virus can carry and express a nonviral gene of <2.5 kb, but it is unlikely to do so unless a very specific recombination takes place. The likelihood of problems with avian viruses can be minimized using line O chickens, which do not encode any endogenous avian viruses.

When performing lineage analysis, there are several signs that can indicate the presence of helper virus within an individual animal. First, if an animal is allowed to survive for long periods of time after inoculation, particularly if embryos or neonates are infected, the animal is likely to acquire a tumor when the helper virus is present. Most naturally occurring replication-competent viruses are leukemogenic, with the disease spectrum being at least in part a property of the viral LTR. Second, if one analyzes either short or long term after inoculation, the clone size, clone number, and spectrum of labeled cells may be indicative of helper virus. For example, the eye of a newborn rat or mouse has mitotic progenitors for retinal neurons, as well as mitotic progenitors for astrocytes and endothelial cells. By targeting the infection to the area of progenitors for retinal neurons, we only rarely see infection of a few blood vessels or astrocytes as their progenitors are outside of the immediate area that is inoculated and they only get infected by leakage of the viral inoculum from the targeted area. However, if the helper virus was present, we would see infection of a high percentage of astrocytes, blood vessels, and eventually, other eye tissues since virus spread would eventually lead to infection of cells outside of the targeted area. A correlation would be expected to exist between the percentage of such nontargeted cells that are infected and the degree to which their progenitors are mitotically active after inoculation, due to the fact that infection requires a mitotic target cell. If tissues other than ocular tissues were examined, one would similarly see evidence of virus spread to cells whose progenitors would be mitotically active during the period of virus spread. In addition, the size and number of "clones" may also appear to be too large for true "clonal" events if the helper virus were present. This interpretation of course relies on some knowledge of the area under study.

Determination of Sibling Relationships

When performing lineage analysis, it is critical to unambiguously define cells as descendents of the same progenitor. This can be relatively straightforward when sibling cells remain rather tightly and reproducibly grouped. An example of such a straightforward case is the rodent retina, where the descendents of a single progenitor migrate to form a coherent radial array.[14,16] In such a system, lineage analysis can be performed using one or two distinct histochemical marker viruses, as described below. For more complex systems, particularly where cell migration is important, many more markers are needed, as described in a later section (see Clonal Analysis Using the PCR/Library Method).

If only a single histochemical marker virus is used, a standard virological titration can be performed in which a particular viral inoculum is serially diluted and applied to tissue. In the retina, the number of radial arrays, their average size, and their cellular composition were analyzed in a series of animals infected with dilutions that covered a 4-log range. The number of arrays was found to have a first order relationship to the dose or virus, whereas the size and cell composition of the radial arrays were independent of the viral dose.[16] Such results indicate that the working definition of a clone, in this case a radial array, fulfilled the statistical criteria expected of a single hit event.

There are several difficulties involved with the use of a single marker virus. First, there must be a wide range of dilutions that can be injected to give countable numbers of events, which is required for determination of a first-order relationship between clone number and viral dose. This type of analysis also relies critically on controlling the exact volume of the injection. These problems may be avoided in cases where it is possible to perform the analysis solely on infections with very small amounts of virus, although generating large amounts of data under these conditions is tedious.

[16] D. L. Turner and C. L. Cepko, *Nature* (*London*) **328,** 131 (1987).

Another potential problem with the dilution approach is that aggregates may form during virus concentration that are not separated by dilution. Thus, even at low dilution, a "clone" may be the result of infection of adjacent cells by two or more members of such an aggregate. In addition, it is difficult to calculate an error rate for the assignment of clonal boundaries, which may make interpretation difficult, particularly of events which occur rarely in the data set.

For these reasons, additional viral vectors with histochemically distinguishable marker genes have been developed for use in mixed infections. In addition to addressing the problems mentioned earlier, use of two marker viruses can often provide a much better initial idea of what clonal boundaries are likely to be by simple visual inspection of the tissue. By using two markers, errors arising from formation of viral aggregates can be visualized since some of these aggregates will include two distinguishable virions. Various criteria for clonal boundaries can be tested, and error rates for each can be computed. Extensive dilutions are not needed.

Two viruses that have been used for this approach encode cytoplasmically localized versus nuclear localized β-Gal. This virus pair is useful when the cytoplasmically localized β-Gal is easily distinguished from the nuclear-localized β-Gal.[17,18] We have found that this is not the case in rodent nervous system cells, as the cytoplasmically localized β-Gal quite often is restricted to neuronal cell bodies and is therefore difficult to distinguish from nuclear-localized β-Gal. To overcome this problem, we created the above-mentioned DAP virus,[13] which is distinctive from the *lacZ*-encoding BAG virus. Another virus that is currently under construction is one encoding the green fluorescent protein.[19] This protein may not only give another easily distinguishable histochemical tag, but may enable studies of clone growth in a live animal.

To perform the two-marker analysis in the rodent retina, a stock containing BAG and DAP was made by growing a $\Psi2$ producer clone for BAG and a $\Psi2$ producer clone for DAP on the same dish.[13,20] The resulting supernatant was concentrated and used to infect rodent retina. The tissue was then analyzed histochemically for the presence of blue (due to BAG infection) and purple (due to DAP infection) radial arrays. If radial arrays were truly clonal, then each one should be only one color. Analysis of 1100 arrays indicated that most were only one color. However, 5 comprised blue cells and purple cells, presumably from infection of adjacent progenitor

[17] S. M. Hughes and H. M. Blau, *Nature* (*London*) **345,** 350 (1990).

[18] D. S. Galileo, G. E. Gray, G. C. Owens, J. Majors, and J. R. Sanes, *Proc. Natl. Acad. Sci. U.S.A.* **87,** 458 (1990).

[19] M. Chalfie, Y. Tu, G. Euskirchen, W. W. Ward, and D. C. Prasher, *Science* **263,** 802 (1994).

[20] A. Halliday and C. L. Cepko, *Neuron* **9,** 15 (1992).

cells with BAG and DAP. (Infection of one cell with both BAG and DAP would not be a problem since the resulting cells would be clonally related.) The 5/1100 figure is an underestimate of the true frequency of incorrect assignment of clonal boundaries as sometimes two BAG or two DAP virions will infect adjacent progenitor cells; the resulting arrays will be a single color, but not clonal. A closer approximation of the true frequency can be obtained using the following formula (for derivation, see Field-Berry *et al.*[13]).

$$\% \text{ error} = \frac{(\text{number of bicolored arrays})[(a + b)^2/2ab]}{\text{number of total arrays}}$$

where a and b are the relative titers of the two viruses in the virus stock. The relative titer of BAG and DAP used in the coinfection was 3:1, and thus the value for percentage errors in clonal assignments in the rodent retina was 1.2%.

Such a low error rate shows that the choice of clonal boundary was reasonable and that viral aggregation during concentration (or at any other step) was not a problem for this viral stock. To take an extreme counter example, if every array were the result of infections by two particles, 6/16 of the arrays would be composed of both blue cells and purple cells. (For a stock with relative viral titers of 1:1, half the arrays would have blue cells and purple cells.) In analyzing the composition of arrays (e.g., by cell type), one need not worry, as one would if using one marker, that rare cell type combinations may be due to occasional errors in assignment of clonal boundaries. If such rare arrays are always of one color (when the error rate is very low), then they are very likely clones.

The error rate being computed here is the rate of "lumping" errors, that is, the frequency with which the criteria defining a clone lumps together daughters of more than one progenitor. However, this computation does not allow assessment of "splitting" errors, where clones that are more spread out (presumably owing to migration) are split by the definition of a clone into two or more subclones. If the criteria used to define clonal relationships are found to generate too many lumping errors, a more restrictive definition can be made, and then this definition can be tested for percentage lumping errors. This process can be done iteratively until the error rate reaches a level that is acceptable relative to the point that is being tested. However, it should be kept in mind that the more restrictive a definition is for clonal relationships, the more it is prone to generate splitting errors. Splitting errors may become obvious in animals injected with very dilute viral stocks; for example, one blue array in a large, otherwise unlabeled area is probably a clone, but it may be divided into several by too restrictive clonal definitions. Thus, combining two-marker analysis with a few dilution experiments may be useful to help balance splitting and lumping errors. If it is necessary to avoid completely both splitting and lumping errors, a much greater number of vectors and more tedious detection method must be used, as detailed below (see Clonal Analysis Using the PCR/Library Method).

The value of 1.2% for lumping errors in assignment of clonal boundaries places an upper limit of 1.2% as the frequency of aggregation for this viral stock since this figure will include errors due to both aggregation and independent virions infecting adjacent progenitors. The presence of helper virus would probably increase the error rate, as it would have the effect of enlarging apparent clone size and creating overlap between clones of different markers. The percentage of errors on injecting this stock in other areas of an animal would depend on the particular circumstances of the injection site and on the multiplicity of infection (MOI, the ratio of infectious virions to target cells). Most of the time the MOI will be quite low (e.g., in the retina it was approximately 0.01 at the highest concentration of virus injected). Injection into a lumen, such as the lateral ventricles, should not promote aggregation or a high local MOI, but injection into solid tissue could present problems since the majority of the inoculum may have access to a limited number of cells at the inoculation site. By coinjecting BAG and DAP, one can monitor the frequency of these events and thus determine if clonal analysis is feasible.

The above analysis, and similar ones using avian vector stocks,[21,22] were performed using viruses that were produced on the same dish and concentrated together. This was done because we felt that the most likely way that two adjacent progenitors might become infected would be through small aggregates of virions. We grew both virions together on the same dish in order for the assay to be sensitive to any aggregation that might occur prior to concentration, but this is probably not necessary in most cases. Aggregation most likely occurs during the concentration step, as one often can see macroscopic aggregates after resuspending pellets of virions. Thus, when the two-marker approach is used to analyze clonal relationships, it is best to coconcentrate the two vectors together in order for the assay to be sensitive to aggregation arising from this aspect of the procedure. (Although aggregation of virions may frequently occur during concentration, it apparently does not frequently lead to problems in lineage analysis, presumably because of the high ratio of noninfectious particles to infectious

[21] E. F. Ryder and C. Cepko, *Neuron* **12,** 1011 (1994).

[22] D. M. Fekete, J. Perez-Miguelsanz, E. F. Ryder, and C. L. Cepko, *Dev. Biol.* **166,** 666 (1994).

particles found in most retrovirus stocks. It is estimated that only 0.1–1.0% of the particles will generate a successful infection. Moreover, most aggregates are probably not efficient as infectious units; it must be difficult for the rare infectious particle(s) within such a clump to gain access to the viral receptors on a target cell.)

Several methods can be used to determine the ratio of two genomes present in a mixed virus stock (e.g., BAG plus DAP). The first two methods are performed *in vitro* and are simply an extension of a titration assay. Any murine virus stock is normally titered on NIH/3T3 cells to determine the amount of virus to inject. The infected NIH/3T3 cells are then either selected for the expression of a selectable marker when the virus encodes such a gene (e.g., *neo* in BAG and DAP) or are stained directly, histochemically, for β-Gal or PLAP activity without prior selection with drugs. If no selection is used, the relative ratio of the two markers can be scored directly by evaluating the number of clones of each color on a dish. Alternatively, selected G418-resistant colonies can be stained histochemically for both enzyme activities and the relative ratio of blue versus purple G418-resistant colonies computed. A third method of evaluating the ratio of the two genomes is to use the values observed from *in vivo* infections. After animals are infected and processed for both histochemical stains, the ratio of the two genomes can be compared by counting the number of clones, or infected cells, of each color.

When all the above methods were applied to lineage analysis in mouse retina[13] and rat striatum,[20] the value obtained for the ratio of G418-resistant colonies scored histochemically was almost identical to the ratio observed *in vivo*. Directly scoring histochemically stained, non-G418-selected NIH/3T3 cells led to an underestimate of the number of BAG-infected colonies, presumably because such cells often are only faint blue, whereas DAP-infected cells are usually an intense purple. *In vivo*, this is not generally the case as BAG-infected cells are usually deep blue.

Regardless of which method is used to score sibling relationships, one further recommendation to aid in the assignments is to choose an injection site that will allow the inoculum to spread. If one injects into a packed tissue, the viral inoculum will most likely infect cells within the injection tract, and it will be very difficult to sort out sibling relationships (i.e., too many lumping errors). In addition, one must inject such that the virus has clear access to the target population; the virus will bind to cells at the injection site and will not gain access to cells that are not directly adjoining that site.

The procedures described below are those that we have used for infection of rodents and chick embryos, histochemical processing of tissue for

β-Gal and PLAP visualization, and preparation and use of a library for the polymerase chain reaction (PCR) method.[23]

II. Procedures

Infection of Rodents

Injection of Virus in Utero

The following protocols may be used with rats or mice. Note that a clean, but not aseptic, technique is used throughout. We routinely soak instruments in 70% ethanol (v/v) before operations, use the sterile materials noted, and include penicillin/streptomycin (final concentrations of 100 units/ml each) in the lavage solution. We have not had difficulty with sepsis using these techniques.

Materials

Ketamine hydrochloride injection (100 mg/ml ketamine)
Xylazine injection (20 mg/ml)
Animal support platform
Depilatory
Scalpel and disposable sterile blades
Cotton swabs and balls, sterile
Tissue retractors
Tissue scissors
Lactated Ringer's solution (LR) containing penicillin/streptomycin
Fiberoptic light source
Virus stock
Automated microinjector
1- to 5-μl micropipettes
3-0 Dexon suture
Tissue stapler

1. Mix ketamine and xylazine 1:1 in a 1-ml syringe with a 27-gauge needle; lift the tail and hindquarters of the animal with one hand and with the other inject 0.05 ml (mice) or 0.18 ml (rats) of anesthetic mixture intraperitoneally.

One or more additional doses of ketamine alone (0.05 ml for mice and 0.10 ml for rats) are usually required to induce and/or maintain anesthesia, particularly if the procedure takes over 1 hr. Respiratory arrest and sponta-

[23] C. Walsh and C. L. Cepko, *Science* **255,** 434 (1992).

neous abortion appear to occur more often if a larger dose of the mixture is given initially or if any additional doses of xylazine are given.

2. Remove hair over entire abdomen using depilatory agent (any commercially available formulation, such as Nair, works well); shaving of remaining hair with a razor may be necessary. Wash skin several times with water, then with 70% ethanol, and allow to dry.

3. Place animal on its back in support apparatus. For this purpose, we find that a slab of styrofoam with two additional slabs glued on top to create a trough works well for this purpose. With the trough appropriately narrow, no additional restraint is then needed to hold the anesthetized animal.

4. Make a midline incision in the skin from xyphoid process to pubis using a scalpel and retract; attaching retractors firmly to the styrofoam support will create a stable working field. Stop any bleeding with cotton swabs before carefully retracting the fascia and peritoneum, and incising them in the midline with scissors (care is required here not to incise the underlying bowel). Continue incision cephalad along the midline of the fascia (where there are few blood vessels) to expose entire abdominal contents. If it is necessary to expose the uterus, gently pack the abdomen with cotton balls or swabs to remove the intestines from the operative field, being careful not to lacerate or obstruct the bowel. Fill the peritoneal cavity with LR, and lavage until clear if the solution turns at all turbid.

Wide exposure is important to allow the later manipulations. During the remainder of the operation, keep the peritoneal cavity moist and free of blood; dehydration or blood around the uterus increases the rate of postoperative abortion.

5. Elevate the embryos one at a time out of the peritoneal cavity and transilluminate with a fiberoptic light source to visualize the structure to be injected. For lateral cerebral ventricular injections, for example, the cerebral venous sinuses serve as landmarks.

When deciding on a structure to inject, keep in mind that free diffusion of virus solution throughout a fluid-filled structure lined with mitotic cells is best for ensuring even distribution of viral infection events throughout the tissue being labeled. The neural tube is an example of such a structure; when virus is injected into one lateral ventricle, it is observed to quickly diffuse throughout the entire ventricular system.

6. Using a heat-drawn glass micropipette attached to an automatic microinjector, penetrate the uterine wall, extraembryonic membranes, and the structure to be infected in one rapid thrust; this minimizes trauma and improves survival. Once the pipette is in place, inject the desired volume of virus solution, usually 0.1–1.0 μl. Coinjection of a dye such as 0.005% (w/v) trypan blue or 0.025% fast green helps determination of the accuracy

of injection and does not appear to impair viral infectivity; coinjection of the polycation Polybrene (80 μg/ml) aids in viral attachment to the cells to be infected.

The type of instrument used to deliver the virus depends on the age of the animal and the tissue to be injected. At early embryonic stages, the small size and easy penetrability of the tissue makes a pneumatic microinjector (such as the Eppendorf 5242) best for delivering a constant amount of virus at a controlled rate with a minimum of trauma. Glass micropipettes should be made empirically to produce a bore size that will allow penetration of the uterine wall and the tissue to be infected. At later ages (late embryonic and postnatal), a Hamilton syringe with a 33-gauge needle works best.

When injecting through the uterine wall, all embryos may potentially be injected except those most proximal to the cervix on each side (injection of these greatly increases the rate of postoperative abortion). In practice, it is often not advisable to inject all possible embryos, if excessive uterine manipulation would be required. At the earliest stages at which this technique is feasible [embryonic day (E) 12 in the mouse or E13 in the rat], virtually any uterine manipulation may cause abortion, so any embryo which cannot be reached easily should not be injected.

7. Once all animals have been injected, lavage the peritoneal cavity until it is clear of all blood and clots, ensure that all cotton balls and swabs have been removed, and move retractors from the abdominal wall/fascia to the skin.

Filling the peritoneal cavity with LR with penicillin/streptomycin before closing increases survival significantly, probably by preventing maternal dehydration during recovery from anesthesia as well as preventing infection.

8. Using 3-0 Dexon or silk suture material on a curved needle, suture the peritoneum, abdominal musculature, and fascia from each side together, using a continuous locking stitch. After closing the fascia, again lavage using LR with penicillin/streptomycin.

9. Close the skin using surgical staples (such as the Clay-Adams Autoclip) placed 0.5 cm apart. Sutures may also be used, but these require much more time (often necessitating further anesthesia, which increases abortion risk) and are frequently chewed off by the animal, resulting in evisceration.

10. Place animal on its back in the cage and allow anesthesia to wear off. Ideally, the animal will wake up within 1 hr of the end of the operation. Increasing time to awakening results in increasing abortion frequency. Food and water on the floor of the cage should be provided for the immediate postoperative period.

11. Mothers may be allowed to deliver progeny vaginally or the offspring may be harvested by cesarean section. Maternal and fetal survival are approximately 60% at early embryonic ages of injection, and increase with gestational age to virtually 100% after postnatal injections.

Injection of Virus Using ex Utero Surgery

Injections into small or delicate structures (such as the eye) require micropipettes that are too fine to penetrate the uterine wall. In addition, it is impossible to precisely target many structures through the rather opaque uterine wall. These problems can be circumvented, although with a considerable increase in technical difficulty and decrease in survival, by use of the *ex utero* technique.[24] The procedure is similar to that just detailed with the following modifications to free the embryos from the uterine cavity:

1. The technique works well in our hands only with outbred, virus-antigen free CD-1 and Swiss-Webster mice, but even these strains may have different embryo survival rates when obtained from different suppliers or different colonies of the same supplier. This variability presumably results from subclinical infections which may render some animals unable to survive the stress of the operation. We have had no success with this technique in rats.

2. After the uterus is exposed and *before* filling the peritoneum with LR, incise the uterus longitudinally along its ventral aspect with sharp microscissors. The uterine muscle will contract away from the embryos, causing them to be fully exposed, surrounded by their extraembryonic membranes.

3. Only two embryos in each uterine horn can be safely injected, apparently because of trauma induced by neighboring embryos touching each other. Thus, all other embryos must be removed. Using a dry sterile cotton swab, scoop out each embryo to be removed, with its placenta and extraembryonic membranes, and press firmly against the uterine wall where the placenta had been attached for 30–40 sec to achieve hemostasis.

It is very important to stop all bleeding before proceeding. From this point on, the embryos must be handled extremely gently, as only the placenta is tethering an embryo to the uterus, and it detaches easily.

4. Fill the peritoneal cavity with LR, and cushion each embryo to be injected with sterile cotton balls soaked in LR. Keeping the embryos submerged throughout the remainder of the procedure is essential to their survival.

5. The injection should then be done with a pneumatic microinjector and heat-pulled glass micropipette. This may usually be done by puncturing

[24] K. Muneoka, N. Wanek, and S. V. Bryant, *J. Exp. Zool.* **239,** 289 (1986).

the extraembryonic membranes first and then the structure to be injected; for some very delicate injections it may be necessary to make an incision in the extraembryonic membranes, which is then closed with 10-0 nylon suture after the injection.

6. At the time of desired fetal harvest or at the latest early on the last day of gestation, sacrifice the mother by cervical dislocation, rapidly incise the abdomen, and deliver the fetuses. If survival to a postnatal time point is desired, it is necessary to foster the pups with another lactating female. This is best done with a mouse that has delivered at the same time as the experimental animal, but we have successfully fostered pups with mothers that delivered several days to a week previously. Attempts to reanesthetize the experimental mother for delivery, thus allowing her to survive and obviating the need for fostering, have been unsuccessful due to poor survival of the pups. This is probably due to the deleterious effect of the anesthesia on the pups, as well as poor milk production by the mother after multiple operations.

Infection of Chick Embryos

Retroviral Injections into Chicken Embryos

Materials

Rotating, wooden egg incubator, optional (Petersime, Gettysburg, OH)
Benchtop egg incubator with plastic egg trays (Kuhl, Flemington, NJ)
Fertilized White Leghorn chicken eggs, specific pathogen free (SPF) (SPAFAS, Inc., Norwich, CT). Stored at 4–12° until needed.
70% ethanol
Small curved scissors, previously soaked in ethanol
Fine forceps, previously soaked in ethanol
20-gauge needle with a 5-ml syringe
Clear packing tape
Egg holder, homemade wooden or plastic rings, 2-inch diameter, one-half inch high
Dissection microscope
Fiber optic light source with green filter (optional)
Micromanipulator
Microinjector with motorized, direct-displacement advance (optional) (Stoelting, Wood Dale, IL)
Evacuated mineral oil to fill tubing of microinjector and micropipettes
Approximately 10 micropipettes (Omego Dot, 1 mm outer diameter, 0.75 mm inner diameter), broken with fine forceps to an outer diameter of 10–20 mm.

Pipette puller
Fast green (0.25% in water), filter sterilized and stored frozen as aliquots
Concentrated virus, stored at −80°, thawed, and placed on ice
Polybrene (0.8–8 mg/ml in water), not required for A-subgroup viruses but may help; titer for your application by performing infections *in vivo* in the site you wish to infect

1. Warm the eggs to room temperature for 1–2 hr and then incubate them on their sides in a humidified, rocking incubator at 37.5–38°. The first 24 hr of incubation is considered embryonic day 0 (E0). A nonrocking (benchtop) incubator is adequate for injections and survival to E19, provided that albumin is removed as detailed in steps 2–4.

2. On E1 or E2, remove a batch of eggs from the incubator, drench with 70% ethanol, and air dry.

3. Lay the egg on its side in an egg holder, and use the 20-gauge needle attached to the 5-ml syringe to jab a small hole in the side of the egg facing up. Jab a second hole in the large end of the egg and insert the needle about 0.5 cm into the albumin at a steep angle to avoid hitting the yolk. Withdraw 1.5 ml of albumin and discard.

4. Seal the holes with tape. Place the egg on its side in a humidified, nonrotating, benchtop incubator.

5. Use small curved scissors to open and enlarge the hole on the top of the egg. Cutting through tape prevents small pieces of shell from falling into the egg. Make a hole approximately 1 cm in diameter, locate the embryo, and then enlarge the hole as necessary (typically about 2 cm wide) to expose the embryo for injection. This step can be done in advance, on E1 or E2, or it can be done immediately before injection. If done in advance, the hole in the shell is resealed with tape.

6. On the day of injection, mix 1 part fast green, 1 part polybrene (we use 1 part 0.8 mg/ml for subgroups E and A), and 8 parts concentrated virus. Spin for approximately 10 sec in a microfuge (10,000 rpm) to pellet large aggregates. Keep the virus stock on ice for the 2–4 hr required to inject a batch of embryos (usually about two to three dozen per experiment).

7. Insert a micropipette into the microinjector tubing that was previously filled with heavy mineral oil, avoiding air bubbles. Attach the micropipette to the micromanipulator, and use gentle pressure to fill the micropipette with oil to the tip.

8. Place 3–5 μl of viral stock on a piece of Parafilm, and lower the tip of the micropipette into the viral stock. Pull back on the Hamilton syringe of the microinjector to fill the pipette with virus. Place the tip of the filled micropipette in sterile saline whenever there is a lag in the procedure.

9. Open the hole in the side of the egg, stabilize the egg with an egg holder, and illuminate the embryo with fiber optics under a dissection microscope. Optimal contrast is obtained by illuminating the embryo from an oblique angle; a green filter can be added to enhance contrast. Embryos younger than E4 can usually be injected directly. Older embryos (E4 and later) usually require some additional preparation because of a thickening and increased vascularization of the embryonic membranes. With fine forceps, tear the serosa, avoiding or diverting the allantois if possible. Otherwise, make an additional tear in the allantois, avoiding the blood vessels as much as possible. Tease open the amnionic membrane and drape the torn edges over the edge of the hole in the shell to stabilize the embryo during injection.

10. Lower the micropipette into the desired structure or cavity of the embryo using the micromanipulator. For neural tube injections, the tip is lowered directly through the overlying amnion and epidermis until it enters the ventricles of the neural tube. The location of the inoculum is readily observed because of the presence of carrier dye.

11. Inject with the motorized microinjector, usually 0.1–1.0 μl; this usually takes 5–50 sec.

12. Reseal the hole with tape and return to the incubator.

Human Placental Alkaline Phosphatase as a Histochemical Marker Gene

Human placental alkaline phosphatase (PLAP) has only recently been adapted for use as a histochemical marker for lineage studies. As such, neither its benefits nor its potential drawbacks have been exhaustively analyzed, and this should be kept in mind as this enzyme begins to enjoy wider use. Our laboratory has undertaken a systematic lineage analysis in the postnatal mouse retina[13] and rat striatum[20] comparing β-Gal and PLAP and has seen very little difference between the two. Thus for these particular combinations of tissue and vectors (BAG and DAP), there appears to be no effect of ectopic PLAP expression on the choice of cell fate or survival. We do not yet know whether this conclusion will hold for other tissues. In the chick retina and cerebellum, for example, clonal analysis does show differences between β-Gal and PLAP in the ease of detection of different cell types.[21,22] It is not yet clear whether these differences reflect cell-specific inactivation of the promotor or the enzymes, differential distribution of the two enzymes within intracellular compartments, or (most worrisome) the perturbation of cell fate or survival.

Human PLAP was initially chosen as a potential histochemical marker for several reasons. Among the variety of isoenzymes of alkaline phosphatase that have been studied, human PLAP is by far the most heat stable, by a factor of about 100.[25] Thus, although many tissues express endogenous alkaline phosphatase(s), it is possible in most cases to greatly reduce this background reactivity by preincubating the tissue at 65° for 30 min. In addition, PLAP is resistant to a variety of substances that act to inhibit other isoenzymes of alkaline phosphatase.[26] We have tested the following inhibitors on mouse and chick neural tissue as recommended by Zoellner and Hunter[26]: 0.5 m*M* levamisole (L[-]-2,3,5,6-tetrahydro-6-phenylimidazo[2,1-*b*]thiazole), 2 m*M* mercuric chloride, 5 m*M* L-leucylglycylglycine, 1 m*M* EDTA, 1 m*M* L-phenylalanine-glycyl-glycine, 0.2 *M* lysine HCl, or 0.3 m*M* sodium arsenate. Levamisole was the most useful in reducing the background staining in brains, although it also reduced PLAP staining slightly in some cases; it was less effective in retinas. None of the inhibitors was nearly as effective as heat treatment in reducing background in the central nervous system. Nonetheless, it is certainly possible that their use may facilitate detection in other tissues. Another benefit of PLAP as a marker is the fact that its activity is probably minimal at normal intracellular pH since this is considerably below the optimal pH for enzyme function. This may suggest that ectopic expression of PLAP during development is not likely to perturb normal physiological processes, but this remains to be tested systematically for different tissues.

Double-Staining of Infected Tissues for β-Galactosidase and Alkaline Phosphatase Activities

The following protocol was adopted for the double staining of β-Gal and PLAP in nervous system tissue. Cells expressing β-Gal will be rendered a bright blue, whereas cells expressing PLAP will be rendered purple due to the presence of NBT in the reaction. The order in which the staining is done is critical since β-Gal is inactivated by the heat treatment that is required to inhibit endogenous alkaline phosphatases. Obviously, if only a single marker is needed, the protocol can be minimally adapted for use with either enzyme individually.

Detecting both enzymes in the same cell is not always possible using the following protocols because the PLAP reaction product is so intense that it usually obscures the β-Gal reaction product. For example, it is difficult to detect cells cotransfected with both genes *in vitro*. This is not a problem when the two enzymatic markers are used in retroviral vectors for lineage mapping studies because individual cells are very unlikely to be infected with both vectors and express both enzymes. (Even if they

[25] R. B. McComb, G. N. Bowers, Jr., and S. Posen, "Alkaline Phosphatase." Plenum, New York, 1979.

[26] H. F. A. Zoellner and N. Hunter, *J. Histochem. Cytochem.* **37,** 1893 (1989).

were, the interpretation of the results would not be affected since one is still observing a clone.) If detection of cells containing both enzymes is necessary, use of different chromogenic substrates might circumvent this difficulty. PLAP can be reacted with naphthol-AS-Bl-phosphate/New Fuchsin, which produces a red stain,[27] and NBT can be added to the β-Gal reaction to produce a purple precipitate. Red/blue or red/purple might be distinguishable from the corresponding single stains; we have not tested these possible combinations.

We have found that performing immunohistochemistry in conjunction with either β-Gal or PLAP histochemical staining has not been very satisfactory. The colored precipitates block fluorescence, thus preventing the use of fluorescently conjugated antibodies. Using HRP-conjugated antibodies, it is difficult to distinguish blue/brown double-labeled cells from brown-labeled ones whereas purple and brown/purple cells would probably look identical. Situations in which the subcellular localization of β-Gal and an antigen are significantly nonoverlapping (e.g., a nuclear antigen) may allow for detection of both.[28] In addition, one can try to limit the intensity of the β-Gal or PLAP reaction product by shortening the time of the reactions to allow for more sensitivity in detecting an antigen within the same cell by HRP-conjugated antibodies.[29] Alternatively, antibodies exist for both β-Gal (rabbit polyclonal, Cappel; mouse monoclonal, Boehringer Mannheim) and PLAP (rabbit polyclonal, Zymed; mouse monoclonal, Medix), although using them is obviously more time-consuming than relying on histochemical reactions.

Different species, tissues, or parts of tissues (i.e., brain regions) can have varying amounts of background labeling.[30-32] This is especially noticeable with alkaline phosphatase reactions, but is also true for β-Gal. We recommend always including negative controls in order to assess the extent of background for a particular tissue or region of interest. Important variables that affect the signal to noise ratio include type of fixative, length of fixation, length of washes, length of heat treatment, prolonged exposure to light, and prolonged storage after staining. For a particularly problematic

[27] E. Harlow and D. Lane, eds., "Antibodies: A Laboratory Manual." Cold Spring Harbor Lab., Cold Spring Harbor, NY, 1988.

[28] E. Y. Snyder, D. L. Deitcher, C. Walsh, S. Arnold-Aldea, E. A. Hartwieg, and C. L. Cepko, *Cell (Cambridge, Mass.)* **68,** 33 (1992).

[29] P. J.-J. Vaysse and J. E. Goldman, *Neuron* **5,** 227 (1990).

[30] B. Pearson, P. L. Wolf, and J. Vazquez, *Lab. Invest.* **12,** 1249 (1963).

[31] E. Robins, H. E. Hirsch, and S. S. Emmons, *J. Biol. Chem.* **213**(16), 4246 (1968).

[32] A. M. Rutenburg, S. H. Rutenburg, B. Monis, R. Teague, and A. M. Seligman, *J. Histochem. Cytochem.* **6,** 122 (1958).

tissue, it may prove advantageous to try different inhibitors of endogenous alkaline phosphatases (see above), as well as different substrates.

Solutions

Phosphate-buffered saline (PBS; 10×): 80 g NaCl, 2 g KCl, 11.5 g Na_2HPO_4, and 2 g KH_2PO_4 in 1 liter H_2O. Adjust to pH 7.2–7.4. Dilute 1:10 before using.

0.5% glutaraldehyde: 25% stock (Sigma) can be stored at −20° and frozen/thawed many times. Make dilution immediately before use.

4% paraformaldehyde: 4 g solid paraformaldehyde, 2 m*M* $MgCl_2$, and 1.25 m*M* EGTA (0.25 ml of a 0.5 *M* EGTA stock, pH 8.0) in 100 ml PBS, pH 7.2–7.4. Heat about 80 ml H_2O to 60° and add paraformaldehyde; add enough NaOH to get paraformaldehyde in solution. Cool to room temperature, add 10 ml 10× PBS, adjust pH with HCl, add $MgCl_2$ and EGTA, and make up to 100 ml with H_2O. The solution can be stored at 4° for 1–2 weeks.

X-Gal detection buffer: 35 m*M* potassium ferrocyanide (can vary from 5 to 35 m*M*), 35 m*M* potassium ferricyanide (can vary from 5 to 35 m*M*), 2 m*M* $MgCl_2$, 0.02% Nonidet P-40 (NP-40) (diluted from 10% stock solution), and 0.01% sodium deoxycholate (diluted from 10% stock solution) in PBS. This buffer can be stored for at least 1 year at room temperature in a foil-covered container.

X-Gal stock (40×): 40 mg/ml X-Gal (5-bromo-4-chloro-3-indolyl-β-D-galactopyranoside) in dimethylformamide. Store at −20° in a glass container covered with foil.

X-Gal reaction mix: Make a 1:40 dilution of X-Gal stock into X-Gal detection buffer. Make dilution immediately before using. Final concentration of X-Gal is 1 mg/ml.

X-P detection buffer (Buffer 3, Genius Kit, Boehringer-Mannheim): 100 m*M* Tris–HCl, pH 9.5, 100 m*M* NaCl, and 50 m*M* $MgCl_2$. Store at room temperature. Tends to precipitate over several weeks. This does not seem to noticeably affect the staining.

X-P stock (100×): 10 mg/ml 5-bromo-4-chloro-3-indolyl phosphate, X-P, also referred to as BCIP in H_2O. Store in the dark as aliquots at −20°. Can be frozen and thawed several times.

NBT stock (50×): 50 mg/ml nitro blue tetrazolium in 70% dimethylformamide, 30% H_2O. Store at −20° in a glass container covered with foil. Does not freeze at this temperature.

X-P reaction mix: 50 μl X-P stock (100×), 100 μl NBT stock (50×), and 50 μl 50 m*M* levamisole (100×) (if desired) in 5 ml X-P detection buffer. Make fresh immediately before using. Final concentrations:

X-P, 0.1 mg/ml; NBT, 1 mg/ml; and levamisole, 0.5 m*M*. Levamisole was ineffective at reducing background in chick cerebellum, while slightly inhibiting PLAP, and was therefore not used in this tissue. Any other desired inhibitor (see introduction) would also be added to the X-P reaction mix.

Gelatin subbing solution for slides: 2 g gelatin and 0.1 g chromium potassium sulfate (chrome alum) in 200 ml H_2O. Heat H_2O to 60°. Dissolve chrome alum, then gradually dissolve gelatin. Filter before use. Can increase or decrease percentage of gelatin. Load slides in racks, dip quickly, and air-dry overnight.

Gelatin/sucrose embedding medium: 7.5% gelatin (porcine skin, Sigma), 15% sucrose, and 0.05% sodium azide in 1× PBS. Dissolve gradually at 60°, with stirring. The medium solidifies at room temperature to a transparent gel. Store at room temperature. Liquefy in microwave with frequent swirling before embedding samples.

Gelvatol mounting medium: Make up according to the protocol given by Rodriguez and Deinhard.[33] Instead of Elvanol, we use Vinol grade 205, polyvinyl alcohol resin (Air Products and Chemicals, Inc., Allentown, PA).

Whole Mount Staining Procedure

The primary protocol shown was worked out for staining of intact mouse retinas that had been double-infected with BAG and DAP.[13] The order in which staining is done is critical since β-Gal is inactivated by the heat treatment that is required to inhibit endogenous alkaline phosphatases. The incubation times for X-Gal and X-P reactions are variables that may require adjustment for different tissues. Longer times may be necessary for large or dense chunks of tissue, although there is a trade off as the background staining of either enzyme intensifies with increasing reaction time. Whole chick embryos (E7) and chick brains (E10) were found to be incompletely reacted in their centers after 4-hr incubation periods. For tissues that are difficult to stain completely as whole mounts, one can stain as sections (see next section). Alternatively, at least for the X-P reaction, one can do a 3- to 4-hr whole mount stain to locate cells of interest, dissect out and section only those areas of interest, and restain the sections for 20–30 min to obtain optimal staining.

1. Dissect the tissue into PBS containing 2 m*M* $MgCl_2$ (PBS + Mg^{2+}) on ice.

[33] J. Rodriguez and F. Deinhardt, *Virology* **12,** 316 (1960).

2. Fix in 0.5% glutaraldehyde in (PBS + Mg^{2+}) for not longer than 45 min on ice. Glutaraldehyde (0.5%) decreases the alkaline phosphatase activity in chick cells stained *in situ* (but not in chicken embryo fibroblasts stained *in vitro*). Therefore, fixation of chick whole mounts is typically done in 4% paraformaldehyde in PBS for 2–4 hr at 4°. In areas where background alkaline phosphatase activity is a problem, increasing the time in 4% paraformaldehyde, even up to several days, can decrease endogenous background without significantly decreasing PLAP activity. However, such long fixation times may decrease β-Gal activity.

3. Rinse in (PBS + Mg^{2+}) five times for 5 min. Rinsing overnight is fine; but waiting for several days at this step may decrease β-Gal activity.

4. Stain in X-Gal reaction mix for 2–4 hr at 37°. (Tris buffer was tried in place of PBS, with no success.)

5. Rinse many times in PBS until the solution no longer turns yellow. This usually takes about five changes. An overnight rinse is fine. It is important to remove X-Gal since residual β-Gal activity in the presence of X-Gal and NBT (added for the following reaction) may enable β-Gal$^+$, PLAP$^-$ cells to turn purple. Chick retinas and cerebella have been kept in PBS at 4° for at least 1 month at this point with no appreciable loss of signal in subsequent X-P staining. X-Gal staining can be easier to examine prior to carrying out the X-P reaction as background alkaline phosphatase staining can obscure X-Gal signal somewhat, particularly in whole mounts.

6. Heat tissue in PBS at 65° for 30 min. This is usually done by floating the dish containing the tissue in a water bath preset to 65°. For staining of embryonic chick diencephalon (one of the areas of the brain with the highest background), this step was increased to 1.5 hr.

7. Preincubate in X-P detection buffer for 15 min. Extending the time of this step results in diffusion of the alkaline phosphatase reaction product.

8. Incubate in X-P reaction mix for 3 hr at room temperature. Since background staining increases in light, cover with foil during and after staining.

9. Rinse in 20 m*M* EDTA in PBS for 2–4 hr. Background can be due to endogenous AP or other reactions which generate hydride ions and thus reduce NBT to form a purple precipitate. We have noted that background staining comes up more slowly in the presence of EDTA. Tissue can be stored in the dark at 4° in PBS + EDTA or 30% sucrose in PBS + EDTA for many months, although the background clearly increases over time.

10. Embed in Paraffin wax using minimum necessary times for the tissue of interest. For mouse retina, which is approximately 250 μm thick, the following procedure was used. Dehydrate through graded ethanols: 50%, 70%, 95%, 100%, 100% for 20 min each. Clear in xylene, 2 × 15 min. Infiltrate with 1:1 mix of xylene and Paraffin, 65° for 30 min. Paraffin,

2 × 15 min. Embed in Paraffin. Clearing and Paraffin embedding is not recommended for tissues fixed with paraformaldehyde. In paraformaldehyde-fixed chick tissues, both the β-Gal and the PLAP reaction products are very sensitive to xylene treatment; even relatively short exposures to xylene causes the reaction product to diffuse into the surrounding tissue. In some cases, this was true in the ethanol washes as well. Frozen sections are a workable alternative (see next section on the preparation of tissue for cryostat sectioning). Other embedding protocols, particularly aqueous-based procedures, may be worth testing. It is worth noting that strong staining in glutaraldehyde-fixed material can even withstand preparation for electron microscopy.[28]

11. Section onto slides coated with gelatin. Silane-treated slides are equally effective.[34]

12. Remove Paraffin with xylene and mount with Permount. For frozen sections, fix sections to slides with 4% paraformaldehyde for 15 min. Rinse with PBS, then mount in Gelvatol (+ EDTA if desired).

Storing slides at −80° helps prevent background staining from increasing.

Protocol for Staining Sections

This protocol was worked out for embryonic rat striatum.[20] Fixation and staining times may need to be altered for other areas of interest.

1. Fix tissue by perfusion and follow with immersion in 4% paraformaldehyde at 4° for 8 hr. Rinse briefly in PBS, then sink in 30% sucrose in PBS containing 2 m*M* $MgCl_2$ (PBS + Mg^{2+}) at 4°. Fixation times will vary with size of tissue. Perfusion may not be necessary for all tissues, especially in embryonic animals. Shorter fixation times may be preferable, as X-Gal staining may be decreased by lengthy fixation.

2. Embed brain in OCT or gelatin/sucrose mounting medium and freeze on liquid N_2. Gelatin/sucrose embedding gives better frozen sections for embryonic tissue than does OCT. Paraffin embedding destroys β-Gal activity, but cells in culture treated as if to Paraffin embed retain PLAP activity.

3. Cut cryostat sections and mount on gelatin-coated slides; air-dry overnight. Sections up to 90 mm thick (the thickest we have tried) have been successfully stained.

4. Fix sections to slides in 4% paraformaldehyde for 10–15 min at 4°.

5. Rinse slides in (PBS + Mg^{2+}) twice, for 10 min each, at 4°.

6. Stain slides in X-Gal reaction mix for 6 hr at 37°.

[34] M. Rentrop, B. Knapp, H. Winter, and J. Schweizer, *Histochem. J.* **19**, 271 (1986).

7. Rinse slides in PBS three times, for 10 min each, or until solution is no longer yellow. Slides can be left in PBS overnight. (See comments in Whole Mount Staining Procedure, step 5.)

8. Transfer slides to preheated PBS at 65° and heat for 30 min.

9. Rinse slides in X-P detection buffer for 10 min.

10. Stain slides in X-P reaction mix for 12 hr. Since background staining increases in light, cover with foil during and after staining.

11. Rinse slides in PBS + 20 m*M* EDTA three times for 10 min. Mount in Gelvatol (+ EDTA if desired). (See comments on background staining in Whole Mount Staining Procedure, step 9. Storing slides at −80° helps prevent background staining from increasing.)

Clonal Analysis Using the PCR/Library Method

As described earlier in the section on determination of clonal boundaries, there are situations where a great deal of migration can lead to splitting errors. In these situations, a larger number of distinguishable viruses can lead to identification of sibling relationships with a greater degree of certainty. We thus devised a method in which a large number of viruses (a "library") could be produced and injected into a single animal.[23] PCR was then used to distinguish the individual library members. In our first libraries, we used insertion of 50–300 irrelevant DNA fragments to create libraries in murine BAG[12] and DAP[13] vectors and in avian RDlac[35] and CHAP[21] vectors. We have also made a library using degenerate oligonucleotides and the PCR products are sequenced.[35a] This library is of far greater complexity (at least 10^4 members) and is still being tested, but looks very promising for use in situations where more certainty is desirable. The protocols used for creating a murine library in BAG and for amplifying the products from single cells are given below.

Preparation of a Retroviral Library for PCR Analysis

In principle, any retroviral plasmid can be used to make a library. We started with the BAG plasmid which contains a unique cloning site (*Xho*I) downstream of the reporter gene. The insertion of DNA into this site does not appear to interfere with expression of the upstream genes. The inserted DNA was genomic DNA from *Arabidopsis thaliana* digested with *Mbo*I. The digested DNA was run on an agarose gel and DNA fragments less than 450 bp in size were used as inserts.[23]

[35] S. T. Reddy, A. W. Stoker, and M. J. Bissell, *Proc. Natl. Acad. Sci. U.S.A.* **88,** 10505 (1991).

[35a] J. Golden, S. Fields-Berry, and C. Cepko, *Proc. Natl. Acad. Sci. U.S.A.* in press (1995).

Before making retroviruses with any of these constructs, approximately 100 constructs were identified whose inserts were conveniently distinguishable by size or by their pattern of digestion using restriction enzymes. PCR products were prepared from bacterial colonies, the products were separated on agarose gels, and the approximate size of each PCR fragment was recorded. PCR products were sorted by size, and each product was digested with a mixture of restriction enzymes with four base recognition sites that were chosen because they were inexpensive and compatible within the same digestion buffer (*Msp*I, *Rsa*I, *Alu*I, *Cfo*I, and *Mse*I). The size and restriction pattern of each tag were recorded, and approximately 100 constructs were chosen that were distinguishable in this standard assay.

Retroviruses were next prepared from the 100 constructs by transfecting minipreparation DNA,[36] purified using crushed glass, into a packaging cell line. The 100 DNAs were transfected[11] pairwise into 50 dishes of the ΨCRIP amphotropic packaging line[4] so that each transfected plate contained a mixture of cells producing two viral constructs. The supernatant of each transfected plate was used to infect a dish of the ecotropic packaging cell line Ψ2. Infected Ψ2 producer cells were selected by growth in medium containing G418 for 7 to 10 days, and the population of resistant colonies was raised to confluence.

Viral supernatants were recovered and titered on NIH/3T3 cells in 6-well tissue culture dishes.[11] DNA from the infected NIH/3T3 cells was amplified using PCR to evaluate the passage of the genetic tags. After the titering reaction, 0.5 ml of a solution of 0.5% Tween 20 and 200 μg/ml proteinase K in 1× PCR buffer was added to each plate of cells. After incubation for 60 min at 65°, the proteinase solution was transferred to a 0.6-ml microfuge tube and incubated at 85° for 20 min, then at 95° for 10 min. The undissolved X-Gal precipitate was pelleted by centrifugation at 10,000 rpm for 10 min, and 10-μl samples of the supernatant were used as templates for 50-μl PCR reactions as described earlier.

The 50 viral supernatants were then mixed in approximately equal quantities and concentrated by centrifugation.[11] The concentrated stock was titered, tested for helper virus, and checked by infection of rodent retina. It was then used to perform experiments in the cerebral cortex.

The method just described can be easily modified in several ways. Any convenient restriction site in a replication-incompetent vector can be used for insertion of tags so long as it does not interfere with expression of the reporter gene or transmission of the virus. Any source of DNA that is not

[36] T. Maniatis, E. F. Fritsch, and J. Sambrook, "Molecular Cloning: A Laboratory Manual." Cold Spring Harbor Lab., Cold Spring Harbor, NY, 1982.

present normally in the host tissue or vector can be used. Any set of enzymes can be chosen for the restriction enzyme characterization of the inserts, and any range of insert sizes can be used if the longer ones are not so long that their size would cut down on the efficiency of the PCR. Finally, it is possible to transfect the entire library as one plasmid preparation into producer cells rather than going through the tedium of making individual stable producers or producer populations. We have used this method in making extremely complex libraries using oligonucleotides as tags.[35a]

Tissue Analysis for PCR-Based Clonal Assignments

Animals are infected and processed histochemically for the appropriate reporter gene as described in the previous sections. Usually sections are made for this analysis, but in special circumstances, whole mounts can be used. Once the histochemistry is complete, the labeled cells are analyzed and their position as well as the outline and hallmarks of the tissue are recorded. The goal is to create a permanent record of the morphology and relative location of the labelled cells since the PCR analysis will destroy the cells and surrounding tissue. A standard microscope and camera lucida device allows morphological details of cells to be drawn at high magnification, whereas cell location can be plotted at low magnification. Alternatively, cells can be photographed or plotted on a computerized system (e.g., CARP[37]). The anatomical analysis is often the most time-consuming aspect of the entire analysis.

PCR of Labeled Cells

After anatomical analysis, PCR amplification of the tags that distinguish the vectors allows clonal analysis. For these experiments, start with two "nested" pairs of oligonucleotides specific for the region of the vector in which the inserts were placed. PCR parameters for each oligonucleotide pair must be optimized, especially the Mg^{2+} concentration in the PCR buffer.[38]

Materials

Histological sections prepared and stained as described in the preceding section, and coverslipped in Gelvatol
Sterile distilled water
1× PCR buffer
Two nested pairs of oligonucleotides (as 20 μM solutions)

[37] C. P. Austin and C. L. Cepko, *Development* (*Berlin*) **110,** 713 (1990).
[38] K. B. Mullis and F. A. Faloona, this series, Vol. 155, p. 335.

Deoxyribonucleotide solution (20 m*M* each in dATP, dTTP, dCTP, dGTP)
Proteinase K, 10 mg/ml solution, in sterile, distilled water
Tween 20, 10% solution, in steriled distilled water
Mineral oil (light), from Sigma
Disposable, breakable razor blades and blade holder (Fine Scientific Tools)
600-μl microfuge tubes (autoclaved and silanized) or 96-well microtiter dishes and lids (e.g., Falcon No. 3911 and 3913)
Automated thermal cycler
Dissecting microscope
50-ml centrifuge tubes

Note: All solutions and containers must be assembled and stored using the most stringent precautions to prevent contamination with DNA that could be amplified in the PCR. A dedicated set of solutions should be handled with dedicated, positive displacement (or other contamination-resistant) pipetting devices in a separate laboratory remote from the thermal cycler. Good discussions of minimizing PCR contamination are available elsewhere.[38]

Dissect Cells and Digest Tissue

1. Prepare a lysis solution. For 50 samples, mix
50 μl, or optimum amount, of 10× PCR buffer
25 μl of 20 μM solution of each of the outermost oligonucleotides
10 μl of 10 mg/ml solution of proteinase K (final concentration is 0.2 mg/ml)
25 μl of 10% Tween 20 solution (final concentration is 0.5%)
375 μl distilled water

2. Soak off coverslips in distilled water in a clean, sterile 50-ml centrifuge tube. After the slide has soaked about 30 min, the coverslip can be carefully pried off with a razor blade. After the coverslip comes off, soak the tissue about 5 min more to remove traces of Gelvatol.

3. Pipette 10 μl of the lysis solution into each 0.6-ml microfuge tube (or well of a 96-well microtiter dish).

4. Break off a fresh fragment (2–5 mm wide at the edge) of the breakable razor blade in the blade holder.

[39] F. M. Ausubel, R. Brent, R. E. Kingston, D. D. Moore, J. G. Seidman, J. A. Smith, and K. Struhl, eds., "Current Protocols in Molecular Biology." Greene Publishing Associates, New York, 1989.

5. Under the dissecting microscope, locate a labeled cell. Well-stained cells can be seen with a low magnification objective (0.8×), but lighter cells may only be seen with a high (5×) magnification objective.

6. Using the razor blade, cut a fragment of tissue that includes the nucleus of the labeled cell. If labeled cells are widely scattered, they can be dissected one at a time in chunks that contain approximately 1000 unlabeled cells. Chunks are typically <500 μm in each dimension, but larger pieces can be used (PCR sensitivity may be less with larger pieces since they do not dissolve as well). If labeled cells are immediately adjacent, dissect several cells in one chunk.

7. Transfer the tissue piece, on the razor blade, to the lysis solution. Confirm under the microscope that the labeled cell is in the lysis solution. It is important to keep careful notes of which cell goes where—drawings are helpful here.

8. Cover the lysis solution with 100 μl of mineral oil, and cap the tube (or cover the microtiter plate).

9. Replace the blade fragment with a fresh one, and dissect the next cell (return to step 5). While dissecting cells, prepare negative controls. Intersperse samples that contain no tissue, or unlabeled tissue, among the positive samples.

10. After all cells have been dissected, and all lysis samples are covered with mineral oil, cover the tubes (or plate).

11. Transfer samples to a thermal cycler. Digest for 2–3 hr at 65°. Inspect a few samples after this time to confirm that the tissue is totally dissolved. If not, digest longer (overnight at 37° is alright). The X-Gal precipitate does not dissolve, but does not interfere with the PCR.

12. Once the tissue is digested, heat it to 85° for 20 min, then 95° for 5 min. This inactivates the proteinase K and denatures the genomic DNA. The samples are now ready for the PCR.

First PCR Reaction

1. Prepare PCR solution. For 50 samples, use
   - 100 μl (or optimum amount) of 10× PCR buffer
   - 50 μl of 20 μ*M* solution of each of the outermost oligonucleotide primers
   - 10 μl of the mixed deoxyribonucleotide solution (20 m*M* of each dNTP)
   - 7.5 μl (37.5 U) of Amplitaq DNA polymerase (Perkin Elmer)
   - 800 μl of distilled water

These reactions can be scaled up or down proportionately. *Taq* polymerase from other manufactures can be used but should be tested first.

2. Put the samples in the thermal cycler and start the PCR reaction, which should begin with an initial denaturation at 92–94° for 3 min. PCR conditions need to be optimized, but will comprise 45 repetitions of denaturation (92–94° for 30–45 sec), annealing (55–70° for 1–3 min), and extension (72° for 1 min).

3. Once the samples have reached >85°, uncover them and add 20 μl of the PCR solution to each sample. The "hot start" enhances the sensitivity and specificity of the PCR. The added PCR solution does not have to be mixed in. It will sink beneath the oil and join the aqueous phase.

4. Cover the samples and allow the PCR reaction to proceed.

Second PCR Reaction

The samples now contain large amounts of amplified target DNA and should not be opened or even transported into the clean laboratory. Since contamination of them is not a concern, they may be handled in the main laboratory. The sensitivity of the second PCR reaction is not critical, and *Taq* polymerase from any manufacturer can be used.

1. Prepare the following solution. For 50 samples, use
   10 μl mixed deoxyribonucleotide solution (20 m*M* of each dNTP)
   200 μl (or optimum amount) of 10× PCR buffer
   50 U *Taq* DNA polymerase
   1800 μl sterile distilled water
   20 μl of each of the internal pair of oligonucleotide primers (final concentration is 0.2 μM)

The lower concentration of primers and deoxyribonucleotides used in the second reaction does not affect the PCR sensitivity or product yield.

2. Pipette 40 μl of this solution into 50 tubes (or microtiter wells).

3. Transfer 2–4 μl of the product of each PCR reaction into each tube (or well). Keep samples carefully labeled, and take note of any pipetting errors.

4. Run the second PCR for 25–35 cycles. Conditions for the second PCR will also have to be optimized, but will likely be similar to those used for the first reaction. If the same thermal cycler is used for both the first and second PCR reactions, be sure not to open any of the samples of the second reaction near the machine.

5. Separate the PCR products on 3% NuSieve/1% Seakem agarose gels. Use 1× TBE as the running buffer, and include pBR/*Msp*I or ΦX/*Hae*III DNA size standards. Anticipate that 40–70% of the samples will produce a band. Suspect a problem if yields are consistently <40%; suspect contamination if many contiguous samples show the same product.

Direct Restriction Enzyme Digestion of PCR Products

DNA inserts may be distinguished by size or sequence analysis. Restriction mapping offers the most convenient assay and is sufficiently specific for libraries with 100–300 different inserts. We chose five restriction enzymes with four-base recognition sequences (*Cfo*I, *Rsa*I, *Alu*I, *Mse*I, *Msp*I) that were active in similar buffer conditions (50–100 m*M* NaCl). The mixture cuts small DNA fragments frequently, allowing them to be easily distinguished. Before loading the diagnostic gel or prior to restriction digestion, sort the PCR products by size. When samples of similar initial size are run side by side on the gel, it allows the most direct comparison of the restriction fragments.

1. Prepare the following solution (for 50 samples):
   250 U of each restriction enzyme
   150 μl of 10× restriction enzyme buffer, appropriate for the enzymes chosen.
   15 μl 10 mg/ml bovine serum albumin
   Sterile distilled water to make a total of 500 μl
2. Pipette 10 μl of this solution into individual wells of a microtiter dish (tubes can also be used).
3. Add 20 μl of one of the PCR reaction products to each well.
4. Cover the microtiter plate and incubate at 37° for 3 hr.
5. Terminate the reactions by adding loading buffer, and separate the products of the reaction on a 3% Nusieve/1% Seakem agarose gel.
6. Record which samples contain the same tags and the overall number of tags seen. Compare this information to the original plots of cell location to get clonal information.

Statistical Analysis

The tentative conclusion from the PCR analysis is that cells containing the same tag are members of the same clone. The confidence of this conclusion rests on (1) the number of clones in a given experiment, k, and (2) the number of tags in the library, n. There is a surprisingly large probability that the same tag will appear in two different clones by coincidence, and this probability needs to be considered in the clonal analysis.

The probability of "coincidental double infections" by one tag can be calculated using binomial theory or computer-simulated using a Monte Carlo simulation. Either method requires the assumption that all tags are present in equal ratios, and thus the ratios of tags in the library must be verified experimentally. Some typical results are shown in Table I, to provide guidelines.

TABLE I

Number of tags (n)	Probability of two clones sharing same tag in k clones			
	$k = 3$	$k = 5$	$k = 10$	$k = 20$
20	0.14	0.40	0.86	0.99
80	0.04	0.12	0.43	0.90
85	0.035	0.11	0.41	0.88
100	0.030	0.10	0.36	0.85
250	0.012	0.04	0.16	0.52

$$\text{Probability that} > 1 \text{ clone shows the same tab} = 1 - \frac{n!}{(n-k)!n^k}.$$

To evaluate clonal data, it is important to evaluate the actual complexity of the viral library (i.e., how many tags are seen in total).[40] The best results demand a highly complex library, with very few clones labeled in each experiment (three to four clones). Under these conditions the probability is maximized that observed patterns faithfully reflect clonal patterns. However, be skeptical of patterns that are seen rarely, as they may only reflect rare coincidental double infections instead of true clonal events.

Reagents and Solutions

10× PCR buffer: 10 m*M* Tris buffer, pH 8.3 (purchase as premixed crystals from Sigma), 50 m*M* KCl, 0.01% gelatin, and 1.5–2.5 m*M* $MgCl_2$.

Mixed deoxyribonucleotide solution: Deoxyribonucleotides may be purchased from Pharmacia as separate 100 m*M* solutions of dATP, dCTP, dGTP, and dTTP. Mix them 1 : 1 : 1 : 1 : 1 with distilled water to make a working mixture that is 20 m*M* in each. Store 10-μl aliquots at −70°.

Proteinase K: Can be purchased from many manufacturers, dissolved in sterile distilled water to make a 10-mg/ml solution, and stored as 20-μl aliquots at −70°.

Oligonucleotide primer solutions: Deprotected oligonucleotides can be passed over a NAP-10 ion-exchange column (Pharmacia) and eluted with sterile distilled water. Adjust the concentration of the effluent to 20 μ*M* by measuring absorbance at 260 nm. We routinely

[40] C. Walsh, C. L. Cepko, E. F. Ryder, G. M. Church, and C. Tabin, *Science* **258,** 317 (1992).

use oligonucleotides without further purification. Oligonucleotides should be stored as 25- to 50-μl aliquots at −20°.

Note: Reagents, instruments, and glass microscope slides should be handled with scrupulous technique and UV-irradiated when needed to destroy contaminating DNA.

[26] *In Situ* Hybridization*

By LINO TESSAROLLO and LUIS F. PARADA

Introduction

Identification of the temporal and spatial patterns of gene expression can provide important clues about gene function. This holds particularly true in embryonic development. *In situ* hybridization makes it possible to localize transcripts directly in cells within the context of their tissues and organs.[1,2] Assuming that a protein is expressed in the cells where the encoding transcripts accumulate, roles for gene products consistent with the timing of appearance and observed sites of expression can be proposed.

In situ hybridization relies on the annealing of labeled nucleic acid probes to DNA or RNA sequences within a tissue section or, in the case of whole mount protocols, within an organ or fragment of tissue.[3] The considerations for successful *in situ* are thus not dissimilar to those of effective membrane nucleic acid hybridization. The signal to noise ratio depends on the specificity, length, complexity, and accessibility of the probe to the hybridization substrate. In the case of *in situ* hybridization, accessibility gains considerable importance, as the tissue section has a thickness that must be penetrated by the probe without compromising histological integrity.

The *in situ* hybridization technique has been adapted effectively for studies in a variety of organisms ranging from *Drosophila* to humans. This method can be employed using DNA or RNA probes to hybridize to a variety of tissue targets including endogenous genes (DNA), viral genes

[1] B. L. M. Hogan, F. Costantini, and E. Lacy, *in* "Manipulating the Mouse Embryo: A Laboratory Manual," Cold Spring Harbor Lab., Cold Spring Harbor, NY, 1986.

[2] D. G. Wilkinson, J. A. Bailes, and P. McMahon, *Cell* (*Cambridge, Mass.*) **50,** 79 (1987).

[3] B. Rosen and R. S. P. Beddington, *Trends Genet.* **9,** 162 (1993).

(DNA or RNA), and endogenous or viral transcripts (RNA). Traditionally, probes have been labeled with radioisotopes. Tritiated probes provide the greatest resolution due to the short length of emission. However, the intensity of signal is low and exposure times are relatively long. ^{32}P-labeled probes can also be employed, although, as for DNA sequencing, the resolution is frequently suboptimal. Moreover, because of half-life decay, these probes must be used immediately after preparation. ^{35}S-labeled probes afford the best compromise. The emission path is sufficiently short so that, in most instances, single cell resolution can be obtained. Furthermore, the exposure times are not significantly different from those of ^{32}P-labeled probes, and when maintained under reducing conditions, the probes are stable for a relatively long period (weeks).

Nonisotopic probes have been in use for some time in *Drosophila* and have gained wider use recently, particularly in the whole-mount *in situ* technique.[3] While affording the obvious advantages in regard to toxicity and probe stability, current indications suggest that these methods do not yet provide the sensitivity obtained with isotopically labeled probes in tissue section hybridization protocols. Thus, while in common use for the detection of abundant viral genomes or transcripts, this method has not yet been widely adopted for the detection of endogenous gene expression.

DNA probes have been successfully employed for *in situ* hybridization. The best application of this strategy is the use of terminally labeled oligonucleotides that enable detection of subsets of transcripts from a gene that expresses multiple distinct transcripts through alternative splicing. However, for general purposes the best results have been obtained using probes generated by transcription of antisense RNAs. This method provides large quantities of cRNA probes with high specific activity.

This chapter focuses on the use of *in situ* hybridization for tissue sections with particular emphasis on paraffin-embedded material. We prefer paraffin because of the excellent histology and the relative ease of obtaining and stably maintaining serial sections. As will be noted, frozen sections can also be employed for this technique with minimal modification. Cryostat sections have some distinct advantages. Sections can be prepared rapidly for analysis and similar or adjacent sections can be employed for immunohistochemistry (paraffin-embedded sections tend to be less widely amenable to antibody staining). The disadvantages are the considerable reduction in quality of histology and the need for storage at $-70°$.

Tissue Preparation

This protocol has been employed successfully in our laboratory for many years for analysis of mouse embryos of all stages and of adult tissues.[4–8] We have modified it over time but always for use on paraffin sections of mouse embryos and adult organs.

Fixation

Complete tissue fixation is crucial. If the fixative does not penetrate the tissue entirely, embedding and sectioning quality are compromised. Cardiac perfusion is advisable for complete fixation of large organs, particularly brain. Tissue dissection is done in cold phosphate-buffered saline (PBS) followed by immersion in a freshly made 4% w/v paraformaldehyde–PBS solution (see materials) at 4°. We routinely use 50-ml conical tubes for this purpose. The rule of thumb is about 10 vol of fixative per volume of tissue [i.e., 1 × 18 days postcoitum (dpc, where day of plug is day 0) mouse embryo, 2 × 16 or 17 dpc embryos, up to 4 × 12 to 15 dpc embryos, in 50 ml of fixative]. An intact 18 dpc mouse embryo is a challenge to fix thoroughly. It may require several days for thorough fixation and incisions in the skin and abdominal cavity will enhance fixative penetration of the embryo. The embryo can be trimmed if sections of the whole embryo are not essential. The formaldehyde solution is unstable and should be changed daily until the embryo is fully cleared.

Inclusion in Paraffin

Wash the tissue using volumes approximately 10-fold that of tissue or more. The sequential washes are meant to remove the fixative by replacement with saline and finally to dehydrate the tissue by gradual replacement of saline with alcohol. These washes are done with occasional agitation, and the timing of each step varies depending on size of the tissue. The times indicated here are for 11–14 dpc embryos. Younger or older embryos can have times decreased or increased, respectively.

1×PBS	4°	30 min
0.85% NaCl	4°	30 min
1 : 1 0.85% NaCl/100% ethanol	Room temperature	15 min
70% ethanol (do twice)	Room temperature	15 min
85% ethanol	Room temperature	30 min

[4] D. Martin-Zanca, M. Barbacid, and L. F. Parada, *Genes Dev.* **4,** 683 (1990).

[5] R. Klein, D. Martin-Zanca, M. Barbacid, and L. F. Parada, *Development* (*Berlin*) **4,** 845 (1990).

[6] E. Keshet, S. D. Lyman, D. E. Williams, D. M. Anderson, N. A. Jenkins, N. G. Copeland, and L. F. Parada, *EMBO J.* **10,** 2425 (1991).

[7] L. Tessarollo, L. Nagarajan, and L. F. Parada, *Development* (*Berlin*) **115,** 11 (1992).

[8] L. Tessarollo, P. Tsoulfas, D. Martin-Zanca, D. Gilbert, N. Jenkins, N. Copeland, and L. F. Parada, *Development* (*Berlin*) **118,** 464 (1993).

95% ethanol	Room temperature	30 min
100% ethanol	Room temperature	30 min

At this point the tissue can be stored at 4° or the inclusion in paraffin can be carried out immediately. The paraffin is melted at 60° and filtered (Whatman filter).

100% xylene (twice in fume hood)	Room temperature	30 min
1:1 xylene/paraffin	60°	45 min
100% paraffin (three times)	60°	20 min

The final paraffin step is performed in the embedding cassette where the wax is allowed to solidify. Heated needles are useful at this step to assist in the proper orientation of the embedded tissue. The paraffin block should then be allowed to solidify at room temperature overnight. The embedded tissue can be stored indefinitely (we routinely store blocks at 4°, although the main aim is to avoid extreme heat). If the tissue shrinks away from the paraffin block, the fixation or processing was not complete and useful sections will not be obtained.

When large tissue is prepared, it is important to prolong all steps. Inclusion of the paraffin can be enhanced by incubation in a vacuum oven.

Tissue Sectioning

Properly embedded tissue or embryos can be serially sectioned with ease at 3–5 μm. This provides the advantage of generating multiple closely related sections (sometimes through the same cell), thus permitting comparative studies using different probes. However, fragile tissues, such as 5 dpc embryos, may need to be cut at 7.5 μm to maintain integrity.

Sections are picked up on glass slides coated with gelatin (see Materials) to enhance tissue adherence. However, other substrates including polylysine or 3-aminopropyltriethoxysilane have been employed by many laboratories. The gelatinized slides are preheated to 40° by placing on a heating tray. Experienced histologists will section serially creating a ribbon of 10 or more sections that is then floated on a 40° waterbath (to decompress the individual sections) prior to being picked up on the slides (by dipping the slide under individual sections). For the inexperienced, however, one section can be taken at a time. A large drop of water (containing 0.2% v/v ethanol) is placed on the slides and, using a toothpick or small brush, the sections are carefully placed on the drops where they are allowed to relax to their original configuration (this usually takes 3–5 min). The water drop is then blotted away with a corner of blotting paper and the section is pressed onto the slide by placing the moist blotting paper over the section and pressing lightly with the fingers. After drying overnight in racks, the slides can be placed in boxes and stored indefinitely at 4°.

Good quality sections are essential. There is no point in performing *in situ* on tissue whose histological features cannot be properly identified, thereby impeding correct interpretation.

Preparation of RNA Probe

cRNA probes are most routinely used. The desired stretch of probe is cloned into an appropriate expression vector in antisense orientation with respect to the bacterial or phage promoter to be employed. The vector is linearized prior to transcription at a convenient restriction site within the probe sequence or immediately downstream. In this way only gene-specific sequences are generated and the polymerase cannot transcribe through the length of the plasmid. The resultant runoff transcripts are antisense strands of the gene to be studied.[9] Promoters such as T3, T7, or SP6 function efficiently *in vitro* and produce probes with high specific activity. Because ^{35}S-labeled radionucleotides are sticky, it is important to use siliconized tubes when possible. To minimize loss of probe due to nonspecific sticking to the Eppendorf tubes, the probes are also maintained in 10 m*M* dithiothreitol (DTT). Total yeast RNA is used as a carrier in several steps, so a substantial clean stock should be maintained. Kits are now available from several suppliers (such as Promega) for the preparation of cRNA transcripts.

The *in vitro* transcription reaction is carried out in a final volume of 20 μl with 300 ng of linear template. The radiolabeled nucleotide of preference is [^{35}S]UTP (>1000 specific activity).

5X transcription buffer	4 μl
10 m*M* NTP mix (ATP, CTP, GTP)	3 μl
0.2 *M* DTT	2 μl
[^{35}S]UTP	8 μl
RNasin (40 units/μl)	0.8 μl
Polymerase (10–20 units/μl)	1 μl
DNA template (300 ng)	1 μl
H_2O	0.2 μl
	20 μl final

Incubate for 1 hr at 37°, add 1 μl of DNase RQ1 (Promega), and then incubate for an additional 30 min at 37° to degrade the DNA template; add 3 μl total yeast RNA as carrier (10 mg/ml) and 25 μl of H_2O. Remove and store 1 μl for gel analysis.

[9] P. A. Krieg and D. A. Melton, this series, Vol. 155, p. 397.

Probe Hydrolysis

In principle, these *in vitro* transcription reactions generate a high percentage of full-length material. While the greatest possible homology of a specific probe to its target sequence is desirable for sensitivity, long probes do not penetrate tissues efficiently. Empirically, we have found a balance between optimal penetration and loss of specificity at 150–200 bases. This is achieved by limited alkaline hydrolysis of the transcription reaction at 60° for a time approximated by the application of the following formula[10]:

$$t(\text{min}) = \frac{L_i - L_f}{kL_iL_f}$$

where L_i is the total transcript length in kilobases (length of transcription template), L_f is final transcript length in kilobases (0.15–0.2), and $k = 0.11$.

Add 50 μl of hydrolysis solution to the transcription reaction and incubate at 60° for the time as determined above. Once the hydrolysis reaction is complete, it is neutralized by the addition of 100 μl of neutralizing solution, and precipitation of the probe is carried out by adding 20 μl of 3 *M* sodium acetate (pH 5.2) and 500 μl of ethanol. After overnight incubation at −70°, the sample is centrifuged for 30 min at 4° and resuspended in 20 to 50 μl of 0.1 *M* DTT. One microliter of the probe is then counted and the stock is diluted in 0.1 *M* DTT to 700–800 × 10^3 counts per minute (cpm)/μl.

Save 4 μl of the neutralized solution for gel analysis. The undigested probe saved previously is used for comparison and allows verification that the transcription reaction was successful as evidenced by the presence of a band of the expected size for a specific transcript. The hydrolyzed probe should give a smear with the bulk of the radioactivity ranging from 150 to 200 bases.

Prehybridization

In this step the tissues are prepared for hybridization by removal of the paraffin, leaving the section intact on the slide. This is followed by a deproteinization step to make the tissue porous to the incoming probe. All of the washes which follow are done at room temperature unless noted.

The sections are deparaffinized by immersing the slides in xylene twice for 10 min (the xylene is recycled about every 60 slides). Next, the slide rack is transferred to 100% ethanol (used for this step only!) for 2 min.

[10] J. B. Lawrence and R. H. Singer, *Nucleic Acids Res.* **13,** 1777 (1985).

Run quickly through the following hydration process: 100% ethanol–95%–85%–70%–50%–30%. (The timing of this hydration is short: 10–15 sec.) After hydration the following washes are performed:

0.85% NaCl	5 min
1×PBS	5 min
4% paraformaldehyde–PBS	30 min
1×PBS	5 min
0.2 *N* HCl	5 min
H_2O	5 min

The slides are placed on a clean (benchcote) surface and individually treated with 1 ml of freshly made proteinase K solution:

20 μg/ml proteinase K in proteinase K digestion buffer	10 min

During proteinase K treatment, place a slide rack in PBS. Drain the proteinase K digestion buffer from the slides and place them into the rack for incubation.

1×PBS	5 min
4% paraformaldehyde–PBS	10 min
dip in sterile doubly distilled H_2O	
0.1 *M* triethanolamine (250 ml)	10 min (stirring)

Once the slide rack is introduced in the triethanolamine solution, add 625 μl of acetic anhydride. Then transfer to:

1×PBS	5 min
0.85% NaCl	5 min

Take the slides quickly (10–20 sec per step) through a dehydration process:
30% ethanol–50%–70%–85%–95%–100% (twice)
Air dry for at least 30 min and use the slides the same day.

Hybridization

Add the probe to the hybridization solution to a final concentration of 7–8 × 10^4 cpm/μl. Heat the solution to 85–90° for 5 min prior to adding to the slides in order to denature any possible secondary structure which might inhibit hybridization. Apply approximately 20 to 40 μl probe/slide and cover with coverslips. Remove any air bubbles that may overlie the tissue and incubate in a slide box containing a paper towel soaked in humidifying solution (see Materials) at 50–52° for at least 20 hr. Make sure that the incubation chamber is carefully sealed to avoid loss of moisture.

Washing

It is important to keep the tissues moist, therefore, when transferring the slides between solutions they must not be allowed to dry. Since DTT is labile, it should be the last item added to the washing solutions just prior to incubation.

The slides with coverslips are transferred from the slide box to a rack and immersed in the following washes: (1) 5× SSC, 10 m*M* DTT for 30 min at 42–50°. This first incubation removes the coverslips. When the slide rack is gently removed from the rack tray, the coverslips should remain at the bottom. (2) Wash for 30 min at 50° in 2× SSC, 10 m*M* DTT. (3) 1× RNase washing solution (20 μg/ml RNase A, 2 μg/ml RNase T1) at 37° for 30–45 min. (4) Formamide solution at 55–58° for 30 min. (5) Wash twice for 30 min at 50° in sodium pyrophosphate washing solution. (6) A dehydration step is now performed, but in the presence of 0.3 *M* ammonium acetate in the ethanol : H_2O solutions (see Materials). Wash quickly in 30%–60%–80%–95% ammonium acetate/ethanol solutions. (7) Wash in 100% ethanol (twice) and air dry.

The precise time and temperature of washes will depend on the properties of the probe and the relative levels of gene expression. Therefore these parameters must be adjusted empirically for each probe and each cell or tissue type.

Autoradiography

Safelight conditions are of extreme importance, therefore, avoid unplugging of switches, phosphorescent watches, and handle the emulsion gently.

Warm the emulsion (i.e., Kodak NTB-2) in a 45° water bath in the dark room (this solution is stored carefully sealed from light at 4°). Pour ~30 ml into a slide mailer. We routinely reuse this stock once and discard. Gently pass a clean, blank slide through the emulsion to remove air bubbles. Dip the slides and allow them to air dry vertically in a rack for 1 hr in the dark.

Store in a light-tight box with Drierite at 4° and isolated from any radioactive source. The emulsion is sensitive to shock so the slides must be treated gently. We routinely expose the emulsion for 7–10 days depending on the level of expression of the gene under investigation and the size of the probe used. A few control slides can be developed beginning at 4 days to estimate the timing required for optimal exposure. When it is time to develop the slides, they are first allowed to come to room temperature prior to opening the container. This avoids condensation which can result in image fogging.

The slides are placed in a rack and developed by immersing twice in D-19 developer for 2.5 min at 16°. After dipping briefly in water the slides are transferred to the fixative solution for 10 min at room temperature twice. These times are somewhat empirical and should be optimized in the darkroom with the specific solutions. After fixation, rinse thoroughly in cold water. Residual emulsion must be scraped off the back of the slide with a single-edged razor blade.

Various histological stains can be employed to visualize the tissue sections (i.e., eosin, methylene blue, and toluidine blue). Stain lightly in a 0.2% toluidine blue/H_2O staining solution for 10–30 min. Rinse in H_2O (0.01 *N* HCl can be used to destain if slides were stained excessively). Dehydrate through a series of alcohols which are solely used for this step: 30%–50%–70%–85%–95%–100% ethanol–xylene (2×). Without allowing the slides to dry, overlay with 2–3 drops of emulsion-compatible mounting media (i.e., Permount) and overlay with long coverslips, taking care to remove all bubbles. Clean off any excess mounting media from edges and underneath and lay flat until the coverslip is immobile.

Frozen Sections

Embryos or tissues can be fixed as described earlier and then processed through the 0.45 *M* sucrose/PBS solution prior to embedding in OCT.[1] The frozen sections are stored at −70°, thawed, and treated as described in prehybridization section starting with the 0.85% NaCl treatment.

Cultured Cells

Grow cells on gelatin- or polylysine-coated slides placed in 10-cm culture dishes. Rinse in PBS; fix for 30 min in 4% paraformaldehyde–PBS, and process as described in the prehybridization section.

Photography

To obtain high-quality photographic images of *in situ* slides, certain factors are essential. First, a research quality, upright microscope with high-quality phase contrast lenses and a dark-field condenser is required. Next, the histological integrity and staining of the sections must be sufficient to provide clear identification of structures and good contrast. We have used various kinds of film with relatively good success. For black and white photography, TMax 100 (Kodak) works well. For color we prefer Ektar ISO 50 (Kodak) for prints or Ektachrome 50 (Kodak) for slides. The precise exposure times for dark-field exposure have to be determined empirically. Always bracket dark-field exposures to ensure getting an acceptable print.

A final point regarding film processing. If your laboratory does not develop its own film and you use a service, warn them about the dark-field images and advise them to use standard developing procedures because otherwise they will overdevelop your prints in trying to bring out anticipated images.

Materials

Tissue Preparation

4% paraformaldehyde–PBS: heat stir 50 ml of water with 4 g of paraformaldehyde to 65° (the solution remains cloudy). Add 5 to 10 μl of 10 *N* NaOH which will make the solution become clear. Filter through a Whatman filter to remove debris. Add 10 ml of 10× PBS, and bring volume to 100 ml final with water.

Slides: the slides must be absolutely clean and devoid of grease. Place slides in metal racks and soak in hot tap water with added detergent. Rinse for 1 hr in hot-running water. Rinse three times in distilled water. Air dry or place in a 40° oven. Boil 1 liter of distilled water. Stir with magnetic stirrer, turn off heat, and add 2 g of gelatin. Add 0.1 g of potassium chrome(III) sulfate $[KCr(SO_4)_2 \cdot 12H_2O]$, and allow temperature to drop to about 35°. Dip the washed slide in and allow 5 min to drain. Dip again and allow to drain at an angle such that the gelatin will run toward the frosted end of the slides. Dry overnight in a 35° oven and store in the original slide boxes for not more than 1 month.

Preparation of the Probe

5× transcription buffer: 200 m*M* Tris-HCl (pH 7.5); 30 m*M* $MgCl_2$; 10 m*M* spermidine; 50 m*M* NaCl. (The RNA transcription kit available from Stratagene is essentially equivalent to that described.)

NTP: mix 1 μl of 10 m*M* CTP, GTP, and ATP.

tRNA, obtained from Boehringer Mannheim, is treated with phenol–chloroform three times, then precipitated several times and resuspended at a final concentration of 10 mg/ml to make a stock solution. The RNA is clean when the yellowish tinge has disappeared.

Hydrolysis solution: this solution is made by mixing 10 μl of 1 *M* DTT, 80 μl of 1 *M* $NaHCO_3$ 120 μl of 1 *M* Na_2CO_3, and 790 μl of H_2O.

Neutralizing solution: this is made by mixing 10 μl of 1 *M* DTT, 200 μl of 1 *M* sodium acetate, 10 μl of acetic acid, and 780 μl of H_2O.

Prehybridization

4% paraformaldehyde–PBS (fresh), 500 ml

1× PBS, 1 liter

0.85% NaCl (autoclaved), 1 liter
Proteinase K; a 20-mg/ml stock solution in water should be made from which aliquots are thawed once or twice and then discarded
Proteinase K digestion buffer: 50 m*M* Tris–Cl (pH 7.5), 5 m*M* EDTA
1 *M* triethanolamine: 16.7 ml in 100 ml diethyl pyrocarbonate (DEPC)-treated H_2O. Store wrapped in aluminum foil at 4° in 50-ml aliquots
Acetic anhydride

Hybridization

Hybridization buffer: 50% formamide, 0.3 *M* NaCl, 20 m*M* Tris–HCl, pH 7.4, 5 m*M* EDTA, 10 m*M* $NaH_2PO_4 \cdot H_2O$, pH 8.0, 10% dextran sulfate, 1× Denhardt's solution, 0.5 mg/ml total yeast RNA. The buffer is stable for several months when stored at −20°. This solution represents nine parts of the complete hybridization solution to which one part of probe will be added. Since the probe contains 100 m*M* DTT, the final hybridization mix will contain 10 m*M* DTT.
Humidifying solution: 50% formamide, 4× SSC
100 m*M* Dithiothreitol

Washes

5× SSC, 10 m*M* DTT solution
2× SSC, 10 m*M* DTT solution
10× RNase washing solution: 4 *M* NaCl, 0.1 *M* Tris HCl, pH 7.5, 50 m*M* EDTA
10 mg/ml boiled RNase A
2 mg/ml RNase T1
Formamide washing solution: 50% formamide, 2× SSC, 10 m*M* DTT
1% (w/v) sodium pyrophosphate
Sodium pyrophosphate washing solution: 2× SSC, 10 m*M* DTT, 0.0625% sodium pyrophosphate
Ammonium acetate ethanol solutions: 9.3 g ammonium acetate/400 ml ethanol : H_2O

Autoradiography

Kodak NTB-2 emulsion
Kodak D-19 developer
Kodak Rapid Fix
0.2% toluidine blue in H_2O

Acknowledgments

We thank Janet Flynn for technical assistance and Cindy Fitzpatrick for manuscript preparation. We also thank Peter Donovan for critical reading of the manuscript. This research is sponsored by the National Cancer Institute, DHHS, under Contract No. NO1-CO-74101 with ABL. The contents of this publication do not necessarily reflect the view or policies of the Department of Health and Human Services, nor does mention of trade names, commercial products, or organizations imply endorsement by the U.S. government.

[27] Monoclonal Antibodies to Oncoproteins

By Jin-Hong Chang, William M. Sutherland, and Sarah J. Parsons

Evidence accumulated over the years has led to the conclusion that proteins encoded by oncogenes and their normal cellular counterparts are involved in multiple signaling pathways that regulate various biological processes such as growth, differentiation, and specialized functions of post-mitotic cells. A variety of reagents have been developed for the study of these proteins in numerous biological paradigms. One of the most powerful of these tools is the monoclonal antibody (MAb), which has proved invaluable for the detection, quantitation, isolation, and characterization of oncogene and protooncogene products. This chapter discusses methods for the production of MAbs, with emphasis on those aspects that we have found to be important for the derivation of successful reagents. In addition, we will detail ways MAbs can be used to increase our understanding of the role of normal and oncogenic proteins in growth control. For alternative methods of MAb production and use, readers are referred to the many reviews and original articles dealing with this subject.[1-4]

Antigen Preparation

The first step in MAb production is the preparation of antigen. Antigen sufficient for both immunization and screening of sera and hybridoma

[1] G. Kohler and C. Milstein, *Nature (London)* **256,** 495 (1975).
[2] S. J. Parsons and J. T. Parsons, *in* "Monoclonal Antibody Production Techniques and Applications," p. 207. Dekker, New York, 1987.
[3] E. Harlow and D. Lane, eds., "Antibodies: A Laboratory Manual." Cold Spring Harbor Lab., Cold Spring Harbor, NY, 1988.
[4] H. Zola, "Monoclonal Antibodies: A Manual of Techniques," p. 147. CRC Press, Boca Raton, FL, 1987.

culture supernatants is required (1–10 mg). Although impure immunogens can be used, pure antigens greatly reduce the time and effort involved in complex screening strategies that invariably accompany immunizing with impure immunogens. This in turn allows for rapid hybridoma selection and cloning, which is essential to a successful hybridoma production.

Immunization and screening strategies should reflect the eventual application of recovered MAbs. If the MAbs will be used to detect the protein in its native conformation (such as in immunoprecipitations, immunofluorescence, or microinjection experiments), native protein is the preferred immunogen, and hybridoma culture supernatants should be screened using those assays (such as immunoprecipitation). On the other hand, if the MAbs will be used for detecting denatured protein (as in Western immunoblotting), denatured protein is the preferred immunogen. In practice, however, MAbs that recognize denatured antigen can be derived from animals immunized with native protein (and vice versa). This is most likely due to the presence of significant amounts of both denatured and native protein in an antigen preparation and the fact that native proteins contain both sequential and combinatorial epitopes.

Oncoproteins and their normal progenitors are structurally conserved throughout both ontogeny and phylogeny, rendering the proteins poorly immunogenic. To overcome this, several strategies have been developed, including coupling of the protein to a highly immunogenic carrier, such as keyhole limpet hemocyanin (KLH), or constructing a bacterial fusion protein. In addition to the nature of the antigen, the method of immunization and the age of the host animal also influence the isotypes, affinities, and specificities of the MAbs produced.

Below, we will discuss several strategies for antigen preparation, including peptide synthesis and coupling, recombinant protein production, and protein purification from cell extracts.

Oncogene Protein Peptide Preparation

Peptides can be synthesized based on amino acid sequences obtained from purified protein or deduced from cloned cDNA. One advantage of using conjugated peptides as immunogens is that specific regions of the molecule can be targeted for antibody production. A decided disadvantage of antipeptide antibodies, in our experience, is their frequently low avidity, which prevents them from being used as immunoprecipitating reagents.

The choice of synthetic peptides for MAb production is largely empirical, although several factors appear to influence the usefulness of the antibodies. These include the presence of a proline residue, hydrophilicity, mobility, and length or mass of the peptide. Tertiary structure analysis has shown that peptides containing an abundance of hydrophilic amino acids, proline residues, or those that are located at the N or C terminus are frequently found on the surface of proteins at the solvent interface.[3] Antibodies generated against such peptides have a greater likelihood of reacting with native protein.

In general, an antigen with a molecular size greater than 10 kDa is considered capable of eliciting a good immune response. Since the molecular mass of synthesized peptides is around 1–2 kDa for every 10 residues, it is advisable that the peptide be increased in mass. There are two ways to accomplish this: one is by coupling the peptide to a large carrier molecule, and the other is by the multiple antigen peptide system (MAPs).[5] KLH and bovine serum albumin (BSA) are preferred choices for carrier proteins, since they are highly immunogenic, and coupling procedures for them have been developed.[3] For MAPs, several peptides are synthesized simultaneously and directly onto a branched lysine core. The resulting macromolecules are generally larger than 10 kDa and can be used directly for immunization without coupling to a carrier.

Recombinant Oncogene Proteins Purified from Escherichia coli and Baculovirus

There are several important advantages of using recombinant proteins as antigens for MAb production: (1) Different regions of the protein can be targeted for antibody development; (2) the coupling step can be omitted since most recombinant proteins are larger than 10 kDa; (3) large amounts of protein can be obtained easily for injection and screening; and (4) fusion proteins between bacterial proteins and proteins of interest can be generated to overcome the poor immunogenicity of most oncoproteins. During the course of recombinant protein production, our experience has been that soluble fusion proteins degrade more extensively than insoluble ones. Therefore, to generate MAbs against pp60$^{c\text{-}src}$, p21^{ras} GTPase activating protein (GAP) and GAP-associated p190, we chose to use β-Gal and *trpE* fusion proteins as immunogens.[6–9] Such fusion proteins tend to be more immunogenic, partly because of their propensity to form aggregates.

[5] D. N. Posnett and J. P. Tam, this series, Vol. 178, p. 739.
[6] S. J. Parsons, D. McCarley, C. Ely, D. Benjamin, and J. T. Parsons, *J. Virol.* **45,** 1190 (1983).
[7] S. J. Parsons, D. McCarley, C. Ely, D. Benjamin, and J. T. Parsons, *J. Virol.* **51,** 545 (1984).
[8] J. Settleman, V. Narasimhan, L. C. Foster, and R. A. Weinberg, *Cell (Cambridge, Mass.)* **69,** 539 (1992).
[9] J.-H. Chang, S. Gill, J. Settleman, and S. J. Parsons, *J. Cell Biol.,* in press (1995).

Protocol for Large-Scale Purification of trpE Fusion Proteins for Immunization

1. Using established recombinant DNA procedures,[10] subclone the gene of interest or a portion of it into a suitable pATH vector in the correct reading frame.[11] Confirm that the appropriately sized protein can be produced by expressing the *trpE* fusion protein in *E. coli* (using a small-scale version of the protocol described below) and analyzing bacterial extracts by Western immunoblotting (utilizing a polyclonal rabbit or mouse antibody specific for the protein of interest or for *trpE*).

2. For large-scale production, transform *E. coli* DH5α with the confirmed pATH plasmid and grow 50 ml of overnight culture in M9 medium (42 m*M* Na_2HPO_4, 22 m*M* KH_2PO_4, 8.5 m*M* NaCl, 18 m*M* NH_4Cl, 2 m*M* $MgSO_4$, 0.2% glucose, 0.1 m*M* $CaCl_2$) plus ampicillin (50 μg/ml) and tryptophan (20 μg/ml). Cultures are incubated at 37° with constant agitation.

3. Transfer the overnight culture to 450 ml complete M9 medium without tryptophan, grow 3 hr with shaking to derepress the *trpE* operator, then add iodoacetic acid (IAA, 10 μg/ml final concentration) for an additional 2 hr. (IAA competes with tryptophan for binding to *trp* repressor protein and allows for maximum induction.)

4. Centrifuge bacteria at 4000*g* for 15 min at 4°.

5. Resuspend the pellet in 15 ml SDS sample buffer to lyse the bacteria.

6. Sonicate (Branson Sonifier 450, VWR Scientific) twice for 30 sec each at 20° to shear the DNA.

7. Centrifuge at 100,000*g* for 30 min at 20° to remove debris.

8. Boil supernatant for 5 min and load the supernatant on eight preparative gels (6% SDS gels, 18 cm × 16 cm × 3 mm). (We usually do a trial run to estimate how much to load onto a preparative gel.)

9. After electrophoresis, stain the gel for 5 min in a solution of 1% Coomassie Brilliant blue/50% methanol and destain for 5 min in deionized water. This treatment serves to localize the major protein bands and does not precipitate proteins within the gel.

10. Excise the band, add an equal volume of phosphate-buffered saline (PBS) (136 m*M* NaCl, 2.6 m*M* KCl, 10 m*M* Na_2HPO_4, 1.7 m*M* KH_2PO_4, pH 7.2), homogenize the gel pieces by passing through syringes connected by a stopcock, and shake overnight at 4°.

11. Collect the supernatant by centrifugation.

[10] J. Sambrook, E. F. Fritsch, and T. Maniatis, eds., "Molecular Cloning: A Laboratory Manual," 2nd ed. Cold Spring Harbor Lab., Cold Spring Harbor, NY, 1989.

[11] T. J. Koerner, J. E. Hill, A. M. Myers, and A. Tzagoloff, this series, Vol. 194, p. 477.

12. Dialyze against 500–1000 ml PBS with three changes to remove excess SDS.

13. Determine protein concentration. The yield is usually 1 mg per 100 ml of culture. The fusion protein is ready to be emulsified in adjuvant and injected.

Production of recombinant oncogene proteins from *E. coli* has proved to be very efficient. However, alternative systems have also been used. For example, recombinant human *Bcl-2* (B cell lymphoma/leukemia-2) has been successfully expressed and purified from a baculovirus system in quantities sufficient for immunization.[12]

Purification of Intact Protein by Affinity Chromatography from Cultured Cells

It is very difficult to purify oncoproteins to homogeneity utilizing conventional chromatography, due largely to the low abundance of the oncoproteins in cultured cells and to the time and expense of growing large quantities of cells. However, several strategies have been devised to enhance the amount and purity of not only oncoproteins, but also proteins which interact with oncoproteins and mediate some of their functions. For example, $pp60^{v\text{-}src}$ was first purified from fibroblasts cultured from tumors of mice injected as neonates with Rous sarcoma virus (RSV). Extracts of these cells were chromatographed over an affinity column made of covalently linked RSV-induced tumor-bearing rabbit antibodies.[6,7] More recently, phosphotyrosyl (p-Tyr)-containing proteins have been purified using p-Tyr antibodies as affinity reagents.[13,14] Exploiting another type of protein/protein interaction, we have isolated GAP-associated p190 by coimmunoprecipitation with GAP using GAP-specific antibodies. The isolated protein has been successfully used as antigen for MAb production.[9]

Purification of GAP-Associated p190

1. Grow 200 dishes (150 mm) of v-*src*-transfected 10T1/2 murine fibroblasts (IV5) to confluency.[15]

2. Wash twice with PBS at room temperature.

[12] J. C. Reed, S. Tanaka, M. Cuddy, D. Cho, J. Smith, R. Kellen, H. U. Saragovi, and T. Torigoe, *Anal. Biochem.* **205,** 70 (1992).

[13] S. B. Kanner, A. B. Reynolds, and J. T. Parsons, *J. Immunol. Methods* **120,** 115 (1989).

[14] J. R. Glenney and L. Zokas, *J. Cell Biol.* **108,** 2401 (1989).

[15] D. K. Luttrell, L. M. Luttrell, and S. J. Parsons, *Mol. Cell. Biol.* **8,** 497 (1988).

3. Lyse with RIPA buffer [150 m*M* NaCl, 50 m*M* Tris–HCl, pH 7.5, 1% (v/v) Nonidet P-40 (NP-40), 0.5% (w/v) sodium deoxycholate, 1 m*M* sodium orthovanadate, 1 m*M* phenylmethylsulfonyl fluoride, 0.5% (w/v) aprotinin, 50 μg/ml leupeptin, 10 μg/ml α_2-macroglobulin, 2 m*M* EGTA] (0.8 ml/dish) at 4°. Steps 3, 4, 6–11 are carried out at 4°.

4. Clarify extract at 100,000*g* for 30 min.

5. Determine the protein concentration of the supernatant. (We obtain ca. 4 g of total cellular lysate in 250 ml RIPA buffer from 200 dishes.)

6. Preclear extract with 1 ml Pansorbin (Calbiochem, La Jolla, CA) for 1 hr at 4°, centrifuge, and collect supernatant.

7. Immunoprecipitate with anti-GAP antibody (using a ratio of 4 g lysate/1 ml pansorbin, preabsorbed with 2.5 mg purified anti-GAP polyclonal antibodies).

8. Mix and incubate for 1.5 hr, centrifuge, and recover pellet.

9. Wash immunocomplexes twice with RIPA buffer and once with PBS.

10. Add SDS sample buffer and follow steps 7–12 of fusion protein purification. (P190 bands are localized by immunoblotting a thin longitudinal slice of the preparative gel with p-Tyr antibody).

11. Estimate protein concentration by gel electrophoresis of a small portion of the sample by staining with Coomassie blue and comparing the intensity to that of stained molecular weight standards (BRL).

12. Dialyze against three changes of 1 liter 0.1× PBS and lyophilize.

The purified proteins are ready to be used as antigen for MAb production. We routinely isolate about 40 μg GAP-associated p190 by this method from 4 g IV5 lysate. We have generated five p190 MAbs using a combination of purified p190 and *trpE* p190 fusion protein as immunogens.[9]

Hybridoma and Monoclonal Antibody Production

Immunization

The vast majority of MAbs specific for oncoproteins have been generated in the mouse; however, other species have also been used.[16] BALB/c mice are usually employed for these studies because most available murine myeloma fusion partners are derived from this species. The BALB/c mouse does not, however, respond well to many immunogens. As an alternative, we have found that immunization of the A/J mouse often generates antibodies to epitopes not represented in the BALB/c repertoire. If an antigen is

[16] M. Furth, L. David, B. Fleurdelys, and E. Scolnick, *J. Virol.* **43,** 294 (1982).

known or is suspected to be a poor immunogen, both strains of mice should be immunized. For production in mice, our strategy has been to immunize two to three young adult animals, using a protocol that increases the probability of generating hybridomas secreting IgG subclass MAbs of high affinity with a broad spectrum of specificities. This is accomplished with multiple injections of antigen given at long intervals (minimum of 4 weeks, preferably 6–8 weeks), followed by an intrasplenic injection delivered just prior to fusion. Mice receive a primary injection of 100 μg purified protein antigen (although as little as 25 μg can be used), emulsified 1:1 with complete Freund's adjuvant (CFA), in a total volume of less than 200 μl. The immunogen is divided equally between two injection sites: subcutaneously at the nape of the neck and intraperitoneally. Four weeks later, the mice are injected again using CFA. If circulating antibody titers (as measured by an ELISA or immunoprecipitation) remain low, additional immunizations can be administered, maintaining the minimum 4-week interval between injections.

Pre- and post-immunization sera are collected and used to develop appropriate assay systems or to test existing assays. Such assays include the ELISA,[4] immunoprecipitation, Western blotting, and immunocytochemistry. In our experience, the most productive fusions result from immunized mice whose sera, collected 2 weeks following a "booster" injection, demonstrate a positive ELISA (an absorbance of at least 1.0) at a dilution of 1:5000 or greater.

Once appropriately high circulating antibody titers have been demonstrated, the mice are again rested for at least 4 weeks. The selected fusion mouse is given a "booster" injection prior to the fusion. In most published protocols, this injection is normally administered without adjuvant as an intraperitoneal or intravenous (tail vein) injection. Since 1984, we have routinely used a direct intrasplenic injection of antigen 4 days prior to the fusion. A simple surgery is performed, essentially according to the procedure described by Spitz[17] and 1–50 μg of antigen in up to 50 μl PBS is injected into the anterior end of the spleen.

The spleen of the fusion mouse is often noticeably larger 4 days following an intrasplenic injection because of an increase in the total lymphocyte population (ca. 60% greater). The greater number of B lymphocytes increases the probability of generating MAbs directed against rare epitopes, weak immunogens, and components of the immunogen that are in relatively low abundance. Parallel fusions with splenocytes from mice immunized by conventional methods either do not produce MAbs to these epitopes or are significantly less productive.

[17] M. Spitz, this series, Vol. 121, p. 33.

Fusion and Cell Culture

Our fusion protocol has been modified from published procedures.[18] Briefly, the spleen is placed in a petri dish containing 5 ml of Iscove's medium without serum (Iscove's MDM, Gibco BRL, containing 0.06 m*M* 2-mercaptoethanol). Two "instruments" are prepared by bending 1.5-inch 19-gauge needles (attached to syringes) to a 45° angle approximately 1 cm from the tip using a sterile hemostat. The ends of the spleen sac are teased open with the point of a needle. Cells are extruded from the sac by firmly pressing down on the center of the spleen with the bent side of one needle while gently rubbing repeatedly (from the center to the open end) with the other. The empty sac is removed and the cell clumps are dispersed by pipetting against the bottom of the dish. Studies have shown this technique to be the most efficient method for capturing the total splenocyte population.

Protocol

1. Filter cell suspension through a 20-μm sterile Nitex filter,[19] add Iscove's medium to 50 ml, and centrifuge at 200*g* for 10 min.
2. Resuspend the cell pellet, wash once with 50 ml medium, and mix with washed Sp2/0-Ag14 myeloma cells[20] at a 5:1 ratio (splenocytes/myeloma) in a total volume of 20 ml. (Sp2/0 cells are maintained in log phase growth, at cell densities less than 6×10^5 cells/ml, for at least 5 days prior to the fusion.)
3. Centrifuge the cells (200*g* for 5 min) and aspirate the medium, then add 0.2 ml 50% polyethylene glycol (PEG) 4000 (Gibco BRL) at 37° every 10 sec while stirring the cells slowly with the tip of the pipette (a total of 1 ml PEG is added over approximately 1 min).
4. Stir the suspension for an additional 1 min, add 5 ml medium dropwise, swirl gently, add another 5 ml medium rapidly, and centrifuge for 5 min at 200*g*.
5. Resuspend the cells gently and wash once in 20 ml HT medium (Iscove's medium, 15% fetal bovine serum, 13.4 m*M* hypoxanthine, 62 μM thymidine).
6. Resuspend the cells in 20 ml HT medium, transfer to a 100-mm petri dish, and incubate at 37° in 5% CO_2/95% air (v/v) for 1–3 hr.
7. Dilute the cells with HT medium, plate into 96-well culture plates at $2–4 \times 10^5$ cells/100 μl/well, and incubate overnight.

[18] M. D. Chapman, W. M. Sutherland, and T. A. E. Platts-Mills, *J. Immunol.* **133,** 2488 (1989)
[19] I. R. Konigsberg, this series, Vol. 57, p. 511.
[20] M. Shulman, C. D. Wilde, and G. Kohler, *Nature* (*London*) **276,** 269 (1978).

8. Add 100 μl HT medium containing 50 μM aminopterin (HAT medium)[21] to each well. On days 3, 6, 8, and 10 (every 2–3 days then as needed), remove 100 μl medium and feed with 100 μl fresh HAT medium. Add medium without the aminopterin after day 14 to facilitate robust cell growth.

9. Feed the cells with HT media for several feedings, then feed with Iscove's medium containing 10% fetal bovine serum.

Initial Screening of Hybridoma Culture Supernatants

Cultures are inspected daily, and culture supernatants are assayed for antibody production when cells cover 50–75% of the bottom of a well (beginning around day 8 or 9 following the fusion). The most common assay employed at this stage is an ELISA which entails coating each well of a 96-well plastic plate with antigen (usually the immunogen), incubating different culture supernatants in each well, washing away unbound antibodies, and detecting those antibodies that bind to antigen with a secondary enzyme-linked antibody. Those wells that capture the secondary antibody (by binding to the primary antibody) are revealed by color development when appropriate enzymatic substrates are added. If the antigen used to coat the plate is pure and contains no carrier or bacterial moiety, a positive ELISA denotes the generation of an antigen-specific antibody. If, however, the antigen in the wells is conjugated to a carrier or constitutes a fusion protein, each supernatant is tested for reactivity to the carrier or bacterial fusion moiety alone in a second, parallel 96-well plate. Those clones reactive with the carrier or fusion moiety are discarded. Cells from specific antibody-positive wells are transferred to 24-well culture dishes and retested. Confirmed antibody-positive parent cultures are cloned at least twice and are cryopreserved at each stage of selection. A significant drop in antibody titer in cultures of cloned hybridomas usually reflects the clonal expansion of a population of cells that cannot produce functional immunoglobulin due to chromosome loss (encoding one of the Ig chains). These cultures should be either recloned or discarded and new cultures established from "banked" cells.

Large-Scale Antibody Production and Purification

Large amounts of MAb can be collected either as culture supernatant or as ascites. Hybridomas are most easily grown as ascites tumors in mice with the same genetic background as the myeloma and immune B cell (MHC compatibility). Ascites is commonly generated in BALB/c mice.

[21] J. W. Littlefield, *Science* **145,** 709 (1964).

However, we routinely generate ascites with hybridomas derived from BALB/c in the F_1 cross between an ICR male and a BALB/c female ($IRCF_1$, available from Hilltop Lab Animals, Scottdale, PA). In our hands, $IRCF_1$ male mice generate approximately 60% more ascites per mouse (with comparable titer) than BALB/c hosts. Hybrid cells derived from A/J mice can also be easily grown as ascites tumors in the isogenic hybrid CAF_1 strain.

Mice (8–10 weeks old) are primed for ascites production by an intraperitoneal injection of 0.5 ml pristane (2,6,10,14-tetramethylpentadecane). Ten days later, hybridoma cells in log phase growth are pelleted and resuspended in Iscove's medium (without serum) and are injected intraperitoneally (2×10^6 cells/0.5 ml/mouse). The mice are carefully monitored and as soon as noticeable abdominal swelling is observed, ascites is collected (by gravity flow) by inserting a 1.5-inch 19-gauge needle through the body wall. Mice may be "tapped" every other day until signs of distress are observed. Mice are then sacrificed and the remaining ascites, collected. Cells are removed from the ascites by centrifugation and the fluid is stored frozen. Antibody is purified by protein G chromatography (Pharmacia/LKB, Piscataway, NJ) according to manufacturer's instructions.

Characterization of Monoclonal Antibodies

Monoclonal antibody characterization will provide the investigator with information that will facilitate use of the antibodies in the most appropriate and efficient ways. Important features to know about each MAb include isotype, epitope specificity, the effect each antibody has on its cognate antigen function, and species and tissue cross-reactivity as well as cross-reactivity with different isoforms of the antigen that could be present in a single cell.

Isotype

Isotype characterization informs the investigator, for example, about the probability of each MAb binding to protein A, a reagent used frequently in several assays described below. The immunoglobulin G_2 (IgG_2) subclass has high affinity for protein A, but the IgG_1 subclass has low affinity for protein A and frequently requires a secondary antibody to be used in immunoprecipitation or Western immunoblotting assays. Several methods for the determination of MAb isotypes have been developed and are based on the ability of secondary antibodies to differentiate between the various heavy (μ, ε, α, δ, $\gamma 1$, $\gamma 2a$, $\gamma 2b$, $\gamma 3$) and light (κ and λ) chains of mouse immunoglobulin.[22] Commercial kits are now available for these determinations.

Epitope Mapping

Localization of the antibody binding site on the antigen can be carried out in several ways. Regions of binding may be deduced from the cross-reactivity profile of an antibody if the primary sequences of the various isoforms or family members of the antigen are known. Alternatively, the reactivity profile of a MAb against a panel of deletion or site-directed mutants will allow localization of the epitope to a specific region or site on the molecule. Although the former method frequently suffers from lack of primary sequence information, both methods share the drawback that alterations at one site (especially deletions) may influence a secondary site, which could be the actual site of MAb binding. With this constraint in mind, we used both methods to map MAbs to pp60^{src} and concluded that (1) the majority of the MAbs were directed to the N-terminal unique domain, but within this domain species differences were evident; (2) one MAb (GD11) was specific for the SH3 domain; and (3) one MAb (R2D2) recognized a combinatorial determinant represented in pp60$^{v\text{-}src}$, pp90$^{v\text{-}yes}$, pp70$^{v\text{-}fgr}$.[6,7,23,24]

Inhibition or Enhancement of Antigen Function

Several oncogene and protooncogene proteins have been demonstrated to exhibit intrinsic enzyme activities, such as protein kinase, GTPase, GTP/GDP exchange activity, and DNA binding. The binding of a specific MAb to its cognate protein may alter the enzymatic activity of the oncoprotein and affect its biological activity. This alteration may take the form of an inhibition or an enhancement, depending on the epitope. For example, a pp60$^{v\text{-}src}$-specific MAb, R2D2, was shown to inhibit the abilities of pp60$^{v\text{-}src}$ and two related tyrosine kinases, pp90$^{v\text{-}yes}$ and pp70$^{v\text{-}fgr}$, to autophosphorylate in an immune complex.[23] Similarly, MAb Y13-259 inhibited the GTPase activity of immunocomplexed p21^{ras} *in vitro*[25,26] and *in vivo* following microinjection.[26] In addition, the binding of some MAbs to the extracellular domain of EGF receptors can effectively block EGF binding and inhibit the growth of carcinoma cells, whereas other MAbs can mimic

[22] I. D. Gardner, *Pathology* **17,** 64 (1985).
[23] D. J. McCarley, J. T. Parsons, D. C. Benjamin, and S. J. Parsons, *J. Virol.* **61,** 1727 (1987).
[24] D. J. McCarley and S. J. Parsons, *Proc. Natl. Acad. Sci. U.S.A.* **84,** 5793 (1987).
[25] S. Hattori, L. Ulsh, K. Halliday, and T. Shih, *Mol. Cell. Biol.* **5,** 1449 (1985).
[26] L. Mulcahy, M. Smith, and D. Stacy, *Nature (London)* **313,** 241 (1985).

EGF binding and activate receptors.[27,28] Furthermore, Klempnauer and Sippel have demonstrated that the DNA binding activity of the v-*myb* protein is inhibited by its cognate mAbs.[29]

Cross-Reactivity

It is not uncommon for a protein to exist in various isoforms or as a member of a functionally related family. Isoforms usually differ from one another by only a few amino acids, whereas family members frequently exhibit more extensive variations. Monoclonal antibodies with specificity for a given epitope can often distinguish between these different forms. For example, MAb 2-17, specific to the unique domain of *src*, can distinguish between pp60$^{c\text{-}src}$ and other related nonreceptor tyrosine kinases. In addition, determining the cross-reactivity profile of each MAb will allow the investigator to choose appropriate MAbs for each experiment. For example, MAb EC10 reacts only with avian pp60$^{c\text{-}src}$, whereas MAbs GD11 and 327 react with both avian and mammalian pp60$^{c\text{-}src}$.[6,7,30] These MAbs have frequently been used to distinguish endogenous vs exogenous pp60$^{c\text{-}src}$ in transfection experiments.[15] Likewise, the specificities of MAbs to phospholipase C have made it possible to differentiate various forms of this enzyme found in different species and tissues.[31]

Monoclonal Antibodies as Immunological Probes for Events of Cellular Transformation and Signal Transduction

Monoclonal antibodies to oncogene proteins have proved to be valuable reagents in dissecting the molecular events leading to cellular transformation and uncovering the roles of protooncogene products in normal cells. These antibodies have been used to detect the presence of oncoproteins and their normal counterparts in transformed and normal cells from a variety of tissues and species, to quantitate the levels of expression in these samples, to determine the intracellular localization of the proteins (which provides clues to their functions), to test for post-translational modifications such as phosphorylation and fatty acid acylation, to assess putative enzymatic activities, and to analyze interacting (coprecipitating) proteins. Such

[27] B. W. Ennis, E. V. Valverius, S. E. Bates, M. E. Lippman, F. Bellot, R. Kris, J. Schlessinger, H. Masui, A. Goldenberg, J. Mendelsohn, and R. B. Dickson, *Mol. Endocrinol.* **3,** 1830 (1989).

[28] J. D. Sato, T. Kawamoto, A. D. Le, J. Mendelsohn, J. Polikoff, and J. H. Sato, *Mol. Biol. Med.* **1,** 511 (1983).

[29] K. H. Klempnauer and A. E. Sippel, *EMBO J.* **6,** 2719 (1987).

[30] L. Lipsich, A. Lewis, and J. Brügge, *J. Virol.* **48,** 352 (1983).

[31] P.-G. Suh, S. H. Ryu, W. C. Choi, K.-W. Lee, and S. G. Rhee, *J. Biol. Chem.* **263,** 14497 (1988).

techniques as immunoprecipitation, immunoblotting, immunocytochemistry, and microinjection have been widely employed to address these issues. A more detailed outline of immunoprecipitation and immunocytochemistry techniques is included in this chapter (see below), while extensive discussions on immunoblotting and microinjection are presented elsewhere in this volume.

Immunoprecipitation

Immunoprecipitation provides the investigator a means of separating a specific protein from a complex mixture and subjecting the isolated protein to independent analysis. Antibodies specific to the protein are added to a cell or tissue extract, and the resulting immune complexes are removed by adsorption on an insoluble matrix, most frequently employing protein A, which has a high affinity for immunoglobulin molecules. Quantitative immunoprecipitation of an antigen may not be achieved by a single MAb. This may be due either to the formation of complexes between the antigen and other cellular proteins or to folding of the antigen, both of which are capable of blocking the MAb epitope. Thus, when quantitative immunoprecipitation is desired, a pool of MAbs, each of which recognizes a different epitope, should be used.

Protocol

All procedures are done at 4°.

1. For experiments aimed at studying post-translational modifications, proteins in intact cells can be labeled with $[^{32}P]P_i$ or ^{3}H- or ^{14}C-labeled fatty acids (e.g., myristic acid, palmitic acid) or sugar precursors (e.g., glucosamine, etc.) prior to extract preparation. For determination of the abundance of a protein, cells are frequently incubated with ^{35}S-, ^{3}H-, or ^{14}C-labeled amino acids. The reader is advised to seek out specific references for conditions of labeling.
2. Prepare cellular lysate using RIPA buffer; in some experiments, RIPA buffer also includes 0.1% SDS to expose the antibody binding sites.
3. Clarify extract by centrifugation at 100,000*g* for 30 min.
4. Determine protein concentration of supernatant.
5. Preclear 500 μg extract with 1–3 μg nonspecific MAb and 25 μl of a 50% slurry of protein A-Sepharose for 30 min.
6. Collect supernatant by centrifugation, then add 1–3 μg of specific MAb for 1 hr. Add 25 μl of protein A-Sepharose for an additional 30 min.
7. Centrifuge immunocomplexes for 2 min in microfuge.
8. Wash beads three times with RIPA buffer.

9. Wash once with TBS (150 mM NaCl, 25 mM Tris–HCl, pH 7.5).

10. Add SDS sample buffer to the pellet, boil for 2 min, and subject eluted proteins to SDS-PAGE and autoradiographic analysis (if cells have been radiolabeled), Western immunoblotting, or protein staining (Coomassie blue or silver).

Immunoblotting

Western immunoblotting is most frequently employed as a method for detecting or quantifying a specific protein in a cell or tissue extract, or in an immunoprecipitate. It is also used to detect proteins that coprecipitate with the cognate antigen. Detailed protocols are outlined by Ely *et al.*[32] and elsewhere in this volume.

Enzymatic or Functional Assays in Immune Complex[7]

The *in vitro* immune complex kinase assay utilized routinely in our laboratory is presented here as an example of the many enzyme or functional assays to which immune complexes can be subjected.

1. Resuspend a pellet of pp60$^{c\text{-}src}$ immunocomplexes captured on protein A-Sepharose in kinase buffer [20 μl of 20 mM PIPES (Sigma, St. Louis, MO), 10 mM $MnCl_2$, pH 7.2].
2. Initiate the reaction by adding 10 μCi [γ-^{32}P]ATP in 10 μl kinase buffer and incubate for 10 min at room temperature.
3. Add 15 μl 3× SDS sample buffer to stop the reaction.
4. Boil 2 min and load onto SDS-PAGE.
5. Dry the gel and expose to Kodak X-AR film.

Cellular Localization Techniques

Monoclonal antibodies to oncogenic products and their normal homologs have been immensely useful in studies aimed at determining the intracellular locations of those proteins in transformed and normal cells. For example, application of subcellular fractionation followed by immunoprecipitation or immunoblotting as well as use of conventional immunofluorescent, confocal immunofluorescent, and immunoelectron microscopic techniques have resulted in the cell surface localization of the pp185^{neu} oncogene product,[33] localization of the *ras*-transforming protein to the inner face of the plasma membrane,[16] and localization of *myc* to the nucleus.[34]

[32] C. M. Ely, S. J. Parsons, and J. T. Parsons, *in* "Neuroprotocols: A Companion to Methods in Neuroscience," (Bertics, ed.) Vol. 1, p. 177. Academic Press, San Diego (1992).

[33] J. Drebin, D. Stern, V. Link, R. Weinberg, and M. Green, *Nature (London)* **312,** 545 (1984).

[34] G. I. Evan, D. C. Hacock, T. Littlewood, and C. D. Pauza, *Ciba Found. Symp.* **119,** 245 (1986).

In addition, MAbs to $pp60^{src}$ have been used to colocalize $pp60^{v\text{-}src}$ with sites of fibronectin degradation in rosette structures of Rous sarcoma virus-transformed chicken embryo fibroblasts[35] and $pp60^{c\text{-}src}$ with secretory granules[36] and endosomes.[37] The results of these studies have provided clues as to the functions of the various proteins. Subcellular fractionation techniques have been thoroughly reviewed elsewhere,[38] and the reader is referred to those sources for protocols. A frequently employed immunofluorescence protocol is outlined here.

Protocol[7,9]

1. Grow cells on coverslips.
2. Wash twice with PBS. (This and all following steps are carried out at room temperature.)
3. Fix with freshly prepared 3% paraformaldehyde in PBS (pH 7.4) for 20 min. [Paraformaldehyde (1.5 g) is added to 45 ml deionized H_2O (prewarmed to 60°). Then 25 μl 2 *N* NaOH is added and stirred vigorously until the paraformaldehyde is dissolved. Finally, 5 ml 10× PBS is added.]
4. Permeabilize fixed cells with 0.4% Triton X-100 in PBS for 4 min.
5. Wash twice with PBS.
6. Incubate with purified, primary (1°) antibodies (10 μg/ml in PBS; MAbs can be purified from ascites or hybridoma culture supernatants by protein G chromatography) for 40 min.
7. Wash three times with PBS. We have found that for low abundance proteins fluorescent signals can be increased with the use of a secondary antibody (2°); otherwise steps 8 and 9 can be omitted.
8. Incubate with purified, 2° goat anti-mouse IgG (10 μg/ml, Jackson ImmunoResearch Laboratories Inc., West Grove, PA) for 40 min.
9. Wash three times with PBS.
10. Block nonspecific binding sites for 30 min with 8% serum. The serum used is homologous to the species from which the chromophore-conjugated antibodies are derived.
11. Incubate with chromophore-conjugated (fluorescein- or Texas Red-conjugated donkey anti-goat IgG) antibody for 40 min in the dark.
12. Wash three times with PBS.
13. Mount the coverslip on a glass slide with a few drops of mounting solution (89% (v/v) glycerol and 1% (v/v) *n*-propyl gallate in PBS).

[35] W.-T. Chen, J.-M. Chen, S. J. Parsons, and J. T. Parsons, *Nature* (*London*) **316,** 156 (1985).
[36] S. J. Parsons and C. E. Creutz, *Biochem. Biophys. Res. Commun.* **134,** 736 (1986).
[37] K. B. Kaplan, J. R. Swedlow, H. E. Varmus, and D. O. Morgan, *J. Cell Biol.* **118,** 321 (1992).
[38] J. G. Krueger, E. A. Garber, and A. R. Goldberg, *Curr. Top. Microbiol. Immunol.* **107,** 51 (1983).

14. Record the data using conventional fluorescence or confocal microscopy. If colocalization studies are to be done, 1°, 2°, and chromophore-conjugated antibodies are added together at steps 6, 8, and 11, respectively. In step 6, care should be taken to use 1° antibodies from different species so that they can be distinguished from one another by employing species-specific antibodies in steps 8 and 11. Secondary antibodies should possess minimal cross-reactivity to serum proteins of species other than that of the 1° antibody.

Concluding Remarks

This chapter describes methods for the generation and utilization of MAbs specific for oncogene products and related proteins. These reagents have proven to be valuable tools for elucidating functions of these proteins in normal and malignant cells. In addition, MAbs have been and continue to be employed in the diagnosis, characterization, and treatment of a number of human tumors.

Acknowledgments

We thank J. T. Parsons, M. E. Cox, J. S. Biscardi, and J. S. Moyers for critical review of the manuscript and S. Gill for technical assistance in generating monoclonal antibodies to GAP and p190. We also thank members of the S. J. Parsons and J. T. Parsons laboratories and the University of Virginia Lymphocyte Culture Center for assistance in these studies. This work is supported by Public Health Service Grants CA23062 and CA36731.

[28] Microinjection of Antibodies

By KARLA KOVARY

Introduction

One of the most effective techniques used in molecular and cellular biology research is the microinjection procedure.[1–3] By using glass needles, miscellaneous molecules like DNA, RNA, and protein can be injected into the cytoplasm or into the nucleus of cells in culture. This procedure,

[1] W. Ansorge, *Exp. Cell Res.* **140,** 31 (1982).

[2] J. E. Celis, *Biochem. J.* **223,** 281 (1984).

[3] M. Graessmann and A. Graessmann, *in* "Microinjection and Organelle Transplantation Techniques" (J. Celis and A. Loyter, eds.), p. 3. Academic Press, London, 1986.

therefore, allows the investigation of the biological role of different molecules directly in their natural cellular environment.

Intracellular microinjection of purified antibodies against different target proteins has been successfully used to inhibit the function of these proteins *in vivo*. Using this technique, useful information has been obtained about the role of several proteins in different biological situations; i.e., the analysis of the dynamic properties of vimentin filaments[4]; involvement of the ras protein in the differentiation of PC12 cells[5]; cooperation between the function of various oncogenes in viral transformation[6]; requirement of the activity of different proteins like p53,[7] c-Ha-ras,[8] Fos proteins (c-Fos, FosB, Fra-1, and Fra-2),[9–11] and Jun proteins (c-Jun, JunB, and JunD)[10,11] for cell cycle progression; dependence of heat-shock proteins for the survival of fibroblasts to thermal stress[12]; modulation of $p34^{cdc2}$ activity by dephosphorylation during cell cycle progression[13]; and requirement of cyclin A activity for both DNA replication and entrance into mitosis.[14,15]

This chapter describes our experience in the use of the microinjection technique as a way to deliver immunopurified antibodies into the cytoplasm of cells in culture. We used this approach to neutralize the regulatory activities of the Fos and Jun proteins in an effort to elucidate the role of these transcription factors in the proliferation of fibroblasts. We investigated the requirement of Fos and Jun proteins for cell cycle progression during different growth conditions by monitoring DNA synthesis of the injected cells. The first part of this chapter describes the generation and immunopurification of the antibodies used in our microinjection experiments. The last part includes descriptions of the microinjection procedure and of the methods used for identifying the injected cells and for DNA synthesis determination.

[4] M. W. Klymkowsky, *Nature (London)* **291,** 249 (1981).

[5] N. Hagag, S. Halegoua, and M. Viola, *Nature (London)* **319,** 680 (1986).

[6] M. R. Smith, S. J. DeGudicibus, and D. W. Stacey, *Nature (London)* **320,** 540 (1986).

[7] W. E. C. Mercer, C. Avignolo, and R. Baserga, *Mol. Cell. Biol.* **4,** 276 (1984).

[8] L. S. Mulcahy, M. R. Smith, and D. W. Stacey, *Nature (London)* **313,** 241 (1985).

[9] K. T. Riabowol, R. J. Vosatka, E. B. Ziff, N. J. Lamb, and J. R. Feramisco, *Mol. Cell. Biol.* **8,** 1670 (1988).

[10] K. Kovary and R. Bravo, *Mol. Cell. Biol.* **11,** 4466 (1991).

[11] K. Kovary and R. Bravo, *Mol. Cell. Biol.* **12,** 5015 (1992).

[12] K. T. Riabowol, L. A. Mizzen, and W. J. Welch, *Science* **242,** 433 (1988).

[13] D. L. Brautigan, J. Sunwoo, J.-C. Labbé, A. Fernandez, and N. J. C. Lamb, *Nature (London)* **344,** 74 (1990).

[14] F. Girard, U. Strausfeld, A. Fernandez, and N. J. C. Lamb, *Cell (Cambridge, Mass.)* **67,** 1169 (1991).

[15] M. Pagano, R. Pepperkok, F. Verde, W. Ansorge, and G. Draetta, *EMBO J.* **11,** 961 (1992).

TABLE I
AMINO ACID SEQUENCE FROM EACH JUN AND FOS PROTEIN USED FOR RAISING ANTIBODIES

Protein	Amino acid sequence[a]	Expressing vector[b]
c-Jun	80 to 334 (334 aa)	pEx34 and pEX2
JunB	46 to 344 (344 aa)	pEX2
JunD	1 to 102 (341 aa)	pEx34
c-Fos	1 to 380 (380 aa)	pEx34
FosB	4 to 338 (338 aa)	pEx34
Fra-1	1 to 276 (276 aa)	pEx34
Fra-2	1 to 326 (326 aa)	pEX1

[a] Number in parentheses is the total number of amino acids (aa) of the original Jun or Fos protein.
[b] Bacterial protein portion of the fusion protein expressed is MS2 polymerase for pEx34 vector and β-galactosidase for pEX1 and pEX2 vectors.

Generation and Immunoaffinity Purification of Antibodies against Fos and Jun Proteins

Antibodies were generated by injecting rabbits with bacterially expressed Fos or Jun proteins fused to bacterial MS2 polymerase or β-galactosidase proteins.[10,11,16] The amino acid region of each Fos and Jun protein used for raising antibodies is shown in Table I. These Fos and Jun protein sequences contain the conserved regions DNA binding domain and leucine zipper, with the exception of JunD. Polyclonal antibodies were therefore raised against large segments of Fos and Jun proteins with the objective to obtain high affinity antibodies recognizing a broad range of different epitopes.

The bacterial expression vectors pEx34 (MS2 polymerase)[17] and pEX1 or pEX2 (β-galactosidase)[18] were used to clone *fos* and *jun* mouse cDNAs. The bacterial host *Escherichia coli* W6γ was used for initial cloning of the target cDNAs into the expression vectors whereas the bacterial host *E. coli* K537 was used to express the recombinant plasmids. The *E. coli* K537 strain contains the temperature-sensitive mutant of the λ repressor gene *cI* (*cIts*857) cloned in a kanamycin-resistant plasmid. Recombinant constructs were amplified in K537 cells at 30° and transient expression was induced by shifting the temperature to 42°.

[16] K. Kovary and R. Bravo, *Mol. Cell. Biol.* **11,** 2451 (1991).
[17] K. Strebel, E. Beck, K. Strohmaier, and H. Schaller, *J. Virol.* **57,** 983 (1986).
[18] K. K. Stanley and J. P. Luzio, *EMBO J.* **3,** 1429 (1984).

Fusion Protein Expression and Purification

The following protocol (adapted from Strebel *et al.*[17]) has been routinely used for fusion protein expression and purification. Transformed K537 cells are grown overnight at 30° in Super Broth medium (32 g of tryptone, 20 g of yeast extract, 5 g of NaCl, and 5 ml of 1 *N* NaOH in 1 liter of water) containing 100 μg per ml of ampicillin and 20 μg per ml of kanamycin. The next day, the cultures are diluted with 1 vol of Super Broth medium warmed up to 60° and are immediately mixed to 2 vol of Super Broth medium that has been previously warmed to 42° in a shaking incubator. Fusion protein expression is induced for 15 min at 42° after which the temperature of the shaking incubator is reduced to 37°. Induction at 37° is then continued with vigorous shaking for another 3 hr. Thereafter, cells from a 400-ml culture are treated in the following way:

1. After sedimented by centrifugation for 10 min at 4000*g* (4°), cells are resuspended in 20 ml of 50 m*M* Tris–HCl buffer, pH 8.0, containing 25% sucrose and 1 mg per ml of lysozyme (w/v), and are incubated at 37° for 30 min.
2. Two milliliters of 250 m*M* EDTA, pH 7.5, and 4 ml of 10% TritonX-100 are added to the resuspended cells and mixed.
3. Cells are disrupted by sonication, and then centrifuged at 20,000*g* for 30 min at 4° to bring down the inclusion bodies that contain most of the fusion protein expressed in an insoluble form.
4. This insoluble material is resuspended by sonication in 10 ml of a 1 *M* urea solution containing 1% TritonX-100 and 0.1% 2-mercapthoethanol, and centrifuged at 15,000*g* for 15 min at 4°.
5. The pellet is then resuspended in 0.5 to 2.0 ml (depending on the size of the pellet) of 7 *M* urea containing 0.1% 2-mercapthoethanol, and centrifuged again. This final 7 *M* urea supernatant contains the fusion protein partially purified.
6. The fusion protein is further purified by preparative sodium dodecyl sulfate–polyacrylamide gel electrophoresis (SDS-PAGE). After electrophoresis, the fusion protein is identified by immersion of the gel in ice-cold 250 m*M* KCl. The band containing the fusion protein is excised and electroeluted in 12.5 m*M* Tris–glycine buffer, pH 7.5, containing 10 m*M* sodium dodecyl sulfate (SDS).

Comments: Although the fusion proteins tend to precipitate in the electroelution buffer, part of it remains soluble. This soluble fraction is dialyzed extensively against 10 m*M* ammonium carbonate solution containing 10 m*M* SDS and is injected into rabbits. However, the quantity of MS2 polymerase or β-galactosidase fusion proteins that precipitates during electroelution depends on the amount of the fusion protein electroeluted and somehow of biochemical characteristics of the mammalian protein fused to the bacterial protein.

Antibody Titer and Cross-Reactivity of Fos and Jun Antisera

Each fusion protein antigen is injected in one or two rabbits and standard protocols are used for the inoculations. Six to eight booster inoculations are given every 4 weeks, and animals are bled 7 to 10 days after each boost. Individual bleeds are tested for the presence of specific antibodies against the antigen injected by the immunoprecipitation of *in vitro*-translated Fos or Jun proteins.[10,11,16] The presence of cross-reactivity in a given antiserum is checked by immunoprecipitation against *in vitro*-translated proteins belonging to a same family.

For *in vitro* translation, recombinant plasmids in pTZ18R (c-*jun*), BlueScript KS (Stratagene, San Diego, CA) (*junB*, *junD*, *fra-1*, and *fra-2*), or pCEM1 (*fosB*), containing the complete coding sequences, are linearized and transcribed *in vitro* by using polymerase T3 or T7 in the presence of m^7 GpppA. The RNAs are then *in vitro* translated using a commercial rabbit reticulocyte lysate (Promega), as directed by the manufacturer.

For immunoprecipitation, the following protocol has been routinely used:

1. Three to 5 μl of the reticulocyte lysate containing the *in vitro*-translated proteins is diluted in 1 ml RIPA buffer (10 m*M* Tris–HCl, pH 7.5, 1% sodium deoxycholate, 1% Nonidet P-40, 150 m*M* NaCl, 0.1% SDS and is incubated with 2 to 5 μl antiserum for 1 hr on ice.
2. Fifteen microliters of protein A–Sepharose CL-4B beads (Pharmacia), previously equilibrated in 0.1 *M* Tris–HCl buffer, pH 7.5, is then added and incubation is extended for 2 to 3 hr more on a roller system at 4°.
3. The immunocomplexes with the protein A–Sepharose beads are centrifuged (maximum speed for 15 sec in a microfuge, at room temperature), washed twice with buffer A (10 m*M* Tris–HCl, pH 7.5, 150 m*M* NaCl, 2 m*M* EDTA, and 0.2% Nonidet P-40), once with buffer B (10 m*M* Tris–HCl, pH 7.5, 500 m*M* NaCl, 2 m*M* EDTA, and 0.2% Nonidet P-40), and once with buffer C (10 m*M* Tris–HCl, pH 7.5).
4. After boiling in Laemmli sample buffer, the samples are run overnight on a 12.5% acrylamide/bisacrylamide (200 : 1) gel, at 12 mA per gel. Fixed gels are incubated twice for 1 hr in dimethyl sulfoxide (DMSO) and once for 3 hr in 20% 2,5-diphenyloxazole (PPO) in DMSO and then washed three times with water, 10 min each washing time. Gels are dried and exposed to Kodak X-Omat AR film at −70°.

Comments: Since most of the Fos and Jun protein fragments used as antigens (with the exception of JunD) contain common conserved sequences like DNA binding domain and leucine zipper particular to each protein family, it was expected to generate antiserum with a high level of cross-reactivity. However, the results obtained by immunoprecipitation screening revealed discrepancies in this respect. Significant variations in cross-reactivity were observed between different bleeds from a same rabbit or when bleeds of different rabbits injected with the same antigen were compared. These variations ranged from bleeds showing a high titer of specific antibodies against the target antigen and undetectable levels of cross-reactivity, to bleeds with high titer of antibodies recognizing indiscriminately all members belonging to a same family (Fos or Jun).

Immunoaffinity Purification of Antibodies against Fos and Jun Proteins

Immunoaffinity Chromatography Matrices Preparation

The resin Affi-Gel 15 (Bio-Rad, Richmond, CA) is used as a matrix for the immunopurification of antibodies. Bacterially expressed fusion proteins outlined in earlier sections were used as antigen source, with the exception of JunD. This protein was used as a recombinant construct expressing the complete sequence of JunD fused to β-galactosidase. Expression and purification of the different fusion proteins as well as nonfused MS2 polymerase and β-galactosidase proteins are performed as described earlier, except that after purification by gel electrophoresis and electroelution, purified proteins are dialyzed against 6 *M* urea.

Purified fusion proteins in a 6 *M* urea solution are covalently coupled to Affi-Gel 15 beads, by gentle overnight shaking of 4 to 10 mg pure protein with 1 to 2 ml resin (previously washed two times with water and two times with 6 *M* urea), in a 12-ml final volume at 4°. After protein coupling, beads are packed into a 1-cm-diameter column, washed with at least 10 vol of 6 *M* urea to remove nonbound bacterially expressed protein, and then equilibrated with phosphate-buffered saline (PBS). Packed columns of fusion protein can be kept at 4° for several months in the presence of sodium azide. Before immunopurification, sodium azide is removed from the columns by extensive washing with PBS.

Comments: The coupling step of fusion proteins to Affi-Gel 15 beads in a 6 *M* urea solution has several advantages. First, as mentioned earlier, concentrated preparations of purified mammalian proteins fused to either β-galactosidase or polymerase are usually insoluble in nondenaturing conditions. On the other hand, they are always soluble in strong denaturants such as 6 *M* urea. Second, in our hands, the coupling of fusion proteins in a 6 *M* urea solution works better with Affi-Gel 15 than with Affi-Gel 10 beads. For Fos or Jun fusion proteins, the coupling yield to Affi-Gel 15 beads in 6 *M* urea ranges among 50 to 90% depending on the fusion protein. The coupling yield to Affi-Gel 10 beads in similar coupling conditions is usually 10 to 50% lower. This observation has practical implications since the coupling yield of proteins to Affi-Gel 10 or 15 in nondenaturing aqueous solutions depends on the pH and salt concentration of the coupling buffer. Therefore, for proteins in a nondenaturing condition it is always advisable to verify experimentally which Affi-Gel resin will give a better yield of protein coupling by testing different combinations of pH and salt concentrations. Similar results are also obtained with other proteins (members from the NF-κB family, for instance) fused to either polymerase or β-galactosidase.

Immunoglobulin Fraction of Antisera

For immunoaffinity purification, 10 to 20 ml of each antiserum (depending on the antibody titer) is precipitated by dropwise addition of saturated ammonium sulfate (pH 7.5) to a final concentration of 33 to 40% and is gently stirred at 4° for 12 hr. The precipitate containing the immunoglobulins is sedimented by centrifugation at 10,000*g* for 30 min, resuspended in 4 ml of PBS, and extensively dialyzed at 4° against PBS.

Immunoaffinity Purification of Jun and Fos Antibodies

For immunopurification, the PBS-dialyzed immunoglobulin fraction of each antiserum is sequentially and repeatedly passed through several columns containing different antigens. The columns are used in the following order: first, the bacterial protein portion of the corresponding fusion protein (β-galactosidase or MS2 polymerase), to eliminate antibodies against this protein; second, the fusion proteins belonging to the same mammalian protein family (either Fos or Jun), to eliminate cross-reactivity; and last, the specific Fos or Jun antigen. Even IgG fractions prepared from antisera showing undetectable levels of cross-reactivity (JunD, for instance) are submitted to the same treatment to eliminate any possible cross-reactivity that might have not been detected by immunoprecipitation screening.

After the columns have been washed extensively with PBS, the bound antibodies are eluted with 0.05 *M* glycine buffer, pH 2.3, containing 0.15 *M* NaCl directly into 0.5 *M* sodium phosphate, pH 7.7 (0.5 ml of sodium phosphate buffer for each 2.5 ml glycine buffer). The presence of protein in the eluted fractions is followed by ultraviolet (UV) absorption at

280 nm. All of these steps are performed at room temperature, using solutions previously sterilized by filtration (0.2 μm) or autoclaving.

The fractions containing the antibodies are then pooled and concentrated by centrifugation in Centricon 30 microconcentrators (Amicon Corp.) at 4°. The protein concentration is determined by colorimetric assay (Bio-Rad kit protein assay). Concentrated antibody preparations (7 to 10 mg per ml) are stored in small aliquots (20 to 50 μl) at −70° whereas the aliquot in use for microinjection is always kept at 4°. Immediately before use, antibodies are centrifuged at 15,000*g* for 20 min and are diluted with PBS to the desired concentration.

A summary of the basic steps involved in the immunopurification of the various Fos and Jun antibodies is shown in Fig. 1, using as an example the immunopurification of anti-c-Fos antibodies.

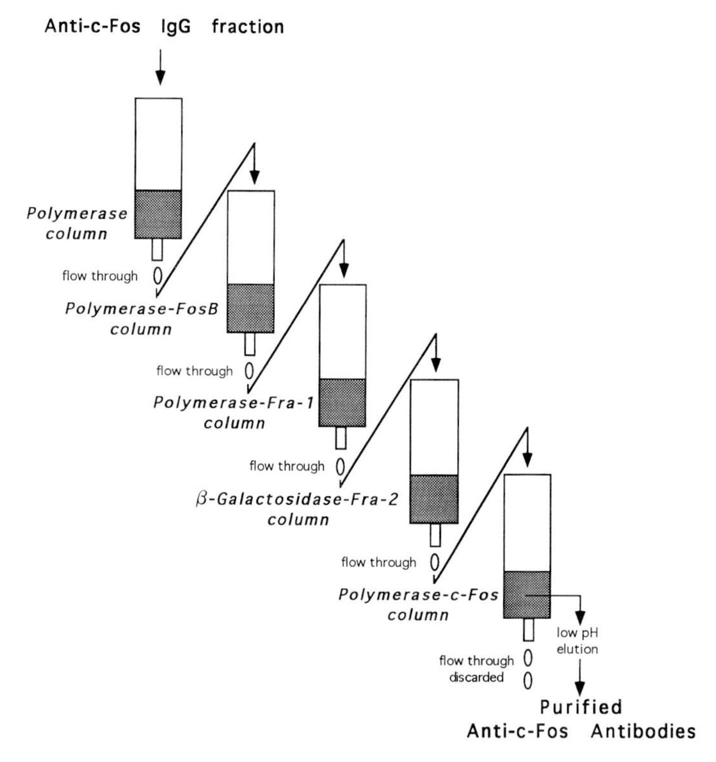

FIG. 1. The primary steps involved in the immunopurification of anti-c-Fos antibodies. The immunoglobulin (IgG) fraction prepared from anti-c-Fos antiserum was passed extensively through several columns containing different antigens. The first column eliminated the antibodies generated against the bacterial protein MS2 polymerase. The following columns eliminated cross-reactivity against the other members of the Fos family (FosB, Fra-1, and Fra-2).

Antibodies that recognize similarly all Fos proteins (anti-Fos family antibodies) are also prepared by using the following procedure. Antisera against c-Fos, FosB, Fra-1, and Fra-2 showing strong cross-reactivity against all Fos proteins are chosen, and the IgG fractions prepared from each of them are passed sequentially over a FosB, Fra-1, and Fra-2; c-Fos, Fra-1, and Fra-2; c-Fos, FosB, and Fra-2, or c-Fos, FosB, and Fra-1 fusion protein columns, respectively. Antibodies absorbed in each column are eluted, pooled, and concentrated. By this procedure, only those antibodies able to recognize common sequences (DNA binding domain and leucine zipper, for instance) present in all Fos molecules are selected.

Anti-Jun family antibodies are prepared by a different procedure. For this purpose, an antiserum raised against the complete c-Jun protein fused to MS2 polymerase was chosen. This antiserum generates antibodies that recognize indiscriminately and with similar efficiency any Jun protein (c-Jun, JunB, and JunD). Anti-Jun family antibodies are immunoaffinity purified on a polymerase–c-Jun fusion protein column.

Control immunoglobulins are purified from normal rabbit serum using the Econo-Pac serum immunoglobulin G purification kit (Bio-Rad).

Each antibody preparation is tested for its capacity to immunoprecipitate *in vitro*-translated proteins. The specificity of a given antibody is checked by immunoprecipitation against the other *in vitro*-translated proteins belonging to the same family as is shown in Fig. 2.

Microinjection of Antibodies

Cell Culture and Microinjection

The cells used in our studies are Swiss 3T3 fibroblasts. Cells are routinely maintained in Dulbecco's modified Eagle's medium (DMEM, GIBCO, Grand Island, NY) supplemented with 10% fetal calf serum (FCS) and antibiotics (100 U of penicillin per ml and 50 μg of streptomycin per ml).

For microinjection, fibroblasts are grown in 60-mm plastic petri dishes (tissue culture grade), on 9 × 9-mm glass coverslips. Cells are microinjected in two growth conditions: quiescence and asynchronous. For the quiescence condition, cells are grown normally in 10% fetal calf serum/Dulbecco's modified Eagle's medium (FCS/DMEM) to 50–70% confluence after which the medium is changed to 2.5% FCS/DMEM. Cells after 48 hr in this low serum medium show a nuclear labeling index that is less than 1% as checked by bromodeoxyuridine incorporation. Cells are microinjected under this condition, and immediately after injection, the medium is changed to 20% FCS/DMEM to induce the transition G_0/G_1 phase. DNA synthesis in these cells is determined 20 to 24 hr after microinjection. For studies during

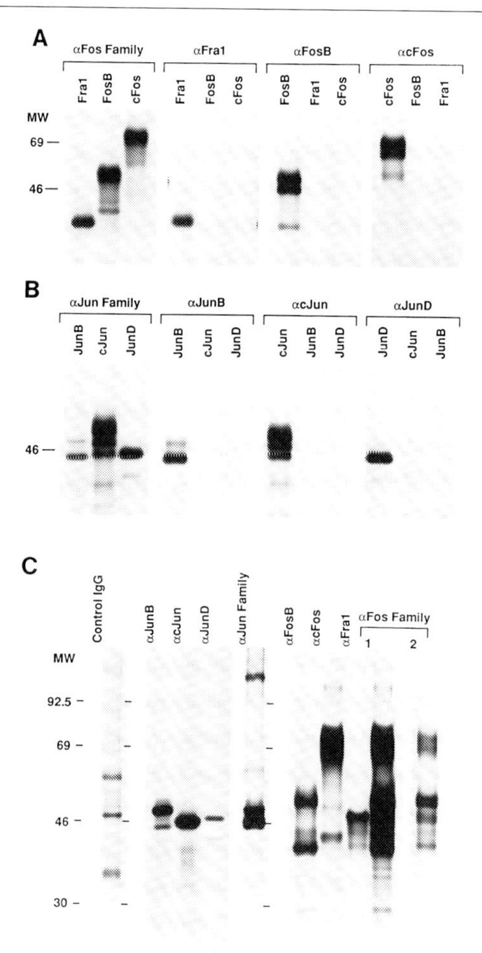

FIG. 2. Specificity of anti-Jun and anti-Fos antibodies. (A) *In vitro*-translated c-Fos, FosB, and Fra-1 immunoprecipitated with anti-Fos family (αFos Family), anti-c-Fos (αcFos), anti-FosB (αFosB), and anti-Fra-1 (αFra1) antibodies. (B) *In vitro*-translated c-Jun, JunB, and JunD immunoprecipitated with anti-Jun family (αJun Family), anti-c-Jun (αcJun), anti-JunB (αJunB), and anti-JunD (αJunD) antibodies. (C) Serum-stimulated cells were labeled for 2 hr with [^{35}S]methionine, lysed under denaturing conditions, and incubated with the respective antibodies. The autoradiograms were exposed for 40 hr except for the anti-Fos family, which was exposed for 4 hr to better visualize the different Fos proteins. Immunoprecipitates were analyzed by (SDS-PAGE). Sizes are shown in kilodaltons on the left.[10]

asynchronous growth, 3T3 cells growing exponentially in 10% FCS/DMEM are microinjected and DNA synthesis is analyzed 30 to 40 hr later.

Microinjections are performed using the AIS automated microinjection system from Zeiss. Briefly, this system consists of a phase-contrast inverted microscope (Axiovert 35M), a B/W CCD TV camera that displays the cells under observation on a RGB TV monitor screen, a micromanipulator that allows exact and safe position of the cells and the glass needle, and a microinjector system (Eppendorf Microinjector 5242). The moving parts of the system, which includes a microscope scanning stage and the micromanipulator, are controlled by a microcomputer (AIS control unit). Detailed information about the AIS system and its use can be found in Pepperkok *et al.*[19-22]

Glass needles are pulled from glass capillaries (10 cm long, 0.78 mm internal diameter, and 1.0 mm external diameter, from Suttler Instrument Co.) using a Suttler P87 pipette puller instrument. No pretreatment has been dispensed to the glass capillaries before pulling them. Each glass capillary originates two pulled glass needles showing high reproducibility in form and dip diameter and very seldom presents problems of clogging. They can be prepared in advance and stored for several months in a dust-free ambient. Needles are filled (1 to 3 μl of sample) from the rear open end using microloaders from Eppendorf that allows the sample to be deposited right at the tip and therefore very little amount of sample is required.

Antibodies are always injected into the cytoplasmic compartment, at concentrations ranging from 0.5 to 10 mg per ml. For each assay, 100 to 200 cells are injected in duplicate and, after injection the medium is always changed. The entire procedure, beginning by removing the cells from the incubator (37°, 5% CO_2), adjusting the cells and the needle position at the microscope, injecting the antibodies, changing the medium after microinjection, and placing the cells back into the incubator, takes around 15 to 20 min. To avoid a rapid pH increase during the time cells are handled outside the incubator, microinjection of quiescent cells is performed in a mixture (without FCS) of 1 vol DMEM to 5 vol MEM-EH (minimal essential medium, Hanks' base). Exponentially growing cells are injected in 5% FCS/DMEM/MEM-EH mixture and the medium is changed to 10% FCS/DMEM after microinjection.

[19] R. Pepperkok, C. Schneider, L. Philipson, and W. Ansorge, *Exp. Cell Res.* **178,** 369 (1988).

[20] W. Ansorge and R. Pepperkok, *J. Biochem. Biophys. Methods* **16,** 283 (1988).

[21] R. Pepperkok, M. Zanetti, R. King, D. Della, W. Ansorge, L. Philipson, and C. Schneider, *Proc. Natl. Acad. Sci. U.S.A.* **85,** 6748 (1988).

[22] R. Pepperkok, S. Herr, P. Lorenz, W. Pyerin, and W. Ansorge, *Exp. Cell Res.* **204,** 278 (1993).

Fluorescent Labeling of Microinjected Cells

Although the AIS system allows the reference coordinates and injection data to be stored and the injected cells can be tracked down several hours after microinjection, a simpler way to find back the microinjected cells is by indirect immunofluorescent staining of the antibodies injected.

DNA synthesis of the injected cells is also detected by immunofluorescent staining. For this, 5-bromodeoxyuridine (BrdU, Sigma), at a final concentration of 100 μM, is added to the medium 6 to 14 hr before the cells are fixed. There is no need to add fluorodeoxyuridine to inhibit thymidylate synthase as is usually recommended when using this high concentration of BrdU. DNA synthesis inhibition is then calculated by determining the percentage of microinjected cells that were not in DNA synthesis after the period of serum stimulation compared to control microinjected cells.

For double indirect immunofluorescence, cells are fixed in cold methanol (4°) for 10 min, rehydrated in PBS, and incubated for 30 min in 1.5 *M* HCl to denature the DNA. After the coverslips have been washed several times with PBS to remove the acid solution, cells are processed for double-immunofluorescent staining using the following protocol:

1. Fifty microliters of a mouse anti-BrdU monoclonal antibody (Becton-Dickinson) diluted 1:50 in PBS is loaded on a sheet of Parafilm stretched onto a flat surface, i.e., the lid of a 24-well tissue culture plate. Coverslips are inverted onto the 50-μl diluted antibody drops and incubated for 30 min at room temperature.
2. After this first antibody incubation, coverslips are removed and placed individually in the wells of the 24-well tissue culture plate containing PBS (cells up). Cells are washed three times with PBS (5 min each washing time, with shaking).
3. Cells are then incubated as described in step 1 of this section, with 50 μl of a mixture of a goat polyclonal anti-rabbit immunoglobulin antibody conjugate to fluorescein isothiocyanate (DAKO) diluted 1:100 in PBS, for the visualization of the injected cells, and a donkey anti-mouse immunoglobulin polyclonal antibody conjugated to Texas Red (Amersham) diluted 1:50 in PBS, to detect DNA synthesis. This incubation is made protected against light.
4. After incubation with the second antibody, cells are washed as described in step 2, protected against light.

Option: After the third washing in PBS, cells can be incubated for an additional 10 min with PBS containing Hoechst dye 33258 (1 to 5 μg per ml, Sigma). This allows easier visualization of the entire population,

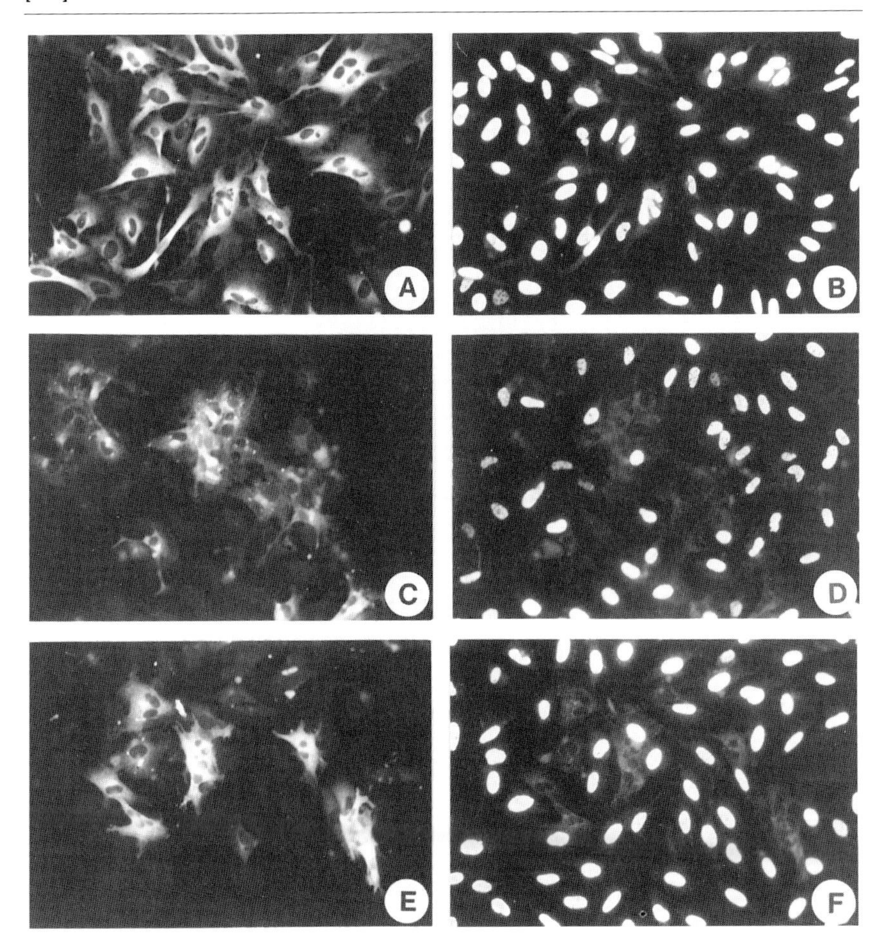

FIG. 3. Microinjection of anti-Fos family and anti-Jun family antibodies. Quiescent cells injected with control immunogloblins (8 mg per ml; A and B) or with anti-Fos family (5 mg per ml; C and D) or anti-Jun family (5 mg per ml; E and F) antibodies were processed for indirect immunofluorescence 20 hr after microinjection to detect the immunoglobulins injected (A, C, and E) and DNA synthesis (B, D, and F), 20 hr following serum stimulation. The microinjected cells were identified by cytoplasmic immunofluorescence of the immunoglobulins injected; cells in DNA synthesis were identified by nuclear immunofluorescence of BrdU incorporated into DNA. BrdU was added to the medium 12 to 14 hr before the cells were fixed.[10]

including injected and noninjected cells. Cells are thereafter rinsed several times with PBS to remove any residual Hoechst dye solution.

5. Coverslips are mounted (cells upside down) on slides with a drop of Fluoromount G (Southern Biotechnic). Coverslips are left for a couple of

hours to dry before immunofluorescence visualization, protected against light. Slides can be kept for several months at 4° in the dark without significant detriment of the immunofluorescent signal. Immunofluorescence has been observed and photographed with a Zeiss Axiophot fluorescence microscope. The double immunofluorescent pattern of cells microinjected with control IgG, anti-Fos family, or anti-Jun family antibodies is shown in Fig. 3.

Concluding remarks: The success in the microinjection of antibodies for neutralization purposes relies mainly on the quality of the antibodies used. Polyclonal antibodies have the advantage of recognizing different epitopes on a same target protein and therefore should have stronger neutralizing activities. Finally, it should be mentioned that antibodies can be injected for reasons other than the neutralization of certain target proteins. Because of the facility of immunofluorescent staining of immunoglobulins, several groups have been injecting inert nonspecific antibodies mixed to different molecules as a tracer to allow easier identification of the injected cells.[13,23-25]

Acknowledgments

The author thanks Dr. Rodrigo Bravo (Bristol-Myers Squibb Pharmaceutical Research Institute, Princeton, NJ) and Dr. Katia Maria da Rocha (São Paulo University, São Paulo, S.P., Brazil) for helpful comments on the manuscript. I also thank Dr. Rolf Peter Ryseck for the initial work in cloning the cDNAs of *fos* and *jun*. This work was carried out in the laboratory of Dr. Rodrigo Bravo. His support is gratefully acknowledged.

[23] D. W. Stacey, L. A. Feig, and J. B. Gibbs, *Mol. Cell. Biol.* **11,** 4053 (1991).
[24] A. S. Alberts, T. Deng, A. Lin, J. L. Meinkoth, A. Schönthal, M. C. Mumby, M. Karin, and J. R. Feramisco, *Mol. Cell. Biol.* **13,** 2104 (1993).
[25] C. Gauthier-Rouvière, A. Fernandez, and N. J. C. Lamb, *EMBO J.* **9,** 171 (1990).

[29] Microinjection into *Xenopus* Oocytes

By WAYNE T. MATTEN and GEORGE F. VANDE WOUDE

Introduction

Much of the progress in understanding the cellular function of oncogenes derives from the ability to infect or transfect somatic cells in culture and to analyze the morphological and biochemical effects of expression of the introduced gene product. The amphibian oocyte, most notably that of

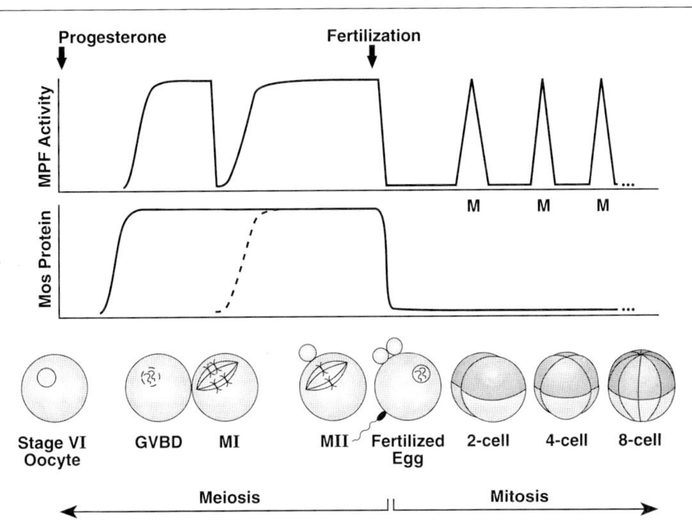

FIG. 1. Meiotic maturation in *Xenopus laevis*. The activity of the maturation promoting factor (MPF) oscillates throughout meiotic and mitotic cell cycles. The Mos protein, a critical element in both initiation of MPF activation and arrest at metaphase II, functions only during meiosis. The dashed line indicates the requirement for *de novo* synthesis of Mos between MI and MII. M, mitosis; GVBD, germinal vesicle (nuclear envelope) breakdown; MI, meiosis I; MII, meiosis II (showing the first polar body resulting from completion of MI).

Xenopus laevis, offers an alternative approach that overcomes many of the disadvantages of using somatic cells. This system uniquely combines a naturally synchronized population of cells; the ability to introduce very small, precisely controlled quantities of DNA, RNA, or protein, whose processing or effects can be rapidly assessed; and the capacity to perform biochemical analyses on a small number of, or even individual, cells. Historically, these oocytes have not been easily amenable to high-resolution immunohistochemical analysis; however, advances in confocal microscopy are allowing direct observation *in situ*,[1] and, like somatic cells, *Xenopus* oocytes have been used very successfully for the transient expression of heterologous membrane receptor proteins and ion channels. It should also be noted that fundamental advances in understanding the meiotic and early embryonic cell cycle in vertebrates were achieved using *Xenopus* oocyte extracts that undergo multiple cell cycles *in vitro*.[2,3]

The major features of *Xenopus* oocyte maturation and early embryonic development are shown schematically in Figure 1. Stage VI oocytes are

[1] D. L. Gard, *Methods Cell Biol.* **38,** 241 (1993).
[2] M. J. Lohka and Y. Masui, *J. Cell Biol.* **98,** 1222 (1984).
[3] A. Murray and M. Kirschner, *Nature* (*London*) **339,** 275 (1989).

naturally arrested in the prophase of meiosis I with no Mos protein or active maturation-promoting factor (MPF), a complex consisting of the $p34^{cdc2}$ serine/threonine kinase and cyclin B.[4] These prophase-arrested oocytes are variously referred to as fully grown oocytes, quiescent oocytes, or immature oocytes. Progesterone release from the surrounding follicle cells, or exposure to numerous inducers *in vitro*, initiates the resumption of meiosis I by stimulating protein synthesis of the Mos serine/threonine protein kinase and activation of MPF. Numerous other signal-transducing molecules are activated as well, either in parallel to or downstream of MPF. After germinal vesicle breakdown (GVBD) and completion of meiosis I, MPF activity drops transiently, then returns to a high level at metaphase II, where a Ca^{2+}-sensitive activity called cytostatic factor (CSF) stabilizes MPF and arrests oocyte development. Fertilization triggers an increase in free cytoplasmic calcium, degradation of Mos and cyclin B, and inactivation of CSF and MPF, all leading to entry into the interphase of the first mitotic division. MPF activity then cycles, increasing at each subsequent M phase.[5-8]

Experiments designed to elucidate the function of the Mos and Ras proteins in *Xenopus* oocyte maturation offer examples of this technique. Mos was shown to be necessary for the initiation of meiotic maturation in Stage VI oocytes by injecting *mos* antisense oligonucleotides into progesterone-treated oocytes and preventing maturation.[9] Moreover, injecting a soluble recombinant Mos protein into oocytes, in the absence of protein synthesis, demonstrated that Mos is sufficient for the induction of MPF activation and GVBD[10] (Fig. 2). The Ras oncoprotein was also shown to induce progesterone-independent meiotic maturation in *Xenopus* oocytes.[11] In addition, Ras, like Mos/CSF, arrests *Xenopus* embryonic cleavage after injection into one blastomere of a two-cell embryo.[12] In fact, this embryonic cleavage arrest assay is a rapid and sensitive test for Ras function.

The protocols described here outline methods for the injection of synthetic mRNAs and purified proteins into oocytes, and include the embryonic cleavage arrest assay. In addition, the basic oocyte preparation and injection techniques should be useful for other studies employing *Xenopus* oocytes.

[4] M. Solomon, *Curr. Opin. Cell Biol.* **5,** 180 (1993).

[5] L. Smith, *Development* (*Berlin*) **107,** 685 (1989).

[6] J. Maller, *Prog. Clin. Biol. Res.* **267,** 259 (1988).

[7] A. Murray and T. Hunt, "The Cell Cycle: An Introduction." Freeman, New York, 1993.

[8] Y. Masui, *Biochem. Cell Biol.* **70,** 920 (1992).

[9] N. Sagata, M. Oskarsson, T. Copeland, J. Brumbaugh, and G. F. Vande Woude, *Nature* (*London*) **335,** 519 (1988).

[10] N. Yew, M. L. Mellini, and G. F. Vande Woude, *Nature* (*London*) **355,** 649 (1992).

[11] C. Birchmeier, D. Broek, and M. Wigler, *Cell* (*Cambridge, Mass.*) **43,** 615 (1985).

[12] I. Daar, A. R. Nebreda, N. Yew, P. Sass, R. S. Paules, E. Santos, M. Wigler, and G. F. Vande Woude, *Science* **253,** 74 (1991).

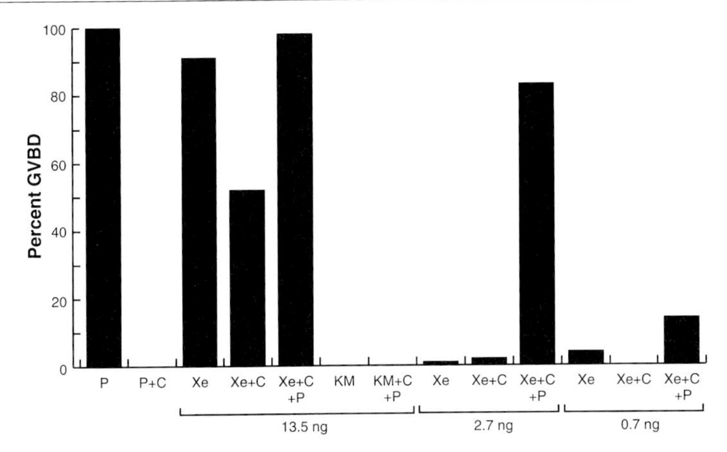

FIG. 2. Induction of GVBD by the maltose binding protein (MBP)-*mos*^{xe} fusion protein in the absence of protein synthesis. Oocytes were injected with the indicated amounts of MBP-*mos*^{xe}, then scored for GVBD 6–8 hr later. Xe, MPB-*mos*^{xe} protein; C, cycloheximide (10 μg/ml, 1 hr before injection and continuously until scoring GVBD); P, progesterone (10 μg/ml); KM, kinase-defective mutant of MBP-*mos*^{xe}. Each percentage is based on the injection of 60 oocytes.[10]

Isolation of Individual Oocytes

Adult frogs were purchased from Xenopus I (Ann Arbor, MI) and kept in a central animal handling facility. The maintenance of *Xenopus* in the laboratory is treated thoroughly by Wu and Gerhart.[13] Briefly, frogs are housed in 19 × 10 × 10 (inches) polycarbonate tanks maintained at 18–22° on a 12-hr light cycle. Feeding (trout chow) and complete water changes occur three times per week. Unfortunately, even these controlled conditions have not precluded the apparently universal phenomenon of lower quality oocytes during the late summer months. Unprimed frogs (not treated with gonadotropins) are anesthetized by immersing in ice–water for 30 min, then euthanized by decapitation. The ovarian tissue is removed and placed in a 100-mm glass or plastic culture dish containing modified Barth's solution (MBS; Table I). Alternatively, if only small numbers of oocytes are required, surgery is performed to remove a small piece of ovary and to allow reuse of the animal.[14]

A sexually mature female can produce 30 cm^3 of ovarian tissue containing oocytes at all stages of development. When several hundred Stage

[13] M. Wu and J. Gerhart, *Methods Cell Biol.* **36,** 3 (1991).

[14] A. Colman, *in* "Transcription and Translation: A Practical Approach" (B. D. Hames and S. J. Higgins, eds.) p. 271. IRL Press, Oxford, 1984.

TABLE I
MODIFIED BARTH'S SOLUTION

Component	Concentration (mM)	10× stock (g/liter)[a]
NaCl	88	51.4
KCl	1.0	0.75
$NaHCO_3$	2.4	2.1
$Ca(NO_3)_2 \cdot 4H_2O$	0.33	0.78
$CaCl_2 \cdot 2H_2O$	0.41	1.0
$MgSO_4 \cdot 7H_2O$	0.82	2.0
HEPES, pH 7.4 (with NaOH)	10.0	23.8

[a] The 10× stock is autoclaved in 500-ml bottles.

VI[15] oocytes are required, we generally use the collagenase method to obtain individual oocytes. Ovaries from one animal are picked apart into ≈ 0.5 cm^3 pieces, rinsed in MBS until the buffer is clear, and gently agitated horizontally for 2 hr at 20–23° in 50 ml of a 2 mg/ml solution of collagenase A (Boehringer-Mannheim; the minimum ratio of buffer to oocytes should be 2:1). Different lots of collagenase should be pretested since they can vary substantially. After thorough rinsing with MBS, approximately 50% have been released as individual oocytes and are largely free of surrounding follicle cells, which generally correlates with a loss of visible blood vessels. It is possible to remove remaining clumps of oocytes, as well as a large number of the smaller oocytes (less than ≈ Stage IV), by gently passing the oocytes consecutively through sieves (Newark Wire Cloth Co., Newark, NJ) with 2.00-mm (No. 10) and 0.71-mm (No. 25) openings. If smaller numbers of oocytes are required or if it is desirable to avoid collagenase treatment, they can be manually dissected[16]; however, be aware that follicle cells are not removed by manual dissection. Collagenase treatment as described here sometimes triggers low levels of spontaneous Mos expression, which we never observe in manually dissected oocytes. We find that the collagenase method produces equal quality oocytes as long as they are allowed sufficient time to recover, generally 12–16 hr at 18–20°.[17]

The last step is to select unblemished, Stage VI oocytes. Careful selection of the healthiest oocytes is necessary to reduce experimental variability. Our somewhat subjective guidelines include a diameter greater than 1.2 mm, a sharp demarcation between the animal (pigmented) and vegetal hemispheres, and even and dense pigment distribution. A slight difference

[15] J. Dumont, *J. Morphol.* **136,** 153 (1972).
[16] L. D. Smith, W. Xu, and R. L. Varnold, *Methods Cell Biol.* **36,** 45 (1991).
[17] R. A. Wallace, D. W. Jared, J. N. Dumont, and M. W. Sega, *J. Exp. Zool.* **184,** 321 (1973).

in color is often the most easily distinguished difference between Stage V (1.0–1.2 mm) and Stage VI oocytes, which are functionally quite different.[12,18] Selected oocytes are maintained in 50% Liebovitz's media with L-glutamine (Gibco; in sterile water, plus 100 U/ml sodium penicillin and 10 μg/ml streptomycin sulfate) at 18° for at least 12 hr before injection. If oocytes are to be radiolabeled (see below), place in MBS with 15 mM HEPES, pH 7.4, and antibiotic. Healthy oocytes should respond to 5 μg/ml progesterone (from a 1000X stock in 100% ethanol) by exhibiting the "white spot" in the animal pole characteristic of GVBD. Although the time required to observe GVBD will vary considerably from batch to batch, 90% of the progesterone-treated oocytes should display white spots in 3–6 hr.

Sample Preparation

Protein solutions are diluted in 1X injection buffer (0.1*M* NaCl, 10 m*M* HEPES, pH 7.4), although injecting 30 nl of a solution containing 1–2 m*M* reducing agent (e.g., dithiothreitol) and less than 0.1% nonionic detergent has not had obvious effects on oocyte maturation.

Messenger RNAs are transcribed *in vitro* by standard protocols using a pTZ18 vector (Pharmacia LKB) altered to contain an 8-mer deoxythymidine tract in the polylinker region. Incorporation of a 5′ terminal cap is essential for efficient translation in the oocyte; we generally use the analog $m^7G(5')ppp(5')G$. We also have used the pSP64T vector successfully.[19] This subject is discussed in detail by Wormington.[20] Synthetic mRNAs are precipitated twice with ethanol and are resuspended in sterile injection buffer (to 1–2 mg/ml) or in DEPC-treated water that has been autoclaved for at least 1 hr to evaporate excess DEPC. A trial *in vitro* translation using the synthetic mRNA is strongly recommended before beginning oocyte studies. To avoid clogged needles, all samples should be centrifuged for 10 sec in a microfuge immediately prior to injection.

Microinjection

We use an Attocyte injector (ATTO Instruments, Rockville, MD) calibrated to deliver 30 nl of reagent; viability is often unaffected by injection of as much as 60 nl of a nontoxic solution. Injection needles are pulled with a Narishige Model No. PB-7 (Narishige Co. Ltd., Tokyo, Japan) from

[18] M. A. Taylor and L. D. Smith, *Dev. Biol.* **121,** 111 (1987).
[19] P. A. Krieg and D. A. Melton, *Nucleic Acids Res.* **12,** 7057 (1984).
[20] M. Wormington, *Methods Cell Biol.* **36,** 167 (1991).

5-inch lengths of glass tubing purchased from Custom Glass Tubing (Drummond Scientific, Broomhall, PA) (0.63 mm o.d.; 0.20 mm i.d.; R-6 glass). For a given delivery volume, manually break off the tip of a pulled needle, using a fine forcep under the dissecting scope, to give a 2- to 5-sec injection time. Our injection chamber is a 60 × 15-mm culture dish, covered on the bottom with a 1-mm pore size support mesh (FlowMesh, Diversified Biotech, Boston, MA) or with sterile gauze with a fine enough weave to exclude the oocytes. A final selection of unblemished oocytes is made before injection. Injections are made into the animal hemisphere, to a depth approximately one-fourth the distance from pole to equator, about 20–30° above the equator.

Metabolic Labeling

Most oocyte proteins are labeled by incubation with 0.1–0.5 mCi/ml [^{35}S]methionine for 8–12 hr, although longer periods are possible. Ten microliters of solution per oocyte is sufficient, but larger volumes will not significantly decrease incorporation. [^{32}P]P_i labeling of proteins is generally performed at 0.2 mCi/ml for 2–4 hr. Direct injection of [^{35}S]methionine or [^{32}P]P_i is reported to increase labeling, but, in our hands, has not produced significantly higher specific activities. RNA labeling is treated in detail elsewhere.[21]

Analysis of Results

Stage VI *Xenopus* oocytes are naturally arrested in the prophase of meiosis I. The easiest way to assess reentry into the meiotic cell cycle is to quantitate the percentage of oocytes that have undergone GVBD. GVBD can be assessed externally as a prominent white spot at the animal pole, due to migration of pigment out of that area. To confirm whether or not GVBD has taken place, fix oocytes in 10% trichloroacetic acid (TCA) for 10 min, then bisect with a scalpel 90° to the equator and look for the absence or presence of an intact nucleus (whitish ball surrounded by yellow yolk protein in the animal hemisphere).

To make whole cell extracts, standard lysis buffers, e.g., RIPA, at 4° are used. In addition to empirically determined protease and phosphatase inhibitors, up to 10 m*M* EGTA is generally added to inhibit numerous Ca^{2+}-sensitive proteases. To homogenize, transfer oocytes to an appropriate tube, remove overlying buffer with a pipette, and triturate with a small

[21] J. B. Gurdon and M. P. Wickens, this series, Vol. 101, p. 370.

pipette tip. For most analyses, 10 μl buffer per oocyte is sufficient. The majority of the yolk protein and pigment are removed by a 5-min centrifugation at $\approx$12,000*g*. Most enzyme activities remain stable after extracts are quickly frozen in liquid nitrogen or dry ice/ethanol, and stored at $-80°$. For immunoprecipitations, the crude extracts are usually centrifuged at 2° for 30–60 min, 10,000*g*, then precleared with protein A–agarose or the appropriate immunoglobulin before the addition of the primary antibody.

Cytostatic Factor Assay

This assay has been useful in determining the effects of many proteins on M phase activity, as mentioned earlier for Mos and Ras. To obtain unfertilized eggs, adult frogs are injected with 100 IU of pregnant mare serum gonadotropin 3–7 days before injecting 600 IU of human chorionic gonadotropin (hCG). Injection is made with a 27-gauge needle into the dorsal lymph sac by subcutaneous injection within a "V"-shaped area at the posterior of the animal. After hCG injection, maintain frogs for 12–14 hr at 16–18° in 1 × MMR (0.1 *M* NaCl, 2 m*M* KCl, 1 m*M* $MgCl_2$, 2 m*M* $CaCl_2$, 0.1 m*M* Na–EDTA, 5 m*M* HEPES, pH 7.8); if possible, incubate frogs separately since different batches of eggs will be of different quality. Wash lain eggs extensively and maintain in 1 × MMR (20 to 23° is adequate) while preparing testes. Testes are surgically removed from one adult male and washed in 100% Ringer's solution (0.1 m*M* NaCl, 1.8 m*M* KCl, 1 m*M* $MgCl_2$, 2 m*M* $CaCl_2$, 4 m*M* $NaHCO_3$), then minced with a scalpel in a 100-mm dish containing the eggs in 0.3 × Ringer's solution. After 15 min, dejelly the eggs in 0.3 × MMR containing 2% cysteine (50 ml of 0.3 × MMR + 200 μl 10 *N* NaOH + 1 g cysteine); gently swirl for 3–5 min, wash several times with 0.3 × MMR, and maintain in 0.3 × MMR containing 5% (w/v) Ficoll 400. Inject the protein of interest into one blastomere of a two-cell embryo and maintain in 0.3 × MMR, 5% (w/v) Ficoll 400 for 4–6 hr and examine for cleavage. Use 30 nl of CSF extract as a positive control. To prepare CSF extract, wash unfertilized, dejellied eggs with extraction buffer (0.15 *M* sucrose, 0.2 *M* NaCl, 2 m*M* $MgSO_4$, 2.5 m*M* Na–EGTA, 20 m*M* HEPES, pH 6.5, 10 m*M* NaF, 80 m*M* sodium β-glycerophosphate, 2 m*M* DTT, 6 m*M* Na–ATP), with a last wash containing protease inhibitors. Centrifuge the eggs at 1000 rpm for about 60 sec to compact them, and remove as much overlying buffer as possible. Crush the eggs by centrifuging at 16°, 10,000*g* for 15 min. The middle (cytoplasmic) layer is removed with a syringe and is clarified by centrifuging at 100,000*g* for 1 hr at 2°.

Acknowledgments

We thank Michelle Reed for assistance in the preparation of this manuscript and Richard Frederickson for the artwork. Research is sponsored in part by the National Cancer Institute, DHHS, under contract No. NO1-CO-74101 with ABL. Additional research funding was provided by the G. Harold and Leila Y. Mathers Charitable Foundation. The contents of this publication do not necessarily reflect the views or policies of the Department of Health and Human Services, nor does mention of trade names, commercial products, or organizations imply endorsement by the U.S. government.

Section IV

Protein–Protein Interactions

[30] Regulatable Chimeric Oncogenes

By AMY K. WALKER and PAULA J. ENRIETTO

I. Introduction

Chimeric proteins constructed using oncogenes and members of the steroid receptor family have been useful tools in studying both biological and biochemical functions of oncogenes. As demonstrated by Picard *et al.*,[1] fusion of the steroid receptor hormone binding domain to a heterologous protein results in a chimeric protein whose activity can be regulated by the addition of the steroid hormones. Application of this technique to the study of oncogenes has provided insight into their biochemical functions as well as to their roles in transformation, cellular differentiation, gene regulation, and enzymatic activity. This chapter focuses on nuclear oncogene–estrogen receptor fusions and discusses methods and considerations when constructing fusion proteins, criteria for functional evaluation, and experimental applications.

II. Background

A. *Steroid Hormone Receptors*

Steroid hormone receptors are members of a large family of proteins including the glucocorticoid (GR), mineralocorticoid (MR), and estrogen receptors (ER). (For a list of receptors and ligands, see Table I.) These receptors can be located in both the nucleus and cytoplasm of cells. Nonligand-bound receptors are thought to be associated with the heat-shock protein 90 (hsp90) which functions to reduce the activity of the nonhormone-bound receptor.[2] On stimulation with hormone, the receptor dimerizes, translocates to the nucleus, and demonstrates increased affinity for DNA. Most of these proteins contain a zinc finger DNA binding domain, a ligand binding domain, and include sequences that can activate transcription (see Fig. 1). In particular, the ER contains two transactivation domains: TAF-1, located in the N terminus of the protein, and TAF-2 that overlaps the hormone binding domain.[2] The domains of steroid receptors can not

[1] D. Picard, S. J. Salser, and K. Yamamoto, *Cell* (*Cambridge, Mass.*) **54,** 1073 (1988).

[2] S. Green and P. Chambon, *in* "Nuclear Hormone Receptors" (M. Parker, ed.), p. 15. Academic Press, New York, 1991.

TABLE I
STEROID RECEPTORS USED IN CHIMERAS

Steroid hormone receptor	Ligand	Activity in mammalian cells	Ref.
Glucocorticoid receptor	Dexamethasone	Agonist	
	Progesterone	Antagonist	24
	RU486	Antagonist	13
Estrogen receptor	β-Estradiol	Agonist	
	Diethylstibesterol	Agonist	
	Tamoxifen	Partial agonist	24
	ICI 164.389	Antagonist	18
Mineralocorticoid receptor	Aldosterone	Agonist	
	Spirolactone	Antagonist	24
Drosophila ecydsone receptor	Ponasterone A	Agonist	
	Muristerone A	Agonist	15

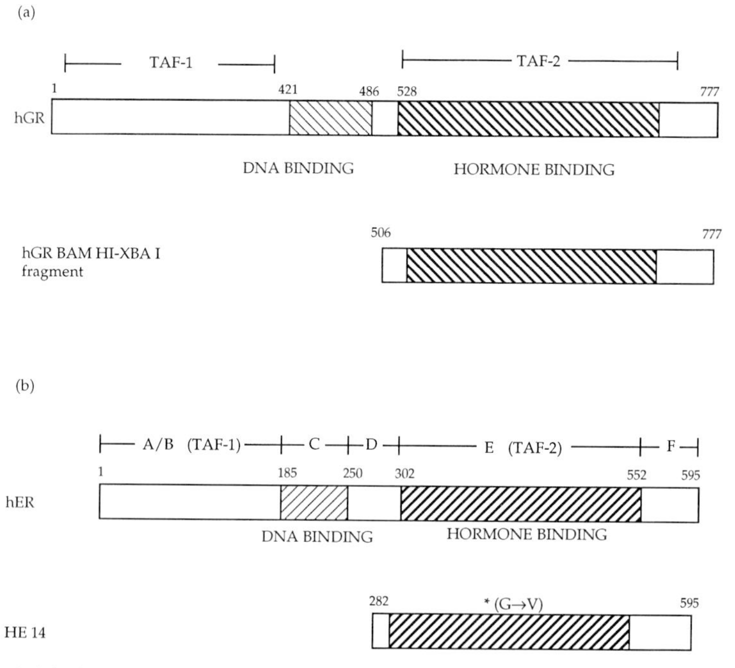

FIG. 1. (a) The human glucocorticoid receptor and (b) the human estrogen receptor. From Green and Chambon[2] and Kumar *et al.*[17]

only function independently but the ligand binding domain can confer hormone regulation on a heterologous protein.[1]

B. Steroid Receptor–Oncogene Fusions

Picard *et al.*[1] demonstrated that the hormone binding domain of a steroid receptor could function as a regulatory element when fused to another protein. They showed that the transcriptional activity of the adenovirus E1A protein became ligand dependent on fusion to the hormone binding domain of the glucocorticoid receptor. In the absence of hormone the E1A-GR fusion was unable to activate transcription from its normal target, the E3 promoter; the addition of dexamethosone restored activity to wild-type E1A levels. This regulation may be mediated by the binding of hsp90 to the receptor in the absence of hormone. It is suggested that the inhibitory effect is relieved by binding of the hormone to the chimera and the concomitant release of hsp90.[1]

Subsequently, this approach has been used to generate hormone-inducible fusions for a variety of diverse proteins: Myc,[3] C/EBP,[4] Fos,[5,6] v-Myb,[7] v-Rel,[8,9] Raf-1,[10] c-Abl,[11] and p53[12] (see Table II). The majority of these chimeras are fusions to the hormone binding domain of the human estrogen receptor.

Although other methods have been used to produce regulatable oncogenes (i.e., isolation of temperature sensitive mutants), oncogene steroid receptor hormone binding domain fusions offer several advantages. First, the addition of hormone is often less "invasive" to a cell than heat shock.[7] Second, it is easier to screen for transformants that grow in the presence of hormone than for cells that are unable to transform at the nonpermissive temperature. Finally, the response to hormone is tightly controlled, relatively fast, and readily reversible.[1,3] Two disadvantages must be considered, however. Chimeric proteins cannot be studied in cell types containing endogenous hormone responsive receptors. In addition, the transactivation domain located

[3] M. Eilers, D. Picard, K. Yamamoto, and J. M. Bishop, *Nature (London)* **340,** 66 (1989).
[4] R. Umek, A. Friedman, and S. McKnight, *Science* **251,** 288 (1991).
[5] G. Superti-Furga, G. Bergers, D. Picard, and M. Busslinger, *Proc. Natl. Acad. Sci. U.S.A.* **88,** 5114 (1991).
[6] M. Schuermann, G. Hennig, and R. Muller, *Oncogene* **8,** 2781 (1993).
[7] O. Burk and K.-H. Klempnauer, *EMBO J.* **10,** 3713 (1991).
[8] G. Boehmelt, A. K. Walker, N. Kabrun, G. Mellitzer, H. Beug, M. Zenke, and P. J. Enrietto, *EMBO J.* **11,** 4641 (1992).
[9] A. J. Capobianco and T. D. Gilmore, *Virology* **193,** 160 (1993).
[10] M. Samuels, M. Weber, J. M. Bishop, and M. McMahon, *Mol. Cell. Biol.* **13,** 6241 (1993).
[11] P. Jackson, D. Baltimore, and D. Picard, *EMBO J* **12,** 2809 (1993).
[12] K. Roemer and T. Friedmann, *Proc. Natl. Acad. Sci. U.S.A.* **90,** 9252 (1993).

TABLE II
ONCOGENE–STEROID RECEPTOR FUSIONS[a]

Protein	Fusion	Hormone-dependent activity	Ref
Adenovirus E1A	E1A GR	Transactivation	1
Human c-Myc	c-Myc ER	Transformation	
	c-Myc GR		3
Rat C/EBP	C/EBP ER	Growth inhibition	
	C/EBP GR		4
Mouse Fos	c-Fos ER	Transformation and cellular	
	c-Fos GR	gene regulation	5
	c-Fos ER	Transformation and cellular	
	v-Fos ER	gene regulation	
	FosB ER		6
Chicken Myb	E26 Myb ER	Transactivation and transformation	7
	AMV Myb ER	Transactivation	19
Turkey v-Rel	v-Rel ER	Transformation and cellular gene regulation	8
	v-Rel ER	Transformation	9
Human Raf-1	Raf-1 ER	Transformation and kinase activity	10
Mouse Abl	v-Abl ER	Transformation and kinase activity	
	c-Abl ER	Transformation and kinase activity	11
Human p53	p53 ER	Growth inhibition	12
Chicken Ets	Ets MR	Transactivation	14

[a] ER, Estrogen receptor; GR, glucocorticoid receptor; MR, mineralocorticoid receptor.

within the hormone binding domain of the receptor may contribute to ligand independent activity and, thus, must be controlled if possible.[13]

III. Initial Considerations

Construction of a hormone-regulatable oncogene requires careful consideration of the type of hormone binding domain to be used and the vector system. Transcription factor oncogenes have been successfully fused to steroid receptor hormone binding domains. Surprisingly, several members of the tyrosine kinase family such as Src, Erb-B, Raf,[10] and Abl,[11] also exhibit ligand-inducible activity when fused to the hormone binding domain of the estrogen receptor. Hormone binding domains from several members

[13] N. Webster, S. Green, J.-R. Jin, and P. Chambon, *Cell (Cambridge, Mass.)* **54,** 199 (1988).

of the steroid receptor family have been used to construct chimeric genes, including the GR,[1,3,4,7] MR,[14] estrogen,[4–12] and *Drosophila* ecdysone receptors.[15] Of these the human estrogen receptor is most commonly used, perhaps because of the availability of an estrogen agonist and antagonist.

As mentioned earlier, the estrogen receptor has two transcriptional transactivation domains, TAF-1 and TAF-2.[16] The fragment of the estrogen receptor most commonly used in these chimeras lacks TAF-1 but contains the hormone binding domain and part of the TAF-2 transactivation domain.[17] Thus, it becomes critical to separate the properties of TAF-2 from the fused heterologous protein, especially if that protein functions as a transcription factor. This can be accomplished using two compounds, ICI 164.384 [*N*-n-butyl-11-(3,17β-dihydroxyestra-1,3,5(10)-trien-7α-yl)-*N*-methylundecamide] and tamoxifen. ICI 164.384, an estrogen antagonist, blocks the activity of the receptor by competing with estrogen for binding.[18] It prevents ligand-induced activation of the receptor, including DNA binding and transcriptional activation.[2] Tamoxifen, on the other hand, functions as a partial agonist of the estrogen receptor and is particularly useful in providing a control for estrogen-specific effects. Estrogen receptor bound to tamoxifen binds DNA; however, it exhibits different properties than the estrogen-bound receptors. Both estrogen- and tamoxifen-bound receptors activate gene expression through the constitutive element, TAF-1, located in the N terminus of the receptor.[2] However, tamoxifen cannot stimulate transcription through the ligand-inducible element, TAF-2, which overlaps the hormone binding domain.[13] Tamoxifen binds to and "switches on" an oncogene–ER protein, but does not elicit additional transactivational effects. Properties of the oncogene–ER constructs that remain constant between estrogen and tamoxifen most likely represent the intrinsic activity of the oncogene and do not stem from an interaction between estrogen and the TAF-2 domain of the ER.

The fragment of the estrogen receptor that is commonly used for these constructs contains the hormone binding domain and has a mutation at aa position 400 (gly → Val) which reduces its affinity for estrogen.[17] Lower affinity in the hormone binding domain helps prevent illegitimate activation of the fusion protein by endogenous estrogens.

[14] C. T. Quang and J. Gysdal, personal communication (1994).

[15] K. Christopherson, M. Mark, V. Bajaj, and P. Godowski, *Proc. Natl. Acad. Sci. U.S.A.* **89,** 6341 (1992).

[16] L. Tora, J. White, C. Brou, D. Tasset, N. Webster, E. Scheer, and P. Chambon, *Cell* (*Cambridge, Mass.*) **59,** 477 (1989).

[17] V. Kumar, S. Green, G. Stack, M. Berry, J.-R. Jin, P. Chambon, *Cell* (*Cambridge, Mass.*) **51,** 941 (1987).

[18] A. E. Wakeling and J. Bowler, *J. Steroid Biochem.* **31,** 645 (1988).

Other fragments of the human estrogen receptor have been used in the construction of chimeras. For example, v-Myb fusions have been created with the D-E fragment of the ER (HE14) and with the D fragment of the receptor. The v-mybER fusion containing only the E fragment of the receptor performs better in transactivation assays than the D–E construct does. This may be due to the presence of a small negative regulatory domain in the D domain.[19]

Many types of vectors have been successfully used to introduce the ER fusion proteins into cells for performing transient assays or for establishing stable cell lines. For example, both replication competent and neomycin selectable retroviruses are compatible with this system.[3,8] Generally, vectors are chosen for compatibility with the cellular system. However, choice of the vector and the cell type can effect the activity of the construct. For example, two v-relER fusions have been published: one expressed by a replication-competent virus (RCAS)[8] in primary bone marrow cells and another expressed by a replication defective virus (SNV) in primary spleen cells.[9] The RCAS v-relER virus transforms well in the presence of estrogen and tamoxifen,[8] whereas the SNV v-relER produces few clones in soft agar assays.[9] The difference in their activity is most likely due to the vector system and is not related to the ability of v-rel to transform these cell types.

IV. Methods

A. Construction and Expression of Chimeric Proteins

Oncogene–steroid receptor fusions are created to maintain a correct reading frame between the two portions of the chimeric protein. Standard polymerase chain reaction and cloning procedures should be followed and details will vary for each particular chimera. Following construction of the fusion protein, the chimera can be cloned into an expression vector of choice.

After introducing the construct into cells it is important to check the expression and stability of the fusion protein. Expression can be assayed by either immunoprecipitation or immunoblotting with an antibody to the oncogene or an antibody to the hormone binding domain of the estrogen receptor.[20] Stability can be measured via pulse–chase experiments where cells are pulse labeled with a radiolabeled amino acid for a short period of time and chased for varying periods with an excess of the corresponding

[19] U. Engelke and J. Lipsick, personal communication (1994).

[20] Obtained from Geoffery Greene, The Ben May Institute, University of Chicago, Chicago IL 60637.

cold amino acid. Samples are then immunoprecipitated with antibodies. In this way the half-life of the wild-type and chimeric protein can be compared.

B. Hormones

The hormones used in these experiments are available from several sources. Dexamethasone, which activates the glucocorticoid receptor, can be obtained from Sigma and can be made up as a 10-μg/ml stock solution. It is active when added to liquid culture media at a concentration of 1–10 μM.[1] 17β-Estradiol, which binds the estrogen receptor, is also available from Sigma. It should be dissolved in ethanol as a 10 mM stock and should be stored at $-20°$. Estrogen should be added to the culture media at concentrations ranging from 10 nM to 2 μM.[4] Diethylstilbesterol, an estrogen agonist, is available from Sigma and should be used at 10 nM concentration in liquid media.[6] ICI 164.384 (molecular weight 525) and 4-hydroxytamoxifen (molecular weight 387), an antagonist and partial agonist of the estrogen receptor, respectively, are available through ICI Pharmaceuticals (Mereside, Alderly Park, Macclesfield, Cheshire SK10 4TG). They should be dissolved in ethanol at 10 mM and should be stored at $-20°$. ICI 164.384 can be added to liquid media at similar concentrations as estrogen. 4-Hydroxytamoxifen (OHT) can be toxic to some cells at concentrations above 1 μM and is usually added to media at concentrations ranging from 1- to 100 nM.[8,10] Tamoxifen citrate can be obtained from Sigma and should be treated in a similar manner as OHT. However, in the case of v-relER, this compound is unable to activate the fusion proteins.[21] All hormones should be applied daily to liquid cultures and every 2–3 days to soft agar assays.

C. Stripped Serum

Fetal calf or newborn calf serum contains low levels of steroid hormones. Therefore, to reduce activation of the chimeric protein by serum, it is advisable to remove the endogenous steroids from the serum used in the tissue culture media. In the following protocol, charcoal is coated with dextran and is used to treat the serum batches.

Preparation of Dextran-Coated Charcoal (*DCC*). Twenty-five grams of Norit A charcoal (Serva) should be washed three times with 10 mM Tris–HCl, pH8, to remove small particles. Centrifuge the washed charcoal at 1000 rpm in a Beckman GS4 rotor. The pellet should be resuspended in 200 ml 10 mM Tris–HCl, pH 8, plus 25 g dextran T70 (Pharmacia), autoclaved for 30 min at 0.5 atm, and stored at 4°.

[21] O. Petrenko and P. J. Enrietto, unpublished observation (1994).

Treatment of Serum. Ten milliliters of DCC should be added to 500 ml of serum and stirred slowly overnight at 4°. The next day remove the charcoal by spinning the mixture at 4000 rpm for 20 min in a Beckman GS 4 rotor. Next, 10 ml of DCC is added to the serum supernatant. This mixture is heated to 50° for 30 min followed by cooling to 4°. After cooling, the serum should be centrifuged at 4000 rpm for 20 min in a Beckman GS4 rotor. Repeat the addition of DCC, heating and cooling one time. After the second cooling the serum should be filtered, first through Whatman 3MM paper and then through a 20-μm Millipore filter to sterilize. It is recommended that the last filtration step be carried out in small (about 100 ml) batches to minimize clogging of the filter. The filtered serum should be divided into aliquots and stored at −20°.[22]

D. *Phenol Red*

In order to assure that the phenol red present in most tissue culture media does not contribute to illegitimate activation of the chimeric protein, cells containing the fusion protein can be propagated in phenol red-free media. However, phenol red does not biochemically activate the v-relER chimera.[22]

V. Evaluation

Several parameters should be tested to ensure that the oncogene–ER construct is working correctly. This section describes additional constructs and alternative hormones that serve as controls as well as suggestions for evaluating the biological and biochemical properties of the constructs.

A. *Control Constructs*

It is advisable to construct several variations of the oncogene hormone binding domain chimera to ensure the highest possible activity of the chimeric protein. In addition, these control constructs serve to rule out artifactual effects caused by the hormone binding domain. First, the ER domain can be fused to either the amino or carboxy terminus of the protein. Eilers *et al.*[3] constructed a 5′ fusion (ERmyc) and the 3′ fusion (MycER) of the hormone binding domain of the estrogen receptor and the oncogene myc. Both were transforming, however, mycER bound hormone less well and produced 10% fewer colonies in soft agar assays. In addition, v-RelER and ER v-Rel constructs were also tested. While the v-RelER fusion trans-

[22] H. Beug, personal communication (1992).

TABLE III
ESTROGEN RECEPTOR LIGANDS

Hormone	Receptor binding	DNA binding	Transactivation	Ref.
β-Estradiol	+	+	TAF-1/TAF-2	16
Tamoxifen	+	+	TAF-1	2
ICI 164.384	+	−	—	18

formed cells in a manner similar to v-Rel, ER v-Rel was not active.[23] v-Rel contains a protein–protein interaction domain in its amino terminus which may have been disturbed by the addition of the estrogen receptor sequences. Thus, the activity of the chimeric protein may vary with the placement of the hormone binding domain, presumably because of structural alterations.

It is critical to be able to ascribe the potential activity of the oncogene–ER fusion to the oncogene portion of the chimera rather than the estrogen receptor. One way to accomplish this is to fuse the ER domain to a nontransforming mutant of the oncogene. Eilers *et al.*,[3] Burk and Klempnauer,[7] and Jackson *et al.*[11] use nontransforming mutants of c-*myc*, v-*myb*, and c-Abl, respectively. The c-*myc* and v-*myb* mutants are still unable to transform when fused to the ER domain, demonstrating that the ER domain by itself is incapable of producing transformation. However, one of the nontransforming c-Abl mutants is capable of transforming when produced as part of an ER fusion. It is not clear what the mechanism is for this activation.

B. Agonists and Antagonists

The steroid hormone receptors respond to an array of naturally occurring and manmade ligands (see Table I). The glucocorticoid receptor was one of the first to be used in fusion proteins and can be activated by dexamethasone. Although progesterone can serve as a antagonist to the GR, it has not been used to control oncogene–GR fusions since it also activates the progesterone receptor. The MR agonist is aldosterone and it can be antagonized by progesterone and spirolactone.[24] However, in experiments using oncogene–GR or –MR fusions, activated cultures are usually compared with cultures from stripped-serum media.[1,3–5,14]

One of the advantages of using the estrogen receptor hormone binding domain rather than other steroid receptor family members is the availability of estrogen analogs for use as controls (see Table III). Two analogs, ICI and tamoxifen, are particularly useful. ICI 146.384 completely inhibits the

[23] G. Boehmelt and P. J. Enrietto, unpublished observation (1992).

[24] J. D. Wilson and D. W. Foster, "Textbook of Endocrinology." Saunders, New York, 1985.

activity of the chimeric protein, providing a negative control, whereas tamoxifen produces an effect similar to estrogen. Biological or biochemical activity induced by estrogen but not by tamoxifen most likely stems from the hormone binding domain and not the oncogene in the fusion.

C. Biological Activity

Measurement of the biological activity of the oncogene–ER fusion protein provides an important indication of its functional status. When treated with estrogen or tamoxifen the oncogene–ER should produce a phenotype very close to the wild-type oncogene. In most cases, the oncogene–ER proteins produce similar numbers of colonies in transformation assays. Fos-B transforms strongly when fused to ER and is activated with DES (an estrogen analog). Since it does not transform in the presence of tamoxifen,[6] the transforming activity observed is most likely the result of estrogen activation of the TAF-2 domain and is not an intrinsic activity of the Fos-B.

D. Biochemical Activity

Many chimeric oncogenes retain most of their biochemical properties when fused to a steroid hormone binding domain. These may include participation in protein–protein interactions, DNA binding, transcriptional activation, or enzymatic activity. However, some caution should be used when measuring the biochemical activities of chimeric proteins. For example, the v-relER construct, when tested in transactivation assays, showed a significant increase in the ability to transactivate gene expression in the presence of estrogen.[8] Tamoxifen does not elicit this response, although it does induce the biological activity of the fusion protein. Thus, the transactivational activity seen with estrogen but not with tamoxifen is most likely due to an interaction between estrogen and the TAF-2 domain of the estrogen receptors. Measurement of transcription from endogenous genes containing the DNA binding sites of interest provides a more accurate estimation of the ability of the oncogene–ER fusion to regulate gene expression.

VI. Applications

A. Biological Assays

Oncogene steroid receptor fusions have been useful in a variety of biological systems, particularly in the study of hematopoietic differentiation. For example, in a series of experiments investigating the differentiation potential of avian hematopoietic cells, v-relER-transformed bone marrow cells were withdrawn from estrogen and allowed to differentiate. Under normal growth conditions, v-relER cells exhibit markers of the myeloid, lymphoid, and erythroid lineages.[25] Manipulation of the culture conditions allowed the differentiation of two different cell types: one which exhibited neutrophilic characteristics and one which resembled dendritic cells. Growth in standard media favored by the outgrowth of the dendritic cells are characterized by their polarized appearance, cytoplasmic protrusions, and vimetin aggregates. Growth in complex media supported the appearance of cells with a lobed nucleus which appear in the neutrophilic lineage. Thus, the v-relER system was able to provide information on the differentiation potential of the hematopoietic target cell for v-*rel* transformation by allowing the replacement of the oncogenic signal with inducers of differentiation.[25]

In another type of experiment, Reichmann *et al.*[26] used a FosER fusion protein to investigate the effect of Fos activity on the differentiation of polarized epithelial cells. Activation of FosER by the addition of estrogen for specific amounts of time at defined stages of epithelial development allowed discrimination between the short-term and long-term effects of Fos expression. Short-term activation of FosER through the addition of estrogen caused the loss of polarity in the epithelial cells, along with other morphological and functional changes. However, upon removal of estrogen these cells were able to reestablish polarity, and completely reverted to their previous phenotype. Long-term activation of FosER, on the other hand, caused the epithelial cells to fully convert to a fibroblastoid phenotype. These cells were unable to regain the phenotype of epithelial cells on inactivation of FosER.[26] Thus the FosER chimera provided a useful tool in investigating the temporal effects of oncogene activation within a defined biological system.

This technology has been useful in investigating the activity of proteins which by themselves initiate differentiation programs (C/EBP)[4] or inhibit cell growth (p53).[12] In this case the cells were propagated when the chimeric protein is turned off, then stimulated with the appropriate hormone to activate the fusion protein and concomitantly the differentiation or cell death program.

B. Molecular/Biochemical Assays

Modulation of the activity of oncogenes has provided insight into their biological effects. In addition, the correlation of these biological changes with the biochemical effects has been possible. Oncogene–ER systems

[25] G. Boehmelt, J. Madruga, P. Dorfler, K. Briegel, H. Schwartz, P. Enrietto, and M. Zenke, *Cell* (*Cambridge, Mass.*) **80,** 341 (1995).

[26] E. Reichmann, H. Schwartz, E. Deiner, I. Leitner, M. Eilers, J. Berger, M. Busslinger, and H. Beug, *Cell* (*Cambridge, Mass.*) **71,** 1103 (1992).

have provided an excellent means for the identification of the target genes regulated by oncogenic transcription factors. The ability to turn the oncogene–ER proteins on and off allows the generation of transformed and nontransformed cells from which subtractive cDNA libraries have been made. In this way, it has been possible to identify genes expressed differentially in the transformed and nontransformed cells. For example, it was shown that v-relER was able to alter the expression of two cellular genes, chicken MHC I and high mobility group protein 14(b) [HMG 14(b)].[8] Comparison of rates of transcription of HMG 14(b) in v-relER cells with and without estrogen by nuclear run-on assays has demonstrated that HMG 14(b) is transcriptionally regulated by v-*rel*. Furthermore, gel shift assays comparing the ability of extracts from v-relER cells grown with or without estrogen to bind to *rel* consensus binding sites in the HMG 14(b) promoter have shown that *rel* binding activity correlates with the activation of v-relER, and the expression of the gene.[27] Thus the v-relER fusion provides a powerful tool for both identifying the cellular targets for v-*rel* regulation and investigating the mechanisms of the regulation.

VII. Conclusions

The regulatable chimeras composed of fusions between steroid receptor hormone binding domains and oncogenes provide an interesting and powerful system for studying how oncogenes work in the context of the cells they transform. Regulatable chimeras offer several advantages over other similar systems such as temperature-sensitive mutants. They are tightly regulated by hormone, respond quickly to stimulation, and provide an easily manipulatable system for controlling the activity of an oncogene. This technique may offer many insights into the biological and biochemical nature of oncogenes.

[27] A. Walker and P. J. Enrietto, unpublished observations (1994).

[31] Use of Tetracycline Operator System to Regulate Oncogene Expression in Mammalian Cells

By Emelyn R. Eldredge, Paul J. Chiao, and Kun Ping Lu

Introduction

Defining the mechanism of oncogene activation is an important step in the understanding and treatment of cancer. However, it is often difficult to ascribe the biochemical, physiological, and morphological phenotypes observed in transformed cells directly to the constitutively activated oncogenes in the cell. It is easier to define effects in a system which allows one to regulate the activity of the suspect protein. One particularly effective method of study is the use of inducible genes, which allow the candidate gene to be activated or repressed in a regulated manner. Introduction or removal of the inducer allows the investigation of events proximal to oncoprotein activation.

Several inducible systems presently exist which use external inducers to regulate gene expression. Inducible eukaryotic promoters include the use of heavy metal ions, hormone induction, and heat shock. These approaches have inherent problems because either the promoters are "leaky" in the uninduced state or the stimuli themselves exert pleiotropic effects on compatible endogenous promoters and proteins. The use of prokaryotic-inducible promoters in eukaryotic systems effectively eliminates these problems by affording the use of nonphysiological inducers. Two inducible prokaryotic systems which have recently received attention are the Lac activator protein, discussed elsewhere in this volume,[1] and the use of the operator of the tetracycline resistance genes, the subject of this chapter.

General Principles of Tetracycline Regulation Expression

In *Escherichia coli*, the tetracycline repressor (*tet*R) binds the tetracycline operator *tet*O to prevent transcription of tetracycline resistance genes in the absence of the drug tetracycline. The *tet*R has high specificity for the *tet*O,[2] whereas tetracycline has high affinity for the *tet*R.[3] In the absence of tetracycline, the *tet*R binds the *tet*O, effectively blocking transcription

[1] M. Labow, this volume [23].

[2] C. Kleinschnidt, K.-H. Tovar, W. Hillen, and D. Porshke, *Biochemistry* **27,** 1094 (1988).

[3] J. Degenkolb, M. Takahashi, G. Ellested, and W. Hillen, *Antimicrob. Agents Chemother.* **36,** 1591 (1991).

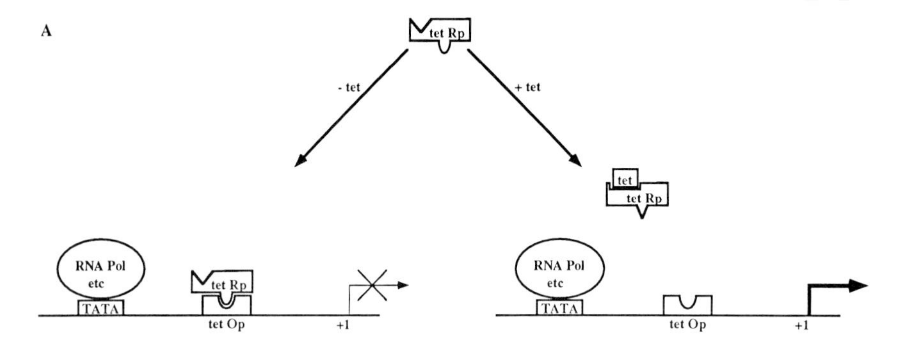

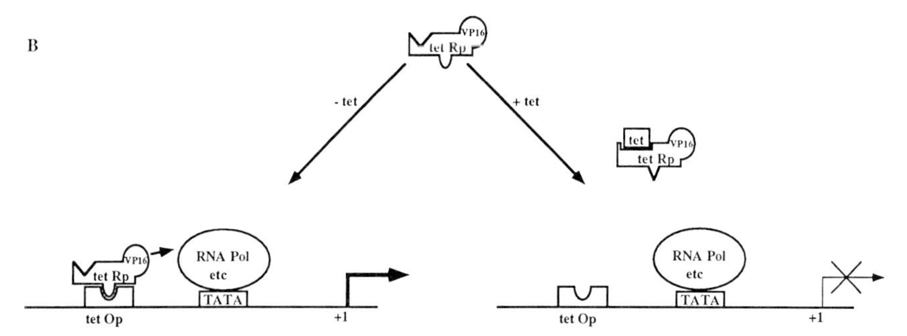

FIG. 1. General principles of the *tet*O system demonstrating repression by the *tet*R (A) and activation by the tTA (B) (see text for details).

by steric interference (Fig. 1A). The binding of tetracycline by the *tet*R, however, abrogates this transcriptional repression as the consequent conformational change in the DNA binding domain of the protein releases the *tet*R from its association with the *tet*O.[4] Although the *tet*R system was developed by Gatz and Quail[5] for regulated gene expression in plant cells, the direct use of the *tet*R to repress transcription and allow tetracycline activation of the target gene has yet to be described in mammalian cells. However, a tetracycline-transactivating system based on the *tet*R has been developed and is described below.

Gossen and Bujard converted the *tet*R into a tetracycline transcriptional activator (tTA) by fusing the full-length *tet*R with the acidic transcriptional activation domain of the herpes simplex viral protein VP16.[6] Unlike the *tet*R

[4] W. Hinrichs, C. Kisker, M. Düvel, A. Müller, K. Tovar, W. Hillen, and W. Saenger, *Science* **264,** 418 (1994).

[5] C. Gatz and P. H. Quail, *Proc. Natl. Acad. Sci. U.S.A.* **85,** 1394 (1988).

[6] M. Gossen and H. Bujard, *Proc. Natl. Acad. Sci. U.S.A.* **89,** 5547 (1992).

system, in which transcription is activated by the addition of tetracycline, in the tTA system, transcription is constitutively activated in the absence of tetracycline (Fig. 1B). When the tTA is associated with its cognate *tet*O DNA sequence, the fused VP16 portion of the protein activates the RNA polymerase complex and stimulates transcription. The addition of tetracycline releases the tTA from the *tet*O, removing the fused transcriptional activator moiety and depressing gene expression.

Experimental Design

Properties of tTA and tetO Constructs

Gatz and Quail[5] reported the likely passive nuclear diffusion of the *tet*R protein as well as the high affinity of tetracycline for the *tet*R as reasons for choosing the *tet*R system for use in regulated gene expression. Indeed, the tTA has been shown to be present in the nucleus of HeLa cells transiently transfected with a tTA-expressing plasmid.[6] Plasmid constructs encoding tTA expression are depicted in Fig. 2.[7–9] It is critical that this transactivator is driven by a strong constitutive promoter to ensure consistent expression throughout the cell cycle and other environmental conditions such as contact inhibition or serum starvation when many promoters are turned off.

The *tet*O must be placed within a minimal promoter in order to observe the "on/off" activation of the tTA reported by Gossen and Bujard.[6] In addition, the element must be positioned upstream of the TATA box. Whereas a single copy of the *tet*O is sufficient for the repressor function of the *tet*R,[10,11] multiple repeats of the *tet*O will increase the induction mediated by the tTA without elevating basal transcription.[6]

Properties of Tetracycline

Tetracycline is an ideal inducer for regulated gene expression, in part because of its low toxicity to mammalian cells and ready absorption across the plasma membrane. It is also relatively inexpensive: 100 mg is enough for 100 liter of culture media. Tetracycline action interrupts prokaryotic

[7] L. Altschmied, R. Baumeister, K. Pfleiderer, and W. Hillen, *EMBO J.* **7,** 4011 (1988).

[8] K. P. Lu and A. R. Means, *EMBO J.* **13,** 2103 (1994).

[9] D. Resnitzky, M. Gossen, H. Bujard, and S. I. Reed, *Mol. Cell. Biol.* **14,** 1669 (1994).

[10] T. Dingermann, U. Frank-Stoll, H. Werner, A. Wissmann, W. Hillen, M. Jacquet, and R. Marschalek, *EMBO J.* **11,** 1487 (1992).

[11] T. Dingermann, H. Werner, A. Schütz, Z. I., K. Nerke, D. Knecht, and R. Marschalek, *Mol. Cell. Biol.* **12,** 4038 (1992).

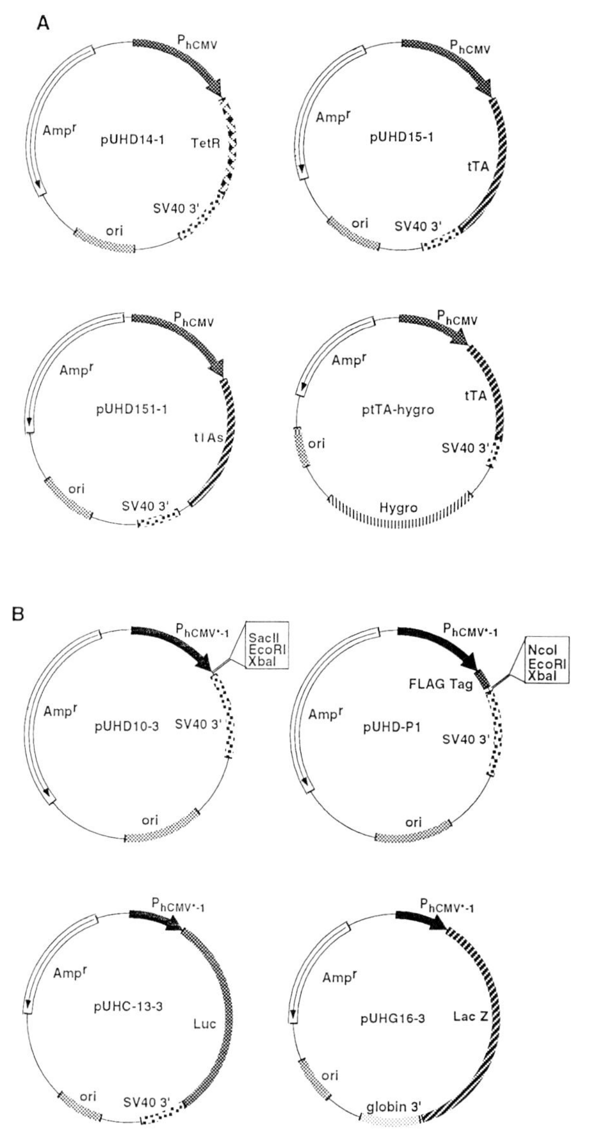

FIG. 2. Vectors used in the tetracycline-responsive system All vectors (not drawn to scale) contain the bacterial replication origin (ori) and β-lactamase-resistant gene conferring ampicillin resistance (Amp^r), both of which are used to maintain these constructs in *E. coli.* Vectors in (A) are used to generate tTA-expressing cells, whereas constructs in (B) are used to test expression of the tTA gene or to express the target genes in the tTA-containing cells. The properties of individual vectors are described as follows. pUHD14-1: The tetracycline repressor sequence (*tet*R) originally from pWH510[7] inserted between the CMV promoter and

polypeptide chain elongation. The only effect tetracycline has on mammalian cells is to inhibit mitochondrial protein synthesis. However, the affinity of tetracycline for the *tet*R is three orders of magnitude greater than its affinity for the 30S ribosome (10^9 M^{-1} versus 10^6 M^{-1}), enabling its use as an inducer metabolite[12] without significant effects on mitochondrial function. The concentration of tetracycline required for regulated expression has been found to be as little as 0.1 μg/ml (0.2 μM) for inactivation of transcription directed by the tTA.[6]

Tetracycline analogs may also be useful in regulating *tet*R activity. For example, anhydrotetracycline, although a poor antibiotic, binds in a slightly different manner[13] and with greater affinity (10^{11} M^{-1}) to the *tet*R. Hence it can be used at a lower concentration than tetracycline with even less of an effect on mitochondrial translation. Gossen and Bujard[6] have shown anhydrotetracycline to be a more potent effector of the tTA.

It must be noted that the tTA system is not ideal for studies which require immediate temporal regulation of candidate genes. Tetracycline is passively taken up by the cell and, in the absence of the tetracycline transporter, it is also passively secreted. Removal of the drug from the culture

[12] M. Takahashi, L. Altschmied, and W. Hillen, *J. Mol. Biol.* **187,** 341 (1986).
[13] B. Hecht, G. Müller, and W. Hillen, *J. Bacteriol.* **175,** 1206 (1993).

---

the 3′ sequence of SV40.[6] pUHD15-1: The 130 amino acid COOH-terminal portion of VP16 fused in frame to the 3′ end of the *tet*R gene in pUHD14-1, resulting in a transactivator, tTA.[6] pUHD151-1: The 97 amino acid COOH-terminal portion of VP16 fused in frame to the 3′ end of the *tet*R gene in pUHD14-1, resulting in a smaller transactivator, tTAs.[6] ptTA-hygro: The 1.8-kb *Xho*I–*Hin*dIII fragment from pUHD15-1 containing the CMV promoter and the tTA gene subcloned into the *Bam*HI site of pSV2hygro, which contains a hygromycin-resistant gene (hygro) (P. Chiao, unpublished data). pUHD10-3: The upstream enhancer region of the CMV promoter in pUHD10-1 is removed and replaced with seven copies of the tetracycline operator O2 sequence of Tn*10*, a 19-bp inverted repeat, resulting in a hybrid promoter, P_{CMV*-1}, which is repressed by tetracycline. The multiple cloning sites, *Sac*II, *Eco*RI, and *Xba*I, are used to insert target genes.[6] pUHD-P1: The 8 amino acid FLAG epitope with a eukaryotic translation initiation signal inserted between P_{CMV*-1} and the SV40 polyadenylation signal in pUHD10-3. The FLAG tag translation frame is ATG-GAC-TAC-AAA-GAC-GAT-GAC-GAT-AAA-GCC-ATG-GCT-CGA-GAA-TTC-GAG-CTC-GGT-ACC-CGG-GGA-TCC-TCT-AGA-GGA-. The underlined *Nco*I, *Eco*RI, and *Xba*I sites are unique for insertion of target genes. The FLAG-specific monoclonal antibodies can be used to detect the expressed proteins.[8] pUHC13-3: A luciferase gene (Luc) inserted between P_{CMV*-1} and the SV40 polyadenylation signal in pUHD 10-3. The resulting construct can be used to test whether transfected cells express the transactivator tTA gene by measuring luciferase activity.[6] pUHD 13-4 is essentially the same as pUHD 13-3, except that there is a slightly diverent spacer between the *tet*O and the minimal CMV promoter. pUHG16-3: P_{CMV*-1} followed by the β-galactosidase gene (LacZ) and a globlin intron/polyadenylation signal. This construct can be also used to test expression of the transactivator tTA gene by measuring β-galactosidase activity.[9]

media merely changes the gradient. Activation to the maximum level of induction can take up to 48 hr after the removal of tetracycline.[6] Repression of transcription on reintroduction of tetracycline can take somewhat longer.

Establishment of Cell Lines Expressing tTA

The establishment of the tTA system must be tailored to individual needs. The general experimental design strategy is depicted in Fig. 3. Initially, cell lines must be developed which stably express the tTA. Those already established were derived from the parental lines HeLa,[6] Rat1,[9] NIH/3T3 (E. R. Eldredge and W. Eckhart, unpublished data), and 208F (P. J. Chiao and I. M. Verma, unpublished data). There are several well-documented methods of eukaryotic cell transfection presently in use includ-

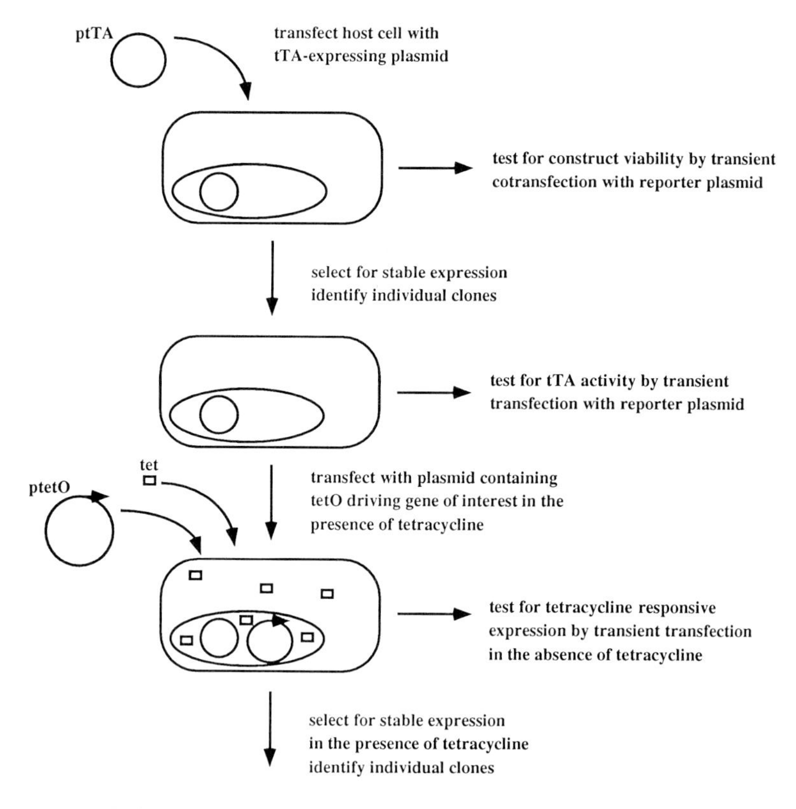

FIG. 3. Experimental design to establish cell lines.

ing DEAE dextran, lipofection, calcium phosphate precipitation, and electroporation.[14]

After transfection it is important to use drug selection to identify stable transformants. In order to do this, the concentration of the selection drug necessary to kill untransfected parental cells, or "kill curve," must be determined. The kill curve varies with the drug used and the cell line transfected. For example, 400 μg/ml of the neomycin analog G418 added to NIH/3T3 cells 48 hr post-transfection will result in the cell death of untransfected cells within 3 days and complete killing within 7 days. The tTA-expressing plasmids pUHD15-1 and pUHD151-1 (see Fig. 2) must each be cotransfected with plasmid containing a gene for drug selection or the appropriate fragment from these vectors, containing the cytomegalovirus (CMV) promoter and the tTA gene, must be subcloned into a selectable marker plasmid like pSV2neo or pSV2hygro, as is the case for ptTA-hygro.

Clonal selection can be quite a lengthy process, depending on the growth characteristics of the parental cells. It is important to harvest clones while they are still spatially distinct on the plates. Individual isolates can be selected through the use of cloning rings or limiting dilution.

Once stably expressing clonal lines have been identified by drug selection, they will need to be evaluated for tTA function. Despite being driven by a strong promoter, the expression and consequently the transactivating activity of the tTA will vary among clones due to integration sites within the genome. To assay the tetracycline inducibility of individual clones, we recommend the use of the reporter genes luciferase and β-galactosidase, which already exist under the influence of the *tet*O (see Fig. 2). Assay protocols for these enzymes are readily available.[14]

Expression of Target Genes in tTA-Containing Cell Lines

Genes of interest, driven by a *tet*O-regulated promoter, are introduced into cell lines which stably express the constitutively active tTA gene, again using standard transfection techniques. Conditional expression and functional studies can be performed by either transient or stable transfection depending on experimental intent. In order to generate cells transfected with, but not actively expressing, target proteins, cells must be preloaded with tetracycline at a final concentration of 1 μg/ml 24–48 hr prior to transfection. End point expression of target genes can be observed 4–48 hr after the replacement of tetracycline-containing media with tetracycline-free media.

[14] F. M. Ausubel, R. Brent, R. E. Kingston, D. D. Moore, J. G. Seidman, J. A. Smith, and K. Sturhl, eds., "Current Protocols in Molecular Biology," Chapter 9. Wiley, New York, 1994.

Transient Transfection

Assaying target gene expression by transient transfection is a relatively rapid procedure and often provides important information regarding the function of candidate genes. Transient transfection assays should be carried out prior to the establishment of stable cell lines to ensure expected expression of target genes. The expressed gene products can be detected through the use of specific antibodies in standard laboratory procedures like immunoblot analysis, immunoprecipitation, and immunocytochemistry or by assaying protein function.

For many newly identified gene products, however, antibodies may not be available and assays may still need to be developed. To facilitate the detection of the expressed proteins, the eight amino acid FLAG epitope has been introduced into pUDH10-3, resulting in the amino-terminal expression of the tag in pUHD-P1.[8] Insertion of this FLAG epitope at the amino terminus of proteins can be accomplished usually without affecting protein function.[8,15] However, characteristics of the target protein, such as complex association and catalytic domains, should be considered when designing epitope fusions.

The commercially available anti FLAG-tag monoclonal antibodies [M1 (hybridoma 4E11 from the American Tissue Type Collection) and M2 (purified antibody available from Kodak/IBI)] can then be used to determine expression of target genes by Western analysis and immunoprecipitation[8] as well as immunocytochemistry.[15] It should be noted that the M1 monoclonal antibody recognizes the FLAG tag in a calcium-dependent manner only when it is inserted at the amino terminus of fusion proteins.[16] The M2 monoclonal antibody, on the other hand, requires neither calcium nor amino-terminal location of the FLAG epitope.[17] Other tags and cognate monoclonal antibodies are available and could be engineered into pUHD10-3.

Stable Cell Lines

Once it has been determined that transiently transfected target genes can be expressed in a tetracycline-responsive manner in the tTA-containing cells, stable cell lines can be established which contain both the tTA and the target gene under the control of the tTA-regulated promoter. Regardless of

[15] K. P. Lu and T. Hunter, *Cell* **81,** in press (1995).

[16] T. P. Hopp, K. S. Prickett, V. Price, R. T. Libby, C. J. March, P. Cerretti, D. L. Urdal, and P. J. Conlon, *Bio/Technology* **6,** 1205 (1988).

[17] B. L. Brizzard, R. G. Chubert, and D. L. Vizard, *BioTechniques* **16,** 730 (1994).

the transfection procedure used, the following considerations should be taken in account.

The first consideration is the selectable marker. Since tTA-containing stable cell lines are already resistant to one selection drug, a second selection drug should be used. G418 (Sigma) and hygromycin (Calbiochem) are eukaryotic selection drugs frequently used to establish doubly stable cell lines. Puromycin is an inexpensive selection drug when compared with G418, but it has a very rapid kill curve, making it problematic for clonal selection. Neither pUHD10-3 nor pUHD-P1 (see Fig. 2) contains a drug-resistance gene, so cotransfection with a second plasmid containing such a marker is necessary to establish doubly stable cell lines. Alternatively, one can subclone the fragment from pUHD10-3 or pUHD-P1 containing the *tet*O promoter and target gene into an established drug selection plasmid, such as pSV2neo or pSV2hygro. Although anecdotal in nature, there is some evidence that cells expressing SV40 large T antigen are difficult to clone out under hygromycin selection. Under these conditions, "superselection" with a single selection marker is also an option. In this process the second construct is transfected with the same selection marker as the first. During the selection process the working concentration of the selection drug is increased (1.5- to 2.0-fold) and those cells containing multiple copies of the resistance gene, from two separate transfection procedures, will grow out (M. Broome, personal communication). Cells can be maintained at the previous drug concentration after selection is complete. Although somewhat unconventional, this method has been shown to work in the selection of G418-resistant *src*$^-$ cells infected with a G418-resistant src plasmid (M. Broome, unpublished data) and in the production of homozygous mutant ES cells targeted with a single construct.[18]

The second factor to consider is the biochemical effect of the target gene on cells. The conditional expression afforded by the tTA system allows stable cells to be established and maintained which contain proteins in the absence of their activity. For example, it can be very difficult to establish stable cell lines if expression of the target gene results in a lethal phenotype, such as cell cycle arrest or cell death. If the gene to be expressed is lethal, the cells should be maintained in tetracycline throughout the selection procedure. The protein can be induced for study at the appropriate time by replacing tetracycline-containing media with tetracycline-free media.

Third, for the same reasons described for the establishment of tTA-expressing cell lines, it is important to obtain individual clones of the doubly stable transformant. It has been shown that expression varies considerably

[18] R. A. Mortensen, D. A. Conner, S. Chao, A. A. T. Geisterfer-Lowrance, and J. G. Seidman, *Mol. Cell. Biol.* **12,** 2391 (1992).

among different cell clones.[6] Again, clonal isolation can be achieved through the use of cloning rings or by plating cells out in multiple-well plates in limiting dilution. The clonal cell lines expressing different levels of transfected genes in the absence of tetracycline are useful in determining the dose-dependent effects of target gene expression.

Finally, the induction ratio of target gene activation in selected clones will have to be determined. In order to do this, the basal level of expression in the presence of tetracycline (1 μg/ml) needs to be compared to maximal expression levels in the absence of the drug. Target gene expression can be attenuated by titrating tetracycline concentration, permitting the investigation of dose–response effects of the target protein on the cell. In addition, it is important to determine the kinetics of induction on drug removal.

Applications of Tetracycline-Regulated Expression

This tTA system has been reported to regulate the activity of *tet*O-driven expression of the reporter protein luciferase over five orders of magnitude.[6] Studies examining the transcriptional activation of physiological proteins have not shown such dramatic induction. A 3- to 10-fold induction of cyclin expression has been demonstrated via immunoblotting.[9,19] We have observed 10- to 20-fold induction of several proteins, including polyoma middle T antigen (E. R. Eldredge and W. Eckhart, unpublished data), the three cell cycle regulated protein kinases NIMA kinase, NLK1, SRPK1,[15] and c-*fos* (P. J. Chiao and I. M. Verma, unpublished data).

Gatz and Quail[5] initially investigated the use of the *tet*R system for regulated gene expression in plant cells. Dingermann and co-workers subsequently utilized the *tet*R system for the regulation of RNA polymerase III genes in both yeast[10] and slime mold.[11] Presently there are no reports of the *tet*R being used for the regulation of oncogene expression. It has been proposed that the *tet*R would be unsuitable for inducible gene expression because repression, being a mechanical and not a biochemical interaction, would require full-time occupancy of the *tet*O.[20] In contrast, the tTA system can achieve a high level of transcription with a much lower occupancy rate because of the strong transactivation properties of the VP16 domain.

The inducible tTA system affords several possible applications for use in the study of oncogenes. The first, and most obvious application, is the directed expression of known (or putative) oncogenes in the tissue culture setting. Conversely, endogenous expression of candidate genes could be repressed by the tTA-inducible expression of antisense RNA. This applica-

[19] A. Wimmel, F. C. Lucibello, A. Sewing, A. Adolph, and R. Müller, *Oncogene* **9,** 995 (1994).
[20] M. Gossen, A. Bonin, and H. Bujard, *Trends Biochem. Sci.* **18,** 471 (1993).

tion would enable the identification of suspected downstream effectors required for the transformation phenotype. The tTA system also seems ideal for the conditional expression of tumor suppressor genes as tive expression of such genes would necessarily be cytostatic. As tetracycline is readily taken up by animals, this system may allow for the development of transgenic animals expressing the tTA and a target gene either out the animal, to examine the effects of oncogene expression on a number of distinct tissues, or under the direction of tissue-specific promoters, for use in examining developmentally regulated genes.

The tTA system presented in this chapter describes methods and applications for the inducible expression of target genes. Specific manipulation of genes in the cell enhances the study of their effects on cell function and eventually the identification of downstream effectors. The development of inducible systems such as this one will greatly advance our understanding of the inner workings of the cell, leading to more effective treatments for all types of disease.

Acknowledgments

We thank Kellie Neet for helpful discussions in the preparation of this manuscript. We also thank Martin Broome, Nik Somia, and Tony Hunter for critically reviewing this manuscript. We are grateful to Hermann Bujard for the generous donation of plasmids and tTA-1 cells.

NOTE ADDED IN PROOF. The generation of transgenic mice expressing the tTA system has been reported since submission of this manuscript. (P. A. Furth, L. St. Onge, H. Böger, P. Gruss, M. Gossen, A. Kistner, H. Bujard, and L. Henninghausen, *Proc. Natl. Acad. Sci. U.S.A.* **91,** 9302–9306.)

[32] Detection of Protein–Protein Interactions by Coimmunoprecipitation and Dimerization

By LYNN J. RANSONE

Introduction

Regulation of transcription is controlled both by general transcription factors and by sequence-specific binding proteins.[1,2] A number of oncogene products can act as transcription factors or cofactors. These proteins may act directly by binding to specific DNA regulatory sequences as exhibited

[1] T. Maniatis, S. Goodburn, and J. A. Fischer, *Science* **236,** 1237 (1987).
[2] P. Mitchell and R. Tjian, *Science* **245,** 371 (1989).

by Jun and ErbA.[3–8] Alternatively, they may act as cofactors like the Fos family members which alone do not bind to DNA, but in association with any of the Jun family members can activate and augment transcription.[9,10] Many of these proteins can be classified into specific families based on common structural motifs, such as the leucine zipper family,[11,12] the helix–loop helix family,[13,14] the homeodomain family,[15] and zinc finger binding proteins.[16,17] There have been several examples demonstrating transcriptional control mediated by homo- and heterodimer formation within the same family of proteins.[9–17] However, there is another level of control that can be mediated by specific protein–protein interactions occurring between transcription factors from different families. This is exemplified by the interaction of Jun, a leucine zipper containing protein, that can interact with the muscle-specific regulatory protein MyoD, a member of the helix–loop helix family,[18] with the DNA binding and ligand binding regions of the glucocorticoid receptor,[19–22] and with the Rel homology domain of NF-

[3] J. Sap, A. Munoz, K. Damm, Y. Goldberg, J. Ghyduel, A. Leutz, H. Beug, and B. Vennström, *Nature (London)* **324,** 635 (1986).

[4] C. Weinberger, C. C. Thompson, E. S. Ong, R. Lebo, D. J. Groul, and R. M. Evans, *Nature (London)* **324,** 641 (1986).

[5] P. Angel, M. Imagawa, R. Chiu, B. Stein, R. J. Imbra, H. J. Rahmsdorf, C. Jonat, P. Herrlich, and M. Karin, *Cell (Cambridge, Mass.)* **49,** 729 (1987).

[6] W. Lee, A. Haslinger, M. Karin, and R. Tjian, *Nature (London)* **325,** 368 (1986).

[7] T. Curran and B. R. Franza, *Cell (Cambridge, Mass.)* **55,** 395 (1988).

[8] L. J. Ransone and I. M. Verma, *Annu. Rev. Cell Biol.* **6,** 539 (1990).

[9] R. Chiu, W. M. Boyle, J. Meek, T. Smeal, T. Hunter, and M. Karin, *Cell (Cambridge, Mass.)* **54,** 541 (1988).

[10] P. Sassone-Corsi, W. M. Lamph, M. Kamps, and I. M. Verma, *Cell (Cambridge, Mass.)* **54,** 553 (1988).

[11] W. Landschultz, P. Johnson, and S. L. McKnight, *Science* **240,** 1759 (1988).

[12] C. R. Vinson, P. B. Sigler, and S. L. McKnight, *Science* **246,** 911 (1989).

[13] C. Murre, P. S. McCaw, and D. Baltimore, *Cell (Cambridge, Mass.)* **56,** 777 (1989).

[14] C. Murre, P. S. McCaw, H. Vaessin, M. Caudy, L. Y. Jan, Y. N. Jan, C. V. Cabrera, J. N. Buskin, S. D. Hauschka, A. B. Lassar, H. Weintraub, and D. Baltimore, *Cell (Cambridge, Mass.)* **58,** 537 (1989).

[15] W. J. Gehring, *Science* **236,** 1245 (1987).

[16] R. M. Evans, *Science* **240,** 889 (1988).

[17] A. Klug and D. Rhoades, *Trends Biochem. Sci.* **12,** 464 (1987).

[18] E. Bengal, L. Ransone, R. Scharfmann, V. J. Dwarki, S. J. Tapscott, H. Weintraub, and I. M. Verma, *Cell (Cambridge, Mass.)* **68,** 507 (1992).

[19] M. I. Diamond, J. N. Miner, S. K. Yoshinaga, and K. R. Yamamoto, *Science* **249,** 1266 (1990).

[20] C. Jonat, H. J. Rahmsdorf, K.-K. Park, A. C. B. Cato, S. Gebel, H. Ponta, and P. Herrlich, *Cell (Cambridge, Mass.)* **62,** 1189 (1990).

[21] R. Schüle, P. Rangarajan, S. Kliewer, L. J. Ransone, J. Bolado, N. Yang, I. M. Verma, and R. M. Evans, *Cell (Cambridge, Mass.)* **62,** 1217 (1990).

[22] H.-F. Yang-Yen, J.-C. Chambard, Y.-L. Sun, T. Smeal, T. J. Schmidt, J. Drouin, and M. Karin, *Cell (Cambridge, Mass.)* **62,** 1205 (1990).

κB p65.[23] Additionally, there have been examples of physical associations between transcriptional regulatory factors such as Jun, Fos, and Rel with members of the basal transcriptional machinery, specifically TFIIB and TATA binding protein.[24,25]

Association between proteins from different classes not only increases the repertoire and complexity of regulatory transcriptional pathways, but also allows for coordination and cross-coupling of these pathways. The challenge then is to develop methods to identify and characterize these specific protein–protein interactions. Several methods have been used to successfully detect the physical associations mentioned earlier. Among these are cross-linking/dimerization studies and coimmunoprecipitation, which are summarized here, as well as GST-fusion chromatography, which is described elsewhere in this volume.

Methods

Homodimerization

Glutaraldehyde Cross-Linking. Glutaraldehyde cross-linking has been used to detect successfully the ability of an *in vitro*-synthesized protein to form multimers. This noncleavable cross-linker is mixed with standardized aliquots of the protein of interest and is subjected to SDS–PAGE analysis. ^{35}S-labeled proteins are translated *in vitro* in the presence of L-[^{35}S]methionine using either rabbit reticulocyte lysate or wheat germ extract according to manufacturers instructions (Promega). Routinely, 5 μl of ^{35}S-labeled *in vitro*-translated protein is incubated with 2 μl 750 m*M* sodium phosphate buffer (pH 7.0), 5 μl 0.004% glutaraldehyde (Sigma, EM grade), and 8 μl H_2O at room temperature for 1 hr. Cross-linked products along with untreated controls are analyzed by SDS–PAGE.[26] In addition to the monomeric form of the protein, a second radioactive species, with an estimated molecular mass approximately twice that of the monomer, will be detected in cases where homodimers form. In some instances, multimeric complexes may be visible. The formation of higher order complexes can be determined through an estimation of the molecular mass of the protein of interest.

[23] B. Stein, A. S. Baldwin, Jr., D. W. Ballard, W. C. Greene, P. Angel, and P. Herrlich, *EMBO J.* **12,** 3879 (1993).

[24] L. J. Ransone, L. D. Kerr, M. J. Schmitt, P. W. Wamsley, and I. M. Verma, *Gene Express.* **1,** 37 (1993).

[25] L. D. Kerr, L. J. Ransone, P. Wamsley, M. Schmitt, T. Boyer, Q. Zhou, A. Berk, and I. M. Verma, *Nature* (*London*) **365,** 412 (1993).

[26] J.-I. Inoue, L. D. Kerr, L. J. Ransone, E. Bengal, T. Hunter, and I. M. Verma, *Proc. Natl. Acad. Sci. U.S.A.* **88,** 3715 (1991).

Coimmunoprecipitation

Coimmunoprecipitation of proteins from cellular extracts is perhaps the most convincing evidence that two proteins physically interact with each other. Unfortunately, this result is not always possible. The availability of antibodies and the normal physiological levels of a given protein affect the ability to carry out this particular type of experiment. The following section summarizes coimmunoprecipitation techniques which can be used to measure the ability of two proteins to interact physically.

Coimmunoprecipitation from Cellular Extracts. Coimmunoprecipitation from cellular extracts is generally performed with ^{35}S-labeled whole cell or fractionated extracts. Metabolic labeling requires that the cells are washed in phosphate-buffered saline and then resuspended in methionine-free medium containing dialyzed serum. The labeling time and induction, if any, are experimentally determined. For example, the detection of a Fos and Jun protein complex requires a period of serum starvation (0.5% serum) for up to 24 hr prior to labeling.[10] To induce c-Fos expression, cells are stimulated with 20% fetal bovine serum or 50 ng/ml phorbol 12-myristate 13-acetate (PMA).[10] At the time of stimulation, L-[^{35}S]methionine is added to a concentration of 1 mCi/ml in 2 ml. The cells are labeled for 90 min and cell extracts are prepared.

For nuclear extracts, cells are washed two times with Tris-buffered saline (TBS: 20 m*M* Tris–HCl, pH 7.5, 100 m*M* NaCl), scraped from the dish into ice-cold TBS, and collected by centrifugation. The cell pellet is resuspended in 0.6 ml of hypotonic buffer [25 m*M* Tris (pH 7.4), 1 m*M* $MgCl_2$, 5 m*M* KCl] and left on ice for 5 min. An equal volume of hypotonic buffer containing 1% Nonidet P-40 (NP-40) is added, and after an additional 5-min incubation on ice, the cell lysate is centrifuged at 5000*g* for 5 min. The supernatant containing the cytoplasmic fraction is removed, and the nuclei are resuspended in 100 μl of nuclear resuspension buffer [20 m*M* HEPES (pH 7.9), 0.4 *M* NaCl, 1 m*M* EDTA, 1 m*M* EGTA, 1 m*M* dithiothreitol (DTT), 1% aprotinin, 1 m*M* phenylmethylsulfonyl fluoride]. The nuclear extracts are incubated with shaking at 4° for 30 min, clarified by centrifugation at 12,000*g* for 10 min, and used for coimmunoprecipitations.

For whole cell extracts, cells are washed two times in TBS and lysed in RIPA buffer (20 m*M* Tris, pH 7.5, 150 m*M* sodium chloride, 2 m*M* EDTA, 1% (w/v) sodium deoxycholate, 1% (v/v) Triton X-100, 0.25% (w/v) SDS). Lysates are clarified by centrifugation at 10,000*g* for 10 min. Normal rabbit serum plus protein A–Sepharose are added to the lysate and the mixture is incubated at 4° for 1 hr. This reduces nonspecific binding of the labeled protein to the protein A–Sepharose. The immunoreaction is then cleared, and the reaction with specific antibody can be carried out.

Initial studies should be carried out titrating the amount of antibody to add to a given nuclear or whole cell extract. Generally a 1:100 dilution of antibody is a good starting point for most experiments. For single cycle coimmunoprecipitations, cell lysate and antibody are incubated overnight at 4° in a total volume of 1 ml of RIPA buffer. Precipitates are collected on 30 μl of a 1:1 mixture of protein A–Sepharose (Pharmacia), washed three to four times in RIPA, boiled in SDS–PAGE loading buffer, and run on a 10–12% SDS–PAGE. Gels are treated with EN^{3}Hance (New England Nuclear), dried, and autoradiographed at −80°.

For two-cycle immunoprecipitations, i.e., detecting Jun proteins from a Fos antibody coimmunoprecipitation,[10] an aliquot equivalent to one-tenth of the total reaction is removed following the anti-Fos immunoprecipitation, to monitor the input proteins of the second immunoprecipitation. The immune complexes are dissociated by heating the immunoprecipitate at 100° in RIPA buffer for 5 min. Protein A–Sepharose and absorbed antibodies are removed by centrifugation. The eluted proteins are then mixed with the antibody of choice, in this case Jun antibody, and the immunoprecipitation is carried out as detailed earlier. The stringency of the precipitation conditions can be diminished by reducing or eliminating the SDS in the RIPA buffer and reducing the deoxycholate concentration. Alternatively, the RIPA buffer can be prepared using a nonionic detergent such as NP-40 at concentrations from 0.2 to 0.5%.[18] In all of these experiments it is important to include a preimmune control to test the specificity of the antibody.

A variation of the coimmunoprecipitation experiments outlined earlier was used to show an *in vivo* association between c-Rel and TATA binding protein (TBP).[25] In this type of experiment, cell extracts are prepared from unlabeled cells and the specific protein–protein interactions are collected first by immunoprecipitation and then detected by Western blot analysis. For example, to detect the c-Rel/TBP association, HeLa cells are treated with 100 ng/ml PMA and 5 ng/ml phytohemagglutinin for 4 hr, or left untreated. Nuclear extracts are prepared as described earlier, and 150 μg of each sample is immunoprecipitated with either preimmune or anti-TBP serum. The immunoprecipitates are treated and washed as indicated earlier. At this point, bound proteins are separated from immunoglobulins by boiling in the presence of 5 m*M* iodoacetamide (Sigma). Proteins are separated by SDS–polyacrylamide electrophoresis and transferred to PVDF Immobilon membrane. A standard Western blot is then carried out by first blocking the membrane with TBS containing 5% dried milk for at least 1 hr, followed by an incubation with c-Rel specific antibody (a 1:250 dilution is a good starting point). After an overnight incubation at 4°, the blot is washed four times with TBS plus 0.05% Tween 20. The immunospecific complexes can then be stained using the Proto-Blot kit (Promega) or a hard copy (autoradiograph) can be obtained using the ECL system (Amersham). The ability to positively identify individual proteins present in a coimmunoprecipitation makes this an attractive assay for the detection of *in vivo* protein–protein associations.

Coimmunoprecipitation Using in Vitro-Translated Proteins. In instances where two proteins cannot be coimmunoprecipitated from cell lysates, one can carry out an *in vitro* protein binding assay. ^{35}S-labeled proteins are cotranslated *in vitro* in the presence of L-[^{35}S]methionine using either rabbit reticulocyte lysate or wheat germ extract according to manufacturers instructions (Promega). Alternatively, equal volumes of separately synthesized ^{35}S-labeled proteins can be incubated at 30° in a buffer containing 20 m*M* HEPES–KOH (pH 7.9), 50 m*M* KCl, 2.5 m*M* $MgCl_2$, 1 m*M* DTT, and 10% glycerol to allow protein heterodimeric complex formation to occur. Depending on the strength of the protein–protein interaction, the immunoprecipitation can be carried out directly at this point[27] or a chemical cross-linker may be added. For interactions such as MyoD-Jun,[18] NF-κB(p65)-Fos/Jun,[23] and Fos-TBP,[24] the chemical cross-linker dithiobis(succinimidyl propionate) (DSP) (Pierce) was used to facilitate complex stability. DSP is added at a final concentration of 2 m*M*, in a volume of 50 μl for 15–30 min at room temperature. The cross-linking reaction is quenched by the addition of ethanolamine to a final concentration of 100 m*M*. The volume of the reaction (cross-linked and untreated) is adjusted to 200 μl with RIPA buffer, and precleared for 1 hr using 30 μl of a 1:1 mixture of protein A–Sepharose as described earlier. At this point the lysate is cleared by centrifugation, the supernatant is moved to a new tube, and specific antibody is added for a minimum 2-hr incubation at 4°. Depending on the antibody, this incubation can go overnight. Protein A–Sepharose is added, and the immunocomplexes are collected as described earlier.

There are several variations of the *in vitro* binding assay. As with the immunoprecipitations from cell lysates, these reactions can be performed under a variety of detergent conditions. Additionally, if the two proteins of interest have approximately the same molecular mass, as with Jun and MyoD,[18] one of the proteins in the reaction can be unlabeled as long as the antibody is directed against this member of the potential complex. Alternatively, bacterially expressed proteins can be used as a source of unlabeled protein. These have the advantage that they can be produced and purified in large quantities. Bacterial expression vectors can be modified to attach an epitope tag such as the influenza virus epitope to the amino-

[27] L. J. Ransone, J. Visvader, P. Sassone-Corsi, and I. M. Verma, *Genes Dev.* **3,** 770 (1989).

terminal portion of the protein. This alleviates the need for protein-specific antibodies since the anti-flu epitope monoclonal antibody 12CA5 is commercially available.

Acknowledgments

I am very grateful to Drs. Timothy C. Fong and Lawrence D. Kerr for their critical reading of the manuscript. Many of the methods described here were worked out in the laboratory of Dr. Inder M. Verma at the Salk Institute, San Diego, California, and were supported by funds from the National Institutes of Health, American Cancer Society, and the Leukemia Society of America.

[33] Epitope Tagging

By YUZURU SHIIO, MICHIYASU ITOH, and JUN-ICHIRO INOUE

Epitope tagging is an experimental strategy in which a specific amino acid sequence that is recognized as an epitope by a particular antibody is incorporated into a recombinantly expressed protein. Advances in molecular biology have made it easier to prepare recombinant protein to be used as an antigen as long as the cDNA for the protein has been cloned. In many cases, proteins can be expressed as a fusion protein with glutathione *S*-transferase (GST: see GST section) in *Escherichia coli.* The protein can be purified easily using glutathione agarose and can be injected into animals to raise antibody. However, in some cases, it is difficult to prepare enough protein due to a low level of expression or to obtain serum with a sufficient amount of antibody. To overcome these difficulties, the epitope tagging technique is of great use. In most cases, the epitope is incorporated into either N or C terminus of the protein. However, the epitope can be incorporated into the center of the molecule as long as the function of the protein is not altered and the epitope is not sterically masked by another region of the molecule. According to our experience, in some cases the position of the tag sequences affects the efficiency of immunoprecipitation of the tagged protein by the antiepitope antibody. The advantages of epitope tagging in protein–protein association experiments are as follows: (1) if a specific antibody that recognizes X protein (X is a protein of your interest) is not available, epitope-tagged recombinant X protein can be expressed and recognized by a specific antibody raised against the epitope. Using this system, the proteins that are associated with X protein can be analyzed *in vitro* or *in vivo.* Two different epitopes can be used in the same set of

TABLE I
COMBINATIONS OF EPITOPE AND ANTIBODY USED FOR EPITOPE TAGGING

Origin	Epitope	Antibody[b]	Ref.
Influenza virus Hemagglutinin protein	YPYDVPDYA	12CA5	1, 2
T7 gene 10 protein[a]	MASMTGGQQMG	α-T7 gene 10	3
FLAG	DYKD	α-FLAG M1	4, 5
Myc	QVFFRNKLLF	9E10	2
Bovine papillomavirus L1 protein	DTYRYI	AU1	6, 7

[a] Expression vectors for gene 10-tagged protein (Novagen) and FLAG-tagged protein (International Biotechnology INC.) are commercially available.
[b] Each antibody is commercially available as: 12CA5 and AU1 (Babco), α-T7 gene 10 (Novagen), α-FLAG (International Biotechnology INC.), and 9E10 (Danova).

experiments when the two proteins have different epitope tags. The association between these proteins could be demonstrated by using antibodies directed against each epitope. (2) Even if the specific antibody against X protein is available, the epitope tagging system has remarkable advantages. When the protein–protein interactions are analyzed in cell culture by cDNA transfection, the X protein expressed from the transfected plasmid can be distinguished from the endogenous X protein. This technique is especially useful when the behavior of the mutant X proteins is analyzed.

Several combinations of the epitope and the corresponding antibody have been successfully used (Table I).[1–7] Epitope tagging could be applied to uses other than protein–protein interaction experiments, such as immunofluorescence staining, electrophoretic mobility shift assays, and cDNA cloning from expression libraries.

Experimental Procedure

Tagging

For expressing tagged protein, oligonucleotides encoding the epitope sequence are linked to the cDNA of the protein by recombinant DNA

[1] J. Field, J. I. Nikawa, D. Broek, B. MacDonald, L. Rodgers, I. A. Wilson, R. A. Lerner, and M. Wigler, *Mol. Cell. Biol.* **8,** 2159 (1988).
[2] S. E. Egan, B. W. Giddings, M. W. Brooks, L. Buday, A. M. Sizeland, and R. A. Weinberg, *Nature (London)* **363,** 45 (1993).
[3] M. Naumann, F. G. Wulczyn, and C. Scheidereit, *EMBO J.* **12,** 213 (1993).
[4] B. Stein, A. S. Ballard, W. C. Greene, P. Angel, and P. Herrlich, *EMBO J.* **12,** 3879 (1993).
[5] M. A. Blanar and W. J. Rutter, *Science* **256,** 1014 (1992).
[6] D. J. Goldstein, T. Andresson, J. J. Sparkowski, and R. Schlegel, *EMBO J.* **11,** 4851 (1992).
[7] D. J. Goldstein, R. Toyama, R. Dhar, and R. Schlegel, *Virology* **190,** 889 (1992).

technology. Expression vectors for FLAG-tagged protein and gene 10-tagged protein are available from International Biotechnologies, Inc. and Novagen, respectively.

Hemagglutinin (HA) Epitope Tagging. To express the protein N-terminally tagged with the HA epitope, we have constructed a vector, pBSHA, shown in Fig. 1 (M. Itoh, unpublished data, 1989). A restriction enzyme site (GTCGAC), which can be digested by *Sal*I, *Acc*I, and *Hin*cII, is present just downstream of the epitope sequence. By using one of the three enzymes followed by Klenow treatment, any translation frame could be chosen for establishing an in frame junction. *In vitro*-translated tagged proteins analyzed so far were efficiently immunoprecipitated by the 12CA5 monoclonal antibody, which recognizes this epitope.

Tagging with Other Epitopes. Synthetic oligonucleotides with an appropriate restriction enzyme site at both ends are incorporated into the cDNA by standard recombinant DNA techniques. Recognition of the tagged protein by the antibody must be checked using *in vitro*-translated protein or the protein expressed by transient transfection into mammalian cells.

Protein–Protein Association in Vitro

A typical experiment is shown in Fig. 2. This experiment demonstrates the association of the Rel transcription factor with its inhibitor pp40(I-κBα) and is performed as follows.

1. The HA epitope was incorporated into the N-terminal end of pp40 by inserting pp40 cDNA into pBSHA (Fig. 1).

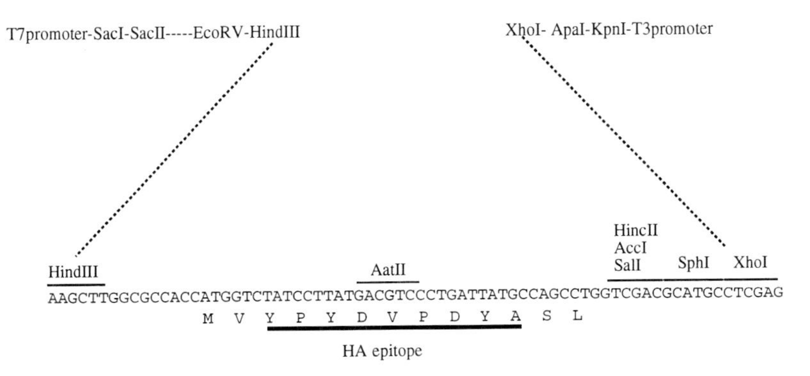

FIG. 1. The structure of pBSHA. Oligonucleotides encoding the HA epitope were inserted between the *Hin*dIII and *Xho*I sites of pBluescript II KS(−) (Stratagene). Only the region between T7 and T3 promoter is shown.

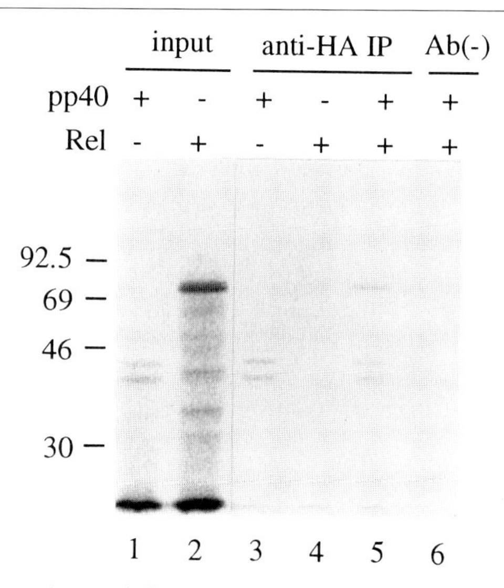

FIG. 2. Protein–protein association experiment using HA epitope tagging. HApp40 and avian Rel proteins were *in vitro* translated in the presence of [^{35}S]methionine. Five microliters of each *in vitro*-translated protein was analyzed by 10% SDS–PAGE (lanes 1 and 2). Five microliters of HApp40 (lane 3), Rel (lane 4), or both (lanes 5 and 6) was incubated at room temperature. The proteins were immunoprecipitated with 12CA5 antibody (lanes 3–5). The control experiment was performed without the addition of antibody (lane 6). After washing, precipitates were analyzed on 10% SDS–polyacrylamide gels.

2. ^{35}S-labeled HApp40 and avian Rel proteins were translated *in vitro* in the presence of L-[^{35}S]methionine by using a wheat germ extract (Promega) and *in vitro*-transcribed RNA from each template (Fig. 2, lanes 1 and 2).

3. Set up four samples as follows:
 sample A: 5 μl of HApp40
 sample B: 5 μl of Rel
 sample C: 5 μl of HApp40 and Rel
 sample D: 5 μl of HApp40 and Rel
Incubate at room temperature for 30 min.

4. Add 1 μl of anti-HA monoclonal antibody (12CA5) to samples A, B, and C and add 1 μl of H_2O to sample D. Incubate on ice for 1 hr.

5. Add 400 μl of binding buffer [20 m*M* Tris–HCl (pH 8.0), 150 m*M* NaCl, 2 m*M* EDTA, and 1% (w/v) NP-40] and 10 μl of protein A–Sepharose [50% (v/v) suspension in binding buffer]. Rotate tubes at 4° for 30 min.

6. Spin down and wash Sepharose beads three times with 800 μl of binding buffer.

7. Final pellet was suspended in 10 μl of sample buffer and analyzed by 10% SDS–PAGE.

HApp40 was immunoprecipitated by the anti-HA monoclonal antibody (Fig. 2, lane 3), whereas the Rel protein was precipitated by the same antibody only in the presence of HApp40 (Fig. 2, lanes 4 and 5). These data clearly demonstrate that pp40 associates Rel *in vitro.*

Protein–Protein Association in Vivo

The same procedure can be used for the protein expressed by DNA transfection into mammalian cells. The monkey kidney cell line Cos-7 and an expression vector driven by SV40 or SRα[8] promoter are usually used since these vectors replicate after transfection because of the presence of T antigen in Cos-7 cells and therefore provide high level protein expression. After Cos-7 cells were transfected with the expression vector for HApp40 and Rel, protein–protein association could be analyzed by the following two methods.

1. Cells are metabolically labeled with [^{35}S]methionine and then lysed with an appropriate buffer. We usually use TNE buffer [10 m*M* Tris–HCl (pH 7.8), 1% w/v nonidet P-40 (NP-40), 150 m*M* NaCl, 1 m*M* EDTA, 10 μg/ml aprotinin]. However, the most suitable composition of the buffer should be determined in each case. The type and concentration of detergent and salt are especially critical for maintaining protein–protein association during immunoprecipitation. For example, RIPA buffer [10 m*M* Tris–HCl (pH 7.4), 0.15 *M* NaCl, 1% NP-40, 0.1% SDS, 0.1% sodium deoxycholate, 1 m*M* EDTA, 10 μg/ml aprotinin] or digitonin buffer[9] (10 m*M* triethanolamine, 150 m*M* NaCl, 1 m*M* EDTA, 10 m*M* iodoacetamide, 1% digitonin, 10 μg/ml aprotinin, pH 7.8) could also be used. The lysate is divided into two tubes and an appropriate amount of the anti-HA antibody is added to one of the tubes (tube A). The other tube (tube B) is supplemented with SDS [final 0.5% (w/v)] and boiled for 5 min to dissociate proteins bound noncovalently. Four volumes of TNE buffer are added to adjust the concentration of SDS down to 0.1%. The same amount of anti-HA antibody used for tube A is added to tube B. Immunoprecipitation is performed by adding protein G–Sepharose, and the final washed immunoprecipitate is analyzed by SDS–PAGE. The pp40-associated Rel protein was observed only in the immunoprecipitate from tube A.

[8] Y. Takebe, M. Seiki, J. Fujisawa, P. Hoy, K. Yokota, K. Arai, M. Yoshida, and N. Arai, *Mol. Cell. Biol.* **8,** 466 (1988).

[9] Y. Yamanashi, T. Kakiuchi, J. Mizuguchi, T. Yamamoto, and K. Toyoshima, *Science* **251,** 192 (1990).

2. Preparation of the lysate and immunoprecipitation are performed as described in step 1 without metabolic labeling. After SDS–PAGE, Western blotting is performed using a PVDF membrane (Millipore) and an anti-Rel antibody. The same result as described in step 1 was obtained.

As described in the introduction, when pp40 mutants are analyzed, the epitope tagging system has a great advantage. Endogenous pp40 protein can be ignored since the anti-HA antibody recognizes only the exogenous mutant pp40 proteins. Thus, one could ask whether different mutant pp40 proteins can associate with Rel protein and which domain of pp40 is required for the association with Rel protein.

Immunofluorescence Staining

The HA epitope and 12CA5 antibody can be used for immunofluorescence staining. Low background staining by the 12CA5 antibody usually leads to reliable results. Subcellular localization of mutant proteins can be easily determined since the antibody does not recognize endogenous proteins. The protocol for staining is:

1. Seed cells onto a coverslip and transfect the appropriate cDNA expression plasmid.
2. After 2 days, wash cells with phosphate-buffered saline (PBS). Add 3% (w/v) paraformaldehyde in PBS and incubate at room temperature for 10 min to fix cells.
3. Wash cells twice with PBS.
4. Add 1% (v/v) Triton X-100 in PBS and incubate at room temperature for 10 min to permeabilize the cell membrane.
5. Wash cells twice with 0.2% (v/v) Tween 20 in PBS.
6. Add 100-fold dilution of 12CA5 antibody and incubate at room temperature for 30 min.
7. Wash cells twice with 0.2% (v/v) Tween 20 in PBS.
8. Add 200-fold dilution of fluorescein isothiocyanate-conjugated anti-mouse IgG and incubate at room temperature for 30 min.
9. Wash cells twice with 0.2% (v/v) Tween 20 in PBS.
10. Add a drop of mounting solution on cells.
11. Observe the sample under the fluorescent microscope.

Since the utility of other antibodies for immunofluorescence staining has not been checked, pilot experiments should be set up to check the antibody.

As described in this section, the epitope tagging system has been proven to be of great use. Since the epitope is identified and the monoclonal antibody has been established, epitope tagging could be applied to novel experimental systems other than those described here. For antibodies whose epitope is identified, one could set up new combinations for epitope tagging.

[34] Biochemical Analysis of SH2 Domain-Mediated Protein Interactions

By GERALD GISH, LOUISE LAROSE, RANDY SHEN, and TONY PAWSON

Introduction

Src homology 2 (SH2) domains are protein modules of approximately 100 amino acids that are found in many cytoplasmic and nuclear polypeptides involved in signal transduction.[1-3] Examples of SH2-containing molecules include proteins involved in phospholipid metabolism (phospholipase C-γ1; phosphatidylinositol 3′-kinase), tyrosine phosphorylation (cytoplasmic tyrosine kinases and cytoplasmic phosphotyrosine phosphatases), regulation of small guanine nucleotide-binding proteins (Ras GTPase-activating protein, Grb2, Shc, *n*-chimerin), control of the cytoskeleton (tensin), and transcription (STAT proteins). These signaling proteins, each of which contains one or two SH2 domains, apparently mediate the activation of biochemical pathways in response to extracellular signals such as growth factors, cytokines, and antigen.

Functionally, SH2 domains bind to phosphotyrosine sites on activated growth factor receptors and on other transmembrane and cytoplasmic signaling proteins.[4-13] In addition to a conserved phosphotyrosine binding

[1] T. Pawson and J. Schlessinger, *Curr. Biol.* **3,** 434 (1993).
[2] T. Pawson and G. D. Gish, *Cell (Cambridge, Mass.)* **72,** 359 (1992).
[3] C. A. Koch, D. Anderson, M. F. Moran, C. Ellis, and T. Pawson, *Science* **252,** 668 (1991).
[4] B. Margolis, N. Li, A. Koch, M. Mohammadi, D. Hurwitz, A. Ullrich, A. Zilberstein, T. Pawson, and J. Schlessinger, *EMBO J.* **9,** 4375 (1990).
[5] A. Kazlauskas, A. Kahishian, J. A. Cooper, and M. Valius, *Mol. Cell. Biol.* **12,** 2534 (1990).
[6] W. J. Fantl, J. A. Escobedo, G. A. Martin, C. W. Turck, M. del Rosario, F. McCormick, and L. T. Williams, *Cell (Cambridge, Mass.)* **69,** 413 (1992).
[7] L. Rönnstrand, S. Mori, A.-K. Arridsson, A. Eriksson, C. Wernstedt, U. Hellman, L. Claesson-Welsh, and C.-H. Heldin, *EMBO J.* **11,** 3911 (1992).
[8] M. Valius, C. Bazenet, and A. Kazlauskas, *Mol. Cell. Biol.* **13,** 133 (1993).
[9] A. Kazlauskas, G.-S. Feng, T. Pawson, and M. Valius, *Proc. Natl. Acad. Sci. U.S.A.* **90,** 6939 (1993).
[10] S. M. Mori, L. Rönnstrand, K. Yokote, A. Engström, S. A. Courtneidge, L. Claesson-Welsh, and C.-H. Heldin, *EMBO J.* **12,** 2257 (1993).
[11] J. McGlade, A. Cheng, G. Pelicci, P. G. Pelicci, and T. Pawson, *Proc. Natl. Acad. Sci. U.S.A.* **89,** 8869 (1992).
[12] A. M. Pendergast, L. A. Quilliam, L. D. Cripe, C. H. Bassing, Z. Dai, N. Li, A. Batzer, K. M. Rabun, C. J. Der, J. Schlessinger, and M. L. Gishizky, *Cell (Cambridge, Mass.)* **75,** 175 (1993).

pocket, SH2 domains have variable binding sites for up to six residues C-terminal to the phosphotyrosine (the +1 to +6 positions), which therefore dictate the specificity of a SH2 domain for a particular phosphotyrosine-containing site.[14] Known SH2 domains can be divided into two main groups.[15] One class typified by the Src kinase SH2 domain preferentially select motifs with phosphotyrosine followed by a hydrophilic residue at the +1 position, a hydrophilic residue at +2, and a hydrophobic residue at +3 from a degenerate phosphopeptide library.[15] The binding surface of the Src SH2 domain has two pockets for the phosphotyrosine and the +3 Ile side chain.[16] The second class, including the SH2 domains of phospholipase C-γ1, phosphatidylinositol 3′-kinase and SH2-containing phosphotyrosine phosphatases, binds phosphotyrosine followed primarily by hydrophobic residues. This class of SH2 domain apparently contains a long hydrophobic groove that contacts up to six residues C-terminal to the bound phosphotyrosine.[14,17] These biochemical and structural data have started to explain how a SH2 domain binds with high affinity only to phosphotyrosine in a specific amino acid sequence context and to reveal the molecular basis for differences in SH2 domain binding specificities.

The binding of a growth factor to a receptor tyrosine kinase can initiate an intracellular response by inducing rapid dimerization and the subsequent intermolecular autophosphorylation of specific tyrosine residues within the cytoplasmic region of the receptor.[18,19] Cytoplasmic proteins containing SH2 domains with appropriate binding properties are thereby localized to the receptor at the membrane. Activation of the cytoplasmic signaling protein can potentially be achieved either as a consequence of its relocalization to the plasma membrane or by allosteric activation mediated directly by phosphotyrosine binding to its SH2 domain(s).[3]

The formation of signaling complexes by phosphotyrosine–SH2 interac-

[13] L. Puil, J. Liu, G. Gish, G. Mbamalu, D. Bowtell, P. G. Pelicci, R. Arlinghaus, and T. Pawson, *EMBO J.* **13,** 764 (1994).

[14] S. M. Pascal, A. U. Singer, G. Gish, T. Yamazaki, S. E. Shoelson, T. Pawson, L. E. Kay, and J. D. Forman-Kay, *Cell* (*Cambridge, Mass.*) **77,** 461 (1994).

[15] Z. Songyang, S. E. Shoelson, M. Chaudhuri, G. Gish, T. Pawson, W. G. Haser, F. King, T. Roberts, S. Ratnofsky, R. J. Lechleider, B. G. Neel, R. B. Birge, J. E. Fajardo, M. M. Chou, H. Hanafusa, B. Schaffhausen, and L. C. Cantley, *Cell* (*Cambridge, Mass.*) **72,** 767 (1993).

[16] G. Waksman, S. E. Shoelson, N. Pant, D. Cowburn, and J. Kuriyan, *Cell* (*Cambridge, Mass.*) **72,** 779 (1993).

[17] C.-H. Lee, D. Kominos, S. Jacques, B. Margolis, J. Schlessinger, S. E. Shoelson, and J. Kuriyan, *Structure* **2,** 423 (1994).

[18] A. Ullrich and J. Schlessinger, *Cell* (*Cambridge, Mass.*) **61,** 203 (1990).

[19] L. C. Cantley, K. R. Auger, G. Carpenter, B. Duckworth, A. Graziani, R. Kapeller, and S. Soltoff, *Cell* (*Cambridge, Mass.*) **64,** 281 (1991).

FIG. 1. Structure of PLC-γ1 showing the locations of SH2 and SH3 domains and the catalytic regions.

tions is a widely used feature of many signal transduction pathways, of which transmembrane receptor tyrosine kinases provide one example. Interactions involving SH2 domains appear important for activation of the Ras pathway by growth factor receptors,[20] the responses of lymphoid cells to antigen,[21] the activation of STAT transcription factors by cytokines,[22] and malignant transformation by cytoplasmic tyrosine kinases such as v-Fps[23] and Bcr-Abl.[12,13]

Here we have used the specific example provided by phospholipase C-γ1 (PLC-γ1) to illustrate techniques for the analysis of SH2 domain function. PLC-γ1 catalyzes the hydrolysis of phosphatidylinositol 4,5-bisphosphate to inositol 1,4,5-trisphosphate and diacylglycerol, two second messengers that regulate intracellular calcium release and protein kinase C activity, respectively.[24] Activated growth factor receptors, including those for epidermal growth factor (EGF),[25] platelet-derived growth factor (PDGF)[7,8] and fibroblast growth factor (FGF),[26] have specific binding sites for the SH2 domains of PLC-γ1. Binding of PLC-γ1 to a phosphotyrosine site in the tail of the β-PDGFR (Tyr-1021) is required for phosphatidylinositol turnover in response to PDGF.[27] Receptor-associated PLC-γ1 becomes phosphorylated on tyrosine, and this modification is also apparently important for its activation.[28,29] The structure of PLC-γ1 is illustrated in Fig. 1.

Three techniques for probing the interactions between SH2 domains and tyrosine-phosphorylated ligands are illustrated. First, the use of SH2 domain fusion proteins to identify proteins, including receptor tyrosine

[20] J. P. Olivier, T. Raabe, M. Henkemeyer, B. Dickson, G. Mbamalu, B. Margolis, J. Schlessinger, E. Hafen, and T. Pawson, *Cell (Cambridge, Mass.)* **73,** 179 (1993).

[21] A. Weiss and D. R. Littman, *Cell (Cambridge, Mass.)* **76,** 263 (1994).

[22] K. Shuai, C. M. Horvath, L. H. Tsai Huang, S. A. Qureshi, D. Cowburn, and J. E. Darnell, Jr., *Cell (Cambridge, Mass.)* **76,** 821 (1994).

[23] I. Sadowski, J. C. Stone, and T. Pawson, *Mol. Cell. Biol.* **6,** 4396 (1984).

[24] S. G. Rhee, P.-G. Suh, S.-H. Ryu, and S. Y. Lee, *Science* **244,** 546 (1989).

[25] Q. C. Vega, C. Cochet, O. Filhol, C.-P. Chang, S. G. Rhee, and G. N. Gill, *Mol. Cell. Biol.* **12,** 128 (1992).

[26] M. Mohammadi, A. M. Honegger, D. Rotin, R. Fischer, F. Bellot, W. Li, C. A. Dionne, M. Jaye, M. Rubinstein, and J. Schlessinger, *Mol. Cell. Biol.* **11,** 5068 (1991).

[27] M. Valius and A. Kazlauskas, *Cell (Cambridge, Mass.)* **73,** 321 (1993).

[28] S. G. Rhee, *Trends Biochem. Sci.* **16,** 297 (1991).

[29] M. I. Wahl, G. A. Jones, S. Nishibe, S. G. Rhee, and G. Carpenter, *J. Biol. Chem.* **267,** 10447 (1992).

kinases, which associate with a particular cytoplasmic SH2-containing protein is discussed. By this procedure we can define some of the components involved in a specific signal transduction pathway. For example, PLC-γ1 associates with the PDGF receptor on stimulation with growth factor.

Second, in order to localize regions on a receptor tyrosine kinase or other phosphoprotein which interact with a particular SH2 domain, we describe methods for the preparation of tyrosine-phosphorylated polypeptides and their use in the analysis of SH2 domain specificity. Using this approach we were able to show that PLC-γ1 binds selectively to the extreme C-terminal region of the PDGF β-receptor at the tyrosine phosphorylation site 1021.[30]

Finally, procedures for more quantitative analysis of these interactions using surface plasmon resonance technology (BiaCore, Pharmacia Biosensor) are described. As PLC-γ1 contains two SH2 domains, this technique allowed us to probe in detail the affinity of each individual domain for a phosphotyrosine-containing peptide composed of the PDGF β-receptor sequence surrounding the amino acid tyrosine 1021.

Strategy for Use of SH2 Fusion Proteins

DNA encoding the SH2 domain of a cytoplasmic protein can be isolated by the polymerase chain reaction (PCR) and subcloned into fusion protein expression vectors such as the pATH bacterial TrpE fusion[31] or pGEX glutathione *S*-transferase (GST) fusion[32] protein systems. The expressed protein is immobilized on beads, then incubated with lysate from various growth factor-stimulated cell lines. The immobilized proteins are then washed, resolved by SDS–PAGE, and transferred to a nitrocellulose membrane. The blots are then probed with antiphosphotyrosine antibodies or antibodies specific for receptors or other phosphotyrosine-containing proteins.

To demonstrate a direct interaction between a SH2 domain and a phosphotyrosine-containing protein, a "Far Western" analysis can be carried out.[33] When the association with a specific phosphotyrosine-containing protein is to be tested, the protein of interest is first immunoprecipitated. Alternatively, to probe for direct binding to unknown proteins, a whole cell lysate is prepared. The material is then resolved by SDS–PAGE and

[30] L. Larose, G. Gish, S. Shoelson, and T. Pawson, *Oncogene* **8,** 2493 (1993).

[31] T. J. Koemer, J. E. Hill, A. M. Myers, and A. Tzagoloff, this series, vol. 194, p. 477.

[32] D. B. Smith and K. S. Johnson, *Gene* **67,** 31 (1991).

[33] C. J. McGlade, C. Ellis, M. Reedijk, D. Anderson, G. Mbamalu, A. D. Reith, G. Panayotou, P. End, A. Bernstein, A. Kazlauska, M. D. Waterfield, and T. Pawson, *Mol. Cell. Biol.* **12,** 991 (1992).

is transferred to a nitrocellulose membrane. SH2 fusion proteins used to probe the blot bind to proteins with appropriate phosphotyrosine-containing sites. These interactions are detected using an antibody specific for the TrpE or GST fusion protein or by exploiting a biotinylated fusion protein.[34] The method relies on the observation that the phosphotyrosine-containing regions of the proteins which act as binding sites for SH2 domains are small and have little secondary structural requirements for activity.

Construction and Expression of SH2 Domain Fusion Proteins

Prior to the advances in the knowledge of SH2 domain structure obtained by X-ray crystallography[16,35,36] and nuclear magnetic resonance[14,17,37,38] studies, guidelines for the construction of the optimal SH2 domain fusion protein were based on the pattern of invariant or highly conserved residues deduced by sequence alignment. The structural studies confirmed that the SH2 domain forms an independently folded modular unit and that truncation at either the N or C termini could have profound effects on domain stability and therefore also on function. Figure 2 shows an alignment of selected SH2 domain amino acid sequences, along with the predicted positions of secondary structure. The absence of one or two residues at either end of a SH2 domain can abrogate its function. For construction of a functional SH2 domain fusion protein, it is therefore prudent to include some residues outside of the N and C termini shown in the alignment to avoid any structural instability.

DNA encoding the SH2 domain from the relevant cDNA is isolated by PCR. The primers used in the reaction, which are typically 27 nucleotides in length, are composed of 12 to 15 bases from the cDNA sequence for selective binding to template and the restriction enzyme recognition site sequence required for subcloning into the expression vector. By including an extra 4 to 6 nucleotides at the 5′ end of the primer, the resulting PCR product is less susceptible to fraying at the ends and efficient restriction enzyme digestion is assured. As stop codons are supplied in both the pATH and pGEX bacterial expression vectors, it is not necessary to include these

[34] B. J. Mayer, P. K. Jackson, and D. Baltimore, *Proc. Natl. Acad. Sci. U.S.A.* **88,** 627 (1991).

[35] G. Waksman, D. Kominos, S. C. Robertson, N. Pant, D. Baltimore, R. B. Birge, D. Cowburn, H. Hanafusa, B. J. Mayer, M. Overduin, M. D. Resh, C. B. Rios, L. Silverman, and J. Kuriyan, *Nature (London)* **358,** 646 (1992).

[36] M. J. Eck, S. E. Shoelson, and S. C. Harrison, *Nature (London)* **362,** 87 (1993).

[37] G. W. Booker, A. L. Breezee, A. K. Downing, G. Panayotou, I. Gout, M. D. Waterfield, and I. D. Campbell, *Nature (London)* **358,** 684 (1992).

[38] M. Overduin, C. B. Rios, B. J. Mayer, D. Baltimore, and D. Cowburn, *Cell (Cambridge, Mass.)* **70,** 697 (1992).

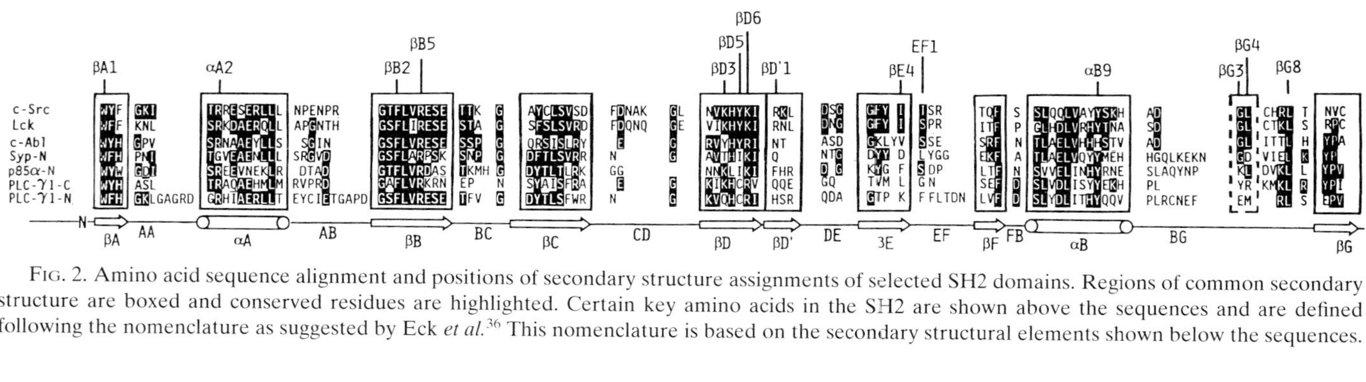

FIG. 2. Amino acid sequence alignment and positions of secondary structure assignments of selected SH2 domains. Regions of common secondary structure are boxed and conserved residues are highlighted. Certain key amino acids in the SH2 are shown above the sequences and are defined following the nomenclature as suggested by Eck *et al.*[36] This nomenclature is based on the secondary structural elements shown below the sequences.

in the primers. Conventional techniques are used for subcloning these fragments into the bacterial expression vector.

For our investigations of PLC-γ1, we chose to study each individual SH2 domain as a fusion protein along with a construct that contains both domains. Using bovine phospholipase C-γ1 cDNA,[39] the more N-terminal SH2 domain (N-SH2) was composed of amino acids from residue 545 to 659, the C-terminal SH2 (C-SH2) from residues 663 to 759, and both domains (N+C SH2s) from residues 545 to 759. The PCR products were cloned into pGEX-KT bacterial vector for expression of GST fusion proteins.

Expression and Isolation of GST–SH2 Domain Fusion Proteins

A wide variety of *Escherichia coli* strains, including DH5α, JM101, XL1-blue, and RR1, can be used for the efficient expression of GST fusion protein from the pGEX expression plasmid. The bacteria are grown by diluting, in a range from 1:10 to 1:20, an overnight culture into 50 ml of fresh LB medium containing 100 μg/ml ampicillin. Subsequent bacterial growth is monitored by optical density (OD) at 600 nm and once a value of 0.6 is achieved, protein expression is induced through the addition of isopropyl-β-D-thiogalactopyranoside (IPTG) to a final concentration from 0.2 to 1 m*M*. The temperature at which the protein induction is performed can be critical for obtaining soluble protein. Although most GST–SH2 domain fusion proteins are soluble when cell growth is continued at 37° following induction, others require that a lower temperature be used. Using freshly transformed bacteria can also aid in obtaining soluble protein. The length of the induction period required for efficient protein expression is temperature dependent and should be increased from 3 to 6 hr at 37° to 6 to 18 hr when performed at room temperature.

The cells are harvested by centrifugation (20 min at 5000 rpm, Sorvall GS-3 rotor) and are washed with 40 m*M* sodium phosphate buffer, pH 7.5, containing 150 m*M* NaCl. At this stage the cells can be stored at −20° or the isolation procedure can be started by resuspending the bacteria in 10 ml of PLC lysis buffer (50 m*M* HEPES, pH 7.5, 150 m*M* NaCl, 10% (v/v) glycerol, 1% (v/v) Triton X-100, 1 m*M* EGTA, 1.5 m*M* $MgCl_2$, 100 m*M* NaF, 10 m*M* sodium pyrophosphate, 1 m*M* sodium orthovanadate, 1 m*M* phenylmethylsulfonyl fluoride (PMSF), 10 μg of aprotinin per ml, 10 μg of leupeptin per ml). Sonication of the resuspended bacteria incubated on ice, using two to three consecutive 30-sec bursts, is used to break open the cells and the lysate is isolated by centrifugation at 15,000*g* for 10 min at

[39] M. L. Stahl, C. R. Ferenz, K. L. Kelleher, R. W. Kriz, and J. L. Knopf, *Nature (London)* **332,** 269 (1988).

4°. The GST fusion proteins are recovered from the supernatant solution through the addition of 200 μl of a 50% (v/v) suspension of glutathione–Sepharose resin (Sigma) equilibrated in PLC lysis buffer, 10–20 min of tumbling at 4°, collection of the resin by centrifugation, and washing with HNTG buffer (20 m*M* HEPES, pH 7.0, 150 m*M* NaCl, 0.1% Triton X-100, 10% glycerol, 1 m*M* sodium orthovanadate, 1 m*M* PMSF, 10 μg of aprotinin per ml, 10 μg of leupeptin per ml). Finally, the beads are resuspended in 1 ml of HNTG buffer, and a portion (20 to 50 μl) is analyzed for yield of GST fusion protein by 10% SDS–PAGE and detection using Coomassie blue staining. The immobilized proteins, as a suspension in HNTG buffer, can be aliquoted, flash frozen, and stored at −80° with little loss of activity over a 3-month period.

Expression and Isolation of TrpE SH2 Domain Fusion Proteins

Efficient expression of TrpE fusion protein is obtained when the pATH expression plasmid is amplified in *E. coli* strain RR1. M9 medium supplemented with 0.5% (w/v) casamino acids, 1 m*M* tryptophan and containing 100 μg/ml ampicillin is used for the growth of bacteria. A 5-ml overnight culture is diluted to 50 ml with fresh medium not containing tryptophan and is grown to an OD of 0.6 (600 nm). Induction for the production of the TrpE fusion protein is carried out through the addition of indole acrylic acid to a final concentration of 40 μM. Following 3 to 6 hr of incubation at 37°, the cells are collected by centrifugation and are washed with 1 ml of 50 m*M* Tris–HCl, pH 7.5/10% (w/v) sucrose. The cell pellet is resuspended in 1 ml of PLC lysis buffer, sonicated twice for 30 sec, and clarified by centrifugation at 15,000*g* for 10 min at 4°. The TrpE fusion proteins are recovered by incubating the clarified lysate with 40 μl of crude anti-trpE serum and 300 μl of a 10% (w/v) protein A–Sepharose bead suspension. Following a 90-min incubation at 4°, the immune complexes are washed three times with HNTG buffer and resuspended in the same buffer. A portion of the immune complex is analyzed for TrpE fusion protein content by 10% SDS–PAGE and immunoblotting with anti-TrpE antiserum. The immune complexes can be divided into aliquots as a suspension in HNTG buffer, flash frozen, and stored at −80° without loss of activity for 1 month.

Detection of Protein Complex Formation

Bacterial SH2 fusion proteins can be used to identify tyrosine-phosphorylated SH2 binding proteins in a cell lysate. The nature of the protein complexes that can be detected depends on the culture cells investigated and the amount of protein present in the cell lysate. For many mammalian cell lines, including Rat2 fibroblasts,[33] Rat2 cells expressing the v-*src* or v-*fps* tyrosine kinase oncogenes,[23,33,40,41] R1hER (Rat1 cells expressing human EGFR), and Cos-1 monkey cells,[33] a density of approximately 5×10^6 cells per ml is required to obtain a sufficiently strong signal. These cell lines are maintained in Dulbecco's modified Eagle's medium supplemented with 10% (v/v) fetal bovine serum (FBS) in a humidified atmosphere containing 5% CO_2. Confluent cells grown in 10-cm-diameter dishes are lysed in 1 ml of PLC lysis buffer and are clarified by centrifugation at 10,000*g* for 10 min at 4°. Depending on the cell line, one plate of confluent cells can yield from 1 to 10×10^6 cells.

To form the complexes, an excess of immobilized fusion protein is added to 1 ml of lysate, typically to a final concentration of 3 to 5 μg/ml. The reaction is mixed with gentle inversion for 90 min at 4°. The complexes are then washed with three 1-ml volumes of HNTG buffer, resuspended in 40 μl of 2× SDS sample buffer, and heated at 100° for 3 min. The proteins are then resolved using SDS–PAGE and are transferred to a nitrocellulose membrane in a semidry blotting apparatus at 0.8 mA/cm^2 for 60 min. Prior to probing with the antiphosphotyrosine antibody, the blots are treated with a blocking solution for at least 1 hr at room temperature in 20 m*M* Tris–HCl, pH 7.5, 150 m*M* NaCl (TBS) containing 3–5% (w/v) bovine serum albumin (Sigma Immunoglobulin-free) and 1% (w/v) ovalbumin. When using other antibodies or crude rabbit sera as probes, blocking is performed with TBS containing 3–5% (w/v) skim milk powder.

When probing the blot with a primary antibody, the optimal antiserum dilution to use and the conditions for washing the blot after incubation are best determined empirically. For affinity-purified rabbit antibodies, a final concentration of 1 μg/ml, in the respective blocking solution, is usually sufficient. The blot is incubated with the primary antibody for 1 hr at room temperature or overnight at 4°. Following this period the blot is washed twice for 5 min with TBS, then twice for 5 min with TBS containing 0.05% (v/v) NP-40 (TBS-N), and finally twice for 5 min with TBS. For primary antibodies other than antiphosphotyrosine, washing with TBS containing 0.05% (v/v) Tween (TBS-T) is used in place of TBS-N.

A variety of different procedures can be used to detect the primary antibody, including secondary antibody–alkaline phosphatase conjugates which are supplied by Sigma or Bio-Rad; ECL which is supplied by Amersham; or by ^{125}I-labeled protein A supplied by NEN (for antiphosphotyrosine primary antibody NEN product NEX-146L is recommended).

[40] C. A. Koch, M. F. Moran, D. Anderson, X. Liu, G. Mbamalu, and T. Pawson, *Mol. Cell. Biol.* **12,** 1366 (1992).

[41] A. Brooks-Wilson, E. Ball, and T. Pawson, *Mol. Cell. Biol.* **9,** 2214 (1989).

Association of PLC-γ1 SH2 Domains with the PDGF Receptor

The following procedure can be used to demonstrate that the PDGF receptor binds to both SH2 domains of PLC-γ1, but only after activation of the receptor following treatment with PDGF. The Rat2 mammalian cell line is used as the source of the PDGF receptor.

1. Twenty 10-cm plates of Rat2 cells are grown to confluence in Dulbecco's modified Eagle's medium supplemented with 10% FBS.
2. Cells are serum starved for 48 hr in 10 ml of medium containing only 0.5% FBS.
3. A 2-ml portion of the medium is removed from each plate, clarified by brief centrifugation, and divided into two equal portions. PDGF is added to a final concentration of 50 ng/ml to one portion whereas the other portion is used as a control and is not stimulated with PDGF.
4. The remaining medium is removed from each plate and the cells are incubated with 1 ml of PDGF-containing medium or 1 ml of the PDGF-free control medium for 5 min at 37°. The plates are then placed on ice, the medium is removed, and the cells are washed with 2.5 ml of 50 m*M* sodium phosphate buffer, pH 7.0, 150 m*M* NaCl (PBS). The cells are then scraped from each set of 10 plates into 3 ml of PLC lysis buffer, transferred to polypropylene tubes, and centrifuged for 15 min at 10,000 rpm in a SS-34 rotor (Sorvall).
5. A 150-μl portion of the clarified cell lysate is diluted with 1 ml PLC lysis buffer supplemented with 10 m*M* DTT and 25 μl of a 50% (v/v) (3–5 μg protein) slurry of GST–SH2 domain immobilized on glutathione agarose beads. The quantity of immobilized GST–SH2 fusion protein used to obtain a good signal is determined empirically. In many instances the original amount of protein immobilized to the beads will be too high. In these cases it is best to dilute the slurry with a resin such as Sephadex G-25 equilibrated in PLC lysis buffer. The cell lysate and immobilized GST–SH2 fusion protein are gently agitated at 4° for 90 min to allow protein–protein association to occur. Following this period, three washes of the beads are carried out using 0.5 ml of HNTG buffer each time.
6. For SDS–PAGE analysis the beads are resuspended in 40 μl of 2× SDS sample buffer and are boiled for 3 min at 100°. Following a brief centrifugation, the solution is applied to an 8% SDS gel to resolve the proteins, being careful not to load any beads onto the gel. The associated proteins are detected using antiphosphotyrosine or more specific antibodies as described earlier. It is often convenient to reprobe the same blot with a different antibody. In order to strip off the first antibody, the nitrocellulose membrane is incubated at 50° for 30 min in 62.5 m*M* Tris–HCl, pH 6.7, 2% SDS, and 100 m*M* 2-mercaptoethanol. The membrane is then washed

twice at room temperature for 10 min in TBS-T buffer. The blot must then be blocked as described earlier and checked for any residual signal before beginning to probe with the new antibody.

Typical results are displayed in Fig. 3. Initial experiments where DTT was omitted during the mixing of the cell lysate with the GST-PLC-γ1 SH2 fusion protein (step 5 above) resulted in only the C-SH2 and N+C SH2 domains associating with the activated PDGF receptor (Fig. 3A). When the experiment was repeated and 10 m*M* DTT was added, as indicated earlier, the N-SH2 domain also associated with the receptor tyrosine kinase (Fig. 3B). This indicates that some SH2 domains must be kept in a reduced environment containing DTT in order to preserve their activity. The C-terminal SH2 domain from the p85 subunit of phosphatidylinositol 3′-kinase also displays this behavior.

SH2 Blotting: Far Western Analysis

When performing a "Far Western" analysis to test for a direct SH2 domain interaction with a specific phosphorylated protein, optimal immunoprecipitation conditions must first be determined, thus assuring that a strong signal will be obtained. This is done empirically and can be aided by using cell lines that overexpress the protein of interest.

If the Far Western analysis is carried out to identify unknown cellular proteins that directly associate with a SH2 domain protein, samples of cell

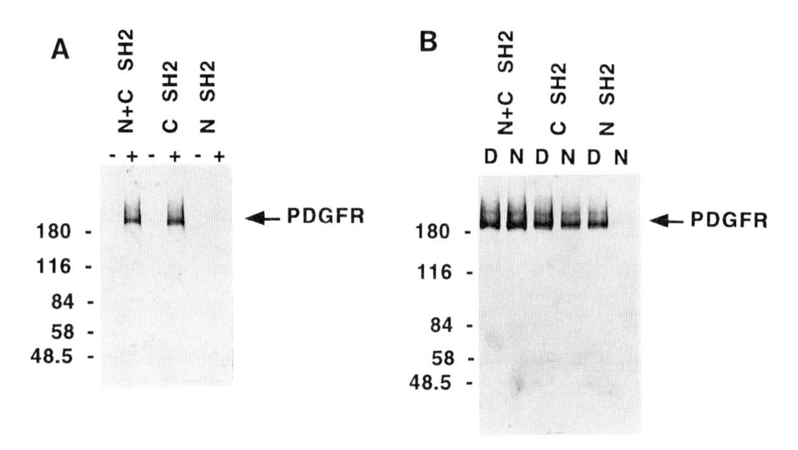

FIG. 3. Binding of GST–PLC-γ1 fusion proteins to PDGFR in lysates of PDGF-stimulated Rat2 cells. (A) Effect of PDGF stimulation on binding (−, unstimulated Rat2 cells; +, stimulated cells). (B) Effect of DTT addition on binding (N, no DTT added; D, 10 m*M* DTT).

lysate that contain at least 10–20 μg of total protein are required to obtain a good signal. The amount of protein in the cell lysate is determined using the bicinchoninic acid protein assay (Pierce), which is compatible with the PLC lysis buffer. Typically, 5×10^6 fibroblast cells, when resuspended in 1 ml of PLC lysis buffer, will contain approximately 1 mg/ml of protein, and a 10- to 20-μl sample will provide a strong signal, although this is, of course, dependent on the cellular expression level of the individual proteins.

The immunoprecipitated or cell lysate proteins are resolved by SDS–PAGE gel, transferred to a nitrocellulose membrane, and blocked overnight at room temperature in TBS-T buffer containing 3% (w/v) skim milk powder. Lysates containing the TrpE–SH2 fusion protein are prepared by sonication of induced bacteria resuspended in 50 m*M* sodium phosphate, pH 7.5, 150 m*M* NaCl, 10 m*M* DTT, 1% (v/v) Triton X-100, 1% (v/v) Tween 20, 10 μg/ml aprotinin, and 10 μg/ml leupeptin. The amount of fusion protein is estimated by Coomassie blue staining of a 10% SDS gel following SDS–PAGE analysis of the bacterial lysates. The blots are probed with 5 μg/ml (approximately 100 n*M*) TrpE–SH2 fusion protein in a blocking solution for 1 to 2 hr at room temperature, then washed three times for 5 min with TBS. To detect bound TrpE–SH2 fusion proteins, the blots are first incubated with mouse monoclonal anti-TrpE antibodies followed by rabbit anti-mouse immunoglobulin G antibodies and finally with ^{125}I-labeled protein A.

Use of Bacterially Expressed Tyrosine-Phosphorylated Polypeptides for Analysis of SH2 Domain-Containing Protein Binding

In order to probe which region of a particular tyrosine-phosphorylated protein, such as a receptor tyrosine kinase, participates in SH2 domain binding and to analyze further the amino acids important for specificity, we sought a convenient method to express selected regions of the protein as tyrosine-phosphorylated fusion proteins in bacteria. With these tools it was possible to investigate the binding of various SH2-containing signaling proteins from several mammalian cell lysates.

The intracellular portion of the PDGF receptor contains three distinct regions that undergo tyrosine phosphorylation following stimulation with PDGF. The first region lies between the transmembrane and the kinase domain, the second is located within the kinase domain as an insert, and the third is located at the extreme C-terminal of the protein. To test if PLC-γ1 bound selectively to the kinase insert or C-terminal of the PDGF receptor, we expressed these regions individually as phosphorylated GST fusion proteins, as described below, and used the immobilized material to probe a mammalian cell lysate that overexpressed PLC-γ1.[30]

Scheme for the Generation of Tyrosine-Phosphorylated Proteins

The region of interest from the receptor tyrosine kinase is expressed in *E. coli* as either a TrpE or a GST fusion protein using the pATH or pGEX plasmids, respectively. The bacteria is then infected with a λ phage (λB1) encoding the cytoplasmic domain of the Elk tyrosine kinase as a LacZ–Elk fusion.[42] Clones containing the λB1 lysogen are identified by duplicate plating and selecting colonies that grow at 30°, but undergo cell lysis at 42°. To obtain protein, these infected cells are induced to produce both the fusion protein of interest and the Elk tyrosine kinase. Phosphorylation occurs during this induction and the tyrosine-phosphorylated fusion protein is isolated by capture with anti-TrpE antibodies, for TrpE fusions,[43] or immobilization of the GST fusion proteins on glutathione–agarose beads.[30]

Preparation of Phage Stocks

The Elk tyrosine kinase utilized in this procedure was obtained from a rat brain λgt11 cDNA expression library probed with antiphosphotyrosine antibody in order to identify clones that encode enzymatically active protein-tyrosine kinase.[42] The resulting bacteriophage expression system encodes 380 amino acids of the Elk tyrosine kinase which encompasses the complete catalytic domain and allows expression of the polypeptide as a lacZ fusion protein. Untransformed bacteria infected with this λElk phage, which can be stored indefinitely as a glycerol stock at −70° or as a stab at room temperature, serve as the source of material for infection. To obtain a working phage stock, a 200 μl of an overnight culture of λElk phage-containing bacteria grown at 30° is added to 3 ml of LB top agar, and the mixture is plated onto a LB plate. Incubation is continued at 42° overnight to ensure that cell lysis will occur. Once lysis is apparent, the phage is eluded by adding 3 ml of SM buffer (50 m*M* Tris–HCl, pH 7.5, 100 m*M* NaCl, 10 m*M* $MgSO_4$, 0.1% gelatin) and gently mixing the solution across the surface of the plate for 1 to 2 hr at 4°. The eluate is then removed and treated with 1 ml of chloroform. Following mixing the eluate is clarified by centrifugation and this solution is analyzed for the phage titer using conventional methods. For successful work the titer of the phage stock should range from 10^9 to 10^{10} plaque-forming units (PFU)/ml.

The phage working stock can be stored in a glass tube protected from light and in the presence of chloroform at 4° for periods up to 1 year.

[42] K. Letwin, S.-P. Yee, and T. Pawson, *Oncogene* **3,** 621 (1988).

[43] M. Reedijk, X. Liu, P. van der Geer, K. Letwin, M. D. Waterfield, T. Hunter, and T. Pawson, *EMBO J.* **11,** 1365 (1992).

Alternatively, the phage stock can be stored as a 7% dimethyl sulfoxide solution at −80°.

λgt11 Elk Infection and Protein Tyrosine Phosphorylation

A variety of bacterial cells, including DH5α, XL-1 Blue, JM101, and C600hfl, can be successfully used for fusion protein expression and infection by the λgt11 Elk phage. For efficient phage infection, 100 μl of 10^{10} PFU/ml phage working stock is added to 200 μl of a overnight culture of bacteria, previously transformed with the protein expression vector, grown in LB+ampicillin medium supplemented with 0.2% maltose and 10 m*M* $MgSO_4$. After a 30-min incubation at 37°, the mixture is streaked onto LB+ampicillin plates and grown overnight at 30°. Isolated colonies are then picked and streaked on duplicate LB+AMP plates. One plate is grown overnight at 42° and the other at 30°. The colonies that do not grow at 42° are lysogenic and contain the Elk-containing λ phage.

For protein phosphorylation the lysogenic-transformed bacterial cells are grown at 30° to an OD (600 nm) from 0.6 to 1.0 and are then incubated for 2 hr in the presence of 1 m*M* IPTG to induce protein expression. Subsequent heat shock of the culture through incubation at 42° for 15 min induces phage replication, thereby enhancing the induction of Elk kinase protein expression and consequent protein phosphorylation. After the heat shock, the cells are incubated at 37° for a further 60 min to maximize phosphorylation levels. This brief period of growth following heat shock is not sufficient to induce the lytic cycle of the phage and bacterial cell lysis is typically not observed.

Prior to using the material generated by this method, it is important to check both the level of fusion protein induction and its phosphorylation. This is conveniently performed by analyzing a portion of the immobilized fusion proteins by SDS–PAGE electrophoresis with detection for protein induction by Coomassie blue staining and protein phosphorylation by Western immunoblotting with an antiphosphotyrosine antibody using ^{125}I-labeled protein A for detection. For this application the use of enhanced chemiluminescent detection can give unreliable results due to a background reaction with the GST portion of the fusion protein.

Localization of PLC-γ1 Binding Site on PDGF Receptor

Figure 4 shows the level of Elk tyrosine kinase-mediated phosphorylation of GST, GST–C-terminal tail (amino acids 953 to 1103 in β-PDGFR), and GST–kinase insert (residues 698 to 797) as monitored by antiphospho-

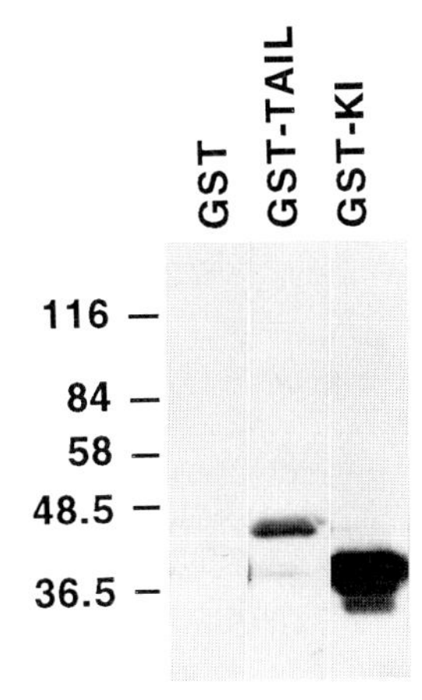

FIG. 4. Elk tyrosine kinase phosphorylation of PDGFR GST fusion proteins. GST and GST fusion proteins containing either the C-terminal region (GST-Tail) or kinase insert (GST-KI) of the PDGF receptor were phosphorylated by the Elk tyrosine kinase as described in the text. Affinity-purified GST fusion proteins were analyzed using antiphosphotyrosine antibodies in a Western blot procedure and detection with ^{125}I-labeled protein A.

tyrosine antibodies. The amount of fusion protein loaded, as determined by Coomassie blue staining, was equivalent. In Fig. 5 the association of PLC-γ1, Ras GTPase-activating protein, or phosphatidylinositol 3′-kinase to the fusion proteins is shown. Whereas Ras GTPase-activating protein and phosphatidylinositol 3′-kinase bound only to the kinase insert, PLC-γ1 interacts specifically with the C-terminal tail region. Further localization of the PLC-γ1 binding site was achieved through the use of GST–C-terminal tail proteins containing specific tyrosine to phenylalanine point mutations, and tyrosine-1021 was identified as the unique site. Based on these results, the phosphopeptide D.N.D.(pY).I.I.P.L.P.D.P.K, corresponding to the region surrounding amino acid 1021, was chemically synthesized and found to inhibit PLC-γ1 binding to the GST fusion protein with high affinity (IC_{50} of 5 n*M*).[30]

Quantitative Analysis of SH2 Domain/Phosphotyrosine-Containing Peptide Interactions Using Surface Plasmon Resonance Technology

Commercial developments in biosensor-based technology have provided a convenient method for quantitative analysis of the affinity SH2

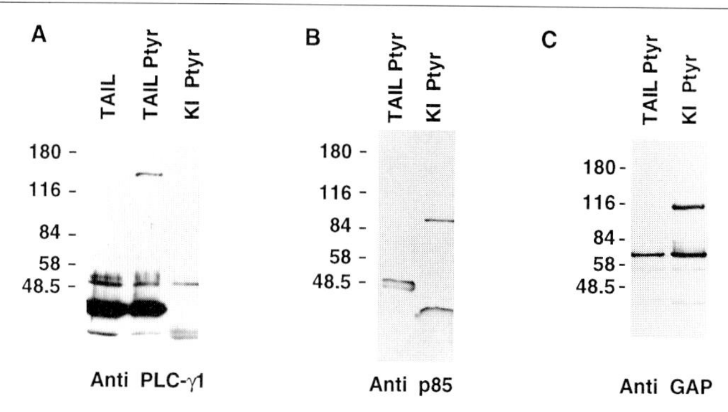

FIG. 5. Association of signaling proteins containing SH2 domains with the kinase insert or C-terminal tail of the PDGFR. A GST fusion protein containing the PDGFR tail, either unphosphorylated (Tail) or phosphorylated (Tail P-Tyr), and a phosphorylated GST fusion protein containing the PDGFR kinase insert (KI P-Tyr) were immobilized on beads and incubated with lysates from Rat2 cells overexpressing PLC-γ1. Association of SH2-containing signal proteins with the GST fusion proteins was detected by Western blot analysis using antibodies to PLC-γ1 (A), p85α (B), or GAP (C), followed by enhanced chemiluminescence (ECL, Amersham).

domains for phosphotyrosine-containing peptides.[44–46] The techniques makes use of surface plasmon resonance (SPR); an optical phenomena characteristic of thin metal films placed at the interface between a glass surface and liquid.[47] The electrons of the metal film interact with the evanescent wave component of light shone on its surface and causes a dip in the intensity of light reflected. The angle that this dip is observed, known as the resonance angle, is affected by the refractive index of liquid solution passing across the opposite surface of the metal film. By monitoring the resonance angle, small changes in the refractive index caused by the concentration of mass near the metal surface can be detected and recorded. An arbitrary unit called the resonance unit (RU) is used to define these changes in the resonance angle. A response of 1000 RU corresponds to a change in surface protein concentration of approximately 1 ng/mm^2.

In the application of this technology, a phosphotyrosine-containing peptide is immobilized to a dextran surface that coats a gold film and is in

[44] S. Felder, M. Zhou, P. Hu, J. Urena, A. Ullrich, M. Chaudhuri, M. White, S. E. Shoelson, and J. Schlessinger, *Mol. Cell. Biol.* **13,** 1449 (1993).

[45] G. Panayotou, G. Gish, P. End, O. Truong, I. Gout, R. Dhand, M. J. Fry, I. Hiles, T. Pawson, and M. D. Waterfield, *Mol. Cell. Biol.* **13,** 3567 (1993).

[46] G. Payne, S. E. Shoelson, G. D. Gish, T. Pawson, and C. T. Walsh, *Proc. Natl. Acad. Sci. U.S.A.* **90,** 4902 (1993).

[47] U. Jönsson, L. Fägerstam, H. Roos, J. Rönnberg, S. Sjölander, E. Stenberg, R. Stahlberg, C. Urbaniczky, H. Östlin, and M. Malmqvist, *BioTechniques,* **11,** 520 (1991).

contact with liquid. Purified SH2 domain-containing proteins flowed over this surface that bind to the immobilized peptide elicit a change in the SPR response. As the SPR response can be monitored as a function of time, the kinetics of SH2 domain binding can be evaluated. This section describes procedures for peptide immobilization and use of the resulting surface for a kinetic evaluation of GST fusion proteins containing either both PLC-γ SH2 domains or the isolated domains for binding to a phosphotyrosine-containing peptide corresponding to the tyrosine-1021 binding site of the β-PDGF receptor.

Peptide Design and Conditions for Immobilization to Biosensor Chip

Two factors are important when designing a peptide for Biocore analysis. First, the peptide must be constructed in such a way that the structural elements important for the interaction being monitored will not be obstructed when immobilized to the chip. Fortunately, for the analysis of most SH2 domain–phosphotyrosine peptide interactions, the minimal requirement for high-affinity binding appears to be inclusion of only 4 to 6 amino acids C-terminal to the phosphotyrosine residue, allowing reasonably small peptides to be used in the assay. In our experiments we used peptides that are 12 amino acids in length with the phosphotyrosine residue located at the fourth position from the N-terminal.

The second important element in peptide design is the inclusion of a lysine at the C-terminal to aid in coupling to the biosensor chip surface. The lysine side chain amine provides a good nucleophilic group for the coupling reaction, and the side chain length can act as a spacer between the phosphopeptide and the surface. As peptides not containing a C-terminal lysine, or other nucleophilic side chains, are only very poorly immobilized, as determined by monitoring the surface for binding antiphosphotyrosine antibody, we have assumed that only a minor fraction of the peptide will be immobilized through the N-terminal amino group.

The dextran surface of the biosensor chip is activated following the manufacture's conditions (Pharmacia BioSensor). A 1:1 (v/v) mixture of *N*-hydroxysuccinimide and *N*-ethyl-*N'*-(3-dimethylaminopropyl)carbodiimide hydrochloride is first flowed over the surface. This is then followed by injection of the phosphopeptide at millimolar concentrations in 50 m*M* HEPES, pH 7.5, 2 *M* NaCl for 8 min at a flow rate of 5 μl/min. Any remaining activated groups on the surface are blocked with the injection of 1.0 *M* ethanolamine. The activity of the surface is monitored by injection of a 100-μg/ml solution of antiphosphotyrosine antibody for 2 min at the same flow rate. A surface suitable for kinetics studies will typically yield values between 1500 and 5000 RU; however, the optimal surface is dependent on the nature of the interaction being monitored and must be determined empirically. A 36-sec injection of 2 *M* guanidium hydrochloride is used to regenerate the surface.

Determination of Kinetic Parameters

Independent experiments are carried out to determine the association and dissociation rate constants. To obtain data for the association rate constant, repetitive injections, across the immobilized phosphopeptide surface, of increasing concentrations of SH2-domain protein are carried out. The SPR response is recorded as a function of time and data are analyzed following procedures recommended by the manufacture.[48] The treatment assumes an approach to equilibrium process described by Eq. (1):

$$dR/dt = k_{ass}\, C\, R_{max} - (k_{ass}\, C + k_{diss})\, R, \tag{1}$$

where k_{ass} and k_{diss} are the association and dissociation rate constants, respectively, C is the initial protein concentration, R_{max} is the maximum binding capacity on the chip surface, and R is the amount of protein bound at time t. The protein concentrations tested are empirically determined so that a sufficiently linear region of the data, as displayed graphically in a plot of dR/dt as a function of R, is obtained. A secondary plot of the slopes from the various runs as a function of the initial protein concentrations yields the k_{ass} value as the slope. Although the k_{diss} rate should also be obtained by this analysis, practice has found that these values are often unreliable.

To obtain the k_{diss} rate constant the phosphopeptide surface of the chip is first loaded with protein. The SPR response is then monitored following the injection of excess (50–100 μM) free phosphopeptide. The k_{diss} is calculated by fitting the data to the first-order kinetic equation

$$\ln R - \ln R_0 = -\, k_{diss}\, t, \tag{2}$$

where R and R_0 are the SPR response at time t and $t = 0$, respectively. The slope of the line in a plot of ln R as a function of t gives k_{diss}. The kinject modification to the operating software can now be obtained from the manufacturer. The kinject function allows for the injection of phosphopeptide immediately following the end of the initial protein injection, thus providing data for both the association and dissociation rate in a single run.

Kinetic Evaluation of GST–PLC-γ1 SH2 Domain Binding

The phosphopeptide corresponding to the region surrounding tyrosine-1021 of the PDGF receptor, D.N.D.(pY).I.I.P.L.P.D.P.K, and found to

[48] R. Karlsson, A. Michaelsson, and L. Mattsson, *J. Immunol. Methods* **145,** 229 (1991).

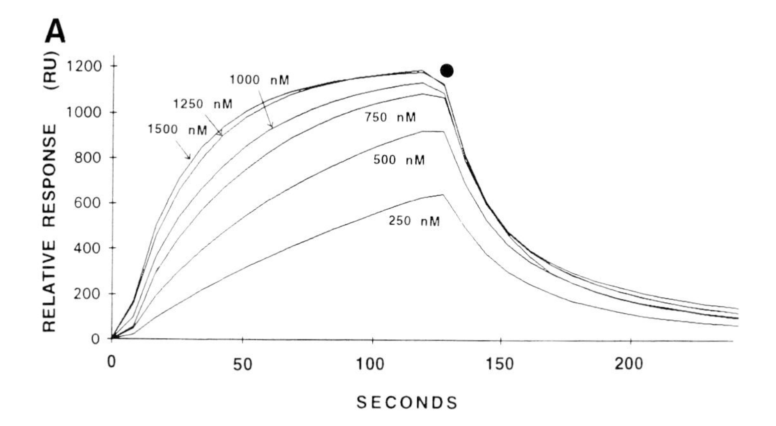

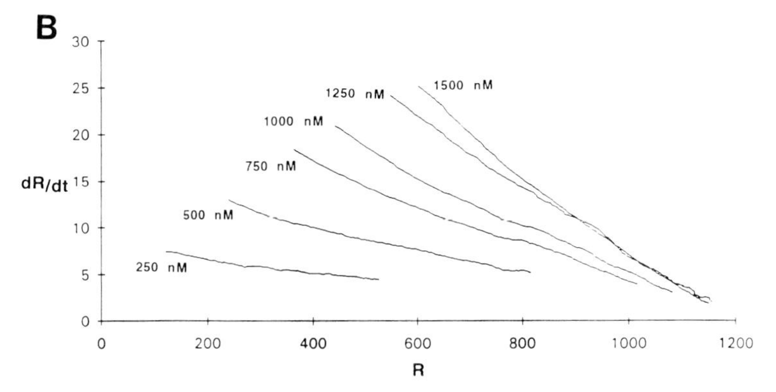

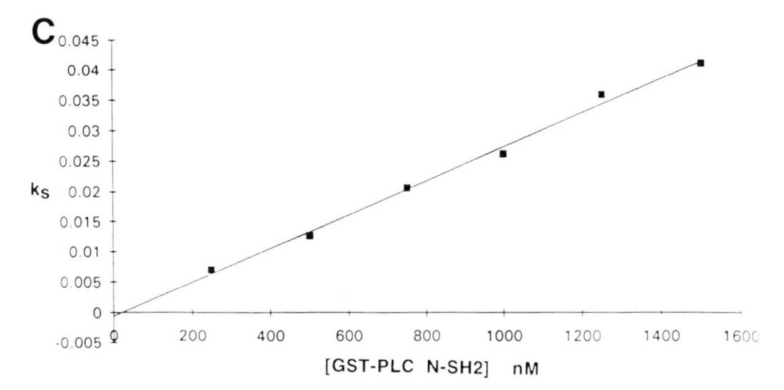

FIG. 6. Biosensor analysis of GST–PLC-γ1 N-SH2 domain binding to the PDGFR Y1021 phosphopeptide. (A) The SPR response following the injection of increasing concentrations of GST–PLC-γ1 N-SH2 from 250 to 1500 n*M*. The dot (●) designates the start of injection of free phosphopeptide across the chip surface. (B) The *dR/dt* vs *R* plots of the data in (A). (C) Replot of the slope of each line in (B) ($k_{ass} \times$ [SH2] + k_{diss}) as a function of the injected protein concentration.

effectively compete with PLC-γ1 binding to the PDGF was immobilized to the biosensor chip as described earlier. A number of different surfaces with varying phosphopeptide densities were prepared using a range of phosphopeptide concentrations, from 40 to 700 μM, in the immobilization. As monitored by the antiphosphotyrosine antibody, the SPR response for these surfaces ranged from 1500 to 5500 RU.

A preliminary kinetic run using the high-density surface (5500 RU) and testing the three GST–PLC-γ1 SH2 proteins containing either both SH2 domains (N+C) or the two individual domains (N or C) from concentrations from 250 to 1500 nM was carried out. In order to evaluate the k_{diss} rate constant, the kinject function was used with a free phosphopeptide concentration of 50 μM. Higher phosphopeptide concentrations tested gave similar results. The results showed that each SH2 protein displayed different kinetic properties. In particular, whereas the GST–PLC-γ1 N-SH2 gave good kinetic data under these conditions, the other GST–PLC-γ1 SH2 proteins bound too quickly. In subsequent experiments the use of a biosensor chip with a lower phosphopeptide surface density (1500 RU) and testing lower GST–PLC-γ1 SH2 concentrations (50 to 300 nM) allowed measurement of the binding kinetics for the N+C and C-SH2 domains.

Figure 6, which displays the results obtained for the binding of GST–PLC-γ1 N-SH2 using the kinject function, illustrates the type of data sought in a SPR experiment. The SPR response for GST–PLC-γ1 N-SH2 concentrations ranging from 250 to 1500 nM is shown in Fig. 6A. The family of curves from time point 0 to 115 sec is the association phase of the binding interaction. The SPR response following injection of excess free phosphopeptide is seen at time points after 140 sec and represents the dissociation phase.

Analysis of the association phase data, in a plot of dR/dt as a function of R [see Eq. (1)], is shown in Fig. 6B. Although only a selected linear portion of the dR/dt vs R data is used in subsequent calculations, the experimental conditions used allowed a majority of the association phase data to be evaluated. In contrast, these conditions were not suitable for

TABLE I
KINETIC CONSTANTS FOR INTERACTION OF PLC-γ1 SH2 DOMAIN FUSION PROTEINS WITH A PDGFR Y1021 PHOSPHOPEPTIDE

GST–PLC	$10^{-4} \times k_{ass}$ [1/(M sec)]	$10^{2} \times k_{diss}$ (1/sec)	K_d (nM)
N-SH2	2.8	3.4	1071
C-SH2	39.0	15.0	385
N+C SH2s	25.2	2.0	80

measurement of N+C SH2 or C-SH2 domain binding as only a small region of the association phase data gave a linear dR/dt vs R plot. To calculate the k_{ass} rate constant, the slope of each line in the dR/dt vs R plot, which is equal to k_{ass} C + k_{diss} as shown in Eq. (1), is plotted as a function of SH2 concentration (Fig. 6C). From the slope of this graph, a k_{ass} rate constant of 2.8×10^4 $M^{-1}\text{sec}^{-1}$ is obtained.

In the evaluation of the k_{diss} rate constant, using the dissociation phases in Fig. 6A, a first-order rate process is assumed. Typically, two phases are observed following the injection of free phosphopeptide across the surface of the chip. The first phase, which is the faster process, causes dissociation of up to 90% of the bound protein from the chip and is independent of the amount of GST–PLC-γ1 SH2 protein bound to the chip surface. It is this portion of the data that is used in calculating the k_{diss} rate constant. With this value the dissociation equilibrium constant (K_d) is obtained using the equation: $K_d = k_{diss}/k_{ass}$.

A summary of the kinetic results obtained using this technique for the analysis of GST–PLC-γ1 N+C SH2, N-SH2, and C-SH2 domain binding to the PDGFR Y1021 phosphopeptide is shown in Table I. The individual N-SH2 domain binds less tightly than either the C-SH2 domain or the N+C SH2 domains. Kinetically, the N-SH2 both associates and dissociates slower from the phosphopeptide than the C-SH2 domain. The higher affinity of the N+C SH2 domains arises from having both a fast association rate and a slow dissociation rate. The relative magnitude of the rate constants suggests how both individual SH2 domains might act to fine tune the specificity of PLC-γ1 binding to an activated growth factor receptor.

[35] SH2 Domain Specificity Determination Using Oriented Phosphopeptide Library

By ZHOU SONGYANG and LEWIS C. CANTLEY

Introduction

Transmission of growth or differentiation signals from the cell membrane to the nucleus requires the collaboration of many signaling molecules. Protein–protein interactions seem to be the basic element for these signaling events. Nature has designed numerous functional domains to ensure the fidelity of protein–protein interactions. One of these domains, the Src homology 2 (SH2) domain, is particularly important because it is an essential component in protein-tyrosine kinase-mediated signaling pathways[1,2] (see also [34] in this volume). A unique property of SH2 domains is their ability to recognize proteins specifically in a phosphotyrosine (pTyr)-dependent manner.[2–4] These interactions can therefore be regulated specifically by tyrosine phosphorylation or dephosphorylation. An additional level of specificity for the SH2 domain recognition is provided by the primary sequences surrounding the phosphotyrosine.[1,5,6]

Knowing the optimal binding sequence for a SH2 domain is particularly helpful because one could quickly identify SH2 domain targets as well as SH2-containing molecules that associate with a given protein.[1] Conventional methods for identifying SH2 domain binding sequences are both time-consuming and expensive, involving various techniques (e.g., phosphopeptide mapping, intensive site-directed mutagenesis, and phosphopeptide competition). In cases where a site can be mapped, the consensus sequence for SH2 domain binding still remains unclear. To circumvent this problem, we have developed an oriented phosphopeptide library technique to expediently evaluate the specificity of SH2 domains.[6] Using this technique, we have demonstrated that different SH2 domains have distinct specificities. The following methods have been very successful in determining optimal binding motifs for more than 30 SH2 domains[6,7] (Z. Songyang and L. C. Cantley, unpublished data, 1994). The identification of these motifs not only provides structural basis for SH2 domain recognition, but also leads to the predictions of likely *in vivo* targets of SH2-containing proteins.[2,8–11]

[1] L. C. Cantley, K. R. Auger, C. L. Carpenter, B. Duckworth, A. Graziani, R. Kapeller, and S. Soltoff, *Cell* (*Cambridge, Mass.*) **64,** 281 (1991).

[2] T. Pawson and J. Schlessinger, *Curr. Biol.* **3,** 434 (1993).

[3] D. Anderson, C. A. Koch, L. Grey, C. Ellis, M. F. Moran, and T. Pawson, *Science* **250,** 979 (1990).

[4] B. J. Mayer, P. K. Jackson, and D. Baltimore, *Proc. Natl. Acad. Sci. U.S.A.* **88,** 627 (1991).

[5] W. J. Fantl, J. A. Escobedo, G. A. Martin, C. W. Turck, M. del Rosario, F. McCormick, and L. T. Williams, *Cell* (*Cambridge, Mass.*) **69,** 413 (1992).

[6] Z. Songyang, S. E. Shoelson, M. Chaudhuri, G. Gish, T. Pawson, W. G. Haser, F. King, T. Roberts, S. Ratnofsky, R. J. Lechleider, B. G. Neel, R. B. Birge, J. E. Fajardo, M. M. Chou, H. Hanafusa, B. Schaffhausen, and L. C. Cantley, *Cell* (*Cambridge, Mass.*) **72,** 767 (1993).

[7] Z. Songyang, S. E. Shoelson, J. McGlade, P. Olivier, T. Pawson, R. X. Bustelo, M. Barbacid, H. Sabe, H. Hanafusa, T. Yi, R. Ren, D. Baltimore, S. Ratnofsky, R. A. Feldman, and L. C. Cantley, *Mol. Cell. Biol.* **14,** 2777 (1994).

[8] M. J. Eck, S. E. Shoelson, and S. C. Harrison, *Nature* (*London*) **362,** 87 (1993).

[9] G. Waksman, S. E. Shoelson, N. Pant, D. Cowburn, and J. Kuriyan, *Cell* (*Cambridge, Mass.*) **72,** 779 (1993).

[10] R. J. Lechleider, S. Sugimoto, A. M. Bennett, A. S. Kashishian, J. A. Cooper, S. E. Shoelson, C. T. Walsh, and B. G. Neel, *J. Biol. Chem.* **268,** 21478 (1993).

[11] K. S. Ravichandran, K. K. Lee, Z. Songyang, L. C. Cantley, P. Burn, and S. J. Burakoff, *Science* **262,** 902 (1993).

Principle of Oriented Peptide Library

The oriented peptide library is similar to peptide libraries used for chaperon BiP (BiP)[12] and HLA class I Antigen (HLA)[13] recognition. A mixture of soluble degenerate peptides is presented to the proteins to be tested. The affinity-purified mixture is then sequenced to determine the binding consensus. However, in contrast to the latter libraries in which peptides are degenerate at all positions, an oriented peptide library has a critical recognition residue locked in (e.g., phosphotyrosine) so that each degenerate position is oriented relative to this critical residue. Therefore, unlike other libraries in which peptides are out of phase, an oriented library is able to compare the relative importance of amino acids at a given degenerate position.

Construction of the Oriented Phosphopeptide Library

The first phosphopeptide library we made had the following composition: Gly-Asp-Gly-pTyr-X-X-X-Ser-Pro-Leu-Leu-Leu, where X indicates all amino acids except Cys and Trp.[6] Cys and Trp were omitted to avoid oxidation and sequencing problems. The theoretical degeneracy of this library is $18^3 = 5832$. The three residues C-terminal to the pTyr were degenerate because these positions have been found to be the major determinates for 1-phosphoinositide (PtdIns) 3-kinase p85 SH2 domain binding.[1,5,14,15] At the time we made the peptide library, there was no evidence that residues N-terminal of pTyr influenced SH2 domain binding. The sequence Gly-Asp-Gly provides a negative charge, frequently found N-terminal to the *in vivo* tyrosine phosphorylation sites.[16] Ser and Pro are present C-terminal to the degenerate positions only because these residues are often located within p85 SH2 domain binding sequences (e.g., IRS-1, see Table II). These residus also give an estimate of how many peptides are being sequenced. All peptides are terminated with three leucine (or lysine for hydrophilic sequencing membrane) residues to prevent washout during sequencing.

Because the above library cannot be used to examine specificity N-terminal to the pTyr, we have designed a second library with three degenerate residues on both sides of the phosphotyrosine (Gly-Ala-X-X-

[12] G. C. Flynn, J. Pohl, M. T. Flocco, and J. E. Rothman, *Nature (London)* **353,** 726 (1991).

[13] T. N. Schumacher, G. M. Van Bleek, M. T. Heemels, K. Deres, K. W. Li, M. Imarai, L. N. Vernie, S. G. Nathenson, and H. L. Ploegh, *Eur. J. Immunol.* **22,** 1405 (1992).

[14] J. A. Escobedo, D. R. Kaplan, W. M. Kavanaugh, C. W. Turck, and L. T. Williams, *Mol. Cell. Biol.* **11,** 1125 (1991).

[15] K. R. Auger, C. L. Carpenter, S. E. Shoelson, W. H. Piwnica, and L. C. Cantley, *J. Biol. Chem.* **267,** 5408 (1992).

[16] T. Hunter and J. A. Cooper, *Adv. Cyclic Nucleotide Protein Phosphorylation Res.* **17,** 443 (1984).

X-pTyr-X-X-X-Ala-Lys-Lys-Lys). For six of the SH2 domains we have studied (i.e., p85, Crk, Nck, SHC, GRB2, and Csk), this library has predicted the same optimal motifs as the original library, and no apparent preference for amino acids at positions N-terminal to the pTyr was observed. In theory, one can make a peptide library as degenerate as possible. However, the more degenerate a library is, the less sensitive sequencing will be. In fact, probably only five to six residues on either end of the pTyr would directly interact with SH2 domains. We are now in the process of making such a library.

Phosphotyrosine Peptide Library Synthesis

N^{α}-Fmoc-based syntheses were conducted on a Milligen/Biosearch 9600 synthesizer using standard BOP/HOBt coupling protocols and protected amino acids, except incorporations following that of the protected phosphotyrosine derivative N^{α}-Fmoc-*O*-(*O*,*O*-dimethoxyphosphoryl)-L-tyrosine were prolonged. For degenerate positions, the resin was deprotected with 30% (v/v) piperidine as usual and was washed extensively with dimethylformamide; the solvent was removed and the moist resin was divided into 18 equivalent amounts by weight. Each aliquot was coupled for 18 hr with four equivalents each of BOP, HOBt, and a different N^{α}-Fmoc-amino acid having the appropriate side-chain protecting group. All aliquots were recombined and the procedure was repeated at each degenerate position. Peptide cleavage and side-chain deprotections were done with trimethylsilyl bromide as described.[17] The peptide mixture was precipitated with diethyl ether (4°) and desalted on a column of Bio-Gel P-2. Sequence analysis of the peptide mixture was performed.[6] The variability in molar amounts of the individual amino acids at a given cycle was generally less than threefold as judged by sequencing (some variability is due to recovery in the sequencer) and less than twofold on the basis of amino acid analysis of the mixture. There might be some decrease in yield with successive cycles; however, since all peptides had the same carboxy-terminal sequence, the washout was similar for all peptides.

Affinity Purification of Phosphopeptides Using GST–SH2 Domains

Preparation of GST–SH2 Domain Fusion Proteins

cDNAs encoding SH2 domains were subcloned into pGEX vectors. As different laboratories usually have different ways to construct expression

[17] S. Domchek, K. Auger, S. Chatterjee, T. Borke, and S. Shoelson, *Biochemistry* **31,** 9865 (1992).

plasmids, we have found it to be beneficial to the activity of the SH2 domain to include six or more residues outside the SH2 domain conserved region.

Overnight cultures of *Escherichia coli* (most commonly DH5α) transformed with various pGEX plasmids were diluted 50-fold in Terrific Broth (TB) with 50 μg/ml ampicillin and shaken at 37°. When the OD_{600} reading is 0.5–0.8 (about 2–3 hr), isopropylthiogalactoside (IPTG) was added to a final concentration of 0.5 m*M* to induce expression at 37° for 4 to 8 hr. Sometimes the expression of certain SH2 domain fusion proteins needed to be done at low temperature (e.g., 30°). After induction, the bacteria were collected for lysis by either of the two methods listed below:

Method 1

1. Resuspend the bacterial pellet in NETN (20 ml/100 ml induction) with 0.5 mg/ml lysozyme, 1 m*M* dithiothreitol (DTT) and 100 μM phenylmethylsulfonyl fluoride (PMSF).
2. After 20 min on ice, sonicate the mixture to get rid of DNA.
3. Spin at 5000*g* for 20 min and save the supernatant.

Method 2

1. Lyse the pellet in lysis buffer (3 ml/g bacteria, 50 m*M* Tris, pH 8.0, 1 m*M* EDTA, 100 m*M* NaCl) with lysozyme (4 mg/g bacteria), 1 m*M* DTT, and 100 μM PMSF.
2. After 20 min on ice, add sodium deoxycholic acid (4 mg/g bacteria) and shake at 25° until the solution becomes very sticky (about 30 min).
3. Add DNase I, spin as in Method 1, and save the supernatant.

The purpose is to purify fusion proteins with high activity. Method 1 has the advantage of being less time-consuming, but also has the risk of denaturing proteins so we therefore recommend Method 2 (see also Molecular Cloning).

To obtain GST–SH2 domain fusion proteins, the supernatant from step 3 was incubated with glutathione beads at 4° for 2 hr. The beads were washed three times with phosphate-buffered saline (PBS: 150 m*M* NaCl, 3 m*M* KCl, 10 m*M* Na_2HPO_4, 2 m*M* KH_2PO_4, pH 7.2) containing 0.5% nonidet P-40 (NP-40) and 1 m*M* DTT, and once with PBS. At this point the beads were ready for use in affinity purification of peptides.

Affinity Purification and Sequencing

For affinity purification (Fig. 1), 150 μl of fusion protein beads (300 μg protein or as concentrated as possible) was packed in a 1-ml syringe used as an affinity column. Beads were washed with 1 ml PBS, and the degenerate peptides (0.3–1 mg) were loaded on the top of the column and allowed to

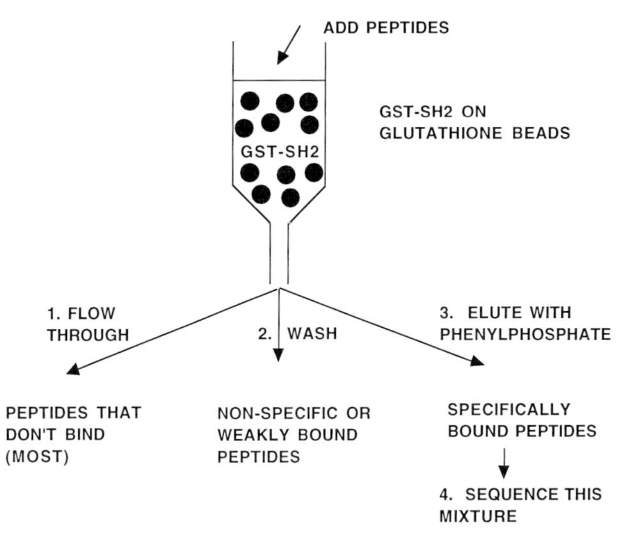

FIG. 1. A scheme representation of the phosphopeptide library technique.

sit at room temperature for 5 to 10 min. During this period, the column flow was stopped. The column was then quickly washed twice with 1 ml ice-cold PBS (containing 10 mg/ml blue dextran and 0.5% NP-40) and once with 1 ml ice-cold PBS in the cold room. Blue dextran and NP-40 help to block nonspecific binding. The color of blue dextran also helps in following the washing. Pressure was added to accelerate the wash so that the entire washing time is less than 1 min to avoid peptide dissociation. Quick washing is critical because phosphopeptide and SH2 domain association exhibits a very fast off rate (about 10 sec).[18] To elute the peptides, 200 μl of 20 m*M* sodium phenyl phosphate solution (pH 7.8) was loaded onto the column at room temperature. The flow through was collected, dried down into a small volume, and sequenced on the Applied Biosystems 477A protein sequencer.

[18] S. Felder, M. Zhou, P. Hu, J. Urena, A. Ullrich, M. Chaudhuri, M. White, S. E. Shoelson, and J. Schlessinger, *Mol. Cell. Biol.* **13,** 1449 (1993).

Sequencing of the peptide mixture was performed for several continuous cycles on the protein sequencer until the residue immediately following the degenerate position was reached. In a typical assay, the total amount of peptides sequenced ranges from 100 pmol to 1 nmol. Occasionally, trace amounts of GST fusion protein would leak out of the column. To avoid sequencing contamination, the peptide mixture can be purified through Centricon 10 (Amicon).

Data Analysis

Because we affinity isolated a mixture of peptides rather than individual peptides, the only way to determine selectivity was to compare the purified mixture to the crude mixture and calculate the relative enrichment value (see below). To explain this process better, we will use the first library as an example. At a given cycle (e.g., cycle 7, Fig. 2), if a particular amino acid is responsible for high affinity binding (in this case Met), peptides with the sequence pTyr-X-X-Met will be prevalent in the mixture. Theoretically, if the sequence of this mixture is compared to that of the crude material, the greatest ratio for Met should be found at cycle 7. However, if the SH2 domain does not have any preference for any amino acids at a degenerate cycle (cycle 6 in Fig. 2), every amino acid should have a similar ratio. If the sum of ratios is normalized to 18 (total number of degenerate amino acids), amino acids not selected by the SH2 domain will have a value of 1 or less. Practically, two changes have been made to the calculation. First, because the rate of washout during sequencing differs from day to day and from sequencer to sequencer, the total amount of picomoles at a degenerate cycle should be normalized to that of the first degenerate cycle. Second, we divided the amount of each amino acid eluted from GST–SH2 beads by that of the control GST column at the same cycle instead of the original peptide mixture. This way, we eliminate variations resulting from the affinity columns and the sequencing. The resulting numbers do not indicate the real enrichment value since background binding to GST has not been subtracted. To obtain the enrichment value, a correction was necessary, i.e., subtracting the background level away and normalizing again.

Using data obtained this way, the order of preference for every amino acid at a given degenerate position can be easily determined. A larger number indicates a stronger selection. In addition, a comparison of enrichment values between different degenerate positions will indicate which position is more selective. A remaining concern is whether the selectivities at different degenerate positions are additive or synergistic. It is possible that a given peptide may have a high affinity for a particular SH2 domain but be underrepresented in this calculation since amino acid X at residue

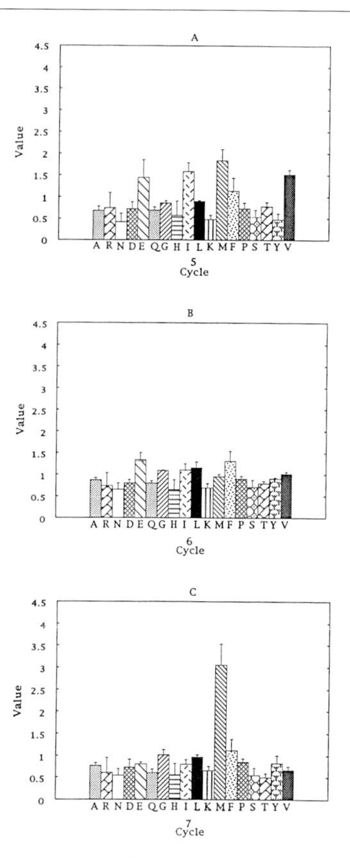

FIG. 2. Specificity of the N-terminal SH2 domain of PtdIns 3-kinase p85 subunit. The degenerate phosphopeptide mixture (Table I) was added to a column containing the N-SH2 domain of p85 as a GST fusion protein. The column was washed, and bound peptides were eluted with phenylphosphate. The eluted peptide mixture was subjected to microsequence analysis and the results were compared to those from the eluate of a control column containing GST alone. A, B, and C are results from the fifth, sixth, and seventh cycle of the sequencing (i.e., the first, second, and third positions after phosphotyrosine). The value represents the ratio of the amount of each amino acid eluted from GST–SH2 bead columns divided by that of the control GST bead columns at the same cycle. Data from three separate experiments were averaged and normalized so that the sum of the values of all amino acids equals the number of amino acids (18). Error bars indicate the standard error ($n = 3$).

+1 will bind only if amino acid Y is located at position +2. Thus, all peptides with only X at +1 or only Y at +2 will fail to bind. Because each peptide is randomly presented to the SH2 domain and the mixture is sequenced, statistically the enrichment value should be position- rather than sequence-dependent. To address this question, the results from less degenerate libraries can be compared with those from highly degenerate libraries. We have made such efforts for p85 SH2 domains, where a less degenerate library (the first amino acid C-terminal to the pTyr was fixed) was used. The order of preference of amino acids at +2 and +3 position C-terminal to the pTyr was found to be very similar to that using the more degenerate library (not shown), indicating that selectivities are sequence-independent. Based on these assumptions, one could use the data in Fig. 2 to determine the relative affinity of individual phosphopeptides. In reality, the product of the enrichment values (called selectivity)[6] at different degenerate positions is a better indicator for affinity. Given the evidence that nonspecific phosphopeptides bind to SH2 domains with roughly 100-fold lower affinity, a 1-fold change in selectivity could result in a 100/[maximum selectivity]-fold (e.g., 5.5-fold in Fig. 2) change in affinity.

Structural Basis for SH2 Domain Recognition

The identification of optimal binding motifs for more than 30 SH2 domains using the oriented phosphopeptide technique has greatly promoted SH2 domain-related research.[2,6,7] Information from the crystal structures of src and lck SH2 domains with high affinity ligand (pTyr-Glu-Glu-Ile, identified using our library) has made it possible to study SH2–phosphopeptide interaction on a structural basis.[8,9] In these structures, the phosphotyrosine is deeply buried in the src and lck SH2 domains. The +1 Glu forms hydrogen bonds with Lys-200 (βD3) and Tyr-202 (βD5) (see also Table I). The +2 Glu makes close contact with Arg-205 (βD′1). The +3 Ile is plugged into a hydrophobic pocket surrounded by Tyr-202 (βD5), Ile-214 (βE4), Thr-215 (EF1), and Leu-237 (BG4). Since the residues of the src SH2 domain backbone are highly conserved among SH2 domains, it is convenient to align SH2 domains and predict which residues may interact with phosphopeptides. As listed in Table I,[7] although the residues that resemble the pTyr binding pocket are highly conserved, those that form the binding sites for amino acids C-terminal to the pTyr vary. This may explain the specificity of SH2 domains.

Based on a comparison of optimal motifs for 13 SH2 domains and differences in their βD5 residues, we have divided the known SH2 domains into four distinct groups (Table I).[6] SH2 domains in Group I have a Tyr or Phe at βD5, and tend to select a general motif of pTyr–hydrophilic–hydrophilic–hydrophobic. On the other hand, SH2 domains in Group III have Ile, Val, or Cys at the βD5 position and select a pTyr–hydrophobic–X–hydrophobic motif. The recently solved structures of the C-terminal SH2 domains of phospholipase C-γ (PLC-γ) and Syp strongly support this

TABLE I
PHOSPHOPEPTIDE MOTIFS FOR SH2 DOMAINS: RESIDUES PREDICTED TO INTERACT WITH SIDE CHAINS OF ASSOCIATED PHOSPHOPEPTIDES[a]

		SRC-SH2			SRC–SH2		SRC-SH2				
SH2 domain	+1	200 βD3	202 βD5	+2	205 βD′1	+3	202 βD5	214 βE4	215 EF1	230 αB9	237 BG4
					Group 1A						
SRC	E	K	Y	E	R	**I**	Y	I	T	Y	L
FYN	**E**	K	Y	**E**	R	**I**	Y	I	T	Y	L
LCK	**E**	K	Y	E	R	**I**	Y	I	S	Y	L
FGR	**E**	K	Y	**E**	R	IV	Y	I	T	Y	L
LYN		K	Y		R		Y	I	S	Y	L
YES		K	Y		R		Y	I	T	Y	L
HCK		K	Y		R		Y	I	S	Y	L
Dsrc		K	Y		K		Y	L	S	Y	L
					Group 1B						
SYK N		H	Y		E		Y	I	S	H	L
*SYK C	QIE	L	Y	eqt	D	**L**	Y	I	P	Y	L
ZAP70 C		Y	Y		S		Y	I	P	L	L
TEC		R	Y		K		Y	L	A	H	L
ATK		R	Y		C		Y	L	A	H	L
ITK		K	Y		K		Y	V	A	H	L
ABL	E	Y	Y	**N**	N	P	Y	V	S	H	L
ARG		Y	Y		N		Y	V	T	H	L
*CSK	**T**	E	Y	n	M	mr	Y	I	D	Y	?
CRK	D	S	Y	H	N	**P**	Y	A	G	Y	T?
NCK	**D**	K	F	**E**	Q	**P**	F	I	G	Y	T?
*fes/fps	**E**	R	F	-	Q	vi	F	R	L	L	G
ZAP70 N		H	F		E		F	I	A	Y	L
SEM5	LV	Q	F	**N**	L	vp	F	L	W	H	R?
*DGBR2	y	Q	F	**N**	L	-	F	L	W	H	R?
*GRB2	qy	Q	F	**N**	L	y	F	L	W	H	R?
GAP C		Q	F		C		F	M	G	Y	I?
GAP N		N	F		I		F	I	G	Y	L?
TENSIN		R	F		T		F	?	?	H	?
*3BP2	**E**	R	Y	**N**	F	-	Y	E	G	Y	P?

notion.[19,20] As more and more SH2 domain binding motifs are mapped, one can predict the specificity of a novel SH2 domain from its primary sequence using both the information in Table I and computer modeling. One can also predict the effect of a mutation on changing the specificity of a SH2 domain. For instance, we have engineered mutations in the binding pockets of src and p85 SH2 domains. The specificities of these SH2 mutants

[19] S. M. Pascal, A. U. Singer, G. Gish, T. Yamazaki, S. E. Shoelson, T. Pawson, L. E. Kay, and J. D. Forman Kay, *Cell* (*Cambridge, Mass.*) **77,** 461 (1994).

[20] C. H. Lee, D. Kominos, S. Jacques, B. Margolis, J. Schlessinger, S. E. Shoelson, and J. Kuriyan, *Structure* **2,** 423 (1994).

TABLE I (*continued*)

		SRC-SH2			SRC–SH2		SRC-SH2				
SH2 domain	+1	200 βD3	202 βD5	+2	205 βD′1	+3	202 βD5	214 βE4	215 EF1	230 αB9	237 BG4
					Group 2						
*VAV	**M**	K	T	E	I	P	T	I	T	Y	?
					Group 3						
p85aN	MIVE	K	I	-	F	**M**	I	F	S	Y	A?
p85bN		K	I		F		I	F	S	Y	A?
p85aC	mli	K	C	-	N	**M**	C	F	A	Y	V?
p85bC		K	C		Y		C	F	A	Y	V?
PLC g1C	VI	K	C	IL	N	**PIV**	C	L	G	Y	Y?
PLC g2C		K	C		Q		C	L	G	Y	Y?
PLC g1N	**LIV**	Q	C	Ed	H	LIV	C	K	F	Y	L?
PLC g2N		Q	C		R		C	K	Y	Y	L?
*SHPTP1 N	F	T	I	-	Q	**F**	I	D	L	Y	V
SHPTP2 N	IV	T	I	-	Q	VI	I	D	L	Y	L
CSW N		T	I		Q		I	D	L	Y	L
SHPTP1 C		T	I		M		I	T	V	F	E
*SHC	EI	K	L	-	V	**ILM**	L	T	K	H	P?
					Group 4						
ShB		M	M		A		M	L	G	Y	?
SHPTP2 C		T	V		R		V	D	V	Y	E
CSW C		T	V		R		V	D	V	Y	E
113 TF		F	A		P		A	L?	S?	Y	?
91 TF		I	S		P		S	L?	Q?	Y	?

[a] Columns +1, +2, and +3 comprise the first, second, and third residue C-terminal to P-tyrosine of the optimal phosphopeptide selected by each SH2 domain (e.g., P-YEEI for src SH2). SRC-SH2 200 and 202 indicate the residues of src (and residues at analogous positions of other SH2 domains) predicted to contact the +1 residue side chain of the associated peptide. SRC-SH2 205 is predicted to be near the +2 side chain, and SRC 202, 214, 215, and 237 are predicted to form a hydrophobic pocket to bind the +3 residue side chain. The alignments were made on the basis of Waksman *et al.*[9] and Lee *et al.*[20] Bold letters indicate strong selection. Uppercase letters without bold indicate medium selection. Lowercase letters indicate weak selection. A hyphen indicates no selection. Motifs not yet determined or not submitted for publication are left blank.

are consistent with our predictions[21] (Z. Songyang and L. C. Cantley, unpublished data), demonstrating the value of Table I.

Search for Potential Targets of SH2 Domains

The existing protein database programs tend to look for sequence homologies rather than preference. This is insufficient in identifying potential

[21] L. E. Marengere, Z. Songyang, G. D. Gish, M. D. Schaller, J. T. Parsons, M. J. Stern, L. C. Cantley, and T. Pawson, *Nature (London)* **369,** 502 (1994).

target proteins. The best way to search for SH2 domain targets is to use the enrichment values as a comparison matrix because the enrichment values most accurately represent relative importance of different amino acids at a given position. We are in the process of writing a new search program. As shown in Table II, many optimal motifs (Table I) have been utilized successfully to predict SH2 domain binding proteins.

Conclusion

We have reviewed an oriented phosphopeptide technique that allows the quick identification of ligand binding specificity for SH2 domains. From

TABLE II

COMPARISON OF ORIENTED LIBRARY PREDICTED OPTIMAL SEQUENCES AND KNOWN *in Vivo* SH2 DOMAIN BINDING SITES[a]

SH2 domains	Optimal motifs	Known *in Vivo* sites	
GRB-2	Y* X **N** X	LPVPE**YIN**QSV	Human EGF receptor
		FDDPS**YVNV**QN	Human SHC
		HVNAT**YVNV**LC	Human HGF receptor
		EGDSS**Y**K**NI**HL	Human CSF-1 receptor
		RKGHE**YTNI**KY	SHPTP2
		AEKPF**YVNV**EF	BCR-ABL
		KSPGE**YVNI**EF	Rat IRS-1
		DAFSD**Y**A**NF**K	RPTP α
SRC family	Y* **E E I**	SETDD**Y**A**EI**ID	Chicken FAK
p85	Y* M X **M**	ESDGG**YM**D**M**SKDESVDYVPMLD	Human PDGFR β
	V	EEEEE**YM**P**M**ED	Mouse Polyoma MT
	I	QGVDT**YV**E**M**RP	Mouse CSF-1 Rec.
	E	DSTNE**YM**D**M**KP	Human c-*kit*
		EELSN**YI**C**M**GG	Rat IRS-1
		VSIEE**Y**TE**M**MP	Rat IRS-1
		HTDDG**YM**P**M**SP	Rat IRS-1
		KGNGD**YM**P**M**SP	Rat IRS-1
		VDPNG**YM**M**M**SP	Rat IRS-1
		PCTGD**YM**N**M**SP	Rat IRS-1
		TGSEE**YM**N**M**DL	Rat IRS-1
		NSRGD**YM**T**M**QI	Rat IRS-1
		LGSQS**YE**D**M**RG	Mouse B cell CD19
		EDADS**YE**N**M**DK	Mouse B cell CD19
		LLHSD**YM**N**M**TP	Human T cell CD28
SHPTP2	Y* I X V	TSSVL**Y**TA**V**QP	Human PDGF receptor β
	V I	KSPGE**YV**N**I**EF	Rat IRS-1
PLC-γ NSH2	Y* **L** X **L**	TXXXE**YL**D**L**XX	Mouse/human bFGF
	Y* **V** I **P**	EGDND**YI**I**P**LP	Human PDGF receptor β
CSH2	I	SADSG**YI**I**P**LP	Human PDGF receptor α

[a] Abbreviations for amino acid residues are A, Ala; D, Asp; E, Glu; F, Phe; G, Gly; H, His; I, Ile; K, Lys; L, Leu; M, Met; N, Asn; P, Pro; Q, Gln; R, Arg; S, Ser; T, Thr; V, Val; Y, Tyr. Y* stands for phosphotyrosine. X indicates any amino acid.

a single experiment, one can get information about the relative importance of almost every amino acid at each location in the SH2 recognition motif. We have also discussed the molecular basis for the SH2 domain–phosphopeptide interaction. Since the residues in the SH2 domains that are responsible for binding to phosphopeptides may be predicted from structural data, the specificity of SH2 domains may be inferred from their primary sequences. This technique also allows us to identify potential *in vivo* targets for various SH2 domains. Finally, the principle of the oriented peptide library can be applied to the study of protein–protein interactions in general.

Acknowledgments

We thank Dr. K. Carraway III and D. Liu for critical reading of this manuscript.

[36] Rapid High-Resolution Western Blotting

By HEIDI OKAMURA, CATHERINE T. SIGAL, LEILA ALLAND, and MARILYN D. RESH

Introduction

Western blotting is a versatile and widely used technique in oncogene research. This method relies on transfer of proteins out of a polyacrylamide gel matrix and onto the surface of a filter.[1,2] Several filter matrices are available which adsorb proteins on their surface, most likely through hydrophobic interaction, without the need for covalent cross-linking. Once immobilized on a solid support, the adsorbed proteins are readily accessible to added ligands and can be used to detect antigen–antibody, protein–protein, or DNA/RNA–protein interactions. Here we present detailed methods for Western blotting, with pertinent examples relevant to oncogene research provided.

Sample Preparation

The first step in a Western blot procedure is typically denaturing SDS–PAGE. Samples for transfer are usually prepared in SDS–PAGE sample buffer.[3] For two-dimensional electrophoresis, samples are prepared

[1] H. Towbin, J. Staehelin, and J. Gordon, *Proc. Natl. Acad. Sci. U.S.A.* **76,** 4350 (1979).
[2] W. N. Burnett, *Anal. Biochem.* **112,** 195 (1981).
[3] U. K. Laemmli, *Nature* (*London*) **227,** 680 (1970).

in sample buffer for isoelectric focusing; the second dimension SDS–PAGE gel is then blotted. Transfer of proteins electrophoresed through native gels has also been reported.[4] Preparation of samples to be blotted from SDS–PAGE gels will be described here.

Whole Cell Lysates

Expression of oncoproteins in tissue culture cells is commonly employed in oncogene research. Here we describe the preparation of samples, typically from NIH/3T3 cell fibroblasts, which can be adapted to most adherent cell cultures. To analyze total cell protein, a confluent plate of tissue culture cells is directly lysed in SDS–PAGE sample buffer in the following manner. The medium is aspirated from a 100-mm dish of cells, and the monolayer is rinsed with 4 ml phosphate-buffered saline (PBS). On aspiration of the PBS wash, 1 ml of 1X sample buffer (70 m*M* Tris, pH 6.8, 11.15% v/v glycerol, 0.0015% v/v bromphenol blue, 3% v/v SDS, 5% v/v 2-mercaptoethanol) is added to the plate. The cells are scraped into the sample buffer using a rubber policeman to ensure complete detachment from the plate. This mixture is transferred to a microcentrifuge tube and sonicated extensively until the viscosity of the mixture is decreased. The sample is heated in a boiling water bath for 3 min to allow complete denaturation and reduction of disulfide bonds, followed by microcentrifugation and SDS–PAGE.

Alternatively, cells can be lysed in other lysis buffers, as long as the protein(s) of interest is soluble in the buffer and the buffer components are compatible with SDS–PAGE. For example, a washed, 100-mm dish of cells can be lysed at 4° in 1 ml RIPA buffer (10 m*M* Tris–HCl, pH 7.4, 150 m*M* NaCl, 1% sodium deoxycholate, 1% Triton X-100, 0.1% SDS, 1 m*M* EDTA), a buffer commonly used for immunoprecipitations. A rubber policeman is used to scrape the cells from the plate into the buffer. The lysed cell mixture is briefly vortexed, followed by centrifugation at 100,000*g* at 4° for 30 min to remove insoluble debris. Then 5X sample buffer (250 μl) is added to the clarified lysate, which is boiled and loaded on the gel.

To work with smaller sample volumes, washed cells may be isolated by centrifugation at 1000*g* as described in steps 1–3 below followed by the addition of the desired amount of lysis buffer. Protein concentrations in lysates are determined using an assay compatible with lysate buffer ingredients (see p. 540).

[4] C. Stacey and M. Merion, *BioChromatography* **3,** 36 (1988).

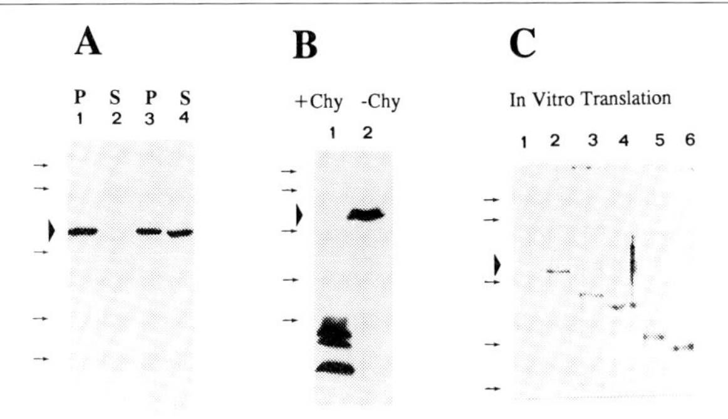

FIG. 1. Western blotting with anti-oncoprotein antibody. (A) Detection of pp60$^{v\text{-}src}$ in membrane fractions. Plasma membrane-enriched fractions were prepared from Rous sarcoma virus-transformed chicken embryo fibroblasts and treated with pH 7 buffer (lanes 1 and 2) or pH 11 buffer (lanes 3 and 4). Membrane-bound (pellet, P) and soluble (supernatant, S) fractions were isolated by ultracentrifugation, solubilized in SDS–PAGE sample buffer, electrophoresed, and Western blotted with anti-p60^{src} antibody. (B) Peptide mapping by Western blotting. Membrane-bound pp60$^{v\text{-}src}$ was incubated with chymotrypsin (lane 1) or with buffer (lane 2), and membrane-bound material was analyzed by Western blotting with anti-p60^{src} antibody. The antibody, which recognizes N-terminal epitopes, can be used to detect peptide fragments deriving from the N terminus of src. (C) Detection of *in vitro* translation products by Western blotting. Rabbit reticulocyte lysates were programmed with dH_2O (lane 1), mRNA encoding full-length pp60$^{v\text{-}src}$ (lane 2), or mRNA encoding v-src protein truncated at amino acid 432 (lane 3), 384 (lane 4), 306 (lane 5), or 258 (lane 6). Aliquots (10 μl) were analyzed by SDS–PAGE and Western blotting with the anti-src antibody just described. The migration positions of the molecular mass markers are indicated with arrows (from top to bottom: 106, 80, 49.5, 32.5, and 27.5 kDa); the arrowhead denotes the position of full-length pp60$^{v\text{-}src}$.

Cell Fractionation

It is often desirable to prepare subcellular fractions prior to protein detection by Western blotting (Fig. 1). A method based on differential centrifugation that is commonly used for fractionation of tissue culture fibroblasts is briefly described here.[5] This protocol is written for fractionation of one 100-mm plate of cells into P1, P100, and S100 fractions. All steps are performed at 4°.

1. Aspirate media from the plate.
2. Rinse plate with 5 ml cold STE (100 m*M* NaCl, 10 m*M* Tris, 1 m*M* EDTA, pH 7.4). Aspirate rinse from plate.
3. Add 5 ml cold STE to plate and scrape cells from plate using a rubber policeman. Centrifuge the cells for 5 min at 1000*g*. Aspirate the wash solution.

[5] M. D. Resh and R. L. Erikson, *J. Cell Biol.* **100,** 409 (1985).

4. Add 0.8 ml hypotonic lysis buffer (10 mM Tris, pH 7.4, 0.2 mM Mg Cl_2) to the pellet. Gently vortex and incubate on ice for 15 min.

5. Break open cells by homogenizing 30 strokes in a Dounce homogenizer.

6. Add 200 μl 5X sucrose (1.25 M sucrose) and 2 μl 0.5 M EDTA. Gently vortex.

7. Spin out nuclei by centrifugation at 1000g for 10 min.

8. Carefully remove the supernatant (S1) to an ultracentrifuge tube.

9. Resuspend the pellet in 1 ml 0.25 M sucrose, 10 mM Tris, pH 7.4, 1 mM EDTA and Dounce homogenize 10 strokes.

10. Centrifuge the pellet suspension at 1000g for 10 min. Combine the supernatant with the S1 supernatant above in an ultracentrifuge tube. The pellet is designated P1.

11. Centrifuge the S1 at 100,000g for 1 hr.

12. Remove the supernatant, designated S100, from the pellet, which is designated P100.

The P1 fraction is enriched with nuclei, whereas the P100 is enriched in plasma membranes and other intracellular membranes. The S100 contains cytosolic proteins. A fraction enriched in mitochondrial membranes can be obtained by centrifuging the S1 at 10,000g prior to the 100,000g spin. P100 membranes can also be further fractionated by centrifugation through sucrose gradients.

To prepare samples for Western blotting, the pellets are resuspended in 1X sample buffer for SDS–PAGE, whereas 5X sample buffer is added in the appropriate amount to the S100. The amounts of sample buffer used can be adjusted so that samples are at constant volume. Alternatively, protein concentrations can be determined prior to adding sample buffer and the final samples loaded according to protein content. Samples are heated in boiling water for 3 min, microcentrifuged, and loaded on a gel.

Immunoprecipitation

Western blotting of immunoprecipitated samples is commonly used to detect specific protein–protein interactions or to detect the presence of phosphotyrosine on specific immunoprecipitated proteins (Fig. 2). Immunoprecipitation can be performed using a variety of lysis buffers. We typically use RIPA buffer. However, to preserve weak protein–protein interactions, milder detergents such as digitonin and octyl glucoside may be more desirable than those used in RIPA buffer. When using antiphosphotyrosine antibody for Western blot detection, 1 mM Na_3VO_4 is added to the lysis buffer as a phosphatase inhibitor.

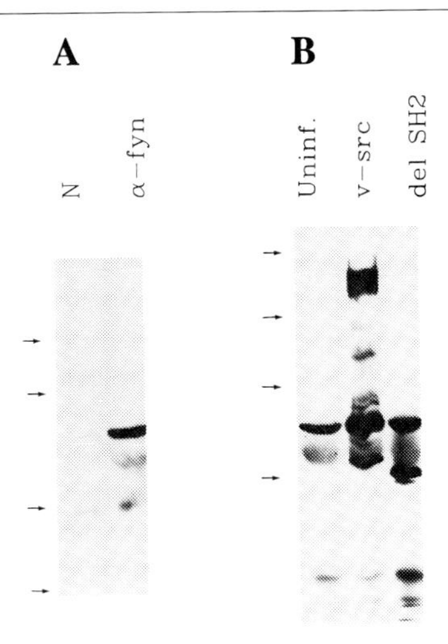

FIG. 2. Western blotting of immunoprecipitated oncoproteins. (A) Human B cell lines were lysed, immunoprecipitated with nonimmune (N) or anti-fyn antibody, and subjected to SDS–PAGE and Western blotting with the anti-pTyr antibody. The migration positions of the molecular mass markers are indicated with arrows (from top to bottom: 106, 80, 49.5, and 32.5 kDa). (B) Chicken embryo fibroblasts which were uninfected, infected with recombinant viruses encoding v-*src*, or a v-*src* mutant with the SH2 domain deleted were lysed and immunoprecipitated with anti-src antibody. The immunoprecipitates were analyzed by Western blotting with anti-pTyr antibody for associated tyrosine phosphorylated proteins. The migration positions of the molecular mass markers are indicated with arrows (from top to bottom: 205, 116, 80, and 49.5 kDa).

The following steps are then performed.

1. Samples in lysis buffer are clarified at 100,000*g*, 4°, for 30 min.
2. To 1 ml of clarified sample, add 2–5 μl of primary antibody, and incubate at 4° for 30 min to 2 hr.
3. Add 50 μl formalin-fixed *Staphylococcus aureus* (prepared as in Kessler[6]), and incubate for 20 min at 4°. Alternatively, immune complexes can be adsorbed to protein A–Sepharose or other similar solid supports.
4. Collect the immunoprecipitates by centrifugation (5 min in a tabletop centrifuge or 1 min in a microfuge). Resuspend the pellet in 1 ml lysis buffer, vortex, and recentrifuge. Repeat this step two to three times.
5. Transfer to a new 1.5-ml microfuge tube, and centrifuge for 2 min in a microfuge. Remove the supernatant.
6. Add 30–50 μl SDS–PAGE sample buffer to the pellet and resuspend. Brief sonication at this point assists in sample resuspension. Boil the suspen-

[6] S. W. Kessler, *J. Immunol.* **115,** 1617 (1975).

sion for 3 min and centrifuge for 2 min in the microfuge. Carefully remove the supernatant containing the solubilized proteins and load the sample on the gel.

It is important to note that the antibody used for immunoprecipitation must not bind the antibodies used for Western blot detection, otherwise the immunoglobulin chains of the immunoprecipitating antibody will react on the Western blot. This problem can be readily avoided by using, for example, a rabbit polyclonal antibody for immunoprecipitation and a mouse monoclonal antibody for Western blotting. The secondary antibody for blot detection would then be an anti-mouse antibody.

Standard or Positive Control

It is advantageous to include a positive control on the gel to be blotted to provide a standard for the subsequent transfer and detection steps. Lysates of cells overexpressing the protein of interest or lysates of cells expressing sufficient quantities of the endogenous protein to be detected can serve as positive controls. If available, enriched fractions or the purified protein can be used for this purpose. If the concentration of the protein of interest is known in the positive control, this can be used to estimate the concentration of the protein in the experimental samples (see the section on Quantitation).

Protein Assay of Samples Containing Detergent

Quantitation of the protein concentration of each sample is necessary if equivalent amounts of protein are to be loaded in each lane of the gel. Protein assay kits that are compatible with up to 1% of a variety of widely used detergents (SDS, Triton X-100, Brij 35, CHAPS, octylglucoside, etc.) are now available. These assays are based on the redox chemistry between proteins and copper ions. Note that reagents that are incompatible with the above assays include certain concentrations of EDTA and DTT. We have successfully used both the Pierce BCA microassay and the Bio-Rad DC assay in our laboratory. Protein concentrations must be in excess of 1–5 μg/ml to be detected in these assays.

Immunoblotting Protocol

There are many considerations to take into account before proceeding to the immunoblotting step, including the use of tank vs semidry blotting apparatus and the type of filter matrix to be used. The choice of electroblotting apparatus depends on the applications required. The use of the tank apparatus is preferable when the antigen is limited and quantitation of protein is important. The tank or "wet" method may also be more efficient at transferring high molecular weight proteins. This method, however, requires large amounts of buffer since the entire gel sandwich is submersed. In contrast, semidry electroblotters use substantially less buffer since only the filter paper sheets of the gel sandwich need to be saturated, and protein transfer may be completed in as little as 15–30 min. As for the choice of filter matrix, nitrocellulose, nylon, and polyvinylidene fluoride (PVDF) membranes are all commonly used, although we find that Immobilon-P (Millipore, Bedford, MA) PVDF membranes give the highest signal to background ratio when used in the following procedure. Likewise, we have optimized our system using a Genie electroblotter (Idea Scientific, Corvallis, OR), but other electroblotting apparatuses are available from a variety of manufacturers. Although the following protocol describes the use of a tank apparatus, all steps subsequent to transfer may be applied to the processing of a blot produced by semidry electroblotting as well.

1. Electrophoretic Blotting

The transfer buffer used was first described by Towbin *et al.*[1] (25 m*M* Tris, 192 m*M* glycine, 20% v/v methanol). The addition of SDS (<0.1%) to the buffer may help the transfer of high molecular weight proteins; however, care must be taken if blotting with Immobilon-PVDF since the use of this detergent may allow proteins to pass through the membrane. If necessary, a lower concentration of SDS (<0.01%) may be used.[7]

The basic gel sandwich for "wet" blotting consists of (1) the cathode plate, (2) foam pad or sponge (Scotch-Brite pads work well), (3) heavy filter paper, (4) gel, (5) nitrocellulose or PVDF membrane, (6) heavy filter paper, (7) sponge, and (8) the anode plate. Make sure that the membrane is on the anode (+) side of the blotter relative to the gel. The sponges should be presoaked, and the entire sandwich should be assembled in transfer buffer to avoid trapping bubbles.

To prepare the gel for blotting, separate the gel plates, leaving the gel adhering to one of the plates. Cut off the stacker and any extra lanes that do not contain sample. Place the gel onto a piece of heavy filter paper (such as Whatman 3MM) that has been wetted with transfer buffer. (It is sometimes easier to place the wet filter paper first on top of the gel, and gently "peel" the gel off the glass plate.) Cut a piece of Immobilon-PVDF slightly larger than the gel. Always use gloves when handling the membrane. It is important to note that unlike nitrocellulose, PVDF membranes must

[7] E. R. Tovey and B. A. Baldo, *J. Biochem. Biophys. Methods* **19,** 169 (1989).

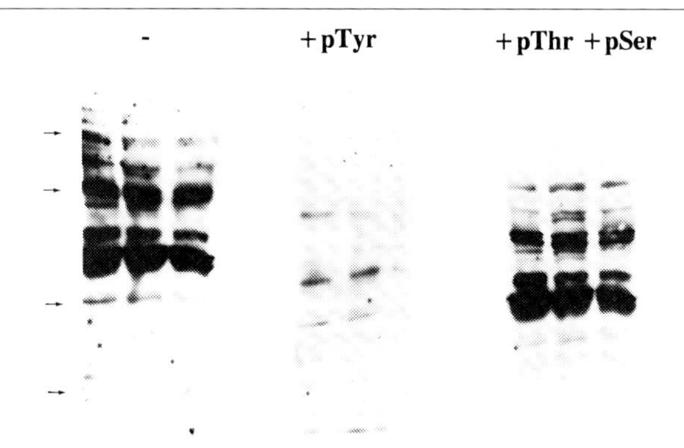

FIG. 3. Detection of phosphotyrosine-containing proteins by Western blotting. Whole cell lysates from three human B cell lines containing 80 μg of protein were separated by SDS–PAGE. Western blotting was performed with anti-pTyr antibody with no addition (−), 1 m*M* phosphotyrosine, or 1 m*M* phosphothreonine + 1 m*M* phosphoserine. The migration positions of the molecular mass markers are indicated with arrows (from top to bottom: 106, 80, 49.5, and 32.5 kDa).

be prewet in alcohol before being used with aqueous solutions. Submerge the PVDF membrane in methanol, then immerse in water for 5 min. Lay the wet membrane on top of the gel, being careful not to leave any air bubbles in between. It may be helpful to cut one corner of the membrane for orientation. Finish assembling the gel sandwich with moistened filter paper and enough sponge to ensure a tight contact between the gel and membrane when placed in the transfer chamber.

Blotting is carried out at 460 mA with a constant current for 1–2 hr at room temperature.

2. Blocking

For most applications the use of 1–5% bovine serum albumin (BSA) (w/v) is adequate for the blocking of nonspecific protein binding sites. We typically use 3% BSA (fraction V) with 10% calf serum in saline solution [0.9% NaCl, 10 m*M* Tris–HCl, (pH 7.4), 0.5 g/liter $MgCl_2$] as a blocking buffer. Other common blocking agents include 2–5% nonfat dry milk, gelatin, ovalbumin, and serum. Small concentrations (0.01–1.0%) of a mild detergent such as Tween 20 or Triton X-100 can also be included.

It is important to note that if antiphosphotyrosine antibodies are to be used as the primary antibody (Fig. 3), nonfat dry milk should not be used as a blocking agent.[8] Also, if protein A is to be used in the detection

[8] M. Kamps, this series, Vol. 201, p. 101.

process, serum should not be used during the blocking step or nonspecific recognition may occur.

After electrophoretic blotting is completed, the membrane should be fully immersed in blocking solution and incubated with gentle agitation for either 1 hr at 37° or overnight at 4°.

3. *Primary Antibody*

The amount of primary antibody to be used will obviously depend on the titer, and the optimal working concentration may have to be determined by trial and error. We typically use a 1:1000 dilution (about 1 μg/ml) of antibody into fresh blocking buffer (3% BSA in saline). The addition of 10% calf serum to the blocking buffer at this step appears to greatly reduce nonspecific binding by the primary antibody, although we do not add the serum if antiphosphotyrosine antibodies are being used.

Incubation of the blot should be done in a small container to minimize the amount of primary antibody used. We have found that best results are obtained by using disposable square petri dishes (Nunc, Naperville, IL) or the plastic cover of a pipette tip rack, where only 10–20 ml of antibody solution is required. Blots are transferred into the primary antibody solution after blocking and are incubated at 4° for 1 hr with very gentle agitation.

4. *Washing*

After incubating in primary antibody, the blot is washed with several changes of buffer to remove excess, unbound antibody. The standard immunoblotting protocol calls for the inclusion of small amounts of detergent (usually 0.05% Tween 20 or NP-40) in some of the washing steps. We have found that the addition of up to 3% Tween 20 in two of the incubations helps to eliminate background problems without affecting the integrity of the antibody–protein interaction. Each incubation is performed for 5 min at room temperature with gentle agitation: two times with saline; two times with saline + 3% Tween 20; and two times with saline.

5. *Secondary Antibody or Protein A*

If the primary antibody is derived from a species that binds protein A or protein G, a labeled form of these proteins can be used in the detection process. Protein A and protein G conjugated to ^{125}I, horseradish peroxidase (HRP), or alkaline phosphatase are all commercially available. Protein A binds well to rabbit, guinea pig, and human immunoglobulin G (IgG), but poorly to mouse, rat, and chicken antibodies. Mouse IgG_1 monoclonal immunoglobulins, as well as goat and sheep antibodies, bind strongly to protein G.

Alternatively, a labeled secondary antibody that is directed against the species of the primary antibody can be used. Again, these antibodies are conjugated to various compounds that enable detection and are commercially available.

We most often use an appropriate secondary antibody conjugated to HRP (anti-rabbit Ig-HRP from donkey or anti-mouse Ig-HRP from sheep; Amersham, Arlington Heights, IL). The secondary antibody is diluted 1:2000 in fresh blocking buffer. Again, if antiphosphotyrosine antibodies are not being used, 10% calf serum is added to the antibody solution. The blot is incubated with the secondary antibody for 30 min with gentle agitation at 4°.

6. Washing

After incubation with the secondary antibody the blot must be washed as in step 4. The washes are again 5 min each at room temperature for a total of 30 min: twice with saline; twice with saline + 3% Tween 20; and twice with saline.

7. Detection

A variety of methods can be utilized for detection, including the use of colorimetric substrates, chemiluminescence, and iodinated antibodies or protein A. Colorimetric methods such as horseradish peroxidase can be limited in their sensitivity and the stained blot may be subject to fading if not stored properly. The use of iodinated antibodies or protein A has the advantage of increased sensitivity and the permanence of recording results by autoradiography, but suffers from the inherent problems associated with the handling and disposal of a radioisotope. In our experience, detection by chemiluminescence gives the highest sensitivity and the most reproducible results for the applications used in our laboratory. Chemiluminescence allows for rapid multiple exposures (typically between 10 sec and 5 min) that can be recorded on X-ray film without the hazards of working with a radioisotope.

A. Chemiluminescence. If a secondary antibody conjugated to horseradish peroxidase was used in step 5, detection can be achieved either by chemiluminescence or by the use of a colorimetric method (see below). Although several chemiluminescence detection reagents are commercially available, we use the enhanced chemiluminescence (ECL) Western blotting kit (Amersham). Equal volumes of the provided reagents are mixed according to the manufacturer's instructions to form the detection solution. We find that the addition of BSA to a concentration of 3% (w/v) in this detection solution greatly improves the appearance of the blot. After incubating in detection solution for 1 min, the liquid is drained, and the blot is placed on Whatman paper, covered with Saran wrap, and exposed to X-ray film. Exposure times can vary from 1 sec to 20 min, depending on the antigen and antibody.

B. Horseradish Peroxidase. The detection of horseradish peroxidase-conjugated antibodies or protein A can also be achieved by utilization of a colorimetric substrate such as chloronaphthol. Make the following solutions up fresh before each use. (1) Dissolve 15 mg of chloronaphthol in 5 ml methanol. (2) Add 25 ml saline [0.9% NaCl, 10 m*M* Tris base (pH 7.4)]. (3) Immediately before use add 16 μl 30% H_2O_2. After removing the last wash, add the substrate solution to the blot. Incubate on a shaker at room temperature for 10–15 min, monitoring carefully. When the bands are sufficiently dark, stop the reaction by removing the substrate and rinsing with several changes of distilled water. Dry the blot between filter paper and store protected from light to prevent fading (wrap in aluminum foil).

C. Alkaline Phosphatase. The following stock solutions should be made and stored at 4°: (1) 0.1 *M* NaCl, 0.01 *M* $MgSO_4$, 0.1 *M* Tris–HCl (pH 9.5); (2) 50 mg nitro blue tetrazolium per ml of 70% dimethylformamide; and (3) 50 mg of 5-bromo-4-chloro-3-indolyl phosphate (BCIP) per ml of 100% dimethylformamide. Add 100 μl of solution 2 and 50 μl of solution 3 to 15 ml of solution 1. Add substrate solution to the blot and monitor carefully while shaking. Stop reaction by removing substrate and washing several times with water.

D. ^{125}I. If iodinated antibody or protein A is used in step 5, the blot is dried between filter paper after the last wash step, then mounted on Whatman paper or cardboard with cellophane tape. The blot is covered with Saran wrap and exposed to X-ray film.

E. General Protein Stains. To visualize the blotted proteins, two methods can be used. A duplicate set of samples can be blotted and stained with 0.1% amido black or 0.2% Coomassie blue R-250 in 40% v/v methanol/10% acetic acid. The blot is stained for 2–3 min, rinsed for 30 sec with water, and destained with three incubations (1 min each) in 90% methanol/2% acetic acid. The blot is rinsed with water for 2 min. Transferred proteins appear as blue/black bands.

Alternatively, proteins can be stained on the same blot after the antibody detection steps are performed.[9] The blot is stained with a solution containing 1 μl/ml black India ink and 0.4% Tween 20 in PBS for 1 hr. Surprisingly, the BSA present in the blocking solution does not react with the ink, and only the transferred proteins are stained.

[9] J. Glenney, *Anal. Biochem.* **156,** 315 (1986).

Rapid Western Blotting: From Gel to Blot in Less Than 6 hr

The following is a modified blotting protocol for high-resolution antigen detection that can be accomplished in as little as 6 hr.[10] This technique has worked well for the detection of phosphotyrosine-containing proteins probed with antibodies to the oncogene product of Rous sarcoma virus (RSV), pp60$^{v\text{-}src}$, as well as other src family members, and antiphosphotyrosine (anti-pTyr) antibodies.

Samples are prepared for SDS–polyacrylamide gel electrophoresis by any of the methods described in the section on sample preparation. The fastest method for preparation of whole cell lysates is to lyse cells directly into SDS/sample buffer. Gel electrophoresis is conducted using a standard apparatus at 25–30 mA for approximately 1.5 hr. Alternatively, a minigel apparatus (Bio-Rad Laboratories, Hercules, CA) can be used, which takes approximately 30 min at the same current.

Blotting is carried out in just 30 min using a PVDF membrane (Immobilon-P, Millipore, Bedford, MA) and a Mini-Genie electroblotter (Idea Scientific). Follow the protocol described in the section on electrophoretic blotting for preparation of the basic gel sandwich. Blotting is carried out at 460 mA with constant current for 30 min at room temperature. The use of prestained molecular weight markers (Bio-Rad Laboratories) during gel electrophoresis allows detection of protein transferred to the blot and obviates the need to stain the gel post-transfer. To complete the protocol in 1 day, blocking is performed by incubating the blot for 1 hr at 37° instead of overnight at 4°.

Incubation of the blot with primary antibody and washing are performed as described earlier. Incubations with antibody can be performed using disposable square petri dishes as described. Alternatively, the blot can be sealed together with the antibody solution in a Seal-A-Meal bag (Dazey, Industrial Airport, KS). This will minimize the volume of the antibody solution and thereby minimize the amount of antibody needed for incubation.

After the washes, the blot is incubated for 30 min with a secondary antibody conjugated to horseradish peroxidase, then washed again (see above section). The ECL Western blotting kit (Amersham) is used for detection. Equal volumes of two reagents are combined, with the addition of BSA to a final concentration of 3% (w/v). The blot is incubated in this detection solution for 1 min, then placed on Whatman paper, covered with Saran wrap, and exposed to X-ray film.

[10] A. Klapper, B. MacKay, and M. D. Resh, *BioTechniques* **12,** 650 (1992).

Blotting for Protein Sequencing

Microgram quantities of purified protein are generally required for sequencing. When the last step in protein purification entails SDS–PAGE, blotting the protein onto a suitable membrane provides an extremely convenient method to prepare purified protein for sequencing. The blotting step removes contaminants such as SDS and acrylamide which would otherwise interfere with the sequencing. Care should be taken that all reagents used for SDS–PAGE and blotting are of the highest quality.

Before committing an entire sample for blotting, it is judicious to perform a pilot experiment to determine blotting yield. Yield is dependent on many factors, such as the molecular weight and isoelectric point of the protein, the percentage of acrylamide used in the gel, the presence or absence of SDS in the transfer buffer, etc. If the protein is radiolabeled, yield can readily be estimated by excising the radioactive band in the blot and comparing this radioactivity with the initial amount. Whether or not the protein is radioactive, it is always advisable to load a standard protein of similar molecular weight on the gel in amounts estimated to be within the range of the amount of protein to be sequenced, e.g., loading samples of the standard ranging from 1 to 20 μg. After blotting, both the blot and the gel are stained to assess the completeness of transfer. The standards allow a reasonable estimate to be made of the amount of purified protein to be submitted for sequencing.

Yield can be optimized in a number of ways. Transfer parameters such as time and current can be adjusted taking into account limitations with respect to heat generation (see below). If protein loss is due to precipitation, adding a low level of SDS to the transfer buffer may be beneficial; however, since SDS also reduces binding to the membrane, the level of SDS must be carefully optimized. The amount of methanol in the buffer may also be adjusted to optimize transfer. Alcohol has the competing effects of reducing pore size in the gel (thus making transfer of high MW proteins more difficult) but increasing binding to the membrane.

Amino-Terminal Sequencing

Blotting for amino-terminal sequencing is performed using a PVDF membrane. To avoid glycine contamination, the transfer buffer devised by Matsudaira[11] is commonly used (10 m*M* CAPS, 10% methanol, pH 11.0). Generally, blotting can be performed according to the blotter manufacturer's recommendations for time and current/voltage. The blot is then stained

[11] P. Matsudaira, *J. Biol. Chem.* **262,** 10035 (1987).

with Coomassie blue[11] as follows: Immerse the blot in staining solution (0.1% Coomassie blue in 50% methanol) for 5 minutes. Destain in 50% methanol/10% acetic acid for 5–10 min. Rinse the blot in distilled H_2O for 5–10 min, air dry, and store at −20°. The protein band to be sequenced is excised from the blot using a clean razor blade or scalpel.

Internal Sequencing

Sequencing via Edman degradation requires a free amino terminus. Since the amino termini of many proteins are blocked, usually by acetylation, sequencing of internal sequences is required. In this case, the purified protein is blotted and the band excised. The blotted protein is digested *in situ* with protease, and the resulting peptides are eluted, purified by HPLC, and sequenced.

The preferred membrane material for internal sequencing is nitrocellulose since it is difficult to elute proteolytic fragments from PVDF using HPLC compatible buffers. The standard Towbin buffer[1] (25 m*M* Tris, 192 m*M* glycine, 20% methanol, pH 8.3) can be used in the transfer, as residual glycine is removed via HPLC. It is important not to "bake" the protein on the blot so that good elution yields are obtained. Therefore, blotting is performed using a cooling system or at 4°. We have had good results blotting at a constant voltage of 15 V overnight in the cold room.

After transfer, the nitrocellulose blot is stained with Ponceau S dye according to the method of Aebersold *et al.*[12] The blot is immersed for 1 min in staining solution (0.1% Ponceau S dye, 1% acetic acid) followed by 1 to 2 min in destaining solution (1% acetic acid). The desired band is excised using a scalpel, taking care to minimize the area that is cut out, as excess membrane can lower the efficiency of the subsequent protease digestion. A blank piece of the blot of similar size is also excised to use as a blank control in the subsequent steps. Each excised nitrocellulose piece is stored in high-purity deionized water at −20° until further processing.

Quantitation

Depending on the detection system used, Western blots can be quantitated using densitometry or phosphor imaging. Blots are developed so that the resulting band intensities are in a linear range. To estimate absolute concentrations, appropriate standards can be electrophoresed and transferred with samples of unknown concentrations. To determine relative

[12] R. H. Aebersold, J. Leavitt, R. A. Saavedra, L. E. Hood, and S. B. H. Kent, *Proc. Natl. Acad. Sci. U.S.A.* **84,** 6970 (1987).

sample concentrations, care should be taken to prepare samples at a constant basis appropriate for the application, such as constant protein concentration or sample volume.

For blots developed using the chemiluminescent technique, the resulting film is scanned and quantitated using a densitometer in the transmittance mode. In our laboratory we have used this method to quantitate the distribution of Src and Src chimeras in cell fractions (P1, P100, and S100). The same procedure applies for autoradiograms resulting from blots developed using radioactive probes. Alternatively, the use of phosphor imaging to detect radioactive probes (^{125}I, ^{32}P, or ^{35}S) provides an extremely convenient means of image printout and quantitation. Blots developed using colorimetric substrates such as those for alkaline phosphatase and horseradish peroxidase are quantitated using reflectance densitometry.

Additional Applications

Besides protein–antibody interactions, other protein–ligand interactions may be exploited for detection. A wide variety of ligands have been used for this purpose, including nucleotides for the detection of proteins such as GTPases; lectins for the detection of glycoproteins; hormones for the detection of receptors; and so forth. The ligand which serves as a probe can be labeled with radioactivity or detected by subsequent antibody recognition.

Although beyond the scope of this chapter, two additional features of Western blotting can be exploited. The first is the ability to "erase" the blot by removing bound antibodies. The blot can then be reprobed with a different antibody.[13] The second feature is the ability to renature some proteins while they are bound to the filter membrane.[14] Enzymatic activity or ligand binding assays can then be performed on the reactivated protein.

Troubleshooting

Here are some of the more commonly experienced problems that we have encountered and some suggested solutions.

1. The signal is too low. (a) If staining of the blot shows inadequate transfer of protein, try increasing the transfer time during blotting or try adding SDS to the blotting solution. (b) Increase the concentration of primary and/or secondary antibody used. (c) Increase the amount of protein initially loaded on the gel.

[13] S. H. Kaufmann, C. M. Ewing, and J. H. Shaper, *Anal. Biochem.* **161,** 89 (1987).

[14] J. E. Ferrell, Jr. and G. S. Martin, this series, Vol. 200, p. 430.

2. The background is too high. (a) Decrease the concentration of primary and/or secondary antibody. (b) Increase the amount and/or duration of the washes. (c) There may be too much or too little agitation during the washes. (d) Change the blocking agent. (e) For chemiluminescence detection, it may be helpful to leave the blot at room temperature for several minutes, prior to exposure to X-ray film. Background signals often decrease during this "rest" period.

3. Missing patterns on blot; blots with smudges or smears. (a) This suggests poor contact between the gel and the membrane during blotting. Use a Pasteur pipette as a rolling pin to roll over the membrane and remove air bubbles between the gel and membrane. (b) As the Scotch Brite pads flatten with age, they need to be replaced to ensure adequate contact between the gel and membrane. (c) If the transfer time is too long, the transfer buffer may overheat and dry out areas of the blot causing missing areas on the blot.

4. Multiple bands appear on the blot. (a) The antigen may have undergone proteolysis. The addition of protease inhibitors during sample preparation may prevent this. (b) When using different antibodies simultaneously on multiple blots, it is important to wash each blot in a separate container to avoid cross-contamination of the antibodies.

[37] Phosphorylation of Transcription Factors

By CHRISTOPHER C. FRANKLIN, VICTOR ADLER, and ANDREW S. KRAFT

Introduction

Post-translational phosphorylation plays an important role in the regulation of transcription factor activity.[1] Certain transcription factors reside in the cell cytoplasm and are translocated into the nucleus in a phosphorylation-dependent manner. This type of regulation is exemplified by the NF-κB family of transcription factors which are complexed in the cytosol by members of the I-κB family.[2] Phosphorylation of the I-κB protein by protein kinase A or protein kinase C releases NF-κB, which is then translocated to the nucleus in an active, DNA-binding state.[3] Alternatively, the nuclear localized transcription factor, CREB, may be regulated by the translocation

[1] T. Hunter and M. Karin, *Cell (Cambridge, Mass.)* **70,** 375 (1992).
[2] F. Shirakawa and S. B. Mizel, *Mol. Cell. Biol.* **9,** 2424 (1989).
[3] S. Ghosh and D. Baltimore, *Nature (London)* **344,** 678 (1990).

METHODS IN ENZYMOLOGY, VOL. 254

of the catalytic subunit of protein kinase A into the nucleus.[4] Within the nucleus, post-translational phosphorylation dramatically affects both DNA-binding activity and transcriptional activity.[1] With few exceptions, phosphorylation of the basic DNA-binding domain of transcription factors inhibits their DNA-binding activity. On the other hand, phosphorylation of the activation domain of a transcription factor generally enhances its transcriptional activity. In some cases, DNA binding and dimerization play a key role in regulating the level of transcription factor phosphorylation.[5-7]

A major focus of our own research has been the regulation of the Jun family of transcription factors. This family of proteins binds the AP-1 enhancer sequence (5′ TGAC/GTCA 3′) as a homo- or heterodimeric complex with other Jun or Fos proteins.[8] Phosphorylation of the c-Jun carboxy-terminal DNA-binding domain by either glycogen synthase kinase 3 or casein kinase II inhibits c-Jun DNA-binding activity.[9,10] Mitogenic stimulation or expression of transforming oncogenes induces both the dephosphorylation of these inhibitory sites and the phosphorylation of the c-Jun amino-terminal activation domain which stimulates c-Jun transcriptional activity.[11,12] The methods used in our laboratory to investigate the regulation of c-Jun phosphorylation can be generally applied to other transcription factors whose phosphorylation status is unknown. The study of transcription factor phosphorylation is aided by the modular nature of these proteins. Distinct regions of these proteins mediate DNA binding, dimerization, and transcriptional activation.[13] This permits discrete functional domains to be fused to other proteins for biochemical analysis. We have exploited this feature to examine the mechanisms regulating c-Jun phosphorylation *in vivo* and *in vitro*, to determine the role of phosphoryla-

[4] M. Hagiwara, P. Brindle, A. Harootunian, R. Armstrong, J. Rivier, W. Vale, R. Tsien, and M. R. Montminy, *Mol. Cell. Biol.* **13,** 4852 (1993).

[5] C. Abate, S. J. Baker, S. P. Lees-Miller, C. W. Anderson, D. R. Marshak, and T. Curran, *Proc. Natl. Acad. Sci. U.S.A.* **90,** 6766 (1993).

[6] A. J. Bannister, T. M. Gottlieb, T. Kouzarides, and S. P. Jackson, *Nucleic Acids Res.* **21,** 1289 (1993).

[7] A. G. Papavassiliou, C. Chavrier, and D. Bohmann, *Proc. Natl. Acad. Sci. U.S.A.* **89,** 11562 (1992).

[8] P. Angel and M. Karin, *Biochim. Biophys. Acta* **1072,** 129 (1992).

[9] W. J. Boyle, T. Smeal, L. H. K. Defize, P. Angel, J. R. Woodgett, M. Karin, and T. Hunter, *Cell (Cambridge, Mass.)* **64,** 573 (1991).

[10] A. Lin, J. Frost, T. Deng, T. Smeal, N. Al-Alawi, U. Kikkawa, T. Hunter, D. Brenner, and M. Karin, *Cell (Cambridge, Mass.)* **70,** 777 (1992).

[11] T. Smeal, B. Binetruy, D. Mercola, A. Grover-Bardwick, G. Heidecker, U. R. Rapp, and M. Karin, *Mol. Cell. Biol.* **12,** 3507 (1992).

[12] B. J. Pulverer, J. M. Kyriakis, J. Avruch, E. Nikolokaki, and J. R. Woodgett, *Nature (London)* **353,** 670 (1991).

[13] M. Ptashne, *Nature (London)* **335,** 683 (1988).

tion in the regulation of c-Jun transcriptional activity, and to identify putative protein kinases that may mediate this phosphorylation.

This chapter provides both *in vivo* and *in vitro* methods to (1) determine whether the transcription factor under study is phosphorylated *in vivo*, (2) analyze the sites of phosphorylation, and (3) examine the nature of the protein kinase(s) that mediates this phosphorylation. These methods require either transcription factor-specific antisera or the cloned cDNA for the protein of interest. Examples of these protocols using the c-Jun transcription factor are described both in the text and with the figures.

In Vivo Labeling of Cultured Cells

A relatively simple method to determine whether a transcription factor is phosphorylated *in vivo* is by analyzing immunoprecipitates from cells metabolically labeled with ortho[^{32}P]phosphate. Post-translational phosphorylation often results in a decrease in the electrophoretic mobility of the protein when analyzed by SDS–PAGE. This mobility shift allows certain phosphorylation events, such as the amino-terminal phosphorylation of c-Jun[12,14,15] and the carboxy-terminal phosphorylation of c-Fos,[16–18] to be monitored by changes in the electrophoretic mobility of proteins immunoprecipitated from cells radiolabeled with [^{35}S]methionine. Consequently, pulse–chase labeling with [^{35}S]methionine can be used to analyze the kinetics of transcription factor phosphorylation.[16–19]

Procedure

1. Wash 5–10 × 10^6 cells twice with Tris-buffered saline (TBS) (15 m*M* Tris–HCl, pH 7.5, at 37°, 150 m*M* NaCl) warmed to 37° and then once with minimum essential medium (MEM) lacking either sodium phosphate (No. 16-227-49; GIBCO, Grand Island, NY) or methionine (No. 16-222-54; GIBCO) and supplemented with 2 m*M* glutamine and 25 m*M* HEPES, pH 7.4. The inclusion of HEPES both buffers the radioisotope and permits cell labeling in the absence of CO_2. If serum is to be present during labeling, it must be dialyzed extensively against TBS to remove trace levels of phosphate and methionine. Preincubate cells for 30 min in media lacking either

[14] C. C. Franklin, V. Sanchez, F. Wagner, J. R. Woodgett, and A. S. Kraft, *Proc. Natl. Acad. Sci. U.S.A.* **89,** 7247 (1992).
[15] E. J. Black, A. J. Street, and D. A. F. Gillespie, *Oncogene* **6,** 1949 (1991).
[16] T. Curran, A. D. Miller, L. Zokas, and I. M. Verma, *Cell (Cambridge, Mass.)* **36,** 259 (1984).
[17] J. R. Barber and I. M. Verma, *Mol. Cell. Biol.* **7,** 2201 (1987).
[18] T. Curran, M. B. Gordon, K. L. Rubino, and L. C. Sambucetti, *Oncogene* **2,** 79 (1987).
[19] C. C. Franklin and A. S. Kraft, unpublished observation (1993).

phosphate or methionine. Label the cells in 1–2 ml fresh media containing either 0.5–2 mCi/ml carrier-free ortho[^{32}P]phosphate (No. PBS.43, 40 mCi/ml; Amersham, Arlington Heights, IL) for 6–12 hr or 0.1 mCi/ml Tran^{35}S-label (No. 51006, 1100 Ci/mmol; ICN, Irvine, CA) for 1–4 hr. Cells grown in monolayers can be labeled in a tissue culture incubator with the appropriate radioactive screening. Radiolabeling of suspended cells with ortho[^{32}P]phosphate can be enhanced by gentle shaking in a 37° water bath.

2. For pulse–chase labeling the cells are washed and preincubated as described earlier. The cells are then pulse labeled in 1 ml of fresh methionine-free MEM containing 0.2 mCi/ml Tran^{35}S-label. After 15 min the cells are washed twice with MEM containing 1 m*M* unlabeled L-methionine and are "chased" in 5 ml of the same media for various periods of time (usually 15 min to 2 hr).

3. Radiolabeled cells are washed twice with ice-cold phosphate-buffered saline (PBS) (138 m*M* NaCl, 2.7 m*M* KCl, 8.1 m*M* $Na_2HPO_4 \cdot 7H_2O$, 1.2 m*M* KH_2PO_4, pH 7.4). Suspended cells are harvested by centrifugation, whereas attached cells are harvested either by scraping into PBS with a rubber policeman or by pipetting up and down in lysis buffer (see below). From this point on it is very important to work on ice and to always have protease and phosphatase inhibitors present to minimize protein degradation and dephosphorylation during manipulation of the sample.

Comments

Serum induces the phosphorylation of many transcription factors. Therefore, it is suggested that the cells are either labeled in the absence of serum or are preincubated in fresh labeling media prior to cell stimulation. The labeling time necessary for sufficient incorporation of ^{32}P into the ATP pool will have to be determined empirically for each cell line examined. To ensure steady-state labeling with ortho[^{32}P]phosphate, cells should be labeled overnight. Similarly, for pulse–chase labeling, the transcription factor of interest must incorporate enough [^{35}S]methionine during the labeling period for analysis and may require pulse times longer than 15 min. This is dependent on the half-life and the rate of synthesis of the transcription factor. The number of cells to be used is dependent on the abundance of the transcription factor and its relative state of phosphorylation. Adequate labeling may be possible using less than 1×10^6 cells per immunoprecipitation.

Immunoprecipitation of *in Vivo*-Labeled Proteins

The radiolabeled transcription factor can now be purified by immunoprecipitation and analyzed by SDS–PAGE and autoradiography. This approach is dependent on the availability of an antibody specific to the transcription factor of interest. Immunoprecipitation with specific antisera can coprecipitate a complex set of proteins. In this section we supply several useful hints to reduce the copurification of nonspecific protein complexes.

Procedure

1. It is important to remember that from this point on all waste is radioactive and should be disposed of properly. Radiolabeled cells are harvested and crude nuclei are prepared by gently resuspending cells by pipetting in 1 ml of ice-cold hypotonic lysis buffer [10 m*M* Tris–HCl, pH 7.4 at 4°, 10 m*M* NaCl, 3 m*M* $MgCl_2$, 0.1% Nonidet P-40 (NP-40)] containing 0.5 m*M* phenylmethylsulfonyl fluoride (PMSF), 0.5 m*M* benzamidine hydrochloride, 1 μg/ml each of pepstatin A, leupeptin, and aprotinin, 50 m*M* NaF, 5 m*M* β-glycerophosphate, and 0.1 m*M* $NaVO_3$, followed by a 5-min incubation on ice.[20] Nuclei are recovered by centrifugation at 500*g* for 5 min at 4°. This step should be omitted if the transcription factor of interest resides solely or partially in the cell cytoplasm.

2. Lysates can be prepared in either a native or a denatured state. Native lysates are prepared by lysis in RIPA buffer (10 m*M* Tris–HCl, pH 7.4, 1% deoxycholate, 1% NP-40, 0.1% SDS, 150 m*M* NaCl, 5 m*M* EDTA) containing protease and phosphatase inhibitors (see above). Intact cells or isolated nuclei are resuspended in 0.5 ml of RIPA and rotated at 4° for 1 hr. Denatured lysates are prepared by boiling the samples for 5 min in 0.1 ml of denaturing lysis buffer [50 m*M* Tris–HCl, pH 8.0, 0.5% SDS, 5 m*M* dithiothreitol (DTT)] followed by a 1:5 dilution with 0.4 ml ice-cold RIPA buffer lacking SDS.

3. Lysates are clarified by centrifugation at 12,000*g* for 15 min, and the supernatant is precleared with 20 μl of a 50% slurry of protein A–Sepharose beads (Pharmacia, Piscataway, NJ). The mixture is rotated for 30 min at 4°, and the supernatant is recovered by centrifugation at 12,000*g* for 1 min.

4. Specific antiserum (1 μl; 1:500 dilution) is added, and samples are rotated at 4° for 1 hr. Immune complexes are collected by the addition of 20 μl of a 50% slurry of protein A–Sepharose beads for 30 min. The immune complexes are washed three times with RIPA buffer, once with 0.5 *M* LiCl, 0.1 *M* Tris–HCl (pH 7.9), and once with RIPA buffer.

5. To determine whether the retarded mobility of a protein on SDS–PAGE is due to protein phosphorylation, the immunopurified protein(s) can be dephosphorylated while complexed to the beads. Resuspend the beads in 50 μl of phosphatase buffer (50 m*M* Tris, pH 8.0, 5 m*M*

[20] M. E. Greenberg and E. B. Ziff, *Nature* (*London*) **311,** 433 (1984).

$MgCl_2$, 100 m*M* NaCl, 5% glycerol) containing 5 U calf intestinal alkaline phosphatase (CIAP) (No. 713 023; Boehringer Mannheim, Indianapolis, IN) and incubate for 1 hr at 37°. Wash the beads twice with PBS.

6. The washed beads are boiled for 5 min in 1× SDS sample buffer [1% (w/v) SDS, 10% (v/v) glycerol, 62 m*M* Tris–HCl, pH 6.8, 10% (v/v) 2-mercaptoethanol], and eluted proteins are resolved on a 10% SDS gel. Gels are fixed for 1 hr in 30% methanol and 10% acetic acid, dried, and labeled proteins are visualized by autoradiography at −70° with intensifying screens. Gels containing ^{35}S-labeled proteins are exposed to En^3Hance (No. NEF 981; New England Nuclear, Boston, MA) for 1 hr and washed with water for 30 min prior to drying. En^3Hance should be saved and can be reused at least five times.

Comments

Although immunoprecipitation of proteins is routinely carried out in many laboratories, analysis of specific ^{32}P-labeled proteins can be difficult because of significant contamination from highly phosphorylated nonspecific proteins. Several modifications of the standard immunoprecipitation protocol can help to reduce this background. Since the majority of transcription factors reside in the nucleus, the isolation of crude nuclei removes a large number of contaminating cytoplasmic proteins prior to immunoprecipitation. However, several transcription factors reside in the cell cytoplasm prior to activation, as is the case for NF-κB.[2] The nuclear isolation protocol described in step 1 may be omitted if the transcription factor is localized even transiently in the cytoplasm.

Extracts prepared using the relatively nondenaturing RIPA buffer (0.1% SDS) often contain transcription factor complexes formed by both specific and nonspecific protein–protein interactions. Depending on the strength of interaction, these proteins will coimmunoprecipitate with the protein of interest. Boiling in denaturing lysis buffer (0.5% SDS) dissociates these protein complexes. The dissociation of such complexes permits the differentiation between specifically coprecipitated proteins, such as Jun/Fos heterodimeric complexes, and post-translationally modified proteins that exhibit heterogeneous electrophoretic mobilities on SDS–PAGE.

When testing a new antibody it is necessary to demonstrate that the protein of interest is immunoprecipitated under the appropriate conditions. ^{35}S labeling used in conjunction with ^{32}P labeling (1) ensures that the protein of interest is actually being immunoprecipitated, (2) ensures that an increase in phosphorylation is due to increased phosphate incorporation and not to an increase in protein synthesis, (3) can be utilized to analyze the kinetics of transcription factor phosphorylation by pulse–chase labeling, and (4) can be used to demonstrate that the reduced electrophoretic mobility of a protein on SDS–PAGE is due to post-translational phosphorylation by the abolishment of the retarded bands by treatment with CIAP (step 5). The washing procedure described earlier is very stringent and is dependent on the specificity and avidity of the antibody. When testing a new antibody it is advisable to utilize less stringent wash conditions, such as a low salt buffer containing 1% NP-40 and lacking SDS.

In Vitro Phosphorylation of Transcription Factors

A number of *in vitro* techniques can be used to analyze transcription factor phosphorylation and to characterize a putative protein kinase(s) that may mediate this phosphorylation *in vivo*. These *in vitro* techniques include (1) analysis of the electrophoretic mobility of transcription factors synthesized by *in vitro* transcription/translation, (2) *in vitro* protein kinase assays using immobilized transcription factor fusion proteins as substrates, and (3) "in gel" protein kinase assays to identify putative protein kinases.

Analysis of Transcription Factor Phosphorylation during Cell-Free *in Vitro* Translation

Proteins can be synthesized *in vitro* by translation of RNA transcribed from the cDNA of a transcription factor using one of several cell-free translation systems. Proteins translated *in vitro* using the rabbit reticulocyte lysate (RRL) translation system often undergo extensive post-translational modifications. Reticulocyte lysate contains a number of protein kinases which can induce protein phosphorylation during translation. For example, c-Jun and c-Fos protein translated in rabbit reticulocyte lysate are highly phosphorylated and retarded in their electrophoretic mobility on SDS–PAGE.[12,15,18] The use of this approach to analyze protein phosphorylation is absolutely dependent on this phosphorylation-induced electrophoretic mobility shift.

Procedure

1. Plasmid DNA (20 μg) containing a SP6, T7, or T3 promoter is linearized with the appropriate restriction endonuclease. Linearized DNA is extracted with phenol/chloroform and precipitated in the presence of 0.1 vol 3 *M* sodium acetate and 2.5 vol 100% ethanol for 1 hr at −70°. The pellet is dried and resuspended in 20 μl TE buffer (10 m*M* Tris, pH 8.0, 1 m*M* EDTA). The transcription reaction is performed according to the manufacturer's instructions (Promega, Madison, WI) in a volume of 50 μl

containing 40 m*M* Tris–HCl, pH 7.5, 6 m*M* $MgCl_2$, 2 m*M* spermidine, 10 m*M* NaCl, 10 m*M* DTT, 50 U RNasin, 0.5 m*M* NTPs, 0.5 m*M* $m^7G(5')ppp(5')G$, 2 μg of linearized plasmid DNA, and 40 U of the appropriate RNA polymerase. After incubation for 1 hr at 37° the DNA template is removed by digestion with RNase-free DNase I (1 U/μg DNA). The amount and integrity of RNA are analyzed by electrophoresis on a denaturing agarose gel.

2. Extract the RNA once with phenol/chloroform and precipitate for 1 hr at −70° in the presence of 0.5 vol 7.5 *M* ammonium acetate and 2.5 vol ethanol. The RNA is pelleted, washed with 70% ethanol, dried, resuspended in 20 μl of TE, and stored in aliquots at −70°.

3. Protein can be translated using nuclease-treated RRL or wheat germ extract (WG) (Promega). The translation reaction (50 μl) contains 35 μl RRL, 50 U RNasin, 20 μ*M* amino acid mixture minus methionine, 40 μCi L-[^{35}S]methionine (No. SJ.204, 1100 Ci/mmol; Amersham), and 2–4 μl RNA and is incubated at 30° for 1 hr. The WG extract translation reaction is performed in a similar manner at 25° for 1 hr using 25 μl WG extract, 80 μ*M* amino acid mixture minus methionine, and 20 m*M* potassium acetate.

4. ^{35}S-labeled translation products are boiled for 5 min in 1× SDS sample buffer and resolved on a 10% SDS minigel. After electrophoresis at 200 V for 45 min at 25°, the gel is exposed to En^3Hance and dried, and proteins are visualized by autoradiography.

Comments

Rabbit reticulocyte lysate contains an unknown protein kinase(s) that phosphorylates the c-Jun amino terminus producing several c-Jun isoforms of reduced electrophoretic mobility on SDS–PAGE (Fig. 1, first lane).[12,15]

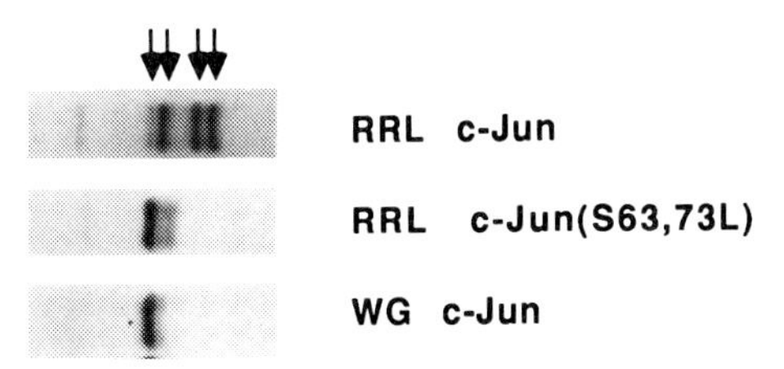

FIG. 1. Synthesis of c-Jun protein by *in vitro* transcription/translation. Wild-type c-Jun RNA or c-Jun RNA containing serine to leucine mutations at positions 63 and 73 (S63,73L) were synthesized by standard *in vitro* transcription protocol as described in the Procedures. RNA transcripts were used to program messenger-dependent rabbit reticulocyte lysate (RRL) or wheat germ extract (WG) in the presence of L-[^{35}S]methionine. The translation products were resolved on a 10% polyacrylamide–SDS gel that was subjected to fluorography and the radiolabeled proteins were visualized by autoradiography. The arrows indicate various post-translationally modified forms of c-Jun.

Mutation of the *in vivo* sites of c-Jun phosphorylation, serines 63 and 73 (second lane),[11,12] or treatment with CIAP[14,15] abolishes the formation of the slower migrating c-Jun species, suggesting that the site-specific phosphorylation of these sites is responsible for this reduced mobility. Wheat germ extracts apparently lack the appropriate protein kinase(s) to induce c-Jun amino-terminal phosphorylation as WG c-Jun appears as a single band (third lane).[12] A similar result has been reported for c-Fos carboxy-terminal phosphorylation in WG extracts.[18] The ability of a transcription factor to be phosphorylated during *in vitro* translation in RRL can be exploited to investigate various parameters that regulate its phosphorylation. A great deal of variability exists between sources of RRL with respect to their ability to phosphorylate c-Jun[19] or c-Fos[18] during *in vitro* translation. It is therefore important to test several batches of RRL.

Expression vectors used for either bacterial or mammalian expression of proteins which contain a SP6, T7, or T3 promoter can be used for this protocol without further manipulation. The use of this approach to analyze transcription factor phosphorylation is dependent on (1) the ability of the system to phosphorylate the transcription factor of interest, *and* (2) the resulting shift in electrophoretic mobility of the phosphorylated forms on SDS–PAGE. To demonstrate that this modification is due to post-translational phosphorylation, the mobility shift must be abolished by either mutation of the site(s) responsible for the phosphorylation (Fig. 1, second lane) or phosphatase treatment of the translation product.[14,15] It must be determined empirically whether the transcription factor of interest is phosphorylated by translation in either of these cell-free systems.

In Vitro Phosphorylation of Immobilized Fusion Proteins

In vitro protein kinase assays are useful both for mapping putative *in vivo* sites of protein phosphorylation and for determining the nature of the protein kinase(s) that mediates this phosphorylation. Because of the modular nature of transcription factors, their distinct functional domains can be expressed as fusion proteins and utilized as substrates for *in vitro* kinase assays. Several expression systems allow for the production of large quantities of highly purified fusion protein (these systems are discussed in more detail elsewhere in this volume). We have used the pGEX expression system (Pharmacia, Piscataway, NJ) to prepare glutathione *S*-transferase (GST) fusion proteins containing various regions of the Jun family of proteins (Fig. 2A).[21] Using these fusion proteins as *in vitro* substrates in conjunction with site-directed mutagenesis of specific serine, threonine, or

[21] D. B. Smith and K. S. Johnson, *Gene* **67,** 31 (1988).

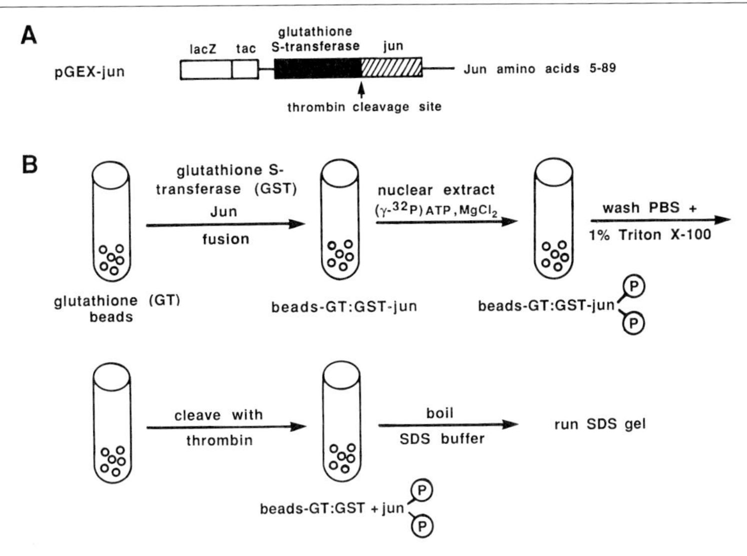

FIG. 2. Procedure for the use of GST fusion proteins as protein kinase substrates. (A) The GST–c-Jun amino-terminal fusion protein construct. (B) Summary of *in vitro* kinase assay using immobilized GST–Jun fusion proteins as substrates.

tyrosine residues permits the mapping of potential phosphorylation sites.[22–24] These immobilized fusion proteins can also be used to develop an affinity purification scheme to identify a putative protein kinase(s) that phosphorylates these sites.[22,23–25]

Procedure

1. Overnight cultures of bacteria transformed with the pGEX–Jun fusion are diluted 1:10 and grown for 1 hr at 37° before being stimulated with 0.4 m*M* IPTG for 4 hr. The bacteria are pelleted for 10 min at 10,000*g*. Bacteria from a 250-ml culture are resuspended in 5 ml of PBSTE (150 m*M* NaCl, 16 m*M* Na_2HPO_4, 4 m*M* NaH_2PO_4, 1% Triton X-100, 2 m*M* EDTA) containing 0.1% 2-mercaptoethanol, 2 m*M* PMSF, 5 m*M* benzamidine hydrochloride and sonicated on ice four times for 30 sec. Cell debris is removed by centrifugation at 12,000*g* for 10 min at 4°. The supernatant is mixed for 1 hr at 4° with 1 ml 50% (v/v) glutathione-agarose beads

[22] V. Adler, C. C. Franklin, and A. S. Kraft, *Proc. Natl. Acad. Sci. U.S.A.* **89,** 5341 (1992).
[23] V. Adler, A. Polotskaya, F. Wagner, and A. S. Kraft, *J. Biol. Chem.* **267,** 17001 (1992).
[24] B. J. Pulverer, K. Hughes, C. C. Franklin, A. S. Kraft, S. J. Leevers, and J. R. Woodgett, *Oncogene* **8,** 407 (1993).
[25] M. Hibi, A. Lin, T. Smeal, A. Minden, and M. Karin, *Genes Dev.* **7,** 2135 (1993).

(Sigma, St. Louis, MO) preswollen in PBSTE. The beads are washed twice with PBSTE containing 2 *M* KCl and twice with PBSTE, and glycerol is added to 50%. The beads can be stored at −20° for 1 year without loss of intact protein.

2. The protein kinase assay is summarized in Fig. 2B. One microgram of immobilized GST–Jun fusion protein is washed once with kinase buffer (20 m*M* HEPES, pH 7.5, 50 m*M* NaCl, 1 m*M* EGTA, 2 m*M* $MnCl_2$, 10 m*M* $MgCl_2$, 1 m*M* DTT, 0.5% NP-40, 0.5 m*M* PMSF, 5 m*M* benzamidine hydrochloride, 5 m*M* NaF, 1 m*M* $NaVO_3$) and resuspended in 50 μl of kinase buffer containing 20 μM ATP and 10 μCi [γ-^{32}P]ATP (3000 Ci/mmol, Amersham).

3. The assay is initiated by the addition of a cell extract or purified protein kinase and is incubated at 30° for 10 min and then washed extensively with ice-cold TBSTE containing 0.1% 2-mercaptoethanol, 0.5 m*M* PMSF, 5 m*M* benzamidine, and 1 m*M* ATP. The phosphorylated GST–Jun fusion is eluted from the beads by boiling in 1× SDS sample buffer and is resolved on a 10% SDS gel. Depending on the purity of the sample to be analyzed, substrate phosphorylation can be readily detected using 0.1–0.5 U of purified protein kinase or approximately 10 μg of crude cell extract depending on the abundance of the protein kinase present in the extract.

4. If the protein fragment is cloned into pGEX-2T and is to be cleaved from the GST fusion prior to analysis, the phosphorylated substrate is washed once with thrombin cleavage buffer (50 m*M* Tris–HCl, pH 7.5, 150 m*M* NaCl, 2.5 m*M* $CaCl_2$, 1% 2-mercaptoethanol) and is incubated with 0.125 U thrombin (Sigma) for 1 hr at 25°. The reaction is stopped by the addition of 20 μl of 2× SDS sample buffer, boiled, and the proteolytic fragments are analyzed by SDS–PAGE. Alternatively, if the fragment is cloned into pGEX-3×, cleavage can be performed by an overnight incubation in 50 m*M* Tris–HCl, pH 7.5, 100 m*M* NaCl, 1 m*M* $CaCl_2$, containing 2 μg human factor Xa (Boehringer Mannheim).

Comments

We have purified a protein kinase that binds and phosphorylates the c-Jun amino-terminal activation domain.[22,23] Various GST–Jun fusion proteins were used as *in vitro* substrates to map the sites of phosphorylation in c-Jun and to determine the substrate specificity of this protein kinase *in vitro*.[22,23] This protein kinase efficiently phosphorylates a full-length (Fig. 3A, lane 4) and amino-terminal (Fig. 3A, lane 2) c-Jun fusion, but not a carboxy-terminal c-Jun fusion containing the basic leucine zipper DNA-binding domain (Fig. 3A, lane 5). Mutation of the *in vivo* sites of c-Jun

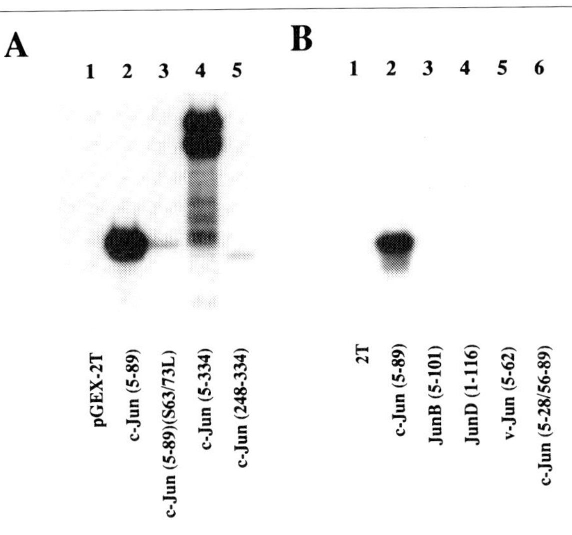

FIG. 3. *In vitro* phosphorylation of GST–Jun fusion proteins. GST fusion proteins containing various regions of either wild-type or mutant Jun proteins were prepared and immobilized on glutathione-agarose beads. The fusion proteins were then used as substrates in an *in vitro* protein kinase assay (see Fig. 2B) using a partially purified c-Jun amino terminal protein kinase as the kinase source.[22] The numbers in parentheses indicate the amino acid residues of the Jun protein that are expressed in the GST fusion. The entire GST–Jun fusion protein was eluted from the beads by boiling in 1× SDS sample buffer and was resolved on a 10% acrylamide–SDS gel. The gel was dried, and radiolabeled proteins were visualized by autoradiography.

amino-terminal phosphorylation (Ser-63 and Ser-73)[11,12] abolishes the amino-terminal phosphorylation (Fig. 3A, lane 3). This protein kinase is highly specific for the amino terminus of c-Jun and not other Jun family members (Fig. 3B, lanes 2–5). The integrity of both the *in vivo* phosphorylation sites (Fig. 3A, lane 3) *and* the c-Jun delta domain (Fig. 3B, lane 6) are necessary for efficient phosphorylation of the c-Jun amino terminus both *in vivo*[14] and by this purified protein kinase *in vitro*.[22]

Substrate level phosphorylation by certain protein kinases may require a specific tertiary structure for substrate recognition. Although short stretches of 5–6 amino acids may be sufficient for protein kinase recognition *in vitro*, there are examples of amino acid sequences that are highly phosphorylated *in vivo* but not *in vitro*. For example, a full-length pp42 MAP kinase–GST fusion functions as an excellent substrate for MAP kinase kinase *in vitro*.[26] However, a 20 or 40 amino acid GST–MAP kinase fusion containing the *in vivo* phosphorylation sites and flanking amino acids did

[26] Y. Zhao and A. S. Kraft, unpublished observation (1993).

not serve as a suitable substrate for *in vitro* phosphorylation.[26] Soluble GST–Jun fusion protein eluted from the beads (see below) may also be used as a substrate for immunocomplex protein kinase assays or be subjected to tryptic phosphopeptide mapping for comparison with the phosphorylation of its corresponding endogenous protein *in vivo*.[27]

Analysis of Phosphorylation by "in Gel" Kinase Assays

The "in gel" kinase assay has been used extensively to analyze various protein kinase signal transduction mechanisms.[28] The "in gel" technique has been utilized for the determination of the apparent molecular weight of a purified c-Jun protein kinase[22,23] and other putative protein kinases that phosphorylate the amino-terminal activation domain of c-Jun.[25] This technique involves the resolution of samples containing a putative protein kinase(s) on an SDS gel polymerized in the presence of a transcription factor substrate. The kinase(s) and substrate are then renatured by removal of the SDS, and phosphorylation of the "in gel" substrate is analyzed *in situ* in the presence of [γ-^{32}P]ATP.

Procedure

1. A 10% polyacrylamide–SDS gel is polymerized in the presence of 60 μg/ml of the appropriate protein substrate. Here we used the amino-terminal GST–c-Jun(5-89) fusion eluted from the glutathione-agarose beads with 20 m*M* reduced glutathione (Sigma) in 50 m*M* Tris–HCl, pH 7.5 final (prepare fresh), for 10 min at 25°. The fusion protein is dialyzed extensively against PBS and concentrated with an Amicon concentrator.
2. Cell extracts are solubilized by boiling for 5 min in an equal volume of 2× SDS sample buffer [2% (w/v) SDS, 20% (v/v) glycerol, 125 m*M* Tris–HCl, pH 6.8, 20% (v/v) 2-mercaptoethanol] and the samples are cleared by centrifugation at 12,000*g* for 5 min.
3. Soluble protein (25–100 μg) is resolved on the minigel by electrophoresis at 200 V for 45 min at 25°. The stacking gel and unused lanes are cut away and discarded. SDS is removed from the minigel by washing four times over a period of 60 min with 20 ml of 40 m*M* HEPES (pH 7.8) containing 10% glycerol.
4. The gel is incubated in 15 ml of "in gel" kinase buffer (20 m*M* HEPES, pH 7.8, 10 m*M* $MgCl_2$, 2 m*M* $MnCl_2$, 1 m*M* EGTA, 0.25 m*M*

[27] E. Nikolakaki, P. J. Coffer, R. Hemelsoet, J. R. Woodgett, and L. H. K. Defize, *Oncogene* **8,** 833 (1993).

[28] J. E. Hutchcroft, M. Anostario Jr., M. L. Harrison, and R. L. Geahlen, this series, Vol. 200, p. 417.

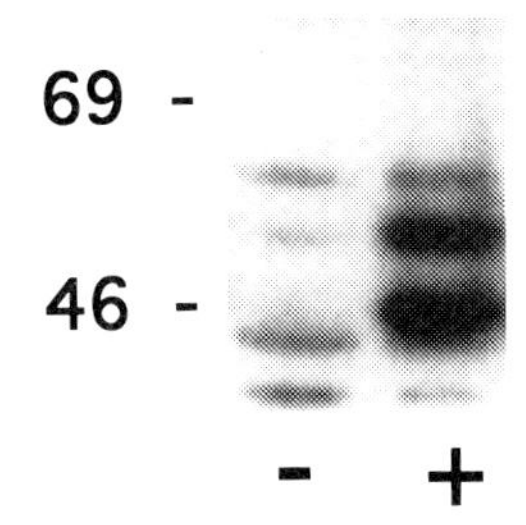

FIG. 4. Analysis of c-Jun protein kinase activity in U937 cell extracts by an "in gel" protein kinase assay. The solubilized GST–c-Jun (5–89) substrate was added to a 10% polyacrylamide–SDS gel prior to polymerization. Whole cell extracts were prepared from either untreated U937 cells (−) or cells treated with 200 n*M* PMA and 100 μM cycloheximide (+) as previously described.[22] Cellular proteins were resolved on the substrate-containing SDS gel and renatured by removal of SDS from the gel. *In situ* substrate phosphorylation was detected by an "in gel" protein kinase assay in the presence of [γ-^{32}P]ATP. The gel was washed free of [γ-^{32}P]ATP and dried, and the radiolabeled substrate was detected by autoradiography.

PMSF, 5 m*M* benzamidine hydrochloride, 40 m*M* KCl, 5 m*M* 2-mercaptoethanol, 0.5% NP-40, 0.5 m*M* $NaVO_3$, 5 m*M* NaF) containing 20 μM ATP and 75 μCi [γ-^{32}P]ATP (3000 Ci/mmol) for 1 hr at 25°.

5. Excess [γ-^{32}P]ATP is removed from the gel by rinsing five times over 2 hr with a 20-ml solution containing 20 m*M* HEPES (pH 7.6) and 10% glycerol. Dialysis tubing containing 1 g of 2 × 8-50 anion-exchange resin is added to the wash solution to bind free [γ-^{32}P]ATP.[28] If the background is still high, wash four times over 2 hr in 20 ml 5% trichloroacetic acid containing 10 m*M* sodium pyrophosphate.

6. The proteins are fixed by a 30-min incubation in 20 ml of 15% methanol and 7% acetic acid, and the gel is dried and exposed to X-ray film.

Comments

We have performed this procedure with a c-Jun amino-terminal GST fusion as the "in gel" substrate and used U937 cell extracts as a source of protein kinase(s). Extracts from untreated U937 cells contain several nonspecific protein kinase activities of variable molecular weight (Fig. 4, lane −). Treatment of cells with PMA and cycloheximide (Fig. 4, lane +), both of which induce c-Jun amino-terminal phosphorylation *in vivo*,[24,29] induces two c-Jun-specific kinase activities of 46 and 54 kDa.[25] These results demonstrate how multiple putative transcription factor protein kinases can be identified using the "in gel" kinase assay.

[29] C. C. Franklin, T. Unlap, V. Adler, and A. S. Kraft, *Cell Growth Differ.* **4,** 377 (1993).

The "in gel" kinase assay depends on the ability of the substrate and protein kinase to renature after removal of SDS from the gel. Protein tyrosine kinases are not readily reconstituted following renaturation in SDS gels.[28] The addition of a denaturation step utilizing either 6 *M* guanidine hydrochloride or 8 *M* urea has been shown to enhance the reconstitution of some, but not all, serine/threonine protein kinases. Members of the MAP kinase family renature quite well under these conditions, and are therefore routinely analyzed by this method. The concentration of substrate incorporated into the gel (between 30 and 60 μg/ml) is critical for the identification of putative transcription factor protein kinases. Excess substrate may lead to the detection of nonspecific protein kinases. Varying the concentrations of substrate may be helpful in differentiating between low and high specific activity protein kinases in the sample. It is important to perform an "in gel" assay in the absence of substrate to differentiate between substrate phosphorylation and autophosphorylation of a protein kinase in the sample. In addition, the specificity and putative phosphorylation sites of various protein kinases in a sample can be determined by running identical "in gel" kinase assays utilizing either wild-type or phosphorylation site mutants as the "in gel" substrate.

Summary

Many transcription factors are regulated by post-translational phosphorylation mechanisms. This chapter described several approaches that have been utilized to examine the phosphorylation of the c-Jun transcription factor. A combination of the techniques described in this chapter can be used to determine whether a transcription factor is phosphorylated *in vivo*, to analyze the sites of phosphorylation *in vitro*, and to permit the identification of putative protein kinases that may mediate this phosphorylation *in vivo*. Proteins labeled with ^{32}P by either *in vivo* or *in vitro* techniques can be further analyzed by tryptic phosphopeptide mapping or phosphoamino acid analysis. These procedures have been described in detail elsewhere.[30]

Acknowledgments

The authors thank Fred Wagner and Vicki McCulloch for assistance in preparing the GST–Jun fusion constructs. This work was supported by National Cancer Institute Grant CA42533 to A.S.K. and American Cancer Society Grant IRG-66-32 to C.C.F.

[30] W. J. Boyle, P. Van Der Geer, and T. Hunter, this series, Vol. 201, p. 110.

[38] Transdominant Negative Mutations

By STEVEN M. SHAMAH and CHARLES D. STILES

Animal cell growth factors regulate gene expression through a signal transduction cascade. The cascade is initiated at the cell surface with the assembly of a multienzyme "signal-generating particle" consisting of an activated tyrosine kinase, proteins with SH2 domains, and small GTP binding proteins (Fig. 1). It terminates in the cell nucleus with the induction of transcription factors such as Fos, Jun, and Myc. In between, this signal transduction cascade is channeled through serine/threonine protein kinases [protein kinase C, Raf, mitogen-activating protein (MAP) kinase], through cytoplasmic tyrosine kinases (Src, Fyn, Yes), or through shuttle proteins such as p91 which are phosphorylated by tyrosine kinases and translocate directly to the nucleus where they function as transcription factors. The role of these proteins in normal and abnormal cell growth can be displayed by a loss-of-function phenotype, most easily achieved through transdominant negative mutations.

This chapter begins by discussing the general properties of transdominant negative mutations. Next, we will review specific strategies and present guidelines for designing transdominant negative mutations that perturb signal transduction at four distinct levels of the mitogenic signaling cascade indicated in Fig. 1. Finally, we will present perspectives for the future of transdominant negative technology.

General Properties

In 1987, Herskowitz[1] drew attention to multiple "experiments of nature" wherein a loss-of-function phenotype was transmitted in a transdominant fashion and detailed the defective intermolecular interactions which culminated in these phenotypes. He suggested that transdominant acting mutations could be experimentally designed and would have general utility—especially in higher organisms which do not lend themselves readily to conventional genetic analysis. As reviewed by Herskowitz,[1] mutated proteins can generate a transdominant negative phenotype by a variety of different mechanisms. As a result, there are no strict guidelines that govern the design of transdominant negative mutations. However, there is a common principle that underlies the vast majority of such mutations and two

[1] I. Herskowitz, *Nature (London)* **329,** 219 (1987).

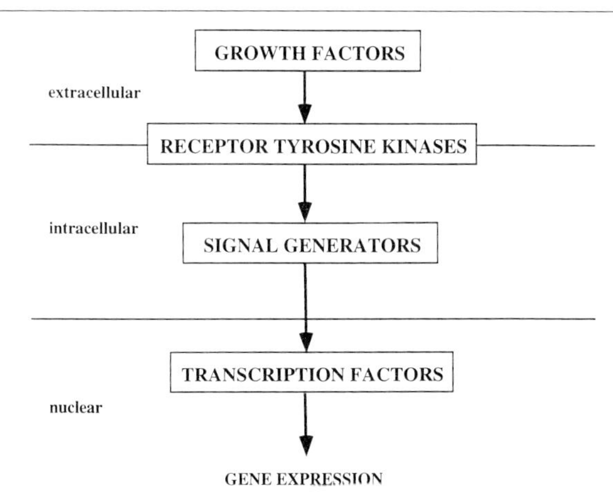

FIG. 1. Four classes of molecules involved in growth factor-initiated mitogenic signaling.

preconditions that must be satisfied for these mutations to generate a transdominant negative phenotype. First, the mutated protein must have at least two domains that are required for function. Second, these domains must be autonomous, i.e., it must be possible to mutate one domain while preserving wild-type activity in the other(s). In this way, mutant proteins can be engineered to lack function yet retain normal protein interaction, resulting in titration of the corresponding wild-type protein or critical cellular substrates. In the following sections, we will show how these two preconditions for transdominant negative mutations are satisfied at four different levels in the signal transduction cascade illustrated in Fig. 1. For some of these classes of proteins the availability of structure–function data facilitates the design of transdominant negative mutations. Thus, we will also describe conserved structural motifs at each of these levels and present general strategies for the generation of transdominant mutations to exploit these opportunities.

Growth Factors

Many polypeptide growth factors are disulfide-linked homodimeric or heterodimeric proteins. These include all members of the platelet-derived growth factor (PDGF) and transforming growth factor-β (TGF-β) superfamilies together with the hematopoietic factors colony-stimulating factor-1 (CSF-1) and stem cell growth factor.[2] All members of the PDGF and

[2] N. Q. McDonald and W. A. Hendrickson, *Cell (Cambridge, Mass.)* **73,** 421 (1993).

TGF-β superfamilies are characterized by a set of phylogenetically conserved cysteine residues that participate in interchain disulfide bonds. In designing transdominant negative mutations for these growth factors, domains required for recognition of wild-type subunits (especially the aforementioned cysteines) should be unperturbed so that a "poisoned" subunit can interact with and titrate the biologically active, wild-type subunits. Functionally autonomous domains that are attractive targets for the generation of transdominant negative mutations include those required for post-translational processing, secretion, protein stability, or receptor recognition.

Processing Mutants

Post-translational processing has been exploited to produce transdominant negative mutations for both PDGFs and TGF-βs.[3,4] The initial transcription/translation product of the PDGF-A gene contains a substantial amount of amino-terminal protein which is normally removed by proteolytic cleavage. If this amino terminal material is not removed, the unprocessed PDGF-A subunit can still interact with and form disulfide-linked dimers with wild-type PDGF-A subunits; however, these heterodimers are incapable of receptor activation. Thus, mutations which perturb post-translational processing of the PDGF-A subunit function as transdominant negative mutations for the PDGF-A gene.[3] A comparable requirement for amino-terminal proteolytic cleavage has been well documented for TGF-βs.[5,6] Transdominant negative mutations of these growth factors have been produced by mutations similar to those described for PDGF-A.[4]

The PDGF-B subunit likewise contains amino-terminal material that is normally removed within the cell.[3] In contrast to the processing requirement for the PDGF-A chain, PDGF-B subunits retain biological activity even without removal of the amino-terminal region.[7] Heterodimers between an unprocessed PDGF-A subunit and a PDGF-B subunit are also biologically active.[3] Thus, amino-terminal processing mutants of PDGF-A are selective transdominant negative mutations with activity on cells that express PDGF-A, but without activity on cells that express PDGF-B.[3]

[3] M. Mercola, P. L. Deininger, S. M. Shamah, J. Porter, C. Y. Wang, and C. D. Stiles, *Genes Dev.* **4,** 2333 (1990).

[4] A. R. Lopez, J. Cook, P. L. Deininger, and R. Derynck, *Mol. Cell. Biol.* **12,** 1674 (1992).

[5] A. M. Gray and A. J. Mason, *Science* **247,** 1328 (1990).

[6] K. Miyazano, A. Olofsson, P. Colosetti, and C.-H. Heldin, *EMBO J.* **10,** 1091 (1991).

[7] M. Hannink and D. Donoghue, *Mol. Cell. Biol.* **6,** 1343 (1986).

Secretion Mutants

Another approach to blocking ligand function is to prevent its secretion and, hence, disrupt access to its cognate receptor. In the case of TGF-β, the amino-terminal propeptide is necessary for secretion and may function as a transport chaperone.[5] Mutations in highly conserved residues of the TGF-β1 proregion that permit dimerization result in an inhibition of wild-type TGF-β secretion and function as transdominant negative mutations for several TGF-β isoforms.[4] The secretion of TGF-β can also be suppressed by modification of conserved N-glycosylation sites, mutations which also exert a transdominant negative effect.[4]

The use of secretion as a point of attack in production of growth factor transdominant negative mutations does present some potential pitfalls. Chief among these is that heterodimers between the mutated and wild-type subunits retain biological activity. These heterodimeric proteins could reach and activate their cognate receptors by leakage or by formation of an intracellular ligand–receptor complex. Several groups have shown that PDGF receptors can become tyrosine phosphorylated in cytoplasmic compartments.[8–11] Furthermore, Deuel and colleagues[12] have demonstrated that addition of an endoplasmic retention signal sequence (KDEL) to v-*sis*, the viral homolog of the PDGF-B chain, efficiently blocks secretion yet permits the intracellular activation PDGF receptors.

Stability Mutants

As noted earlier, the cysteine residues involved in interchain disulfide bonds are well conserved in the dimeric growth factors. However, other cysteines essential for intrachain disulfide bridge formation are equally well conserved.[2] Amino acid substitutions for these cysteines in the PDGF-B chain reduce the biological activity of the growth factor.[13,14] A substitution for an equivalent cysteine residue in the PDGF-A subunit has produced a transdominant negative mutation that works equally well on the PDGF-A and -B gene products.[3] It appears that the mutated PDGF-A subunit is less stable than the wild-type subunits and that this instability dominates in mutant/wild-type heterodimers.[3]

[8] T. Fleming, T. Matsui, C. Molloy, K. Robbins, and S. Aaronson, *Proc. Natl. Acad. Sci. U.S.A.* **86,** 8063 (1989).

[9] M. T. Keating and L. T. Williams, *Science* **239,** 914 (1988).

[10] B. E. Bejcek, R. M. Hoffman, D. Lipps, D. Y. Li, C. A. Mitchell, P. W. Majerus, and T. F. Deuel, *J. Biol. Chem.* **267,** 3289 (1992).

[11] V. B. Lokeshwar, S. S. Huang, and J. S. Huang, *J. Biol. Chem.* **265,** 1665 (1990).

[12] B. Bejcek, D. Y. Li, and T. F. Deuel, *Science* **245,** 1496 (1989).

[13] M. K. Sauer and D. J. Donoghue, *Mol. Cell. Biol.* **8,** 1011 (1988).

[14] N. A. Giese, K. C. Robbins, and S. A. Aaronson, *Science* **236,** 1315 (1987).

Receptor Recognition Mutants

Structure–function analysis of PDGF has revealed small regions implicated in receptor binding.[15,16] Amino acid substitutions in one of these domains of the PDGF-A chain gene yield mutated PDGF subunits that recognize and form stable heterodimers with wild-type PDGF subunits; however, the resulting heterodimeric molecules are not well recognized by PDGF receptors.[17] Thus, these mutations titrate PDGF ligands from within the cell and exert a transdominant negative effect.

Receptor Tyrosine Kinases

Receptor tyrosine kinases (RTKs) can be grouped into subfamilies on the basis of ligand affinity, amino acid sequence similarity, and tertiary structural homology. The larger and better characterized subfamilies include the epidermal growth factor (EGF), insulin, PDGF, fibroblast growth factor (FGF), and nerve growth factor (NGF) receptors.[18] The RTKs (with the exception of the insulin receptor subfamily members) exist as monomeric subunits in the plasma membrane. Formation of the ligand–receptor complex induces a noncovalent dimerization of receptor subunits that is essential for activation of the receptor tyrosine kinase domain.[19] The ligand binding, oligomerization, and tyrosine kinase domains are functionally autonomous.[18] In both laboratory experiments and "experiments of nature," this autonomy has been exploited to produce transdominant negative mutations.

In laboratory studies, the entire cytoplasmic portion of the receptor, including the tyrosine kinase domain, has been deleted to produce transdominant negative mutations of the EGF, FGF, PDGF, and insulin-like growth factor I (IGF-I) receptors.[20–23] Point mutations at a highly conserved lysine residue in the ATP binding site have also yielded a transdominant

[15] W. J. LaRochelle, J. H. Pierce, M. May-Siroff, N. Giese, and S. A. Aaronson, *J. Biol. Chem.* **267,** 17074 (1992).

[16] A. Ostman, M. Andersson, U. Hellman, and C.-H. Heldin, *J. Biol. Chem.* **266,** 10073 (1991).

[17] F. S. Vassbotn, M. Andersson, B. Westermark, C. H. Heldin, and A. Ostman, *Mol. Cell. Biol.* **13,** 4066 (1993).

[18] A. Ullrich and J. Schlessinger, *Cell (Cambridge, Mass.)* **61,** 203 (1990).

[19] J. Schlessinger and A. Ullrich, *Neuron* **9,** 383 (1992).

[20] E. Amaya, T. J. Musci, and M. W. Kirschner, *Cell (Cambridge, Mass.)* **66,** 257 (1991).

[21] O. Kashles, Y. Yarden, R. Fischer, A. Ullrich, and J. Schlessinger, *Mol. Cell. Biol.* **11,** 1454 (1991).

[22] D. Prager, H. Yamasaki, M. M. Weber, S. Gebremedhin, and S. Melmed, *J. Clin. Invest.* **90,** 2117 (1992).

[23] H. Ueno, J. A. Escobedo, and L. T. Williams, *J. Biol. Chem.* **268,** 22814 (1993).

negative mutation of the receptor for colony-stimulating factor-1 (CSF-1), a member of the PDGF receptor family.[24] These mutants titrate functional receptors by forming ligand-induced, nonfunctional oligomers with wild-type receptor subunits. Collectively, RTK mutants have been shown to block mesoderm induction in *Xenopus* embryos (FGF), keratinocyte organization and differentiation (FGF), growth hormone suppression (IGF-1), mitogenic responses (EGF and PDGF), and transformation (PDGF and CSF-1).[17,20–26]

In nature, tyrosine kinase-negative variants of FGF and neurotrophin receptors are produced by alternative splicing.[27,28] These splice variants may function as negative regulators of cell growth and differentiation by a transdominant negative action (although there is no supporting data at this time). Naturally occurring RTK variants that account for transdominant negative animal phenotypes have been documented for the insulin and c-*kit* receptors. Here, point mutations in the cytoplasmic domain result in tyrosine kinase-deficient receptor variants which culminate in an insulin-resistant diabetic state and developmental defects (the *W* mouse phenotype), respectively.[29,30] These naturally occurring mutants might be especially instructive in the design of other mutations that will function well in transgenic animal disease models (see below).

Signal Generators

The next stage in the mitogenic signal transduction cascade is the interaction of activated RTKs with intracellular signal generators (Fig. 1). Transdominant negative mutations for Ras, a small GTP binding protein, for Raf, a serine/threonine kinase, and for the Src family of tyrosine kinases have been utilized to delineate the order of events in a critical downstream pathway of RTK mitogenic signaling. These mutants function by competing with their wild-type counterparts for substrates that are present in limiting amounts within the cell.

[24] A. D. Reith, C. Ellis, N. Maroc, T. Pawson, A. Bernstein, and P. Dubreuil, *Oncogene* **8,** 45 (1993).

[25] S. Werner, W. Weinberg, X. Liao, K. G. Peters, M. Blessing, S. H. Yuspa, R. L. Weiner, and L. T. Williams, *EMBO J.* **12,** 2635 (1993).

[26] S. M. Shamah, C. D. Stiles, and A. Guha, *Mol. Cell. Biol.* **13,** 7203 (1993).

[27] M. V. Chao, *Neuron* **9,** 583 (1992).

[28] M. Jaye, J. Schlessinger, and C. A. Dionne, *Biochim. Biophys. Acta* **1135,** 185 (1992).

[29] K. Nocka, J. C. Tan, E. Chiu, T. Y. Chu, P. Ray, P. Traktman, and P. Besmer, *EMBO J.* **9,** 1805 (1990).

[30] A. Cama, M. J. Quon, M. de la Luz Sierra, and S. I. Taylor, *J. Biol. Chem.* **267,** 8383 (1992).

The Ha-*ras* gene product is a 21-kDa protein that binds GDP in its resting state and GTP when activated by extracellular stimuli. A serine-to-asparagine substitution at amino acid position 17 in Ha-*ras* creates a mutant protein with preferential affinity for GDP.[31] When expressed at high levels, a Ha-*ras* Asn-17 mutant forms inactive Ras–GDP complexes. These complexes titrate limiting upstream regulators, preventing them from interacting with wild-type Ras proteins. The Ha-*ras* Asn-17 mutant mutation has been used extensively to probe the contribution of Ras activation to growth factor mitogenic signaling.[32] Interestingly, *RAS* genes with a functionally equivalent mutation act as transdominant negative mutations in *Saccharomyces cerevisiae* by blocking the interaction of wild-type Ras with its upstream regulator CDC25.[33] These experiments have led to a model suggesting that CDC25 proteins interact with Ras–GDP molecules and catalyze nucleotide exchange by stabilizing the intermediate nucleotide-free state of Ras.[33]

The Raf-1 protooncogene product is a serine/threonine kinase. Following RTK activation, Raf-1 becomes activated resulting in the phosphorylation of substrate factors and the induction of transcription factors implicated in growth control (Fig. 1).[34] Raf-1 is a member of a family of related serine-threonine kinases that share three highly conserved regions: CR1, CR2, and CR3.[34] The amino-terminal CR1 region contains a cysteine finger motif, the CR2 region consists of numerous serine and threonine residues, and the carboxy-terminal CR3 region contains the serine/threonine kinase domain. Mutant Raf-1 proteins with deletions in the CR1 region are constitutively activated suggesting that this region functions as a negative regulatory domain.[35] An upstream regulator may interact with the CR1 region via the zinc finger to relieve the inhibitory effect and activate the CR3 kinase domain.[34] The CR1 and CR3 domains are functionally autonomous. Thus, mutations that inactivate the CR3 kinase domain exhibit transdominant negative activity.[34] These mutants, like the Ha-*ras* Asn-17 mutant, may titrate the upstream activator of Raf-1 by forming inactive complexes through the intact CR1 regions. These transdominant negative mutations have been used to demonstrate that Raf-1 is an essential mitogenic signal transducer downstream of Ras.[36]

Three members of the Src family of cytoplasmic tyrosine kinases (Src,

[31] L. A. Feig and G. M. Cooper, *Mol. Cell. Biol.* **8,** 3235 (1988).

[32] K. W. Wood and T. M. Roberts, *Biochim. Biophys. Acta* **1155,** 133 (1993).

[33] S. Powers, K. O'Neill, and M. Wigler, *Mol. Cell. Biol.* **9,** 390 (1989).

[34] J. T. Bruder, G. Heidecker, and U. R. Rapp, *Genes Dev.* **6,** 545 (1992).

[35] V. P. Stanton, D. W. Nichos, A. P. Laudano, and G. M. Cooper, *Mol. Cell. Biol.* **9,** 639 (1989).

[36] W. Kolch, G. Heidecker, P. Lloyd, and U. R. Rapp, *Nature* (*London*) **349,** 426 (1991).

Fyn, and Yes) associate with activated RTKs.[37] These signaling enzymes compete for specific RTK phosphotyrosine binding sites and become activated upon binding.[37–39] Thus, overexpression of one family member can prevent others from associating with activated PDGF receptors.[37] Mutations in Fyn or Src that render these enzymes kinase inactive function as transdominant negative mutants when overexpressed by preventing the formation of functional Src kinase/RTK complexes.[40] Studies with these transdominant negative mutations have demonstrated that the Src family tyrosine kinases are required for PDGF-mediated mitogenic signaling.[40]

Transcription Factors

Signal transduction concludes with induction of the immediate early gene set. Many of these genes encode transcription factors that function within homo- or heterodimeric protein complexes. Functionally autonomous motifs within these complexes mediate dimerization, DNA binding, and transactivation of transcription.[33] This molecular architecture presents opportunities for designing transdominant negative mutations. The Fos and Jun immediate early gene products provide a useful example.

The Fos and Jun proteins associate noncovalently to form the AP1 complex which binds to a specific response element found in a variety of different genes.[33] Fos and Jun are characterized by two distinct structural domains: a basic motif which is rich in lysine and arginine residues, and a "leucine zipper" which consists of a series of regularly spaced leucine residues consistent with the formation of an amphipathic α helix. Structure–function analysis of these domains has demonstrated that the leucine zipper is necessary for Fos/Jun heterodimer formation and DNA binding. The basic motif is required only for DNA binding.[33] Point mutations that substitute valine for critical basic residues or that truncate the basic motif entirely can abolish DNA binding yet preserve heterodimerization activity. Coexpression of Jun mutants with wild-type Fos and Jun proteins results in a transdominant negative effect on DNA binding activity of the wild-type AP1 complex.[41] The mutant proteins need not be overexpressed relative to

[37] R. M. Krypta, Y. Goldberg, E. T. Ulug, and S. A. Courtneidge, *Cell (Cambridge, Mass.)* **62,** 481 (1990).

[38] S. Mori, L. Ronnstrand, K. Yokote, A. Engström, S. A. Courtneidge, L. Claesson-Welsh, and C.-H. Heldin, *EMBO J.* **12,** 2257 (1993).

[39] G. M. Twamley, B. Hall, R. Kypta, and S. A. Courtneidge, *Oncogene* **7,** 1893 (1992).

[40] G. M. Twamley-Stein, R. Pepperkok, W. Ansorge, and S. A. Courtneidge, *Proc. Natl. Acad. Sci. U.S.A.* **90,** 7696 (1993).

[41] L. J. Ransone, J. Visvader, P. Wamsley, and I. M. Verma, *Proc. Natl. Acad. Sci. U.S.A.* **87,** 3806 (1990).

wild-type protein to exert the transdominant negative effect, suggesting that the mutants have a higher affinity for heterodimerization than do wild-type Fos and Jun proteins. In fact, Jun transdominant negative mutations with truncated basic motifs can disrupt preformed Fos/Jun complexes. Similar mutations in the Fos protein yield a transdominant negative mutant which must be overexpressed at least twofold compared to wild-type Fos and Jun in order to disrupt AP1 binding. Fos mutations which delete the basic motif cannot interfere with preformed Fos/Jun heterodimers and thus display a weaker transdominant negative phenotype than the corresponding Jun mutants.[41]

Another immediate-early transcription factor is encoded by the Myc protooncogene. The Myc protein associates with another intranuclear protein, Max, to generate the functional unit. The amino-terminal regions of Myc and Max are essential for transcription activity. The carboxy-terminal regions contain three conserved motifs: a basic region, a helix–loop–helix motif, and a leucine zipper. Collectively, these three regions are referred to as "bHLH-Zip."[42] The basic regions of bHLH-Zip are necessary for DNA binding whereas the HLH-Zip regions are important for noncovalent heterodimerization. A mutated version of Max which lacks the basic DNA binding region functions as a transdominant negative mutation for Myc by forming nonfunctional heterodimers.[42]

The receptors for estrogen, thyroid hormone, and retinoic acid function as ligand-activated transcription factors.[43,44] As with Fos, Jun, Myc, and Max, these receptors share distinct structural domains that are involved in ligand binding, DNA binding, dimer formation, and transactivation of transcription. A centrally located zinc finger motif mediates DNA binding for these receptors whereas ligand binding, dimerization, and transcription activities are mediated by overlapping carboxy-terminal regions.

Transdominant negative mutations for the steroid/nuclear receptors have been modeled after an oncogenic permutation of the thyroid-hormone receptor (TR), v-erbA.[45] The v-erbA protein contains several amino acid changes and a nine amino acid deletion in the carboxy terminus which, collectively, ablate the ability to bind thyroid hormone.[46,47] Because the zinc finger region is unaffected by these mutations, the v-erbA protein is fully capable of DNA binding. When coexpressed with TR, v-erbA re-

[42] M. Billaud, K. J. Isselbacher, and R. Bernards, *Proc. Natl. Acad. Sci. U.S.A.* **90,** 2739 (1993).
[43] M. Beato, *Cell (Cambridge, Mass.)* **56,** 335 (1989).
[44] R. M. Evans, *Science* **240,** 889 (1988).
[45] K. Damm, C. C. Thompson, and R. M. Evans, *Nature (London)* **339,** 593 (1989).
[46] J. Sap, A. Munoz, K. Damm, Y. Goldberg, J. Ghysdael, A. Leutz, H. Beug, and B. Vennström, *Nature (London)* **324,** 635 (1986).
[47] A. Munoz, M. Zenke, U. Gehring, J. Sap, H. Beug, and B. Vennström, *EMBO J.* **7,** 155 (1988).

presses TR responsiveness in a transdominant fashion.[45,48] The mechanism of this transdominant negative effect is unclear. The TR/DNA interaction was thought to require ligand binding. Moreover, deletion of the ligand binding domain from the glucocorticoid, estrogen, and progesterone receptors causes them to function as constitutive activators of transcription.[44] Since the DNA binding domain is essential for its transdominant negative function, v-erbA may compete with TR for response element templates and, once bound, remain in an inactivated state.[45] However, these same studies demonstrated that a 1:1 ratio of v-erbA to TR suppresses the wild-type receptor by approximately 90%, suggesting that perhaps an additional mechanism is involved.

Retinoic acid receptor (RAR) and estrogen receptor (ER) variants with mutations modeled after the v-erbA protein have been shown to also exert transdominant negative effects on the parent receptors.[49,50] Carboxy-terminal truncations in RAR α, β, or γ resembling the nine residue deletion of v-erbA preserve DNA binding and reduce ligand binding of the receptor but fail to transactivate and, thus, block wild-type RAR function.[50] Point mutations in the carboxy-terminal region of the ER involved in hormone binding (residues 516–524) or transactivation (residues 538–552) result in a failure to activate transcription in response to estrogen.[49] However, only the transactivation mutants display a transdominant negative effect on wild-type ERs. As with v-erbA, RAR and ER transdominant negative mutations are extremely potent and the mechanism of action of both remains unclear. The formation of inactive mutant/wild-type heterodimers, competition for response elements, and/or titration of a limiting cellular factor remain as possible modes of action for these mutants.

Future Considerations

The signal generating pathways that relay information from growth factors at the cell surface to transcription factors in the nucleus are complex and still not fully understood. Transdominant negative mutations for the emerging superfamily of Src homology-2 (SH2) containing signal generators would be invaluable tools for dissecting these networks. SH2 domains are structural domains (approximately 100 amino acids in length) that function

[48] J. Sap, A. Munoz, J. Schmitt, H. Stunnenberg, and B. Vennström, *Nature* (*London*) **340,** 242 (1989).

[49] B. A. Ince, Y. Zhuang, C. K. Wrenn, D. J. Shapiro, and B. S. Katzenellenbogen, *J. Biol. Chem.* **268,** 14026 (1993).

[50] K. Damm, R. A. Heyman, K. Umesono, and R. M. Evans, *Proc. Natl. Acad. Sci. U.S.A.* **90,** 2989 (1993).

as highly specific recognition sites for phosphorylated tyrosine residues.[51] Through these highly conserved domains, signal generators bind to activated RTKs and subsequently become tyrosine phosphorylated and enzymatically active. The SH2 domains of the various RTK signal generators possess a high degree of specificity for individual phosphotyrosine residues. Thus, signal transducing enzymes such as phospholipase C-γ, phosphatidylinositol 3-kinase, and GTPase activating protein each bind to different phosphotyrosine sites on the cytoplasmic surface of RTKs.[51]

In general, the SH2 domains of these enzymes are structurally distinct from the functional catalytic domains, making possible nonoverlapping mutations in these two domains. Mutations in these proteins that eliminate enzymatic activity but retain SH2-mediated RTK binding should exert a transdominant negative effect by competing for specific RTK phosphotyrosine residues with wild-type enzymes. In fact, individual SH2 domains, truncated from the remainder of the protein, might also function as transdominant negative mutations for specific signal transducing enzymes by the same mechanism.

Another prominent application for transdominant negative mutations will be in disease modeling. The animal cell growth factors and the signal transduction apparatus regulate essential events in hematopoiesis, wound healing, immune surveillance, and central nervous system function. For PDGF receptor, c-*kit* receptor, stem cell growth factor, and CSF-1, *in vivo* biological functions are displayed through loss-of-function mutations in mice such as patch (*Ph*), white spotting (*W*), steel (*Sl*), and osteopetrosis (*op/op*).[52–58] However, given the large number of growth factors and the multiple links in the signal transduction cascade, spontaneous loss-of-function mutations seem to be rare events. Functional redundancy precludes

[51] T. Pawson and G. D. Gish, *Cell (Cambridge, Mass.)* **71,** 359 (1992).

[52] E. N. Geissler, M. A. Ryan, and D. E. Housman, *Cell (Cambridge, Mass.)* **55,** 185 (1988).

[53] D. A. Stephenson, M. Mercola, E. Anderson, C. Wang, C. D. Stiles, D. F. Bowen-Pope, and V. M. Chapman, *Proc. Natl. Acad. Sci. U.S.A.* **88,** 6 (1991).

[54] K. M. Zsebo, D. A. Williams, E. N. Geissler, V. C. Broudy, F. H. Martin, H. L. Atkins, R.-Y. Hsu, N. C. Birkett, K. H. Okino, D. C. Murdock, F. W. Jacobsen, K. E. Langley, K. A. Smith, T. Takeishi, B. M. Cattanach, S. J. Galli, and S. V. Suggs, *Cell (Cambridge, Mass.)* **63,** 213 (1990).

[55] E. Huang, K. Nocka, D. R. Beier, T.-Y. Chu, J. Buck, H.-W. Lahm, D. Wellner, P. Leder, and P. Besmer, *Cell (Cambridge, Mass.)* **63,** 225 (1990).

[56] H. Yoshida, S.-I. Hayashi, T. Kunisada, M. Ogawa, S. Nishidawa, H. Okamura, T. Sudo, L. D. Shultz, and S.-I. Nishikawa, *Nature (London)* **345,** 442 (1990).

[57] W. Wiktor-Jedrzejczak, A. Bartocci, A. W. Ferrante, Jr., A. Ahmed-Ansari, K. W. Sell, J. W. Pollard, and R. E. Stanley, *Proc. Natl. Acad. Sci. U.S.A.* **87,** 4828 (1990).

[58] N. G. Copeland, D. J. Gilbert, B. C. Cho, P. J. Donovan, N. A. Jenkins, D. Cosman, D. Anderson, S. D. Lyman, and D. E. Williams, *Cell (Cambridge, Mass.)* **63,** 175 (1990).

some interesting disease states and more are surely obscured by embryonic lethal mutations. Transdominant negative mutations should provide useful tools for physiologic analysis of the growth factors/signal transduction apparatus. The incorporation of new genes into a mouse can be easier than the "knock-out" of endogenous genes. Moreover, with the use of tissue-specific promoter elements, the loss of function can be localized to obtain insights not possible with homologous recombination experiments.

Section V

Protein–DNA Interactions

[39] Mapping DNA–Protein Interactions in Cells and Nuclei: Genomic Sequencing by Template Purification

By JOVAN MIRKOVITCH

Understanding the molecular mechanisms of gene expression in eukaryotes necessitates a precise knowledge of DNA–protein interactions at regulatory sequences in DNA. Most convenient approaches rely on determining *in vitro* which DNA sequences bind proteins derived from either unfractionated or fractionated cellular extracts, or cloned genes. However, a number of procedures aimed at identifying at the nucleotide level DNA–protein interactions on genomic DNA, either in intact cells or in isolated nuclei, have been reported.[1–8] This chapter presents in detail a procedure for analyzing DNA–protein interactions on genomic DNA using template purification and genomic sequencing. First, the principles of the purification method and genomic sequencing are discussed. Then, precise protocols for the treatment of cells with dimethyl sulfate (DMS) or $KMnO_4$, and the isolation of nuclei for DNase I footprinting are presented. Finally, protocols for the purification of a specific sequence from whole genomic DNA followed by detection of a sequence ladder using primer extension are described in detail.

Genomic Sequencing Procedures

A number of genomic sequencing procedures, principally aimed at determining at the nucleotide level DNA–protein interactions on genomic DNA, have been developed.[2–9] Although these sequencing procedures depend on different approaches, they all are difficult to carry out. The principal limitation is that a sequencing signal must be generated from a single-copy sequence which, for man, or mouse, is present in a total of about 3×10^9 bp of genomic DNA. The original protocol developed by Church and

[1] A. Ephrussi, G. M. Church, S. Tonegawa, and W. Gilbert, *Science* **227,** 134 (1985).
[2] P. D. Jackson and G. Felsenfeld, *Proc. Natl. Acad. Sci. U.S.A.* **82,** 2296 (1985).
[3] S. B. Selleck and J. E. Majors, *Mol. Cell. Biol.* **7,** 3260 (1987).
[4] M. M. Becker, Z. Wang, G. Grossman, and K. A. Becherer, *Proc. Natl. Acad. Sci. U.S.A.* **86,** 5315 (1989).
[5] P. R. Mueller and B. Wold, *Science* **246,** 780 (1989).
[6] H. Saluz and J.-P. Jost, *Proc. Natl. Acad. Sci. U.S.A.* **86,** 2602 (1989).
[7] L. Zhang and J. D. Gralla, *Genes Dev.* **3,** 1814 (1989).
[8] J. Mirkovitch and J. E. Darnell, *Genes Dev.* **5,** 83 (1991).
[9] G. M. Church and W. Gilbert, *Proc. Natl. Acad. Sci. U.S.A.* **81,** 1991 (1984).

Gilbert[9] is based on the transfer of genomic DNA, which has undergone electrophoresis on a sequencing gel, to a nylon membrane followed by hybridization with a single-stranded probe. Detailed protocols have been published,[10,11] but until the latter part of the 1980s genomic sequencing was performed in only a few laboratories.

The large excess of nonrelevant DNA over the sequence of interest has two major practical consequences. First, a large amount of genomic DNA has to be processed to obtain a distinct signal. Second, a highly specific detection procedure must be used in which the nonrelevant DNA produces a minimal background. To circumvent these limitations, two genomic sequencing techniques have been developed. First, a polymerase chain reaction (PCR)-based technique, called ligation-mediated PCR (LM-PCR), relies on the enzymatic amplification of DNA fragments from a specific sequence. This technique circumvents both major limitations set by the genome organization. The PCR procedure amplifies only the specific sequence, which is then highly enriched over the rest of the genomic DNA, and can produce a distinct signal with different detection protocols.[5,12] This procedure has been used in a number of different systems and detailed protocols for LM-PCR have been published.[13,14]

The second genomic sequencing technique, described in detail in this chapter, is based on the rapid purification from genomic DNA of the sequence of interest.[8] Initially, genomic DNA in intact cells, permeabilized cells, or isolated nuclei is modified by chemical probes or by enzymes (Fig. 1, step 1). As a control, naked genomic DNA is subjected to the same procedure. Genomic DNA is purified by proteinase K digestion followed by phenol extraction and ethanol precipitation. The genomic DNA is then cleaved with one or more restriction enzymes (Fig. 1, step 2). This digestion is necessary in order to define the boundaries of the sequence to be analyzed, as well as to obtain DNA fragments short enough to permit hybridization to the genomic DNA. Usually at least one 4-bp cutter enzyme is included. Omitting this step produces an intractable viscous gel after hybridization since most fragments will contain two or more repetitive sequences. At this stage, genomic DNA modified at G and A residues by dimethyl sulfate

[10] P. B. Becker and G. Schutz, *in* "Genetic Engineering, Principles and Methods" (J. K. Setlow, ed.), p. 1. Plenum, New York, 1988.

[11] H. P. Saluz and J. P. Jost, *Anal. Biochem.* **176,** 201 (1989).

[12] G. P. Pfeifer, S. D. Steigerwald, P. R. Mueller, B. Wold, and A. D. Riggs, *Science* **246,** 810 (1989).

[13] P. R. Mueller and B. Wold, *in* "Current Protocols in Molecular Biology" (F. M. Ausubel, R. Brent, R. E. Kingston, D. D. Moore, J. G. Seidman, J. A. Smith, and K. Struhl, eds.), p. 15. Wiley (Interscience), 1991.

[14] I. K. Hornstra and T. P. Yang, *Anal. Biochem.* **213,** 179 (1993).

1
Treat cells or nuclei with DNA modifying or cleaving reagents

2
Isolate genomic DNA and digest with restriction enzymes

3
If necessary, induce cleavage at modified residues

4
Denature genomic DNA and hybridize with gene-specific biotinylated RNA probe.

5
Immobilize hybrids on streptavidin beads and eliminate bulk of genomic DNA

6
Recover immobilized genomic sequences by alkali treatment with concomitant hydrolysis of RNA probe

7
Detect cleavage or modified sites by primer extension or ligation-mediated PCR

FIG. 1. Procedure for template purification genomic sequencing.

(DMS) or oxidized at other residues by $KMnO_4$ can be cleaved by a piperidine treatment (Fig. 1, step 3). Subsequently, a biotinylated RNA corresponding to the DNA sequence of interest is hybridized in solution with whole genomic DNA and drives the hybridization reaction (Fig. 1, step 4). The hybrids between the biotinylated RNA and the target genomic molecules are then immobilized on streptavidin beads (Fig. 1, step 5) and the bulk of the genomic DNA is washed away. The purified genomic sequences are recovered from the beads by an alkali treatment (Fig. 1, step 6) which also hydrolyzes the biotinylated RNA which otherwise may interfere with detection procedures. This enrichment protocol allows the detection of the target DNA sequences by a single round of primer extension using a highly radioactive primer (Fig. 1, step 7). Alternatively, the purified material can be used as starting material for LM-PCR. We have consistently observed better results with LM-PCR using purified sequences instead of whole genomic DNA.[15]

[15] J. Mirkovitch, unpublished observations.

This template purification procedure presents two main advantages over ligation-mediated PCR. First it is rapid and necessitates only a single enzymatic step. This prevents the appearance of artifactual bands and the loss of real bands during some LM-PCR steps. The second advantage of the template purification technique is that the original genomic DNA is used as a template for the primer extension, without an amplification step. In this case, modifications to the DNA that create an arrest point for DNA polymerase can be detected without the necessity of breaking the sugar–phosphate backbone. This turns out to be convenient for the detection of single-stranded regions *in vivo* by $KMnO_4$.[16] Breakage of the sugar–phosphate backbone is required for the ligation-mediated PCR procedure.[5,12] While working on genuine genomic DNA may be an advantage, it also presents the major limitation of the template purification procedure. Typically about 200 μg of mouse or human genomic DNA is required to obtain a clear signal by biotinylated RNA purification and primer extension with a highly labeled primer. Template purification is fast and effective when working with most tissue culture cells. However, if the amount of material is a limiting factor, LM-PCR may be more convenient. As mentioned earlier, prior purification of specific sequences, even from small amounts of genomic DNA (1 μg or less), greatly facilitates the LM-PCR procedure.[15]

Before planning to undertake an extensive analysis of *in vivo* DNA–protein interactions, it is wise to consider the type of data generated by this experimental approach. As footprints appear as the reduction of a sequencing signal, only stable DNA–protein interactions can be detected. As another consequence, it is important to work with a biological material that is as homogenous as possible. If a sequence is transiently occupied *in vivo* during, for example, the cell cycle or a transcription initiation event, no clear footprint will appear. However, a protein extract from these cells may well show stable binding as assessed by gel mobility shift assays. As a general rule, the picture obtained by *in vivo* footprinting is more dynamic and appears different from that derived from *in vitro* DNA–protein procedures. The investigator should expect that some sequences binding proteins *in vitro* do not appear occupied *in vivo*, and sequences that were considered as unimportant *in vitro* may appear well protected in living cells. The different *in vivo* picture may turn out to be most informative. For example, we have demonstrated *in vivo* the transient presence of a binding protein on a DNA regulatory element for which no gel mobility shift could be detected using nuclear extracts.[17] Although the *in vivo* binding was tran-

[16] J. Mirkovitch and J. E. Darnell, *Mol. Biol. Cell* **3,** 1085 (1992).

[17] J. Mirkovitch, T. Decker, and J. E. Darnell, *Mol. Cell. Biol.* **12,** 1 (1992).

sient, i.e., appearing for only a short time during gene activation, *in vitro* exonuclease III assays identified this binding activity in nuclear extracts even at times after the *in vivo* footprint was lost. In conclusion, *in vivo* footprinting results provide information that assesses the relevance of *in vitro* binding assays and transient transfections, and as such are well worth the time invested to obtain them. The protocols presented here are adapted from previously published work[8,16–18] and are aimed at helping investigators that desire to probe protein interactions with genomic DNA in isolated nuclei or intact cells.

Cells, Nuclei, and Isolation of Genomic DNA

A number of approaches can be used to determine the occupancy of a specific nucleotide sequence in whole cells or nuclei. The most informative procedure is the probing of methylation at G residues when intact cells are treated with DMS. Indeed, DMS penetrates the cells easily and is highly reactive at room temperature or at 37°. Sequences complexed with proteins *in vivo* can protect G or A residues from methylation. Alternatively, DNA–protein interactions can create hydrophobic pockets or a DNA structure that can show a hypersensitivity to DMS modification. Although DMS modification seems to be the best *in vivo* approach, some DNA–protein interactions may not be observed. Indeed, DMS usually produces narrow footprints involving only a few residues. This is because the small size of the DMS molecule (126 Da) allows reaction with residues that are not tightly bound to proteins, it reacts mostly with guanines, and shows lower reactivity with adenines.

In contrast to DMS, DNase I digestion usually produces large protected regions (typically 15 to 30 bases) that are easily detected. The main drawback to DNase I is that the enzyme does not penetrate cells unless the membrane has been disrupted, conditions that cannot be considered as *in vivo*. However, DNase I footprints are usually easier to obtain than methylation protection data, and nuclear DNase I footprints can be conducted as a first step to determine which regions will have to be further analyzed by DMS. Two approaches are possible to obtain DNase I digestion of nuclear DNA. First, nuclei are isolated from disrupted cells. We favor the isolation of nuclei in a polyamine–EDTA buffer containing a mixture of protease inhibitors (described below). The principal advantage of this type of buffer is the absence of Mg^{2+} ions, thereby preventing DNA degradation by endogenous DNases during the isolation procedure. The poly-

[18] J. Mirkovitch, *in* "*In vivo* Footprinting" (I. Cartwright, ed.), JAI Press, Greenwich, CT, in press.

amine buffer is suitable for nuclei isolation from tissue culture cells[17] as well as from animal tissues.[8] However, different DNase I footprints can be obtained depending on the isolation procedure. An alternative to nuclei isolation is to allow the entry of DNase I into cells by permeabilizing them with lysolecithin.[7,19] Although this procedure appears closer to the *in vivo* situation and can result in clear footprints,[7,19] certain cell lines may contain high levels of endogenous nucleases which results in the digestion of genomic DNA even in the absence of added DNase I.[15]

1. Treatment of Cells with DMS

Extreme care must be used when handling the very toxic and odorless DMS. All manipulations should take place in a fume hood and waste processed as described.[20] Cells are treated with DMS at a final concentration of 0.2% in their medium for 2 min at about 37°.

Cells Growing on Plates. Volumes here are for 150-cm^2 plates. The medium (20 ml) is taken up in a tube which can be closed tightly as a 50-ml Corning or Falcon polypropylene tube. DMS is added to the medium at a final concentration of 0.2% and is vigorously shaken for a few seconds to mix the hydrophobic DMS. The medium is immediately placed back in the plates and left to react with the cells for 2 min. The medium is then aspirated, and the cells are rapidly washed with 10 ml of ice-cold phosphate-buffered saline (PBS) and lysed in 4 ml of SDS–proteinase K buffer (25 m*M* Tris–HCl, pH 8.0, 250 m*M* NaCl, 5 m*M* Na-EDTA, 0.4% SDS, 0.1 mg/ml proteinase K) supplemented with 0.2% v/v 2-mercaptoethanol.

Cells Growing in Suspension. For cells like HeLa or leukocytes, cells are resuspended at about 3×10^7 cell/ml in their growth medium to which DMS is added to a final concentration of 0.2% and the mixture is vigorously shaken for a few seconds. After 2 min, an equal volume of ice-cold buffer W (see below, isolation of nuclei) supplemented with 1% v/v 2-mercaptoethanol and 0.05% v/v Nonidet P-40 is used to lyse the cells which are sedimented at 2000*g* for 5 min. The nuclear pellet is resuspended in 2 ml of supplemented W buffer (for about 3×10^8 cells) to which 10 ml of SDS–proteinase K buffer is added.

Dimethyl sulfate and piperidine treatment of plasmid DNA and naked genomic DNA are as described,[20] except that after piperidine treatment, the DNA is ethanol precipitated, resuspended in TE with 200 m*M* NaCl, extracted with PCI (phenol : $CHCl_3$: isoamyl alcohol, 25 : 24 : 1, v/v), and ethanol precipitated in new silanized microtubes. Under these conditions

[19] G. P. Pfeifer and A. D. Riggs, *Genes Dev.* **5,** 1102 (1991).
[20] A. M. Maxam and W. Gilbert, this series, Vol. 65, p. 499.

it is not necessary to dry the samples thoroughly in vacuum to remove traces of piperidine.

2. *Treatment of Cells with $KMnO_4$*

The $KMnO_4$ treatment of cells can take place either on ice ($KMnO_4$ final concentration about 20 m*M*) or at 37° ($KMnO_4$ final concentration about 7 m*M*) with similar results. $KMnO_4$ is freshly prepared as a 150 m*M* solution in water and added to PBS just before use.

Cells Growing on Plates. Cells on 150-cm^2 dishes are washed with 10 ml of PBS and excess PBS is carefully removed. The monolayers are covered with 10 ml/dish of PBS containing $KMnO_4$ and left to react for 1 min. The monolayers are then washed with 10 ml of ice-cold PBS containing 2 m*M* Na-EDTA and the cells are lysed with 4 ml SDS–proteinase K buffer supplemented with 0.2% v/v 2-mercaptoethanol.

Cells Growing in Suspension. Cells in suspension are sedimented and resuspended in one-tenth volume of PBS. $KMnO_4$ from a 0.15 *M* freshly dissolved stock is added to the desired final concentration and left to react for 1 min. Reactions are stopped by the addition of 2-mercaptoethanol and EDTA to final concentrations of 0.2% v/v and 50 m*M*, respectively, and cells are lysed by the addition of 1 vol of solution 2× SK (50 m*M* Tris-HCl, pH 8.0, 400 m*M* NaCl, 5 m*M* EDTA, 0.4% v/v SDS, and 0.2 mg/ml proteinase K). Under these conditions the manganese ions remain soluble during further PCI extractions and ethanol precipitation.

Treatment of Nuclei and Naked DNA. Nuclei in buffer W are treated with $KMnO_4$ on ice as for the cells, except that 3 vol of 2× SK buffer are added to lyse the nuclei. Naked genomic DNA or plasmid DNA are treated with $KMnO_4$ in TE buffer as for cells or nuclei, except that after quenching, the samples are extracted once with PCI and ethanol precipitated at −20°.

3. *Isolation of Nuclei and DNase I Treatment*

Cells Growing on Plates. Cells on 150-cm^2 dishes are washed with 10 ml of ice-cold PBS and excess PBS is carefully removed. The monolayers are covered with 5 ml/dish of ice-cold lysis buffer and left to swell for 5 min on ice [lysis buffer is 3.75 m*M* Tris–HCl, pH 7.4, 0.05 m*M* spermine, 0.125 m*M* spermidine, 20 m*M* KCl, 1 m*M* K-EDTA, 1 m*M* dithiothreitol (DTT), 0.5 m*M* phenylmethylsulfonyl fluoride (PMSF), 3 μg/ml aprotinin, 0.5 μg/ml leupeptin, 1 μg/ml pepstatin]. Cells are then scraped from the plates and lysed in a Dounce homogenizer with 10 strokes of a tight-fitting pestle. Nuclei are pelleted in a swing-out rotor at 1000*g* for 5 min without braking. Nuclei are resuspended in the same volume (5 ml/150-cm^2 dish) of lysis buffer containing 20% glycerol and spun exactly as before. The nuclei are resuspended in 1 ml buffer W/150-cm^2 dish (15 mM Tris–HCl, pH 7.4, 5% glycerol, 0.2 mM spermine, 0.5 mM spermidine, 80 mM KCl, 2 mM K-EDTA, 1 mM DTT, and protease inhibitors as in lysis buffer).

Cells Growing in Suspension. About 3×10^8 cells are sedimented and resuspended in 40 ml of ice-cold lysis buffer and left to swell for 5 min. Cells are lysed with 10 strokes of a tight-fitting pestle in a Dounce homogenizer and are spun at 1000g for 5 min without braking. Nuclei are resuspended in the same volume of lysis buffer with 20% glycerol and spun as before. Nuclei are resuspended in 6 ml of buffer W and treated with DNase I as for nuclei from cells growing on plates.

DNase I Digestion. Nuclei at about 10^8/ml are preheated at 25° for 2 min and the same volume of preheated buffer W supplemented with 20 mM $MgCl_2$ and DNase I is added for 1 min. Digestion is stopped by the addition of 20 mM EDTA and then of 3 vol of SDS–proteinase K solution. The concentration of DNase I that gives an adequate sequencing signal varies among sequences and cell lines. Concentrations from 2 to 20 μg/ml should be tested for regions that are in DNase I hypersensitive sites such as promoters and enhancers of active genes. For other regions, concentrations up to 100 μg/ml may be necessary.

4. Isolation of Genomic DNA

1. Samples of genomic DNA are left in SDS–proteinase K buffer overnight at 37° or for 2 to 4 hr at 50°.

2. Samples are then extracted twice with PCI at room temperature. This is most conveniently done in screw-capped tubes of 15- or 50-ml polypropylene tubes (Falcon or Corning) in which the aqueous and phenol phases can be vigorously mixed. All PCI extractions are first heated for a few minutes to warm up the PCI that has been stored at 4°, followed by mixing and centrifugation. The limited DNA shearing resulting from the vigorous mixing will not interfere with genomic sequencing and facilitates pipetting. Polypropylene tubes with PCI can be safely centrifuged at 2000g to separate the phases, and the PCI is removed with a pipette.

3. DNA is recovered after precipitation with 2.5 vol of ethanol, usually in Falcon 15-ml snap cap polypropylene tubes. For DMS- or $KMnO_4$-treated samples, a precipitate must be visible and is recovered immediately by a 5-min centrifugation at 1000g. This results in the quantitative recovery of DNA while most of the RNA remains in the supernatant. If the genomic DNA is fragmented (as after DNase I treatment), the samples are left overnight at −20° and the DNA is recovered after a 20-min 5000g centrifugation (SS34 or similar rotor).

4. After centrifugation, genomic DNA precipitates are dried under vacuum. Samples are then resuspended in TE. The DNA should dissolve in 1–2 hr at room temperature with gentle mixing and occasional vigorous shaking.

5. Samples derived from whole cells are further treated for 1 hr at 37° with 10 μg/ml of pancreatic RNase A. Then NaCl is added to 200 m*M*, and the samples are extracted once with PCI, precipitated with ethanol, and resuspended as before. DNA concentration is determined by measuring the OD at 260 nm and the samples are now ready for digestion with restriction enzymes followed by purification of specific sequences.

Purification of Specific Genomic Sequences

1. Synthesis of Biotinylated RNA Probes

Biotinylated ribonucleotides can be synthesized by conventional procedures using SP6 and T7 RNA polymerases[21] in the presence of biotin-11–UTP. Biotin-11–UTP can be obtained from Enzo (Cat. No. 42815) or can be purchased from Sigma. Biotin-16–UTP available from other sources has not been tested in our laboratory. We use a 3 : 1 ratio of UTP : biotin–UTP as the biotinylated nucleotide is expensive, and a single biotin molecule per nucleic acid is in theory sufficient to bind it to streptavidin. We have consistently obtained better yields of whole length RNA with SP6 polymerase when compared to T7. We have not tested T3 polymerase. However, the RNA product does not need to be full-length for efficient purification, and "smeary" preparations work as well. The biotinylated RNA does not need to be longer than 50 nucleotides but, for optimal results, should encompass the sequence that anneals to the primer. Photobiotinylation after synthesis of RNA with regular nucleotides has not been tested.

During the preparation of the biotinylated RNA, two aspects are critical for purification of genomic sequences. The RNA preparation should be free of the template plasmid DNA. Any of this DNA left will copurify with genomic sequences and yield artifactual bands. RNA is therefore extensively digested with DNase I. As a control, an aliquot of the RNA preparation is used in a primer extension and should not give any signal. Another critical step is the elimination of unincorporated biotinylated UTP. This is achieved by selective ethanol precipitation in the presence of 0.75 *M* NH_4CH_3COOH with 2.5 vol of ethanol at room temperature. A protocol that yields large amounts of biotinylated RNA is presented below.

[21] P. A. Krieg and D. A. Melton, this series, Vol. 115, p. 397.

Alternatively, the RiboMAX kit sold by Promega works as well. If LM-PCR is the main procedure of detection, lower amounts of biotinylated RNA are required and the RNA preparation can be reduced in size. It should be noted that for SP6 and T7 protocols, enzymes that produce 3′ overhangs should not be used to linearize the plasmid template unless the overhangs are eliminated.

1. The reaction mix is prepared in the following order at room temperature to avoid precipitation of the DNA template by the polyamine-containing buffer:
   - 20 μg of a plasmid (linearized and purified by PCI and ethanol precipitation)
   - H_2O to make a final reaction volume of 100 μl
   - 2 μl of 1 *M* DTT
   - 10 μl of each ATP, CTP, and GTP at 10 m*M*
   - 7.5 μl of UTP at 10 m*M*
   - 1.25 μl of biotin-11–UTP at 20 m*M*
   - 20 μl of 5× buffer (1× is 20 m*M* Tris–HCl, pH 7.4, 10 m*M* NaCl, 6 m*M* $MgCl_2$, 1 m*M* spermidine, for both SP6 and T7)
   - 10 μl of RNasin 40 units/microliter and 150 units of SP6 or 300 units of T7

2. The reaction is left to proceed at 40° (SP6) or at 37° (T7) for 1 hr. Then the same amounts of polymerase and RNasin are added and incubation is continued for another hour.

3. Two hundred units of DNase I ("RNase-free," from BRL or Boehringer Mannheim) is then added together with 200 units of RNasin and the samples are incubated at 37° for 1–2 hr.

4. The reaction is stopped by the addition of 170 μl or HE, 30 μl of 7.5 *M* NH_4CH_3COOH, and 0.8 ml of PCI. The samples are vortexed, spun for 1 min in a microfuge at room temperature, and the PCI layer is removed from the bottom of the microfuge tube. The interphase is left with the aqueous phase as biotinylated nucleic acids tend to partition there.

5. After the addition of 0.75 ml of ethanol, the interphase should dissolve. Samples in ethanol are left for 10 min at room temperature and are then spun in a microfuge for 10 min at room temperature. The ethanol is carefully removed and the visible pellets are left to dry on the bench for a few minutes.

6. Pellets are resuspended in 170 μl of HE. Twenty microliters of Boehringer-Mannheim 10× MS buffer (1× is 10 m*M* Tris–HCl, pH 7.4, 10 m*M* $MgCl_2$, 50 m*M* NaCl, 1 m*M* dithioerythritol), 5 μl of 1 *M* DTT, 10 μl of

RNasin 40 U/μl, and 200 units of DNase I are added in this order. Digestion is left to proceed for 1–2 hr at 37°.

7. The addition of 20 μl of 7.5 M NH_4CH_3COOH and 0.8 ml of PCI stops the reaction, and 500 μl of ethanol is added to the aqueous phase and interphase (that may not appear at this point). The RNA is precipitated for 10 min at room temperature and is collected by centrifugation for another 10 min.

8. The biotinylated RNA is resuspended in 200 μl of HE, transferred to a new tube with 20 μl of 7.5 M NH_4CH_3COOH, and the RNA is again precipitated at room temperature by the addition of 0.5 ml of ethanol. This last resuspension and precipitation step is then repeated. The successive precipitations are aimed at removing the unincorporated biotin-11–UTP.

9. The biotinylated RNA is resuspended in a final volume of 200 μl of HE, and 2 μl is used for UV quantification. Another 2 μl can be used for analysis on a small denaturing polyacrylamide gel and visualized with ethidium bromide. About 20 to 100 μg of biotinylated RNA should be obtained using this protocol.

2. *Hybridization with Genomic DNA*

For hybridization with a biotinylated probe, genomic DNA has to be digested with at least one 4-bp cutter restriction enzyme and, if necessary, other enzymes to determine the extremities of the sequence to be analyzed. Between 200 and 500 μg of genomic DNA is needed for one primer extension with a radiolabeled primer. If LM-PCR is used, 1 to 10 μg is sufficient and results, after PCR, in unlimited material for extensions with a kinased primer. For large amounts of genomic DNA, samples are digested overnight with about 1 unit of each restriction enzyme for every 4 to 5 μg of genomic DNA. For economical reasons, it is best to use 4-bp cutters like *Hae*III, *Hin*fI, *Msp*I, *Rsa*I, or *Taq*I (in this case 4 hr at 65°). Also, one or two inexpensive 6-bp cutters are included to define the ends of the sequence to be purified. Digestions are stopped by EDTA, and the DNA samples are extracted once with PCI, precipitated with ethanol at −20°, dried well in vacuum, and resuspended in RNase-free HE (20 mM HEPES–NaOH, pH 7.4, 1 mM Na-EDTA) at a final DNA concentration of 2 to 5 mg/ml. For DNA samples of a few hundred micrograms, usually 200 μl is used when samples for LM-PCR (1 to 10 μg) are resuspended in 20 μl of HE. All subsequent manipulations are carried out with silanized microtubes and tips under RNase-free conditions.

The following hybridization protocol was devised for 400-μg to 1-mg samples of DNA that will be analyzed by primer extension with a highly labeled probe. DNA samples for LM-PCR are processed in one-tenth the volume during hybridization and in the specified volumes later on.

1. To genomic DNA dissolved in 200 μl of HE, 100 μl of 0.6 *M* NaOH is added and the samples are heated at 65° for 5 min.

2. The samples are cooled for 2 min at room temperature. Thirty microliters of 1 *M* HEPES–NaOH, pH 7.4, and 100 μl of 0.6 *M* HCl are added to each tube in this order, and the tube is immediately vortexed.

3. Seventy microliters of 5× HB containing freshly added biotinylated RNA is then mixed with the denatured genomic samples (5× is 100 m*M* HEPES–NaOH, pH 7.4, 3*M* NaCl, 5 m*M* Na-EDTA, 0.5 mg/ml yeast tRNA, and 0.25% Tween 20). The final concentration of the biotinylated probe used in the hybridization is 0.1 to 0.3 μg/ml. If the biotinylated RNA spans only the region to be purified, 0.1 μg/ml should be sufficient.

4. Tubes are placed in containers (such as NEN lead pigs) that are submerged in a 65° water bath. This prevents the formation of condensation in the microtubes. Hybridization is left to proceed overnight, up to a maximum of 16 hr. Samples can be conveniently frozen after hybridization.

3. *Recovery of Specific Genomic Sequences*

DNA–RNA hybrids are immobilized on streptavidin–agarose (Sigma) and the bulk of the genomic DNA is washed away. To prevent nonspecific absorption on beads, the Sigma preparation is diluted five times with 1× HB (packed volume is about 10% of the slurry volume) and is washed four or five times before addition to the hybridization mix. Streptavidin magnetic beads available from several suppliers were not tested but may work as well.

1. Washed streptavidin–agarose diluted five times in 1× HB (pipetted with wide mouth tips) is added to the hybridization mix (50 μl for large samples, 10 μl for LM-PCR samples) and the hybrids are left to bind for 30 to 60 min at 37° with slow agitation.

2. Beads are sedimented by a short spin in a microfuge (ideally a swing-out microfuge) and the supernatant containing most of the genomic DNA is removed. This supernatant can be used, after PCI extraction (leave the interphase behind) and ethanol precipitation, for the purification of other sequences (such as the other DNA strand). We have successfully purified three different sequences from a single genomic DNA preparation.

3. The beads are washed once (twice for LM-PCR samples) with 1 ml of HE. With some brands of microtubes the pellet of agarose beads may be difficult to see and the supernatant should be removed carefully.

4. Genomic DNA is recovered from the beads by alkali treatment using 180 μl of 0.2 *M* NaOH (90 μl for LM-PCR samples). For the recovery of the immobilized genomic sequences, it is important to work with silanized plastic ware. After the addition of NaOH to the beads, the tubes are briefly vortexed and left for 2 to 5 min on the bench.

5. The tubes are then centrifuged briefly and the supernatant containing the released DNA is carefully recovered in a silanized microtube using a silanized tip. The samples are then placed at 70° for 20 min to hydrolyze the RNA.

6. Tubes are chilled on ice, and 30 μl (15 for LM-PCR) of 1 *M* Tris–HCl, pH 7.4, containing 50 μg/ml yeast RNA is added. Then 90 μl of 0.4 *M* HCl is added (45 μl for LM-PCR) and the tubes are immediately vortexed.

7. The samples are extracted with 0.5 ml of PCI and the aqueous phase is placed into another silanized microtube containing 0.9 ml (0.45 ml for LM-PCR) of ethanol. The tubes are usually left overnight at −20°, but can be chilled twice on dry ice before centrifugation for 15 min in a microfuge. The visible small pellet is dried on the bench and is resuspended in 3 or 6 μl of HE. The ethanol should be readily pipetted out from silanized tubes, otherwise a short spin to bring down the ethanol should be performed before drying the DNA pellet.

Primer Extension Using High Specific Activity Primers

Purified sequences can be detected either by primer extension with a highly labeled probe or by LM-PCR. In both protocols a similar primer extension is carried out on the purified material. Described here are the synthesis of a highly radioactive primer and the conditions for performing primer extension to obtain accurate sequencing ladders. It is optimal that the primer anneals to the extremity of a purified sequence. However, satisfactory results have been obtained with primers that anneal in the middle of the purified DNA fragments. The primer extension is the most difficult part of the genomic sequencing procedure. It is worth testing the conditions using plasmid DNA that has been treated with DMS and piperidine (G-ladder) to determine the optimal conditions before samples derived from genomic DNA are used. Plasmids can also be used conveniently to test the template purification.

1. Synthesis of Radioactive Primers

A 25- to 30-mer primer containing 10 or more radioactive nucleotides (specific activity about 30,000 Ci/mmol) is prepared essentially according to Saluz and Jost.[6] A 9-mer primer precursor anneals to a 29-mer to produce a 4-bp 3′ overhang. Fill in of 16 nucleotides produces a 25-mer-labeled primer that is then separated from the 29-mer template in a polyacrylamide-urea gel. Both starting oligonucleotides were previously purified on polyacrylamide denaturing gels before annealing. The 9-mer/29-mer duplex is produced after an overhight annealing of 200 pmol of 29-mer to 500 pmol of 9-mer primer in a final volume of 50 μl in 8× MS buffer (from Boehringer-Mannheim, 1× is 10 m*M* Tris–HCl, pH 7.4, 10 m*M* $MgCl_2$, 50 m*M* NaCl, 1 m*M* dithioerythritol) which yields 5 pmol template/μl in 8× MS buffer.

1. Radioactive nucleotides (5 μl of each one at 3000 Ci/mmol, equivalent to 17 pmol) and different cold nucleotides (250 pmol) are dried in a silanized tube. Synthesis of the 25-mer radiolabeled primer is carried out in a volume of 4 μl. The number of hot nucleotides is not important; one or all four nucleotides can be radioactive and a full-length primer should be obtained. For economy, it is usually convenient to use one or two radioactive nucleotides chosen to introduce 10 or more radioactive residues in the primer.
2. The dried nucleotides are resuspended in 3 μl of H_2O, then 0.5 μl of annealed oligonucleotides (2.5 pmol of template) and 0.5 μl of Klenow enzyme at 2 units/μl are added in this order. The reaction is carried out for 10 min at room temperature and is stopped by the addition of 4 μl of sequencing gel-loading buffer.
3. To separate the radioactive primer from the template, the sample is electrophorized on a 15-cm 12% acrylamide–urea gel. The position of the highly radioactive 25-mer primer is detected by a few seconds exposure to a X-ray film. The region of the gel that contains the primer is cut out and crushed with a small pestle before the addition of extraction buffer.
4. After crushing, 300 μl of elution buffer is added (25 m*M* Tris–HCl, pH 8.0, 400 m*M* NH_4CH_3COOH, 2 m*M* Na-EDTA, 0.1% SDS). The radioactive primer is left to diffuse from the gel for 1 hr with vigorous mixing (a bacterial shaker at 300 rpm is adequate). The gel pieces are sedimented (5 min in a microfuge), and the supernatant is recovered. Another 200 μl of extraction buffer is added, and the crushed acrylamide is extracted for another hour and is then recentrifuged.
5. The two supernatants (corresponding to about 80% of the probe as estimated by monitor) are pooled and 3 μg of yeast RNA is added as a carrier. The mixture is extracted once with PCI and is precipitated with 3 to 4 vol of ethanol overnight at −20° or by chilling twice on dry ice.
6. After precipitation, the probe is resuspended in 20 μl of HE to yield an estimated concentration of 0.1 pmol/μl. The primer must be used within a few days as older degraded primers tend to produce high backgrounds.

2. Primer Extension

A number of parameters should be tested to obtain a sequencing ladder that truly represents the population of modified molecules. These parameters include annealing temperature and time as well as concentrations of primer, but the most important appears to be the type of DNA polymerase. The protocol presented here works well for Klenow and *Taq* polymerase, conditions under which different results can be obtained with the different enzymes. Conditions for Vent and T4 DNA polymerase are also discussed. Pilot experiments are most easily carried out on plasmid DNA that has been cleaved by a Maxam–Gilbert sequencing reaction, for example, a G-ladder. One nanogram of plasmid of about 3 kb is equivalent to sequences purified from 1 mg of human or rodent genomic DNA. When using double-stranded plasmids for control reactions, it is important to use short annealing times as the two strands can renature and inhibit primer extension. A single-stranded template can be obtained from plasmids by selection with biotinylated RNA. In this case, the amount of biotinylated RNA should be in excess over the plasmid DNA.

1. The primer extension starts in a silanized microtube with a template resuspended in 3 μl of HE.

2. One microliter of 0.2 *M* NaOH and 1 μl of primer are added (as 2 μl of a freshly prepared mix) along the side of the tube close to the bottom and left to fall at the bottom. This results in adequate mixing.

3. Samples are placed for 5 min at 65° for denaturation and are then chilled on ice. The amount of primer necessary is variable depending of the sequence; 0.01 to 0.02 pmol of radioactive primer (0.1 to 0.2 μl of a highly labeled primer preparation, about 10^5 cpm) produces clear results. More primer may increase the specific signal but also the background. A background smear may not appear with plasmid DNA controls but will certainly be present with genomic samples and is usually proportional to the primer amounts.

4. After denaturation and chilling on ice, 2.5 μl of a freshly prepared mix of 1 μl of 0.2 *M* HCl and 1.5 μl of 10× extension buffer are added along the side of the tube and left to fall into the sample as for the NaOH primer mix (1× extension buffer is 10 m*M* Tris–HCl, pH 8.5, at room temperature, 25 m*M* KCl, and 3.5 m*M* $MgCl_2$).

5. Annealing is typically for 10 min at 60°, but temperature may vary between 55 and 65° depending of the sequence and length of the primers. For longer annealing times, samples are put into NEN lead containers that are submerged into a water bath.

6. After annealing, samples are placed on ice, and 12.5 μl of nucleotides and enzyme mix is added. The samples are vortexed and briefly centrifuged to bring down the liquid. Extension is carried out for 10 min at 70° for *Taq*

polymerase or at 50° for Klenow.[7] The enzyme–nucleotide mix is made of 0.5 μl of 10× buffer, 0.8 μl of dNTPs stock (each at 10 m*M*, final concentration in reaction is 0.4 m*M*), and polymerase to obtain a final concentration in the reaction of 10 units/ml (*Taq* or Amplitaq) or 25 units/ml (Klenow). Low amounts of polymerase are important as extension with high concentrations often results in unreadable sequencing ladders. For *Taq* or Amplitaq (Boehringer or Perkin-elmer) it is necessary to leave the extension to proceed for 10 min. These enzymes have a terminal deoxytransferase activity and it is important that every extended product has an extra added residue. Enzymes such as Vent (New England Biolabs) or T4 DNA polymerase can work as well. T4 DNA polymerase seems to be particularly suitable for substrates where the primer anneals on the template at some distance from the ends.[15] Another advantage of T4 DNA polymerase is that it hydrolyzcs most of the free primer which results in lower backgrounds. However, a drawback of the T4 polymerase is that it works mostly at 37° and is more susceptible under these conditions to artifactual pause sites.[15] Vent is used with the buffer supplied by New England Biolabs, whereas the T4 DNA polymerase is used in a final buffer consisting of 50 m*M* Tris–HCl, pH 8.5, 15 m*M* NH_4CH_3COOH, 25 m*M* NaCl, and 7 m*M* $MgCl_2$. BSA and 2-mercaptoethanol are added in the 12.5-μl nucleotide–enzyme mixture to obtain final concentrations of 0.1 mg/ml and 0.1% v/v, respectively. Vent and T4 DNA polymerase are both used at a final concentration of 25 units/ml.

7. After extension, reactions are stopped by the addition of 170 μl of 25 m*M* Tris–HCl, pH 7.4, 1 m*M* Na-EDTA, 0.2% SDS, and 5 μg/ml denatured salmon sperm DNA. Tubes are briefly vortexed, microfuged, and incubated at 80° for about 20 min.

8. Samples are cooled to room temperature and 10 μl of 5 *M* NaCl and 300 μl of PCI are added. After vortexing and centrifugation, the aqueous phase is put in a new silanized tube containing 0.9 ml of ethanol and is chilled twice on dry ice.

9. After a 10- to 15-min centrifugation, the supernatant is carefully removed. A small pellet should be visible. If ethanol remains along the tube walls, the samples are briefly spun and the ethanol is pipetted from the pellet. Radioactivity may be still present on the sides of the silanized tube, but the pellet contains most of the extended products.

10. The pellet is dried for a few minutes on the bench and is resuspended in 3 μl of 90% formamide containing 0.5× TBE, 0.2% bromphenol blue, and 0.2% xylene cyanol. The samples are heated at 80° for 5 min prior to electrophoresis.

11. Sequencing gels have been poured the day before and prerun for 3 to 4 hr (usually the time required to make the primer extension). Sharktooth

combs are not suitable for footprinting experiments. Unincorporated primer is run out of the gel which is then fixed in 10% methanol, 10% acetic acid. Gels can be dried on cellophane sheets for exposure between two intensifying screens, but Whatman 3MM in place of cellophane is usually adequate. Exposures are for 12 hr to 3 days at $-80°$.

Acknowledgments

Most of these procedures were developed during a postdoctoral fellowship in the laboratory of James E. Darnell, Jr. I am grateful to Melanie Price and Lewis Pizer for critical reading of the manuscript. Research at ISREC was supported by the Swiss National Science Foundation, the Swiss League against Cancer, and by ISREC.

[40] DNA Affinity Chromatography

By TADASHI WADA, HAJIME WATANABE, HARUMA KAWAGUCHI, and HIROSHI HANDA

Introduction

DNA affinity chromatography for purification of sequence-specific DNA-binding proteins was originally developed with multimerized synthetic oligonucleotides covalently coupled to agarose resins.[1] A variety of DNA affinity resins have also been developed.[2-4]

This chapter describes the preparation and use of DNA affinity beads that consist of multimerized synthetic oligonucleotides covalently coupled to latex beads. The beads have several advantages for purification of sequence-specific DNA-binding proteins. First, the beads are able to immobilize relatively large amounts of DNA because beads with an extremely small diameter of 0.2 μm on average have a larger surface area for total DNA coupling than agarose beads. Second, DNAs coupled on the beads are more accessible to DNA-binding proteins because the beads are free to move in the binding mixture. Third, the property of the bead with a hydrophilic surface lowers nonspecific proteins binding on it. Fourth, the

[1] J. T. Kadonaga and R. Tjian, *Proc. Natl. Acad. Sci. U.S.A.* **83,** 5889 (1986).

[2] J. T. Kadonaga, this series, Vol. 208, p. 10.

[3] L. A. Kerrigan and J. T. Kadonaga, *in* "Current Protocols in Molecular Biology" (F. M. Ausubel, R. Brent, R. E. Kingston, D. D. Moore, J. G. Seidman, J. A. Smith, and K. Struhl, eds.), p. 1, 10, 12. Wiley (Interscience), New York, 1993.

[4] H. Kawaguchi, A. Asai, Y. Ohtsuka, H. Watanabe, T. Wada, and H. Handa, *Nucleic Acids Res.* **17,** 6229 (1989).

batchwise purification procedure reduces the contamination of nonspecific proteins because the volume of the beads precipitated after a brief centrifugation is extremely small. These advantages enable us to directly purify sequence-specific DNA-binding proteins from crude cell extracts within a few hours. This particular method has been successfully employed in the purification of several transcription factors.[4–8]

A brief description of the procedure is as follows. Latex beads composed of a styrene (St) core and a glycidyl methacrylate (GMA) surface are prepared. Complementary chemically synthesized oligonucleotides that contain a recognition site for a sequence-specific DNA-binding protein are annealed and ligated to give oligomers with protruding ends. The DNA oligomers are then covalently coupled to latex beads by simply mixing them at 50° for more than 3 hr. Epoxy groups of the beads mainly react with amino groups at the protruding ends of DNA. Crude cell extracts are incubated with the affinity beads in the presence of nonspecific DNA such as single-stranded DNA or poly(dI-dC). The desired sequence-specific DNA-binding protein that binds to the recognition sites at the affinity beads is batchwisely purified. The sequence-specific protein precipitated with the beads by brief centrifugation is then eluted from the affinity beads with a high salt solution. A typical protein is purified more than 1000-fold with a 70% yield. If necessary, it is possible to repeat the batchwise step for further purification. This procedure is not only effective, but is also simple and straightforward to perform. A DNA affinity chromatography experiment can be carried out within a few hours.

Materials and Methods

Preparation of Latex Beads

Latex beads are prepared as described.[6] Briefly, GMA, St, and divinylbenzene (DVB) are used as the monomers. To prepare the latex beads, mix 1.8 g of GMA, 1.2 g of St, 0.04 g of DVB, and 110-ml distilled water in a 200-ml three-necked round-bottom flask equipped with a stirrer, a nitrogen gas inlet, and a condenser. To purge oxygen in the mixture, bubble

[5] K. Morohashi, S. Honda, Y. Inomata, H. Handa, and T. Omura, *J. Biol. Chem.* **267,** 17913 (1992).

[6] Y. Inomata, H. Kawaguchi, M. Hiramoto, T. Wada, and H. Handa, *Anal. Biochem.* **206,** 109 (1992).

[7] F. Hirose, M. Yamaguchi, H. Handa, Y. Inomata, and A. Matsukage, *J. Biol. Chem.* **268,** 2092 (1993).

[8] H. Watanabe, J. Sawada, K. Yano, K. Yamaguchi, M. Goto, and H. Handa, *Mol. Cell. Biol.* **13,** 1385 (1993).

nitrogen gas into it, and keep the mixture at 70° in a water bath. Then add 10 ml of distilled water containing 0.06 g of azobisamidinopropane dihydrochloride to initiate soap-free emulsion polymerization. Since the resulting GMA–St copolymer has a partially hydrophobic surface due to exposure to polystyrene microdomains, add 0.3 g of GMA to the mixture 2 hr after the initiation of polymerization, then allow the mixture to stand for 24 hr until the whole surface of the copolymer is covered with poly-GMA. Collect the beads with epoxy groups derived from GMA on the surface by centrifugation, wash three times with water, and suspend in 10 vol of water. The beads can be stored at 4°.

Solutions

Storage Buffer. 10 m*M* Tris–HCl (pH 8.0), 0.3 *M* KCl, 1 m*M* EDTA, and 0.02% NaN_3.

TGEN. 50 m*M* Tris–HCl (pH 7.9), 20% (v/v) glycerol, 1 m*M* EDTA, 0.1% (v/v) Nonidet P-40 (NP-40), and 1 m*M* dithiothreitol (DTT).

10× Ligation Buffer. 660 m*M* Tris–HCl (pH 7.4), 100 m*M* $MgCl_2$, 10 m*M* spermidine, 150 m*M* DTT, and 2 mg/ml bovine serum albumin (BSA).

10× Exobuffer. 330 m*M* Tris–acetate (pH 7.9), 660 m*M* CH_3COOK, 100 m*M* $(CH_3COO)_2Mg$, and 5 m*M* DTT.

Methods

Preparation of DNA Oligomers. The core sequences to which proteins of interest bind specifically can be determined using DNase I footprinting,[9] methylation interference,[10] or other methods.[11–13] Precise determination of the binding site is necessary for efficient purification. Two complementary oligonucleotides are designed to produce 3′-protruding ends with 4 or 5 bases. Oligonucleotides are synthesized on an ABI Model 392 DNA synthesizer by the phosphoamidite method in a 40-nmol scale using the "trityl off" program. The oligonucleotides are cleaved from the columns, incubated at 55° for more than 8 hr, and dried under vacuum according to the manufacturer's directions. In the case of the synthesis of 20 bases, about 200 μg of oligonucleotides can be recovered. Resuspend the dried oligonucleotides in water (400 μl), and use an aliquot to determine the concentration using a spectrophotometer at a wavelength of 260 nm (an

[9] D. Galas and A. Schmitz, *Nucleic Acids Res.* **5,** 3157 (1978).
[10] U. Siebenlist and W. Gilbert, *Proc. Natl. Acad. Sci. U.S.A.* **77,** 122 (1980).
[11] M. Fried and D. M. Crothers, *Nucleic Acids Res.* **9,** 6505 (1981).
[12] T. D. Tullius, B. A. Dombroski, M. E. A. Churchill, and L. Kam, this series, Vol. 155, p. 537.
[13] R. P. Hertzberg and P. B. Dervan, *J. Am. Chem. Soc.* **104,** 313 (1982).

OD 1 corresponds to 20 μg/ml). Mix approximately 200 μl of each oligonucleotide (100 μg) and boil at 100° for 10 min in a water bath. Then allow over 1 hr for the mixture to cool to room temperature. After brief centrifugation, divide the mixture into five microcentrifuge tubes (80 μl each). Add 20 μl of 10× ligation buffer, 2 μl of 100 m*M* ATP, 96 μl of water, and 2 μl of T4 polynucleotide kinase (15 units/μl) to each tube. One of the tubes contains 1 μCi of [γ-^{32}P]ATP for monitoring the DNA-coupling efficiency. After incubation at 37° for 60 min, add 4 μl of T4 DNA ligase (2.8 units/μl) and incubate at 4° overnight. For checking the length of the ligated DNA oligomers, aliquots (10 μl) of the mixture are subjected to a 10% polyacrylamide gel electrophoresis (PAGE). The ligated DNA with 5- to 15-mers should be produced under these conditions. If monomers are not ligated as expected, repeat phosphorylation and ligation again after phenol–chloroform extraction and ethanol precipitation. Rinse the pellet with 1 ml of 70% ethanol and centrifuge again for a few minutes followed by drying the pellet under vacuum. Resuspend the pellet with 80 μl of water and repeat the phosphorylation and ligation steps.

Preparation of 5′-Protruding Ends for Efficient Immobilization. The latex beads are composed of a styrene core and a polyglycidyl methacrylate surface, to which DNA oligomers are immobilized by means of covalent bonds formed between epoxy and amino groups.[6] In our studies, DNA oligomers with protruding ends were about five times more efficiently bound to the beads than those with blunt ends at either strand, indicating that the amino groups at the protruding ends possibly react on the epoxy groups. Also, a DNA fragment with about 15 nucleotides at the 5′-protruding ends gave the most efficient purification of the transcription factor ATF/E4TF3.[6] However, longer protruding ends may decrease both the coupling and purification efficiency because amino groups on one protruding end react with several epoxy groups or one DNA molecule is immobilized at both ends. The ligated DNA oligomers are subjected to phenol–chloroform extraction followed by ethanol precipitation. Suspend the DNA oligomers in 40 μl of water per tube and combine them (total: 200 μl). Add 40 μl of 10× exobuffer, 120 μl of water, and 40 μl of T4 DNA polymerase (5 units/μl). Incubate at 37° for 30–45 sec to digest the DNA oligomers by a 3′ → 5′-exonuclease activity of T4 DNA polymerase. To stop the reaction, add 4 μl of 0.5 *M* ETDA and 400 μl of phenol–chloroform simultaneously and vortex for 1 min. Centrifuge at 15,000 rpm for 5 min, transfer the aqueous phase to another microcentrifuge tube followed by ethanol precipitation, and resuspend in 40 μl of water. Load half of the sample (20 μl; up to 100 μg) on a NICK column (Pharmacia) that is equilibrated with water and add 400 μl of water. Further add 400 μl of water three times (second to fourth fraction). Collect all the fractions and check the concentration. Usually the ligated DNA oligomers are collected in the second fraction. Even if the DNA is eluted in the third fraction, it is not recommended for use in the coupling reaction. Measure the concentration of the DNA oligomers in the second fraction at the 260-nm wavelength using a spectrophotometer (an OD of 1 corresponds to 50 μg/ml). If the concentration of the DNA oligomers is less than 0.2 μg/μl, precipitate the oligomers with ethanol and adjust the concentration to more than 0.2 μg/μl with H_2O. Add 1 *M* potassium phosphate buffer (pH 8.0) to a final concentration of 10 m*M* and use the DNA solution immediately for the coupling reaction.

Immobilization of DNA on Latex Beads. The beads corresponding to 10 mg are transferred with a yellow 200-μl tip to a 1.5-ml microcentrifuge tube with a screw cap and are collected by centrifugation at 15,000 rpm for a few minutes at room temperature until the supernatant becomes clear. Add 400 μl of 10 m*M* potassium phosphate buffer and vortex for at least 2 min until aggregation of the beads disappears. Centrifuge and discard the cleared supernatant. Repeat this wash twice. Before addition of the DNA to the beads, count the radioactivity of the DNA to estimate the DNA coupling efficiency. Mix 10 mg of the beads and 100 μg of the DNA and incubate at 50°. Occasional mixing during the incubation is likely to increase the DNA coupling efficiency. The coupling reaction usually requires more than 3 hr. The beads bearing the DNA oligomers are collected by centrifugation at 15,000 rpm for a few minutes at room temperature. The uncoupled DNA oligomers included in the supernatant can be reused for another coupling reaction after phenol–chloroform extraction and ethanol precipitation. Wash the beads two times with 500 μl of 2.5 *M* NaCl, suspended in 1 ml of ethanolamine hydrochloride (pH 8.0), and incubate at room temperature for 24 hr to inactivate unreacted epoxy groups on the beads' surfaces. Collect the beads and wash once with water (500 μl) and three times with 500 μl of storage buffer. Resuspend the beads in 200 μl of storage buffer and count radioactivity of the beads for the calculation of the DNA coupling efficiency. Usually 20 to 30% of input DNA is coupled to the beads. The beads can be stored at 4°.

Purification of DNA-Binding Protein. One of the advantages of using the DNA affinity latex beads is the possibility of one-step purification of DNA-binding proteins from a crude nuclear extract.[6] However, the following points must be taken into consideration for the direct purification. First, contamination of nuclease activity affects the purification efficiency. The starting materials must be examined for the presence of nuclease activity. To test for the nuclease activity, the labeled DNA probe is usually used and incubation is done in the conditions of the gel retardation assay in the presence of competitor DNAs. If the crude extracts contain little nuclease activity, such as nuclear extracts from HeLa cells, the DNA-binding proteins can be purified directly from the crude nuclear extracts. If the starting extract contains high nuclease activities, partial purification, using conventional column chromatography such as DEAE-Sepharose, DEAE-cellulose, phosphocellulose, Mono Q, Mono S, or gel filtration, is necessary. Heparin–agarose is also useful because it could separate DNA-binding proteins from DNase activity.

Second, optimal binding conditions should be established. The effects of temperature, pH, Mg^{2+} concentration, and requirement for metallic ion are some of the things that should be examined by a small-scale purification or a gel retardation assay.

Third, the effect of the addition of nonspecific DNA to the starting fraction must be examined. Single-stranded DNA prepared from salmon sperm or calf thymus DNA is usually used. The addition of poly(dI-dC) or poly(dG-dC) also has different effects to reduce the background noise. To determine type and concentration of the nonspecific DNAs, the data from a small-scale purification and a gel retardation assay are necessary. In the case of ATF/E4TF3 purification, an equal concentration of single-stranded DNA such as that from the gel shift assay gives good results.[6] The concentration of competitor DNA affects the specific activity of the eluted protein. The larger the amount of nonspecific DNA, the higher the specific activity obtained and the less the DNA-binding protein recovered.

Here we present a typical purification procedure carried out for the purification of ATF/E4TF3. Transfer the beads bearing about 5 μg of the DNA oligomers from the storage solution to a 1.5-ml siliconized microcentrifuge tube with a screw cap using a yellow 200-μl tip and centrifuge at 4° for 2 min. After discarding the supernatant, wash the beads three times with 100 μl of TGEN containing 0.1 *M* KCl and keep it on ice until use. Add 1 ml of nuclear extracts[14] (dialyzed in TGED containing 0.1 *M* KCl) into a 1.5-ml microcentrifuge tube containing appropriate amounts of nonspecific DNA and mix well. After incubating the mixture on ice for 15 min, transfer the mixture to the tube containing the beads and mix well. Allow the mixture to stand for 30 min. During this step, the color of the solution is white because of a stable distribution of the beads in the solution. Centrifuge at 15,000 rpm for a few minutes to clarify the supernatant. The length of the centrifugation time depends on the buffer and the proteins. Transfer the supernatant to another microcentrifuge tube, quickly freeze using liquid nitrogen, and store at −80°. Wash the beads three times with 200 μl of TGEN containing 0.1 *M* KCl. Then wash three times with 200 μl of TGEN containing 0.3 *M* KCl and add 20 μl of the buffer containing 1.0 *M* KCl. Vortex well and collect the suspension to the bottom by brief centrifugation.

[14] J. D. Dignam, R. M. Lebovitz, and R. G. Roeder, *Nucleic Acids Res.* **11,** 1475 (1983).

Stand on ice for a few minutes, centrifuge for 5 min, and collect the supernatant. About 70% of the bound proteins is recovered in one step. Repeat the elution step three times to obtain about 95% recovery. At least 10 μl of the elution buffer is required for a good recovery. Affinity-purified proteins can be characterized by SDS–PAGE. Usually, the protein concentration is so low that silver staining is necessary to identify the purified proteins. The purified proteins should be tested for the sequence-specific DNA-binding activity by the gel retardation assay. The used latex beads can be regenerated after washing once with 2.5 M KCl and three times with storage buffer. The latex beads can be recycled at least 10 times when a HeLa nuclear extract is used as a starting material.

Comparison of Latex Beads with Sepharose Resins. The DNA–Sepharose resins were originally used in the chromatography method developed by Kadonaga and Tjian.[1] Typically, a crude nuclear extract is prepared, and the desired protein is partially purified by the column chromatography step and is then applied to the affinity Sepharose resins.[2] However, our developed latex beads enabled us to purify DNA-binding proteins directly from a crude extract.[6] Here we compared the two methods by a direct purification from HeLa cell crude nuclear extracts. DNA-binding proteins are usually purified by column chromatography on DNA–Sepharose. A batchwise purification procedure as well as column chromatography was used to compare the affinity latex beads with the DNA–Sepharose resins. The latex beads were too small to be packed in a column. As seen in Fig. 1A, several distinct bands were observed in the purified fraction using the latex beads (lane 1) but not using the Sepharose resins (lane 2). All the proteins except the 39-kDa protein were able to specifically bind to the ATF/E4TF3 sites. These results indicate that the latex beads are superior to the Sepharose resins for purifying members of the ATF/E4TF3 family directly from the crude cell extracts. We also examined the DNA-binding activities of the purified fractions. The procedure with latex beads gave a DNA-binding activity higher than that obtained with the two procedures of the Sepharose resins (Fig. 1B, lanes 1 to 3).

Troubleshooting

Low Coupling Efficiency of DNA. When the DNA couples the latex beads, epoxy groups of the beads react with amino groups of the DNA and this reaction is easily inhibited by the presence of other amino groups. To couple the DNA to the latex efficiently, it is necessary to eliminate other amino groups. It is recommended that DNA is gel filtrated before a coupling reaction with H_2O instead of 10 mM phosphate buffer and coupled to the beads in H_2O. Furthermore, before a masking reaction, the beads

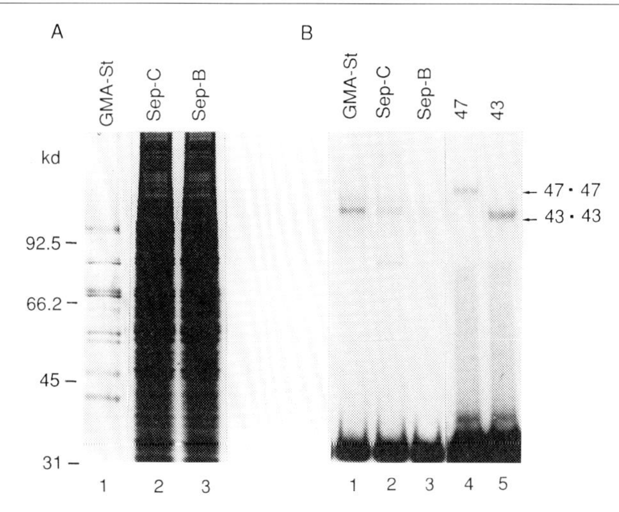

FIG. 1. Comparison of affinity latex beads with affinity Sepharose resins. (A) Silver staining of a SDS-polyacrylamide gel of affinity-purified ATF/E4TF3. Lane 1, 1/30 volume of eluate from the batchwise procedure (started from 1 ml of HeLa cell extracts) with the affinity latex beads (GMA-St); lane 2, 1/50 volume of eluate from column chromatography with the DNA affinity Sepharose resins (Sep-C); lane 3, 1/50 volume of eluate from batchwise procedure with the DNA affinity Sepharose resins (Sep-B). (B) DNA-binding activity of ATF/E4TF3. Gel retardation assays were performed. The DNA probe bearing the ATF/E4TF3-binding sites was incubated with a 1/600 volume of eluate from the batchwise procedure with the affinity latex beads (lane 1) and with the DNA–Sepharose (lane 3), and from the column chromatography with the DNA–Sepharose (lane 2). Lanes 4 and 5 show the complexes of the DNA probe and a homodimer of the renatured CREB (47 kDa) and ATF-1 (43 kDa) proteins, respectively (from Inomata *et al.*[6]). Molecular mass markers in kilodaltons.

should be washed well with H_2O. Some reagents containing amino groups such as EDTA and nucleotides derived from digestion of DNA usually remain even after ethanol precipitation.

Low Recovery of Purified Protein. There are three main reasons for failure to recover DNA-binding activity. (1) Improper binding conditions. Optimal conditions should be examined again. The reduction of nonspecific DNA concentration sometimes improves the recovery. (2) Nonspecific adsorption to tips and tubes. This can be reduced by the addition of 0.01–0.1% (v/v) of Nonidet P-40 in the buffer. Siliconization of plastic ware is also effective, although it is not always necessary. (3) Inefficient coupling reaction or improper storage of the latex beads. If the latex beads have been used several times, the DNase contained in the input fraction might digest the coupled DNA.

In the case of ATF/E4TF3, approximately 70% of the DNA-binding activity is trapped to the DNA affinity latex beads and is eluted by the buffer containing 1 *M* KCl, and flow through the fraction hardly contains specific DNA-binding activity. Sometimes the DNA-binding activities cannot be detected after the purification due to denaturation of the protein during the DNA-binding assay. The addition of BSA (final concentration: 0.1–1.0 mg/ml) to the reaction mixture sometimes gives good results.

Low Quality of Purified Protein. Sometimes multiple bands are observed when the purified fraction is subjected to SDS–PAGE. Such low purity is sometimes observed when the washing of the resin is insufficient. Because the beads easily stick together, latex beads must be suspended well in a washing buffer with vigorous agitation. A middle washing buffer, containing 0.2 to 0.3 *M* NaCl or KCl, is also effective in reducing the background noise. This is dependent on the character of the DNA-binding factor so be sure to confirm that the DNA-binding protein is not eluted by the middle washing buffer. For determining the best washing condition, carry out the small-scale purification. Prepare five tubes containing 100 μl of cell extracts. Pretreat with appropriate amounts of nonspecific DNA, mix with the beads, and wash three times with TGEN containing 0.1 *M* KCl. Wash each sample three more times with TGEN containing various concentrations of salt (usually from 0.1 to 0.5 *M*), respectively, and then bound proteins are eluted with TGEN containing 1.0 *M* KCl. Sometimes a lower salt concentration of the elution buffer contributes to the high purity of the purified protein. The optimum salt concentration of the elution buffer can be determined by a small-scale purification experiment as described earlier. When multiple bands are reproducibly detected, all proteins should be tested for their DNA-binding activities. To test for the DNA-binding activities of each polypeptide included in the purified fraction, renature the polypeptides after fractionation by a SDS–polyacrylamide gel.[15,16] Polypeptides of interest are cut out from the gel, denatured, and renatured as described previously. Renatured proteins are tested for DNA-binding activity by the gel retardation assay. If the DNA-binding protein forms a complex, all components of the protein can be identified by a combination of these renatured proteins.[16] In special cases, DNA-binding activities are observed only when all the proteins are mixed at the denatured condition and renatured together.

Some nonspecific DNA-binding proteins reported previously, such as ADP ribosylase and Ku, are also purified by these DNA affinity latex beads.[2,3] When only several bands are visible, all the proteins should be tested for their DNA-binding activities. Starting from nuclear extracts or whole cell extracts, all the members of the family of the DNA-binding

[15] D. A. Harger and R. R. Burgess, *Anal. Biochem.* **109,** 76 (1980).

[16] H. Watanabe, T. Wada, and H. Handa, *EMBO J.* **9,** 841 (1990).

proteins can be purified at one time.[6] In this case, all the proteins must be denatured and renatured for good characterization. When we purified ATF/E4TF3 directly from the HeLa cell nuclear extracts, at least eight proteins with different molecular weights were copurified (Fig. 1A). Each protein could bind to the ATF/E4TF3-binding sites specifically, indicating the ATF/E4TF3 family.

Using DNA affinity beads, some associated proteins that form a complex with the DNA-binding proteins can also be identified.[17]

Acknowledgments

This work was supported by a Grant-in-Aid for Scientific Research on Priority Areas from the Ministry of Education, Science, and Culture of Japan and by a HFSPO grant to H.H. and was supported by Fellowships from the Japan Society for the Promotion of Science for Japanese Junior Scientists to T.W.

[17] T. Wada, T. Takagi, Y. Yamaguchi, H. Kawase, M. Hiramoto, Ferdous, M. Takayama, K. A. W. Lee, H. C. Hurst, and H. Handa, submitted for publication.

[41] Selection of Protein Binding Sites from Random Nucleic Acid Sequences

By T. KEITH BLACKWELL

Introduction

A variety of strategies have been developed by which specific DNA or RNA sequences that can be bound by a protein, or by a protein complex, can be isolated by selecting these molecules *in vitro* from libraries of random sequences. In principle, these protocols make it possible to evaluate the capability of any protein to recognize specific nucleic acid sequences, thus providing information which can be invaluable for investigations of how that protein functions. These techniques have a number of experimental applications that can be useful for study of various proteins that have been implicated in oncogenesis.

Either a biological[1] or a biochemical[2] selection scheme can be employed to isolate protein binding sites from a random sequence nucleic acid library, and each strategy can potentially yield different information. This chapter

[1] M. S. Z. Horwitz and L. A. Loeb, *Proc. Natl. Acad. Sci. U.S.A.* **83,** 7405 (1986).
[2] A. R. Oliphant, C. J. Brandl, and K. Struhl, *Mol. Cell. Biol.* **9,** 2944 (1989).

is restricted to discussion of biochemical selection strategies, which have been exploited and developed to a much greater extent. The power of current random sequence selection techniques derives from a central concept: that the selection procedure itself is coupled with a step in which the selected molecules are amplified by the polymerase chain reaction (PCR). The amplification step makes it possible to perform multiple sequential rounds of selection and amplification to isolate specific binding sequences. We have referred to this general approach as the selection and amplification of binding sites (SAAB) technique,[3] and it has been given various different names by others.[4–8] These various "*in vitro* selection" techniques, which have been reviewed in Szostak,[9] all have in common the general strategy shown in Fig. 1, which depicts selection from a library of random sequence oligonucleotides. This basic strategy has also been employed to select binding sites *in vitro* from pools of genomic DNA fragments[10] and to isolate specific RNA sequences which bind to a protein of interest, using libraries of RNA transcribed from random sequences *in vitro*.[5] In RNA selection protocols, the selected RNA is reverse-transcribed into cDNA, which after PCR amplification is itself then transcribed *in vitro* to generate a RNA template for a subsequent round of selection.[5,6] Since these protocols make it possible to perform multiple rounds of selection, they can be employed to isolate specific binding sequences even under circumstances in which these sequences are extremely rare or when very small amounts of protein are used, as long as the affinity of the sequence-specific interaction exceeds that of nonspecific binding.

With an *in vitro* selection scheme, it is possible in principle to identify binding sites for any protein that binds to specific DNA or RNA sequences, even if no candidate *in vivo* targets for that protein are known. For example, with such a technique it was first demonstrated that c-Myc is a sequence-specific DNA binding protein.[11] These protocols have a variety of additional important applications. For example, they can be used to identify sequences that are bound by multiprotein complexes that are present in cell extracts, if the selection is performed using an antibody or affinity column that binds

[3] T. K. Blackwell and H. Weintraub, *Science* **250,** 1104 (1990).

[4] H. J. Thiesen and C. Bach, *Nucleic Acids Res.* **18,** 3203 (1990).

[5] C. Tuerk and L. Gold, *Science* **249,** 505 (1990).

[6] A. D. Ellington and J. W. Szostak, *Nature (London)* **346,** 818 (1990).

[7] R. Pollock and R. Treisman, *Nucleic Acids Res.* **18,** 6197 (1990).

[8] G. Mavrothalassitis, G. Beal, and T. Papas, *DNA Cell Biol.* **9,** 783 (1990).

[9] J. W. Szostak, *Trends Biochem. Sci.* **17,** 89 (1992).

[10] K. W. Kinzler and B. Vogelstein, *Nucleic Acids Res.* **17,** 3645 (1989).

[11] T. K. Blackwell, L. Kretzner, E. M. Blackwood, R. N. Eisenman, and H. Weintraub, *Science* **250,** 1149 (1990).

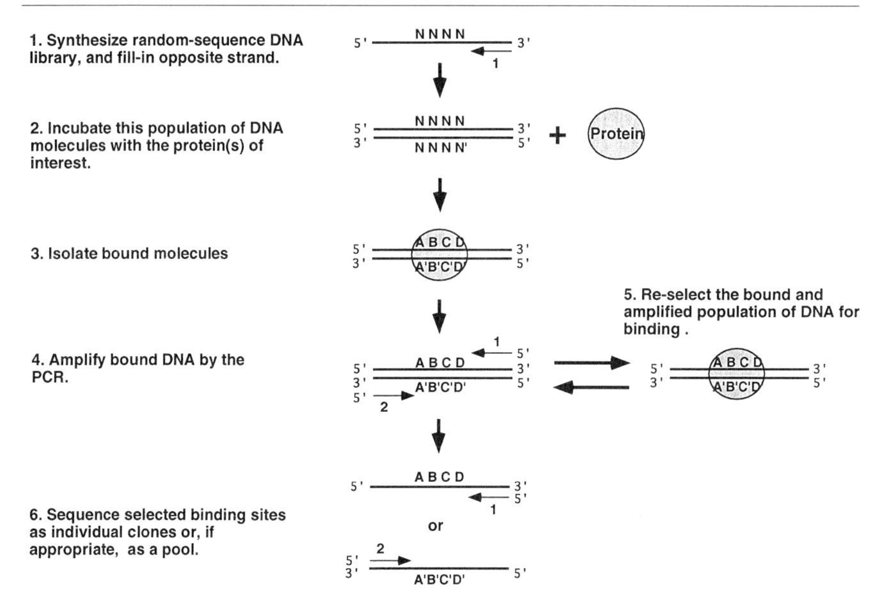

FIG. 1. General scheme for *in vitro* selection of DNA binding sites. A pool of random sequence DNA oligonucleotides is synthesized (1) so that the randomized bases are flanked by fixed sequences that allow amplification by the polymerase chain reaction (PCR) using primers 1 and 2. This population is converted to a double-stranded DNA sequence library by either a fill-in or PCR reaction. The random sequence library is incubated with the protein complex of interest (2), then the bound sequences are isolated (3) by one of a variety of methods such as an electrophoretic mobility shift assay (EMSA), a column, or a filter binding assay. The selected pool is then amplified by the PCR (4), allowing selection for binding to be performed again (5). After an appropriate number of rounds of selection and amplification, the selected population is analyzed by nucleotide sequencing (6). If a portion of the binding site had been specified in the original library, the selected population can be sequenced as a pool. Otherwise, individual molecules are cloned and sequenced. N designates random nucleotides, and ABCD indicates specific selected sequences.

to one of the proteins in the complex.[7,12] With such a strategy, it was demonstrated that the retinoblastoma (Rb) protein is associated with another cellular protein that binds to E2F sites, providing one of the first demonstrations that Rb and E2F interact with each other.[13] If only part of a binding site is fixed in the starting oligonucleotide library, the binding sites that are selected *in vitro* can be sequenced as a pool.[3,5] This pool sequence assay can detect very subtle differences in binding sequence preferences, and it thus greatly augments the sensitivity of the *in vitro* selection

[12] W. Funk and W. Wright, *Proc. Natl. Acad. Sci. U.S.A.* **89,** 9484 (1992).

[13] T. Chittenden, D. M. Livingston, and W. G. Kaelin, *Cell (Cambridge, Mass.)* **65,** 1073 (1991).

experiment. This latter approach is especially useful for studying the DNA recognition properties of different protein family members, all of which bind to related DNA sequences.[3,11,14,15] However, perhaps its greatest value is as a tool for investigating how the binding sequence specificity of a particular protein is determined, because the pool sequence assay is an especially sensitive means for evaluating how mutation of particular amino acid residues affects DNA recognition.[14,16]

Through *in vitro* selection, it is also possible to identify RNA and single-stranded DNA molecules (termed "aptamers") which form complex structures that are capable of highly specific molecular recognition.[6,17] In such experiments, which are beyond the scope of this chapter, aptamers have been identified which bind specifically to small molecule ligands[6,17] or to proteins that do not normally bind to nucleic acids.[18,19] The uses and limitations of this application of *in vitro* selection are still being defined, but it potentially offers the capability of selecting molecules that will interact specifically with almost any kind of biochemical target.

This chapter consists of an overview and discussion of various methods for performing *in vitro* binding site selection experiments. It focuses primarily on the selection of DNA recognition sequences from oligonucleotide libraries and places particular emphasis on use of the electrophoretic mobility shift assay (EMSA) to select sequences bound by a protein of interest, and on analysis of the selected molecules by means of the pool sequencing assay.

Binding Site Selection *in Vitro*

Preparation of Random Sequence Library

To produce a random sequence DNA library, first an oligonucleotide is synthesized so that a stretch of random sequence (equivalent amounts of the nucleotides A, C, G, and T) is flanked by fixed sequences that permit PCR amplification.[3] In general, a length of 16 bases should be sufficient for each PCR primer. These fixed sequences should contain restriction

[14] S. C. Ekker, D. P. von Kessler, and P. A. Beachy, *EMBO J.* **11,** 4059 (1992).

[15] A. Ma, T. Moroy, R. Lollum, H. Weintraub, F. W. Alt, and T. K. Blackwell, *Oncogene* **8,** 1093 (1993).

[16] T. K. Blackwell, J. Huang, A. Ma, L. Kretzner, F. W. Alt, R. N. Eisenman, and H. Weintraub, *Mol. Cell. Biol.* **13,** 5216 (1993).

[17] A. D. Ellington and J. W. Szostak, *Nature (London)* **355,** 850 (1992).

[18] L. C. Bock, L. C. Griffin, J. A. Latham, E. H. Vermaas, and J. J. Toole, *Nature (London)* **355,** 564 (1992).

[19] D. Jellinek, C. K. Lynott, D. B. Rifkin, and N. Janjic, *Proc. Natl. Acad. Sci. U.S.A.* **90,** 11227 (1993).

enzyme sites that permit cloning of individual selected molecules and that are placed close enough to the ends of the DNA to allow PCR amplification after subcloning. The length of random sequence should depend on the experiment. In general, 15–20 randomized bp should be more than adequate for most DNA binding proteins. However, for selections involving multiprotein complexes or for proteins that might bind cooperatively to two adjacent sites, larger regions of random sequence (>30 bp) will be necessary to allow recognition of multiple sites on the DNA, especially if it is required that these sites be separated by spacer regions.[12] If several different library templates are to be used side by side, as may be the case in pool sequencing experiments (see below), the danger of cross-contamination might be particularly high. It may then be advisable to insert a small number of additional fixed bases into each template as a "tag" that unambiguously identifies the library from which the selected site or sites are derived.

The double-stranded library can be produced by PCR amplification, using procedures described below for amplification of DNA molecules that are isolated during the selection. However, it may be advantageous instead to perform a large-scale synthesis of the second strand using the Klenow fragment of DNA polymerase. In this strategy, microgram amounts of the library oligonucleotide are annealed to a 10-fold excess of the appropriate PCR primer (Fig. 1), which is extended in a large-scale polymerization reaction by standard protocols[20] to complete the second strand. The resulting double-stranded library is then run on a high percentage (14%) polyacrylamide gel to permit isolation of full-length molecules, which can be eluted as described below. This technique for synthesizing the double-stranded library yields a large-scale stock which can be used repeatedly, and can also be used to produce libraries that include modified bases, are restricted to a subset of bases, or have part of the site defined for use in the pool sequencing assay.

Separation of Bound and Free Nucleic Acid Sequences by EMSA

If the protein preparation to be assayed is relatively pure, an EMSA has many advantages as a method for selecting specific DNA binding sites.[3] In this method, the random sequence library is labeled and incubated with the protein of interest, then the reaction is loaded onto a nondenaturing polyacrylamide gel. Separation is achieved because DNA that is bound by the protein migrates through the gel more slowly than free DNA. The bound complexes are visualized by autoradiography and are excised from the gel, and the selected DNA is amplified following elution from the

[20] J. Sambrook, T. Maniatis, and E. F. Fritsch, eds., "Molecular Cloning: A Laboratory Manual," 2nd ed. Cold Spring Harbor Lab., Cold Spring Harbor, NY, 1989.

gel slice. The most significant advantage of this protocol is that it allows visualization of each selection step, making apparent the relative ratios of bound and free DNA, and of specific and nonspecific binding (Fig. 2). It thus indicates the stringency of selection and reveals whether selection for specific binding sites has occurred during the course of an experiment.

In formulating the initial DNA binding reaction, it is important to

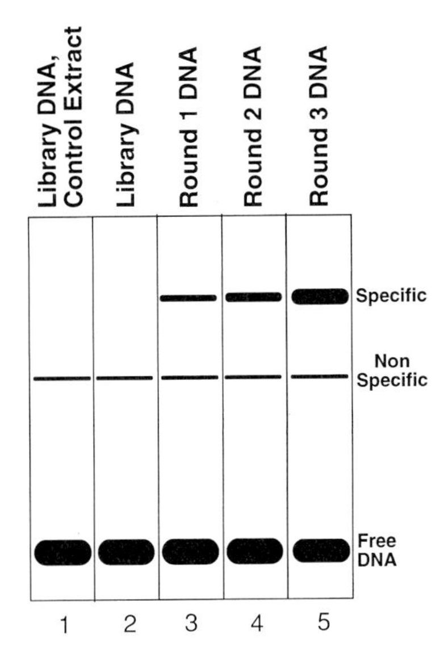

FIG. 2. Enrichment for specific binding sequences by EMSA selection. A hypothetical EMSA of binding of a protein preparation to the random sequence library and to the products of successive rounds of selection for binding by that protein. In many cases, complexes that correspond to sequence-specific binding will not be visible in an assay of binding to the input library, and those that are apparent might correspond to nonspecific interactions (lane 2), especially if the protein preparation is not pure. However, if *all* complexes with retarded mobility are isolated in the first selection round (see text), the enrichment for specific binding sequences might be sufficient to allow detection of sequence-specific complexes in the second round (lane 3). Alternatively, further selection for all complexes of decreased mobility can be performed. If a control extract that does not contain the protein of interest is available, it might be useful for identification of some complexes as background. For example, in this gel the nonspecific complex is present in an assay of the control extract (lane 1). By definition, sequence-specific interactions of interest should be of higher affinity than nonspecific interactions and should be enriched for during successive rounds of *in vitro* selection, so that the corresponding complex becomes increasingly prominent (lanes 4 and 5). The stringency of selection can then be increased by decreasing the input protein concentration. When the relative representation of specific complexes no longer increases with further selection, it is likely that the highest affinity sites have been selected for. The input DNA in each EMSA sample is indicated above the corresponding lane. All lanes except 1 show assays of binding by the protein preparation used for selection.

include enough library DNA so that its sequence complexity encompasses all of the potential specific binding sites that are being selected for. For example, DNA binding proteins usually recognize fewer than 15 specific bp. To cover all possibilities for 15 completely random bases would require $4^{15}/2$ (because either orientation suffices) or about 5.4×10^8 molecules. This amount is equivalent to 9.0×10^{-16} mol (or 3.2×10^{-11} g of a 55-bp library template), which for a 20-μl DNA binding reaction corresponds to a concentration of 4.5×10^{-11} M. However, in an experimental situation, a higher concentration would be required to ensure a high probability that all possibilities were represented. For example, to have a 99% probability that each of these sequences were represented in a 20-μl binding reaction would require a concentration of about 2×10^{-10} M (or 1.4×10^{-10} g of a 55-bp library template).[21] In selections that involve multiprotein complexes and require large library templates, it may not be feasible to have a complete representation of the library present in the reaction. However, a complete representation may not be required because the recognition elements themselves would constitute only a fraction of the total template length.[12]

For most DNA binding site selections, a more practical consideration is that it is necessary to include enough input DNA in each round so that only a fraction will be bound in the selection EMSA. The smaller the bound fraction, the more the selection is "driven" by the higher affinity of sequence-specific protein–DNA interactions relative to nonspecific binding. Under some circumstances, the appropriate amount of input protein (and DNA) can be estimated from the predicted binding affinity. For example, a high-affinity DNA binding protein such as a homeodomain would be expected to bind to its cognate site with an affinity in the nanomolar range[22] and thus could be used to select binding sites at a concentration of 10 nm or less, with the DNA library added to a higher concentration. In other situations the binding affinity cannot be predicted and the amount of protein added must be determined empirically. For example, if the protein must form a dimer or a higher order oligomer to bind DNA, the dimerization affinity will influence the amount of DNA binding oligomers that will be present. During selection of binding sites for the basic helix–loop–helix (bHLH) protein MyoD, which requires dimer formation to bind DNA, it was necessary for the protein concentration to be higher than that of the DNA,[3] apparently because MyoD forms DNA binding homodimers relatively inefficiently.

[21] L. Clarke and J. Carbon, *Cell (Cambridge, Mass.)*, **9,** 91 (1976).

[22] M. Affolter, A. Percival-Smith, M. Muller, W. Leupin, and W. J. Gehring, *Proc. Natl. Acad. Sci. U.S.A.* **87,** 4093 (1990).

For the first round of selection, the random sequence library is end labeled by a kinase reaction. In general, the DNA binding reaction can be performed under standard conditions, such as 50 m*M* Na^+ or K^+, 20 m*M* HEPES, 1 m*M* dithiothreitol 3 m*M* $MgCl_2$, 1 m*M* EDTA, and 5% glycerol.[3] For proteins produced by overexpression in bacteria, it may be useful to add the detergent NP-40 to 0.5% to inhibit aggregation.[23] A nonspecific competitor like poly(dI : dC) can be added to increase the stringency of selection, but is appropriate only if it does not compete significantly with sequence-specific interactions by the protein complex of interest. Incubation at room temperature for 20 min prior to electrophoresis is usually sufficient for achieving equilibrium conditions. The EMSA gel can then be run according to any of a variety of protocols, which have been described elsewhere.[3,24] It is usually appropriate to run the gel at room temperature, although it should always be considered a possibility that some protein–DNA complexes might require a lower temperature for stability. In general, except for the initial rounds of selection (see below), these gels are run until the free library DNA nearly reaches the bottom of the gel, then they are dried and autoradiographed.

Recovery, Amplification, and Reselection of Bound Sequences

It is possible that protein–DNA complexes would be apparent on the autoradiogram of the first-round EMSA gel and that they would contain a population of specific binding sites. Alternatively, this autoradiogram might show either multiple complexes, some of which could derive from nonspecific interactions, or no obvious complexes at all.[11,16] It is therefore often advantageous to run the first-round selection gel for a very short time (under standard conditions for about 20 min[11,16]) so that the free fraction of DNA runs only about 1 cm into the gel. The entire upper 0.75 cm of the lane, which includes all DNA that has migrated at reduced mobility relative to the free fraction, is then excised from the dried gel (including the paper backing) for recovery and amplification of the DNA within. In the second selection round, this batch selection can be repeated along with, or instead of, a full-length EMSA gel. As this procedure selects for specific sequences that bind with affinities that are higher than that of nonspecific DNA, specific complexes can be identified because they become increasingly abundant in the EMSA as the selection progresses (Fig. 2). DNA within these specific protein–DNA complexes can then be isolated from full-length gels by excising them in slices that encompass the center two-

[23] A. B. Lassar, J. N. Buskin, D. Lockshon, R. L. Davis, S. Apone, S. D. Hauschka, and H. Weintraub, *Cell (Cambridge, Mass.)* **58,** 823 (1989).

[24] J. Carey, this series, Vol. 208, p. 103.

thirds of the lane and are 0.3 cm wide. By this procedure any DNA that might be trailing up the sides of the lane (and migrating with mobility that is artifactually low) is left behind.

DNA is purified from these gel slices by incubation at 37° for 3–4 hr in 0.5 ml of 0.5 *M* ammonium acetate, 10 m*M* $MgCl_2$, 1 m*M* EDTA, and 0.1% SDS.[3] Alternatively, comparable yields (measured by recovered cpm of labeled DNA; usually about 20%) can generally be obtained with 0.5 *M* ammonium acetate alone. An overnight incubation increases DNA recovery (to approximately 50%), but makes it more likely that PCR artifacts such as multimers will occur subsequently. After debris are spun out and 5 μg of tRNA carrier is added to the eluate, it is extracted twice each with phenol and chloroform/isoamyl alcohol (24:1 v/v), and precipitated with ethanol. These samples are then resuspended in 0.3 *M* sodium acetate and reprecipitated with ethanol. Approximately one-fifth of the resuspended sample is then amplified by 35 cycles of the PCR in a 100-μl reaction. The PCR is performed under standard conditions following optimization of Mg concentration, with extreme care taken to avoid cross-contamination of samples.[3] These conditions yield about 100 ng of DNA product and allow recovery from as few as 50 cpm of EMSA-isolated template. The products of these reactions are electrophoresed on a 14% polyacrylamide gel, which is then stained with ethidium bromide. The sample is excised from the wet gel, and eluted and purified as above.

For subsequent selection rounds, after amplification and purification these selected DNA pools are labeled by incorporation using the PCR. Approximately 5 ng (as estimated from the preparative gel) of the purified amplified template is labeled for *one* cycle in a 20-μl reaction that contains 30 μCi of [^{32}P]dTTP, 50 m*M* each of dATP, dGTP, and dCTP, and 100 ng of each primer in the standard PCR reaction buffer.[3] It is crucial to add a large excess of primers to ensure that synthesis occurs on all templates in the reaction because denatured templates that reanneal are likely to form heteroduplexes. Following the PCR, unincorporated label is removed with a G50 spin column (Boehringer). With this labeling protocol all of the labeled DNA derives from *de novo* synthesis, thus eliminating the danger of heteroduplex formation and allowing conformation that the synthesis was complete. The binding reaction and EMSA are then performed as above, using an appropriate amount of the PCR-labeled template pool (usually 1–2 ng).

As an alternative to the one-cycle labeling protocol, the selected DNA can be labeled during the amplification process. This latter labeling protocol is simpler, but to prevent heteroduplex formation during this procedure it is necessary to maximize the probability that complete synthesis occurs on the last round of amplification, a goal which requires that the efficiency of amplification be titrated and remain consistent.[7,25] Two additional disadvantages of labeling during the amplification step are that this protocol does not yield a purified nonlabeled amplified stock, which can be used repeatedly, and that it omits the gel purification step, which is helpful for decreasing the likelihood that PCR artifacts such as multimers or primer bands might accumulate.

An important issue to keep in mind in these experiments is the danger of contamination of a reagent or a DNA sample by template DNA. Only trace amounts of a template oligonucleotide are required for PCR amplification, and because of its low molecular weight the likelihood of contamination occurring is much greater than with plasmid DNA. This issue is of particular concern if multiple library templates and proteins are being used in simultaneous selection experiments. It is advisable when handling these samples to use plugged pipette tips and to employ a separate set of pipetters for working with DNA eluted from EMSA gels (to avoid contamination from large-scale stocks). It may be worthwhile to obtain primers and random sequence library DNA from separate sources because primers can become contaminated through use of the same oligonucleotide synthesis machine or speed-vac for production of primer and library DNA, even if the primers are made several days after the library. Alternatively, elution of the oligonucleotides from the synthesis column by hand can prevent cross-contamination deriving from the machine port.

Other Selection Methods

As is discussed earlier, the EMSA can be advantageous for *in vitro* binding site selections that involve protein preparations that are nearly or partially pure. However, a variety of other methods can also be very useful for isolating nucleic acids that are bound by a protein of interest, and can thus be used for *in vitro* binding site selections. These selection methods include a filter-binding assay[4,5] and various types of affinity columns. For an affinity column selection, the protein of interest can be engineered so that it includes a selectable "tag," such as glutathione S-transferase.[13] In the selection step, the random sequence library is incubated with the tagged protein in a DNA binding cocktail, which is then incubated with the affinity column. DNA that is not bound to the protein is washed away, and then the bound DNA is eluted from the column either with a high salt buffer, which disrupts protein–DNA interactions, or by elution of the protein itself. The selected DNA can then be amplified for further selection. By an analogous affinity column strategy, an antibody that is specific to the protein

[25] W. Wright, M. Binder, and W. Funk, *Mol. Cell. Biol.* **11,** 4104 (1991).

of interest or to an epitope with which that protein is tagged can also be used to isolate specific sequences that bind to complexes which contain that protein.[7,25]

Affinity column (and antibody selection) strategies share the advantage that the protein used for selection need not be pure because it is isolated from other proteins in the mixture by the column selection step. For example, although an EMSA can be used for selections involving proteins expressed in *in vitro* translation mixes,[3] in general an affinity column selection is essential for isolation of specific sequences that are bound by complexes present in extracts from eukaryotic cells because of the heterogeneity of these extract preparations.[7,12] These selection methods also have the advantage that under many circumstances they are simpler to perform than an EMSA selection. However, when an affinity column strategy is employed, before the selected products are cloned and sequenced, it is advisable to determine whether selection of specific binding sites has actually occurred. By labeling the input DNA at each selection step, it is possible to obtain an estimate of the relative fraction of DNA that binds to the protein on the column. Nevertheless, as compared to the EMSA, this procedure does not give as clear or direct an indication of the relative proportions of bound and free DNA. Perhaps of most significance, with column selection methods it is not as apparent to what extent nonspecific binding, such as sticking to the column matrix itself, might be interfering with the selection. For these reasons, it may be useful to employ an analytical EMSA (Fig. 2) as a complement to affinity column selections[7] to evaluate the extent to which each selection step enriches for specific binding sequences.

Analysis of Selected DNA Molecules

Sequence Analysis of Selected Sites

After it has been established that the population of selected DNA molecules is enriched with specific binding sequences (Fig. 2), individual selected molecules can be cloned using the restriction sites within the primer sequences. The nucleotide sequences of these individual molecules can then be determined. The presence of a sequence motif in a high percentage of selected molecules is helpful as an indication of a potential binding site consensus, but this criterion misses sites that conform only partially to the consensus.[16] It is therefore essential to assay binding to individual selected sequences, which can be recovered from plasmid clones and labeled by PCR amplification. This analysis discriminates those selected molecules that actually contain specific binding sequences from any that might be present in the selected population because of nonspecific binding.[16] However, to identify specific binding sites unequivocally, it is necessary to perform biochemical footprinting analyses of binding to selected DNA molecules[16,26] or to test binding to individual synthesized oligonucleotides with sequences that are based on these molecules.

Pool Sequence Assay

To analyze the DNA or RNA binding specificity of a protein complex in detail, it can be useful to perform *in vitro* selection experiments in which only certain positions have been randomized within an otherwise defined binding site. Under these circumstances, it is possible to analyze the selected sites using the pool sequencing assay.[3,5,11] Sequencing of the selected sites as a pool gives not only a "read out" of the sequence preferences of the protein at each selected position, but also provides a visual image of its binding which is akin to a footprint (Fig. 3).[27] This assay provides an exquisitely sensitive means for detecting binding sequence preferences, whether they derive from direct contact or from secondary steric effects.

For *in vitro* selection experiments that are to be analyzed by pool sequencing, the random sequence library template is synthesized as described earlier, except that a portion of the binding site is fixed in sequence so that the protein complex will bind at a defined location on the template DNA. A total template size of about 55 bp is optimal for the sequencing protocol, which is described below. After the selection process has been completed, the resulting sequence pool is amplified by the PCR and purified as in each selection round, except that it is eluted from the gel overnight to increase the yield for sequencing. As a further strategy for maximizing the yield of eluted DNA, it may be helpful to increase the relative efficiency of elution by expanding the size of the molecules that contain the selected sequences. This can be done by perfoming the last amplification step using primers that each overhang the ends of the starting library in the 5′ direction. For example, by such a strategy a 55-bp template has been increased in size to 124 bp prior to sequencing.[16]

For the sequencing protocol, one of the two original PCR primers is labeled with ^{32}P by a kinase reaction, and unincorporated label is removed with a G25 or G50 spin column (Boehringer). Because these primers are small (15–16 bases), it is generally necessary to reload the column with buffer and spin a second time to recover most of the labeled DNA. The labeled primer eluted in this second spin is usually used for sequencing

[26] T. K. Blackwell, B. Bowerman, J. Priess, and H. Weintraub, *Science* **266,** 621, (1994).

[27] A. R. Ferre-D'Amare, G. C. Prendergast, E. B. Ziff, and S. K. Burley, *Nature (London)* **363,** 38 (1993).

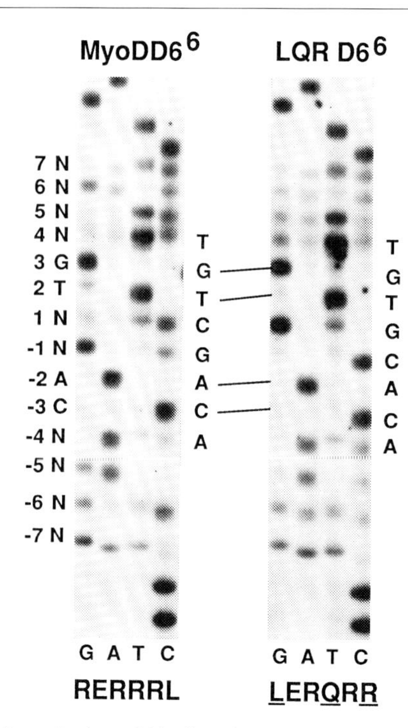

FIG. 3. Pool sequencing of selected binding sites. This experiment taken from Blackwell *et al.*[16] provides an example of pool sequencing data. The indicated sequence pools (MyoDD6[6] and LQRD6[6]) are respectively derived from six rounds of *in vitro* selection for binding to the bHLH protein MyoD and to a mutant MyoD (LQR) into which has been substituted the corresponding residues from the bHLH protein cMyc. They have been selected from the library D6, which contains the sequence NNNNCANNTGNNNN, in which the bHLH protein binding site consensus CA -- TG[3] is specified, and in which N signifies completely random sequence at that position. Specification of this consensus within the library sequences has fixed the location at which these two proteins bind. Substitution of these c-Myc residues into MyoD has changed its preferred recognition sequence at the central bases (-1 and 1) from GC to CG, but has left its preferences at flanking sequences, particularly at ± 4, largely intact. Further experiments have indicated that this change in internal sequence preferences is mediated by the introduced c-Myc R residue,[16] and crystallographic studies have suggested that within each dimer subunit, this R residue directly contacts the G nucleotide in each of these central base pairs.[27] Selected sequences are indicated to the right of each gel, and corresponding amino acid residues in MyoD (RERRRL) and c-Myc (LERQRR) are designated below the gels, with the substituted residues underlined.

because it is generally free of a labeled artifactual species, which is sometimes present in the first elution and can obscure some bands in the sequencing gel. The sequencing protocol itself is an adaptation of the termination step of the Sequenase procedure [United States Biochemical (USB)]. Labeled primer (10 ng) is mixed with approximately 5 ng of a purified amplified template pool in a 12-μl sample, which includes 1 μl of Sequenase manganese (Mn) buffer (to increase the uniformity of band intensities) and 2 μl of 5× Sequenase buffer. This mixture is placed at 95° for 5 min, then allowed to cool at room temperature for 1 min, during which it must be quick-spun to bring down condensation. The sample is placed on ice, and then 1 μl of 0.1 *M* dithiothreitol and 2 μl of diluted Sequenase enzyme (1:8 in ice-cold TE, pH 7.4) are added. A 3.5-μl aliquot of this mix is then added to 2.5 μl of each of the four Sequenase dGTP termination mixes (USB). After incubation at 45° for 4 min, these reactions are stopped by the addition of 4 μl of Sequenase stop solution. dITP termination mixes can be employed, but do not give data of the same high quality.[3] This protocol yields sequences up to about 80 bp from the 5′ end of the primer, at which point the extent of termination by dideoxynucleotide incorporation is virtually complete. A 1.5-μl aliquot of each reaction is run on a 14% sequencing gel containing 8 *M* urea in 1 × TBE. Prior to autoradiography, the gel is fixed in 10% acetic acid, 10% methanol, but only after the unreacted primer (which is present in vast excess of incorporated product) has been cut away to prevent its diffusion.

Perspectives

The development of *in vitro* selection strategies has made it possible to apply the power of "genetic" selections to biochemical investigations of interactions between proteins and nucleic acids. With these approaches, it is possible to "screen" for specific molecules that bind at high affinity to a particular protein complex instead of examining large numbers of individual molecules separately. These procedures can in principle be performed with any type of protein preparation, as long as it is possible to isolate the sequences that are bound to the protein. They can also be performed with libraries of any mixture of nucleic acids, or of modified nucleic acids, as long as these molecules can be copied by the PCR.

This strategy is very useful in experimental situations in which it is suspected that a protein has nucleic acid binding activity, but in which no candidate targets have been identified. Although it does not yield actual physiological genomic regulatory sequences, *in vitro* selection does identify binding sites that will be helpful in a search for such sequences and that provide a basis for further investigations of the function of that protein. It can also reveal the complete spectrum of sequences that can be recognized by any protein, information that can be useful when a limited number of targets might have been identified. When this approach is applied to studies of multiprotein–DNA complexes, it can provide evidence of previously unidentified factors which interact with a known protein by yielding specific sequences that are bound by those factors. This technique is also very useful for investigations of how particular proteins bind to specific DNA sequences. Simply by comparing the sequences that are selected by two different proteins, it is possible to detect differences in their recognition properties that might not be as readily demonstrated by testing their binding to several individual sequences. The sensitivity of *in vitro* selection experiments is significantly augmented by the pool sequencing assay, which allows screening of the selected pool simultaneously, and can reveal very subtle aspects of nucleic acid binding preferences. When this assay is coupled with mutagenesis of protein coding sequences, it becomes a powerful means of elucidating how particular amino acids might be involved in DNA (or RNA) recognition. As more structures are determined for representative members of DNA binding protein families, this strategy should become increasingly useful for exploring how different family members recognize their respective cognate sites.

It appears likely that *in vitro* selection schemes in the future will be employed in an increasing variety of contexts. Biologically based selections may be used more commonly as a strategy for identifying nucleic acid sequences that can perform a particular function. These experiments could take the form of intracellular selections, such as screens for sequences that allow activation of a particular promoter.[1] This strategy offers the possibility of identifying sequences that are bound by proteins that can function in a particular context in a cell. Other types of functional selections could be performed *in vitro*, such as selections for molecules that activate or inhibit a biological process or interaction. Sequences have already been isolated *in vitro* that function as ribozymes.[28,29] Finally, the promise of "aptamer" technology is just beginning to be explored. If it is possible to isolate high-affinity specific aptamers for a wide range of molecular targets, then this strategy could prove extremely useful for generating recognition molecules for research purposes and could have almost unlimited potential as a tool for drug design.[30]

Acknowledgments

Preparation of this chapter was supported in part by a grant from the National Institutes of Health (GM50900) to T.K.B. I am indebted to the late Harold Weintraub for his support and advice, and for the example he set as a scientist and mentor.

[28] R. Green and J. W. Szostak, *Science* **258,** 1910 (1992).
[29] D. P. Bartel and J. W. Szostak, *Science* **261,** 1411 (1993).
[30] M. L. Riordan and J. C. Martin, *Nature (London)* **350,** 442 (1991).

[42] Electrophoretic Mobility Shift Assay

By LAWRENCE D. KERR

Introduction

Molecular approaches to the study of gene regulation and expression have demonstrated that the transcriptional activity of a given gene is regulated, in part, by protein complexes associated with specific DNA sequences within that gene. This observation has been facilitated greatly by the addition of techniques which analyze specific protein–DNA complexes. Although a pattern of gene expression may be extremely complex, the ability to analyze defined regions of DNA sequence with respect to the protein(s) bound by those sequences remains a relevant and useful measure for the binding of these elements and the subsequent analysis on transcription. This chapter focuses on the practical use of the electrophoretic mobility shift assay (EMSA) in the analysis of specific protein–DNA complex formation and the use of this technique in the characterization of these important interactions.

Practical Considerations of EMSA

General

The "gel retardation" or EMSA was first described by Fried and Crothers[1] and Garner and Revzin[2] in their analyses of RNA polymerase/CAP interaction with the *lac* promoter and has been reviewed previously.[3–5] The technique exploits the simple rationale that as DNA travels through a nondenaturing matrix, it is retarded in this matrix by the presence of the protein associated with it (Fig. 1). In contrast to the footprinting assay described in [6] in this volume, in which one detects the absence of DNA sequences in the presence of bound protein, the EMSA reveals a protein–DNA complex that is detected as separate from the unbound DNA. It is therefore quite common to employ both types of assay in the ultimate

[1] M. Fried and D. M. Crothers, *Nucleic Acids Res.* **9,** 6506 (1981).
[2] M. M. Garner and A. Revzin, *Nucleic Acids Res.* **9,** 3047 (1981).
[3] A. Revzin, *Bio Techniques* **7,** 346 (1989).
[4] C. L. Dent and D. S. Latchman, "Transcription Factors: A Practical Approach," p. 1. IRL Press, Oxford, 1992.
[5] M. J. Garabedian, J. LaBaer, W.-H. Liu, and J. R. Thomas, "Gene Transcription: A Practical Approach," p. 243. IRL Press, Oxford, 1992.

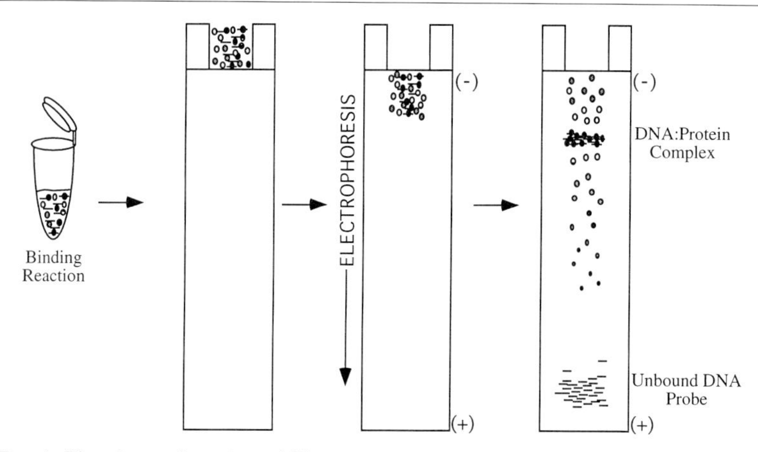

FIG. 1. The electrophoretic mobility shift assay. The incubation reaction containing the protein extract (●) and the DNA probe (−) are combined. The reaction is loaded onto a nondenaturing polyacrylamide gel for electrophoretic fractionation of DNA–protein complexes from unbound DNA. Only those complexes containing the radiolabeled probe are visualized during autoradiography.

analysis of the interactions of specific DNA binding proteins on a promoter. While one must constantly recognize the caveat that these techniques do represent *in vitro* assays and that genes in the context of chromatin *in vivo* may require a far more complex scenario of regulation, these techniques are invaluable in the approach to initial studies of gene regulation.

Preparation of Extracts

The source of protein used in EMSA is of critical importance and requires at least some educated guesses about the nature of the protein one believes might bind to a given DNA sequence. This section reviews the preparation of extracts ranging from the easiest yet crudest form, the whole cell extract, to the analysis of DNA binding proteins whose coding sequence have been cloned.

The preparation of whole cell extracts is easy, relatively fast, and the quantity of protein achieved per mass of starting material is usually very good. In addition, the protein components of the entire cell can be sampled for its DNA binding potential. At times, whole cell extracts are the only choice when faced with a DNA binding complex of completely unknown subcellular origin or when extracts are required from samples that have been frozen (especially those frozen at −20° or below). Freezing samples damages both cytoplasmic and nuclear membranes and makes the recovery of isolated, intact protein fractions from the cytoplasm and nucleus very difficult. The method of preparation of whole cell extracts used most commonly is a modification of the Dignam *et al.*[6] protocol. This preparation is highly effective for the isolation of protein extracts from tissue samples or cell lines. The disadvantage to such a preparation lies in its crudeness. Whole extracts generally require either more potent or more stable protease inhibitors than do more purified preparations, particularly in the final dialysis buffer in which the extract is to be stored. In addition, whole cell extracts generally degrade faster than purer preparations due to the presence of cellular proteases, and one might anticipate the loss of binding activity between a few weeks to a couple of years, depending on the particular nature of the protein of interest and the conditions of storage utilized. As a general rule, if two or more different DNA binding probes for ubiquitous DNA binding proteins [e.g., SP1, TATA binding protein (TBP), etc.] fail in EMSA, the extract might be "dead." One additional disadvantage to whole cell extracts is that for many low abundant nuclear proteins (e.g., some transcription factors), the dilution that occurs with the cytoplasmic proteins in a whole cell extract may render the concentration too low to be detected with a reasonable amount of whole cell extract in the EMSA. This problem might be corrected by isolating cytoplasmic proteins apart from nuclear extracts.[6,7]

One footnote to the preparation of extracts from tissue samples. Tissue samples can be very difficult to disperse and may require separation first with scissors, followed by processing with a macerator, then a homogenizer, and then a Dounce homogenizer. Failure to disperse clumps of cells in the early stages of the preparation will lead to lower yields in the final extract. In addition, if fewer than 10^5 cells are to be extracted, the preparation of extracts by microextraction should be used.[7,8]

The efficient separation of cytoplasmic proteins from nuclear extracts requires patience and practice. The time-tested protocol described by Dignam *et al.*[6] remains one of the best for achieving such separations. This protocol is particularly effective for isolation of extracts from cells ranging in numbers from 1×10^7 to 1×10^9. Once the binding component of interest has been identified from an extract, it is possible then, through empirical trial, to change the conditions to maximize the extraction of the protein.

Under conditions in which the supply of tissue or cells is limiting, a miniextraction procedure can be utilized.[7,8] This procedure employs the use of a hypotonic lysis solution and Nonidet P-40 (NP-40) to extract nuclear

[6] J. D. Dignam, R. M. Lebovitz, and R. G. Roeder, *Nucleic Acids Res.* **11,** 1475 (1983).
[7] R. J. Roy, P. Gosselin, and S. L. Guérin, *Bio Techniques* **11,** 770 (1991).
[8] E. Schreiber, P. Matthias, M. M. Muller, and W. Schaffner, *Nucleic Acids Res.* **17,** 6419 (1989).

proteins without the need for homogenization. It is especially useful when multiple samples are being prepared from a relatively small number of cells ($0.5–1 \times 10^6$). Similar to whole cell extracts, however, micronuclear extracts are not as pure as when prepared by the large-scale method involving dialysis of the extract. Micronuclear extracts therefore require additional or more stable protease inhibitors and may proteolytically degrade at a faster rate than large-scale nuclear extracts.

The molecular cloning of numerous cDNAs encoding DNA binding proteins affords investigators the opportunity to produce pure proteins in relatively large quantities. *In vitro* transcription of cDNAs into mRNA followed by *in vitro* translation of the mRNA allows us to synthesize a specific DNA binding protein of interest for use in EMSA.[9,10] Although only picogram quantities of protein are synthesized in a single reaction, this is more than sufficient for detection for the EMSA. Thus, many samples of a mutagenized cDNA can be assayed easily and quickly using *in vitro* transcription/*in vitro* translation of the mutant cDNA. This is a very popular method for the mapping of the DNA binding domain(s). The disadvantages of producing protein by this method are few but significant. One difficulty in this method is that *in vitro*-translated proteins used in EMSA cannot be synthesized as radiolabeled proteins for analysis by sodium dodecyl sulfate–polyacrylamide gel electrophoresis (SDS–PAGE). Our laboratory routinely sets up an *in vitro* translation reaction in which the sample is divided into two tubes: one contains radiolabeled methionine, the other with unlabeled methionine. An aliquot of the radiolabeled sample is analyzed by SDS–PAGE and autoradiography. Approximately equal amounts of protein are loaded into the EMSA reaction based on results of the radiolabeled translation. One additional caveat to the use of *in vitro* transcription/*in vitro* translation applies to scenarios where a heterodimeric complex of factors is required for association with DNA. Although some heterodimeric proteins work exceptionally well in this system (e.g., c-Fos/c-Jun), other complexes are not formed at all (e.g., NF-κB p65/p50 heterodimers) due most probably to inappropriate post-translational modifications. Despite these disadvantages, the ability to translate *in vitro* individual DNA binding protein cRNA has greatly facilitated our analysis of protein–DNA interactions.

Certain applications require large quantities of a specific DNA binding protein. Several techniques have arisen over the recent past for the production of these proteins, including (1) expression in bacteria, (2) baculovi-

[9] D. Melton *et al.*, *Nucleic Acids Res.* **12,** 7035 (1984).
[10] P. A. Krieg and D. A. Melton, this series, Vol. 155, p. 397.

rus[11,12] and vaccinia virus,[13] expression, and (3) expression in mammalian cells. The expression of cloned cDNAs in bacteria is rapidly becoming one of the most popular methods for the large-scale production of cloned proteins. Routinely, a cDNA encoding a protein of interest is cloned into one of several expression vectors resulting in either a fused or unfused cDNA. The production of cloned proteins as fusion proteins has been reviewed elsewhere in this volume and will be discussed briefly here. A common technique employed by our laboratory utilizes unfused proteins produced in bacteria that are partially purified by column chromatography using heparin-agarose (Table I). We have successfully utilized this procedure for several different families of DNA binding proteins and have found the proteins useful in EMSA, coimmunoprecipitation assays, *in vitro* association assays, and in antibody production. One disadvantage to the production of certain proteins in bacteria is the lack of certain post-translational modifications. We have found, for example, that while the c-Rel protooncogene protein produced in *Escherichia coli* is able to bind DNA efficiently, it is unable to activate transcription due to improper phosphorylation of its transactivation domain.

The expression of cloned DNA sequences in mammalian cells for the purpose of analyzing the protein products in EMSA is used widely. Two approaches are routine and of particular relevance in this chapter. Both techniques produce a desired protein in mammalian cells under conditions in which the protein is appropriate modified (i.e., by phosphorylation, glycosylation, isoprenylation, etc.).[14,15] In the first approach, cDNA encoding the protein of interest is placed in a mammalian expression vector. The plasmid is then introduced into a cell line of interest (e.g., COS-7) by standard methods including calcium phosphate precipitation, DEAE-dextran precipitation, electroporation, or retroviral-mediated gene transfer. By way of example, transfection of an expression vector bearing an SV40 (simian virus 40) origin of replication sequence into the T-antigen of SV40-expressing cell line COS-7 will result in amplification of the transfected plasmid and subsequent overexpression of the encoded protein. Micronuclear extraction of the transfected cells will yield nuclear extracts which contain abundant levels of the protein of interest.[14] In the second

[11] P. A. Kitts *et al.*, *Nucleic Acids Res.* **18,** 5667 (1990).

[12] D. R. O'Reilly *et al.*, "Baculovirus Expression Vectors: A Laboratory Manual." Freeman, New York, 1992.

[13] T. R. Fuerst, E. G. Niles, W. Studier, and B. Moss, *Proc. Natl. Acad. Sci. U.S.A.* **83,** 8122 (1986).

[14] B. Stein and A. S. Baldwin, Jr., *Mol. Cell. Biol.* **13,** 7191 (1993).

[15] Q. Zhou, P. M. Lieberman, T. G. Boyer, and A. J. Berk, *Genes Dev.* **6,** 1964 (1992).

TABLE I
PRODUCTION AND PURIFICATION OF DNA BINDING PROTEINS FROM *Escherichia coli*

1. 80 ml Luria broth containing 50–100 μg/ml ampicillin is inoculated with a single colony of a high expressing clone of BL21 bacteria containing a bacterial expression clone (i.e., pET-, pT7/7-, pGST-, pHis-Tag, etc.).
2. Culture is shaken vigorously (>250 rpm) overnight at 32–37°.
3. Add 800 ml Luria broth with ampicillin and shake vigorously for 1 hr.
4. Add 800 μl 250 mg/ml isopropylthiogalactoside and continue shaking for 3 hr at 32°.
5. Collect cells by centrifugation (2000*g*, 10 min, 4°).
6. Resuspend pellet in 8 ml MTPBS [phosphate buffer saline (pH 7.4, autoclaved), 1% NP-40, 1% Triton X-100, protease inhibitor cocktail [0.1 m*M* phenylmethylsulfonyl fluoride, 1 μg/ml aprotinin, 10 μg/ml leupeptin, 10 μg/ml pepstatin].
7. Sonicate for 1 min on ice.
8. Spin down debris at 8000*g* for 30 min at 4°.
9. Isolate supernatant and dialyze overnight in 1 liter of buffer A [20 m*M* Tris–HCl (pH 7.9), 0.5 *M* NaCl, 1 m*M* EDTA, 10% glycerol, protease inhibitors as above].
10. Dilute dialyzate so that the salt concentration is 0.1 *M* (add 40 ml TM0*M*[a])
11. Load 5 ml heparin-agarose in a disposable column and wash with 5 ml TM2*M*.
12. Load sample onto column and resolve by gravity. Save the flow through.
13. After the sample has flowed through, wash column with 20 column volumes (100 ml) TM0.1*M*. Save the flow-through fractions.
14. Elute protein with a step gradient, collecting 3 × 5-ml fractions at each step. Elute with final KCl concentrations in TM of 0.2 *M*, 0.4 *M*, 0.6 *M*, 0.8 *M*, and 1 *M*. Most DNA binding proteins elute between 0.4 and 0.6 *M*.
15. 50 μl of each sample tube is analyzed by SDS–PAGE and in EMSA.
16. Positive samples are dialyzed against TM0.5*M*, aliquoted, and stored at −80°.

[a] TM0*M*: 50 m*M* Tris–HCl, pH 7.9, 1 m*M* EDTA, 12.5 m*M* $MgCl_2$, 1 m*M* DTT, and 20% glycerol (v/v) and protease inhibitor cocktail. For example, TM0.1*M* indicates the TM buffer containing 0.1 *M* KCl.

approach, a large number of cells in suspension (e.g., HeLa, lymphocyte lines, etc.) are infected with a retrovirus expressing a cloned sequence for the desired protein.[15] We have exploited this system even further by expressing our proteins fused at their amino termini to an epitope from either the influenza HA protein or the human c-myc protein.[16] The expressed protein can then be rapidly purified over Sepharose-coupled monoclonal antibodies to either epitope. After extensive washing, the desired protein is removed by elution with the epitope peptide and the protein is dialyzed into the appropriate storage buffer. Protein purified in this manner appears to undergo proper post-translational modification and is highly useful in gel shift analyses.

The aforementioned techniques for generation of protein to be used in EMSA are just examples of ways in which DNA binding proteins can

[16] L. D. Kerr *et al.*, *Nature* (*London*) **365,** 412 (1993).

be isolated for characterization. The EMSA is useful not only for the characterization of a specific protein–DNA interaction but has been used successfully in the purification of DNA binding protein prior to microsequencing and molecular cloning. The quality of the protein extract will affect the conditions utilized during the DNA binding reaction (i.e., for the addition of competitors of nonspecific DNA–protein interactions) and will therefore be discussed in other sections.

Preparation of Probe

The ability to radiolabel DNA fragments to high specific activity allows investigators to use either DNA fragments containing a binding site(s) of interest (usually 50 to 400 bp in length) or double-stranded synthetic oligonucleotides containing the desired binding site(s). The procedures described for the production of high specific activity probes utilize either the property of the Klenow fragment of DNA polymerase I to "fill in" the 3′ end of a recessed 3′ end restriction endonuclease-cleaved DNA site,[17] or the ability of T4 polynucleotide kinase to catalyze the transfer of a radiolabeled γ-phosphate from ATP to the 5′-hydroxyl group of DNA.[18] In order to efficiently transfer this phosphate moiety, the 5′-phosphate groups normally present on DNA must be removed prior to the kinase reaction.[19]

Purification of Probe

The quality of the radiolabeled probe, that is the extent of removal of the unincorporated nucleotides and contaminating proteins, is essential to a clean gel shift pattern. Several different approaches can be used ranging from simple ethanol precipitation or spin-column chromatography to electrophoresis through low melting agarose or polyacrylamide gels. The choice of which protocol to use is something of an empirical one, many double-stranded oligonucleotide binding site probes can be used after two rounds of spin-column chromatography, affording a cheap, simple, and fast method of purification. Alternatively, many longer probes (DNA fragments >150 bp in length) require purification by gel electrophoresis and subsequent purification from the gel matrix. The following protocols are recommended according to the type of DNA fragment used in the generation of the EMSA probe.

[17] T. Maniatis, E. F. Fritsch, and J. Sambrook, eds., "Molecular Cloning: A Laboratory Manual," p. 113. Cold Spring Harbor, Lab., Cold Spring Harbor, NY, 1992.

[18] T. Maniatis, E. F. Fritsch, and J. Sambrook, eds., "Molecular Cloning: A Laboratory Manual," p. 122. Cold Spring Harbor Lab., Cold Spring Harbor, NY, 1992.

[19] P. H. Seeburg, J. Shine, J. A. Marshall, J. D. Baxter, and H. M. Goodman, *Nature (London)* **220,** 486 (1982).

For the removal of unincorporated nucleotides and proteins from a double-stranded oligonucleotide probe <75 bp in length, spin-column chromatography through Sephadex G-50 is highly effective.[20] One concern with probes longer than 75 bp yet smaller than 200 bp is that during the various manipulations separation of the labeled double-stranded DNA fragments might have occurred. Radiolabeled single strands can result in the binding of nonspecific single-strand binding proteins which can interfere with the interpretation of authentic double-stranded DNA–protein complexes. Purification of the double-stranded fragments eliminates this concern and can be accomplished using a high-resolution nondenaturing polyacrylamide gel.[21] For DNA probes greater than 200 bp, the lower resolving power of agarose gel electrophoresis is highly effective at separating the double-stranded probe from the unincorporated nucleotides.

Gel Types

Many different types of gel matrices have been used for analysis of specific protein–DNA interactions. This section will be used to review the choice of gel types, the percentage for particular applications, the choice of buffers used in EMSA, and the conditions of electrophoresis (Table II).

The most commonly used type of gel electrophoresis for gel retardation involves a 4–6% polyacrylamide gel. Gel matrices in this range are highly effective in the separation of proteins from approximately 15 to 500 kDa and DNA fragments ranging from 30 to 200 bp. The acrylamide : bisacrylamide concentration can range from 19 : 1 to 29 : 1, and our laboratory routinely maintains a working stock solution of 40% (v/v) acrylamide (19 : 1 acrylamide : bisacrylamide) that is stable for several weeks at 4°.

Agarose gel matrices are particularly useful in the separation of protein–DNA complexes from free DNA when large fragments of DNA (>200 bp) or large protein complexes are expected to bind the probe DNA. Molecular biology grade agarose in the appropriate buffer (0.5–1× TBE[15]) is used to pour a horizontal gel ranging from 0.7 to 1.5% agarose. Agarose gels must not be allowed to heat up during electrophoresis as disassociation of protein–DNA complexes may occur. For studies in which dimer, trimer, and oligomer interactions of proteins complexes are bound to DNA simultaneously, a mixed polyacrylamide : agarose gel (1–2.5% acrylamide : 0.5–0.7% agarose) can be utilized to better separate the various monomeric to oligomeric protein–DNA interactions.[3]

[20] P. J. Mason, T. Enver, D. Wilkinson, and J. G. Williams, *in* "Gene Transcription: A Practical Approach" (B. D. Hames and S. J. Higgins, eds.), p. 10. IRL Press, Oxford, 1992.

[21] P. J. Mason, T. Enver, D. Wilkinson, and J. G. Williams, *in* "Gene Transcription: A Practical Approach" (B. D. Hames and S. J. Higgins, eds.), p. 29. IRL Press, Oxford, 1992.

TABLE II
EMSA GELS AND RUNNING CONDITIONS

Gel composition	Running conditions
EMSA gel 1	
4% polyacrylamide (30:2)	Gel prerun at 10 V/cm, 0.5 hr
0.25X TBE[a]	Gel run at 10 V/cm, 2.5 hr
2.5% glycerol	Ambient temperature
EMSA Gel 2	
5% polyacrylamide (29:1)	Gel rerun at 10 V/cm, 0.5 hr
0.5X TG[b]	Regular run at 10 V/cm, 2.5 hr
	Ambient temperature
EMSA Gel 3	
5% polyacrylamide (30:1)	Polymerize in cold room
0.5X TBE	Gel prerun at 15 V/cm, 0.3 hr
5% glycerol	Gel run at 15 V/cm, 2.5 hr
0.025% NP-40	4°
EMSA Gel 4	
1.4% agarose	Gel run at 6 V/cm, 3 hr
0.5X TBE	Ambient temperature

[a] 1X TBE: 0.045 *M* Tris, 0.045 *M* boric acid (pH 8.3), 1 m*M* EDTA.
[b] 1X TG: 25 m*M* Tris (pH 7.9), 200 m*M* glycine.

The choice of buffers that can be used in an EMSA are as varied as the types of gels (Table II). Buffers for EMSA gels are generally low in ionic strength and in certain circumstances can be the buffer used in the protein–DNA binding reaction. Medium salt concentrations in buffers stabilize the interactions between proteins and DNA. High salt concentrations not only disrupt the stability of protein–DNA complexes but can interfere in the movement of these molecules into the gel matrices; in addition, high salt buffers experience significant heating during electrophoresis which can lead to disassociation of the protein from the DNA probe. Low salt buffers are equally detrimental; too low of an ionic concentration can lead not only to disassociation of the protein from the DNA probe but separation of the double-stranded DNA probe template. In addition to salt(s) and buffering agents (Tris, glycine), many buffers contain accessory molecules which aid in stabilizing the protein–DNA complex formation, including divalent cations (Mg^{2+}, Mn^{2+}, or Zn^{2+}), low concentrations of nonionic detergents (NP-40), carrier cofactors (cAMP, insulin, or bovine serum albumin), reducing agents [dithiothreitol (DTT), 2-mercaptoethanol), or glycerol. The presence of each agent or even additional factors must be determined for each protein–DNA complex although we have included the commonly used buffer systems below (Table II).

One final consideration which can dramatically affect the outcome of results in EMSA is the temperature under which the gel is run. For some protein–DNA complexes, the rate of disassociation versus the rate of association is slowed by lowering the temperature. By way of example, our TATA-binding protein (TBP)–DNA binding reactions are performed at 30°. Once the binding reaction is complete, the products are separated and analyzed on a 5% polyacrylamide gel precooled to 4° and electrophoresis proceeds for 1.5 hr at 4° at 10 V/cm.[16] Once again, trial and error will determine whether electrophoresis at ambient temperature or cooler conditions will facilitate the particular protein–DNA complex.

We have described various conditions which alter the stability and analysis of protein–DNA complexes *in vitro*. Modification of gel percentages, cross-linking reagents, pH, ionic strength, the presence of accessory factors, temperature, and running conditions can all either positively or adversely affect the interaction between protein and DNA.

Binding Buffers

A variety of reagents and conditions affect whether a protein will associate with its DNA binding site in solution or not. This section reviews the protein–DNA binding buffer, the addition of nonspecific competitor DNA to the binding reaction, the time and temperature of incubation, and the preparation for loading the binding reaction (Table III).

The binding buffer used in an individual EMSA can be as unique as the protein–DNA complex formed. Each binding buffer starts with a basis of a buffering agent and a salt. The necessity of these agents are intuitive or have been discussed in the preceding section. The addition of cofactors or accessory reagents is completely dependent on the specific nature of the DNA binding protein complex. We have supplied several examples for commonly used transcription factor binding buffers in Table III and some commonly used accessory reagents and their recommended concentration range. Some notable examples might be useful in designing a specific protein–DNA binding buffer. The association of the heterodimeric activator protein −1 (AP-1) complex containing the products of the protooncogenes c-*fos* and c-*jun* with its DNA binding site, the TPA-responsive element (TRE), requires the presence of a relatively high concentration of reducing agent in order to observe complex formation. The presence of a high concentration of EDTA in the binding buffer of the Zn^{2+} finger Sp1 transcription factor will abrogate binding of the complex to DNA. The addition of 0.025% NP-40 to the binding buffer of NF-κB aids in the association of this transcriptional factor complex to the κB site probe.

TABLE III
BINDING BUFFERS AND COMPONENTS

1X κB binding buffer	1X AP-1 binding buffer
10 mM Tris–HCl (pH 7.9)	10 mM HEPES (pH 8.0)
50 mM NaCl	20 mM NaCl
0.5 mM EDTA	4 mM $MgCl_2$
10% glycerol	0.1 mM EDTA
1 mM DTT	2 mM spermidine
	17.5% glycerol
	2 mM DTT
1X Sp1 binding buffer	1X GATA binding buffer
50 mM Tris–HCl (pH 7.9)	20 mM HEPES (pH 7.9)
100 mM KCl	50 mM NaCl
12.5 mM $MgCl_2$	5 mM $MgCl_2$
1 mM EDTA	0.1 mg/ml BSA
1 mM DTT	6% glycerol
20% glycerol	0.025% NP-40
0.1% NP-40	1 mM DTT

Component	Final concentration ranges
Tris–HCl or HEPES	10–75 mM
NaCl or KCl	25–250 mM
DTT	1–6 mM
$MgCl_2$	5–15 mM
$ZnCl_2$	0.1–0.5 μM
NP-40	0.1–0.3%
Glycerol	15–25%
BSA	2–25 μg

Binding Reaction

Analysis of protein–DNA interactions via EMSA requires that a stable protein–DNA complex be formed *in vitro* prior to separation from unbound nucleotides. Intuitively the goal is to identify specific interactions between the input proteins and the DNA binding site as a probe. In a typical experiment, a binding buffer is mixed with the protein(s) of interest in the presence of nonspecific DNA which competes with many proteins in a crude extract that can bind nonspecifically to DNA and prevent the probe from entering the gel. Commonly used sources for this nonspecific competitor DNA include *E. coli* DNA, salmon or human sperm DNA, calf thymus DNA, or any of a series of synthetic oligonucleotide strands of which the most common is poly(dI-dC) or poly (dI-dC). The amount of nonspecific competitor DNA can range from none at all (under conditions in which purified proteins are used in the assay), to a few nanograms, to a few micrograms (routinely 1–3 μg) when using crude cellular extracts. These amounts must be determined empirically and are specific to each individual protein–DNA complex. After incubation of this mixture of binding buffer, protein, and nonspecific DNA, the purified probe is added and the reaction is again incubated at the appropriate temperature and for the appropriate amount of time; both parameters are best determined empirically. By way of example, a standard EMSA for the association of NF-κB with the kB site,[22] the association of AP-1 with the TPA-responsive element,[23] the Sp1–DNA association conditions,[24] and the GATA binding buffer we also use for the TBP–TATA site association assay[25] are provided.

After the incubation of the protein extract with the probe, the reaction is prepared for loading onto the gel in one of two ways. If the glycerol content of the final binding reaction is 4% or greater, the sample can be loaded directly onto the gel without further additions. In the absence of any glycerol, either glycerol to approximately 5% or Ficoll to 4% of the total binding reaction volume can be added. This will ensure that the sample will not float away in the buffer. Bromphenol blue and/or xylene cyanole dyes can be added in a probe alone control lane or to the samples as desired. This is particularly useful in estimating the migration of the probe from the loading origin. It is also considered good practice to leave the unbound probe on the bottom of the gel in order to show the consistency of loading and the integrity of the probe itself.

Characterization of a DNA–protein interaction can involve a series of separate gel shifts which address the specificity of association of the protein complex with the DNA binding site and subsequent analyses to identify the binding protein(s). Specificity is determined partially by adding a 100-fold excess of the unlabeled oligonucleotide or DNA fragment to the incubation at the same time that the radiolabeled probe is added. This establishes a competition between the unlabeled DNA binding site and the probe for association with a protein. This experiment should be run in contrast to an identical assay in which a 100-fold excess of an unlabeled oligonucleotide or DNA fragment is mutated in such a way that it no longer binds the protein complex of interest. Where the addition of the direct competitor DNA should reduce or abolish binding at a given concentration, the addition of the mutant DNA, at this same concentration, should have no affect

[22] A. S. Baldwin, Jr., J. C. Azazkhan, D. E. Jenson, A. A. Beg, and L. R. Coodley, *Mol. Cell. Biol.* **11,** 4943 (1991).

[23] Y. Nakabeppu, K. Ryder, and D. Nathans, *Cell (Cambridge, Mass.)* **55,** 907 (1988).

[24] J. T. Kadanaga, *et al.*, *Cell (Cambridge, Mass.)* **51,** 1079 (1987).

[25] T. C. Fong and B. M. Emerson, *Genes Dev.* **6,** 521 (1992).

on the specific protein–DNA complex. This type of assay can be pursued further by introducing competitor DNA which contains subtle mutations both in the DNA binding site and the flanking sequences. Similar sequences can be compared in such a manner to determine the extent of competition. A high affinity binding site would be able to bind the protein complex more efficiently than a low affinity binding site; that is if one sequence requires a 25-fold excess of competitor DNA to completely abolish binding of a given probe, whereas a second sequence requires a 100-fold excess of competitor to abolish binding off the same probe, the first sequence constitutes a higher affinity binding site for the protein complex than the second sequence.

One modification of the EMSA routinely used to establish the molecular mass of the DNA binding protein complex involves the use of ultraviolet light (UV) cross-linking of photoreactive-radiolabeled probes in the gel shift analysis.[26] In the production of the probe for EMSA, bromodeoxyuridine (BrdU) is incorporated into one strand whereas the opposing strand serves as a template for the polymerase reaction. Our laboratory utilizes three radiolabeled nucleotide triphosphates in addition to BrdU to produce a very high specific activity probe which, when exposed to UV light in the presence of protein, will cross-link any protein associated with BrdU residues. The *in situ* reaction can be analyzed directly by SDS–PAGE, or a gel shift analysis can be performed with the cross-linked reaction, the retarded bands isolated individually, and these gel fragments separated by SDS–PAGE. Only those proteins covalently attached to the radiolabeled probe will be visualized by autoradiography, reflecting the size of the protein subunit directly contacting the DNA binding site.

The production and commercial availability of antisera to numerous DNA binding proteins have broadened the ability of the investigator to examine the identity of the DNA binding protein complex. Purified antisera (polyclonal or monoclonal) can be added directly to the protein incubation stage prior to the addition of the gel shift probe. The response of the protein–DNA complex depends on the specific nature of the antibody generated. Some antisera will bind protein domains not associated with the DNA binding domain due to the resulting increase in the size of the antibody–protein–DNA complex, the affected complex will experience a "supershift." If the antibody recognizes the DNA binding portion of the protein as its epitope, the association of the antibody–protein complex will thus inhibit the association with DNA and the complex will not be visualized. Specificity of an antibody–protein complex is extremely important to determine as even normal rabbit serum from some bleeds can cause a

[26] S. Miyamoto, K. Cauley, and I. M. Verma, this volume [43].

nonspecific supershift to be seen. Antipeptide antisera are particularly useful as supershift reagents because the preincubation of the antibody and its immunogenic peptide can be done prior to the addition of the protein extract as an additional control. While not all antisera to DNA binding proteins are useful as EMSA reagents, the availability of antisera to numerous transcription factors and basal regulatory factors have facilitated greatly our ability to identify several DNA binding proteins in a fast, efficient, and cost-effective manner.

Acknowledgments

We thank Dr. Lynn Ransone and Dr. Sarah Lilly for critical reading of the chapter. The protocols outlined in this chapter are the cumulative work of individuals in Dr. Inder Verma's laboratory at the Salk Institute to whom we are grateful: Penny Wamsley, Mark Schmitt, and Terry Leveillard. This work was supported in part by Grants R01-GM51249 from the National Institutes of Health.

[43] Ultraviolet Cross-Linking of DNA Binding Proteins

By SHIGEKI MIYAMOTO, KEITH CAULEY, and INDER M. VERMA

I. Introduction

To understand the regulation of transcription, it is of great importance to identify proteins that bind to DNA on the regulatory elements of the genes of interest. In general, DNA binding and transcription factor complexes can be resolved by an electrophoretic mobility shift assay (EMSA).[1] EMSA resolves protein complexes that bind to a radiolabeled DNA fragment or double-stranded oligonucleotide probe representing the promoter or enhancer region of a gene. Although this assay provides information on the specificity and number of various complexes formed on a given DNA sequence, it fails to give further information such as the molecular weights of proteins that are directly contacting the DNA. Knowing the size of DNA binding proteins can greatly aid in the purification and cloning of the genes encoding them. In contrast to EMSA, ultraviolet (UV) radiation can bring about the formation of a covalent linkage between DNA and contacting protein, hence knowledge of specific DNA binding proteins within a protein complex can be obtained by UV cross-linking.[1]

[1] F. M. Ausubel, R. Brent, R. E. Kingston, D. D. Moore, J. G. Seidman, J. A. Smith, and K. Struhl, eds., "Current Protocols in Molecular Biology," Chapter 12, and references therein. Wiley (Interscience), New York, 1988.

Smith[3] first reported that UV radiation brings about the covalent cross-linking of protein to nucleic acids *in vivo*[2] and *in vitro*. Smith[3] characterized the reactivity of each amino acid to uracil on UV light exposure and demonstrated that cysteine, phenylalanine, and tyrosine were more reactive than other amino acids. In 1974, Lin and Riggs[4] demonstrated that a substitution of thymidine with bromodeoxyuridine (BrdU) significantly increased the efficiency of UV cross-linking between DNA and protein.

The exact molecular mechanisms of covalent bond formation between protein and DNA are not known. When UV light energy exceeds the ionization potential of the bases in solution, purine and pyrimidine cation radicals are formed. BrdU-substituted DNA is more reactive than unsubstituted DNA in that input energy readily results in debromination and consequent production of highly reactive uracil radicals.[4] These reactive radicals can form covalent bonds with proximal amino acid residues of DNA-bound proteins. Since no cross-linking agent is involved, the bond formed between the protein and nucleic acid is referred to as "zero length."[5] As the oligonucleotide can be radiolabeled, the UV cross-linked products are also radiolabeled and can be resolved as specific DNA binding proteins by sodium dodecyl sulfate–polyacrylamide gel electrophoresis (SDS–PAGE). This procedure has been used in the characterization of a number of DNA and RNA binding protein complexes[6–9] and has found productive application in the characterization of protein complexes that bind the κB enhancer motif in the immunoglobulin κ light chain gene.[10] This chapter attempts to describe the method of UV cross-linking using κB enhancer binding complexes as an example.

II. Mechanics of Ultraviolet Cross-Linking

A. Nuclear Extract Preparation

Many protocols exist for the preparation of cellular or nuclear extracts. Generally, extracts have been prepared from cultured cells by a large-

[2] K. C. Smith, *Biochem. Biophys. Res. Commun.* **8,** 157 (1962).

[3] K. C. Smith, *Biochem. Biophys. Res. Commun.* **34,** 354 (1969).

[4] S.-Y. Lin and A. D. Riggs, *Proc. Natl. Acad. Sci. U.S.A.* **71,** 947 (1974).

[5] J. N. Hockensmith, W. L. Kubasek, W. R. Voracheck, E. M. Evertsz, and P. H. von Hippel, this series, Vol. 208, p. 211.

[6] I. Sagami, S. Y. Tsai, H. Wang, M.-J. Tsai, and B. O'Malley, *Mol. Cell. Biol.* **6,** 4259 (1986).

[7] T. D. Allen, K. L. Wick, and K. S. Mathews, *J. Biol. Chem.* **266,** 6113 (1991).

[8] L. A. Chodosh, R. W. Carthew, and P. A. Sharp, *Mol. Cell. Biol.* **6,** 4723 (1986).

[9] J. M. Gott, M. C. Willis, T. H. Koch, and O. C. Uhlenbeck, *Biochemistry* **30,** 6290 (1991).

[10] S. Miyamoto, M. J. Schmitt, and I. M. Verma, *Proc. Natl. Acad. Sci. U.S.A.* **91,** 5056 (1994).

scale protocol[11] involving a dialysis step. We and others have found very satisfactory results using a simple, small-scale extract procedure.[12] Extracts prepared in this way are generally stable at −70° for extended periods, and can be thawed and refrozen, although stability through the freeze–thaw cycle may vary for specific factors. The presence of protease inhibitors is essential for stability of the transcription factors. Phenylmethylsulfonyl fluoride (PMSF), aprotinin, pepstatin A, leupeptin, etc., may be added to block protease activity. Extracts can also be prepared from animal tissues, although results are generally less reliable. We have had good results using the protocol of Sonnenberg *et al.*[13] for extracts of brain tissue.[14] Apparently, because of high protease content, these extracts are not as stable as those prepared from cultured cells, and we have used them for only a week or so following preparation and storage at −70°. Other protocols for nuclear extract preparation involve sucrose gradient centrifugation and yield high quality, transcriptionally active extracts.[15,16] These protocols generally require larger amounts of initial starting material.

B. Bromodeoxyuridine-Substituted Oligonucleotide Probe Preparation

The double-stranded oligonucleotide probe for UV cross-linking requires the substitution of BrdU for thymidine, as well as the incorporation of radiolabeled nucleotides for radioactive tagging of the cross-linked protein. BrdU can be substituted for thymine in a DNA polymerase I (Klenow fragment) reaction (Fig. 1). As an empirical consensus, incorporation of BrdU for thymidine in the oligonucleotide probe does not qualitatively effect the binding of proteins to their target DNA sites,[4] although this should be verified in each case. Quantitatively, the affinity of protein binding can be effected by BrdU incorporation, however. Dissociation of the *lac* repressor is reported to be slower from BrdU-substituted DNA than from unsubstituted DNA.[4]

The overall success of BrdU-mediated UV cross-linking is determined by the abundance of the DNA binding protein, the exact sequence of the binding site, and the proximity of the incorporated BrdU to a protein–DNA contact point. As the overall efficiency of DNA cross-linking is low, oligonucleotide probes labeled to a high specific activity are used. Although it is possible to incorporate BrdU into both strands of the oligonucleotide,

[11] J. D. Dignam, R. M. Lebovitz, and R. G. Roeder, *Nucleic Acids Res.* **11,** 1475 (1991).
[12] N. C. Andrews and D. V. Faller, *Nucleic Acids Res.* **19,** 2499 (1991).
[13] J. L. Sonnenberg, P. F. Macgregor-Leon, T. Curran, and J. Morgan, *Neuron* **3,** 359 (1989).
[14] K. Cauley and I. Verma, *Proc. Natl. Acad. Sci. U.S.A.* **91,** 390 (1994).
[15] B. J. Graves, P. F. Johnson, and S. L. McKnight, *Cell (Cambridge, Mass.)* **44,** 565 (1986).
[16] K. Gorski, M. Carneiro, and U. Schibler, *Cell (Cambridge, Mass.)* **47,** 767 (1986).

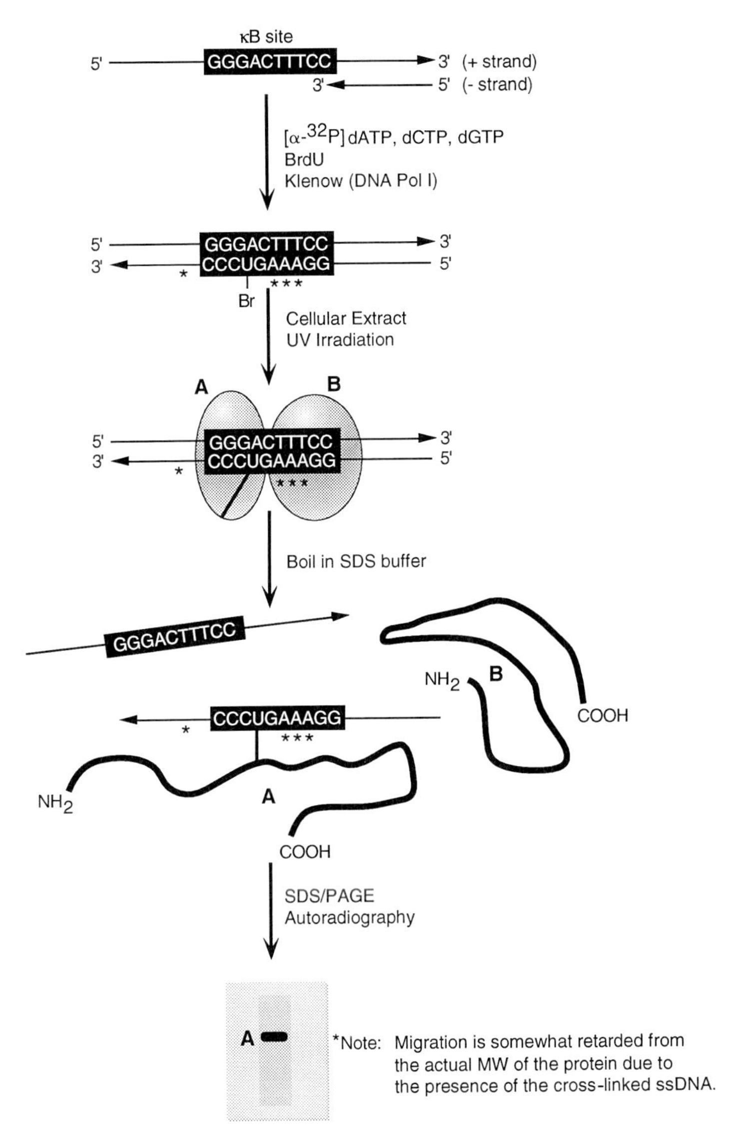

FIG. 1. The UV cross-linking procedure. One oligonucleotide containing the DNA binding site of interest (in this case a κB site) is synthesized along with an 8- to 10-mer oligonucleotide which anneals to the extreme 3′ end. The second strand is synthesized by the Klenow fragment of the DNA polymerase I in the presence of BrdU and radioactive nucleotides. The cellular extract is added to the BrdU-substituted radioactive probe and is UV irradiated to form a covalent linkage between specific DNA binding proteins and the probe. Following boiling in Laemmli SDS sample buffer, the DNA binding proteins can be separated by SDS–PAGE and visualized by autoradiography. The cross-linked proteins will migrate slower than their actual molecular weights due to the presence of a covalently attached single-stranded oligonucleotide probe. If the binding site is not palindromic in nature, the "plus" and "minus" strands may yield different sets of cross-linked proteins when labeled individually.

generation of one reactive strand is generally simpler to perform and the cross-linking results are easier to interpret. As such, it is often worthwhile to perform separate cross-linking experiments where each strand of the probe includes BrdU.

Protocol: BrdU Incorporation

Synthesize one strand of the 20- to 30-mer oligonucleotide including the DNA binding site. Synthesize a second oligonucleotide of 8–10 bp in length which will anneal to the extreme 3′ end of the first oligonucleotide. BrdU is incorporated by a Klenow extension, and so the exact oligonucleotide sequence should be chosen based on the optimal incorporation of BrdU into the enzymatically synthesized strand. The short annealing primer can be designed for use against different 30-mers with various mutations so that only one new oligonucleotide need be generated for each additional desired mutation. A single thymidine residue in the DNA binding site may be sufficient for a successful UV cross-linking. One strand of incorporated BrdU may be more effective in cross-linking protein than the other, and this should be tested experimentally (see Fig. 1). It is also important to realize that multiple proteins may be contacting DNA, and so the results of cross-linking of one BrdU-labeled DNA strand may be different from those where the opposing oligonucleotide strand is labeled. In theory, if a palindromic sequence is used, the same set of proteins should be cross-linked.

Oligonucleotide Annealing. Mix 0.5 μg 30-mer, 0.2 μg 10-mer, 10 μl 10× annealing buffer [100 m*M* Tris (pH 7.4), 60 m*M* $MgCl_2$, 500 m*M* NaCl, 60 m*M* 2-mercaptoethanol, and 2 mg/ml gelatin], and H_2O to a final volume of 100 μl. Heat sample to 90° and cool slowly to room temperature.

Extension Reaction. Mix 100 μl of reannealed oligonucleotide with 25 μl each of [α-^{32}P]dATP (800 Ci/m*M*), [α-^{32}P]dGTP (800 Ci/m*M*), and [α-^{32}P]dCTP (800 Ci/m*M*), 4 μl 50× dNTP/BrdU (250 μM dATP, 250 μM dGTP, 250 μM dCTP, and 2.5 m*M* BrdUTP), 10 μl 10× annealing buffer, 5 μl Klenow (5 U/μl), and 6 μl H_2O. Incubate sample at 16° for 90 min. Add 200 μl TE. Pass through a Sephadex G-50 spun column.

If cross-linking is very efficient for a given oligonucleotide, the number of radioactive nucleotides may be reduced. It is also possible to ^{32}P end label nonradioactive BrdU-substituted probes using T4 polynucleotide kinase, although we have had less success in cross-linking using end-labeled probes, possibly because UV cleavage of DNA can result in the loss of the terminal label or presence of phosphatases in cellular extracts.

C. Ultraviolet Cross-Linking in Solution

DNA binding complexes are generally characterized by EMSA before UV cross-linking is undertaken. Probes are compared and characterized by EMSA. Cross-linking is linear with time,[8,16] (Fig. 2A), but the heat dissipated by UV irradiation and evaporation of the sample limit cross-linking times. The effect of the UV dose on the protein extract, and on the oligonucleotide probe alone, can be monitored by EMSA. Following UV exposure, cross-linked products can be analyzed by SDS–PAGE directly. It is not necessary to digest the oligonucleotide from the protein in order to determine the molecular weight as the small molecular weight of the oligonucleotide can be subtracted from the protein molecular weight.

It is often the case that solution-phase UV cross-linking yields a large number of labeled proteins on SDS–PAGE. Many or most such proteins are cross-linked nonspecifically to the probe, and one must then determine which proteins represent specific interactions with the DNA. The specificity can be assessed by competition analysis with excess unlabeled wild-type and mutant oligonucleotide (Fig. 2C). Choice and titration of competitor oligonucleotides can be determined by EMSA (for example, see Fig. 3A). If EMSA demonstrates that DNA binding of the complex of interest changes in the course of a treatment, such as stimulation of a second messenger pathway, cross-linked nuclear extracts from stimulated and unstimulated cells might identify specifically labeled DNA binding proteins. Such an analysis would yield more or less cross-linked products because UV cross-linking is semiquantitative in that if more specific DNA binding proteins are present, more cross-linked products can be recovered (Fig. 2B). If satisfactory results are not achieved by UV cross-linking in solution, cross-linking in the EMSA gel should be attempted (discussed below).

Protocol: Cross-Linking in Solution

Set up standard binding reaction as for EMSA with a 10^6 cpm BrdU-substituted probe and 10–50 μg extract in a 50-μl reaction volume in a 1.5-ml round-bottom vial, such as a NUNC cryogenic vial. Cover the vials with a small piece of plastic wrap held tightly in place with a strip of Parafilm. Place the vial in a test tube rack. Place a UV transilluminator inverted over the test tube rack, supported so that it is stable. The illuminator should be within 5 cm of the top of the vial. The UV source should be 305 nm and 7000 $\mu W/cm^2$ intensity. We have had success with the Fotodyne transilluminator. Plastic wrap is UV transparent and helps to prevent evaporation of the sample during UV exposure. Cross-link (using Fotodyne transilluminator set on "Analytical" setting) for 60 min at room temperature. After cross-linking, add an equal volume of 2× Laemmli SDS sample buffer and boil for 5 min. Run a SDS–PAGE. The free probe will run at the dye front so it could be removed following drying before exposure to film. If a lower intensity UV source is used, or the distance from the source to the sample vial is increased, time of exposure should be increased. If the binding activity of interest appears to be reduced following cross-linking, as judged

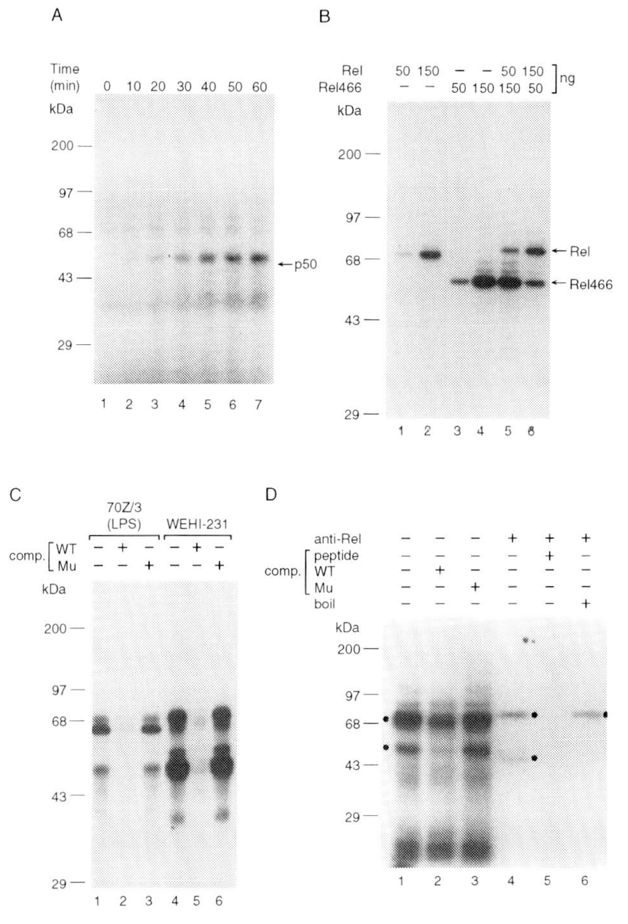

FIG. 2. UV cross-linking in solution. (A) Time course of UV cross-linking using a double-stranded Igκ-κB site (5′-CTCAACAGAGGGGACTTTCCGAGAGGCCAT-3′) with a nuclear extract prepared from the WEHI231 murine B cell line. When the "minus" strand of the Igκ-κB site is labeled, the p50 protein is predominantly cross-linked. As can be seen (compare lanes 1–7), the longer the UV exposure, the more cross-linked p50 protein can be recovered. (B) UV cross-linking is semiquantitative with an excess probe. Two different amounts of bacterially prepared full-length murine c-Rel (Rel) or C-terminally truncated c-Rel (Rel466) are cross-linked using a double-stranded palindromic κB site (5′-CAACGGCAGGGGAATTCCCCTCTCCTT-3′). More proteins were cross-linked when more DNA binding proteins were added (compare lanes 1 and 2 to 3 and 4). When they were mixed (lanes 5 and 6), cross-linking efficiency was correspondingly semiquantitative for both proteins. (C) Specificity of UV cross-linking. The specificity of UV cross-linking is tested by competition with 50-fold excess double-stranded wild-type (5′-CAACGGCAGGGGAATTCCCC-TCTCCTT-3′, lanes 2 and 5) or double-stranded mutant (5′-CAACGGCAGATCTAT-CTCCCTCTCCTT-3′, lanes 3 and 6) oligonucleotides. Lanes 1–3, nuclear extract prepared

by EMSA, it may be worthwhile reducing the temperature of the sample during the cross-linking procedure. This can be done by standing the sample test tube rack in wet ice during cross-linking.

D. Immunoprecipitation of Cross-Linked Proteins

If the EMSA complex is suspected to contain a known factor, the supershift assay can be performed using a specific antibody to determine its presence.[17] Alternatively, the products of solution phase UV cross-linking can be analyzed by immunoprecipitation using antibodies. Controls for antibody specificity are the same as those described earlier for competition with wild-type and mutant unlabeled oligonucleotides. Preimmune serum or competition with an epitope peptide also offers an important control (Fig. 2D, compare lanes 4 and 5). If more than one protein binds DNA simultaneously, immunoprecipitation can be performed with cross-linked samples that are previously boiled in high percentage SDS (0.5%) to disrupt multimeric complexes. This way, the heteromeric nature of DNA binding complexes can be directly examined (Fig. 2D, compare lanes 4 and 6; also see Miyamoto *et al.*[10]).

Protocol: Immunoprecipitation of Ultraviolet Cross-Linked Proteins

Set up UV cross-linking reactions as described earlier. Immunoprecipitate using a standard RIPA buffer and antibodies (see Sambrook *et al.*[18] for detailed protocols on immunoprecipitation). When disruption of the protein complex is preferred, add SDS to 0.5%, boil for 5 min, and dilute with RIPA buffer to a final SDS concentration of 0.1%. The antibody–protein

[17] T. M. Kristie and B. Roizman, *Proc. Natl. Acad. Sci. U.S.A.* **83,** 3218 (1986).

[18] J. Sambrook, E. F. Fritsch, and T. Maniatis, eds., "Molecular Cloning: A Laboratory Manual," 2nd ed., Chapter 18, and references therein. Cold Spring Harbor Lab., Cold Spring Harbor, New York, 1989.

from the bacterial lipopolysaccharide-stimulated 70Z/3 murine pre-B cell line. Lanes 4–6, nuclear extract prepared from unstimulated WEHI231 cells. (D) Immunoprecipitation analysis following UV cross-linking in solution to identify the nature of κB-binding proteins in WHEI231 cells. κB-binding proteins present in the nuclear extract of WEHI231 cells were UV cross-linked using the palindromic κB site as in Fig. 2B and resolved by SDS–PAGE. Lanes 2 and 3 show the specificity test by excess wild-type or mutant competitor. A portion of the cross-linked products were immunoprecipitated using murine c-Rel antiserum 5075 without (lanes 4 and 5) or with (lane 6) boiling in SDS to disrupt protein complexes. Lane 5 shows the competition of immunoprecipitation with an excess epitope peptide. The lower band in lane 4 migrates somewhat faster and is diffused due to the presence of the immunoglobulin heavy chain. For more description, see Miyamoto *et al.*[10] Dots show the specific κB-binding proteins (c-Rel and p50).

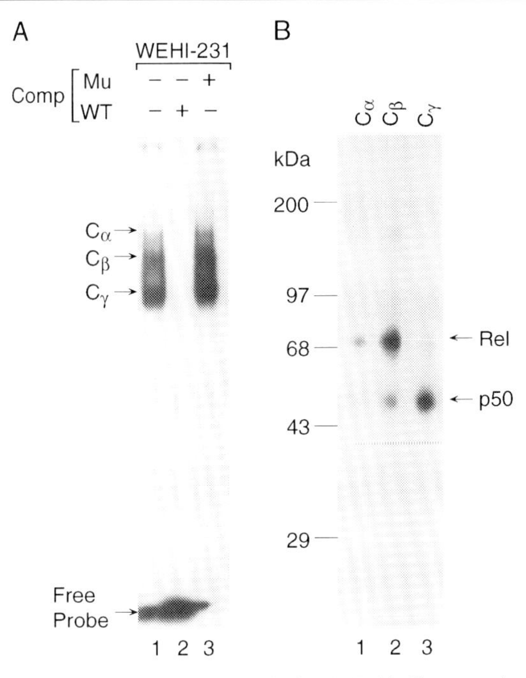

FIG. 3. In-gel cross-linking. (A) EMSA analysis of κB binding complexes present in the nuclear extract isolated from WEHI231 cells. Two major (Cβ and Cγ) complexes and a minor (Cα) complex can be detected by the palindromic κB site (lane 1). Lanes 2 and 3 show the specificity of binding complexes by competition with a 50-fold excess wild-type (lane 2) or mutant (lane 3) oligonucleotide. (B) A similar EMSA gel as in A was irradiated with UV and each complex was cut out, minced, boiled in SDS sample buffer, and resolved by SDS–PAGE. Complex Cα consists of a single protein with a molecular mass of ~80 kDa (lane 1) whereas Cβ and Cγ consist of proteins with molecular masses of ~55 and ~80 kDa (lane 2) and ~55 kDa only (lane 3), respectively. From the studies shown here and elsewhere,[10] the Cα, Cβ, and Cγ are concluded to be a Rel homodimer, a p50-Rel heterodimer, and a p50 homodimer, respectively.

complex can be isolated by protein A– or G–Sepharose. Run SDS–PAGE to resolve cross-linked proteins.

E. Ultraviolet Cross-Linking in the EMSA Gel

When multiple complexes bind a probe, it is useful to find out which of the DNA binding proteins identified by solution-phase UV cross-linking constitute each protein complex. For this purpose, in-gel cross-linking becomes invaluable. Following electrophoresis, a wet EMSA gel is exposed to the UV light source, after which the gel is exposed to film (Fig. 3A). Complexes are visualized by autoradiography, excised from the gel, and analyzed by SDS–PAGE (Fig. 3B). The specific DNA binding proteins in each EMSA complex can be directly identified (Figs. 3A and 3B; complexes Cα, Cβ, and Cγ). Depending on the source of cellular extracts, this protocol may yield results that are more readily interpreted than those seen with cross-linking in solution, as the EMSA serves as a partial purification of the DNA binding complexes. In addition, all of the binding specificity controls can be carried out in the EMSA.

Protocol: In-Gel Cross-Linking

Carry out standard EMSA using a BrdU-labeled probe. If necessary, scale up with wider EMSA gel wells. Remove one glass plate and cover the gel with plastic wrap. Place gel side down on the UV source. Cross-link (using Fotodyne transilluminator set on "Analytical" setting) for 60 min (at room temperature or 4°). Expose gel to identify cross-linked EMSA bands. Recover bands, mince, and boil in 2× Laemmli SDS buffer. Load gelatinous material into the lanes of SDS–PAGE and electrophorese. Fix, dry, and expose the gel to film with intensifying screens.

III. Concluding Remarks

Covalent cross-linking between DNA and protein can serve to identify specific DNA binding proteins. This procedure is particularly useful when the nature of the DNA binding protein is unknown to the investigator. This method can generate the number and approximate molecular weights of different proteins that can specifically bind to a given DNA sequence, such as that in the regulatory region of a gene of interest. Furthermore, since different proteins may contact DNA at different points, BrdU labeling of the "plus" strand may identify a set of proteins different from those seen with the "minus" strand. However, other DNA binding proteins that cannot be cross-linked, as well as other non-DNA binding proteins in a EMSA complex, will not be identified by this procedure.

Author Index

Numbers in parentheses are footnote reference numbers and indicate that an author's work is referred to although the name is not cited in the text.

A

C

D

E

F

G

H

I

J

K

L

M

N

O

P

S

T

U

V

W

X

Y

Z

Subject Index

A

B

E

F

G

H

I

K

L

M

O

V

W

X

Y

ISBN 0-12-182155-2